Analytic Number Theory

American Mathematical Society

Colloquium Publications
Volume 53

Analytic Number Theory

Henryk Iwaniec
Emmanual Kowalski

AMERICAN MATHEMATICAL SOCIETY

2000 *Mathematics Subject Classification.* Primary 11Fxx, 11Lxx, 11Mxx, 11Nxx, 11T23, 11T24, 11R42.

For additional information and updates on this book, visit
www.ams.org/bookpages/coll-53

Library of Congress Cataloging-in-Publication Data

Iwaniec, Henryk.
Analytic number theory / Henryk Iwaniec, Emmanuel Kowalski.
p. cm. — (Colloquium publications, ISSN 0065-9258 ; v. 53)
Includes bibliographical references and index.
ISBN 0-8218-3633-1 (acid-free paper)
Softcover ISBN 978-1-4704-6770-8

1. Number theory. I. Kowalski, Emmanuel, 1969– II. Title. III. Colloquium publications (American Mathematical Society) ; v. 53.

QA241.I85 2004
512.7′3-dc22 2004045081

Printed in the United States of America.
♾ The paper used in this book is acid-free and falls within the guidelines
established to ensure permanence and durability.
Visit the AMS home page at http://www.ams.org/

13 12 11 10 9 8 7 6 5   26 25 24 23 22 21

Contents

PREFACE

This book shows the scope of analytic number theory both in classical and modern directions. There are no division lines; in fact our intent is to demonstrate, particularly for newcomers, the fascinating countless interrelations. Of course, our picture of analytic number theory is by no means complete, but we tried to frame the material into a portrait of a reasonable size, yet providing a self-contained presentation.

We were writing this book in a period of time during and after teaching courses and working with graduate students in Rutgers University, Bordeaux University and Courant Institute. We thank these institutions for providing conditions for both of us to work together. We shared ideas on what this book should be about with many of our colleagues, who gave us critical suggestions. Among them we would like to mention Étienne Fouvry, John Friedlander, Philippe Michel and Peter Sarnak. During a long process of typing and preparation of this book for publication, we received stimulating encouragement and technical advice from Sergei Gelfand, for all of his help we express our gratitude. Carol Hamer helped to polish some of our English phrases while her little boys tried to destroy the TeX files without success. We thank them all for the output.

Henryk Iwaniec
Emmanuel Kowalski
15 December, 2003

INTRODUCTION

Analytic Number Theory distinguishes itself by the variety of tools it employs to establish results, many of which belong to the main streams of arithmetic. It is not part of analysis nor of any particular discipline of mathematics, however it does interact indeed with various fields. Therefore everybody seems to view the subject differently. This vast diversity of concepts of analytic number theory is its great attraction. Our desire in this book is to exhibit the wealth and prospects of the theory, its charming theorems and powerful techniques. However it is not our primary objective to give proofs of the strongest results, although in many cases we come quite close to the best possible ones. Rather we favor a reasonable balance between clarity, completeness and generality. The book was conceived with graduate students in mind so the reader will often find that our emphasis is on reasoning throughout the arguments. Of course our presentation is subjective, and in retrospect may lose its meaning. Certainly we do not always follow the original lines of discovery, but occasionally we do draw brief historical perspectives.

Leonard Euler must get credit for the first use of analytical arguments for the purpose of studying properties of integers, specifically by constructing generating power series. Euler's proof of the infinity of prime numbers makes use of the divergence of the zeta function and the corresponding product over primes, which is named after him. This was the beginning of analytic number theory. Next came P. G. L. Dirichlet whose creation of the theory of L-functions for characters, resulting in the proof of the infinity of primes in arithmetic progressions, makes him the true father of analytic number theory. From these early days to modern times the distribution of prime numbers constitutes the core of the subject. This will be apparent in the course of our book. The first two chapters cover questions of primes up to the elementary methods of P. Tchebychev.

Chapter 3 provides definitions and basic properties of Dirichlet characters and the Gauss sums. Characters on ideals of imaginary quadratic fields are also introduced, not only because they play a supporting role in subsequent chapters but to show a bit of analytic number theory beyond the traditional domain of the rational integers as well; there will be other examples throughout the book, for instance, elliptic curves.

Poisson summation for number theory is what a car is for people in modern communities – it transports things to other places and it takes you back home when applied next time – one cannot live without it. Chapter 4 presents a classical account of this basic technology. Many readers do realize now, others will figure out later, that we are already talking about ideas of modular forms. But we continue our considerations along traditional lines (both classical and more recent ones) before the concept of modularity takes the leading position.

The celebrated memoir of B. Riemann on the zeta function is embedded in the context of abstract L-functions in Chapter 5. It is not our style to consider things in terms more general than necessary, so defining a class of L-functions which suits minimum requirements of our forthcoming applications was not without difficulty and hesitation. In this way we could convey to dedicated researchers that generalizations are not always straightforward. For instance, to establish the zero-free region for L-functions of degree > 1 one cannot rely on the same principles as for the Dirichlet L-functions. The key ingredient is the Rankin-Selberg convolution. On the other hand, the problem of exceptional zero is resolved for many automorphic L-functions of degree > 1 (not without clever constructions) while it remains open for the L-functions with real characters. Furthermore in Chapter 5 a message is sent that a better life exists in the world of automorphic forms than in the zoo of degree one L-functions.

Analytic number theory does not mean non-elementary. The first author recalls that his first serious encounter with analytic number theory started by reading the lovely book of A.O. Gelfond and Yu.V. Linnik, "Elementary Methods of Analytic Number Theory". When an ambitious beginner starts from there her/his love of the subject is sealed forever. Try it yourself! One is instantly captured by sieve methods. In this book we do not have space to give justice to this marvelous idea, nevertheless Chapter 6 should suffice for basic applications.

Next comes the "Large Sieve", which is not a sieve but a name for other things. Yes, it did originate from a short paper by Linnik on a sieve problem, but it took time to recognize the true nature of these ideas. In Chapter 7 we reveal our viewpoint and the crucial attributes (spectral completeness, orthogonality), then we demonstrate the amazing power of the large sieve on selected old and new problems. Other features of the large sieve are scrutinized showing the good and the bad sides. For example, the approach using the duality principle is fruitful for harmonics of degree one (characters) while producing poor results for harmonics of larger degree (like for example the eigenvalues of Hecke operators). The controversy over the proper place of the large sieve is academic. Simply speaking the large sieve inequalities are parts of bilinear forms theory.

Estimates for exponential sums are the first tools which deeply penetrate the problems of analytic number theory beyond natural structures. These cannot be grasped by harmonic analysis alone. See what clever use people made of the property that a shifted interval is another interval, that adding an integer to a set of integers yields again a set of integers (sorry, primes are not preserved!). We challenge algebraists to find a structural explanation of the power of such arguments! They should read Chapter 8 to find what H. Weyl built out of these observations. Van der Corput and Vinogradov are also the main figures from the early stages of that discipline. A lot of work and talking went into our presentation of Vinogradov's method, because it is not quite correctly explained in numerous publications. At some point Vinogradov departs from the Weyl differencing process and treats multi-dimensional exponential sums as bilinear forms (this is the way we think of it anyway).

The next two chapters show more recent technology which was developed to replace the unproven Riemann hypothesis in applications to the distribution of prime numbers. We are talking about estimates for the number of zeros of L-functions in vertical strips which are positively distanced from the critical line. Hopefully in a future one will say we were wasting time on studying the empty set. Great ideas are camouflaged there in arguments of enormous complexity, so this might not be enjoyable for everyone at first. However if you think the Riemann hypothesis is not provable in your lifetime, please read and admire these unconditional substitutes. Special mention goes to Hugh Montgomery, Martin Huxley and Matti Jutila for the most original contributions.

Although we are primary interested in rational integers one can learn and benefit a lot from arithmetic of other fields. Not only from the number fields or p-adic fields, but indirectly from the fields of finite characteristic as well. Particularly fruitful are the methods of exponential sums over the finite fields. In Chapter 11 we prove (among other things) the Riemann hypothesis for special curves which yields the celebrated estimate of Weil for Kloosterman sums. The Kloosterman sums have been employed to solve various problems of analytic number theory from the beginning of their creation in 1926. We also mention briefly the state of the knowledge of exponential and character sums over algebraic varieties. Applications of these are harder to make, yet there is a handful of examples in the literature. It was a painful decision to exclude all but the simplest from presentation in this book. Otherwise to do full justice for these highly sophisticated ideas we would have to choose the most complicated application for which we have no room. It suffices to say that a preparation of a given problem of analytic number theory to an estimate for character sum over varieties can be the state-of-the-art in its own right, never mind that the final argument is powered by the outside forces of algebraic geometry.

Dirichlet characters are already discussed in Chapter 3 and we return to them in Chapter 12 to treat very short character sums. Again one must be inventive to break limits of natural structures. Burgess theorem is a fine example.

Sums over primes are treated in the next chapter. When Vinogradov succeeded in estimating sums over primes of additive characters, which he needed for a solution to the ternary Goldbach problem in conjunction with the circle method, it was a shocking result. Before him the Grand Riemann Hypothesis could do the job, but keep in mind that the Riemann hypothesis is still not established. The original ideas of Vinogradov were borrowed from combinatorial sieve and were rather complicated. Recently developed identities offer much simpler treatments of more general sums over primes. As they share the same fundamental principle (reducing the sum to bilinear forms) the results are pretty much the same, so the choice of the method is a matter of taste and technical convenience. To capture the key elements in Chapter 13 we develop more than one identity.

A popular criterion for analytic number theory is that complex variable analysis is being used. Perhaps it is better to say harmonic analysis, since the action of the latter is more profound. For a long time, analytic number theory flourished exclusively from abelian harmonic analysis, that is to say from the Fourier transform in $\mathbb{R}^n$. There is still a great potential in this classical analysis to be explored. However much stronger fertilizers began to act on the soil of analytic number theory in recent times. These are automorphic functions. Of course, modular forms have been driving algebraic aspects of number theory much longer, but in a limited scope (confined to holomorphic forms). New resources of automorphic theory are found in the spectral analysis, the foundation of which was led by H. Maass and A. Selberg at the turn of the $1940's$ (real-analytic cusp forms, Eisenstein series, trace formula). In simple terms a non-abelian harmonic analysis found its role in analytic number theory. Truly effective expansion of spectral methods into analytic number theory began about twenty five years ago, changing the face of either subject irrevocably.

This book barely addresses the fascinating issues of the new direction throughout Chapters 14, 15, and 16. Our featured application is to estimation for sums of Kloosterman sums. This is a good choice (if no more can be accommodated), because the reader can appreciate the new tools by comparing with the earlier results derived in Chapter 11 by algebraic considerations. Another application of the spectral theory of automorphic forms to an arithmetical question is presented in Chapter 21, that is to the equidistribution of roots of quadratic congruences of prime moduli. The spectral theory continues to grow extensively, so it would be premature to wrap it up here or in any other book. For further reading we recommend [**I3**], [**Sa3**].

Although the spectral methods of automorphic forms predominate current research in analytic number theory, the traditional problems continue getting our attention with respectful intensity throughout the remaining chapters. Great treasures of the subject mustn't be buried in the past. First of all a newcomer should learn the stories of primes in arithmetic progressions to large moduli. In Chapter 17 she/he will find how E. Bombieri and A.I. Vinogradov bypassed the Riemann hypothesis to establish (by the large sieve and other means) unconditional results with applications as good as one can get from the RH itself. Of course our arguments are not identical with the original ones (of 1965) since we take advantage of later simplification, in great measure due to P.X. Gallagher.

Chapter 18 goes further back to 1944 when Linnik gave an extraordinary bound for the least prime in an arithmetic progression. For a long time this bound was considered as the most difficult theorem in analytic number theory. And yes, it is still hard by today's standards, and one can still learn a lot from the technology applied! Faced with the obstacle of the exceptional zero, Linnik brings the repulsion effect (he calls it Deuring-Heilbronn phenomenon) to a new level; amazingly enough he turns the problem to his advantage! This is a fascinating development in the history of analytic number theory which we recommend one should master for a better understanding of the status of the exceptional zero today.

Once upon a time the famous Goldbach problem was worth a million dollar prize award. For applications the problem (representations of even integers by the sums of two primes) has no great merit, but as an intellectual challenge one would be proud to crack it. Probably something new about prime numbers would be revealed then. Read Chapter 19 to improve your chances.

Chapter 20 is serious. Here analytic methods storm the domain of diophantine equations, which from ancient Greeks was exclusively a business of arithmetic. Started by Hardy - Ramanujan, continued by Hardy - Littlewood and developed substantially further by Kloosterman, the circle method uses orthogonality of additive characters to detect equations, not only to solve algebraic equations but a large class of additive problems over special integers as well. The toughest are the binary additive problems. They are not completely solved by the Kloosterman method, but at least we get a very reliable picture of what the true asymptotic for the number of solutions should be. Kloosterman sums which we covered in the preceding chapters are instrumental in the circle methods. After classical ideas we propose a more direct variant which in principle should produce the same results, however without employing Kloosterman sums. One should read Chapter 20 with an open mind, separate technical (still attractive) elements from conceptual devices to see clearly the connections with modular forms. Certainly Kloosterman and Rademacher were aware of these intrinsic connections, while they are overlooked by some specialists in the circle method.

Equidistribution problems for sequences of special integers, lattice points in various domains, solution to diophantine equations, etc, constitute a heavy industry over the analytic number theory. We regret there is no space to run this industry in full capacity in the book. The book of M.N. Huxley [**Hu4**] treats only the lattice point problems, however quite deeply. In Chapter 21 we are dealing with the problem of distribution of roots of a quadratic equation reduced modulo prime. As the prime modulus tends to infinity we show that the roots are uniformly distributed. The arguments include almost everything that we developed in the book so far, thus showing that the industry is robust.

Because of failure of the unique factorization of algebraic integers, the arithmetic of number fields is not as easy as for rational numbers and sometimes perplexing. The complexity is measured by the order of the ideal class group. Naturally the case of imaginary quadratic fields received the first and the most attention because units do not interfere. We do know that the class number grows to infinity (so there is only a finite number of imaginary quadratic fields with a fixed class number), but the serious issue is to estimate the class number effectively. Chapter 22 describes the problem thoroughly and prepares the ground for the advances in Chapter 23. The effective lower bound for the class number (due to D. Goldfeld) may not appear strong for demanding researchers, yet it is deep with respect to results taken from other sources. First of all it uses the Gross-Zagier formula for L-functions of elliptic curves at the central point. We do provide a substantial overview of the involved arguments from elliptic curves, although these are more geometric than analytic. The analytic arguments themselves are quite delicate. Actually they came first, the L-functions of elliptic curves being supplementary. Indeed we worked out an effective lower bound for the class number which depends on the order of vanishing of general L-functions of degree two, suspecting that the requirements are satisfied by quite a few of them.

In Chapter 24 we prove a very classical result of Selberg that a positive proportion of zeros of the Riemann zeta function lies on the critical line. This is a good place to learn about the mollification techniques (a kind of smoothing), which is used in many works today and will reappear in Chapter 26.

Assuming the Riemann hypothesis, H.L. Montgomery revealed in 1974 that the distribution of zeros of $\zeta(s)$ follows the behavior of eigenvalues of certain "ensembles" of unitary matrices. More recently physicists joined the team of workers in number theory, creating a new excitement and hope for finding a path to a proof of the Riemann hypothesis. This is the main objective of the so-called random matrix theory, one of the most popular subject and driving forces of current analytic number theory. It offers reliable models for predicting the behavior of arithmetical quantities which for a long time were shrouded in mystery. The consistency of the random matrix theory with the harmony of integers still seems quite surprising. Whatever the future of this enterprise will be, due to the current cooperation, analysis is closer to arithmetic than ever before. A subject of such magnitude cannot be fully presented in a short space. Therefore in Chapter 25 we stick to the original theme of the correlation of zeros of $\zeta(s)$ and its variations on the zeros of families of automorphic L-functions which are near the central point. We leave it for the reader to judge whether the ideas of random matrix theory are realistic for launching an attack on the Riemann Hypothesis.

In recent investigations the central values of L-functions appear in a variety of formulas with vanishing or non-vanishing assumptions. Take for example [**IS2**] where an effective lower bound for the class number of imaginary quadratic fields is derived essentially from the non-vanishing of central values of families of L-functions, to the contrary of the vanishing requirements in the previous investigations. Another example is the formula of T. Watson [**Wa**] by means of which the quantum-ergodicity conjecture (that is the equidistribution of Maass cusp forms) is reduced to a subconvexity bound for certain L-functions of degree four. We consider in detail one non-vanishing statement which has applications to arithmetic geometry.

We hope this book will show the picture of analytic number theory in plenty of colors. However we must say that a lot of significant topics are left out. Missing are the dispersion method, the amplification method (see [**M2**]), some analytic techniques from diophantine approximations and transcendence. Moreover probability arguments are barely exposed, and we didn't touch ergodic theory either, whose impact on number theory has been felt strongly in the last years.

We also try to show some details of the powerful theories which are developing as the most useful new tools for analytic number theory, in particular the theory of higher-degree automorphic forms and their L-functions, and algebraic geometry; young researchers in particular should be encouraged to develop expertise in these subjects. It is certain that spectacular applications have only begun and more will be open to those who understand both sides. Dually, arithmetic geometry and algebraic number theory also give and promise a wealth of new questions, or new aspects of old ones, where the skills and techniques of analytic number theory will be tested to the utmost. Hopefully they will bring rich rewards to those who will try to come to these open fields... We barely mention some questions related to elliptic curves but we believe that there is much more to discover. The deep conjectures of Lang and Trotter [**LT**] are already quite popular, and a few other challenging problems may be found in [**Ko1**].

The exercises inside each section serve a dual purpose, some are to improve the reader's skill, the others serve as additional information about the subject. Historical remarks are brief, to give some orientation in the development of the matter, rather than to credit exhaustively the inventors. The only advice we offer to new researchers is read! read! read! many papers with complete proofs. Knowing a result in analytic number theory is only the first step to liking it; more important and rewarding is to understand the arguments of its proof. Our viewpoint is that making mathematics should not be rated like breaking sport records. Sometimes the strongest result is boring while a slightly weaker one generates great pleasure.

Formal prerequisites for much of the book are rather slight, not going beyond differential calculus, complex analysis and integration, especially Fourier series and integrals. It is more important for the reader to have or acquire a good understanding of how to manipulate inequalities and not simple identities.

In later chapters automorphic forms become important. We have included two survey chapters, yet we expect that many readers will have already some knowledge of this important topic, or will study it independently.

In some sections (for instance Sections 5.13 and 5.14), which are intended as convenient references for certain facts and results which are hard to locate in proper form in the literature, we assume that the reader has some familiarity with other topics, such as representations of groups and algebraic geometry.

Sections of this book were written over a period of time, therefore readers will notice a slight change of style and repetition. We think that a small redundancy is helpful for reading long arguments. Occasionally the same object is introduced again in a different chapter in local terminology which should be more familiar in a particular context. We believe this flexibility is justified for comfort, even at the expense of losing uniqueness.

Our notations are mostly standard. But since inequalities with unspecified constants are the lifeblood of analytic number theory, and since there are sometimes controversies on this subject, we spell out the meaning of the various comparison symbols $O()$, $o()$, $\sim$, $\asymp$ or $\ll$. Most important, we use Landau's $f = O(g)$ and Vinogradov's $f \ll g$ as synonyms; thus $f(x) \ll g(x)$ for $x \in X$ (where X must be specified either explicitly or implicitly) means that $|f(x)| \leqslant Cg(x)$ for all $x \in X$ and some constant $C \geqslant 0$. Any value of C for which this holds is called an implied constant. Since a constant is most often a function looking for a variable, the "implied constant" will sometimes depend on other parameters, which we explicitly mention at the most important points (but sometimes it is clear from context). If there is no other dependency, we speak of "absolute constants". This usage means that our $O()$, for instance, is not the same as that of Landau or Bourbaki. We use $f \asymp g$ to mean that both relations $f \ll g$ and $g \ll f$ hold, of course with possibly different implied constants.

However $f = o(g)$ for $x \to x_0$ means that for any $\varepsilon > 0$ there exists some (unspecified) neighborhood U_ε of x_0 such that $|f(x)| \leqslant \varepsilon g(x)$ for $x \in U_\varepsilon$. Then $h \sim g$ means $h = g + o(g)$. Those are the same as in Landau or Bourbaki.

Among the few notation which may be unfamiliar to a beginner, we mention that $p^k \| m$, where p is prime and k an integer, means that p^k divides m exactly (i.e. p^{k+1} does not divide m). The integral part $[x]$ is the integer n such that $n \leqslant x < n + 1$.

We sometimes use the notation $\sum^\star$ to denote sums restricted to a subset of "primitive" objects, which will be indicated in each case, and $\sum^\flat$ to denote a sum restricted to squarefree numbers.

CHAPTER 1

ARITHMETIC FUNCTIONS

1.1. Notation and definitions.

Throughout this book $\mathbb{Z}, \mathbb{Q}, \mathbb{R}, \mathbb{C}$ denote the sets of integers, rationals, real and complex numbers respectively. The addition of complex numbers makes $\mathbb{C}$ a group, and $\mathbb{Z} \subset \mathbb{Q} \subset \mathbb{R} \subset \mathbb{C}$ is the natural inclusion of subgroups. The Lebesque measure dx on the additive group $\mathbb{R}$ is the invariant measure (with respect to additive translations). The subsets $\mathbb{Q}^\star \subset \mathbb{R}^\star \subset \mathbb{C}^\star$ of the non-zero numbers are closed under multiplication, they are considered as multiplicative groups. In $\mathbb{R}^\star$ we distinguish the subgroup of positive real numbers $\mathbb{R}^+$ which is endowed with the invariant measure $x^{-1}dx$ (with respect to multiplicative translations).

The positive integers are called natural numbers. The set of natural numbers $\mathbb{N} = \{1, 2, 3, \dots\}$ contains the set of primes $\mathbb{P} = \{2, 3, 5, 7, \dots\}$. These are fundamental elements of arithmetic since every natural number factors uniquely (up to a permutation) into powers of distinct primes. The distribution of primes is one of basic problems of analytic number theory. Prime numbers are often denoted by p.

First of all, to cultivate analytic number theory one must acquire a considerable skill for operating with arithmetic functions. We begin with a few elementary considerations.

A complex valued function f defined on $\mathbb{N}$ is called an arithmetic function. In some contexts an arithmetic function $f : \mathbb{N} \to \mathbb{C}$ is better seen as the sequence $\mathcal{A} = (a_n)$ with $a_n = f(n)$. Very often in such context $\mathcal{A}$ is supported on numbers of primary interest, and a_n is just a multiplicity, or some sort of weight, which is introduced when counting such numbers. A different job is assigned to certain functions $f : \mathbb{N} \to \mathbb{C}$ (such as Dirichlet characters or Hecke eigenvalues) which we call "arithmetic harmonics." These play an instrumental role in analyzing the primary sequence $\mathcal{A} = (a_n)$. Essentially the arithmetic harmonics are employed with the primary sequence $\mathcal{A} = (a_n)$ to produce a family of twisted sequences $\mathcal{A}_f = (a_n f(n))$. The twisted sequences with appropriate harmonics are capable of selecting a special subsequence of $\mathcal{A}$ which is our target (think of Dirichlet characters being employed to detect primes in arithmetic progressions).

Thanks to the additive and multiplicative structures of integers the two important classes of arithmetic functions are distinguished. A function $f : \mathbb{N} \to \mathbb{C}$ is an additive function if it satisfies

$$f(mn) = f(m) + f(n) \tag{1.1}$$

for m, n relatively prime. If this property holds for all m, n, then f is said to be completely additive. For example $f(n) = \log n$ is completely additive. Similarly, a function $f : \mathbb{N} \to \mathbb{C}$ is multiplicative if it satisfies

$$f(mn) = f(m)f(n) \tag{1.2}$$

for m, n relatively prime, and f is completely multiplicative if (1.2) holds for all m, n. For example $f(n) = n^{-s}$ with $s \in \mathbb{C}$ is completely multiplicative. Obviously, the additive and the multiplicative functions are determined by their values at prime powers. We have $f(1) = 0$ if f is additive, $f(1) = 1$ if f is multiplicative not identically zero.

1.2. Generating series.

To an arithmetic function f we shall attach the two most natural infinite series

$$E_f(z) = \sum_n f(n)z^n, \tag{1.3}$$

$$D_f(s) = \sum_n f(n)n^{-s}, \tag{1.4}$$

where z, s are complex variables (in the case of the power series $E_f(z)$ we may also include $n = 0$ if $f(0)$ is defined). These are called the generating series, or functions, of f. Other kinds of generating functions can be attractive as well to capture distinct properties of f. For many arithmetic functions the corresponding generating series converges absolutely in a small domain, and it is an interesting question how far the series can be analytically continued? If the generating series extends beyond the range of absolute convergence, then this property usually manifests some group structure in the coefficients $f(n)$ ("random" series cannot be continued). For example the analytic continuation of Artin L-functions is intimately related to the reciprocity laws in number fields (in the abelian case at any rate), the analytic continuation of Hasse-Weil L-functions is a major step towards the modularity of the corresponding elliptic curves (over $\mathbb{Q}$), and so on.

The generating power series were introduced by L. Euler (1707–1783) for studying special additive problems (when doing so he was the first to mix analysis with arithmetic). Let $E_f(z)$ and $E_g(z)$ be the series for f and g. Then the product

$$E_f(z)E_g(z) = \sum_n h(n)z^n \tag{1.5}$$

yields the power series for the function h given by (the additive convolution)

$$h(n) = \sum_{\ell+m=n} f(\ell)g(m). \tag{1.6}$$

Applying Cauchy's theorem one expresses the coefficient $h(n)$ by the contour integral

$$h(n) = \frac{1}{2\pi i}\int_{|z|=r} E_f(z)E_g(z)z^{-n-1}dz. \tag{1.7}$$

Hence, given reasonable analytic properties of $E_f(z)E_g(z)$ one can deduce a good estimate for $h(n)$, or even an asymptotic formula as n tends to ∞. Of course, Euler did not know Cauchy's theorem, so his ideas were limited to the power series for which one has a secondary expression, giving an exact formula for $f(n)$ rather than an approximation (still not a tautology). We give three cute examples. First is the identity

$$\sum_0^\infty z^n = \prod_1^\infty (1 + z^{2^m}) = (1 - z)^{-1}$$

which is nothing but an analytic statement of the uniqueness of binary expansion of natural numbers. Here is another identity:

$$\sum_0^\infty p(n) z^n = \prod_1^\infty (1 - z^m)^{-1}$$

where $p(n)$ is the partition function. Less obvious is the following formula of Jacobi:

$$\sum_{-\infty}^\infty z^{n^2} = \prod_1^\infty (1 - z^m)(1 - z^{m+1/2})^2. \tag{1.8}$$

About ninety years ago the integral representation (1.7) for the additive equation (1.6) gave birth to the circle method. These ideas will be developed in Chapter 20.

By changing the variable z into

$$e(z) = e^{2\pi i z} \tag{1.9}$$

the power series (1.3) is seen as a Fourier series, which is a common practice in modern analytic number theory.

The generating series $D_f(s)$ is called the Dirichlet series after being used in Dirichlet's fundamental works on primes in arithmetic progressions, though the special case

$$\zeta(s) = \sum_1^\infty n^{-s} \tag{1.10}$$

was already considered by Euler.

The Dirichlet series is particularly attractive for multiplicative functions (but not limited to this case). For, if f is multiplicative, then

$$D_f(s) = \prod_p (1 + f(p)p^{-s} + f(p^2)p^{-2s} + \cdots) \tag{1.11}$$

provided the involved series in prime powers and the product over primes (called the Euler product) converge absolutely. In particular, we have

$$\zeta(s) = \prod_p (1 - p^{-s})^{-1} \tag{1.12}$$

if $\mathrm{Re}\,(s) > 1$. This representation of $\zeta(s)$ as an Euler product is an expression in analytic language of the unique factorization of natural numbers into distinct prime powers.

1.3. Dirichlet convolution.

Assuming that the Dirichlet series $D_f(s), D_g(s)$ converge absolutely it follows that the product $D_f(s)D_g(s)$ is also given by a Dirichlet series, namely we have

$$D_f(s)D_g(s) = \sum_1^\infty h(n)n^{-s}$$

with the coefficients

$$h(n) = \sum_{d|n} f(d)g\Big(\frac{n}{d}\Big). \tag{1.13}$$

The arithmetic function h defined by (1.13) (never mind the convergence of generating series) is called the Dirichlet (or multiplicative) convolution of f and g, and it is denoted by $f \star g$.

The set of all arithmetic functions with the usual addition $+$ and the operation $\star$ is a commutative ring. The function $\delta : \mathbb{N} \to \mathbb{C}$ whose Dirichlet series is $D_\delta(s) = 1$ is the unit element of this ring, i.e.,

$$\delta(n) = \begin{cases} 1 & \text{if } n = 1, \\ 0 & \text{if } n > 1. \end{cases} \tag{1.14}$$

The function $\zeta(s)$ defined in $\mathrm{Re}\,(s) > 1$ by (1.10) is called the Riemann zeta function (after his seminal memoir [**Rie**]), its Dirichlet coefficients make the constant function $1(n) = 1$ for all $n \geqslant 1$. By virtue of (1.12) the inverse of $\zeta(s)$ has also a Dirichlet series expansion, namely

$$\frac{1}{\zeta(s)} = \prod_p \Big(1 - \frac{1}{p^s}\Big) = \sum_1^\infty \frac{\mu(m)}{m^s}, \tag{1.15}$$

say, with coefficients

$$\mu(m) = \begin{cases} (-1)^r & \text{if } m = p_1 \dots p_r \text{ with } p_1, \dots, p_r \text{ distinct}, \\ 0 & \text{otherwise}. \end{cases} \tag{1.16}$$

The function $\mu(m)$ was introduced in 1832 by A. F. Möbius, and it bears his name ever since. Moreover, by the Euler product (1.12) the logarithm of $\zeta(s)$ has the Dirichlet series expansion

$$\log \zeta(s) = \sum_{\ell=1}^\infty \sum_p \ell^{-1} p^{-\ell s}. \tag{1.17}$$

Recall that the Dirichlet convolution $\star$ corresponds to the multiplication of the generating Dirichlet series, therefore the identity $\zeta(s) \cdot \zeta^{-1}(s) = 1$ reads as

$$\delta(m) = \sum_{d|m} \mu(d) = \begin{cases} 1 & \text{if } m = 1, \\ 0 & \text{if } m > 1. \end{cases} \tag{1.18}$$

Using this formula one obtains the following

MÖBIUS INVERSION. *For any $f, g : \mathbb{N} \to \mathbb{C}$ the following two relations are equivalent:*

$$g(n) = \sum_{d|n} f(d), \tag{1.19}$$

$$f(n) = \sum_{d|n} \mu(d) g\Big(\frac{n}{d}\Big). \tag{1.20}$$

REMARK. To be accurate with the history the inversion formulas in the above form were stated only in 1857 by R. Dedekind. The original version stated by Möbius was somewhat different, it amounts to saying that for any real variable functions $F, G : [1, x] \to \mathbb{C}$ the following two relations are equivalent:

$$G(x) = \sum_{n \leqslant x} F(x/n), \tag{1.21}$$

$$F(x) = \sum_{m \leqslant x} \mu(m) G(x/m). \tag{1.22}$$

The Möbius function is multiplicative. If f, g are multiplicative, then so are $f \cdot g$ and $f \star g$. If g is multiplicative, then

$$\sum_{d|n} \mu(d) g(d) = \prod_{p|n} (1 - g(p)). \tag{1.23}$$

This product admits a probabilistic interpretation. Viewing $g(p)$ as a probability of some independent events which may occur at p one can think of (1.23) as the probability that none of the events associated with prime divisors of n occur.

1.4. Examples.

Now we give a large sample of arithmetic functions which one encounters in analytic number theory. Many other functions will be introduced in due course.

The divisor function $\tau(n)$ is the number of positive divisors of n, so we have

$$\zeta^2(s) = \sum_{1}^{\infty} \tau(n) n^{-s}. \tag{1.24}$$

More generally $\tau_k(n)$ denotes the number of representations of n as the product of k natural numbers, so its Dirichlet series is $\zeta^k(s)$. Explicitly

$$\tau_k(n) = \binom{a_1 + k - 1}{k - 1} \cdots \binom{a_r + k - 1}{k - 1} \quad \text{if } n = p_1^{a_1} \ldots p_r^{a_r}. \tag{1.25}$$

For any $\nu \in \mathbb{C}$ we define $\sigma_\nu(n)$ by

$$\zeta(s)\zeta(s - \nu) = \sum_{1}^{\infty} \sigma_\nu(n) n^{-s}, \tag{1.26}$$

so it is the sum of powers of divisors

$$\sigma_\nu(n) = \sum_{d|n} d^\nu. \tag{1.27}$$

We have the Ramanujan formula

$$\zeta(s)\zeta(s-\alpha)\zeta(s-\beta)\zeta(s-\alpha-\beta)\zeta^{-1}(2s-\alpha-\beta)=\sum_1^\infty\sigma_\alpha(n)\sigma_\beta(n)n^{-1}. \tag{1.28}$$

In particular,

$$\zeta^4(s)\zeta^{-1}(2s)=\sum_1^\infty\tau(n)^2n^{-s}. \tag{1.29}$$

One can recognize the above Ramanujan series as a special case of the Rankin-Selberg convolution L-functions attached to modular forms (see Section 5.1). Dividing (1.29) by $\zeta(s)$ one gets

$$\zeta^3(s)\zeta^{-1}(2s)=\sum_1^\infty\tau(n^2)n^{-s}, \tag{1.30}$$

hence multiplying by $\zeta(2s)$

$$\zeta^3(s)=\sum_1^\infty\Big(\sum_{d^2m=n}\tau(m^2)\Big)n^{-s}.$$

This is the symmetric square L-function associated with $\zeta^2(s)$. Dividing again by $\zeta(s)$ one gets

$$\zeta^2(s)\zeta^{-1}(2s)=\sum_1^\infty 2^{\omega(n)}n^{-s} \tag{1.31}$$

where $\omega(n)$ denotes the number of distinct prime factors of n. Next we get

$$\zeta(s)\zeta^{-1}(2s)=\sum_1^\infty|\mu(n)|n^{-s}. \tag{1.32}$$

Note that $|\mu(n)|=\mu^2(n)$ is the characteristic function of squarefree numbers. By (1.32) we have

$$\mu^2(n)=\sum_{d^2|n}\mu(d). \tag{1.33}$$

Inverting (1.32) we obtain another Dirichlet series,

$$\zeta^{-1}(s)\zeta(2s)=\sum_1^\infty\lambda(n)n^{-s}$$

with coefficients

$$\lambda(n)=(-1)^{\Omega(n)} \tag{1.34}$$

where $\Omega(n)$ denotes the total number of prime factors of n (counted with multiplicity). Thus $\lambda(n)$, which is called the Liouville function, is completely multiplicative and it agrees with $\mu(n)$ on squarefree numbers. The Euler function $\varphi(n)$ is defined by the series

$$\frac{\zeta(s-1)}{\zeta(s)} = \sum_{1}^{\infty} \varphi(n) n^{-s}, \tag{1.35}$$

so it is also given by

$$\varphi(n) = n \prod_{p|n} \Big(1 - \frac{1}{p}\Big) = n \sum_{d|n} \frac{\mu(d)}{d}. \tag{1.36}$$

One may think of $\varphi(n)/n$ as the probability that a randomly chosen integer m is coprime with n. In other words, $\varphi(n)$ represents the number of residue classes $(a \bmod n)$ with $(a, n) = 1$.

Very important arithmetic functions emerge by differentiation. We begin with

$$-\zeta'(s) = \sum_{1}^{\infty} (\log n) n^{-s},$$

which gives us the logarithm function

$$L(n) = \log n. \tag{1.37}$$

Since $L(n)$ is additive, it follows that $L \cdot (f \star g) = (L \cdot f) \star g + f \star (L \cdot g)$. This says that the multiplication by L is a derivation in the Dirichlet ring of arithmetic functions.

Next the function $\Lambda(n)$ can be defined as coefficients in the Dirichlet series for the logarithmic derivative $-(\log \zeta(s))' = -\zeta'(s)\zeta(s)^{-1}$, so

$$-\frac{\zeta'}{\zeta}(s) = \sum_{1}^{\infty} \Lambda(n) n^{-s}. \tag{1.38}$$

By the Euler product (1.12) one computes that

$$\Lambda(n) = \begin{cases} \log p & \text{if } n = p^{\alpha}, \alpha \geqslant 1, \\ 0 & \text{otherwise.} \end{cases} \tag{1.39}$$

Thus $\Lambda(n)$ is supported on prime powers. One can read (1.38) as $\Lambda = \mu \star L$. More explicitly,

$$\Lambda(n) = \sum_{d|n} \mu(d) \log \frac{n}{d} = -\sum_{d|n} \mu(d) \log d. \tag{1.40}$$

By Möbius inversion we obtain $L = 1 \star \Lambda$, or equivalently,

$$\log n = \sum_{d|n} \Lambda(d). \tag{1.41}$$

The function $\Lambda(n)$ was known and effectively used by P. Tchebyshev, nevertheless it is called the von Mangoldt function. Here is another identity which is useful in elementary prime number theory:

$$\Lambda - 1 = \mu \star (L - \tau). \tag{1.42}$$

The von Mangoldt function Λ has been generalized to the higher von Mangoldt function of any degree $k \geqslant 0$ by $\Lambda_k = \mu \star L^k$, i.e.,

$$\Lambda_k(n) = \sum_{d|n} \mu(d)\Big(\log \frac{n}{d}\Big)^k \tag{1.43}$$

(notice that $\Lambda_0 = \delta$). This satisfies the recurrence

$$\Lambda_{k+1} = L\Lambda_k + \Lambda \star \Lambda_k \tag{1.44}$$

from which it follows by induction that $\Lambda_k(n)$ is supported on numbers having at most k distinct prime factors, i.e., $\Lambda_k(n) = 0 \quad \text{if } \omega(n) > k$. Moreover, we have

$$0 \leqslant \Lambda_k(n) \leqslant (\log n)^k, \tag{1.45}$$

the lower bound following by induction from (1.44) and the upper bound following from the formula $L^k = 1 \star \Lambda_k$, or explicitly,

$$(\log n)^k = \sum_{d|n} \Lambda_k(d), \tag{1.46}$$

which comes from (1.43) by Möbius inversion. Of course, Λ_k is not multiplicative, but we have the following handy formula (see (1.44))

$$\Lambda_k(mn) = \sum_{0 \leqslant j \leqslant k} \binom{k}{j} \Lambda_j(m)\Lambda_{k-j}(n), \qquad \text{if } (m,n) = 1.$$

EXERCISE 1. For any integer $k \geqslant 0$ and a real number $x > 0$, define

$$\Lambda_k(n,x) = \sum_{d|n} \mu(d) \left(\log \frac{x}{d}\right)^k. \tag{1.47}$$

Note that $\Lambda_k(n,x)$ depends only on the squarefree kernel of n. Prove that

$$\Lambda_k(n,x) = \sum_{0 \leqslant j \leqslant k} \binom{k}{j} \Lambda_j(n) \left(\log \frac{x}{n}\right)^{k-j}.$$

Then, using $\Lambda_j \leqslant L^{j-i}\Lambda_k$, derive that for $n \leqslant x$

$$\Lambda_k(n,x)\left(\frac{\log n}{\log x}\right)^k \leqslant \Lambda_k(n) \leqslant \Lambda_k(n,x). \tag{1.48}$$

Therefore $\Lambda_k(n,x)$ is supported on integers n having at most k distinct prime divisors, and it satisfies $0 \leqslant \Lambda_k(n,x) \leqslant (\log x)^k$ if $n \leqslant x$.

Next show that if $(m,n) = 1$, then

$$\Lambda_k(mn,x) = \sum_{0 \leqslant j \leqslant k} \binom{k}{j} \Lambda_j(n)\Lambda_{k-j}\Big(m, \frac{x}{n}\Big).$$

Hence derive that for $n = p_1^{a_1} \dots p_r^{a_r}$ with $p_1, \dots, p_2$ distinct primes

$$\Lambda_k(n,x) \leqslant r! \binom{k}{r} (\log x)^{k-r}(\log p_1)\dots(\log p_r). \tag{1.49}$$

One can associate a Möbius and von Mangoldt function to any multiplicative function. If $D_f(s)$ is the generating Dirichlet series for f which has Euler product, then μ_f and Λ_f are defined as coefficients in

$$\begin{aligned} \frac{1}{D_f(s)} &= \sum_1^\infty \mu_f(m)m^{-s}, \\ -\frac{D_f'(s)}{D_f(s)} &= \sum_1^\infty \Lambda_f(n)n^{-s}. \end{aligned} \tag{1.50}$$

Hence $f\star\Lambda_f = f\cdot L$ and $\Lambda_f = \mu_f\star(f\cdot L)$. Clearly Λ_f is supported on prime powers. If f is completely multiplicative, then $\mu_f(n) = \mu(n)f(n)$ and $\Lambda_f(n) = \Lambda(n)f(n)$.

There is a variety of truly interesting arithmetic functions. The theory of modular forms is a basic source for multiplicative functions. As a simple model we pick up the function $r(n)$ which is the number of solutions to $a^2+b^2 = n$ in integers a, b. The generating Dirichlet series for $r(n)$ is equal to $4\zeta_K(s)$, where

$$\zeta_K(s) = \sum_{\mathfrak{a}} (N\mathfrak{a})^{-s}$$

is the zeta function of the imaginary quadratic field $K = \mathbb{Q}(\sqrt{-1})$ while the factor 4 is the number of units in K. Here $\mathfrak{a}$ runs over non-zero integral ideals of K. All these are principal ideals generated by the Gaussian integers $\alpha = a+bi \in \mathbb{Z}[\sqrt{-1}]$, $\alpha \neq 0$, and if $\mathfrak{a} = (\alpha)$, then $N\mathfrak{a} = a^2+b^2$. Hence, indeed

$$4\zeta_K(s) = \sum_{n\geqslant 1} r(n)n^{-s}.$$

Clearly $\frac{1}{4}r(n)$ is multiplicative. The prime numbers which are represented as the sum of two squares are characterized in a beautiful theorem of Fermat. It is easy to see that no prime $p \equiv -1 (\bmod 4)$ can be so written and Fermat proved that all other primes can be (uniquely up to the sign and order of a, b). In modern language this theorem of Fermat is just the factorization law in the Gaussian domain $\mathbb{Z}[\sqrt{-1}]$; it asserts that $p = 2$ is square of a prime ideal, $p \equiv 1 (\bmod 4)$ splits into a product of two distinct (complex conjugate) prime ideals, $p \equiv -1 (\bmod 4)$ remains prime. Using the factorization law one can show by verification of local factors that

$$\zeta_K(s) = \zeta(s)L(s,\chi_4)$$

where χ_4 is the non-trivial character to modulus 4 (we have $\chi_4(n) = \sin\frac{\pi n}{2}$) and $L(s,\chi_4)$ is the associated Dirichlet L-function

$$L(s,\chi_4) = \sum_1^\infty \chi_4(n)n^{-s}.$$

Hence

$$r(n) = 4\sum_{d|n}\chi_4(d). \tag{1.51}$$

Besides the Dirichlet series there is great interest in studying the Fourier series for $r(n)$ due to the additive nature of the equation $a^2+b^2=n$. We have

$$\sum_0^\infty r(n)e(nz)=\theta^2(z)$$

(recall (1.9) that $e(z)=e^{2\pi iz}$), where $\theta(z)$ is the theta function

$$\theta(z)=\sum_{-\infty}^{\infty}e(n^2z). \tag{1.52}$$

This series converges absolutely for $\operatorname{Im}(z)>0$. More generally, letting $r_k(n)$ be the number of representations of n as the sum of k squares we have

$$\sum_0^\infty r_k(n)e(nz)=\theta^k(z).$$

On one hand, the theta function $\theta(z)$ is recognized in analytic number theory for its ability to select squares. On the other hand, $\theta(z)$ is usable due to the following transformation rule:

$$\theta\left(\frac{az+b}{cz+d}\right)=\nu(c,d)(cz+d)^{\frac12}\theta(z) \tag{1.53}$$

which holds for any z with $\operatorname{Im}(z)>0$ and any integers a,b,c,d with $ad-bc=1$, $c\equiv 0 \pmod 4$, where $\nu(c,d)$ depends only on c,d. We have $\nu^2(c,d)=\chi_4(d)$, so $\nu(c,d)$ takes only four values $\pm1,\pm i$ (see (3.42) for the exact description). In other words, $\theta(z)$ is a modular form of weight $\frac12$ with multiplier $\nu(c,d)$ while $\theta^2(z)$ is a modular form of weight one and character χ_4 on $\Gamma_0(4)$. In addition to the modular relations (1.53) the theta function satisfies the following involution equation:

$$\theta\Big(\frac{-1}{2z}\Big)=\Big(\frac{z}{i}\Big)^{\frac12}\theta\Big(\frac{z}{2}\Big). \tag{1.54}$$

Note that $\theta(z)$ does not vanish by virtue of the Jacobi product (1.8).

More generally one associates a theta function with any positive definite quadratic form in several variables and corresponding harmonic polynomials (see Section 14.3).

Yet other types of arithmetic functions are given by exponential sums over $\mathbb{Z}/m\mathbb{Z}$ or $(\mathbb{Z}/m\mathbb{Z})^\star$. The most important of these for analytic number theory are Gauss sums, Ramanujan sums and Kloosterman sums. We mention here only the quadratic Gauss sum (for general Gauss sums, see Section 3.4)

$$G(m)=\sum_{x(\bmod m)}e\Big(\frac{x^2}{m}\Big).$$

They were evaluated exactly by Gauss as a consequence of the Quadratic Reciprocity Law (see Theorem 3.3):

$$\bar{G}(m)=\frac{1+i^m}{1+i}\sqrt{m}. \tag{1.55}$$

The Kloosterman sums $S(a,b;c)$ encompass Ramanujan sums. They are defined by

$$S(a,b;c) = \sum_{x \pmod c}^{\star} e\Big(\frac{ax+b\bar{x}}{c}\Big) \tag{1.56}$$

for integers a, b and $c \geqslant 1$. The symbol $\sum^{\star}$ restricts the sum to x coprime with c and $\bar{x}$ denotes the multiplicative inverse of x modulo c, i.e. $x\bar{x} \equiv 1 \pmod c$. Note that $S(a,b;c)$ is real, and has the following symmetries:

$$S(a,b;c) = S(b,a;c), \tag{1.57}$$

$$S(aa',b;c) = S(a,ba';c) \text{ if } (a',c) = 1. \tag{1.58}$$

It also has multiplicative properties (by Chinese Remainder Theorem)

$$S(a,b;cd) = S(a\bar{c},b\bar{c};d)S(a\bar{d},b\bar{d};c) \text{ if } (c,d) = 1. \tag{1.59}$$

Kloosterman sums will make many appearances in this book. They are crucial to modern analytic number theory through links with spectral theory of automorphic forms. Though no "simpler" formula exists (comparable to (1.55) for quadratic Gauss sums), a sharp bound for individual Kloosterman sums is known, due to A. Weil:

$$|S(a,b;c)| \leqslant \tau(c)(a,b,c)^{1/2}\sqrt{c}. \tag{1.60}$$

We will prove this in Chapter 11 (see Corollary 11.12) and use it many times.

When $b=0$ (or $n=0$, see (1.57)), the Kloosterman sum $S(n,0;m)$ is called a Ramanujan sum. It is also sometimes denoted $c_m(n)$. For these there is an explicit formula, see (3.2) and (3.3).

1.5. Arithmetic functions on average.

In the course of analytic number theory the averages of arithmetic functions of various kinds appear all over the place. The first task is to establish estimates for finite sums of type

$$\mathcal{M}_f(x) = \sum_{n\leqslant x} f(n). \tag{1.61}$$

In this section we show a few elementary ideas.

The first common tool is summation by parts, which is the formula

$$M_{fg}(x) = \sum_{n\leqslant x} f(n)g(n) = M_f(x)g(x) - \int_1^x M_f(t)g'(t)dt$$

and its obvious variants. This formula often allows us to evaluate $\mathcal{M}_{fg}(x)$ if $\mathcal{M}_f$ is known and g is smooth and doesn't oscillate.

If f is defined on $[1,x]$, monotonic and continuous, then $\mathcal{M}_f(x)$ is well approximated by a corresponding integral, precisely

$$\mathcal{M}_f(x) = \int_1^x f(y)dy + O(|f(x)| + |f(1)|). \tag{1.62}$$

For example, we get

$$\sum_{n\leqslant x}(\log n)^k = xP_k(\log x) + O((\log x)^k) \tag{1.63}$$

where $P_k(X)$ is a polynomial of degree k

$$P_k(X) = \sum_{0\leqslant \ell\leqslant k}(-1)^{k-\ell}\frac{k!}{\ell!}X^\ell,$$

and

$$\sum_{n\leqslant x}(\log \frac{x}{n})^k = k!x + O((\log x)^k). \tag{1.64}$$

If $f(y)$ is continuously decreasing to zero, one can do better than (1.62). Indeed, putting

$$[x] = \max\{\ell\in\mathbb{Z} : \ell\leqslant x\} \tag{1.65}$$

(the integral part of x) and

$$\{x\} = x - [x] \tag{1.66}$$

(the fractional part of x) the average of f is given by the Stieltjes integral

$$\mathcal{M}_f(x) = \int_1^x f(y)d[y] = \int_1^x f(y)dy - \int_1^x f(y)d\{y\}.$$

Hence

$$\mathcal{M}_f(x) = \int_1^x f(y)dy + \gamma_f + O(f(x)) \tag{1.67}$$

where the constant γ_t is given by

$$\gamma_f = -\int_1^\infty f(y)d\{y\} = f(1) + \int_1^\infty \{y\}df(y).$$

For example, we get

$$\sum_{n\leqslant x}\frac{\log^k n}{n} = \frac{1}{k+1}(\log x)^{k+1} + (-1)^k k!\gamma_k + O\Big(\frac{\log^k x}{x}\Big) \tag{1.68}$$

where γ_k is called the Stieltjes constant. For $k=0$ we get the Euler constant

$$\gamma = 1 - \int_1^\infty \{y\}y^{-2}dy = 0.577215\ldots \tag{1.69}$$

Similarly one can evaluate averages of smooth functions in several variables by repeated applications of the integral approximation method. In this way one can derive Gauss formula for the lattice points in a circle

$$\sum_{n\leqslant x} r(n) = \pi x + O(\sqrt{x}). \tag{1.70}$$

Actually, Gauss' argument was geometrical in nature, he derived (1.70) by packing the circle with a unit square (which amounts to the same thing as the integration). In the same way one obtains

$$\sum_{n\leqslant x} r_k(n) = \rho_k x^{\frac{k}{2}} + O(x^{\frac{k-1}{2}}) \tag{1.71}$$

where $\rho_k = \pi^{k/2}/\Gamma(\frac{k}{2}+1)$ is the volume of the k-dimensional unit ball.

Most interesting arithmetic functions f are not defined on all real numbers in which case it makes no sense to approximate $\mathcal{M}_f(x)$ by the integral (1.67). Yet, if f is given by a convolution of two functions one of which is smooth on real numbers and the other vanishes rapidly on natural numbers, it is still possible to evaluate $\mathcal{M}_f(x)$ by means of integrals and convergent series. Specifically, if $f = g\star h$, where h is monotonic and bounded, we arrange $\mathcal{M}_f(x)$ as

$$\mathcal{M}_f(x) = \sum_{m\leqslant x} g(m)\mathcal{M}_h\Big(\frac{x}{m}\Big),$$

and, after replacing $\mathcal{M}_h(y)$ by $\int_0^y h(t)dt + O(1)$, we get

$$\mathcal{M}_f(x) = \int_0^x \sum_{m\leqslant x} \frac{g(m)}{m} h\Big(\frac{y}{m}\Big)dy + O\Big(\sum_{m\leqslant x} |g(m)|\Big).$$

Estimating the partial sum over $y < m \leqslant x$ trivially we get

$$\mathcal{M}_f(x) = \int_0^x F(y)dy + O(\mathcal{M}_{|g|}(x)) \tag{1.72}$$

where

$$F(y) = \sum_{m\leqslant y} \frac{g(m)}{m} h\Big(\frac{y}{m}\Big). \tag{1.73}$$

If we assume that the series $\sum g(m)m^{-1}$ converges absolutely, then this formula is a slight modification of a theorem of A. Wintner. As an example we apply (1.72) for $f(m) = \varphi(n)n^{-1}$ which is given by (1.36) getting

$$\sum_{n\leqslant x} \frac{\varphi(n)}{n} = \frac{x}{\zeta(2)} + O(\log x). \tag{1.74}$$

For the divisor function $\tau(n)$ the sum $\mathcal{M}_\tau(x)$ has geometric interpretation, namely it represents the number of lattice points $(m,n) \in \mathbb{N}\times\mathbb{N}$ under the hyperbola $mn \leqslant x$. In this case our general formula (1.72) yields

$$\mathcal{M}_\tau(x) = \sum_{n\leqslant x} \tau(n) = \sum_{n\leqslant x} \Big[\frac{x}{n}\Big] = x\log x + O(x)$$

showing that the average value of $\tau(n)$ is $\log n$. In order to improve the error term in the divisor problem Dirichlet applied a simple, yet powerful trick with switching divisors. First by symmetry of the equation $m_1 m_2 = n$ we figure that

$$\mathcal{M}_\tau(x) = 2\sum_{m\leqslant\sqrt{x}} \Big[\frac{x}{m}\Big] - [\sqrt{x}]^2.$$

Here we have fewer terms of summation than in the straightforward arrangement. This trick is called the hyperbola method. Now dropping the integral part symbols we make an error $O(\sqrt{x})$, and it follows by (1.69) that

$$\sum_{n\leqslant x}\tau(n)=x\log x+(2\gamma-1)x+O(\sqrt{x}). \tag{1.75}$$

EXERCISE 2. Prove by Dirichlet's method that

$$\sum_{n\leqslant x}\tau_k(n)=xP_k(\log x)+O(x^{1-\frac{1}{k}})$$

where P_k is a polynomial of degree $k-1$ ($P_k(x)$ is the residue of $\zeta(s)^k x^s s^{-1}$ at $s=1$).

Dirichlet's hyperbola method works nicely for the lattice points in a ball of dimension $k\geqslant 4$. Lagrange proved that every natural number can be represented as the sum of four squares, i.e., $r_4(n)>0$, and Jacobi established the exact formula for the number of representations

$$r_4(n)=8(2+(-1)^n)\sum_{d|n,d\text{ odd}}d.$$

Hence we derive

$$\begin{aligned}\sum_{n\leqslant x}r_4(n)&=8\sum_{m\leqslant x}(2+(-1)^m)\sum_{dm\leqslant x,d\text{ odd}}d\\&=8\sum_{m\leqslant x}(2+(-1)^m)\Big(\frac{x^2}{4m^2}+O(\frac{x}{m})\Big)\\&=2x^2\sum_1^\infty(2+(-1)^m)m^{-2}+O(x\log x).\\&=3\zeta(2)x^2+O(x\log x)=\frac{1}{2}(\pi x)^2+O(x\log x).\end{aligned}$$

This result extends easily for any $k\geqslant 4$ (write r_k as the additive convolution of r_4 and r_{k-4}, apply the above result for r_4 and execute the summation over the remaining $k-4$ squares by integration)

$$\sum_{n\leqslant x}r_k(n)=\frac{(\pi x)^{\frac{k}{2}}}{\Gamma(\frac{k}{2}+1)}+O(x^{\frac{k}{2}-1}\log x). \tag{1.76}$$

Notice that this improves the formula (1.71) which was obtained by the method of packing with a unit square. The exponent $\frac{k}{2}-1$ in (1.76) is best possible because the individual terms of summation can be as large as the error term (apart from $\log x$), indeed for $k=4$ we have $r_4(n)\geqslant 16n$ if n is odd by the Jacobi formula. The only cases of the lattice point problem for a ball which are not yet solved (i.e., the best possible error terms are not yet established) are for the circle ($k=2$) and the sphere ($k=3$). We shall address these fascinating problems on other occasions.

EXERCISE 3. Prove by the hyperbola method that

$$\sum_{n\leqslant x}\tau(n^2+1)=\frac{3}{\pi}x\log x+O(x).$$

1.6. Sums of multiplicative functions.

Throughout f is a multiplicative function. Since f is determined by its values at prime powers it is possible to estimate $\mathcal{M}_f(x)$ in terms of the local sums

$$\sigma_p(f)=\sum_{\nu=0}^{\infty}f(p^\nu)p^{-\nu}. \tag{1.77}$$

First we give simple upper bounds when f is non-negative. The following estimates need no explanation

$$M_f(x)\leqslant x\sum_{n\leqslant x}f(n)n^{-1}\leqslant x\prod_{p\leqslant x}\sigma_p(f). \tag{1.78}$$

One can do better if f is non-decreasing on prime powers. In this case $h=\mu\star f$ is non-negative, indeed $h(p^\nu)=f(p^\nu)-f(p^{\nu-1})\geqslant 0$ if $\nu\geqslant 1$. Writing $f=1\star h$ we get

$$\begin{aligned}\mathcal{M}_f(x)&=\sum_{m\leqslant x}h(m)\Big[\frac{x}{m}\Big]\leqslant x\sum_{m\leqslant x}h(m)m^{-1}\\&\leqslant x\prod_{p\leqslant x}\sigma_p(h)=x\prod_{p\leqslant x}\sigma_p(f)\Big(1-\frac{1}{p}\Big).\end{aligned} \tag{1.79}$$

Here is a crude but useful application of (1.79) for $f(m)=\tau_k(m)^\ell$. We have $f(p)=k^\ell$ and

$$\sigma_f(p)\leqslant\Big(1+\frac{1}{p}\Big)^{k^\ell}\Big(1+\frac{1}{p^2}\Big)^c$$

where $c=c(k,\ell)$ is a positive constant. We have also the elementary bound

$$\prod_{p\leqslant x}\Big(1+\frac{1}{p}\Big)\ll\log x.$$

for any $x\geqslant 2$ (see (2.15)). Hence (1.79) yields

$$\sum_{n\leqslant x}\tau_k(n)^\ell\ll x(\log x)^{k^\ell-1} \tag{1.80}$$

where the implied constant depends on k,ℓ. This is a crude bound, but of the correct order of magnitude. Notice that it implies that

$$\tau_k(n)\ll n^\varepsilon \tag{1.81}$$

for any $\varepsilon>0$, the implied constant depending on ε and k. We shall be using these bounds for the divisor function $\tau_k(n)$ often without mention.

If one takes the average of $f(n)$ over square free numbers, or what amounts to the same thing one assumes f is supported on squarefree numbers, then the above arguments yield

$$\mathcal{M}_f(x) \leqslant x \prod_{p \leqslant x} \left(1 + \frac{f(p)-1}{p}\right) \tag{1.82}$$

provided $f(p) \geqslant 1$ for all p. We may require $f(p) \geqslant 1$ to hold true for primes only in a certain set. Then the result is

$$\mathcal{M}_f(x) \leqslant x \prod_{\substack{p \leqslant x \\ f(p)<1}} \left(1 + \frac{f(p)}{p}\right) \prod_{\substack{p \leqslant x \\ f(p)>1}} \left(1 + \frac{f(p)-1}{p}\right). \tag{1.83}$$

Here is another elementary approach. We only assume that f is non-negative and completely sub-multiplicative, i.e., $f(mn) \leqslant f(m)f(n)$ for all $m, n \geqslant 1$. Moreover, suppose that $f(p)$ is bounded on average,

$$\sum_{m \leqslant x} f(m)\Lambda(m) \leqslant cx \tag{1.84}$$

for any $x \geqslant 2$, where $c \geqslant 1$ is a constant (see the Tchebyshev bound (2.12)). Then we get

$$\sum_{n \leqslant x} f(n) \log n = \sum_{nm \leqslant x} f(nm)\Lambda(m) \leqslant cx \sum_{n \leqslant x} f(n) n^{-1}.$$

Hence by partial summation

$$\mathcal{M}_f(x) \leqslant \frac{3cx}{\log x} \sum_{n \leqslant x} \frac{f(n)}{n} \leqslant \frac{3cdx}{\log x} \prod_{p \leqslant x} \left(1 + \frac{f(p)}{p}\right) \tag{1.85}$$

where

$$d = \sum_{n \leqslant x} \left(\frac{f(n)}{n}\right)^2.$$

To illustrate this result we take the multiplicative function $f(n) = b(n)$ which is the characteristic function of numbers representable as a sum of two squares. We have $f(p) = 0$ if $p \equiv -1 \pmod 4$ and $f(p) = 1$ otherwise. In this case (see (2.30))

$$\prod_{p \leqslant x} \left(1 + \frac{f(p)}{p}\right) \ll (\log x)^{\frac{1}{2}}$$

for any $x \geqslant 2$. Hence one deduces by (1.85) that

$$B(x) = |\{n \leqslant x : n = a^2 + b^2\}| \ll x(\log x)^{-\frac{1}{2}}. \tag{1.86}$$

Here every number which admits a representation as the sum of two squares is counted once. For this reason the upper bound (1.86) is slightly smaller than the area of the circle $a^2 + b^2 \leqslant x$ (see (1.70)). Using analytic methods E. Landau [**La1**] established the asymptotic formula

$$B(x) = \sum_{n \leqslant x} b(n) \sim Cx(\log x)^{-\frac{1}{2}}, \tag{1.87}$$

as $x \to \infty$, where C is a positive constant. Landau's formula also follows by an elementary method with C given by (1.102).

Very differently, still elementarily, the formula (1.87) is proved in [**I8**] by applying a half-dimensional sieve. The sieve method has the advantage of producing results of great uniformity. For example, one gets

$$B(x+y) - B(x) < (1+\varepsilon)B(y)$$

for any $\varepsilon > 0$ provided y is sufficiently large in terms of ε only, where x is any positive number. If x is very large compared to y, then analytic arguments produce rather poor results for $B(x+y) - B(x)$.

Next we present an elementary method which yields an asymptotic formula for $\mathcal{M}_f(x)$ for a large class of multiplicative functions. This requires some regularity in the distribution of f at prime powers. Letting Λ_f be the von Mangoldt function associated with f (see (1.50)) we assume that

$$\sum_{n \leqslant x} \Lambda_f(n) = \kappa \log x + O(1) \tag{1.88}$$

where κ is a constant. This condition means $f(p)$ is about κp^{-1} on average whereas it was κ previously. For many multiplicative functions in practice (1.88) can be established by elementary methods (this condition is an analogue of the Mertens formula (2.14), and it is weaker than the Prime Number Theorem). With some applications in mind we allow $f(p)$ to be slightly negative. Precisely the method works if (1.88) holds with $\kappa > -\frac{1}{2}$. In addition to this we need a crude estimate

$$\sum_{n \leqslant x} |f(n)| \ll (\log x)^{|\kappa|}. \tag{1.89}$$

The method begins with evaluation of the logarithmically smoothed sum

$$\sum_{n \leqslant x} f(n) \log \frac{x}{n} = \int_1^x \mathcal{M}_f(y) y^{-1} dy. \tag{1.90}$$

Since $f \cdot L = f \star \Lambda_f$ we derive by (1.88) that

$$\begin{aligned}\sum_{n \leqslant x} f(n) \log n &= \sum_{d \leqslant x} f(d) \sum_{m \leqslant x/d} \Lambda_f(m) \\ &= \sum_{d \leqslant x} f(d) \Big(\kappa \log \frac{x}{d} + O(1)\Big).\end{aligned}$$

Estimating the sum of error terms by means of (1.89) we get

$$(\kappa + 1) \sum_{n \leqslant x} f(n) \log n = \kappa \mathcal{M}_f(x) \log x + O((\log x)^{|\kappa|}).$$

Inserting this into (1.90) we show that the difference

$$\Delta(x) = \mathcal{M}_f(x) \log x - (\kappa + 1) \int_2^x \mathcal{M}_f(y) y^{-1} dy \tag{1.91}$$

is relatively small, namely

$$\Delta(x) \ll (\log x)^{|\kappa|}. \tag{1.92}$$

Next we divide (1.91) by $x(\log x)^{-\kappa-2}$ and integrate getting

$$\int_2^x \Delta(y)y^{-1}(\log y)^{-\kappa-2}dy = \int_2^x \mathcal{M}_f(y)y^{-1}(\log y)^{-\kappa-1}dy$$
$$-(\kappa+1)\int_2^x y^{-1}(\log y)^{-\kappa-2}\Big(\int_2^y \mathcal{M}_f(u)u^{-1}du\Big)dy.$$

Changing the order of integration in the multiple integral we get

$$\int_2^x \mathcal{M}_f(y)y^{-1}(\log y)^{-\kappa-1}dy + \int_2^x [(\log x)^{-\kappa-1}y - (\log u)^{-\kappa-1}]\mathcal{M}_f(u)u^{-1}du.$$

The first integral cancels out with the second part of the last one, so we are left with

$$\int_2^x \Delta(y)y^{-1}(\log y)^{-\kappa-2}dy = (\log x)^{-\kappa-1}\int_2^x \mathcal{M}_f(u)u^{-1}du.$$

Combining this with (1.91) we arrive at the following identity:

$$\mathcal{M}_f(x) = (\log x)^\kappa \int_2^x -\Delta(y)d(\log y)^{-\kappa-1} + \Delta(x)(\log x)^{-1}.$$

Since the above integral converges absolutely, we can extend its range to infinity getting the identity

$$\mathcal{M}_f(x) = \{c_f + r_f(x)\}(\log x)^\kappa \tag{1.93}$$

where c_f is a constant given by the definite integral

$$c_f = -\int_2^\infty \Delta(y)d(\log y)^{-\kappa-1} \tag{1.94}$$

and $r_f(x)$ is a function given by

$$r_f(x) = \int_x^\infty (\Delta(y) - \Delta(x))d(\log y)^{-\kappa-1}. \tag{1.95}$$

This tends to zero as $x \to \infty$, indeed by (1.92)

$$r_f(x) \ll (\log x)^{|\kappa|-\kappa-1}. \tag{1.96}$$

However, the constant (1.94) does not look natural.

Now having established the asymptotic formula (1.93) with the error term (1.96) we shall use these results to derive another expression for c_f by an appeal to the zeta function for f,

$$D_f(s) = \sum_1^\infty f(n)n^{-s}.$$

The series converges absolutely for $s > 0$ by virtue of (1.89). We compute by partial summation that

$$D_f(x) = \int_1^\infty y^{-s}d\mathcal{M}_f(y) = -\int_1^\infty \mathcal{M}_f(y)dy^{-s}$$
$$= -\int_0^\infty \mathcal{M}_f(e^t)de^{-st} = -\int_0^\infty \{c_f + O(t^{-\varepsilon})\}t^\kappa de^{-st}$$
$$= \{c_f + O(s^\varepsilon)\}s^{-\kappa}\Gamma(\kappa+1)$$

as $s \to 0+$. Comparing this with the κ-th power of the Riemann zeta function we get

$$\zeta(s+1)^{-\kappa} D_f(s) \sim c_f \Gamma(\kappa+1).$$

On the other hand, we have the Euler product

$$\zeta(s+1)^{-\kappa} D_f(s) = \prod_p (1-p^{-s-1})^\kappa \left(\sum_{\nu=0}^{\infty} f(p^\nu) p^{-\nu s} \right)$$

which converges absolutely for $s > 0$ and it has limit as $s \to 0$ by virtue of (1.88). Hence the constant c_f is equal to

$$c_f = \frac{1}{\Gamma(\kappa+1)} \prod_p \Big(1 - \frac{1}{p}\Big)^\kappa (1 + f(p) + f(p^2) + \cdots). \tag{1.97}$$

We have established the following formula (if $\kappa \geqslant 0$, this is due to E. Wirsing [**Wi1**]).

THEOREM 1.1. *Suppose f is a multiplicative function which satisfies (1.88) and (1.89) with $\kappa > -\frac{1}{2}$. Then*

$$\sum_{n \leqslant x} f(n) = c_f (\log x)^\kappa + O((\log x)^{|\kappa|-1}) \tag{1.98}$$

where c_f is a constant given by (1.97).

We illustrate the applicability of this formula by solving a simple sieve problem. Let $\mathcal{P}$ be a set of primes. Our question is how many numbers $n \leqslant x$ have no prime divisors in $\mathcal{P}$? Let f be the completely multiplicative function with

$$f(p) = \begin{cases} 1 & \text{if } p \in \mathcal{P}, \\ 0 & \text{otherwise.} \end{cases}$$

Then

$$g(n) = \prod_{p|n} (1 - f(p))$$

is the characteristic function of numbers in question. Therefore we are asking to evaluate

$$S(\mathcal{P}, x) = \sum_{n \leqslant x} g(n).$$

Opening the convolution $g = \mu f \star 1$ we obtain

$$S(\mathcal{P}, x) = \sum_{d \leqslant x} \mu(d) f(d) \Big[\frac{x}{d}\Big] = x \sum_{d \leqslant x} \mu(d) \frac{f(d)}{d} + O(\mathcal{M}_f(x)).$$

To proceed further we are going to assume that the proportion of primes in $\mathcal{P}$ is less than the proportion of those not in $\mathcal{P}$. More precisely we need the condition

$$\sum_{p \leqslant x} \frac{f(p)}{p} \log p = \kappa \log x + O(1) \tag{1.99}$$

with a constant $\kappa < \frac{1}{2}$. Then we derive $\mathcal{M}_f(x) \ll x(\log x)^{\kappa-1}$ by (1.85) and

$$\sum_{d \leqslant x} \mu(d) \frac{f(d)}{d} = c(\mathcal{P})(\log x)^{-\kappa} + O((\log x)^{\kappa-1})$$

by (1.98), where $c(\mathcal{P})$ is a constant depending on the set $\mathcal{P}$,

$$c(\mathcal{P}) = \frac{1}{\Gamma(1-\kappa)} \prod_p \Big(1 - \frac{f(p)}{p}\Big)\Big(1 - \frac{1}{p}\Big)^{-\kappa}. \tag{1.100}$$

Gathering the above results we conclude

COROLLARY 1.2. *Let $\mathcal{P}$ be a set of primes of density $\kappa < \frac{1}{2}$ (that is (1.99) holds). Then the number of integers $1 \leqslant n \leqslant x$ having no prime divisors in $\mathcal{P}$ satisfies*

$$S(\mathcal{P}, x) = c(\mathcal{P}) x (\log x)^{-\kappa} \{1 + O((\log x)^{2\kappa - 1})\} \tag{1.101}$$

where $c(\mathcal{P})$ is the constant given by (1.100).

EXERCISE 4. Apply (1.98) to derive Landau's formula (1.87) with the constant

$$C = \frac{1}{\sqrt{2}} \prod_{p \equiv -1 (\mathrm{mod}\ 4)} \Big(1 - \frac{1}{p^2}\Big)^{-\frac{1}{2}}. \tag{1.102}$$

1.7. Distribution of additive functions.

Throughout f is an additive function, so f is determined by its values at prime powers, precisely

$$f(n) = \sum_{p^\alpha \| n} f(p^\alpha).$$

Hence the average of f is given by

$$\mathcal{M}_f(x) = \sum_{p^\alpha \leqslant x} f(p^\alpha) \left(\left[\frac{x}{p^\alpha}\right] - \left[\frac{x}{p^{\alpha+1}}\right]\right).$$

Dropping the integral part symbols we arrive at

$$\mathcal{M}_f(x) = xE(x) + O(x^{\frac{1}{2}} D(x)), \tag{1.103}$$

where

$$E(x) = \sum_{p^\alpha \leqslant x} f(p^\alpha) p^{-\alpha} (1 - p^{-1}) \tag{1.104}$$

and

$$D^2(x) = \sum_{p^\alpha \leqslant x} |f(p^\alpha)|^2 p^{-\alpha}. \tag{1.105}$$

Actually the error term in (1.103) is $\sum |f(p^\alpha)|$ which we estimated by $x^{\frac{1}{2}} D(x)$ by applying Cauchy's inequality and that $p^\alpha \leqslant x$. The result reveals that $E(x)$ is approximately equal to the mean value of f.

Similarly we evaluate the average of f^2. We have

$$\mathcal{M}_{f^2}(x) = \sum_{p^\alpha} \sum_{q^\beta} f(p^\alpha) f(q^\beta) \sum_{\substack{n \leqslant x \\ p^\alpha \| n, q^\beta \| n}} 1$$

where p, q run over primes. If $p \neq q$, the inner sum is equal to

$$\Big[\frac{x}{p^\alpha q^\beta}\Big] - \Big[\frac{x}{p^{\alpha+1} q^\beta}\Big] - \Big[\frac{x}{p^\alpha q^{\beta+1}}\Big] + \Big[\frac{x}{p^{\alpha+1} q^{\beta+1}}\Big] = \frac{x}{p^\alpha q^\beta}\Big(1-\frac{1}{p}\Big)\Big(1-\frac{1}{q}\Big) + O(1),$$

and for $p = q$ we must have $\alpha = \beta$ getting

$$\Big[\frac{x}{p^\alpha}\Big] - \Big[\frac{x}{p^{\alpha+1}}\Big] = \frac{x}{p^\alpha}\Big(1-\frac{1}{p}\Big) + O(1).$$

Hence we derive that

$$\begin{aligned}\mathcal{M}_f^2(x) = x \sum_{\substack{p^\alpha, q^\beta \leqslant x \\ p \neq q}}\sum f(p^\alpha) f(q^\beta) p^{-\alpha} q^{-\beta} (1-p^{-1})(1-q^{-1}) \\ + x \sum_{p^\alpha \leqslant x} f^2(p^\alpha) p^{-\alpha}(1-p^{-1}) + O\Big(\sum_{p^\alpha, q^\beta \leqslant x}\sum |f(p^\alpha) f(p^\beta)|\Big).\end{aligned}$$

Here we drop the condition $p \neq q$ and estimate the additional terms with $p = q$ using $|f(p^\alpha)f(p^\beta)| \leqslant |f(p^\alpha)|^2 + |f(p^\beta)|^2$. We arrive at

$$\mathcal{M}_{f^2}(x) = xE^2(x) + O(xD^2(x)). \tag{1.106}$$

This result says that $E^2(x)$ is approximately equal to the mean value of f^2.

Both approximations (1.103) and (1.106) suggest that the individual values $f(n)$ for $n \leqslant x$ should be reasonably well approximated by $E(x)$. Indeed this is true in the following sense:

THEOREM 1.3. *For any additive function $f : \mathbb{N} \to \mathbb{C}$ we have*

$$\sum_{n \leqslant x} |f(n) - E(x)|^2 \leqslant cxD^2(x) \tag{1.107}$$

where c is an absolute constant.

PROOF. We can assume that f is real since the real and the imaginary parts of f are additive and they can be treated separately. We can also assume that x is a positive integer. First we evaluate the variance

$$V(x) = \sum_{n \leqslant x} (f(n) - x^{-1}\mathcal{M}_f(x))^2 = \mathcal{M}_{f^2}(x) - x^{-1}\mathcal{M}_f^2(x).$$

By (1.103) and (1.106) we get

$$V(x) \ll x^{\frac{1}{2}} |E(x)| D(x) + xD^2(x) \ll xD^2(x).$$

Then replacing $x^{-1}\mathcal{M}_f(x)$ by $E(x)$ we make an admissible error, and (1.107) is established. □

Theorem 1.3 is due to P. Túran [**Tu1**] and J. Kubilius [**Kub1**]. The expected value $E(x)$, if one prefers, can be reduced to

$$A(x) = \sum_{p \leqslant x} f(p) p^{-1} \tag{1.108}$$

at the expense of the constant c in (1.107). Our arguments above yield $c = 4$ if x is sufficiently large, however, J. Kubilius [**Kub2**] showed (after having received suggestions from H.L. Montgomery) that (1.107) holds with $c = c(x) \sim \frac{3}{2}$ as $x \to \infty$. He and others also showed that the $\frac{3}{2}$ cannot be replaced by a smaller constant (cf. A. Hildebrand [**Hi1**]).

As an example we choose the additive function $\omega(n)$ which counts the number of distinct prime divisors of n. Therefore $\omega(p^\alpha) = 1$, $E(x) = \log\log x + O(1)$ and $D^2(x) = \log\log x + O(1)$ by (2.15), so the Túran-Kubilius theorem gives

$$\sum_{n \leqslant x} (\omega(n) - \log\log x)^2 \ll x \log\log x. \tag{1.109}$$

Hence it follows that $\omega(n) \sim \log\log n$ for almost all n (a theorem of Hardy and Ramanujan). Precisely we have

COROLLARY 1.4. *For any $x \geqslant 3$ and $z \geqslant 1$,*

$$|\{n \leqslant x : |\omega(n) - \log\log x| > z(\log\log x)^{\frac{1}{2}}\}| \ll z^{-2} x \tag{1.110}$$

where the implied constant is absolute.

Furthermore one can ask if a given additive function f can be renormalized by $A(x)$ and some $B(x)$ so that the frequencies

$$\nu(x; z) = \frac{1}{x} \left| \left\{ n \leqslant x : \frac{f(n) - A(x)}{B(x)} \leqslant z \right\} \right| \tag{1.111}$$

have a limit for any $z \in \mathbb{R}$ as x tends to ∞. This question is addressed in the probabilistic theory of numbers. We shall only give a glimpse of central results. The very first one was established by P. Erdös and M. Kac [**EK**].

THEOREM 1.5. *Let $f(n)$ be an additive function with $-1 \leqslant f(p^\alpha) = f(p) \leqslant 1$ such that*

$$B(x) = \sum_{p \leqslant x} f^2(p) p^{-1} \to \infty \tag{1.112}$$

as $x \to \infty$. Then $\nu(x; z) \sim G(z)$ for any $z \in \mathbb{R}$, where $G(z)$ is the Gaussian distribution function

$$G(z) = \frac{1}{\sqrt{2\pi}} \int_{-\infty}^{z} e^{-t^2/2} dt. \tag{1.113}$$

In particular, it follows from the Erdös-Kac theorem that for any real z

$$|\{n \leqslant x : \omega(n) - \log\log x \leqslant z(\log\log x)^{\frac{1}{2}}\}| \sim G(z) x$$

as x tends to ∞. Many other arithmetic functions (not necessarily additive) follow the Gaussian limiting distribution law. For example, H. Halberstam [**Hal**] proved that the number of prime divisors of shifted primes satisfy

$$|\{p \leqslant x : \omega(p+1) - \log\log x \leqslant z(\log\log x)^{\frac{1}{2}}\}| \sim G(z)\pi(x) \tag{1.114}$$

for any fixed $z \in \mathbb{R}$ as $x \to \infty$.

CHAPTER 2

ELEMENTARY THEORY OF PRIME NUMBERS

2.1. The Prime Number Theorem.

It was observed nearly 200 years ago by Legendre and Gauss (independently) that the density of primes $p \leqslant x$ is $(\log x)^{-1}$, precisely they postulated the following

PRIME NUMBER THEOREM. *Let $\pi(x)$ denote the number of primes $p \leqslant x$. As $x \to \infty$, we have*

$$\pi(x) \sim \frac{x}{\log x}. \tag{2.1}$$

Gauss observed that an even better approximation to $\pi(x)$ is given by the singular integral

$$\mathrm{Li}(x) = \int_0^x \frac{dy}{\log y} = \int_1^x \Big(1 - \frac{1}{y}\Big)\frac{dy}{\log y} + \log\log x + \gamma. \tag{2.2}$$

For $x > 1$ this satisfies the asymptotic expansion

$$\mathrm{Li}(x) = \frac{x}{\log x}\left\{ \sum_{0 \leqslant \ell < m} \ell!(\log x)^{-\ell} + O((\log x)^{-m}) \right\}. \tag{2.3}$$

Some people use $\ell i(x) = \mathrm{Li}(x) - \mathrm{Li}(2)$ in place of $\mathrm{Li}(x)$ for $x \geqslant 2$. Later, Tchebyshev realized it is simpler to count primes p with the weight $\log p$, so he investigated the sum

$$\theta(x) = \sum_{p \leqslant x} \log p$$

rather than $\pi(x)$. It is still more convenient to evaluate the average of the von Mangoldt function

$$\psi(x) = \sum_{n \leqslant x} \Lambda(n). \tag{2.4}$$

Note that (2.1) is equivalent to each of the asymptotic formulas

$$\theta(x) \sim x, \qquad \psi(x) \sim x \tag{2.5}$$

by partial summation and trivial estimation of the contribution of $n = p^\alpha \leqslant x$ with $\alpha \geqslant 2$.

Although Λ is given by the convolution (1.40) the formula (1.72) falls short of yielding (2.5) because the error term exceeds the main term. Applying Dirichlet's trick of switching divisors (the hyperbola method) one can reduce the number of error terms, but it does not complete the work. This only leads to another problem of estimating sums of the Möbius function. Indeed, one can show by the hyperbola method that the Prime Number Theorem is equivalent to the estimate

$$M(x) = \sum_{m\leqslant x} \mu(m) = o(x), \quad \text{as } x \to \infty. \tag{2.6}$$

To derive (2.5) from (2.6) we consider

$$\Delta(x) = \sum_{n\leqslant x} (\log n - \tau(n) + 2\gamma).$$

Subtracting (1.75) from (1.63) for $k = 1$ we get

$$\Delta(x) \ll \sqrt{x}. \tag{2.7}$$

Now applying (1.42) we get

$$\begin{aligned} \psi(x) - x + 2\gamma &= \sum\sum_{dk\leqslant x} \mu(d)(\log k - \tau(k) + 2\gamma) \\ &= \sum_{k\leqslant K} (\log k - \tau(k) + 2\gamma) M\Big(\frac{x}{k}\Big) + \sum_{d\leqslant xK^{-1}} \mu(d)\Big(\Delta(\frac{x}{d}) - \Delta(K)\Big) \end{aligned} \tag{2.8}$$

for any $1 \leqslant K \leqslant x$. The last sum is $O(xK^{-1/2})$ by virtue of (2.7) while for any fixed K the preceding sum is $o(x)$ by the hypothesis (2.6). Since K can be arbitrarily large, this shows that (2.6) implies (2.5). To establish the converse we use the identity

$$\sum_{n\leqslant x} \mu(n) \log \frac{x}{n} = M(x) \log x + \sum\sum_{mn\leqslant x} \mu(m)\Lambda(n)$$

which follows from $\Big(\dfrac{1}{\zeta}\Big)' = -\dfrac{\zeta'}{\zeta}\dfrac{1}{\zeta}$. The left side is trivially $O(x)$ (see (1.64) for $k = 1$), so

$$M(x) \log x = -\sum_{m\leqslant x} \mu(m)\psi\Big(\frac{x}{m}\Big) + O(x).$$

Applying (2.18) we arrive at

$$M(x) \log x = \sum_{m\leqslant x} \mu(m)\Big(\frac{x}{m} - \psi\Big(\frac{x}{m}\Big)\Big) + O(x).$$

Now using the hypothesis (2.5) this equation implies (2.6).

2.2. Tchebyshev method.

The first significant attempt for proving the Prime Number Theorem was made by Tchebyshev in 1848–1852. Retrospectively one may say that his approach is based essentially on the product formula

$$N = \prod_p p^{\nu_p(N)}$$

meaning that the archimedean valuation equals the product of p-adic valuations. This obvious formula is particularly fruitful for the factorial numbers $N = 1\cdot 2 \cdots n$.

Like Tchebyshev we present the arguments in a more friendly analytic format. First we have

$$L(x) = \sum_{n \leqslant x} \log n = x \log x - x + O(\log x). \tag{2.9}$$

On the other hand, using (1.41) we get

$$L(x) = \sum_{d \leqslant x} \Lambda(d) \left[\frac{x}{d}\right] = \sum_{m \leqslant x} \psi\Big(\frac{x}{m}\Big). \tag{2.10}$$

Hence evaluating $L(x) - 2L(\frac{x}{2})$ in two ways one gets

$$\psi(x) - \psi\Big(\frac{x}{2}\Big) + \psi\Big(\frac{x}{3}\Big) - \psi\Big(\frac{x}{4}\Big) + \cdots = x \log 2 + O(\log x). \tag{2.11}$$

Using the monotonicity of $\psi(y)$ one derives from (2.11) the first Tchebyshev estimates

$$x \log 2 + O(\log x) < \psi(x) < x \log 4 + O(\log x). \tag{2.12}$$

Tchebyshev improved these estimates several times using more involved linear combinations of $L(\frac{x}{m})$. For example let us consider $\psi_f(x) = L(x) - L(\frac{x}{2}) - L(\frac{x}{3}) - L(\frac{x}{5}) + L(\frac{x}{30})$. By (2.9)

$$\psi_f(x) = \alpha x + O(\log x)$$

where $\alpha = \frac{1}{2}\log 2 + \frac{1}{3}\log 3 + \frac{1}{5}\log 5 - \frac{1}{30}\log 30 = 0.9212\ldots$ On the other hand,

$$\psi_f(x) = \sum_{n \leqslant x} \Lambda(n) f\Big(\frac{x}{n}\Big)$$

where

$$f(x) = [x] - \left[\frac{x}{2}\right] - \left[\frac{x}{3}\right] - \left[\frac{x}{5}\right] + \left[\frac{x}{30}\right].$$

Since $1 - \frac{1}{2} - \frac{1}{3} - \frac{1}{5} + \frac{1}{30} = 0$, the function $f(x)$ is periodic of period 30. One can check that $f(x)$ takes only two values $0, 1$ and $f(x) = 1$ if $1 \leqslant x < 6$. Therefore

$$\psi_f(x) \leqslant \psi(x) \leqslant \psi_f(x) + \psi\Big(\frac{x}{6}\Big).$$

Hence one derives

$$\alpha x + O(\log x) < \psi(x) < \tfrac{6}{5}\alpha x + O(\log x). \tag{2.13}$$

Since the difference between the upper and the lower bounds in (2.13) is quite small, one can deduce from (2.13) that any dyadic interval $[x, 2x]$ contains primes provided x is sufficiently large (actually Tchebyshev proved this for all $x \geqslant 1$, which was postulated by Bertrand). Tchebyshev failed to prove the Prime Number Theorem, however, he succeeded to show that if $\lim_{x\to\infty} \psi(x)/x$ exists, then this limit is equal to one (this follows instantly from (2.11) because $1 - \frac{1}{2} + \frac{1}{3} - \frac{1}{4} + \cdots = \log 2$).

Interesting results were derived from Tchebyshev works by Mertens. First approximating $[x/d]$ in (2.10) by x/d one gets

$$\sum_{n \leqslant x} \frac{\Lambda(n)}{n} = \log x + O(1). \tag{2.14}$$

Next applying partial summation one derives from (2.14) that

$$\sum_{p\leqslant x}\frac{1}{p}=\log\log x+\beta+O\Big(\frac{1}{\log x}\Big) \tag{2.15}$$

where β is a constant. Finally using $\log(1-p^{-1})=-p^{-1}+O(p^{-2})$, we have the Mertens formula

$$\prod_{p\leqslant x}\Big(1-\frac{1}{p}\Big)=\frac{e^{-\gamma}}{\log x}\Big\{1+O\Big(\frac{1}{\log x}\Big)\Big\}. \tag{2.16}$$

Mertens showed that the exponent γ in (2.16) coincides with the Euler constant (1.69).

Similarly, using $\delta=1\star\mu$ in place of $L=1\star\Lambda$, one derives a handful of results for the Möbius function. First for any $x\geqslant 1$ one gets

$$\sum_{m\leqslant x}\mu(m)\Big[\frac{x}{m}\Big]=\sum_{d\leqslant x}M\Big(\frac{x}{d}\Big)=1 \tag{2.17}$$

in place of (2.10). Hence approximating $[x/m]$ by x/m one deduces that

$$\Big|\sum_{m\leqslant x}\frac{\mu(m)}{m}\Big|\leqslant 1. \tag{2.18}$$

2.3. Primes in arithmetic progressions.

Every natural number $n\equiv -1(\mathrm{mod}\ 3)$ has a prime divisor $p\equiv -1\ (\mathrm{mod}\ 3)$. Hence there are infinitely many primes $p\equiv -1(\mathrm{mod}\ 3)$. In the same way one shows that for $q=4,5$ there are infinitely many primes $p\equiv -1(\mathrm{mod}\ q)$.

Let q be any integer >1. Consider the cyclotomic polynomial of degree $\varphi(q)$,

$$\Phi_q(x)=\prod_{\substack{a(\mathrm{mod}\ q)\\(a,q)=1}}\Big(x-e\Big(\frac{a}{q}\Big)\Big)\in\mathbb{Z}[x].$$

One can show (quite easily by using elementary properties of the multiplicative group $(\mathbb{Z}/q\mathbb{Z})^\star$) that every prime divisor of $\Phi_q(n)$ coprime with q satisfies $p\equiv 1$ (mod q). If q is prime this property is very simple. Indeed in this case we have

$$\Phi_q(x)=\frac{x^q-1}{x-1}=1+x+\ldots+x^{q-1}.$$

Hence if p divides $\Phi_q(x)$, then $x^q\equiv 1\ (\mathrm{mod}\ p)$, and comparing with the little Fermat theorem $x^{p-1}\equiv 1\ (\mathrm{mod}\ p)$ we deduce that q divides $p-1$. Now arguing like Euclid we can deduce that there are infinitely many primes $p\equiv 1\ (\mathrm{mod}\ q)$.

Naturally every class $a(\mathrm{mod}\ q)$ with $(a,q)=1$ should have infinitely many primes. It was for the proof of this theorem that L. Dirichlet (1837) invented multiplicative characters $\chi(\mathrm{mod}\ q)$ (see Chapter 3). To every such character he associated the Dirichlet series

$$L(s,\chi)=\sum_{1}^{\infty}\chi(n)n^{-s}. \tag{2.19}$$

As in the case of the zeta function the Dirichlet L-function has the Euler product

$$L(s,\chi)=\prod_p(1-\chi(p)p^{-s})^{-1} \tag{2.20}$$

which converges absolutely for $s>1$. Hence $L(s,\chi)$ does not vanish if $s>1$. Taking logarithms we get

$$\log L(s,\chi)=\sum_p\chi(p)p^{-s}+O(1). \tag{2.21}$$

Hence by the orthogonality of characters (see Section 3.2)

$$\sum_{p\equiv a(\mathrm{mod}\ q)}p^{-s}=\frac{1}{\varphi(q)}\sum_{\chi(\mathrm{mod}\ q)}\overline{\chi}(a)\log L(s,\chi)+O(1), \tag{2.22}$$

where the error term $O(1)$ is bounded in terms of s uniformly for $s>1$. For the principal character we can relate $L(s,\chi_0)$ to the zeta function

$$L(s,\chi_0)=\zeta(s)\prod_{p|q}(1-p^{-s}).$$

Since

$$\zeta(s)=\sum_1^\infty n^{-s}=\int_1^\infty x^{-s}dx+O(1)=\frac{1}{s-1}+O(1),$$

we infer that the principal character contributes to (2.22)

$$\frac{1}{\varphi(q)}\log L(s,\chi_0)=\frac{1}{\varphi(q)}\log\frac{1}{s-1}+O(1). \tag{2.23}$$

Next we consider $\chi\neq\chi_0$. Since χ is periodic of period q and the sums over any complete set of residue classes vanish, it follows that the character sum over any segment is bounded, precisely

$$\Big|\sum_{x<n\leqslant y}\chi(n)\Big|<q.$$

Therefore the series (2.19) converges for $s>0$ and $|L(s,\chi)|<q$, by partial summation. If we know that

$$L(1,\chi)\neq 0, \tag{2.24}$$

then $\log L(s,\chi)$ has a finite limit as $s\to 1^+$. Hence we can deduce that for $s>1$,

$$\sum_{p\equiv a(\mathrm{mod}\ q)}p^{-s}=\frac{1}{\varphi(q)}\log\frac{1}{s-1}+O(1). \tag{2.25}$$

In this way Dirichlet proved there are infinitely many primes in any primitive residue class. Actually (2.25) shows that prime numbers are equidistributed (in a certain sense) among the primitive residue classes $a(\mathrm{mod}\ q)$.

It remains to show the non-vanishing of $L(1,\chi)$ for every $\chi\neq\chi_0$, which is the heart of the matter. We do not follow exactly the original lines of Dirichlet but rather mix his ideas with more familiar transformations of Tchebyshev. We begin with

$$\begin{aligned}\sum_{n\leqslant x}\chi(n)n^{-1}\log n &= \sum_{n\leqslant x}\chi(n)n^{-1}\sum_{d|n}\Lambda(d)\\ &= \sum_{d\leqslant x}\chi(d)\Lambda(d)d^{-1}\sum_{m\leqslant\frac{x}{d}}\chi(m)m^{-1}\\ &= \sum_{d\leqslant x}\chi(d)\Lambda(d)d^{-1}\Big(L(1,\chi)+O(d/x)\Big)\\ &= L(1,\chi)\sum_{d\leqslant x}\chi(d)\Lambda(d)d^{-1}+O(1)\end{aligned}$$

where the error term $O(1)$ is bounded in terms of x (but not of q). The left side is also bounded (by Abel's criterion) because the average of $\chi(n)$ over any segment is bounded. Therefore we proved that

$$\sum_{d\leqslant x}\chi(d)\frac{\Lambda(d)}{d}\ll 1 \tag{2.26}$$

provided $L(1,\chi)\neq 0$. Similarly we obtain

$$\begin{aligned}\log x+\sum_{n\leqslant x}\chi(n)\frac{\Lambda(n)}{n} &= \sum_{n\leqslant x}\frac{\chi(n)}{n}\sum_{d|n}\mu(d)\log\frac{x}{d}\\ &= \sum_{d\leqslant x}\mu(d)\frac{\chi(d)}{d}\left(\log\frac{x}{d}\right)\sum_{m\leqslant\frac{x}{d}}\frac{\chi(m)}{m}\\ &= L(1,\chi)\sum_{d\leqslant x}\mu(d)\frac{\chi(d)}{d}\log\frac{x}{d}+O(1).\end{aligned}$$

Hence, if $L(1,\chi)=0$, we deduce that

$$\sum_{n\leqslant x}\chi(n)\frac{\Lambda(n)}{n}=-\log x+O(1). \tag{2.27}$$

In view of the Mertens formula (2.14) the above conditional formula can be written as

$$\sum_{p\leqslant x}(1+\chi(p))\frac{\log p}{p}\ll 1.$$

This shows that $\chi(p)=-1$ for almost all primes. In other words, if $L(1,\chi)$ vanishes, then the character $\chi(n)$ on squarefree numbers mimics the Möbius function $\mu(n)$, which is not likely to be true (both functions are multiplicative but $\mu(n)$ is not periodic). In any case, we have shown rigorously that

$$\sum_{n\leqslant x}\chi(n)\frac{\Lambda(n)}{n}=\delta_\chi\log x+O(1) \tag{2.28}$$

where $\delta_\chi = -1, 0$ according to $L(1,\chi) = 0$ or $L(1,\chi) \neq 0$ if $\chi \neq \chi_0$, and $\delta_\chi = 1$ if $\chi = \chi_0$. Summing (2.28) over all characters we get

$$\sum_{\substack{n \leqslant x \\ n \equiv 1 \pmod q}} \frac{\Lambda(n)}{n} = \frac{1}{\varphi(q)} \Big(\sum_{\chi \pmod q} \delta_\chi \Big) \log x + O(1).$$

Hence $\sum_\chi \delta_\chi \geqslant 0$. This shows that there is at most one character $\chi \neq \chi_0$ with $L(1,\chi) = 0$. If such an exceptional character exists, it must be real because $L(1,\chi) = 0$ implies $L(1,\overline{\chi}) = 0$ (the complex conjugation is a continuous automorphism of $\mathbb{C}$).

The non-vanishing of $L(1,\chi)$ for a real character $\chi \neq \chi_0$ requires a different argument. To this end we consider

$$T(x) = \sum_{n \leqslant x} \tau(n,\chi) n^{-\frac{1}{2}}$$

where

$$\tau(n,\chi) = \sum_{d|n} \chi(d). \tag{2.29}$$

Note that

$$\tau(n,\chi) = \prod_{p^\alpha \| n} (1 + \chi(p) + \cdots + \chi(p^\alpha)) \geqslant 0$$

for all n, and $\tau(m^2,\chi) \geqslant 1$. Therefore

$$T(x) \geqslant \sum_{m \leqslant \sqrt{x}} m^{-1} > \tfrac{1}{2} \log x.$$

On the other hand, opening the convolution (2.29) we evaluate $T(x)$ in the same fashion as Dirichlet dealt with the divisor problem (1.75) (the hyperbola method). We obtain

$$\begin{aligned}
T(x) &= \sum\sum_{mn \leqslant x} \chi(m)(mn)^{-\frac{1}{2}} \\
&= \sum_{m \leqslant \sqrt{x}} \chi(m) m^{-\frac{1}{2}} \sum_{n \leqslant \frac{x}{m}} n^{-\frac{1}{2}} + \sum_{n < \sqrt{x}} n^{-\frac{1}{2}} \sum_{\sqrt{x} < m \leqslant \frac{x}{n}} \chi(m) m^{-\frac{1}{2}} \\
&= \sum_{m \leqslant \sqrt{x}} \chi(m) m^{-\frac{1}{2}} \Big\{ \tfrac{1}{2} \Big(\frac{x}{m}\Big)^{\frac{1}{2}} + c + O\Big(\Big(\frac{m}{x}\Big)^{\frac{1}{2}}\Big)\Big\} + O\Big(\sum_{n < \sqrt{x}} n^{-\frac{1}{2}} x^{-\frac{1}{4}} \Big) \\
&= \tfrac{1}{2} L(1,\chi) x^{\frac{1}{2}} + O(1).
\end{aligned}$$

Comparing both estimates we obtain $\log x < L(1,\chi) x^{\frac{1}{2}} + O(1)$, whence $L(1,\chi) \neq 0$ by letting x tend to ∞.

THEOREM 2.1 (DIRICHLET). *For every non-principal character* $\chi \pmod q$ *we have* $L(1,\chi) \neq 0$.

By Theorem 2.1 we complete the proof of (2.25). Moreover, we proved that (2.26) holds for any $\chi \neq \chi_0$. Summing over characters we deduce

THEOREM 2.2. *For any a with $(a,q)=1$ we have*

$$\sum_{\substack{n\leqslant x\\ n\equiv a(\mathrm{mod}\ q)}} \frac{\Lambda(n)}{n} = \frac{\log x}{\varphi(q)} + O(1) \tag{2.30}$$

where the error term $O(1)$ depends only on q.

REMARK. The original proof of Dirichlet that $L(1,\chi)\neq 0$ for a real character χ is a simple consequence of his Class Number Formula: let $K=\mathbb{Q}(\sqrt{D})$ be a quadratic field with $D\equiv 0,1 \pmod 4$ so D is the discriminant of K. There is an associated real primitive character χ such that

$$\zeta_K(s)=\sum_{\mathfrak{a}}(N\mathfrak{a})^{-s}=\zeta(s)L(s,\chi)$$

and every primitive real character modulo $q\neq 1$ arises in this way from $K=\mathbb{Q}(\sqrt{\chi(-1)q})$. Dirichlet showed that

$$L(1,\chi)=\begin{cases}\frac{2\pi h}{w\sqrt{q}} & \text{if } \chi(-1)=-1,\\ \frac{2h\log\varepsilon}{\sqrt{q}} & \text{if } \chi(-1)=1,\end{cases} \tag{2.31}$$

where h is the class number of K, w is the number of units in the ring of integers of K if $\chi(-1)=-1$, and ε is the fundamental unit of K if $\chi(-1)=1$. Hence $L(1,\chi)>\frac{1}{\sqrt{q}}$. (See (22.59) for odd characters).

In Chapters 5, 22, 23, we develop lower bounds for $L(1,\chi)$ using techniques of increasing sophistication. See also Section 15.9 for some results about class numbers of real quadratic fields.

2.4. Reflections on elementary proofs of the prime number theorem.

The Prime Number Theorem was proved in 1896 independently by Hadamard and de la Valleé Poussin by analytic methods. The first elementary proofs were found about fifty years later by Erdös and Selberg. At the heart of their proofs lies the formula (due to Selberg)

$$\sum_{p\leqslant x}(\log p)^2+\sum_{pq\leqslant x}\sum(\log p)(\log q)=2x\log x+O(x). \tag{2.32}$$

To prove this we begin by

$$\begin{aligned}\sum_{n\leqslant x}\log^2 n &= x(\log^2 x-2\log x+2)+O(\log^2 x)\\ &= x(\log x)\Big(\sum_{k\leqslant x}k^{-1}\Big)-\sum_{k\leqslant x}[\gamma+(\gamma+2)\log k]+O(\log^2 x).\end{aligned}$$

where γ is the Euler constant. Hence opening the convolution $\Lambda_2 = \mu\star L^2$ we derive

$$\begin{aligned}\sum_{n\leqslant x}\Lambda_2(n) &= \sum\sum_{mn\leqslant x}\mu(m)\log^2 n\\ &= \sum_{l\leqslant x}\frac{x}{l}\Big(\sum_{m|l}\mu(m)\log\frac{x}{m}\Big)\\ &\quad-\sum_{l\leqslant x}\sum_{m|l}\mu(m)\Big[\gamma+(\gamma+2)\log\frac{l}{m}\Big]+O(x)\\ &= \sum_{n\leqslant x}\frac{x}{n}(\delta(n)\log x+\Lambda(n))+\sum_{n\leqslant x}(\gamma\delta(n)+(\gamma+2)\Lambda(n))+O(x)\\ &= x\log x+x\log x+O(x)=2x\log x+O(x)\end{aligned}$$

by Mertens' formula (2.14) and Tchebyshev's upper bound (2.12) (recall (1.18)).

EXERCISE. Derive from (2.32) by induction on $k\geqslant 2$ (use the recurrence (1.47)) that

$$\sum_{n\leqslant x}\Lambda_k(n) = kx(\log x)^{k-1}\Big\{1+O\Big(\frac{1}{\log x}\Big)\Big\}. \tag{2.33}$$

Using Selberg's formula (2.32) Postnikov and Romanov [**PR**] derived the elegant inequality

$$|M(x)|\log x < \sum_{n\leqslant x}\Big|M\Big(\frac{x}{n}\Big)\Big| + O(x\log\log 3x). \tag{2.34}$$

Compare this with the second part of (2.17). Now the Prime Number Theorem can be derived in the form $\psi(x)\sim x$ from (2.32), or in the form $M(x)=o(x)$ from (2.34) by elementary, however lengthy, tauberian type arguments.

E. Bombieri [**Bo1**] and E. Wirsing [**Wi2**] refined the Erdös-Selberg method to show

$$\psi(x) = x + O(x(\log x)^{-A}) \tag{2.35}$$

for any $A > 0$, the implied constant depending on A. Along the same lines H. Diamond and J. Steinig [**DS**] succeeded in showing that

$$|\psi(x)-x| \leqslant x\exp\Big\{-(\log x)^{\frac{1}{7}}(\log\log x)^{-2}\Big\} \tag{2.36}$$

for all $x\geqslant e^{e^{100}}$. The arguments of Erdös–Selberg are truly elementary (no complex numbers are harmed in the course of the proof) nevertheless we express our mixed feelings, because the earlier analytic methods based on contour integration inside the zero-free region of the Riemann zeta function are considerably simpler and the results are superior (see Sections 5.4 and 5.6). In 1899 de la Vallée Poussin proved quite easily that

$$\psi(x) = x + O(x\exp(-c\sqrt{\log x})) \tag{2.37}$$

where $c>0$ and the implied constant are absolute. The subsequent improvements of the PNT came from sharper upper bounds for $\zeta(s)$ near the line $\mathrm{Re}\,(s)=1$, particularly by Vinogradov's estimate for exponential sums (see Chapter 8). However, we would like to say that Vinogradov's method of exponential sums should not be regarded as a new technique for non-vanishing of the zeta function. In fact, all the known methods of converting the upper bounds into the lower bounds, from which to draw a zero-free region, remain in principle the same one as originated by Hadamard and de la Vallée Poussin. Yet, some variations are more sensitive to the input than the others, giving closer connections between estimates and zeros. For example the Borel-Caratheodory theorem in complex function theory does the job very well. The best error term in (2.37) known today is due to Korobov [**Kor**] and Vinogradov [**V1**] (see Corollary 8.31).

The asymptotic formula $\psi(x) \sim x$ is equivalent to the non-vanishing of $\zeta(s)$ on the line $\mathrm{Re}\,(s) = 1$, and the latter can be established relatively easily without appealing to analytic properties of $\zeta(s)$.

Here is how one can arrange the analytic ideas for a proof of the Prime Number Theorem in the form (2.35) which a broad-minded reader may still find elementary. We avoid complex variable analysis, although for clarity we do not hesitate to employ $\zeta(s)$ at complex numbers $s = \sigma + it$, but only with $\sigma > 1$ where the Dirichlet series and the Euler product converge absolutely. We begin by considering the function $G(s) = (-1)^k (1/\zeta(s))^{(k)}$. This is given by the series

$$G(s) = \sum_m \frac{\mu(m)}{m^s} (\log m)^k. \tag{2.38}$$

From here we go to the finite sum

$$F(x) = \sum_{m \leqslant x} \mu(m) (\log m)^k \log \frac{x}{m} \tag{2.39}$$

where the factor $\log(x/m)$ plays a task of smoothing at the cut-off point. This particular cutting is produced by means of the integral formula

$$\log^+ y = \frac{1}{2\pi i} \int_{(\sigma)} y^s s^{-2} ds$$

where $\sigma > 1$, giving

$$F(x) = \frac{1}{2\pi i} \int_{(\sigma)} x^s G(s) s^{-2} ds. \tag{2.40}$$

For the purpose of estimating $G(s)$ we consider $\zeta^*(s) = (s-1)\zeta(s)$. Note that

$$(1/\zeta(s))^{(k)} = (s-1)(1/\zeta^*(s))^{(k)} + k(1/\zeta^*(s))^{(k-1)}. \tag{2.41}$$

Next we appeal to the formula from differential calculus

$$\left(\frac{1}{f}\right)^{(k)} = \frac{1}{f} \sum\sum_{a_1 + 2a_2 + \cdots = k} \frac{(a_1 + a_2 + \cdots)!}{a_1! a_2! \cdots} \left(\frac{-f'}{1! f}\right)^{a_1} \left(\frac{-f''}{2! f}\right)^{a_2} \cdots \tag{2.42}$$

where $a_1, a_2, \ldots$ run over non-negative integers. This formula for $f(s) = \zeta^*(s)$ reduces the problem to upper bounds for the derivatives of $\zeta^*(s)$ and a lower bound for $|\zeta^*(s)|$. First we derive by partial summation that for $\mathrm{Re}\,(s) > 1$,

$$\begin{aligned}(-1)^\ell \zeta^{(\ell)}(s) &= \sum_1^X n^{-s}(\log n)^\ell + \int_X^\infty x^{-s}(\log x)^\ell dx + O\Big(\frac{|s|}{X}(\log X)^{\ell+1}\Big)\\ &= \int_1^\infty x^{-s}(\log x)^\ell dx + O\Big(\Big(1+\frac{|s|}{X}\Big)(\log X)^{\ell+1}\Big).\end{aligned}$$

Taking $X = 2|s|$ we get

$$(2.43)\qquad (-1)^\ell \zeta^{(\ell)}(s) = \frac{\ell!}{(s-1)^{\ell+1}} + O((\log 2|s|)^{\ell+1}).$$

Hence

$$(2.44)\qquad (\zeta^*(s))^{(\ell)} = (s-1)\zeta^{(l)}(s) + l\zeta^{(l-1)}(s) \ll |s|(\log 2|s|)^{\ell+1}.$$

Next we infer from the Euler product by positivity that

$$\begin{aligned}1 &\leqslant \prod_p \Big(1 + (1+p^{it}+p^{-it})^2 p^{-\sigma}\Big)\\ &= \prod_p \Big(1 + (3+2p^{it}+2p^{-it}+p^{2it}+p^{-2it})p^{-\sigma}\Big)\\ &\asymp \zeta^3(\sigma)|\zeta(\sigma+it)|^4|\zeta(\sigma+2it)|^2\end{aligned}$$

where the implied constant is absolute (a similar formula is also present in the arguments of Hadamard and de la Vallée Poussin, see Theorem 5.26). Hence and by (2.43) we get the lower bound

$$(2.45)\qquad |\zeta^*(s)| \gg (\sigma-1)^{\frac34}|s|(\log 2|s|)^{-\frac12}.$$

From (2.44) and (2.45) we derive using (2.42) that

$$(1/\zeta^*(s))^{(k)} \ll (\sigma-1)^{-\frac34(k+1)}|s|^{-1}(\log 2|s|)^\kappa$$

where κ and the implied constant depend only on k (in fact, $\kappa = (5k+1)/2$ comes out by the above arguments, but it does not matter for application). Hence by (2.41)

$$G(s) \ll (\sigma-1)^{-\frac34(k+1)}(\log 2|s|)^\kappa.$$

Inserting this bound to (2.40) and integrating trivially we get

$$F(x) \ll x^\sigma(\sigma-1)^{-\frac34(k+1)}.$$

Finally taking $\sigma = 1 + (\log x)^{-1}$ we conclude the following estimate for the sum (2.39)

$$(2.46)\qquad F(x) \ll x(\log x)^{\frac34(k+1)}.$$

Note that this is non-trivial if $k > 3$.

Now it is a completely elementary exercise to derive the PNT from (2.46). First we do the sum

$$(2.47)\qquad H(x) = \sum_{n\leqslant x}\mu(m)(\log m)^k.$$

which is obtained by differencing $F(x)$ as follows

$$F(x+y)-F(x)=H(x)\log\frac{x+y}{x}+\sum_{x<m\leqslant x+y}\mu(m)(\log m)^k\log\frac{x+y}{m}$$
$$=\Big[H(x)+O(y(\log x)^k)\Big]\log\frac{x+y}{x}.$$

Hence

$$H(x)\ll y(\log x)^k+y^{-1}x^2(\log x)^{\frac{3}{4}(k+1)}=2x(\log x)^{k-A}$$

by choosing $y=x(\log x)^{-A}$ with $A=(k-3)/8$. Finally we get by partial summation

$$\sum_{m\leqslant x}\mu(m)\ll x(\log x)^{-A} \tag{2.48}$$

where the implied constant depends only on k. Since k is arbitrary so is A. This proves the PNT for the Möbius function. Hence the PNT for the von Mangoldt function follows by the formula (2.8).

CHAPTER 3

CHARACTERS

3.1. Introduction.

The characters on residue classes, both additive and multiplicative, play instrumental parts in analytic number theory. Other characters such as the Hecke characters of a number field, or the characters of a finite field are also indispensable in the modern theory. We shall introduce these in due course. However, to avoid redundancy we start here by giving basic definitions in a general context of a finite abelian group. (Characters for non-abelian groups occur also, but more naturally as Galois groups and are best seen from the automorphic perspective in analytic number theory).

Let G be a finite abelian group. A homomorphism $\chi : G \to \mathbb{C}^\star$ is called a character of G. Therefore (in the multiplicative notation) χ has the properties

$$\chi(xy) = \chi(x)\chi(y) \qquad \text{for all } x, y \in G,$$

$\chi(1) = 1$ and $\chi(x)^m = 1$, where $m = |G|$ is the order of G, therefore $\chi(x)$ is a root of unity.

The characters of G form a group $\hat{G}$ with multiplication given by

$$(\chi_1\chi_2)(x) = \chi_1(x)\chi_2(x) \qquad \text{for all } x \in G.$$

$\hat{G}$ is called the dual group, its identity element is the trivial character

$$\chi_0(x) = 1 \qquad \text{for all } x \in G.$$

Throughout $\overline{\chi}$ denotes the complex conjugate, hence also the inverse. If G is cyclic of order m and g is a generator of G, then every character of G is of type

$$\chi_a(x) = e^{2\pi i a y/m}, \qquad \text{if } x = g^y$$

for some fixed residue class $a (\bmod m)$. These are distinct characters, therefore $\hat{G}$ is also cyclic of order m, so $\hat{G}$ is isomorphic to G. The isomorphism $\hat{G} \simeq G$ can be established for any finite abelian group by writing G as the direct product of cyclic groups. There is a canonical isomorphism between G and $\hat{\hat{G}}$ given by $x \mapsto \hat{\hat{x}}$ where

$$\hat{\hat{x}}(\chi) = \chi(x) \qquad \text{for all } \chi \in \hat{G}.$$

The raison d'être of these characters is found in the following orthogonality relations

$$\sum_{x\in G}\chi(x)=\begin{cases}|G| & \text{if } \chi=\chi_0,\\ 0 & \text{if } \chi\neq\chi_0,\end{cases}$$

$$\sum_{\chi\in\hat{G}}\chi(x)=\begin{cases}|\hat{G}| & \text{if } x=1,\\ 0 & \text{if } x\neq 1,\end{cases}$$

which allows us to detect the group identity element.

Suppose d divides the order of G. An element $g\in G$ is said to have exponent d if g^d is the identity. The order of g is its smallest exponent. The characters of exponent d are exactly these which are trivial on the subgroup

$$G^d=\{x^d : x\in G\},$$

therefore they may be viewed as characters of the factor group G/G^d. The orthogonality relations become

$$\sum_{\chi^d=\chi_0}\chi(y)=\begin{cases}[G:G^d] & \text{if } y\in G^d,\\ 0 & \text{if } y\notin G^d.\end{cases}$$

This will allow us to detect the d-th powers of G.

3.2. Dirichlet characters.

First we consider the characters on the additive group $\mathbb{Z}/m\mathbb{Z}$ of residue classes modulo m. They are given by

$$\psi_a(n)=e\Big(\frac{an}{m}\Big)$$

where $e(z)=e^{2\pi iz}$. This formula makes an additive character a function on $\mathbb{Z}$, periodic of period m. The orthogonality property becomes

$$\sum_{a(\mathrm{mod}\ m)}e\Big(\frac{an}{m}\Big)=\begin{cases}m & \text{if } n\equiv 0(\mathrm{mod}\ m),\\ 0 & \text{otherwise.}\end{cases}$$

We shall call $\psi_a(n)=e(an/m)$ primitive if $(a,m)=1$. Adding all the additive primitive characters we obtain the Ramanujan sum

$$S(n,0;m)=c_m(n)=\sideset{}{^\star}\sum_{a(\mathrm{mod}\ m)}e\Big(\frac{an}{m}\Big). \tag{3.1}$$

(see (1.56)). Recall that throughout this book $\sum^\star$ restricts the summation to "primitive" elements, in the above case among additive characters modulo m. One shows by Möbius inversion that

$$c_m(n)=\sum_{d|(m,n)}d\mu\Big(\frac{m}{d}\Big). \tag{3.2}$$

Hence

$$c_m(n)=\mu\Big(\frac{m}{(m,n)}\Big)\frac{\varphi(m)}{\varphi(m/(m,n))}. \tag{3.3}$$

In particular,

$$c_m(n) = \mu(m) \qquad \text{if } (m,n) = 1. \tag{3.4}$$

In general,

$$|c_m(n)| \leqslant (m,n). \tag{3.5}$$

Next we consider characters on the multiplicative group of residue classes $a(\text{mod } m)$ with $(a,m) = 1$,

$$\chi : (\mathbb{Z}/m\mathbb{Z})^\star \to \mathbb{C}.$$

As with the additive characters, we wish to consider χ as a function on $\mathbb{Z}$; we do so by setting $\chi(n) = 0$ whenever n is not prime to m. This makes χ a Dirichlet character, a function on $\mathbb{Z}$, periodic modulo m and completely multiplicative. The corresponding extension of the trivial character $\chi_0(\text{mod } m)$ is called the principal character to modulus m. The group $(\mathbb{Z}/m\mathbb{Z})^\star$ and its dual have $\varphi(m)$ elements. The orthogonality relations become

$$\sum_{a(\text{mod } m)} \chi(a) = \begin{cases} \varphi(m) & \text{if } \chi = \chi_0, \\ 0 & \text{otherwise}, \end{cases}$$

$$\sum_{\chi(\text{mod } m)} \chi(a) = \begin{cases} \varphi(m) & \text{if } a \equiv 1(\text{mod } m), \\ 0 & \text{otherwise}, \end{cases}$$

If $m = m_1 m_2$ with $(m_1, m_2) = 1$, then $(\mathbb{Z}/m\mathbb{Z})^\star \simeq (\mathbb{Z}/m_1\mathbb{Z})^\star \times (\mathbb{Z}/m_2\mathbb{Z})^\star$, so every multiplicative character $\chi(\text{mod } m)$ is a product $\chi_1\chi_2$ of multiplicative characters $\chi_1(\text{mod } m_1)$ and $\chi_2(\text{mod } m_2)$.

The Dirichlet series associated to Dirichlet characters are of paramount importance in analytic number theory. They are called Dirichlet L-functions and denoted $L(s,\chi)$, or $L(\chi, s)$ depending on emphasis:

$$L(s,\chi) = \sum_{n\geqslant 1} \chi(n) n^{-s} = \prod_p (1 - \chi(p)p^{-s})^{-1} \; ; \tag{3.6}$$

the series and the Euler product are absolutely convergent for $\mathrm{Re}\,(s) > 1$. As is well known, these functions can be analytically continued to $\mathbb{C}$. See Section 4.6 for a proof. Many other analytic properties, applications and generalizations of Dirichlet L-functions will be considered throughout the book.

Note that by total multiplicativity, (1.15), (1.38) give

$$\frac{1}{L(s,\chi)} = \sum_{n\geqslant 1} \mu(n)\chi(n)n^{-s}, \quad -\frac{L'}{L}(s,\chi) = \sum_{n\geqslant 1} \Lambda(n)\chi(n)n^{-s}$$

for $\mathrm{Re}\,(s) > 1$.

3.3. Primitive characters.

Associated to each character χ, in addition to its modulus m, is a natural number $m^\star$, its conductor. The conductor is the smallest divisor of m such that χ may be written as $\chi = \chi_0\chi^\star$ where χ_0 is the principal character to modulus m and $\chi^\star$ is a character to modulus $m^\star$. For some characters the conductor is equal to the modulus. Such characters are called primitive. In the above factorization $\chi^\star$ is a primitive character uniquely determined by χ. We shall say that χ is induced by $\chi^\star$ or that $\chi^\star$ induces χ. We have

$$\chi(a) = \chi^\star(a) \qquad \text{if } (a, m) = 1.$$

The number of primitive characters to modulus m is given by

$$\varphi^\star(m) = m \prod_{p \| m}\Big(1 - \frac{2}{p}\Big) \prod_{p^2 | m}\Big(1 - \frac{1}{p}\Big)^2. \tag{3.7}$$

The proof is a simple exercise in Möbius inversion. Indeed, we have $\varphi = 1 \star \varphi^\star$, whence $\varphi^\star = \mu \star \varphi$, i.e.,

$$\varphi^\star(m) = \sum_{d|m} \mu(d)\varphi\Big(\frac{m}{d}\Big)$$

giving (3.7) by multiplicativity. In the same way using the orthogonality of characters we infer a more general result

$$\sideset{}{^\star}\sum_{\chi(\text{mod } m)} \chi(a) = \sum_{d|(a-1,m)} \varphi(d)\mu\Big(\frac{m}{d}\Big) \qquad \text{if } (a, m) = 1. \tag{3.8}$$

Incidentally, the formula (3.7) shows that the primitive characters $\chi(\text{mod } m)$ exist precisely when $m \not\equiv 2(\text{mod } 4)$.

The primitive characters are pleasant to deal with. For example, we have a convenient formula (which is not valid for χ not primitive)

$$\frac{1}{m} \sum_{c(\text{mod } m)} \chi(ac + b) = \begin{cases} \chi(b) & \text{if } m \mid a, \\ 0 & \text{if } m \nmid a. \end{cases} \tag{3.9}$$

Indeed, letting S be the above sum we obtain $\chi(1 + m_1 x)S = S$ for any x, where $m_1 = m(a, m)/(a^2, m)$. If $S \neq 0$, this yields $\chi(1+m_1x) = 1$ for any x, which implies that χ is periodic of period m_1. Since χ is primitive, we must have $m_1 = m$, i.e., $m \mid a$. Thus $S = 0$ if $m \nmid a$, and (3.9) is obvious otherwise.

The Dirichlet characters of exponent two are just the real characters (real-valued). In a number of ways they play a special role, particularly they are fundamental in the theory of quadratic forms. Below we give a complete list of real primitive characters. If $m = p$ is an odd prime, then there exists exactly one real primitive character of conductor p, namely the quadratic residue symbol (the Legendre symbol)

$$\chi_p(n) = \Big(\frac{n}{p}\Big).$$

This can be defined by saying that $1 + \chi_p(n)$ is the number of solutions $x(\text{mod } p)$ to $x^2 \equiv n(\text{mod } p)$.

For $m = 4$ we have exactly one primitive character defined by

$$\chi_4(n) = (-1)^{\frac{n-1}{2}} \qquad \text{if } 2 \nmid n.$$

If $m = 8$, we have two primitive characters

$$\chi_8(n) = (-1)^{\frac{1}{8}(n-1)(n+1)} \quad \text{if } 2 \nmid n,$$

$$\chi_4\chi_8(n) = (-1)^{\frac{1}{8}(n-1)(n+5)} \qquad \text{if } 2 \nmid n.$$

If m is prime power, there are no real primitive characters of conductor m other than $\chi_4, \chi_8, \chi_4\chi_8$ and χ_p. Every real primitive character $\chi(\text{mod } m)$ is obtained as the product of the above ones. Therefore the conductor of a real primitive character is a number of type $1, k, 4k, 8k$ where k is a positive odd and squarefree integer.

3.4. Gauss sums.

Let us come back for a moment to a general finite abelian group G. The characters of G form a complete orthogonal system so that any function $f : G \to \mathbb{C}$ has the Fourier expansion

$$f = \frac{1}{|G|} \sum_{\psi \in \hat{G}} \langle f, \psi \rangle \psi$$

with coefficients

$$\langle f, \psi \rangle = \sum_{g \in G} f(g)\overline{\psi}(g).$$

EXERCISE 1 (DUE TO DEDEKIND). Prove that

$$\prod_{\psi \in \hat{G}} \langle f, \psi \rangle = \det_{g,h \in G} (f(gh^{-1})).$$

In the case of functions on residue classes, and on a finite field, both additive and multiplicative characters are present and it is often necessary to transform the Fourier expansion in one of these systems to the other. In doing so we encounter Gauss sums $\langle \chi, \psi \rangle$ for any pair of multiplicative and additive characters χ, ψ respectively. We shall present the Gauss sums for finite fields in due time.

Now we consider Gauss sums associated with characters on residue classes, say modulo m. For any multiplicative character $\chi(\text{mod } m)$ we put

$$\tau(\chi) = \sum_{b (\text{mod } m)} \chi(b) e\Big(\frac{b}{m}\Big). \tag{3.10}$$

Multiplying (3.10) by $\overline{\chi}(a)$ and summing over χ we derive by orthogonality

$$e\Big(\frac{a}{m}\Big) = \frac{1}{\varphi(m)} \sum_{\chi (\text{mod } m)} \overline{\chi}(a)\tau(\chi) \qquad \text{if } (a,m) = 1. \tag{3.11}$$

This is the Fourier expansion of additive characters in terms of the multiplicative ones. Similarly

$$\chi(a)\tau(\overline{\chi}) = \sum_{b (\text{mod } m)} \overline{\chi}(b) e\Big(\frac{ab}{m}\Big) \qquad \text{if } (a,m) = 1 \tag{3.12}$$

which gives the Fourier expansion of χ in terms of additive characters provided $\tau(\overline{\chi}) \neq 0$. Note that the condition $(a,m) = 1$ in (3.12) can be dropped if χ is primitive because both sides vanish if $(a,m) \neq 1$ (apply (3.9)).

LEMMA 3.1. *Suppose the character χ modulo m is induced by the primitive character $\chi^\star$ modulo $m^\star$. Then*

$$\tau(\chi) = \mu\Big(\frac{m}{m^\star}\Big)\chi^\star\Big(\frac{m}{m^\star}\Big)\tau(\chi^\star). \tag{3.13}$$

If χ (mod m) is primitive, then

$$|\tau(\chi)| = \sqrt{m}. \tag{3.14}$$

PROOF. We have

$$\tau(\chi) = \sideset{}{^\star}\sum_{a (\mathrm{mod}\ m)} \chi^\star(a)e\Big(\frac{a}{m}\Big) = \sum_{d|m}\mu(d)\chi^\star(d) \sum_{a (\mathrm{mod}\ \frac{m}{d})} \chi^\star(a)e\Big(\frac{ad}{m}\Big).$$

The innermost sum vanishes unless $d = m/m^\star$ giving (3.13). To prove (3.14) we write

$$\begin{aligned} |\tau(\chi)|^2 &= \sum\sum_{a,b (\mathrm{mod}\ m)} \chi(a)\overline{\chi}(b)e\Big(\frac{a-b}{m}\Big) \\ &= \sum_{a (\mathrm{mod}\ m)} \chi(a) \sideset{}{^\star}\sum_{b (\mathrm{mod}\ m)} e\Big(\frac{(a-1)b}{m}\Big). \end{aligned}$$

The inner sum is the Ramanujan sum (which is also the Gauss sum for the principal character). Inserting (3.2) we get

$$|\tau(\chi)|^2 = \sum_{d|m} d\mu\Big(\frac{m}{d}\Big) \sum_{\substack{a (\mathrm{mod}\ m)\\ a\equiv 1 (\mathrm{mod}\ d)}} \chi(a).$$

If χ is primitive of conductor m, then the last sum vanishes unless $d = m$ (apply (3.9)), giving (3.14). □

One can see by Lemma 3.1 that $\tau(\chi)$ does not vanish exactly when $m/m^\star$ is squarefree and prime to $m^\star$. In this case (3.12) becomes a true formula for $\chi(a)$.

We now consider the more general sum

$$\tau(\chi, \psi_a) = \sum_{b (\mathrm{mod}\ m)} \chi(b)\psi_a(b)$$

where $\psi_a(b) = e(ab/m)$ is an additive character.

LEMMA 3.2. *Let χ modulo m be a non-principal character induced by the primitive character $\chi^\star$ modulo $m^\star$. Let $a \geqslant 1$. We have*

$$\tau(\chi, \psi_a) = \tau(\chi) \sum_{d|(a,m/m^\star)} d\overline{\chi}^\star(a/d)\mu(m/dm^\star),$$

in particular,

$$\tau(\chi, \psi_a) = \overline{\chi}(a)\tau(\chi)$$

if $(a, m) = 1$, i.e. if ψ_a is a primitive additive character.

Note that $\overline{\tau}(\chi) = \chi(-1)\tau(\overline{\chi})$, hence if χ(mod m) is primitive, then (3.14) can be written as

$$\tau(\chi)\tau(\overline{\chi}) = \chi(-1)m. \tag{3.15}$$

For any characters $\chi_1(\text{mod } m_1)$ and $\chi_2(\text{mod } m_2)$ with $(m_1, m_2) = 1$ we have

$$\tau(\chi_1\chi_2) = \chi_1(m_2)\chi_2(m_1)\tau(\chi_1)\tau(\chi_2). \tag{3.16}$$

This multiplication rule fails if the moduli are not relatively prime. For characters to the same modulus the correction factor is the Jacobi sum

$$J(\chi_1, \chi_2) = \sum_{a(\text{mod } m)} \chi_1(a)\chi_2(1-a). \tag{3.17}$$

Precisely if χ_1, χ_2 are two characters modulo m such that $\chi_1\chi_2$ is primitive, then

$$\tau(\chi_1)\tau(\chi_2) = J(\chi_1, \chi_2)\tau(\chi_1\chi_2). \tag{3.18}$$

Hence if all $\chi_1, \chi_2, \chi_1\chi_2$ are primitive of modulus m, then

$$|J(\chi_1, \chi_2)| = \sqrt{m}. \tag{3.19}$$

If $\chi(\text{mod } m)$ is primitive, then

$$J(\chi, \overline{\chi}) = \chi(-1)\mu(m). \tag{3.20}$$

3.5. Real characters.

For a primitive character $\chi(\text{mod } m)$ we know that $|\tau(\chi)| = \sqrt{m}$, however, the determination of the argument of $\tau(\chi)$ is a difficult problem. Using Deligne's estimate for multiple Kloosterman sums one can show that $\tau(\chi)/\sqrt{m}$ are asymptotically equidistributed on the unit circle as $\chi(\text{mod } m)$ ranges over all primitive characters and m tends to infinity over primes (see Chapter 21).

In the case of real characters $\tau(\chi)$ were evaluated completely by Gauss. It is easy to see that $\tau(\chi_4) = 2i, \tau(\chi_8) = 2\sqrt{2}$ and $\tau(\chi_4\chi_8) = 2\sqrt{2}i$.

THEOREM 3.3 (GAUSS). *For m odd squarefree*

$$\tau(\chi) = \varepsilon_m\sqrt{m} \tag{3.21}$$

where

$$\varepsilon_m = \begin{cases} 1 & \text{if } m \equiv 1 \ (\text{mod } 4), \\ i & \text{if } m \equiv -1 \ (\text{mod } 4). \end{cases} \tag{3.22}$$

We shall give an analytic proof of (3.21). But first we evaluate other Gauss sums of type

$$G(m) = \sum_{n(\text{mod } m)} e\Big(\frac{n^2}{m}\Big) \tag{3.23}$$

for any integer $m \geqslant 1$. If m is even we derive by shifting n to $n + \frac{m}{2}$ that $G(m) = i^m G(m)$, because $\frac{1}{m}(n + \frac{m}{2})^2 \equiv \frac{n^2}{m} + \frac{m}{4}(\text{mod } 1)$. Hence

$$G(m) = 0, \qquad \text{if } m \equiv 2(\text{mod } 4). \tag{3.24}$$

Next we show that

$$G(m^3) = mG(m), \qquad \text{if } m \not\equiv 2(\text{mod } 4). \tag{3.25}$$

This follows by splitting the summation in $G(m^3)$ into $n \equiv a + bm^2(\text{mod } m^3)$ if $2 \nmid m$ or $n \equiv a + b\frac{m^2}{2}(\text{mod } m^3)$ if $4 \mid m$. Now we are ready to prove

THEOREM 3.4 (DIRICHLET). *For any $m \in \mathbb{N}$,*

$$\overline{G}(m) = \frac{1+i^m}{1+i}\sqrt{m}. \tag{3.26}$$

PROOF. We have

$$G(m) = 2\sum_{0<n<\frac{m}{2}} e\Big(\frac{n^2}{m}\Big) + O(1).$$

In the segment $\frac{m}{4} < n < \frac{m}{2}$ we change n into $[\frac{m}{2}] - n = \frac{m}{2} - \{\frac{m}{2}\} - n$. Since $\frac{1}{m}([\frac{m}{2}]-n)^2 = \frac{1}{m}(n+\{\frac{m}{2}\})^2 - \frac{m}{4} + [\frac{m}{2}] - n \equiv \frac{1}{m}(n+\{\frac{m}{2}\})^2 - \frac{m}{4}(\mathrm{mod}\ 1)$, we obtain

$$G(m) = 2\sum_{0<n<\frac{m}{4}} e\Big(\frac{n^2}{m}\Big) + 2i^{-m}\sum_{0<n<\frac{m}{4}} e\Big(\frac{(n+\{m/2\})^2}{m}\Big) + O(1).$$

Here, and as before, the error term $O(1)$ accounts for the terms which are not covered in the displayed summation (at most two of them). By Lemma 8.8 each of the two short sums above is approximated (up to a bounded error term) by the integral

$$\int_0^{m/4} e\Big(\frac{x^2}{m}\Big)dx = \int_0^{\infty} e\Big(\frac{x^2}{m}\Big)dx + O(1) = \frac{1+i}{4}\sqrt{m} + O(1).$$

Hence adding up these results we get

$$G(m) = \frac{1+i^{-m}}{1+i^{-1}}\sqrt{m} + O(1).$$

It remains to show that the last error term $O(1)$ vanishes. This follows by iterated application of (3.25) if $m \not\equiv 2(\mathrm{mod}\ 4)$; otherwise (3.26) is obvious because both sides vanish. □

We generalize $G(m)$ by setting

$$G\Big(\frac{a}{m}\Big) = \sum_{n(\mathrm{mod}\ m)} e\Big(\frac{an^2}{m}\Big) \tag{3.27}$$

for any a with $(a,m) = 1$. In particular, $G(\frac{1}{m}) = G(m)$. If $m = m_1m_2$ with $(m_1, m_2) = 1$, we obtain

$$G\Big(\frac{a}{m_1m_2}\Big) = G\Big(\frac{am_2}{m_1}\Big)G\Big(\frac{am_1}{m_2}\Big) \tag{3.28}$$

by writing $n = n_1m_2 + n_2m_1$ with n_1 and n_2 ranging modulo m_1 and m_2 respectively. This formula reduces the problem of evaluating generalized Gauss sum $G(a/m)$ to that of a prime power modulus. If $m = p$ is an odd prime we can write

$$G\Big(\frac{a}{p}\Big) = \sum_{b(\mathrm{mod}\ p)} \Big(1 + \Big(\frac{b}{p}\Big)\Big)e\Big(\frac{ab}{p}\Big)$$

since $1 + (\frac{b}{p})$ is the number of solutions to $n^2 \equiv b(\mathrm{mod}\ p)$. Therefore

$$G\Big(\frac{a}{p}\Big) = \sum_{b(\mathrm{mod}\ p)} \Big(\frac{b}{p}\Big)e\Big(\frac{ab}{p}\Big) = \Big(\frac{a}{p}\Big)G\Big(\frac{1}{p}\Big) = \Big(\frac{a}{p}\Big)\varepsilon_p\sqrt{p} \tag{3.29}$$

by changing the variable and using (3.26). Next for distinct odd primes p, q we derive by (3.26),(3.28),(3.29) that

$$\varepsilon_{pq}\sqrt{pq} = G\Big(\frac{1}{pq}\Big) = G\Big(\frac{p}{q}\Big)G\Big(\frac{q}{p}\Big) = \Big(\frac{p}{q}\Big)\Big(\frac{q}{p}\Big)\varepsilon_p\varepsilon_q\sqrt{pq}.$$

Note also that

$$\varepsilon_{pq} = \varepsilon_p\varepsilon_q(-1)^{\frac{p-1}{2}\frac{q-1}{2}}. \tag{3.30}$$

Combining these formulas we deduce

THEOREM 3.5 (QUADRATIC RECIPROCITY LAW). *For any odd primes $p \neq q$ we have*

$$\Big(\frac{p}{q}\Big)\Big(\frac{q}{p}\Big) = (-1)^{\frac{p-1}{2}\frac{q-1}{2}}. \tag{3.31}$$

Since $G(-1/p) = \overline{G}(1/p)$ it follows from (3.29) that for odd p

$$\Big(\frac{-1}{p}\Big) = \varepsilon_p^2 = (-1)^{\frac{p-1}{2}} = \chi_4(p). \tag{3.32}$$

EXERCISE 2. Prove that for odd p,

$$\Big(\frac{2}{p}\Big) = (-1)^{\frac{1}{8}(p^2-1)} = \chi_8(p). \tag{3.33}$$

Now (3.21) follows by $\tau(\chi) = G(p) = \varepsilon_p\sqrt{p}$ if $m = p$ is odd prime, and in general by the multiplicativity on using (3.16),(3.30) and (3.31).

Note we have proved that

$$\sum_{x(\text{mod } m)} \Big(\frac{x}{m}\Big)e\Big(\frac{x}{m}\Big) = \sum_{x(\text{mod } m)} e\Big(\frac{x^2}{m}\Big)$$

if m is odd squarefree. If m is odd but not squarefree, the left side vanishes while the right side does not.

For convenience we extend the Legendre symbol $(\frac{a}{p})$ to $p = 2$ by setting

$$\Big(\frac{a}{2}\Big) = \begin{cases} 1 & \text{if } 2 \nmid a, \\ 0 & \text{if } 2 \mid a. \end{cases} \tag{3.34}$$

Then for any $b > 0$ we define

$$\Big(\frac{a}{b}\Big) = \prod_{p^\nu \| b} \Big(\frac{a}{p}\Big)^\nu. \tag{3.35}$$

If b is odd, this symbol $(\frac{a}{b})$ was introduced by Jacobi.

EXERCISE 3. Derive from (3.31) that for any odd integers a, b relatively prime

$$\left(\frac{a}{|b|}\right)\left(\frac{b}{|a|}\right) = (-1)^{\frac{a-1}{2}\frac{b-1}{2}}(a,b)_\infty. \tag{3.36}$$

Here $(x,y)_\infty$ is the Hilbert symbol defined for $xy \neq 0$ by

$$(x,y)_\infty = \begin{cases} -1 & \text{if } x<0 \text{ and } y<0, \\ 1 & \text{otherwise.} \end{cases} \tag{3.37}$$

Notice that $2(x,y)_\infty = 1 + \text{sign } x + \text{sign } y - \text{sign } xy$.

EXERCISE 4. Prove that for any a,m with $(2a, m) = 1$,

$$G\left(\frac{a}{m}\right) = \left(\frac{a}{m}\right)\varepsilon_m\sqrt{m}. \tag{3.38}$$

Finally we extend the symbol $(\frac{a}{b})$ to all integers a, b except for $a = b = 0$. If $ab \neq 0$, we set

$$\left(\frac{a}{b}\right) = \left(\frac{a}{|b|}\right)(a,b)_\infty. \tag{3.39}$$

Then we set

$$\left(\frac{1}{0}\right) = \left(\frac{0}{1}\right) = \left(\frac{0}{-1}\right) = -\left(\frac{-1}{0}\right) = 1, \tag{3.40}$$

and

$$\left(\frac{a}{b}\right) = 0 \qquad \text{if } (a,b) \neq 1. \tag{3.41}$$

For $b \neq 0$ this symbol was introduced by Shimura [**Sh1**] in the context of modular forms of half-integral weight. He showed that (1.53) holds with

$$\nu(c,d) = \varepsilon_d\left(\frac{c}{d}\right). \tag{3.42}$$

EXERCISE 5. Check the consistency of the above settings. Prove that for any $b \geqslant 1$ the map $a \mapsto (\frac{a}{b})$ is a Dirichlet character modulo b. Prove that for any $a \neq 0$ the map $b \mapsto (\frac{a}{b})$ is a Dirichlet character of conductor $a^\star \mid 4a$.

The above extension of $(\frac{a}{b})$ will be called the Jacobi symbol.

Any integer $\Delta \neq 0$ with $\Delta \equiv 0, 1 (\text{mod } 4)$ is called a discriminant. If $\Delta = 1$ or Δ is the discriminant of a quadratic field, then Δ is said to be fundamental. Any discriminant can be written uniquely as $\Delta = e^2 D$ with D fundamental and $e \geqslant 1$. A prime discriminant is a fundamental discriminant having exactly one prime factor, thus it is a number of type $-4, -8, 8$ and $\chi_4(p)p$ with $p > 2$. Every fundamental discriminant factors into prime discriminants.

To any discriminant Δ one associates the Kronecker symbol $\left(\frac{\Delta}{c}\right)_K$ which is defined for any $c \neq 0$ by means of the Jacobi symbol as follows

$$\left(\frac{\Delta}{c}\right)_K = \left(\frac{2^\nu}{\Delta}\right)\left(\frac{\Delta}{b}\right) \tag{3.43}$$

where $c = 2^\nu b$ with b odd. Note that $(\frac{\Delta}{2})_K = (\frac{2}{\Delta}) = \chi_8(\Delta)$, explicitly

$$\left(\frac{\Delta}{2}\right)_K = \begin{cases} 1 & \text{if } \Delta \equiv 1 (\text{mod } 8), \\ -1 & \text{if } \Delta \equiv 5 (\text{mod } 8), \\ 0 & \text{if } \Delta \equiv 0 (\text{mod } 4). \end{cases} \tag{3.44}$$

We also extend the definition of the Kronecker symbol for $c = 0$ by setting

$$\left(\frac{\Delta}{0}\right)_K = \begin{cases} 1 & \text{if } \Delta = 1, \\ 0 & \text{otherwise.} \end{cases} \tag{3.45}$$

Therefore $\left(\frac{\Delta}{c}\right)_K$ is defined for all integers c and $\Delta \neq 0$, $\Delta \equiv 0, 1 (\text{mod } 4)$.

We shall drop the subscript K in the Kronecker symbol notation and explain in any relevant case that we are dealing with the Kronecker symbol. Remember the Kronecker symbol is not defined for Δ's other than discriminants. When the symbol $\left(\frac{\Delta}{c}\right)$ appears without comments it stands for the Jacobi symbol.

EXERCISE 6. Prove that for a fundamental discriminant Δ the Kronecker symbol $(\frac{\Delta}{c})$ is a primitive character of conductor $|\Delta|$.

3.6. The quartic residue symbol.

Next we construct certain characters of order four. These are associated with $\mathbb{Q}(i)$, the imaginary quadratic field of discriminant $D = -4$. The ring of integers of $\mathbb{Q}(i)$ is

$$\mathbb{Z}[i] = \{z = a + bi : a, b \in \mathbb{Z}\}$$

and the group of units of $\mathbb{Z}[i]$ is $\{1, i, i^2, i^3\}$. The irreducible elements of $\mathbb{Z}[i]$ are (up to the units); $1 + i$ with $N(1 + i) = 2$, the rational primes $q \equiv -1 (\text{mod } 4)$ with $Nq = q^2$ and the complex numbers $\pi = a + bi$ with

$$N\pi = \pi\overline{\pi} = a^2 + b^2 = p \equiv 1 (\text{mod } 4). \tag{3.46}$$

Note that π and $\overline{\pi}$ are coprime, these are called Gaussian primes. Every element of $\mathbb{Z}[i]$ different from zero factors uniquely (up to the units and permutations) into powers of irreducible elements.

For every Gaussian prime π we define a map (the quartic residue symbol)

$$\left(\frac{\alpha}{\pi}\right) : \mathbb{Z}[i] \to \{0, 1, i, i^2 i^3\} \tag{3.47}$$

which satisfies

$$\left(\frac{\alpha}{\pi}\right) \equiv \alpha^{\frac{p-1}{4}} (\text{mod } \pi). \tag{3.48}$$

Note that $(\frac{\alpha}{\pi}) = 0$ if $\pi \mid \alpha$. If $\pi \nmid \alpha$, then $\alpha^{p-1} \equiv 1 (\text{mod } \pi)$ because the residue class ring $\mathbb{Z}[i]/\pi\mathbb{Z}[i]$ is a finite field with $p = N\pi$ elements. This property (the little theorem of Fermat) implies the existence and the uniqueness of solutions to (3.48) with $(\frac{\alpha}{\pi}) = i^m$ for some $0 \leqslant m < 4$. In particular, we have

$$\left(\frac{i}{\pi}\right) = i^{\frac{p-1}{4}}, \quad \text{if} \quad p = \pi\overline{\pi} \equiv 1 (\text{mod } 4). \tag{3.49}$$

EXERCISE 7. Prove the following properties of the quartic residue symbol:

$$\left(\frac{\alpha\beta}{\pi}\right) = \left(\frac{\alpha}{\pi}\right)\left(\frac{\beta}{\pi}\right), \tag{3.50}$$

$$\alpha \equiv \beta (\mathrm{mod}\ \pi) \Rightarrow \left(\frac{\alpha}{\pi}\right) = \left(\frac{\beta}{\pi}\right), \tag{3.51}$$

$$\left(\frac{\alpha}{\pi'}\right) = \left(\frac{\alpha}{\pi}\right) \quad \text{if} \quad \pi' = \pi i^m, \tag{3.52}$$

$$\left(\frac{\overline{\alpha}}{\overline{\pi}}\right) = \overline{\left(\frac{\alpha}{\pi}\right)}, \tag{3.53}$$

$$\left(\frac{\alpha}{\pi}\right) = 1 \text{ if and only if } z^4 \equiv \alpha (\mathrm{mod}\ \pi) \text{ has a solution in } \mathbb{Z}[i]^\star. \tag{3.54}$$

If $\gamma = \pi_1 \cdots \pi_r$ is the product of Gaussian primes (not necessarily distinct), then we set the symbol

$$\left(\frac{\alpha}{\gamma}\right) = \left(\frac{\alpha}{\pi_1}\right) \cdots \left(\frac{\alpha}{\pi_r}\right). \tag{3.55}$$

Clearly the properties (3.50)–(3.53) hold true with π replaced by γ. In particular, (3.53) gives

$$\left(\frac{\alpha}{p}\right) = 1 \quad \text{if} \quad (\alpha, p) = 1.$$

Note that $(1+i)^2 = 2i$. We say that $\alpha \in \mathbb{Z}[i]$ is odd if $(1+i) \nmid \alpha$ and primary if $\alpha \equiv 1 (\mathrm{mod}\ 2(1+i))$. Every odd number in $\mathbb{Z}[i]$ is associated with exactly one primary element, i.e., $\varepsilon\alpha \equiv 1 (\mathrm{mod}\ 2(1+i))$ for exactly one unit $\varepsilon = i^m$. A primary element can be written as the product of primary irreducibles uniquely up to permutations. We have the following

THEOREM 3.6 (THE LAW OF QUARTIC RECIPROCITY). *If π_1, π_2 are distinct primary Gaussian primes, then*

$$\left(\frac{\pi_1}{\pi_2}\right)\left(\frac{\pi_2}{\pi_1}\right) = (-1)^{\frac{p_1-1}{4}\frac{p_2-1}{4}} \tag{3.56}$$

where $p_1 = N\pi_1$ and $p_2 = N\pi_2$. If $\pi = a + bi$ is primary, then

$$\left(\frac{i}{\pi}\right) = i^{(1-a)/2}, \quad \left(\frac{1+i}{\pi}\right) = i^{(a-1-b-b^2)/4}, \quad \left(\frac{2}{\pi}\right) = i^{-b/2}.$$

The quartic residue symbol $(\frac{\alpha}{\pi})$ is a multiplicative character of the finite field $\mathbb{F}_p \simeq \mathbb{Z}[i]/\pi\mathbb{Z}[i]$. The Gauss sum of this character is

$$g(\pi) = \sum_{a (\mathrm{mod}\ p)} \left(\frac{\alpha}{\pi}\right) e\left(\frac{a}{p}\right) \tag{3.57}$$

where a runs over the rational residue classes modulo p and α is the representative of a in $\mathbb{Z}[i]$ modulo π. If π is primary, then

$$g^2(\pi) = -(-1)^{\frac{p-1}{4}} \pi \sqrt{p}. \tag{3.58}$$

We shall be interested in the quartic residue symbol at rational integers

$$\chi_\pi(n) = \left(\frac{n}{\pi}\right), \qquad \text{for } n \in \mathbb{Z}. \tag{3.59}$$

This is a Dirichlet character of conductor $p = \pi\overline{\pi}$ and of order four. Since $(\pi, \overline{\pi}) = 1$, we have $\chi_\pi^2(n) \equiv n^{\frac{p-1}{2}} \pmod p$. Therefore

$$\chi_\pi^2(n) = \left(\frac{n}{p}\right) \tag{3.60}$$

is the quadratic residue symbol. Clearly $\chi_\pi(n)$ and $\chi_{\overline{\pi}}(n)$ are distinct Dirichlet characters but their squares yield the same quadratic character. More generally, if $q = p_1 \cdots p_r$ is the product of distinct primes $p_j \equiv 1 (\bmod\ 4)$, then there are 2^r distinct Dirichlet characters $\chi (\bmod\ q)$ satisfying $\chi^2 = \chi_q$, namely $\chi = \chi_{\pi_1} \cdots \chi_{\pi_r} = \chi_\gamma$, say, for any $\gamma = \pi_1 \cdots \pi_r$ with $\gamma\overline{\gamma} = q$.

3.7. The Jacobi-Dirichlet and the Jacobi-Kubota symbols.

Let q be the product of primes $\equiv 1 (\bmod\ 4)$ (not necessarily distinct primes). The Jacobi symbol $\chi_q(n) = (\frac{n}{q})$ can be extended to Gaussian domain $\mathbb{Z}[i]$ in several ways corresponding to representations of q as the sum of two squares

$$q = u^2 + v^2 \quad \text{with } (u, v) = 1. \tag{3.61}$$

By requiring $w = u + iv$ to be primary we distinguish u from v and we fix the sign of u. Note that $u \equiv 1 (\bmod\ 2)$ and $v \equiv u - 1 (\bmod\ 4)$. A Gaussian integer $w = u + iv$ is said to be primitive if $(u, v) = 1$, thus an odd w is primitive if $(w, \overline{w}) = 1$. For any primary primitive w we define the symbol $(\frac{z}{w}) : \mathbb{Z}[i] \to \{0, 1, -1\}$ by

$$\left(\frac{z}{w}\right) = \left(\frac{\operatorname{Re} wz}{|w|^2}\right) \tag{3.62}$$

where on the right side is the Jacobi symbol. More explicitly we have

$$\left(\frac{z}{w}\right) = \left(\frac{ur - vs}{q}\right), \qquad \text{if } z = r + is \tag{3.63}$$

where $q = w\overline{w}$. If q is prime $\equiv 1 (\bmod\ 4)$, we get two different symbols $(\frac{z}{w})$ and $(\frac{z}{\overline{w}})$ which were considered by Dirichlet [**Dir**]. Throughout we call $(\frac{z}{w})$ the Jacobi-Dirichlet symbol whenever w is primary and primitive.

We could introduce the Jacobi-Dirichlet symbols using roots of quadratic congruences: there is one-to-one correspondence between the roots of

$$\omega^2 + 1 \equiv 0 (\bmod\ q) \tag{3.64}$$

and the factorizations $q = w\overline{w}$ with $w = u + iv$ primary which is given by

$$\omega \equiv -\overline{u}v (\bmod\ q). \tag{3.65}$$

Here $\overline{u}$ denotes the multiplicative inverse of u modulo q. We obtain

$$\left(\frac{z}{w}\right) = \left(\frac{r + \omega s}{q}\right) \quad \text{if } z = r + is \tag{3.66}$$

by $(\frac{u}{q}) = (\frac{|u|}{q}) = (\frac{q}{|u|}) = (\frac{v^2}{|u|}) = 1$. Hence it is clear that

$$\left(\frac{r}{w}\right) = \left(\frac{r}{q}\right) \quad \text{if } r \in \mathbb{Z}. \tag{3.67}$$

For $z = i$ we have

$$\left(\frac{i}{w}\right) = i^{\frac{p-1}{2}}. \tag{3.68}$$

Clearly $(\frac{z}{w})$ is periodic in z of period q, and it is multiplicative as well since

$$(r_1 + \omega s_1)(r_2 + \omega s_2) \equiv r_1 r_2 - s_1 s_2 + \omega(r_1 s_2 + r_2 s_1) \pmod q.$$

EXERCISE 8. Derive from the quadratic reciprocity law (3.31) the following reciprocity law for the Jacobi-Dirichlet symbol

$$\left(\frac{z}{w}\right) = \left(\frac{w}{z}\right) \tag{3.69}$$

for any z and w which are primary and primitive.

For Gaussian integers $z = r + is \equiv 1 \pmod 2$, we define

$$[z] = i^{\frac{r-1}{2}} \left(\frac{s}{|r|}\right) \tag{3.70}$$

where $(\frac{s}{|r|})$ stands for the Jacobi symbol. Note that $[z]$ vanishes if z is not primitive. We are interested in the multiplicative structure of $[z]$ within the Gaussian ring $\mathbb{Z}[i]$ rather than with respect to the co-ordinates $(r, s) \in \mathbb{Z} \times \mathbb{Z}$. For this reason we refer to $[z]$ as the Jacobi-Kubota symbol. Indeed, this symbol is relevant to the Kubota homomorphism $SL(2, \mathbb{Z}) \to \{\pm 1\}$ which has much to do with metaplectic modular forms. Of course, $[z]$ is not multiplicative in the strict sense, yet it is nearly so, up to a factor which is the Jacobi-Dirichlet symbol.

EXERCISE 9. Prove that if w is primary primitive and $z \equiv 1 \pmod 2$, then

$$[wz] = \varepsilon [w][z] \left(\frac{z}{w}\right) \tag{3.71}$$

with $\varepsilon = \pm 1$ depending only on the quadrants to which z, w and wz belong (see the details in [**FI1**]).

Both symbols of Jacobi-Dirichlet and Jacobi-Kubota are utilized in the proof of the asymptotic formula for primes $p = X^2 + Y^4$. They play a role through the estimation for general bilinear forms in $[wz]$ (see Proposition 21.4 of [**FI1**]).

3.8. Hecke characters.

E. Hecke [**Hec1**] generalized the Dirichlet characters to any number field. His characters are multiplicative functions on ideals. We restrict our considerations to imaginary quadratic fields. Every such field is obtained by an extension of rationals by $\sqrt{d}$, where d is a negative, squarefree integer. Denote $K = \mathbb{Q}(\sqrt{d})$. The ring of integers of K is a free $\mathbb{Z}$-module $\mathcal{O} = \mathbb{Z} + \omega\mathbb{Z}$, where

$$\omega = \begin{cases} \sqrt{d} & \text{if} \quad d \equiv 2 (\mathrm{mod}\ 4), \\ \frac{1}{2}(1 + \sqrt{d}) & \text{if} \quad d \equiv 1 (\mathrm{mod}\ 4). \end{cases}$$

The discriminant of $K = \mathbb{Q}(\sqrt{d})$ is

$$D = \left(\det \begin{pmatrix} 1 & \omega \\ 1 & \overline{\omega} \end{pmatrix}\right)^2 = \begin{cases} 4d & \text{if } d \equiv 2, 3 (\mathrm{mod}\ 4), \\ d & \text{if } d \equiv 1 (\mathrm{mod}\ 4). \end{cases}$$

Note that in any case $K = \mathbb{Q}(\sqrt{D})$ and $\mathcal{O} = \mathbb{Z} + \frac{1}{2}(D + \sqrt{D})\mathbb{Z}$. The extension $K/\mathbb{Q}$ has two automorphisms, the identity and the complex conjugation. The group of units $U \subset \mathcal{O}$ is finite, cyclic, generated by the root of unity $\zeta_w = e^{2\pi i/w}$ of order $w = 4, 6, 2$, precisely

$$(3.72)\qquad \begin{aligned} U &= \{\pm 1, \pm i\} &&\text{if } d = -1,\\ U &= \{\pm 1, \pm\omega, \pm\omega^2\} &&\text{if } d = -3,\\ U &= \{\pm 1\} &&\text{if } d < -3. \end{aligned}$$

The Kronecker symbol $\chi_D(n) = (\frac{D}{n})$ is a real primitive character of conductor $-D$, we call it the field character. Notice that $\chi_D(-1) = -1$ (see (3.39)). Every ideal $\mathfrak{a} \neq 0$ in $\mathcal{O}$ is a free $\mathbb{Z}$-module, and the ring of residue classes $\mathcal{O}/\mathfrak{a}$ is finite, the number of its elements being the norm of $\mathfrak{a}$

$$(3.73)\qquad N\mathfrak{a} = |\{\alpha (\mathrm{mod}\ \mathfrak{a}) : \alpha \in \mathcal{O}\}|.$$

The norm is multiplicative. If $\mathfrak{a} = (\alpha)$ is a principal ideal, then $N\mathfrak{a} = N\alpha = |\alpha|^2$. Every ideal $\mathfrak{a} \neq 0$ factors uniquely into powers of prime ideals. Every prime ideal of K appears as a factor of a rational prime. The law of factorization of rational primes into prime ideals in K asserts the following:

- if $\chi_D(p) = 0$, then p ramifies; $(p) = \mathfrak{p}^2$ with $N\mathfrak{p} = p$,
- if $\chi_D(p) = 1$, then p splits; $(p) = \mathfrak{p}\bar{\mathfrak{p}}$ with $\mathfrak{p} \neq \bar{\mathfrak{p}}$ and $N\mathfrak{p} = p$,
- if $\chi_D(p) = -1$, then p is inert; $(p) = \mathfrak{p}$ with $N\mathfrak{p} = p^2$.

An integral ideal $\mathfrak{a} \neq 0$ which has no rational integral divisors other than ± 1 is said to be primitive. Every primitive ideal can be written uniquely as the $\mathbb{Z}$-module

$$(3.74)\qquad \mathfrak{a} = a\mathbb{Z} + \tfrac{1}{2}(b + \sqrt{D})\mathbb{Z} = [a, \tfrac{1}{2}(b + \sqrt{D})]$$

with $a = N\mathfrak{a}$ and b the integer determined by $-a < b \leqslant a$, $b^2 \equiv D (\mathrm{mod}\ 4a)$. Note that we have $b^2 - 4ac = D$ with $(a, b, c) = 1$. This is clear from the following computation:

$$\begin{aligned} (a) = (\mathfrak{a}\bar{\mathfrak{a}}) &= [a, \tfrac{1}{2}(b + \sqrt{D})][a, \tfrac{1}{2}(b - \sqrt{D})]\\ &= [a^2, \frac{a}{2}(b + \sqrt{D}), \frac{a}{2}(b - \sqrt{D}), ac]\\ &\subset [a^2, ab, ac] = a[a, b, c] \subset (a). \end{aligned}$$

For a primitive ideal $\mathfrak{a}$ we set the point in the upper half-plane

$$(3.75)\qquad z_{\mathfrak{a}} = \frac{b + \sqrt{D}}{2a} \in \mathbb{H}.$$

The inverse ideal $\mathfrak{a}^{-1}$ is the free $\mathbb{Z}$-module generated by 1 and $\bar{z}_{\mathfrak{a}}$,

$$(3.76)\qquad \mathfrak{a}^{-1} = \mathbb{Z} + \frac{b - \sqrt{D}}{2a}\mathbb{Z}.$$

Let $\mathfrak{d}$ denote the different of K, so by definition $\mathfrak{d}^{-1}$ is the fractional ideal

$$\mathfrak{d}^{-1} = \{\alpha \in K : Tr\ \alpha \in \mathbb{Z}\}.$$

In our case the different of the quadratic field $K = \mathbb{Q}(\sqrt{D})$ is the principal ideal $\mathfrak{d} = (\sqrt{D})$. Note that $\bar{\mathfrak{d}} = \mathfrak{d}$ and $N\mathfrak{d} = |D|$.

Let I be the group of all non-zero fractional ideals

$$I = \left\{ \frac{\mathfrak{a}_1}{\mathfrak{a}_2} : \mathfrak{a}_1, \mathfrak{a}_2 \subset \mathcal{O}, \mathfrak{a}_1\mathfrak{a}_2 \neq 0 \right\},$$

and $P \subset I$ the subgroup of principal ideals $(\alpha) = \alpha\mathcal{O}$ with $\alpha \in K^\star$. The factor group $H = I/P$ is called the class group of K. It is a finite abelian group of order $h = |H| = [I : P]$ which is called the class number of K (also of the discriminant D). Every class $\mathcal{A}$ is represented by a unique primitive ideal $\mathfrak{a}$ with $z_{\mathfrak{a}}$ in the standard fundamental domain (such ideal is called reduced) $F = F^- \cup F^+$, where

$$\begin{aligned} F^- &= \left\{ z = x + iy \in \mathbb{H} : -\tfrac{1}{2} < x < 0 \text{ and } |z| > 1 \right\} \\ F^+ &= \left\{ z = x + iy \in \mathbb{H} : 0 \leqslant x \leqslant \tfrac{1}{2} \text{ and } |z| = 1 \right\}. \end{aligned} \tag{3.77}$$

Therefore the class number $h = h(D)$ is equal to the number of primitive reduced ideals. For a primitive ideal written as $\mathfrak{a} = [a, \frac{1}{2}(b + \sqrt{D})]$ with $b^2 - 4ac = D$ we have $|z_{\mathfrak{a}}|^2 = \frac{c}{a}$, whence for $\mathfrak{a}$ to be reduced it says that either $-a < b \leqslant a < c$ or $0 \leqslant b \leqslant a = c$. Conversely any integral solution to $b^2 - 4ac = D$ satisfying the above inequalities yields a primitive reduced ideal defined by (3.74).

Now we go to characters. First we consider Dirichlet characters on $\mathcal{O}$. Let $\mathfrak{m}$ be a non-zero integral ideal in $\mathcal{O}$. The residue classes $\alpha \pmod{\mathfrak{m}}$, $\alpha \in \mathcal{O}$, $(\alpha, \mathfrak{m}) = 1$ form a multiplicative group $(\mathcal{O}/\mathfrak{m})^\star$. Any homomorphism

$$\chi : (\mathcal{O}/\mathfrak{m})^\star \to \mathbb{C}^\star$$

is called a Dirichlet character modulo $\mathfrak{m}$. Naturally we extend χ to $\mathcal{O}$ by setting $\chi(\alpha) = 0$ if $(\alpha, \mathfrak{m}) \neq 1$. If χ is a character to modulus $\mathfrak{m}$ then (by natural restrictions) it can be also viewed as a character to any modulus $\mathfrak{n} \subset \mathfrak{m}$ (recall that this inclusion of ideals means $\mathfrak{m}$ divides $\mathfrak{n}$). For any $\chi \pmod{\mathfrak{m}}$ and any $\beta \in (\mathfrak{d}\mathfrak{m})^{-1}$ we define the Gauss sum

$$\tau_\chi(\beta) = \sum_{\alpha \pmod{\mathfrak{m}}} \chi(\alpha) e(\operatorname{Tr}(\alpha\beta)) \tag{3.78}$$

This is well defined, because if $\alpha_1 \equiv \alpha_2 \pmod{\mathfrak{m}}$ then $\beta(\alpha_1 - \alpha_2) \in (\mathfrak{d}\mathfrak{m})^{-1}\mathfrak{m} = \mathfrak{d}^{-1}$ and $\operatorname{Tr}(\beta(\alpha_1 - \alpha_2)) \in \mathbb{Z}$. The same argument shows that $\tau_\chi(\beta)$ depends only on β $\pmod{\mathfrak{d}^{-1}}$, i.e. $\tau_\chi(\gamma) = \tau_\chi(\beta)$ if $\gamma - \beta \in \mathfrak{d}^{-1}$. Clearly

$$\tau_\chi(\alpha\beta) = \bar{\chi}(\alpha)\tau_\chi(\beta) \text{ if } (\alpha, \mathfrak{m}) = 1.$$

The conductor of $\chi \pmod{\mathfrak{m}}$ is the largest ideal $\mathfrak{f} \supset \mathfrak{m}$ (that is the smallest divisor $\mathfrak{f}$ of $\mathfrak{m}$) such that χ factors through $(\mathcal{O}/\mathfrak{f})^\star$. If $\mathfrak{f} = \mathfrak{m}$ then the character is said to be primitive.

PROPOSITION 3.7. *If $\chi \pmod{\mathfrak{m}}$ is primitive, then*

$$|\tau_\chi(\beta)|^2 = \begin{cases} N\mathfrak{m} & \text{if } (\beta\mathfrak{d}, \mathfrak{m}) = 1, \\ 0 & \text{otherwise.} \end{cases} \tag{3.79}$$

After having introduced characters of $(\mathcal{O}/\mathfrak{m})^\star$, we look for (unitary) characters of $\mathbb{C}^\star$. By definition these are continuous homomorphisms of $\mathbb{C}^\star$ into the unit circle

$$S^1 = \{\alpha \in \mathbb{C} \mid |\alpha| = 1\}.$$

Any such character $\chi_\infty : \mathbb{C}^\star \to S^1$ is given by

$$\chi_\infty(\alpha) = \left(\frac{\alpha}{|\alpha|}\right)^\ell$$

for some $\ell \in \mathbb{Z}$. We call ℓ the frequency of χ_∞.

Next we turn to characters on ideal classes. Recall we have already introduced the following groups:

- I the multiplicative group of fractional ideals $\neq 0$.
- P the subgroup of principal ideals.
- $H = I/P$ the ideal class group.

Now for any integral ideal $\mathfrak{m} \subset \mathcal{O}$, $\mathfrak{m} \neq \mathcal{O}$, we set the subgroups

$$I_\mathfrak{m} = \{\mathfrak{a} \in I \mid (\mathfrak{a}, \mathfrak{m}) = 1\} \subset I,$$
$$P_\mathfrak{m} = \{(\alpha) \in P \mid (\alpha, \mathfrak{m}) = 1\} \subset P.$$

Here $(\mathfrak{a}, \mathfrak{m}) = 1$ means $\mathfrak{a} = \mathfrak{b}\mathfrak{c}^{-1}$ for some $\mathfrak{b}$, $\mathfrak{c} \subset \mathcal{O}$, with $(\mathfrak{b}\mathfrak{c}, \mathfrak{m}) = 1$.

EXERCISE 10. Show that $(\alpha, \mathfrak{m}) = 1$ means $\alpha = \beta\gamma^{-1}$ for some β, $\gamma \in \mathcal{O}$, $(\beta\gamma, \mathfrak{m}) = 1$.

The factor groups $I_\mathfrak{m}/P_\mathfrak{m}$ are isomorphic to $I/P = H$. Any homomorphism of H into S^1 is called a class group character. There are exactly $h = |H| = |\hat{H}|$ class group characters.

Finally we introduce the Hecke characters ("Grössencharakteren"). A Hecke character modulo $\mathfrak{m}$ is a continuous homomorphism $\psi : I_\mathfrak{m} \to S^1$ for which there exist two characters

$$\chi : (\mathcal{O}/\mathfrak{m})^\star \to S^1$$
$$\chi_\infty : \mathbb{C}^\star \to S^1$$

such that

$$\psi((\alpha)) = \chi(\alpha)\chi_\infty(\alpha)$$

for every $\alpha \in \mathcal{O}$, $(\alpha, \mathfrak{m}) = 1$. Note that χ (mod $\mathfrak{m}$) and χ_∞ are determined uniquely by ψ. Indeed we have

$$\chi_\infty(\alpha) = \psi((\alpha)) \quad \text{if} \quad \alpha \equiv 1 \pmod{\mathfrak{m}}.$$

Moreover, the group $\{\alpha \in K^\star \mid \alpha \equiv 1 \pmod{\mathfrak{m}}\}$ is dense in $\mathbb{C}^\star$ so the above formula defines χ_∞ uniquely on $\mathbb{C}^\star$ by continuity. After having determined χ_∞ we find χ by

$$\chi(\alpha) = \psi((\alpha))/\chi_\infty(\alpha).$$

Not every pair of characters χ (mod $\mathfrak{m}$) and χ_∞ come from a Hecke character ψ (mod $\mathfrak{m}$). Indeed, if $\varepsilon \in U$, then we have $\chi(\varepsilon)\chi_\infty(\varepsilon) = \psi((\varepsilon)) = 1$. One can show that the condition

$$\chi(\varepsilon)\chi_\infty(\varepsilon) = 1 \text{ for all } \varepsilon \in U \tag{3.80}$$

is sufficient for the pair of χ (mod $\mathfrak{m}$) and χ_∞ to result from a Hecke character ψ (mod $\mathfrak{m}$). However, a pair of characters χ (mod $\mathfrak{m}$) and χ_∞ satisfying the above "units consistency condition" determine ψ (mod $\mathfrak{m}$) only up to a class group character, i.e. it induces exactly h Hecke characters.

In our case the group U is cyclic of order $w = 4$, 6 or 2 generated by the root of unity $\zeta = e^{2\pi i/w}$. Therefore the condition (3.80) reduces to that for $\varepsilon = \zeta$, which becomes

$$\chi(\zeta) = \zeta^{-\ell}. \tag{3.81}$$

In other words, this is a condition for $\ell \pmod w$. If $D < -4$, we have $\zeta = -1$, so the units consistency condition becomes

$$\ell = \tfrac{1}{2}(\chi(-1) - 1) \pmod 2. \tag{3.82}$$

The conductor of a Hecke character $\psi \pmod{\mathfrak{m}}$ is the smallest divisor $\mathfrak{f}$ of $\mathfrak{m}$ such that ψ is the restriction of a Hecke character to modulus $\mathfrak{f}$. This is the same as the conductor of $\chi \pmod{\mathfrak{m}}$. If $\mathfrak{f} = \mathfrak{m}$, then ψ is said to be primitive. Therefore $\psi \pmod m$ is primitive if and only if the corresponding $\chi \pmod{\mathfrak{m}}$ is primitive. The Hecke characters of modulus $\mathfrak{m} = (1)$ and of the frequency $\ell = 0$ correspond exactly to the class group characters. Hence there can be many primitive characters of conductor one.

Let ψ be a primitive Hecke character of conductor $\mathfrak{m}$ and of frequency ℓ. We consider ψ as a multiplicative function on all non-zero integral ideals by setting $\psi(\mathfrak{a}) = 0$ if $(\mathfrak{a}, \mathfrak{m}) \neq 1$. To ψ we associate the Hecke L-function

$$L(s, \psi) = \sum_{0 \neq \mathfrak{a} \subset \mathcal{O}} \psi(\mathfrak{a})(N\mathfrak{a})^{-s}.$$

This series converges absolutely for $\mathrm{Re}\,(s) > 1$ and has an Euler product over prime ideals

$$L(s, \psi) = \prod_{\mathfrak{p}} (1 - \psi(\mathfrak{p})(N\mathfrak{p})^{-s})^{-1}.$$

We also introduce the local factor at the infinite place

$$L_\infty(s, \psi) = (|D|N\mathfrak{m})^{s/2}(2\pi)^{-s}\Gamma(s + \tfrac{1}{2}|\ell|). \tag{3.83}$$

THEOREM 3.8 (HECKE). *Let $\psi \pmod{\mathfrak{m}}$ be primitive. Then the complete product $\Lambda(s, \psi) = L_\infty(s, \psi)L(s, \psi)$ has analytic continuation to the whole complex s-plane except for a simple pole at $s = 1$ when $\psi = \psi_0$ is the trivial character. In this case $\mathfrak{m} = (1)$, $\ell = 0$ and $L(s, \psi_0) = \zeta_K(s) = \zeta(s)L(s, \chi_D)$ is just the Dedekind zeta function of K, so*

$$\operatorname*{res}_{s=1} \Lambda(s, \psi_0) = hw^{-1}.$$

Moreover, $\Lambda(s, \psi)$ satisfies the functional equation

$$\Lambda(s, \psi) = W(\psi)\Lambda(1 - s, \bar{\psi}) \tag{3.84}$$

with the root number $W(\psi)$ given by the corresponding Gauss sum

$$W(\psi) = i^{-\ell}\tau(\psi)(N\mathfrak{m})^{-1/2}. \tag{3.85}$$

The Gauss sum $\tau(\psi)$ associated with $\psi \pmod{\mathfrak{m}}$ is defined by

$$\tau(\psi) = \frac{\chi_\infty(\gamma)}{\psi(\mathfrak{c})} \sum_{\alpha \in \mathfrak{c}/\mathfrak{cm}} \chi(\alpha) e\Big(\mathrm{Tr}\,\frac{\alpha}{\gamma}\Big) \tag{3.86}$$

where $\gamma \in \mathcal{O}$ and $\mathfrak{c} \subset \mathcal{O}$ are such that $(\mathfrak{c}, \mathfrak{m}) = 1$, $\mathfrak{c}\mathfrak{d}\mathfrak{m} = (\gamma)$. Note that the definition of $\tau(\psi)$ does not depend on the choice of γ and $\mathfrak{c}$ due to the units consistency condition (3.80).

EXERCISE 11. Show that for a primitive Hecke character ψ (mod $\mathfrak{m}$) whose Dirichlet component is χ (mod $\mathfrak{m}$) we have

$$\tau(\psi) = \psi(\mathfrak{a})\tau_{\chi}(\beta) \tag{3.87}$$

for some $(\mathfrak{a}, \mathfrak{m}) = 1$ and $\beta \in (\mathfrak{d}\mathfrak{m})^{-1}$.

EXERCISE 12. Prove that for a primitive Hecke character ψ (mod $\mathfrak{m}$),

$$|\tau(\psi)| = (N\mathfrak{m})^{1/2}. \tag{3.88}$$

EXERCISE 13. If ψ modulo $\mathfrak{m}$ and ξ modulo $\mathfrak{n}$ are primitive Hecke characters with $(\mathfrak{m}, \mathfrak{n}) = 1$, then

$$\tau(\psi\xi) = \psi(\mathfrak{n})\xi(\mathfrak{m})\tau(\psi)\tau(\xi).$$

REMARK. If ψ is a primitive Hecke character of conductor $\mathfrak{m} = (1)$ and frequency ℓ, then $\tau(\psi) = i^{\ell}$. In this case the functional equation (3.84) has the root number $W(\psi) = 1$. Since $\bar{\psi}(\mathfrak{a}) = \psi(\bar{\mathfrak{a}})$ we find that $L(s, \psi) = L(s, \bar{\psi})$ by changing $\mathfrak{a}$ into $\bar{\mathfrak{a}}$ in the Dirichlet series.

If ψ(mod $\mathfrak{m}$) has frequency ℓ, then its complex conjugate $\overline{\psi}$(mod $\overline{\mathfrak{m}}$) has frequency $-\ell$. Therefore, without loss of generality, we can assume that ψ(mod $\mathfrak{m}$) has frequency $\ell \geqslant 0$. Then the function $f : \mathbb{H} \to \mathbb{C}$ given by

$$f(z) = \sum_{\mathfrak{a}} \psi(\mathfrak{a})(N\mathfrak{a})^{\frac{\ell}{2}} e(zN\mathfrak{a}) \tag{3.89}$$

is an automorphic form of weight $k = \ell + 1$ on $\Gamma_0(N)$ of level $N = |D|N\mathfrak{m}$ and character χ(mod N) defined by

$$\chi(n) = \chi_D(n)\psi(n) \quad \text{if } n \in \mathbb{Z}. \tag{3.90}$$

If $\ell > 0$, then f is a cusp form, which is primitive if the character ψ(mod $\mathfrak{m}$) is primitive. In particular, if $\mathfrak{m} = (1)$, then $N = |D|$ and $\chi = \chi_D$.

REMARKS. We say a few words about the proofs of the above results. First the Fourier series (3.89) splits into a finite sum over ideal classes. In each class we get an infinite series over equivalent ideals which is a theta function for a positive definite binary quadratic form twisted by a suitable harmonic polynomial (see Section 14.3). Applying the Poisson summation formula one shows that these individual theta functions are automorphic forms of the same type (cf. Chapter 10 of [**I4**]). Hence the f, being a linear combination of theta functions, is also an automorphic form of the same type. Another application of Poisson summation shows that the theta functions satisfy a Jacobi type involution formula. Summing over the classes these formulas yield

$$z^k f\Big(\frac{z}{\sqrt{N}}\Big) = \frac{i\tau(\psi)}{\sqrt{N\mathfrak{m}}} \bar{f}\Big(\frac{-1}{z\sqrt{N}}\Big)$$

(at this point it is crucial that ψ(mod $\mathfrak{m}$) is primitive), where $\tau(\psi)$ is the Gauss sum and $N = |D|N\mathfrak{m}$. From this, one derives the functional equation (3.84) by taking the Mellin transform with respect to $z = iy$. See also Chapter 14.

We end this section by giving specific examples of Hecke characters on imaginary quadratic fields.

EXAMPLE 1. For $K = \mathbb{Q}(i)$ the discriminant is $D = -4$, the ring of integers is $\mathcal{O} = \mathbb{Z}[i]$, the group of units is $U = \{\pm 1, \pm i\}$ and the class number is $h = 1$, thus every ideal is principal. Fix an integer ℓ. Put $\mathfrak{m} = (1)$ or $\mathfrak{m} = 2(1+i)$ according to $4 \mid \ell$ or $4 \nmid \ell$. Define $\xi : I_{\mathfrak{m}} \to \mathbb{C}^\star$ by

$$\xi(\mathfrak{a}) = \Big(\frac{\alpha}{|\alpha|}\Big)^\ell \tag{3.91}$$

where α is the unique primary element which generates $\mathfrak{a}$, with $\alpha \equiv 1 (\mathrm{mod}\ 2(1+i))$. Then ξ is a primitive character of frequency ℓ and conductor $\mathfrak{m}$.

EXAMPLE 2. Let $K = \mathbb{Q}(i)$ be as above. Take a positive integer $q \equiv 0 (\mathrm{mod}\ 4)$ and $\chi : (\mathbb{Z}[i]/q\mathbb{Z}[i])^\star \to \mathbb{C}^\star$ a character on the multiplicative group of primitive residue classes modulo q in $\mathbb{Z}[i]$. Define $\xi : I_q \to \mathbb{C}^\star$ by

$$\xi(\mathfrak{a}) = \chi(\alpha)\Big(\frac{\alpha}{|\alpha|}\Big)^\ell$$

where α is the unique primary element which generates $\mathfrak{a}$. Then ξ is a Hecke character (not necessarily primitive) of frequency ℓ and modulus q.

EXAMPLE 3. Let $K = \mathbb{Q}(\sqrt{D})$ have discriminant $D < -3$, $D \equiv 1 (\mathrm{mod}\ 4)$. Then the ring of integers is

$$\mathcal{O} = \Big\{\tfrac{1}{2}(m + n\sqrt{D}) : m, n \in \mathbb{Z},\ \ m \equiv n (\mathrm{mod}\ 2)\Big\}, \tag{3.92}$$

the group of units is $U = \{\pm 1\}$ while the class number $h = h(D)$ can be arbitrarily large. Since the principal ideal $(\sqrt{D})$ has norm $|D|$ the ring inclusion $\mathbb{Z} \subset \mathcal{O}$ defines an isomorphism

$$\mu \begin{cases} \mathcal{O}/(\sqrt{D}) \to \mathbb{Z}/|D|\mathbb{Z} \\ \dfrac{m + n\sqrt{D}}{2} \mapsto \dfrac{m}{2}\ (\mathrm{mod}\ |D|). \end{cases} \tag{3.93}$$

Composing μ with the Jacobi symbol we get a quadratic character $\varepsilon : \mathcal{O} \to \{0, \pm 1\}$, explicitly

$$\varepsilon\Big(\frac{m + n\sqrt{D}}{2}\Big) = \Big(\frac{2m}{|D|}\Big). \tag{3.94}$$

Note that ε is odd, i.e., $\varepsilon(-\alpha) = -\varepsilon(\alpha)$ because $|D| \equiv -1 (\mathrm{mod}\ 4)$. Given ℓ odd, we have h Hecke characters $\xi : I \to \mathbb{C}$ such that on principal ideals $\mathfrak{a} = (a)$,

$$\xi(\mathfrak{a}) = \varepsilon(\alpha)\Big(\frac{\alpha}{|\alpha|}\Big)^\ell. \tag{3.95}$$

These are primitive characters of frequency ℓ and conductor $\mathfrak{m} = (\sqrt{D})$.

EXAMPLE 4. Let $K = \mathbb{Q}(\sqrt{D})$ be any imaginary quadratic field with w units. Given $\ell \equiv 0 (\mathrm{mod}\ w)$, we have $h = h(D)$ Hecke characters $\xi : I \to \mathbb{C}^\star$ such that $\xi(\mathfrak{a}) = \alpha/|\alpha|)^\ell$ on principal ideals $\mathfrak{a} = (\alpha)$. These are primitive characters of frequency ℓ and conductor $\mathfrak{m} = (1)$. Among these there is a special character of frequency $\ell = hw$ defined by $\xi(\mathfrak{a}) = (\alpha/|\alpha|)^w$ for any $\mathfrak{a} \in I$, where $(\alpha) = \mathfrak{a}^h$.

EXAMPLE 5. Let $K = \mathbb{Q}(\sqrt{D})$ be any imaginary quadratic field of discriminant D, and $\chi(\mathrm{mod}\ q)$ a primitive Dirichlet character of conductor q with $(q, D) = 1$. Define the homomorphism $\chi \circ N : I_q \to \mathbb{C}^\star$ by

$$(\chi \circ N)(\mathfrak{a}) = \chi(N\mathfrak{a}). \tag{3.96}$$

Then $\xi = \chi \circ N$ is a Hecke character of frequency zero which is primitive of conductor $\mathfrak{m} = (q)$. In this case

$$\tau(\xi) = \chi_D(q)\chi(|D|)\tau^2(\chi) \tag{3.97}$$

where $\tau(\chi)$ is the Gauss sum for the Dirichlet character $\chi(\mathrm{mod}\ q)$.

CHAPTER 4

SUMMATION FORMULAS

4.1. Introduction.

Series of arithmetic functions often undergo some kind of involutary transformations. These are derived not by re-grouping and changing the order of terms but deeper changes are made by applying Fourier analysis. For example, the equation

$$\sum_{m\in\mathbb{Z}} e^{-\pi m^2/y} = \sqrt{y}\sum_{n\in\mathbb{Z}} e^{-\pi n^2 y} \tag{4.1}$$

cannot be verified simply by a combinatorial argument. A rich source of relations of this kind lies in automorphic theory. For instance, for a classical cusp form $f(z)$ of weight k on the modular group $\Gamma = SL_2(\mathbb{Z})$ whose Fourier series is

$$f(z) = \sum_1^\infty \lambda(n) n^{\frac{k-1}{2}} e(nz) \tag{4.2}$$

the automorphy equation $f(z) = z^{-k} f(-1/z)$ yields (by integration along the vertical line $z = iy$) the formula

$$\sum_1^\infty \lambda(m) g(m) = \sum_1^\infty \lambda(n) h(n) \tag{4.3}$$

where $g(x)$ is any smooth, compactly supported function on $\mathbb{R}^+$ and

$$h(y) = 2\pi i^k \int_0^\infty J_{k-1}(4\pi\sqrt{xy}) g(x) dx. \tag{4.4}$$

See more general results in (4.71) and (4.72).

Modular transformations are typical but there are plenty of other important cases. For example the formulas (15.30) (the Selberg trace formula) or (16.34) (the Kuznetsov formula for sums of Kloosterman sums) do not come from automorphy equations. Naturally an equation connecting one series of an arithmetic function (weighted by a test function of certain class) with another is called a summation formula. In this chapter we develop a few such formulas and give standard applications.

4.2. The Euler-Maclaurin formula.

We begin by considering a function defined on a segment of real numbers. Suppose $a, b \in \mathbb{Z}$ and f is continuous in $[a, b]$. Then the sum of $f(n)$ can be written as the Stieltjes integral

$$\sum_{a<n\leqslant b} f(n) = \int_a^b f(x)d[x] \tag{4.5}$$

where $[x]$ denotes the integral part of x (see (1.65)). Putting

$$\psi(x) = x - [x] - \tfrac{1}{2} \tag{4.6}$$

we derive by partial integration the following Euler-Maclaurin formula.

LEMMA 4.1. *For f of class C^1 on $[a, b]$ with $a < b, a, b \in \mathbb{Z}$ we have*

$$\sum_{a<n\leqslant b} f(n) = \int_a^b (f(x) + \psi(x)f'(x))dx + \tfrac{1}{2}(f(b) - f(a)). \tag{4.7}$$

This formula produces good results when $f'(x)$ is relatively small. If the higher order derivatives of f exist and are very small, it can be profitable to continue integrating by parts. To this end we need the Bernoulli polynomials $B_k(X)$.

We define $B_k(X) \in \mathbb{Q}[X]$ by the following recurrence conditions:

$$B_0(X) = 1, \tag{4.8}$$

$$B_k'(X) = kB_{k-1}(X), \tag{4.9}$$

$$\int_0^1 B_k(x)dx = 0, \tag{4.10}$$

if $k \geqslant 1$. The condition (4.9) defines $B_k'(X)$ in terms of $B_{k-1}(X)$ up to a constant while the last condition (4.10) (the orthogonality of $B_k(X)$ to constants) determines the constant. The monomial X^k satisfies the first two conditions but it fails the third one. We find that $B_k(X) = X^k - \frac{k}{2}X^{k-1} + \dots$. In particular, we have $B_1(X) = X - \frac{1}{2}$ and $B_2(X) = X^2 - X + \frac{1}{6}$. The generating power series for the Bernoulli polynomials $B_k(X)$ is

$$F(t, X) = \sum_0^\infty B_k(X)\frac{t^k}{k!} = \frac{te^{tX}}{e^t - 1}. \tag{4.11}$$

The last equality follows by noticing that

$$\frac{\partial}{\partial X}F(t, X) = \sum_1^\infty B_{k-1}(X)\frac{t^k}{(k-1)!} = tF(t, X).$$

EXERCISE 1. Prove that

$$B_k(1-X) = (-1)^k B_k(X)$$
$$B_k(X+1) - B_k(X) = kX^{k-1}$$
$$\sum_{0\leqslant a<q} B_k\Big(X+\frac{a}{q}\Big) = q^{1-k}B_k(qX).$$

Next we define the Bernoulli numbers as constant terms of the Bernoulli polynomials

$$B_k = B_k(0). \tag{4.12}$$

The generating power series for Bernoulli numbers is

$$F(t) = \sum_0^\infty B_k\frac{t^k}{k!} = \frac{t}{e^t-1}. \tag{4.13}$$

Since $F(-t) = t + F(t)$, it follows that $B_k = 0$ if $k > 1$, k odd.

EXERCISE 2. Prove that

$$B_k(X) = \sum_{\ell=0}^k \binom{k}{\ell} B_\ell X^{k-\ell}. \tag{4.14}$$

REMARK. The Bernoulli polynomials and numbers have been generalized with respect to a character $\chi(\text{mod } q)$ as follows:

$$\sum_0^\infty B_{k,\chi}(X)\frac{t^k}{k!} = \sum_{1\leqslant\leqslant q}\chi(a)\frac{te^{at}}{e^{qt}-1}.$$

The constant term of $B_{k,\chi}(X)$ is the generalized Bernoulli number, namely

$$B_{k,\chi} = B_{k,\chi}(0) = q^{k-1}\sum_{0\leqslant a<q}\chi(a)B_k\Big(\frac{a}{q}\Big).$$

The generalized Bernoulli numbers occur in values of L-functions at special points. They truly belong to the theory of p-adic L-functions (due to Leopoldt, see e.g. [**W**]).

For any $k \geqslant 0$ we put

$$\psi_k(x) = B_k(\{x\}) \tag{4.15}$$

where $\{x\} = x - [x]$ is the fractional part of x. These functions are periodic of period one, so is the generating function

$$\sum_{k=0}^\infty \psi_k(x)\frac{t^k}{k!} = \frac{te^{t\{x\}}}{e^t-1}.$$

Therefore they can be represented by Fourier series. The n-th Fourier coefficient of the generating function is

$$\int_0^1 \Big(\sum_{k=0}^{\infty} \psi_k(x)\frac{t^k}{k!}\Big)e(-nx)dx = \frac{t}{e^t-1}\int_0^1 e^{(t-2\pi in)x}dx$$
$$= \frac{t}{t-2\pi in} = -\sum_{k=1}^{\infty}\Big(\frac{t}{2\pi in}\Big)^k.$$

Hence the n-th Fourier coefficient of $\psi_k(x)$ is $-k!(2\pi in)^{-k}$ giving the expansion

$$\psi_k(x) = -k!\sum_{n\neq 0}(2\pi in)^{-k}e(nx) \tag{4.16}$$

(there is no term $n=0$ by (4.10)). This series converges absolutely if $k \geqslant 2$. For $k=1$ we arrange the terms symmetrically getting the series

$$\psi(x) = -\sum_1^{\infty}(\pi n)^{-1}\sin(2\pi nx) \tag{4.17}$$

which converges pointwise and boundedly except for $x \in \mathbb{Z}$.

EXERCISE 3. Prove that for any $N \geqslant 1$ and $x \in \mathbb{R}$

$$\psi(x) = -\sum_1^{N}(\pi n)^{-1}\sin(2\pi nx) + O((1+\|x\|N)^{-1}) \tag{4.18}$$

where $\|x\|$ is the distance of x to the nearest integer

$$\|x\| = \min\{|x-m| : m \in \mathbb{Z}\}. \tag{4.19}$$

REMARK. Evaluating (4.16) at $x=0$ we get

$$\zeta(2m) = \frac{-(2\pi i)^{2m}}{(2m)!}B_{2m}$$

for $m \geqslant 1$. This turns out to be also true for $m=0$ by the functional equation (4.75). Moreover, by the functional equation (4.75) it follows that for any $m \geqslant 1$,

$$\zeta(1-2m) = -\frac{1}{2m}B_{2m}.$$

Furthermore, for a primitive character χ of conductor $q>1$ one has

$$L(m,\chi) = \frac{-1}{2m!}\Big(\frac{-2\pi i}{q}\Big)^m \tau(\chi)B_{m,\overline{\chi}}$$
$$L(1-m,\chi) = \frac{-1}{m}B_{m,\chi}$$

for any $m \geqslant 1$ with $m \equiv \frac{1}{2}(1-\chi(-1)) \pmod 2$ (see e.g. [**IR**], [**W**]).

Integrating by parts $k-1$ times in (4.7) and using (4.9) one derives the Euler-Maclaurin formula of order k.

THEOREM 4.2. *Suppose $f \in C^k([a,b])$ with $a < b, a, b \in \mathbb{Z}$. Then*

$$\sum_{a<n\leqslant b} f(n) = \int_a^b \left(f(x) - \frac{(-1)^k}{k!}\psi_k(x) f^{(k)}(x) \right) dx + \sum_{\ell=1}^{k} \frac{(-1)^\ell}{\ell!} (f^{(\ell-1)}(b) - f^{(\ell-1)}(a)) B_\ell. \tag{4.20}$$

Very often the first order Euler-Maclaurin formula (4.7) is as good as (4.20). Inserting (4.18) into (4.7) and integrating by parts we get

COROLLARY 4.3. *For $f \in C^1([a,b])$ with $a < b, a, b \in \mathbb{Z}$ we have*

$$\sum_{a\leqslant n\leqslant b}{}' f(n) = \sum_{|n|\leqslant N} \int_a^b f(x) e(nx) dx + O\Big(\int_a^b \frac{|f'(x)| dx}{1 + N\|x\|} \Big) \tag{4.21}$$

where N is any positive integer and the implied constant is absolute. Here and hereafter the dash indicates that the terms at the end-points of summation are taken with half values.

4.3. The Poisson summation formula.

This section is based on classical Fourier analysis as described in the Appendix. Recall that for any function $f \in L^1(\mathbb{R})$ its Fourier transform is defined by

$$\hat{f}(y) = \int_{\mathbb{R}} f(x) e(-xy) dx. \tag{4.22}$$

THEOREM 4.4. *Suppose that both $f, \hat{f}$ are in $L^1(\mathbb{R})$ and have bounded variation. Then*

$$\sum_{m\in\mathbb{Z}} f(m) = \sum_{n\in\mathbb{Z}} \hat{f}(n) \tag{4.23}$$

where both series converge absolutely.

PROOF. Consider the function

$$F(x) = \sum_{m\in\mathbb{Z}} f(x+m)$$

which is periodic of period one. This has the absolutely convergent Fourier series expansion

$$F(x) = \sum_{n\in\mathbb{Z}} c_F(n) e(nx)$$

with coefficients given by

$$c_F(n) = \int_0^1 F(t) e(-nt) dt = \int_{-\infty}^{\infty} f(t) e(-nt) dt = \hat{f}(n).$$

Taking $F(0)$ we get the Poisson summation formula (4.23). □

EXERCISE 4. Derive (4.23) from (4.21).

Changing $f(x)$ into $f(vx+u)$ with $v \in \mathbb{R}^+$ and $u \in \mathbb{R}$ the formula (4.23) generalizes to

$$\sum_{m\in\mathbb{Z}} f(vm+u) = \frac{1}{v}\sum_{n\in\mathbb{Z}} \hat{f}\Big(\frac{n}{v}\Big)e\Big(\frac{un}{v}\Big). \tag{4.24}$$

By the Fourier inversion $\hat{\hat{f}}(x) = f(-x)$ this can also be written as

$$\sum_{n\in\mathbb{Z}} f\Big(\frac{n}{v}\Big)e\Big(\frac{un}{v}\Big) = v\sum_{m\in\mathbb{Z}} \hat{f}(vm-u). \tag{4.25}$$

EXERCISE 5. Using (4.24) prove that for a primitive character $\chi \pmod q$,

$$\sum_{m\in\mathbb{Z}} f(m)\chi(m) = \frac{\tau(\chi)}{q}\sum_{n\in\mathbb{Z}} \hat{f}\Big(\frac{n}{q}\Big)\overline{\chi}(n) \tag{4.26}$$

where $\tau(\chi)$ is the Gauss sum.

EXAMPLES. From the Fourier pairs (4.83), (4.84), (4.85) we obtain the following equations:

$$\sum_{|n|\leqslant y}\Big(1-\frac{|n|}{y}\Big)e(nx) = y\sum_m\Big(\frac{\sin\pi y(m+x)}{\pi y(m+x)}\Big)^2, \tag{4.27}$$

$$\sum_n e(nx)e^{-2\pi|n|y} = (\pi y)^{-1}\sum_m |m+x+iy|^{-2}, \tag{4.28}$$

$$\sum_n e(nx)e^{-\pi n^2/y} = \sqrt{y}\sum_m e^{-\pi(m+x)^2y}. \tag{4.29}$$

for any $x \in \mathbb{R}$ and $y \in \mathbb{R}^+$. The third equation generalizes (4.1). The second equation has a geometric series on its left side, so by a direct summation

$$\sum_n e(nx)e^{-2\pi|n|y} = \frac{\sinh 2\pi y}{\cosh 2\pi y - \cos 2\pi x}.$$

The first sum can be also executed by using geometric series; it yields

$$\sum_{n\leqslant y}\Big(1-\frac{|n|}{y}\Big)e(nx) = \frac{1}{y}\Big(\frac{\sin\pi x[y]}{\sin\pi x}\Big)^2 + \frac{\{y\}}{y}\frac{\sin\pi x(2[y]+1)}{\sin\pi x}.$$

If y is a positive integer this reduces to $y^{-1}(\sin\pi xy/\sin\pi x)^2$.

The same method (averaging of integral translations) works in several variables giving

THEOREM 4.5. *Suppose f is in the Schwartz class $\mathcal{S}(\mathbb{R}^\ell)$. Then*

$$\sum_{m\in\mathbb{Z}^\ell} f(m) = \sum_{n\in\mathbb{Z}^\ell} \hat{f}(n). \tag{4.30}$$

The formulas (4.24) and (4.25) extend to several variables in a similar fashion. The assumption that f is a Schwartz function can be weakened considerably. In two dimensions the Poisson formula (4.30) is just the trace formula for the Laplace operator and an invariant integral operator on the torus $\mathbb{R}^2/\mathbb{Z}^2$, see [**I4**].

4.4. Summation formulas for the ball.

If f is radial, then so is $\hat{f}$ (see Lemma 4.17). In this case the Poisson formula (4.30) becomes a summation formula for the arithmetic function

$$r_\ell(m) = |\{m_1, \dots, m_\ell \in \mathbb{Z} : m_1^2 + \dots + m_\ell^2 = m\}|.$$

Precisely, if $g(x)$ is smooth and compactly supported on $\mathbb{R}^+$, then

$$\sum_0^\infty r_{2k}(m)g(m)m^{\frac{1}{2}-\frac{k}{2}} = \sum_0^\infty r_{2k}(n)h(n)n^{\frac{1}{2}-\frac{k}{2}} \tag{4.31}$$

where $h(y)$ is the Hankel transform of $g(x)$ given by (4.103). Throughout we assume that $\ell = 2k$ is an integer > 1, so k is a half integer.

REMARKS. The condition that g is compactly supported can be considerably weakened. It suffices that both g and h are of type (α, β) with $\alpha < \frac{k}{2} + \frac{1}{2} < \beta$ (see (4.104)). To this end notice that

$$r_{2k}(m) \ll m^{k-1+\varepsilon}$$

for $m \geqslant 1$, the implied constant depending on ε and k (if $k \geqslant 2$, then this holds without ε).

Although there is no contribution from $m = 0$ on the left side of (4.31) there is still a contribution from $n = 0$ on the right side of (4.31) because $h(y)$ is not compactly supported in $\mathbb{R}^+$ even if $g(x)$ is. In practice this contribution yields a main term, so it is desired to single out the zero-th term. We have $r_{2k}(0) = 1$ and

$$h(y) \sim \frac{\pi^k}{\Gamma(k)} \int_0^\infty g(x)(xy)^{\frac{k-1}{2}}\,dx$$

as $y \to 0$ by the asymptotic $J_\nu(x) \sim x^\nu / 2^\nu \Gamma(\nu+1)$. Hence (4.31) can be written as

THEOREM 4.6 (SUMMATION FORMULA FOR A BALL). *Suppose g is smooth and compactly supported on $\mathbb{R}^+$. Then*

$$\sum_1^\infty r_{2k}(m)g(m)m^{\frac{1}{2}-\frac{k}{2}} = \frac{\pi^k}{\Gamma(k)} M(g) + \sum_1^\infty r_{2k}(n)h(n)n^{\frac{1}{2}-\frac{k}{2}}, \tag{4.32}$$

where $M(g)$ is the Mellin transform of g at $s = \frac{k+1}{2}$, i.e.,

$$M(g) = \int_0^\infty g(x)x^{\frac{k-1}{2}}\,dx, \tag{4.33}$$

and $h(y)$ is the Hankel type transform of g

$$h(y) = \pi \int_0^\infty g(x) J_{k-1}(2\pi\sqrt{xy})dx. \tag{4.34}$$

COROLLARY 4.7 (SUMMATION FORMULA FOR A CIRCLE). *Suppose g is smooth and compactly supported on $\mathbb{R}^+$. Then*

$$\sum_1^\infty r(m)g(m) = \pi \int_0^\infty g(x)dx + \sum_1^\infty r(m)h(m) \tag{4.35}$$

where

$$h(y) = \pi \int_0^\infty g(x) J_0(2\pi\sqrt{xy})dx, \tag{4.36}$$

and both series converge absolutely.

Summation formulas for the circle were established in one form or another by Hardy-Landau [**HaLa**] and Voronoi [**Vor**]. There are plenty of expressions for the Bessel function in the kernel of (4.36). Here we select a few particularly useful integral representations:

$$\begin{aligned}\pi J_0(z) &= \int_0^\pi \cos(z\sin\theta)d\theta \\ &= 2\int_1^\infty (t^2-1)^{-\frac12}\sin(zt)dt \\ &= 2\int_0^\infty \sin(z\,\mathrm{ch}\,t)dt = 2\int_0^\infty \sin\Big(\frac{zw}{2}\Big)\cos\Big(\frac{z}{2w}\Big)\frac{dw}{w}.\end{aligned}$$

Moreover, we have the following asymptotic expansion (see (23.451.1) of [**GR**])

$$\pi J_0(z) = \Big(\frac{2\pi}{z}\Big)^{\frac12}\Big\{\cos(z-\tfrac{\pi}{4}) + \frac{1}{8z}\sin(z-\tfrac{\pi}{4}) + O\Big(\frac{1}{z^2}\Big)\Big\} \tag{4.37}$$

valid for $z > 0$. Inserting this into (4.36) we get

$$\begin{aligned}h(y) = &\int_0^\infty (xy)^{-\frac14} g(x)\cos(2\pi\sqrt{xy}-\tfrac{\pi}{4})dx \\ &+ \frac{1}{16\pi}\int_0^\infty (xy)^{-\frac34} g(x)\sin(2\pi\sqrt{xy}-\tfrac{\pi}{4})dx + O\Big(\int_0^\infty (xy)^{-\frac54}|g(x)|dx\Big).\end{aligned}$$

Integrating by parts in the second integral we obtain

$$h(y) = \int_0^\infty (xy)^{-\frac14} g(x)\cos(2\pi\sqrt{xy}-\tfrac{\pi}{4})dx + O(R(y)) \tag{4.38}$$

where

$$R(y) = \int_0^\infty (xy)^{-\frac54}(|g(x)| + x|g'(x)|)dx. \tag{4.39}$$

Next integrating by parts in the first integral we obtain

$$h(y) = -\frac{1}{\pi y}\int_0^\infty (xy)^{\frac14} g'(x)\sin(2\pi\sqrt{xy}-\tfrac{\pi}{4})dx + O(R(y)). \tag{4.40}$$

In the case of sphere ($\ell = 3, k = \frac{3}{2}$) we encounter the Bessel function of order $\frac{1}{2}$, which is an elementary function, namely $J_{\frac{1}{2}}(z) = (2/\pi z)^{\frac{1}{2}} \sin z$, so (4.34) simplifies to

$$h(y) = \int_0^\infty (xy)^{-\frac{1}{4}} g(x) \sin(2\pi\sqrt{xy})dx.$$

Setting $G(x) = g(x)x^{-\frac{1}{4}}$ and $H(y) = h(y)y^{\frac{1}{4}}$ the formula (4.32) with $k = \frac{3}{2}$ becomes

COROLLARY 4.8 (SUMMATION FORMULA FOR A SPHERE). *Suppose G is smooth and compactly supported on $\mathbb{R}^+$. Then*

$$\sum_1^\infty r_3(m)G(m) = 2\pi \int_0^\infty x^{\frac{1}{2}} G(x)dx + \sum_1^\infty r_3(n)n^{-\frac{1}{2}} H(n) \tag{4.41}$$

where

$$H(y) = \int_0^\infty G(x) \sin(2\pi\sqrt{xy})dx, \tag{4.42}$$

and both series converge absolutely.

To illustrate the results we apply the summation formula (4.35) to improve the error term in the Gauss circle problem (see (1.70)).

COROLLARY 4.9. *We have*

$$\sum_{m \leqslant X} r(m) = \pi X + O(X^{\frac{1}{3}}). \tag{4.43}$$

PROOF. We shall establish an upper bound for the sum in question, the lower bound can be derived by applying similar arguments. In what follows it is crucial that $r(n)$ is nonnegative, so we can smooth out the sum by enlarging its range. To this end we choose the test function $g(x) \geqslant 0$ supported on $0 \leqslant x \leqslant X+Y$ in which segment $g(x) = \min\{x, 1, (X+Y-x)Y^{-1}\}$. This is not smooth, nevertheless (4.35) holds true. Here Y is a parameter at our disposal to be chosen later to minimize the resulting upper bound. We assume that $1 \leqslant Y \leqslant X^{\frac{1}{2}}$. By (4.35) we obtain

$$\sum_{m \leqslant X} r(m) \leqslant \sum_m r(m)g(m) = \pi\Big(X + \frac{Y+1}{2}\Big) + \sum_n r(n)h(n),$$

where $h(n)$ is the integral transform of g given by (4.36). Using (4.40) and (4.39) one shows that

$$h(y) \ll y^{-\frac{3}{4}} X^{\frac{1}{4}} (1 + y/Z)^{-\frac{1}{2}} \tag{4.44}$$

where $Z = XY^{-2}$. Hence

$$\sum_1^\infty r(n)h(n) = \sum_a \sum_b h(a^2 + b^2) \ll (XZ)^{\frac{1}{4}} = (X/Y)^{\frac{1}{2}}.$$

Therefore

$$\sum_{m \leqslant X} r(m) \leqslant \pi X + O(Y + X^{\frac{1}{2}} Y^{-\frac{1}{2}}).$$

Choosing $Y = X^{\frac{1}{3}}$ we balance the error terms to $O(X^{\frac{1}{3}})$. By a similar argument we establish a lower bound with the same main and error terms. This completes the proof of (4.43). □

EXERCISE 6. Prove that

$$\sum_{m\leqslant x} r(m)\Big(1-\frac{m}{x}\Big)=\frac{\pi}{2}x+\sum_{1}^{\infty}\frac{r(n)}{\pi n}J_2(2\pi\sqrt{nx})=\frac{\pi}{2}x+O(x^{-\frac{1}{4}}). \tag{4.45}$$

Note the error term is very small due to smoothing.

Similarly to (4.43) one can derive from (4.41) the following approximate formula:

$$\sum_{m\leqslant X} r_3(m)=\frac{4\pi}{3}X^{\frac{3}{2}}+O(X^{\frac{3}{4}}). \tag{4.46}$$

The strongest error term here was obtained by Chamizo – Iwaniec [**ChI**] and Heath-Brown [**HB5**]. Interestingly enough in either paper character sums were brought into play independently; they obtained the exponents $\frac{29}{44}$ and $\frac{21}{32}$ in place of $\frac{3}{4}$ respectively.

4.5. Summation formulas for the hyperbola.

First we establish summation formulas for the divisor function $\tau(n)$. They are essentially due to Voronoi [**Vor**] except for our arrangements of the integral transforms.

THEOREM 4.10. *Suppose $g(x)$ is smooth and compactly supported on $\mathbb{R}^+$. Let $ad\equiv 1 \pmod c$. Then*

(4.47)

$$\sum_{1}^{\infty}\tau(m)g\Big(\frac{m}{c}\Big)\cos\Big(\frac{2\pi am}{c}\Big)=\int_0^\infty\Big(\log\frac{x}{c}+2\gamma\Big)g(x)dx + \sum_{1}^{\infty}\tau(n)p\Big(\frac{n}{c}\Big)\cos\Big(\frac{2\pi dn}{c}\Big),$$

(4.48)

$$\sum_{1}^{\infty}\tau(m)g\Big(\frac{m}{c}\Big)\sin\Big(\frac{2\pi am}{c}\Big)=\sum_{1}^{\infty}\tau(n)q\Big(\frac{n}{c}\Big)\sin\Big(\frac{2\pi dn}{c}\Big),$$

where $p(y)$ and $q(y)$ are the integral transforms

$$p(y)=\int_0^\infty C(2\pi\sqrt{xy})g(x)dx,$$
$$q(y)=\int_0^\infty S(2\pi\sqrt{xy})g(x)dx,$$

with kernels

$$C(z)=4\int_0^\infty\cos(zw)\cos\Big(\frac{z}{w}\Big)\frac{dw}{w},$$
$$S(z)=4\int_0^\infty\sin(zw)\sin\Big(\frac{z}{w}\Big)\frac{dw}{w}.$$

These kernels are also given by (see (3.864.1) and (3.864.2) of [**GR**])

$$C(z) = 4K_0(2z) - 2\pi Y_0(2z),$$
$$S(z) = 4K_0(2z) + 2\pi Y_0(2z);$$

therefore both results can also be written in the following single formula:

$$\begin{aligned}\sum_1^\infty \tau(m)e\Big(\frac{am}{c}\Big)g(m) &= \frac{1}{c}\int_0^\infty (\log x + 2\gamma - 2\log c)g(x)dx \\ &\quad - \frac{2\pi}{c}\sum_1^\infty \tau(n)e\Big(-\frac{dn}{c}\Big)\int_0^\infty Y_0\Big(\frac{4\pi}{c}\sqrt{nx}\Big)g(x)dx \\ &\quad + \frac{4}{c}\sum_1^\infty \tau(n)e\Big(\frac{dn}{c}\Big)\int_0^\infty K_0\Big(\frac{4\pi}{c}\sqrt{nx}\Big)g(x)dx.\end{aligned} \tag{4.49}$$

We shall derive (4.49) from a still more general summation formula for the tempered divisor function

$$\tau_g(m) = \sum_{m_1m_2=m} g(m_1, m_2). \tag{4.50}$$

Here $g(x,y)$ is a function of class C^1, compactly supported on $\mathbb{R}^2$. By the way, the arithmetic function $\tau_g(m)$ is a prototype of Fourier coefficients of automorphic forms in the space of continuous spectrum.

PROPOSITION 4.11. *Suppose g is a compactly supported function of class C^1 on $\mathbb{R}^2$. Let $ad \equiv 1 \pmod c$. Then*

$$\sum_{m\in\mathbb{Z}} \tau_g(m)e\Big(\frac{am}{c}\Big) = \sum_{n\in\mathbb{Z}} \tau_h(n)e\Big(-\frac{dn}{c}\Big) \tag{4.51}$$

where h is given by the Fourier transform of g, namely

$$h(x,y) = \frac{1}{c}\hat{g}\Big(\frac{x}{c}, \frac{y}{c}\Big). \tag{4.52}$$

For $n = 0$ we have

$$\tau_h(0) = \int_{\mathbb{R}^2}\Big(\frac{1}{c} + \Big\{\frac{x}{c}\Big\}\frac{\partial}{\partial x} + \Big\{\frac{y}{c}\Big\}\frac{\partial}{\partial y}\Big)g(x,y)dxdy \tag{4.53}$$

$$= -\int_{\mathbb{R}^2}\Big(\frac{1}{c} + \Big[\frac{x}{c}\Big]\frac{\partial}{\partial x} + \Big[\frac{y}{c}\Big]\frac{\partial}{\partial y}\Big)g(x,y)dxdy. \tag{4.54}$$

PROOF. First we open $\tau_g(m)$ on the left side of (4.51) as in (4.50) and we split the summation into residue classes $(m_1, m_2) \equiv (u_1, u_2) \pmod c$. Then in each class we apply the Poisson summation formula (4.24) getting

$$\begin{aligned}\sum_m \tau_g(m)e\Big(\frac{am}{c}\Big) &= \sum\sum_{u_1,u_2 \pmod c} e\Big(\frac{a}{c}u_1u_2\Big)\sum_{v_1}\sum_{v_2} g(u_1 + cv_1, u_2 + cv_2) \\ &= c^{-2}\sum_{n_1}\sum_{n_2}\sum\sum_{u_1,u_2 \pmod c} e_c(au_1u_2 + n_1u_1 + n_2u_2)\hat{g}\Big(\frac{n_1}{c}, \frac{n_2}{c}\Big) \\ &= c^{-1}\sum_{n_1}\sum_{n_2} e\Big(-\frac{d}{c}n_1n_2\Big)\hat{g}\Big(\frac{n_1}{c}, \frac{n_2}{c}\Big) = \sum_n \tau_h(n)e\Big(-\frac{dn}{c}\Big).\end{aligned}$$

The term $n = 0$ is special since there are infinitely many divisors of zero. We compute $\tau_h(0)$ by reversing the above analysis (of course, this analysis could be avoided altogether, nevertheless, for the economy of presentation sometimes a repeated application of an involution is justified). First we arrange the summation over n_1, n_2 as follows:

$$\begin{aligned}\tau_h(0) &= \frac{1}{c}\sum_{n_2 n_2 = 0} \hat{g}\Big(\frac{n_1}{c}, \frac{n_2}{c}\Big)\\ &= -\frac{1}{c}\hat{g}(0,0) + \frac{1}{c}\sum_n \hat{g}\Big(0, \frac{n}{c}\Big) + \frac{1}{c}\sum_n \hat{g}\Big(\frac{n}{c}, 0\Big).\end{aligned}$$

Then we apply Poisson's summation and the Euler-Maclaurin formula (4.7) getting

$$\begin{aligned}\frac{1}{c}\sum_n \hat{g}\Big(0, \frac{n}{c}\Big) &= \int \Big(\sum_m g(x, cm)\Big)dx\\ &= \int\int \Big(\frac{1}{c} + \Big\{\frac{y}{c}\Big\}\frac{\partial}{\partial y}\Big)g(x,y)dxdy.\end{aligned}$$

Similarly we execute the summation of $\hat{g}(\frac{n}{c}, 0)$. Gathering the results we obtain (4.53). $\square$

PROOF OF (4.49). One would like to apply (4.51) directly for $g(x, y) = g(xy)$, but this is not possible because such a function does not have compact support in $\mathbb{R}^2$ even if $g(t)$ does have compact support in $\mathbb{R}$. For this reason we attach to $g(m_1 m_2)$ redundant factors $\eta(m_1), \eta(m_2)$ to localize the real variables x, y in compact segments of $\mathbb{R}^+$. Precisely we set $g(x, y) = \eta(x)\eta(y)g(xy)$ where $\eta(t) = \min\{t/\varepsilon, 1\}$ with $0 < \varepsilon < 1$ so that $\tau_g(m) = \tau(m)g(m)$. Note that $\eta'(t)$ has discontinuity at $t = \varepsilon$, but this is allowed for application of Proposition 4.11 as can be seen from the proof by inspection (we could choose $\eta(t)$ smooth, however, the involved computations would have been less explicit). For the above choice we have by (4.53)

$$\begin{aligned}\tau_h(0) &= \int\int \Big(\frac{1}{c} + \Big\{\frac{x}{c}\Big\}\frac{\partial}{\partial x} + \Big\{\frac{y}{c}\Big\}\frac{\partial}{\partial y}\Big)\eta(x)\eta(y)g(xy)dxdy\\ &= \frac{1}{c}\int g(y)L(y)dy,\end{aligned}$$

where

$$L(y) = \int \Big(\frac{\eta(x)}{x}\eta\Big(\frac{y}{x}\Big) + 2c\Big\{\frac{x}{c}\Big\}\frac{\partial}{\partial x}\frac{\eta(x)}{x}\eta\Big(\frac{y}{x}\Big)\Big)\,dx.$$

Since y is bounded from below and above, and ε is small, we find by examining the support of the involved functions the following representation:

$$L(y) = \int \Big(1 - 2\frac{c}{x}\Big\{\frac{x}{c}\Big\}\Big)\eta(x)\eta\Big(\frac{y}{x}\Big)\frac{dx}{x} - 2cy\int \Big\{\frac{x}{c}\Big\}\eta'\Big(\frac{y}{x}\Big)\frac{dx}{x^3}.$$

The last integral ranges over $x > \varepsilon^{-1}y$, so it is bounded trivially by $O(\varepsilon c/y)$. Then we have

$$\int_0^\infty \eta(x)\eta\Big(\frac{y}{x}\Big)\frac{dx}{x} = 2\int_0^{\sqrt{y}} \eta(x)\frac{dx}{x} = 2 + \log y - 2\log\varepsilon,$$

$$\int_0^\infty \frac{c}{x}\Big\{\frac{x}{c}\Big\}\eta(x)\eta\Big(\frac{y}{x}\Big)\frac{dx}{x} = 1 + \int_\varepsilon^\infty \frac{c}{x}\Big\{\frac{x}{c}\Big\}\eta\Big(\frac{y}{x}\Big)\frac{dx}{x}$$
$$= 1 + \int_{\varepsilon/c}^\infty \{x\}\eta\Big(\frac{y}{cx}\Big)\frac{dx}{x^2} = 1 + \int_{\varepsilon/c}^\infty \{x\}\frac{dx}{x^2} + O\Big(\frac{\varepsilon c}{y}\Big).$$

The part of the last integral with $\frac{\varepsilon}{c} < x < 1$ equals $\log\frac{c}{\varepsilon}$ and the remaining part equals $1-\gamma$, where γ is the Euler constant (see (1.69)). From these evaluations we get

$$L(y) = \log y + 2\gamma - 2\log c + O\Big(\frac{\varepsilon c}{y}\Big).$$

Hence

$$\tau_h(0) = \frac{1}{c}\int g(y)(\log y + 2\gamma - 2\log c)dy + O(\varepsilon). \tag{4.55}$$

As ε tends to zero this gives the leading term on the right side of (4.49).

Now we compute $\tau_h(n)$ for $n \neq 0$. Since the factorization $n_1n_2 = n$ permits the sign change of both n_1, n_2 we can replace $\hat{g}$ in (4.52) by the cosine-Fourier transform, i.e., we may take

$$h(u,v) = \frac{2}{c}\iint \eta(x)\eta(y)g(xy)\cos\frac{2\pi}{c}(ux+vy)dxdy.$$

Changing variables we write this as

$$h(u,v) = \frac{2}{c}\int g(y)\Big(\int \eta(x)\eta\Big(\frac{y}{x}\Big)\cos\Big(\frac{2\pi}{c}\Big(ux+\frac{vy}{x}\Big)\Big)\frac{dx}{x}\Big)dy.$$

If we omit $\eta(x)\eta(y/x)$, then the pure cosine integral equals

$$\int_0^\infty \cos\Big(\frac{2\pi}{c}\Big(ux+\frac{vy}{x}\Big)\Big)\frac{dx}{x} = \begin{cases} -\pi Y_0(\frac{4\pi}{c}\sqrt{uvy}) & \text{if } uv > 0,\\ 2K_0(\frac{4\pi}{c}\sqrt{|uv|y}) & \text{if } uv < 0\end{cases}$$

(see (3.871.2) and (3.871.4) of [**GR**], respectively, or our Appendix). This yields exactly the remaining terms on the right side of (4.49).

Now we only need to show that the distortion introduced by $\eta(x)\eta(y/x)$ is negligible. We denote the difference by $h_0(u,v) = h_1(u,v) + h_1(v,u)$, where

$$h_1(u,v) = \frac{2}{c}\int_0^\varepsilon \Big(1-\frac{x}{\varepsilon}\Big)\int_0^\infty g(xy)\cos\frac{2\pi}{c}(ux+vy)dydx.$$

Integrating by parts twice with respect to y and then applying Fubini's theorem we get

$$h_1(u,v) = \frac{e}{2\pi^2v^2}\int_0^\infty\int_0^\varepsilon \Big(1-\frac{x}{\varepsilon}\Big)x^2g''(xy)\cos\frac{2\pi}{c}(ux+xy)dxdy.$$

Now integrating by parts twice with respect to x we estimate the inner integral by $O(\varepsilon u^{-2})$. The inner integral is also estimated trivially by

$$\int_0^\infty x^2|g''(xy)|dx \ll y^{-3}.$$

Combining both estimates we arrive at

$$h_1(u,v) \ll v^{-2}\int_0^\infty \min(\varepsilon u^{-2}, y^{-3})dy = \tfrac{3}{2}\varepsilon^{\frac{2}{3}}|u|^{-\frac{4}{3}}v^{-2}.$$

Next, adding the corresponding bound for $h_1(v,u)$ we obtain $h_0(u,v) \ll \varepsilon^{\frac{2}{3}}|uv|^{-\frac{4}{3}}$. Therefore the total distortion of the right side of (4.51), which was caused by introduction of the localizing factors $\eta(x)\eta(y)$ to the left side, is bounded by $O(\varepsilon^{2/3})$ since the series $\sum \tau(n)n^{-4/3}$ converges. Letting ε tend to zero the distortion vanishes and we complete the proof of (4.49), hence also of Theorem 4.10. □

The summation formula (4.51) for the tempered divisor function (4.50) was established in [**DI**]. The special case (4.49) is less flexible, but it is quite effective when applicable (a different, rather indirect, proof of (4.49) was given by M. Jutila [**Ju1**]). We proceed to compose some variations on (4.49).

First, we get a summation formula for certain series of Kloosterman sums: if $(c,q)=1$, then

$$\begin{aligned}\sum_{m\geqslant 1}\tau(m)S(h,m;c)g(m) &= \frac{2}{c}S(h,0;c)\int_0^\infty\Big(\log\frac{\sqrt{x}}{c}+\gamma\Big)g(x)dx\\ &-\frac{2\pi}{c}\sum_{n\geqslant 1}\tau(n)S(h-n,0;c)\int_0^\infty Y_0\Big(\frac{4\pi}{c}\sqrt{nx}\Big)g(x)dx\\ &+\frac{4}{c}\sum_{n\geqslant 1}\tau(n)S(h+n,0;c)\int_0^\infty K_0\Big(\frac{4\pi}{c}\sqrt{nx}\Big)g(x)dx.\end{aligned} \tag{4.56}$$

for any integer h. To prove this, we open the Kloosterman sum (1.56)

$$S(h,m;c) = \sideset{}{^\star}\sum_{a \ (\mathrm{mod}\ c)} e\Big(\frac{h\bar{a}+ma}{c}\Big),$$

then for each a, we apply (4.49) and then exchange the order of summation again. Note that the Kloosterman sums on the left side collapsed to the Ramanujan sums on the left side. This feature is significant in applications, see e.g. [**DFI2**], [**KM1**], [**KMV1**].

Next, playing with additive characters we derive from (4.49) a summation formula for the divisor function in an arithmetic progression.

COROLLARY 4.12. *Let $(q,r)=1$, and g be smooth, compactly supported on $\mathbb{R}^+$. Then*

(4.57)

$$\begin{aligned}\sum_{m\equiv r (\mathrm{mod}\ q)}\tau(m)g(m) &= \frac{\varphi(q)}{q^2}\int_0^\infty(\log x+2\gamma-2\eta(q))g(x)dx\\ &+\frac{1}{q}\sum_{c|q}\frac{1}{c}\sum_n\tau(n)\Big\{S(r,n;c)T^+\Big(\frac{n}{c^2}\Big)+S(r,-n;c)T^-\Big(\frac{n}{c^2}\Big)\Big\}\end{aligned}$$

where $\eta(q)$ is the additive function

$$\eta(q) = \sum_{p|q} \frac{\log p}{p-1}, \tag{4.58}$$

$S(r, n; c)$ is the Kloosterman sum, and $T^+(y)$, $T^-(y)$ are the integral transforms

$$T^+(y) = -2\pi \int_0^\infty Y_0(4\pi\sqrt{xy})g(x)dx, \tag{4.59}$$

$$T^-(y) = 4\int_0^\infty K_0(4\pi\sqrt{xy})g(x)dx. \tag{4.60}$$

REMARKS. Since $Y_0(z)$ has the asymptotic expansion (4.37) but with cos and sin interchanged (see (23.451.2) of [**GR**]), we get the approximate formula

$$T^+(y) = \frac{2}{\pi y}\int_0^\infty (xy)^{\frac{1}{4}} g'(x)\cos(4\pi\sqrt{xy} - \frac{\pi}{4})dx + O(R(y))$$

by the same argument which led us to (4.40), where $R(y)$ is given by (4.39). Similarly one derives

$$T^-(y) = \frac{2}{\pi y}\int_0^\infty (xy)^{\frac{1}{4}} g'(x)e^{-4\pi\sqrt{xy}}dx + O(R(y)).$$

EXERCISE 7. Using the Weil bound for Kloosterman sums (11.16) and the test function $g(x)$ as in the proof of (4.42), derive from (4.57) that for $(q, r) = 1$ and $X \geqslant 2$,

$$\sum_{\substack{m\leqslant X\\ m\equiv r \pmod q}} \tau(m) = \frac{\varphi(q)}{q^2}X\{\log X + 2\gamma - 1 - 2\eta(q)\} + O(\tau^2(q)(q^{\frac{1}{2}} + X^{\frac{1}{3}})\log X) \tag{4.61}$$

where the implied constant is absolute. In particular,

$$\sum_{m\leqslant X} \tau(m) = X(\log X + 2\gamma - 1) + O(X^{\frac{1}{3}}\log X). \tag{4.62}$$

This improves the error term in the Dirichlet divisor problem (1.75).

EXERCISE 8. Prove that for any primitive character $\chi(\text{mod } q)$ with $q > 1$, and any smooth function $g(x)$ compactly supported on $\mathbb{R}^+$ we have

$$\sum_1^\infty \tau(m)\chi(m)g(m) = \tau^2(\chi)q^{-2}\sum_1^\infty \tau(n)\overline{\chi}(n)h(nq^{-2}) \tag{4.63}$$

where $\tau(\chi)$ is the Gauss sum, and $h(y)$ is the integral transform

$$h(y) = \int_0^\infty K(2\pi\sqrt{xy})g(x)dx \tag{4.64}$$

with kernel

$$K(z) = 4\chi(-1)K_0(2z) - 2\pi Y_0(2z). \tag{4.65}$$

[**Hint:** Expand χ into additive characters (see (3.12)).]

Next we derive a summation formula for

$$\tau_\nu(n,\chi) = \sum_{n_1 n_2 = n} \chi(n_1)\Big(\frac{n_1}{n_2}\Big)^\nu \tag{4.66}$$

where $\chi(\bmod q)$ is a primitive character of conductor $q > 1$ and ν is a fixed complex number. These are eigenvalues of the Hecke operators T_n^χ in the space of Eisenstein series for $\Gamma_0(q)$ and character χ. In principle the method that we used for $\tau(n)$ applies for $\tau_\nu(n,\chi)$, though some arithmetical arguments need to be revised and a few modifications in integral transforms are required. However, the analysis of convergence is the same, so here we do not repeat it. We treat only the two extremal cases $q|c$ and $(q,c)=1$. See also [**KMV1**] for other similar formulas.

THEOREM 4.13. *Let $ad \equiv 1 \pmod c$ and χ be a primitive character of conductor $q > 1$ with $q|c$. Then for any smooth, compactly supported function g on $\mathbb{R}^+$ we have*

$$\begin{aligned}\sum_1^\infty \tau_\nu(m,\chi) e\Big(\frac{am}{c}\Big) g(m) &= \frac{\chi(d)}{c}\Big(\frac{q}{c}\Big)^{2\nu}\tau(\chi)L(1+2\nu,\overline{\chi})\int_0^\infty g(x)x^\nu\,dx \\ &+ \frac{\chi(d)}{c}\sum_1^\infty \tau_\nu(n,\chi)e\Big(-\frac{dn}{c}\Big)\int_0^\infty g(x)J_{2\nu}^\pm\Big(\frac{4\pi}{c}\sqrt{nx}\Big)dx \\ &+ \frac{\chi(d)}{c}\sum_1^\infty \tau_\nu(n,\chi)e\Big(\frac{dn}{c}\Big)\int_0^\infty g(x)K_{2\nu}^\pm\Big(\frac{4\pi}{c}\sqrt{nx}\Big)dx\end{aligned} \tag{4.67}$$

where $\tau(\chi)$ is the Gauss sum, $L(1+2\nu,\overline{\chi})$ is the Dirichlet L-function, and

$$\begin{aligned} J_{2\nu}^+(z) &= \frac{-\pi}{\sin\pi\nu}(J_{2\nu}(z) - J_{-2\nu}(z)), \quad \text{if } \chi(-1)=1, \\ J_{2\nu}^-(z) &= \frac{\pi i}{\cos\pi\nu}(J_{2\nu}(z) + J_{-2\nu}(z)), \quad \text{if } \chi(-1)=-1, \\ K_{2\nu}^+(z) &= 4\cos(\pi\nu)K_{2\nu}(z), \quad \text{if } \chi(-1)=1, \\ K_{2\nu}^-(z) &= 4i\sin(\pi\nu)K_{2\nu}(z), \quad \text{if } \chi(-1)=-1. \end{aligned}$$

REMARK. For $\nu = 0$ we have

$$\tau(n,\chi) = \sum_{d|n}\chi(d). \tag{4.68}$$

If χ is odd (the case of modular forms of weight one), our formula reduces to

$$\begin{aligned}\sum_1^\infty \tau(m,\chi)e\Big(\frac{am}{c}\Big)g(m) &= \frac{\chi(d)}{c}\tau(\chi)L(1,\overline{\chi})\int_0^\infty g(x)\,dx \\ &+ 2\pi i\frac{\chi(d)}{c}\sum_1^\infty \tau(n,\chi)e\Big(-\frac{dn}{c}\Big)\int_0^\infty g(x)J_0\Big(\frac{4\pi}{c}\sqrt{nx}\Big)dx.\end{aligned}$$

PROOF. On the left side of (4.67) we split the summation into residue classes $(m_1, m_2) \equiv (u_1, u_2) \pmod{c}$ and for each class we apply the Poisson formula (4.24) getting

$$\frac{1}{c^2}\sum_{n_1}\sum_{n_2}\sum\sum_{u_1,u_2 \pmod{c}} \chi(u_1)e_c(au_1u_2 - n_1u_1 - n_2u_2)I(n_1,n_2)$$

where $I(n_1, n_2)$ is the Fourier integral

$$I(n_1,n_2) = \int_0^\infty\int_0^\infty \Big(\frac{x}{y}\Big)^\nu g(xy)e_c(xn_1+yn_2)dxdy.$$

The sum over $u_2 \pmod{c}$ vanishes unless $au_1 \equiv n_2 \pmod{c}$ in which case it equals c. Therefore we showed that the left side of (4.67) is equal to

$$\frac{\overline{\chi}(a)}{c}\sum_{n_1}\sum_{n_2}\chi(n_2)e\Big(-\frac{d}{c}n_1n_2\Big)I(n_1,n_2).$$

Note that $\overline{\chi}(a) = \chi(d)$. From $n_1n_2 = 0$ we get

$$\begin{aligned}&\frac{\chi(d)}{c}\sum_1^\infty \chi(n)[I(0,n)+\chi(-1)I(0,-n)]\\ &= \frac{\chi(d)}{c}\Big(\frac{2\pi}{c}\Big)^{2\nu}L(-2\nu,\chi)\Big(\int_0^\infty g(x)x^\nu dx\Big)\int_0^\infty (e^{iy}+\chi(-1)e^{-iy})y^{-1-2\nu}dy.\end{aligned}$$

The last integral is equal to $2\Gamma(-2\nu)\cos\pi\nu$ or $-2i\Gamma(-2\nu)\sin\pi\nu$ according to $\chi(-1) = 1$ or $\chi(-1) = -1$ by (4.108) and (4.109) respectively. Here we transform $L(-2\nu, \chi)$ into $L(1 + 2\nu, \overline{\chi})$ by the functional equation

$$L(-2\nu,\chi) = \frac{\tau(\chi)}{\sqrt{\pi}}\Big(\frac{q}{\pi}\Big)^{2\nu}L(1+2\nu,\overline{\chi})\begin{cases}\Gamma(\frac{1}{2}+\nu)/\Gamma(-\nu), & \text{if } \chi(-1)=1,\\ -i\Gamma(1+\nu)/\Gamma(\frac{1}{2}-\nu), & \text{if } \chi(-1)=-1,\end{cases}$$

see (4.73). In the case of even character we encounter

$$\frac{2\Gamma(\frac{1}{2}+\nu)}{\Gamma(-\nu)}\Gamma(-2\nu)\cos\pi\nu = 2^{-2\nu}\sqrt{\pi}$$

and in the case of odd character we encounter

$$-\frac{-2\Gamma(1+\nu)}{\Gamma(\frac{1}{2}+\nu)}\Gamma(-2\nu)\sin\pi\nu = 2^{-2\nu}\sqrt{\pi}.$$

Hence we conclude that in both cases the terms with $n_1n_2 = 0$ contribute the same amount which is exactly the first term on the right side of (4.67). From $n_1n_2 \neq 0$ we get

$$\begin{aligned}&\frac{\chi(d)}{c}\sum_1^\infty\sum_1^\infty \chi(n_2)e\Big(-\frac{d}{c}n_1n_2\Big)[I(n_1,n_2)+\chi(-1)I(-n_1,-n_2)]\\ &+\frac{\chi(d)}{c}\sum_1^\infty\sum_1^\infty \chi(n_2)e\Big(\frac{d}{c}n_1n_2\Big)[I(-n_1,n_2)+\chi(-1)I(n_1,-n_2)].\end{aligned}$$

Changing the variables $(x,y)\to(u\sqrt{vn_2/n_1},u^{-1}\sqrt{vn_1/n_2})$ we get

$$I(\pm n_1,\pm n_2)=\Big(\frac{n_2}{n_1}\Big)^{\nu}\int_0^\infty g(v)\Big(\int_0^\infty e_c(\sqrt{vn_1n_2}(\pm u\pm u^{-1}))u^{2\nu-1}du\Big)\,dv.$$

Hence the remaining terms on the right side of (4.67) emerge by applying (4.112)–(4.115). □

REMARK. In the proof of Theorem 4.13 we assumed that $-\nu$ is a small positive number, but the result extends to all complex ν by analytic continuation.

THEOREM 4.14. *Let $ad\equiv 1 \pmod c$ and $\chi \pmod q$ be a primitive character of conductor $q>1$ with $(c,q)=1$. Then for any smooth, compactly supported function g on $\mathbb{R}^+$ we have*

$$\begin{aligned}\sum_1^\infty \tau_\nu(m,\chi)e\Big(\frac{am}{c}\Big)g(m)&=\chi(c)c^{2\nu-1}L(1-2\nu,\chi)\int_0^\infty g(x)x^{-\nu}dx\\ &+\frac{\chi(-c)}{c}\tau(\chi)q^{\nu-1}\sum_1^\infty \tau_{-\nu}(n,\overline{\chi})e\Big(-\frac{d\overline{q}n}{c}\Big)\int_0^\infty g(x)J_{2\nu}^{\pm}\Big(\frac{4\pi\sqrt{nx}}{c\sqrt{q}}\Big)dx\\ &+\frac{\chi(c)}{c}\tau(\chi)q^{\nu-1}\sum_1^\infty \tau_{-\nu}(n,\overline{\chi})e\Big(\frac{d\overline{q}n}{c}\Big)\int_0^\infty g(x)K_{2\nu}^{\pm}\Big(\frac{4\pi\sqrt{nx}}{c\sqrt{q}}\Big)dx\end{aligned}\tag{4.69}$$

where $\overline{q}q\equiv 1 \pmod c$, and $J_\nu^\pm, K_\nu^\pm$ are the same as in Theorem 4.13.

REMARK. For $\nu=0$ and χ odd our formula reduces to

$$\begin{aligned}\sum_1^\infty \tau(m,\chi)e\Big(\frac{am}{c}\Big)g(m)&=\frac{\chi(c)}{c}L(1,\chi)\int_0^\infty g(x)dx\\ &-2\pi\frac{\chi(c)}{c}\frac{\tau(\chi)}{q}\sum_1^\infty \tau(n,\overline{\chi})e\Big(-\frac{d\overline{q}n}{c}\Big)\int_0^\infty g(x)J_0\Big(\frac{4\pi\sqrt{nx}}{c\sqrt{q}}\Big)dx.\end{aligned}\tag{4.70}$$

PROOF. On the left side of (4.69) we split the summation into classes $m_1\equiv u_1 (\mathrm{mod}\ cq)$, $m_2\equiv u_2 \pmod c$ and for each class we apply the Poisson formula (4.24) getting

$$\frac{1}{c^2q}\sum_{n_1}\sum_{n_2}\sum_{\substack{u_1 (\mathrm{mod}\ cq)\\ u_2\ (\mathrm{mod}\ c)}}\sum \chi(u_1)e_c\Big(au_1u_2-\frac{n_1}{q}u_1-n_2u_2\Big)I_\nu\Big(\frac{n_1}{q},n_2\Big)$$

where $I_\nu(z_1,z_2)$ is the same integral as in the proof of Theorem 4.13. Note that

$$I_\nu\Big(\frac{n_1}{q},n_2\Big)=q^\nu I_\nu\Big(\frac{n_1}{\sqrt{q}},\frac{n_2}{\sqrt{q}}\Big)=q^\nu I_{-\nu}\Big(\frac{n_2}{\sqrt{q}},\frac{n_1}{\sqrt{q}}\Big).$$

The sum over $u_2 \pmod c$ vanishes unless $au_1\equiv n_2 \pmod c$ in which case it equals c. Putting $u_1=dn_2q\overline{q}+ct$, where t ranges modulo q we obtain

$$\sum_{u_1}\sum_{u_2}=ce\Big(-\frac{d\overline{q}}{c}n_1n_2\Big)\chi(-c\overline{n}_1)\tau(\chi).$$

Hence the left side of (4.69) is equal to

$$\frac{\chi(-c)}{c}\tau(\chi)q^{\nu-1}\sum_{n_1}\sum_{n_2}\overline{\chi}(n_1)e\Big(-\frac{d\overline{q}}{c}n_1n_2\Big)I_{-\nu}\Big(\frac{n_2}{\sqrt{q}},\frac{n_1}{\sqrt{q}}\Big).$$

This looks very similar to the sum we have already dealt with in the proof of Theorem 4.13. Therefore (4.69) is derived from (4.67) by making the substitution $(\chi,\nu,d)\to(\overline{\chi},-\nu,d\overline{q})$ in appropriate places. □

Theorems 4.6, 4.10, 4.13 and 4.14 are special cases of summation formulas for the Fourier coefficients of modular forms. Actually, these results confirm in themselves the modularity of relevant forms. As a matter of fact we were able to establish such modular relations directly by applying the two-dimensional Poisson summation only because in each case our arithmetic function appears in the Fourier coefficients of a GL_2 form which is lifted from GL_1 forms, i.e., the corresponding L-function factors into two Dirichlet L-functions. For genuine cusp form coefficients one cannot apply the ordinary Poisson's summation formula. If one knows that a Fourier series

$$f(z)=\sum_n\lambda_f(n)n^{\frac{k-1}{2}}e(nz)$$

is a modular form, say $f\in S_k(q,\chi)$ (see Chapter 14 for notation and survey of holomorphic modular forms), then the corresponding summation formula for the coefficients $\lambda_f(n)$ would be just another expression for the equation

$$f\Big(\frac{az+b}{cz+d}\Big)=\chi(d)(cz+d)^kf(z).$$

EXERCISE 9. Suppose f is a cusp form of weight $k\geqslant 1$, level $q\geqslant 1$ and character χ modulo q. Let $c\geqslant 1$, $c\equiv 0(\mathrm{mod}\ q)$ and $ad\equiv 1\ (\mathrm{mod}\ c)$. Prove that the Fourier coefficients of f satisfy

$$\sum_1^\infty\lambda_f(m)e\Big(\frac{am}{c}\Big)g(m)=\frac{\chi(d)}{c}\sum_1^\infty\lambda_f(n)e\Big(-\frac{dn}{c}\Big)h(n)\tag{4.71}$$

for any smooth, compactly supported function g on $\mathbb{R}^+$, where

$$h(y)=2\pi i^k\int_0^\infty g(x)J_{k-1}\Big(\frac{4\pi}{c}\sqrt{xy}\Big)dx.$$

[**Hint.**] Apply the modular equation for $z=-\frac{d}{c}+\frac{i}{cy}$ and integrate the resulting Fourier series on both sides with respect to y against a suitable test function $G(y)$.

If f is primitive (see Section 14.7), so that it satisfies the Fricke involution equation

$$f\Big(\frac{-1}{qz}\Big)=\eta_fq^{\frac{k}{2}}z^k\overline{f}(z)$$

with $|\eta_f|=1$ (Proposition 14.14), then we also have

$$\sum_1^\infty\lambda_f(m)g(m)=\eta_fq^{-1/2}\sum_1^\infty\overline{\lambda}_f(n)h(n)\tag{4.72}$$

where h is the same Hankel-type integral transform of g as above with $c = \sqrt{q}$ (apply (4.71) for

$$\begin{pmatrix} a & b \\ c & d \end{pmatrix} = \begin{pmatrix} 0 & -1/\sqrt{q} \\ \sqrt{q} & 0 \end{pmatrix}$$

and change $\chi(d)$ into η_f to get (4.70)).

In modern analytic number theory there is a great demand for summation formulas for arithmetic functions of type $\gamma(m) = \alpha(m)\beta(m+h)$ where $\alpha(m), \beta(n)$ are essentially modular forms coefficients and h is a fixed integer (but a uniformity in h is crucial). These kind of problems are often encountered when evaluating power-moments of families of L-functions on the critical line. See, for instance, [**DFI1**], [**Sa1**], [**M1**].

4.6. Functional equations of Dirichlet L-functions.

A summation formula for arithmetic functions often has its realization as a functional equation for the corresponding generating Dirichlet series. Conversely, a functional equation connecting two Dirichlet series leads (by way of taking the inverse of Mellin transform) to a summation formula for the coefficient sequences. In this section we establish this correspondence in full detail for the Dirichlet L-functions (3.6) using the original method of Riemann. These ideas were developed further by E. Hecke [**Hec1**] for L-functions of number fields and for automorphic L-functions, and later put in the general adélic setting in Tate's Thesis [**Ta1**]. Furthermore the connections between functional equations for twisted L-functions and modularity via summation formulas is the essence of the so-called converse theorems in the theory of automorphic forms in general. There are delicate issues of compatibility which we do not discuss here; see for example the Weil converse theorem for GL_2 forms in Theorem 14.21, or [**I4**].

Let $q \geqslant 1$ and let χ be a primitive character modulo q. We let $\delta(\chi) = 0$ if $q \neq 1$ and $\delta(\chi) = 1$ if $q = 1$, in which case $\chi = 1$ and $L(\chi, s) = \zeta(s)$. Also we put $\kappa = \frac{1}{2}(1 - \chi(-1))$, so $\kappa = 0$ if χ is even and $\kappa = 1$ if χ is odd.

THEOREM 4.15. *With the above notation, $L(s, \chi)$ extends to a meromorphic function on $\mathbb{C}$ which is entire if $\chi \neq 1$ and otherwise admits a unique simple pole with residue 1 at $s = 1$. The completed L-function*

$$\Lambda(s, \chi) = \left(\frac{q}{\pi}\right)^{s/2} \Gamma\left(\frac{s+\kappa}{2}\right) L(s, \chi)$$

is entire if $q \neq 1$ and has simple poles with residue 1 at $s = 0$ and $s = 1$ otherwise. Moreover, it satisfies the functional equation

$$\Lambda(s, \chi) = \varepsilon(\chi) \Lambda(1 - s, \bar{\chi}) \tag{4.73}$$

where

$$\varepsilon(\chi) = i^{-\kappa} \frac{\tau(\chi)}{\sqrt{q}}. \tag{4.74}$$

Recall that the Gauss sum $\tau(\chi)$ is of modulus $\sqrt{q}$ (see (3.10), (3.14)) for χ primitive, so that the "root number" $\varepsilon(\chi)$ is of modulus 1. Theorem 3.3 of Gauss shows that $\varepsilon(\chi) = 1$ for any real primitive character χ.

For the Riemann zeta function, the functional equation is therefore

$$\Lambda(s) = \pi^{-s/2}\Gamma(s/2)\zeta(s) = \Lambda(1-s). \tag{4.75}$$

PROOF. We show the arguments for $q \neq 1$, the case of $\zeta(s)$ is only slightly different because of the pole at $s = 1$. Let $\theta(y, \chi)$ be the theta function

$$\theta(y, \chi) = \sum_{n \in \mathbb{Z}} \chi(n) n^{\kappa} e^{-\pi n^2 y/q}$$

for $y > 0$. By splitting into progressions modulo q, we have

$$\theta(y, \chi) = \sum_{a \pmod q} \chi(a) \theta(y; q, a),$$

where

$$\theta(y; q, a) = \sum_{n \equiv a \pmod q} n^{\kappa} e^{-\pi n^2 y/q}.$$

We apply the Poisson formula (see (4.24)) to this sum. The function $e^{-\pi x^2}$ is its own Fourier transform (4.29). Differentiating the corresponding Fourier integral, it follows that the Fourier transform of $f_y(x) = x^{\kappa} e^{-\pi x^2 y}$ is given by

$$\hat{f}_y(t) = i^{-\kappa} y^{-\kappa - \frac{1}{2}} f_{y^{-1}}(t).$$

By (4.24) we obtain

$$\theta(y; q, a) = i^{-\kappa} y^{-\kappa - \frac{1}{2}} q^{-\frac{1}{2}} \sum_{b \pmod q} e\Big(\frac{ab}{q}\Big) \theta\Big(\frac{1}{y}; q, b\Big)$$

after splitting again in progressions. Using the formula (3.12) for a primitive character it follows that

$$\theta(y, \chi) = \varepsilon(\chi) y^{-\kappa - \frac{1}{2}} \quad \theta\Big(\frac{1}{y}, \bar{\chi}\Big). \tag{4.76}$$

Now we appeal to the Mellin transform formula (see (4.107))

$$\Big(\frac{q}{\pi}\Big)^{(s+\kappa)/2} \Gamma\Big(\frac{s+\kappa}{2}\Big) n^{-s} = \int_0^{+\infty} n^{\kappa} e^{-\pi n^2 y/q} y^{(s+\kappa)/2} \frac{dy}{y}$$

to derive

$$\Big(\frac{q}{\pi}\Big)^{\kappa/2} \Lambda(s, \chi) = \frac{1}{2} \int_0^{+\infty} \theta(y, \chi) y^{(s+\kappa)/2} \frac{dy}{y}$$

for $\mathrm{Re}\,(s) > 1$. Split the integral at $y = 1$ and apply (4.76) for $0 < y < 1$ getting

$$\Big(\frac{q}{\pi}\Big)^{\kappa/2} \Lambda(s, \chi) = \frac{1}{2} \int_1^{+\infty} \Big\{\varepsilon(\chi)\theta(y, \bar{\chi}) y^{(1-s+\kappa)/2} + \theta(y, \chi) y^{(s+\kappa)/2}\Big\} \frac{dy}{y} \tag{4.77}$$

from which both the analytic continuation of $\Lambda(s, \chi)$ (hence that of $L(s, \chi)$) and the functional equation (4.73) follow since $\theta(y, \chi)$ decays exponentially fast at ∞.□

REMARK. This proof should be compared with that of Theorem 14.7, the functional equation of Hecke L-functions.

Quite often in applications a formula for the sum of an arithmetic function $\lambda(n)$ with a sharp cut $n \leqslant x$ is more convenient than that with smooth ending. In this situation it is better to work with an approximate formula having a good error term rather than with an exact infinite series which is only conditionally convergent. Here we state a classical result for a twisted divisor function which can be derived from the functional equation for the Dirichlet L-series (see [**FI2**]).

THEOREM 4.16. *Let $\chi_1, \ldots, \chi_d$ be primitive characters to moduli $q_1, \ldots, q_d$ respectively. Put*

$$\lambda(n) = \sum_{n_1 \cdots n_d = n} \chi_1(n_1) \cdots \chi_d(n_d).$$

For any $1 \leqslant y \leqslant x$ we have

$$\sum_{1 \leqslant n \leqslant x} \lambda(n) = R(x) + w\sqrt{\frac{x}{\pi d}} \sum_{1 \leqslant n \leqslant y} \frac{\bar{\lambda}(n)}{\sqrt{n}} \Big(\frac{q}{nx}\Big)^{\frac{1}{2d}} \cos\Big(2\pi d\Big(\frac{nx}{q}\Big)^{\frac{1}{d}} + \frac{\pi u}{4}\Big) + O\Big(\Big(\frac{q}{xy}\Big)^{\frac{1}{d}} x^{1+\varepsilon}\Big).$$

Here $q = q_1 \cdots q_d$, $u = d - 3 - 2(\kappa_1 + \cdots + \kappa_d)$ and w is the root number of the L-function $L(s) = L(s, \chi_1) \cdots L(s, \chi_d)$ (w is a product of the normalized Gauss sums). The main term $R(x)$ is equal to the residue of $s^{-1}x^s L(s)$ at $s = 1$, so $R(x) = xP(\log x)$, where $P(x)$ is a polynomial of degree $\leqslant d$. The implied constant depends only on d and ε.

Estimating the dual sum trivially and choosing $y = q^{\frac{1}{d+1}} x^{\frac{d-1}{d+1}}$ we get

$$\sum_{1 \leqslant n \leqslant x} \lambda(n) = R(x) + O(q^{\frac{1}{d+1}} x^{\frac{d-1}{d+1} + \varepsilon})$$

where the implied constant depends only on d and ε. This, of course, is not the best error term. Using exponential sums methods one can improve the result slightly. The best error term with respect to x which one can hope for is $O(x^{\frac{1}{2} - \frac{1}{2d} + \varepsilon})$. Indeed, the first term of the dual sum is about as large. Note that the Exponent Pair Hypothesis (see Chapter 8) does yield the best estimate, while the Riemann Hypothesis does not!

4.A. Appendix: Fourier integrals and series.

Let $L^1(\mathbb{R})$ denote the space of Lebesgue integrable functions on $\mathbb{R}$. The Fourier transform of $f \in L^1(\mathbb{R})$ is defined by

$$\mathcal{F}f(y) = \hat{f}(y) = \int_{\mathbb{R}} f(x) e(-xy) dx. \tag{4.78}$$

Though $f(x)$ is not necessarily continuous its Fourier transform $\hat{f}(y)$ is uniformly continuous and $\hat{f}(y) \to 0$ as $|y| \to \infty$ (the Riemann-Lebesgue lemma). The convolution of two functions $f, g \in L^1(\mathbb{R})$ is defined by

$$(f \star g)(x) = \int_{\mathbb{R}} f(x-y)g(y)dy \tag{4.79}$$

for almost all x, and it belongs to $L^1(\mathbb{R})$, indeed the L^1-norm satisfies $\|f \star g\|_1 \leqslant \|f\|_1 \|g\|_1$. The convolution is a smoothing operator, precisely, if g has the first j derivatives in $L^1(\mathbb{R})$, then so does $f \star g$ for any f in $L^1(\mathbb{R})$.

The Fourier transform $\hat{f}(y)$ determines its original $f(x)$. Precisely, if $f \in L^1(\mathbb{R})$, then

$$\int_{-Y}^{Y} \Big(1 - \frac{|y|}{Y}\Big) \hat{f}(y) e(xy) dy \to f(x)$$

as $Y \to \infty$ in the L^1-norm. If both $f, \hat{f}$ are in $L^1(\mathbb{R})$, then the above integral converges uniformly in x giving

$$f(x) = \int_{\mathbb{R}} \hat{f}(y) e(xy) dy. \tag{4.80}$$

In other words, $\hat{\hat{f}}(x) = f(-x)$. This is the inversion formula for the Fourier transform. The Fourier transform of $\rho(Y) = \max\{1 - \frac{|y|}{Y}, 0\}$ is called the Fejér kernel

$$\varphi(x) = \int_{-Y}^{Y} \Big(1 - \frac{|y|}{Y}\Big) e(xy) dy = \Big(\frac{\sin \pi x Y}{\pi x Y}\Big)^2. \tag{4.81}$$

We have $\widehat{f \star g} = \hat{f} \cdot \hat{g}$, in particular, $\hat{\varphi} \star f = \rho \cdot \hat{f}$, which shows that the linear space of functions in $L^1(\mathbb{R})$ having compactly supported Fourier transform is dense.

Here are basic Fourier pairs:

$$f(x) = \begin{cases} 1 \text{ if } |x| < 1, \\ \frac{1}{2} \text{ if } |x| = \frac{1}{2}, \\ 0 \text{ if } |x| > 1, \end{cases} \qquad \hat{f}(y) = \frac{\sin 2\pi y}{\pi y}, \tag{4.82}$$

$$f(x) = \max\{1 - |x|, 0\}, \qquad \hat{f}(y) = \Big(\frac{\sin \pi y}{\pi y}\Big)^2, \tag{4.83}$$

$$f(x) = e^{-2\pi|x|}, \qquad \hat{f}(y) = \pi^{-1}(1+y^2)^{-1}, \tag{4.84}$$

$$f(x) = e^{-\pi x^2}, \qquad \hat{f}(y) = e^{-\pi y^2}, \tag{4.85}$$

$$f(x) = \frac{1}{\operatorname{ch} \pi x}, \qquad \hat{f}(y) = \frac{1}{\operatorname{ch} \pi y}. \tag{4.86}$$

There is a corresponding Fourier analysis of functions on the circle $T = \mathbb{R}/\mathbb{Z}$. These are regarded as functions on $\mathbb{R}$ which are periodic of period one. The Fourier coefficient of a function $f \in L^1(T)$ is defined by

$$c_n(f) = \int_0^1 f(x) e(-nx) dx. \tag{4.87}$$

They form the Fourier series for f

$$f(x) \sim \sum_{n \in \mathbb{Z}} c_n(f) e(nx) \tag{4.88}$$

which is nothing but a formal expression as long as the convergence is not established. The Riemann-Lebesgue lemma asserts that $c_n(f) \to 0$ as $|n| \to \infty$, and, of course, this is not sufficient for the convergence of (4.88). Nevertheless, the Fourier coefficients $c_n(f)$ determine f, precisely, if $c_n(f) = 0$ for all n, then $f = 0$ almost everywhere.

The convolution

$$(f \star g)(x) = \int_0^1 f(x-y) g(y) dy \tag{4.89}$$

is defined almost everywhere, and has its Fourier coefficients equal to the product

$$c_n(f \star g) = c_n(f) c_n(g).$$

In particular, convolving $f \in L^1(T)$ with a trigonometric polynomial

$$P_N(x) = \sum_{|n| \leqslant N} c_n(P) e(nx)$$

gives a trigonometric polynomial

$$(f \star P_N)(x) = \sum_{|n| \leqslant N} c_n(f) c_n(P_N) e(nx).$$

This trick solves many summability problems. The most popular is the Fejér kernel

$$F_N(x) = \sum_{|n| \leqslant N} \Big(1 - \frac{|n|}{N}\Big) e(nx) = N \Big(\frac{\sin \pi N x}{\pi N x}\Big)^2. \tag{4.90}$$

This produces the Fejér partial sums

$$\sum_{|n| \leqslant N} \Big(1 - \frac{|n|}{N}\Big) c_n(f) e(nx) \tag{4.91}$$

which do converge to $f(x)$ in the L^1-norm on T. The exact partial sums of the Fourier series for f can be tailored similarly by convolving f against the Dirichlet kernel

$$D_N(x) = \sum_{|n| \leqslant N} e(nx) = \frac{\sin \pi(2N+1)x}{\sin \pi x} \tag{4.92}$$

for any positive integer N.

Concerning a pointwise convergence for symmetric partial sums we have

$$\sum_{|n| \leqslant N} c_n(f) e(nx) \to \tfrac{1}{2}(f(x+0) + f(x-0)) \tag{4.93}$$

for all $x \in T$ provided $f \in L^1(T)$ has bounded variation. The convergence is uniform on closed intervals of continuity of f.

In many ways the problem of representing a periodic function by its Fourier series simplifies if f is square integrable because $L^2(T)$ is a Hilbert space, the inner product being

$$\langle f, g\rangle = \int_T f(x)\bar{g}(x)dx,$$

and the additive characters $e_n(x) = e(nx)$ form a complete orthonormal system. Note that $\|f\|_1 \leqslant \|f\|_2$ by Cauchy-Schwarz inequality, so $L^2(T) \subset L^1(T)$. For any f, g in $L^2(T)$ we have the Parseval formula

$$\sum_n c_n(f)\bar{c}_n(g) = \langle f, g\rangle. \tag{4.94}$$

Hence the Fourier coefficients are square summable. The symmetric partial sums of the Fourier series converge to the function in the L^2-norm.

Next we consider functions in several variables $x = (x_1, \dots, x_k) \in \mathbb{R}^k$. Say $f : \mathbb{R}^k \to \mathbb{C}$ is a Schwartz function if

$$f^{(a)}(x) \ll |x|^{-A} \tag{4.95}$$

for any $a = (a_1, \dots, a_k)$ and $A \geqslant 0$, where $|x|^2 = x_1^2 + \cdots + x_k^2$. Let $\mathcal{S}(\mathbb{R}^k)$ denote the class of Schwartz functions. For any $f \in \mathcal{S}(\mathbb{R}^k)$ we set the Fourier transform

$$\hat{f}(y) = \int_{\mathbb{R}^k} f(x)e(-x\cdot y)dx \tag{4.96}$$

where $x \cdot y = x_1y_1 + \cdots + x_ky_k$ is the scalar product in $\mathbb{R}^k$. The Fourier transform maps the space $\mathcal{S}(\mathbb{R}^k)$ into itself (check (4.95) for $\hat{f}(y)$ by partial integration), and it has the following properties: $\hat{\hat{f}}(x) = f(-x)$, $\langle \hat{f}, g\rangle = \langle f, \hat{g}\rangle$, explicitly,

$$f(x) = \int_{\mathbb{R}^k} \hat{f}(y)e(x\cdot y)dy, \tag{4.97}$$

$$\int_{\mathbb{R}^k} \hat{f}(x)g(x)dx = \int_{\mathbb{R}^k} f(y)\hat{g}(y)dy. \tag{4.98}$$

In other words, the Fourier transform is an isometry on $\mathcal{S}(\mathbb{R}^k)$. The convolution

$$(f \star g)(x) = \int_{\mathbb{R}^k} f(x-y)g(y)dy \tag{4.99}$$

is turned into the product $\mathcal{F}(f \star g) = \hat{f}\cdot\hat{g}$.

One may think of $\mathbb{R}^k$ as a homogeneous space acted on by the group of translations $G = \mathbb{R}^k$. In addition to the translations the orthogonal group $SO(\mathbb{R}^k)$ acts on $\mathbb{R}^k$ by rotations making $\mathbb{R}^k$ the euclidean space. The riemannian metric on $\mathbb{R}^k$ (of curvature zero) is given by the differential $ds^2 = (dx_1)^2 + \cdots + (dx_k)^2$, and the corresponding Laplace operator is

$$D = \frac{\partial^2}{\partial x_1^2} + \cdots + \frac{\partial^2}{\partial x_k^2}. \tag{4.100}$$

The exponential functions $\varphi(x) = e(x\cdot y)$ for $y \in \mathbb{R}^k$ are eigenfunctions of D, precisely $(D + \lambda(\varphi))\varphi = 0$, with eigenvalue $\lambda(\varphi) = 4\pi^2|y|^2$. An important point is that the Laplace operator is rotation invariant (i.e., the rotations are isometries of

$\mathbb{R}^k$), therefore the Fourier transform of a radial function is also radial. Precisely, employing the Fourier inversion (4.97) one shows

LEMMA 4.17. *Let $k \geqslant 2$. Put $\nu = \frac{k}{2} - 1$. Suppose*

$$f(x) = g(|x|^2)|x|^{-\nu} \tag{4.101}$$

where g is a smooth compactly supported function on $\mathbb{R}^+$. Then

$$\hat{f}(y) = h(|y|^2)|y|^{-\nu} \tag{4.102}$$

with

$$h(y) = \pi \int_0^\infty J_\nu(2\pi\sqrt{xy})g(x)dx \tag{4.103}$$

where $J_\nu(x)$ is the Bessel function of order ν.

If P is a homogeneous polynomial in k variables and complex coefficients such that $DP = 0$ (it is called a spherical harmonic), then we have the Fourier pair

$$F(x) = P(x)e^{-\pi|x|^2}, \qquad \hat{F}(y) = i^d P(y)e^{-\pi|y|^2}$$

where $d = \deg P$. This self-duality (due to Hecke) actually characterizes spherical harmonics. In principle it says that the multiplication by spherical harmonics has little effect on Fourier transform. For example, Lemma 4.17 generalizes to the following equation (due to Bochner)

$$\mathcal{F}(P(x)g(|x|^2)|x|^{-\nu}) = P(y)h(|y|^2)|y|^{-\nu}$$

where h is given by (4.103) with the Bessel function of order increased from ν to $\nu + d$ (recall that $\nu = \frac{k}{2} - 1$). These facts appear naturally in the context of unitary representations.

We end this survey of classical Fourier analysis by comments on the Mellin transform on $\mathbb{R}^+$. Say $f : \mathbb{R}^+ \to \mathbb{C}$ is of type (α, β) if

$$f(y)y^{s-1} \in L^1(\mathbb{R}^+) \quad \text{for } \alpha < \text{Re } s < \beta. \tag{4.104}$$

For such f the Mellin transform

$$M(f)(s) = \int_0^\infty f(y)y^{s-1}dy \tag{4.105}$$

is defined for complex variable s in the vertical strip $\alpha < \text{Re } s < \beta$. Clearly $M(f)(s)$ is holomorphic in this strip. Since the Mellin transform on $\mathbb{R}^+$ is merely a version of the Fourier transform on $\mathbb{R}$ derived by the change of variables $y = e^x$, $s = it$, the results for the latter translate appropriately for the former. For example if f is continuous of type (α, β) and has bounded variation, then the Mellin inversion formula holds:

$$f(y) = \frac{1}{2\pi i} \int_{(\sigma)} M(f)(s)y^{-s}ds \tag{4.106}$$

where (σ) indicates that the integration is on the vertical line $\text{Re}\,(s) = \sigma$.

Here is a selection of Mellin transforms from Chapter 3 of [**GR**] (formulas (381.4), (761.9), (761.4), (411.3), (523.3), (222.2), (293.3), (871.4), (871.3), (871.2), (871.1) respectively)

$$\int_0^\infty e^{-y}y^{s-1}dy = \Gamma(s), \qquad \sigma > 0, \tag{4.107}$$

$$\int_0^\infty (\cos y)y^{s-1}dy = \Gamma(s)\cos\frac{\pi s}{2}, \qquad 0 < \sigma < 1, \tag{4.108}$$

$$\int_0^\infty (\sin y)y^{s-1}dy = \Gamma(s)\sin\frac{\pi s}{2}, \qquad -1 < \sigma < 1, \tag{4.109}$$

$$\int_0^\infty (1+y)^{-1}y^{s-1}dy = \frac{\pi}{\sin \pi s}, \qquad 0 < \sigma < 1, \tag{4.110}$$

$$\int_0^\infty \log(1+y)y^{s-1}dy = \frac{\pi}{s\sin \pi s}, \qquad -1 < \sigma < 0. \tag{4.111}$$

For $x > 0$ we have

$$\int_0^\infty \cos\Big(\frac{x}{2}\Big(y-\frac{1}{y}\Big)\Big)y^{s-1}dy = 2K_s(x)\cos\frac{\pi s}{2}, \tag{4.112}$$

$$\int_0^\infty \sin\Big(\frac{x}{2}\Big(y-\frac{1}{y}\Big)\Big)y^{s-1}dy = 2K_s(x)\sin\frac{\pi s}{2}, \tag{4.113}$$

$$\int_0^\infty \cos\Big(\frac{x}{2}\Big(y+\frac{1}{y}\Big)\Big)y^{s-1}dy = -\pi J_s(x)\sin\frac{\pi s}{2} - \pi Y_s(x)\cos\frac{\pi s}{2}, \tag{4.114}$$

$$\int_0^\infty \sin\Big(\frac{x}{2}\Big(y+\frac{1}{y}\Big)\Big)y^{s-1}dy = \pi J_s(x)\cos\frac{\pi s}{2} - \pi Y_s(s)\sin\frac{\pi s}{2}. \tag{4.115}$$

In the last four formulas $-1 < \sigma < 1$ and J_s, K_s, Y_s are the standard Bessel functions.

In the last two formulas we also have

$$J_s(x)\sin\frac{\pi s}{2} + Y_s(x)\cos\frac{\pi s}{2} = \frac{J_s(x) - J_{-s}(x)}{2\sin\frac{\pi s}{2}}, \tag{4.116}$$

and

$$J_s(x)\cos\frac{\pi s}{2} - Y_s(x)\sin\frac{\pi s}{2} = \frac{J_s(x) + J_{-s}(x)}{2\cos\frac{\pi s}{2}}. \tag{4.117}$$

CHAPTER 5

CLASSICAL ANALYTIC THEORY OF L-FUNCTIONS

We will now discuss a very classical part of analytic number theory. Indeed, most of the results discussed in this chapter can be traced back in some form to Riemann's Memoir on the zeta function. They were subsequently extended to Dirichlet characters, with two essentially new phenomenon appearing. One is that uniformity of estimates with respect to the conductor is vital. The other, not unrelated, is the stubborn appearance of the so-called exceptional zeros (or Landau-Siegel zeros) for quadratic characters. Ruling out the existence of such zeros remains one of the main problems of number theory.

Nowadays more refined L-functions occur frequently (see Chapters 14 and 15 for a survey of automorphic forms, one of the sources for L-functions), but it is well-known that the classical analytic results remain valid in great generality. Finding explicit statements in the literature, uniform in all parameters, is however sometimes difficult. For this reason we prove all main results in a general abstract context. Then, starting from Section 5.9 , we survey the most important classes of L-functions that arise in practice:

(1) Dirichlet L-functions; see Section 5.9.

(2) Dedekind zeta functions and L-functions of Hecke Grössencharakteren of number fields; see Section 5.10.

(3) The automorphic L-functions of classical cusp forms (that is to say of holomorphic and Maass forms on $GL(2)$); see Section 5.11.

(4) General automorphic L-functions on $GL(m)$, $m \geqslant 1$; see Section 5.12.

(5) The L-functions of algebraic varieties and Galois representations (although most of those are only conjectured to fit the framework discussed in this chapter); see Sections 5.13 and 5.14.

In each of these cases we spell out the concrete meaning of the abstract results, and give a few additional results.

5.1. Definitions and preliminaries.

In order to treat all the L-functions that interest us in one stroke, we introduce some abstract definitions, although we are not trying to create axioms. The alternative is to introduce automorphic L-functions at the outset (see Section 5.12), but some translation would still be required for the most classical cases (and Rankin-Selberg convolutions would require a separate treatment). We try to balance familiarity and ease of use with sufficient generality. The reader unfamiliar with L-functions, or acquainted only with Dirichlet L-functions, should read the introductory paragraphs of Section 5.9 first, then read what follows with this important case in mind. Basic ideas are already present in the context of Dirichlet L-functions. We then encourage him or her to get acquainted with automorphic

L-functions (see [**BG**] for instance). The analogy with Dirichlet or Hecke characters on ideals in a number field do not really reveal the beauty and the depth of the genuine modular forms. This is especially true when Rankin-Selberg L-functions occur, which are instrumental for establishing zero-free regions.

We denote L-functions by $L(f,s)$, $L(g,s)$, although the symbols f, g carry no specific meaning until Section 5.9. They are used merely for the suggestion that L-functions usually arise as attached to some interesting arithmetic object.

We say that $L(f,s)$ is an L-function if we have the following data and conditions:

(1) A Dirichlet series with Euler product of degree $d \geqslant 1$,

$$L(f,s) = \sum_{n\geqslant 1} \lambda_f(n) n^{-s} = \prod_p (1-\alpha_1(p)p^{-s})^{-1} \cdots (1-\alpha_d(p)p^{-s})^{-1} \tag{5.1}$$

with $\lambda_f(1) = 1$, $\lambda_f(n) \in \mathbb{C}$, $\alpha_i(p) \in \mathbb{C}$. The series and Euler products must be absolutely convergent for $\mathrm{Re}\,(s) > 1$. The $\alpha_i(p)$, $1 \leqslant i \leqslant d$, are called the local roots or local parameters of $L(f,s)$ at p, and they satisfy

$$|\alpha_i(p)| < p \text{ for all } p. \tag{5.2}$$

(2) A gamma factor

$$\gamma(f,s) = \pi^{-ds/2} \prod_{j=1}^{d} \Gamma\Big(\frac{s+\kappa_j}{2}\Big) \tag{5.3}$$

where the numbers $\kappa_j \in \mathbb{C}$ are called the local parameters of $L(f,s)$ at infinity. We assume these numbers are either real or come in conjugate pairs. Moreover, $\mathrm{Re}\,(\kappa_j) > -1$. This last condition tells us that $\gamma(f,s)$ has no zero in $\mathbb{C}$ and no pole for $\mathrm{Re}\,(s) \geqslant 1$.

(3) An integer $q(f) \geqslant 1$, called the conductor of $L(f,s)$, such that $\alpha_i(p) \neq 0$ for $p \nmid q(f)$ and $1 \leqslant i \leqslant d$. A prime $p \nmid q(f)$ is said to be unramified.

From these, the so-called complete L-function

$$\Lambda(f,s) = q(f)^{\frac{s}{2}} \gamma(f,s) L(f,s) \tag{5.4}$$

is defined. Clearly, it is holomorphic in the half-plane $\mathrm{Re}\,(s) > 1$, yet it must admit analytic continuation to a meromorphic function for $s \in \mathbb{C}$ of order 1 (see the Appendix), with at most poles at $s = 0$ and $s = 1$. Moreover, it must satisfy the functional equation

$$\Lambda(f,s) = \varepsilon(f)\Lambda(\bar{f}, 1-s). \tag{5.5}$$

where $\bar{f}$ is an object associated with f (the dual of f) for which $\lambda_{\bar{f}}(n) = \bar{\lambda}_f(n)$, $\gamma(\bar{f},s) = \gamma(f,s)$, $q(\bar{f}) = q(f)$ and $\varepsilon(f)$ is a complex number of absolute value 1, called the "root number" of $L(f,s)$.

We let $r(f)$ denote the order of the pole (if positive) or zero (if negative) of $\Lambda(f,s)$ at $s=0$ and $s=1$, equal because of (5.5). Since $\gamma(f,1) \neq 0, \infty$, $r(f)$ is also the order of the pole or zero of $L(f,s)$ at $s=1$. Later we shall show in all concrete cases that $L(f,s)$ does not vanish at $s=1$, so $r(f) \geqslant 0$.

We seek uniform estimates for various analytic quantities related to $L(f,s)$. For an individual L-function, the only parameter is $s \in \mathbb{C}$, but when $L(f,s)$ varies, we also have to deal with the degree, conductor, local parameters, and each of these or a combination may be under investigation. It turns out that most results for $L(f,s)$ are expressed conveniently in terms of the *analytic conductor* . First put

$$\mathfrak{q}_\infty(s) = \prod_{j=1}^{d} (|s+\kappa_j|+3). \tag{5.6}$$

Multiplying this by $q(f)$ we get the analytic conductor

$$\mathfrak{q}(f,s) = q(f)\mathfrak{q}_\infty(s) = q(f)\prod_{j=1}^{d} (|s+\kappa_j|+3). \tag{5.7}$$

We also denote

$$\mathfrak{q}(f) = \mathfrak{q}(f,0) = q(f)\prod_{j=1}^{d} (|\kappa_j|+3).$$

Note that $\mathfrak{q}(f) \geqslant 3^d q(f)$, so $d < \log \mathfrak{q}(f)$, and that

$$\mathfrak{q}(f,s) \leqslant \mathfrak{q}(f)(|s|+3)^d, \tag{5.8}$$

so estimates could be performed in terms of this last quantity without compromising much of strength.

From the definition, we see that if $L(f,s)$ and $L(g,s)$ are L-functions, then $L(f,s)L(g,s)$ is one with conductor $q(f)q(g)$ and analytic conductor $\mathfrak{q}(f,s)\mathfrak{q}(g,s)$, gamma factor $\gamma(f,s)\gamma(g,s)$, root number $\varepsilon(f)\varepsilon(g)$. Moreover, $L(\bar{f},s)$ is also an L-function by construction, with same degree, conductor, gamma factor and with $\varepsilon(\bar{f}) = \bar{\varepsilon}(f)$. Also if $L(f,s)$ is entire, then for any fixed $t \in \mathbb{R}$, the shifted L-function $L(g,s) = L(f,s+it)L(\bar{f},s-it)$ is another L-function, with gamma factor $\gamma(g,s) = \gamma(f,s+it)\gamma(f,s-it)$, conductor $q(g) = q(f)^2$, root number $\varepsilon(g) = 1$. The analytic conductor is $\mathfrak{q}(g,s) = \mathfrak{q}(f,s+it)\mathfrak{q}(f,s-it)$. The condition that $L(f,s)$ be entire is required because we are not allowed to shift the poles. We take the product of two shifted L-functions by it and $-it$ to maintain the local parameters coming in complex pairs. Of course, these are only cosmetic arrangements.

In the sequel we often simplify the above notation by not displaying the dependence on f, s; we write $q = q(f)$, $\mathfrak{q}_\infty = \mathfrak{q}_\infty(s)$, $\mathfrak{q} = \mathfrak{q}(f,s)$, $r = r(f)$.

If $\bar{f} = f$, then $L(f,s)$ is said to be self-dual. This means that the Dirichlet series of the L-function has real coefficients. For a self-dual L-function, the root number is real, hence $\varepsilon(f) = \pm 1$. It is then called the sign of the functional equation. The following observation is originally due to Shimura; despite its simplicity, it is significant in a number of applications (see Section 23.6 and Chapter 26 for instance):

PROPOSITION 5.1. *Let $L(f,s)$ be self-dual with $\varepsilon(f) = -1$. Then $L(f,\frac{1}{2}) = 0$.*

PROOF. By definition we have $\gamma(f,\frac{1}{2}) \neq 0$, and the functional equation applied to the central point $s = \frac{1}{2}$ yields $L(f,\frac{1}{2}) = -L(f,\frac{1}{2})$, hence $L(f,\frac{1}{2}) = 0$. □

The local roots $\alpha_i(p)$ are well defined up to permutation of the indices. If for any i we have $|\alpha_i(p)| = 1$ for all $p \nmid q(f)$ and $|\alpha_i(p)| \leqslant 1$ otherwise, then $L(f,s)$ is said to satisfy the Ramanujan-Petersson conjecture. This implies, in particular, that $|\lambda_f(n)| \leqslant \tau_d(n)$. Similarly, if for any j we have $\mathrm{Re}(\kappa_j) \geqslant 0$, then $L(f,s)$ is said to satisfy the (generalized) Selberg Conjecture, or the Ramanujan-Petersson conjecture at the infinite place. In this case $\gamma(f,s)$ has no poles for $\mathrm{Re}(s) > 0$.

Because $\Lambda(f,s) = \gamma(f,s)L(f,s)$ is assumed to be holomorphic, except possibly at $s = 0$ and $s = 1$, it follows that poles of $\gamma(f,s)$ at $s \neq 0$ are zeros of $L(f,s)$. These are called the "trivial zeros". They are located exactly at the points $-2m - \kappa_j \neq 0$, $1 \leqslant j \leqslant d$, where $m \geqslant 0$ is an integer. For example, the trivial zeros of $\zeta(s)$ are at $s = -2, -4, -6, \ldots$ while $\zeta(0) = -\frac{1}{2}$. The other zeros of $L(f,s)$ are called the non-trivial zeros. They are located in the critical strip $0 \leqslant \mathrm{Re}(s) \leqslant 1$. Presumably some trivial zeros may lay in the critical strip as well, however, one conjectures they are never in the interior.

As mentioned in the introduction, we will discuss examples starting in Section 5.9. It is important to note that many interesting L-series exist beyond our context, either because of lack of strength in our fingers to prove the required assumptions (for instance Artin L-functions for non-trivial irreducible characters are not known to be entire in general, and general L-functions of varieties are not even known to be meromorphic, see Sections 5.13 and 5.14), or because some of the conditions fail. For instance, in the second class come Dirichlet L-functions of non-primitive characters (they have Euler product but the functional equation has extra factors), L-functions of general modular forms (see Theorem 14.7 for instance) which have no Euler products and a functional equation relating $L(f,s)$ with some function not directly related to $L(\bar{f},s)$. Still interesting are partial zeta functions of ideal classes of number fields (see (22.55) for imaginary quadratic fields) which have functional equations (relating an ideal class $\mathfrak{a}$ with $\mathfrak{d} - \mathfrak{a}$, where $\mathfrak{d}$ is the different), but no Euler product if the class number is > 1. However, these partial zeta functions can be reconstructed as linear combinations of the full L-functions for the class group characters.

The modern analytic theory of L-functions is much influenced by two concepts: that of a family of L-functions, and that of (Langlands) functoriality. The former lacks a formal definition but is a strong guiding principle (see [**M2**]). In this book we will indicate when examples of families appear. Note that in this context, varying the imaginary part t of s should count as varying in a family. The main parameters in a family of L-functions are encoded in the analytic conductor.

Functoriality, on the other hand, reflects the principle that sometimes new L-functions are built out of old ones by making some operation (simple looking, yet quite subtle) on their coefficients $\lambda_f(n)$, such as raising to some power ($\lambda_f(n^k)$ or $\lambda_f(n)^k$). A proper formalization requires quite major notation and setup, and the naïve ideas must usually be changed; moreover, much of the global setting is still conjectural. We will do here with a simple-minded definition of Rankin-Selberg type L-functions, and mention elsewhere the symmetric square and other symmetric power L-functions (see Section 5.12 and the discussion of the Sato-Tate conjecture in Chapter 21).

Let $L(f,s)$ and $L(g,s)$ be L-functions of degrees d and e, with local components at infinity κ_i and ν_j, and local roots $(\alpha_i(p))$ and $(\beta_j(p))$ respectively. For $p \nmid q(f)q(g)$, let

$$L_p(f \otimes g, s) = \prod_{i,j} (1 - \alpha_i(p)\beta_j(p)p^{-s})^{-1}. \tag{5.9}$$

We say that f and g have a Rankin-Selberg convolution if there exists an L-function $L(f \otimes g, s)$ of degree de such that

$$L(f \otimes g, s) = \prod_{p \nmid q(f)q(g)} L_p(f \otimes g, s) \prod_{p \mid q(f)q(g)} H_p(p^{-s})$$

where

$$H_p(p^{-s}) = \prod_{j=1}^{de} (1 - \gamma_j(p)p^{-s})^{-1} \quad \text{with} \quad |\gamma_j(p)| < p. \tag{5.10}$$

The gamma factor must be written as

$$\gamma(f \otimes g, s) = \pi^{-des/2} \prod_{i,j} \Gamma\Big(\frac{s + \mu_{i,j}}{2}\Big)$$

with $\mathrm{Re}\,(\mu_{i,j}) \leqslant \mathrm{Re}\,(\kappa_i + \nu_j)$ and $|\mu_{i,j}| \leqslant |\kappa_i| + |\nu_j|$, where κ_i and ν_j are the local components at infinity of f and g. In addition, the conductor of $f \otimes g$ must divide $q(f)^e q(g)^d$ and $L(f \otimes g, s)$ must have a pole at $s = 1$ if $g = \bar{f}$. In fact, unless $L(f, s)$ or $L(g, s)$ factor, we will see that in concrete cases $L(f \otimes g, s)$ is entire if $g \neq \bar{f}$.

We call $L(f \otimes g, s)$ the Rankin-Selberg L-function or convolution of f and g, or the Rankin-Selberg square if $g = \bar{f}$.

Note that if $L(f \otimes f, s)$ or $L(f \otimes \bar{f}, s)$ exists, then the estimates on the local roots and on κ_j can be improved to $|\alpha_i(p)| < \sqrt{p}$ and $\mathrm{Re}\,(\kappa_j) > -\frac{1}{2}$.

The conditions also imply the inequality for the analytic conductor

$$\mathfrak{q}(f \otimes g, s) \leqslant \mathfrak{q}(f)^e \mathfrak{q}(g)^d (|s| + 3)^{de} \tag{5.11}$$

(see (5.8)). Although it is by no means obvious that $L(f \otimes g, s)$ should exist, it is the case when $L(f, s)$ and $L(g, s)$ are automorphic L-functions, as discussed in Section 5.12. This covers most cases of importance in analytic number theory (including the L-functions of number fields). Notice that the dual L-function is $L(\bar{f} \otimes \bar{g}, s)$.

5.2. Approximations to L-functions.

We begin with a crude result which will be used extensively, and improved later.

LEMMA 5.2. *Any L-function $L(f, s)$ is polynomially bounded in vertical strips $s = \sigma + it$ with $a \leqslant \sigma \leqslant b$, $|t| \geqslant 1$.*

PROOF. The L-function is bounded in the half-plane $\sigma \geqslant 1 + \epsilon$ by the absolute convergence of the Dirichlet series. Hence we deduce a polynomial bound for $L(f, s)$ in the half-plane $\sigma < -\epsilon$ by the functional equation and the Stirling formula (5.114). Then the polynomial bound in the remaining strip follows by the Phragmen-Lindelöf principle (see Theorem 5.53). □

We now state an exact formula, known as the "approximate functional equation", which gives analytically convenient expressions for $L(f, s)$ in the critical strip where the series does not converge absolutely.

THEOREM 5.3. *Let $L(f, s)$ be an L-function. Let $G(u)$ be any function which is holomorphic and bounded in the strip $-4 < \mathrm{Re}(u) < 4$, even, and normalized by $G(0) = 1$. Let $X > 0$. Then for s in the strip $0 \leqslant \sigma \leqslant 1$ we have*

$$L(f,s) = \sum_n \frac{\lambda_f(n)}{n^s} V_s\Big(\frac{n}{X\sqrt{q}}\Big) + \varepsilon(f,s) \sum_n \frac{\overline{\lambda_f(n)}}{n^{1-s}} V_{1-s}\Big(\frac{nX}{\sqrt{q}}\Big) + R \tag{5.12}$$

where $V_s(y)$ is a smooth function defined by

$$V_s(y) = \frac{1}{2\pi i} \int_{(3)} y^{-u} G(u) \frac{\gamma(f, s+u)}{\gamma(f,s)} \frac{du}{u} \tag{5.13}$$

and

$$\varepsilon(f,s) = \varepsilon(f) q(f)^{\frac{1}{2}-s} \frac{\gamma(f, 1-s)}{\gamma(f,s)}. \tag{5.14}$$

The last term $R = 0$ if $\Lambda(f, s)$ is entire, otherwise

$$R = \Big(\operatorname*{res}_{u=1-s} + \operatorname*{res}_{u=-s}\Big) \frac{\Lambda(f, s+u)}{q^{s/2}\gamma(f,s)} \frac{G(u)}{u} X^u.$$

PROOF. Consider the integral

$$I(X, f, s) = \frac{1}{2\pi i} \int_{(3)} X^u \Lambda(f, s+u) G(u) \frac{du}{u}.$$

This integral exists because $\Lambda(f, s)$ decays rapidly as $t \to +\infty$ for fixed σ by Stirling's formula (see (5.113)). For the same reason one can move the integration to $\mathrm{Re}(u) = -3$ by Lemma 5.2. Applying the functional equation there yields

$$\Lambda(f,s) = I(X, f, s) + \varepsilon(f) I(X^{-1}, \bar{f}, 1-s) + R q^{s/2} \gamma(f,s) \tag{5.15}$$

where $\Lambda(f, s)$ comes from the simple pole of $u^{-1}G(u)$ at $u = 0$ and the last term R comes from possible residues at $u = 1 - s$ and $u = -s$ if $\Lambda(f, s)$ is not entire.

By expanding into absolutely convergent Dirichlet series we have

$$\begin{aligned} I(X, f, s) &= q^{s/2} \sum_{n \geqslant 1} \lambda_f(n) n^{-s} \frac{1}{2\pi i} \int_{(3)} \gamma(f, s+u) \Big(\frac{X\sqrt{q}}{n}\Big)^u G(u) \frac{du}{u} \\ &= q^{s/2} \gamma(f,s) \sum_{n \geqslant 1} \frac{\lambda_f(n)}{n^s} V_s\Big(\frac{n}{X\sqrt{q}}\Big). \end{aligned}$$

We do the same for $I(X^{-1}, \bar{f}, 1-s)$ and combine them with (5.15). Dividing both sides by $q^{s/2}\gamma(f, s)$ yields (5.12). □

EXERCISE 1. Let $L(f, s)$ be an L-function which is entire. For a smooth, compactly supported function F on $\mathbb{R}^+$ with Mellin transform $\hat{F}$, prove that

$$\sum_{n\geqslant 1} \lambda_f(n)F(n) = \frac{\varepsilon(f)}{\sqrt{q}} \sum_{n\geqslant 1} \bar{\lambda}_f(n)H(n) \tag{5.16}$$

where

$$H(y) = \frac{1}{2\pi i} \int_{(3)} \hat{F}(1-s) \frac{\gamma(f, s)}{\gamma(f, 1-s)} y^{-s} ds.$$

(More explicit formulas, in the case of classical automorphic forms, are described in Section 4.5; in particular, see the formula (4.71) in which the test function goes through the Hankel transform).

Show that if $L(f, s) = \zeta(s)$ is the Riemann zeta function, (5.16) holds with an additional term

$$R = \operatorname*{res}_{s=1} \zeta(s)\hat{F}(s) = \int_0^{+\infty} F(y)dy$$

on the right side. Prove that in this case

$$H(y) = 2\int_0^{+\infty} F(y)\cos(2\pi xy)dy,$$

and (5.16) is the Poisson summation formula.

EXERCISE 2. Let $L(f, s)$ be the L-function associated with a holomorphic primitive cusp form f of weight $k = 2$ and level q (think of the Hasse-Weil zeta function of an elliptic curve). Derive from (5.12) the following formula for the central value

$$L(f, \tfrac{1}{2}) = \sum_n \frac{\lambda_f(n)}{\sqrt{n}} \exp\Big(-\frac{2\pi n}{X\sqrt{q}}\Big) + \varepsilon(f)\sum_n \frac{\overline{\lambda_f}(n)}{\sqrt{n}} \exp\Big(-\frac{2\pi nX}{\sqrt{q}}\Big),$$

where X is any positive number [**Hint:** take $G(u) = 1$.] Differentiating with respect to X this yields the summation formula

$$\sum_n \frac{\lambda_f(n)}{\sqrt{n}} \exp\Big(-\frac{2\pi n}{X\sqrt{q}}\Big) = X^2\varepsilon(f)\sum_n \frac{\overline{\lambda_f}(n)}{\sqrt{n}} \exp\Big(-\frac{2\pi nX}{\sqrt{q}}\Big).$$

Taking $X = 1$ we get for self-dual L-function with root number $\varepsilon(f) = 1$,

$$L(f, \tfrac{1}{2}) = 2\sum_n \frac{\lambda_f(n)}{\sqrt{n}} \exp\Big(-\frac{2\pi n}{\sqrt{q}}\Big).$$

For suitable test functions $G(u)$, both sums in (5.12) are effectively limited to the terms with $n \ll \sqrt{\mathfrak{q}(f, s)}$. We shall see this clearly for a particular choice of $G(u)$,

$$G(u) = \Big(\cos\frac{\pi u}{4A}\Big)^{-4dA}$$

where A is a positive integer.

PROPOSITION 5.4. *Suppose* $\mathrm{Re}\,(s+\kappa_j) \geqslant 3\alpha > 0$ *for* $1 \leqslant j \leqslant d$. *Then the derivatives of* $V_s(y)$ *satisfy*

$$y^a V_s^{(a)}(y) \ll \Big(1 + \frac{y}{\sqrt{\mathfrak{q}_\infty}}\Big)^{-A}, \tag{5.17}$$

$$y^a V_s^{(a)}(y) = \delta_a + O\Big(\Big(\frac{y}{\sqrt{\mathfrak{q}_\infty}}\Big)^{\alpha}\Big) \tag{5.18}$$

where $\delta_0 = 1$, $\delta_a = 0$ *if* $a > 0$ *and the implied constants depend only on* α, a, A *and* d. *Recall that* $\mathfrak{q}_\infty = \mathfrak{q}_\infty(s)$ *is given by (5.6).*

PROOF. For s and u with $\mathrm{Re}\,(s) = \sigma > 0$ and $\mathrm{Re}\,(u) = \beta > -\sigma$ we derive by Stirling's formula (5.112)

$$\begin{aligned}\frac{\Gamma(s+u)}{\Gamma(s)} &\ll \frac{|s+u|^{\sigma+\beta-\frac{1}{2}}}{|s|^{\sigma-\frac{1}{2}}} \exp\Big(\frac{\pi}{2}(|s| - |s+u|)\Big)\\ &\ll (|s|+3)^{\beta} \exp(\frac{\pi}{2}|u|)\end{aligned}$$

where the implied constant depends only on σ and β. Hence

$$\frac{\gamma(f,s+u)}{\gamma(f,s)} \ll \mathfrak{q}_\infty^{\beta/2} \exp\Big(\frac{\pi d}{2}|u|\Big).$$

We have

$$y^a V_s^{(a)}(y) = \frac{1}{2\pi i} \int_{(3)} y^{-u} G(u)(-u)^a \frac{\gamma(f,s+u)}{\gamma(f,s)} \frac{du}{u}.$$

Moving the integration to the line $\mathrm{Re}\,(u) = \beta = -\alpha$ we derive (5.17) while moving to the line $\mathrm{Re}\,(u) = A$ we derive the bound $O((\sqrt{\mathfrak{q}_\infty}/y)^A)$. This combined with (5.17) yields (5.18). □

The formula (5.12) expresses $L(f,s)$ in the critical strip, essentially, as two partial sums of the Dirichlet series $L(f,s)$ and the dual one, each of length $\sqrt{\mathfrak{q}(f,s)}$. Hence one can easily derive estimates in the critical strip. We give here the individual estimate, and the uniform version for L-functions satisfying the Ramanujan-Petersson conjecture. See Section 5.12 for stronger results in the case of automorphic L-functions.

Because $L(f,s)$ can have a pole or zero at $s = 1$ of order r, it is appropriate to kill this singularity before estimating. Therefore we shall often estimate $p_r(s)L(f,s)$ rather than $L(f,s)$, where

$$p_r(s) = \Big(\frac{s-1}{s+1}\Big)^r. \tag{5.19}$$

EXERCISE 3. Prove that for s with $0 \leqslant \sigma \leqslant 1$ we have

$$p_r(s)L(f,s) \ll \mathfrak{q}(f,s)^{(1-\sigma)/2+\varepsilon} \tag{5.20}$$

for any $\varepsilon > 0$, with the implied constant depending on ε and f. If $L(f,s)$ satisfies the Ramanujan-Petersson conjecture, then the implied constant depends only on ε and the degree of the L-function. [**Hint:** Use the absolute convergence of $L(f,s)$ in $\sigma > 1$ (or the bound $|\lambda_f(n)| \leqslant \tau_d(n)$, respectively), the functional equation,

Stirling's estimate for the gamma function and the Phragmen-Lindelöf convexity principle.]

In particular, (5.20) yields the following bound on the critical line

$$L(f,s) \ll \mathfrak{q}(f,s)^{\frac{1}{4}+\varepsilon} \tag{5.21}$$

which is known as the convexity bound. Using the approximate functional equation rather than the convexity principle, and the Ramanujan-Petersson conjecture, one can derive a slightly better estimate

$$L(f,s) \ll \mathfrak{q}(f,s)^{\frac{1}{4}}(\log \mathfrak{q}(f,s))^{d-1} \tag{5.22}$$

where the implied constant depends only on the degree d of $L(f,s)$. It is conjectured that

$$L(f,s) \ll \mathfrak{q}(f,s)^{\varepsilon}$$

for $\operatorname{Re}(s) = \frac{1}{2}$ with any $\varepsilon > 0$, the implied constant depending only on ε. This is the so-called Lindelöf Hypothesis, which is one of the consequences of the Grand Riemann Hypothesis (see Corollary 5.20). However, in many applications the major step is to improve on the convexity bound for relevant L-functions, in the sense of reducing the exponent $\frac{1}{4}$ by a positive number, however small that number may be. Such results go back to H. Weyl for the Riemann zeta function, namely

$$\zeta(s) \ll |s|^{\frac{1}{6}+\varepsilon}$$

(see (8.22)). For Dirichlet characters, one derives from Burgess's estimate for short character sums that

$$L(\chi,s) \ll |s|q^{\frac{3}{16}+\varepsilon}$$

(see Theorem 12.9), which is worse than (5.21) in t-aspect, but much better in q-aspect: the latter is often the most important (e.g. for applications to algebraic number theory).

In view of the proof of (5.22) by the formula (5.12) it is clear that any subconvexity estimate amounts to proving that there is considerable cancellation when summing the oscillating values $\lambda_f(n)n^{-\frac{1}{2}-it}$, and this proves to be an arithmetical problem.

Currently, it is known that a subconvexity bound holds in t-aspect and q-aspect separately (but not yet for both jointly) for any L-function attached to classical modular forms (see the surveys [**M2**], [**IS1**], and the papers [**DFI2**], [**DFI3**], [**KMV2**], [**PSa**], [**M1**], among others).

5.3. Counting zeros of L-functions.

One of the deepest subjects of the theory of L-functions is the distribution of the zeros of $L(f,s)$. We discuss first what can be done using basic properties of holomorphic functions through the method of Hadamard and de la Vallée Poussin.

LEMMA 5.5. *Let $L(f,s)$ be an L-function. All zeros ρ of $\Lambda(f,s)$ are in the critical strip $0 \leqslant \sigma \leqslant 1$. For any $\varepsilon > 0$, we have*

$$\sum_{\rho \neq 0,1} |\rho|^{-1-\varepsilon} < +\infty.$$

PROOF. Since the Euler product expansion of $L(f,s)$ is absolutely convergent and $\gamma(f,s)$ does not vanish for $\mathrm{Re}(s) > 1$, there are no zeros of $\Lambda(f,s)$ in this region. By the functional equation, the same holds for $\mathrm{Re}(s) < 0$. The second statement is a general result of complex analysis for entire functions of order 1. □

Notice that if $\rho = \beta + i\gamma$ is a zero of $\Lambda(f,s)$, then $\bar\rho$ is a zero of the dual $\Lambda(\bar f,s)$. Hence, by the functional equation, the point reflected in the critical line $\rho^* = 1-\bar\rho = 1-\beta+i\gamma$ is also a zero of $\Lambda(f,s)$ (with corresponding multiplicity). Of course, there is no symmetry among the trivial zeros of $L(f,s)$ which come from the poles of $\Gamma(\frac{s+\kappa_j}{2})$.

THEOREM 5.6. *Let $L(f,s)$ be an L-function. There exist constants $a = a(f)$ and $b = b(f)$ such that*

$$(s(1-s))^r\Lambda(f,s) = e^{a+bs}\prod_{\rho\neq 0,1}\Big(1-\frac{s}{\rho}\Big)e^{s/\rho} \tag{5.23}$$

where ρ ranges over all zeros of $\Lambda(f,s)$ different from $0,1$. Hence

$$-\frac{L'}{L}(f,s) = \frac{1}{2}\log q + \frac{\gamma'}{\gamma}(f,s) - b + \frac{r}{s} + \frac{r}{s-1} - \sum_{\rho\neq 0,1}\Big(\frac{1}{s-\rho}+\frac{1}{\rho}\Big). \tag{5.24}$$

both expressions being uniformly absolutely convergent in compact subsets which have no zeros or poles (recall that r is the order of the pole or zero of $L(f,s)$ at $s=1$).

PROOF. The expansion (5.23) is simply the application of the Hadamard factorization theorem of entire functions of finite order, and (5.24) follows by taking the logarithmic derivative. □

We denote

$$-\frac{L'}{L}(f,s) = \sum_{n\geqslant 1}\Lambda_f(n)n^{-s} \tag{5.25}$$

the expansion of the logarithmic derivative of an L-function in Dirichlet series supported on prime powers. In terms of the local roots $\alpha_i(p)$ of the Euler product (5.7) we have

$$\Lambda_f(p^k) = \sum_{j=1}^{d}\alpha_j(p)^k\log p. \tag{5.26}$$

For Dirichlet characters, $\Lambda_\chi(n) = \chi(n)\Lambda(n)$, and in general $\Lambda_f(p) = \lambda_f(p)\log p$ for p prime. Note that $\Lambda_{\bar f}(n) = \overline{\Lambda_f(n)}$.

PROPOSITION 5.7. *Let $L(f,s)$ be an L-function of degree $d\geqslant 1$ and let ρ denote the zeros of $\Lambda(f,s)$ different from $0,1$.*

(1) The number of zeros $\rho=\beta+i\gamma$ such that $|\gamma-T|\leqslant 1$, say $m(T,f)$, satisfies

$$m(T,f) \ll \log \mathfrak{q}(f,iT) \tag{5.27}$$

with an absolute implied constant.

(2) For any s in the strip $-\frac{1}{2} \leqslant \sigma \leqslant 2$ we have

$$\frac{L'}{L}(f,s) + \frac{r}{s} + \frac{r}{s-1} - \sum_{|s+\kappa_j|<1} \frac{1}{s+\kappa_j} - \sum_{|s-\rho|<1} \frac{1}{s-\rho} \ll \log \mathfrak{q}(f,s), \tag{5.28}$$

with an absolute implied constant.

(3) The constant $b(f)$ in (5.23) satisfies

$$\mathrm{Re}\,(b(f)) = -\sum_{\rho} \mathrm{Re}\,(\rho^{-1}). \tag{5.29}$$

PROOF. First we prove (5.29). Taking the logarithmic derivative of the functional equation $(s(1-s))^r \Lambda(f,s) = \varepsilon(f)(s(1-s))^r \Lambda(\bar{f}, 1-s)$, we get from (5.23)

$$2\mathrm{Re}\,(b(f)) = b(f) + b(\bar{f}) = -\sum_{\rho \neq 0,1} \Big(\frac{1}{s-\rho} + \frac{1}{1-s-\bar{\rho}} + \frac{1}{\rho} + \frac{1}{\bar{\rho}}\Big). \tag{5.30}$$

We have $(s-\rho)^{-1} - (1-s-\bar{\rho})^{-1} \ll |\rho|^{-2}$ for $s \neq \rho$, the implied constant depending on s, and similarly $\rho^{-1} + \bar{\rho}^{-1} \ll |\rho|^{-2}$ so the series

$$\sum_{\rho \neq 0,1} \Big(\frac{1}{s-\rho} + \frac{1}{1-s-\bar{\rho}}\Big) \quad \text{and} \quad \sum_{\rho \neq 0,1} \Big(\frac{1}{\rho} + \frac{1}{\bar{\rho}}\Big)$$

are absolutely convergent by Lemma 5.5. So we can separate them in (5.30) and the first one vanishes (the terms cancel out since ρ and $1-\bar{\rho}$ are zeros of $\Lambda(f,s)$), giving (5.29).

Let $T \geqslant 2$ and $s = 3 + iT$. By (5.26) we have $|\Lambda_f(n)| \leqslant dn \log n$. Hence

$$\Big|\frac{L'}{L}(f,s)\Big| \leqslant d\zeta'(2) \ll \log \mathfrak{q}(f). \tag{5.31}$$

By Stirling's formula (5.116) we have $\frac{1}{2}\log q + \frac{\gamma'}{\gamma}(f,s) \ll \log \mathfrak{q}(f,s)$. Observe also that for any zero $\rho = \beta + i\gamma$ we have

$$\frac{2}{9+(T-\gamma)^2} < \mathrm{Re}\,\Big(\frac{1}{s-\rho}\Big) < \frac{3}{4+(T-\gamma)^2}.$$

Hence we can take the real part in (5.24) and rearrange the resulting absolutely convergent series to derive using (5.29) that

$$\sum_{\rho} \frac{1}{1+(T-\gamma)^2} \ll \log \mathfrak{q}(f, iT). \tag{5.32}$$

This implies (5.27). To derive (5.28), we write $s = \sigma + it$ and

$$-\frac{L'}{L}(f,s) = -\frac{L'}{L}(f,s) + \frac{L'}{L}(f, 3+it) + +O(\log \mathfrak{q}(f,s)).$$

by (5.31), and we get by (5.24) and (5.116) again

$$-\frac{L'}{L}(f,s) = \frac{\gamma'}{\gamma}(f,s) + \frac{r}{s} + \frac{r}{s-1} - \sum_{\rho} \Big(\frac{1}{s-\rho} - \frac{1}{3+it-\rho}\Big) + O(\log \mathfrak{q}(f,s)).$$

In the series, keep the zeros with $|s-\rho|<1$ and estimate the remainder by $\log \mathfrak{q}(f,s)$ using

$$\Big|\frac{1}{s-\rho}-\frac{1}{3+it-\rho}\Big|\leqslant\frac{3}{1+(T-\gamma)^2}$$

and (5.32). Moreover we have (see (5.116))

$$\begin{aligned}\frac{\gamma'}{\gamma}(f,s)&=-\frac{d}{2}-\sum_j\frac{1}{s+\kappa_j}+\sum_j\frac{\Gamma'}{\Gamma}\Big(1+\frac{s+\kappa_j}{2}\Big)\\&=-\sum_{|s+\kappa_j|<1}\frac{1}{s+\kappa_j}+O(\log\mathfrak{q}_\infty(s)).\end{aligned}$$

Thus (5.28) follows. □

THEOREM 5.8. *Let $L(f,s)$ be an L-function of degree d. Let $N(T,f)$ be the number of zeros $\rho=\beta+i\gamma$ of $L(f,s)$ such that $0\leqslant\beta\leqslant 1$ and $|\gamma|\leqslant T$. We have*

$$N(T,f)=\frac{T}{\pi}\log\frac{qT^d}{(2\pi e)^d}+O(\log\mathfrak{q}(f,iT)) \tag{5.33}$$

for $T\geqslant 1$ with an absolute implied constant.

PROOF. Let $N'(T,f)$ be the number of zeros ρ of $\Lambda(f,s)$ with $0\leqslant\beta\leqslant 1$ and $0<\gamma\leqslant T$. We have

$$N(T,f)=N'(T,f)+N'(T,\bar f)+O(\log\mathfrak{q}(f)) \tag{5.34}$$

where the error term accounts for possible real and trivial zeros of $L(f,s)$ in $0\leqslant\sigma<1$. By (5.27) we can assume that $\Lambda(f,s)$ does not vanish on $\mathrm{Im}\,(s)=T$ by adding to T an arbitrarily small number if necessary, modifying $N(T,f)$ by a quantity $\ll\log\mathfrak{q}(f,iT)$. Choosing small $\delta>0$ such that $\Lambda(f,s)\neq 0$ in $-\delta\leqslant\mathrm{Im}\,(s)<0$, we have $N(T,f)=I(T)+O(\log\mathfrak{q}(f,iT))$ with

$$I(T)=\frac{1}{2\pi i}\int_{\mathcal{C}}\frac{\Lambda'}{\Lambda}(f,s)ds$$

where $\mathcal{C}$ is the rectangle with vertices $3-i\delta$, $3+iT$, $-2+iT$ and $-2-i\delta$. We cut this rectangle symmetrically at the points $\frac12-i\delta$ and $\frac12+iT$. By the functional equation, the integral on the left part of the contour equals that on the right part. Therefore $I(T)$ equals twice the integral over the right part of $\mathcal{C}$ which we are going to estimate by observing the variation of arguments of each factor in

$$\Lambda(f,s)=q^{s/2}\gamma(f,s)L(f,s)=\pi^{-ds/2}q^{s/2}\prod_{j=1}^d\Gamma\Big(\frac{s+\kappa_j}{2}\Big)L(f,s).$$

The variation of the argument of $\pi^{-ds/2}q^{s/2}$ gives a contribution to $I(T)$ equal to

$$\frac{T}{4\pi}\log\frac{q}{\pi^d}+O(1). \tag{5.35}$$

By the Stirling formula (5.113), the argument of $\Gamma(\sigma+it)$ is $t\log t-t+O(1)$ if $t\geqslant 1$, so the contribution of gamma factors to $I(T)$ is

$$\frac{1}{2\pi}\Big(\frac{dT}{2}\log\frac{T}{2}-\frac{dT}{2}\Big)+O(\log\mathfrak{q}(f)). \tag{5.36}$$

For $L(f,s)$ on the vertical segment from $3-i\delta$ to $3+iT$, we estimate

$$\log L(f,s) = -\sum_n \frac{\Lambda_f(n)}{\log n} n^{-s} \ll \log \mathfrak{q}(f)$$

as in (5.31), so the integral on this segment is $\ll \log \mathfrak{q}(f)$. Then on the remaining horizontal parts from $\frac{1}{2}-i\delta$ to $3-i\delta$ and $3+iT$ to $\frac{1}{2}+iT$, we appeal to (5.28) and (5.27) to get the estimate $O(\log \mathfrak{q}(f,iT))$.

By (5.34), (5.35) and (5.36) applied to $L(f,s)$ and $L(\bar{f},s)$, we get (5.33). □

REMARKS. The main term of (5.33) comes out from our computation of the variation in the argument of $q^{s/2}\gamma(f.s)$ along certain (right) segments, the contribution from the L-function itself being small (well, only from the relevant segments). This illustrates how the L-function is intrinsically connected with the companion at the infinite place.

5.4. The zero-free region.

Analytic applications of L-functions make use of complex integration in a region free of zeros of $L(f,s)$ for various f. The results are stronger if the zero-free region spreads deeper into the critical strip. The basic method to cut one is still that of Hadamard and de la Vallée Poussin. It is remarkable that this method also reappears in Deligne's second proof of the Riemann Hypothesis for varieties over finite fields [**De2**] (the basic trigonometric inequality (5.69) is misprinted there). We present the traditional arguments, but in somewhat more elegant fashion, starting with a useful general lemma of Goldfeld, Hoffstein and Liehman [**GHL**].

LEMMA 5.9. *Let $L(f,s)$ be an L-function of degree d with* $\mathrm{Re}\,(\Lambda_f(n)) \geqslant 0$ *for $(n,q(f))=1$. Suppose that at ramified primes $|\alpha_j(p)| \leqslant p/2$. Then $L(f,1)\neq 0$. Let $r\geqslant 0$ be the order of the pole of $L(f,s)$ at $s=1$. There exists an absolute constant $c>0$ such that $L(f,s)$ has at most r real zeros in the interval*

$$s \geqslant 1 - \frac{c}{d(r+1)\log \mathfrak{q}(f)}.$$

PROOF. Let β_j be zeros of $L(f,s)$ in the segment $\frac{1}{2}\leqslant \beta_j \leqslant 1$. By (5.28) for $s=\sigma>1$ we get

$$\sum_j \frac{1}{\sigma-\beta_j} < \frac{r}{\sigma-1} + \mathrm{Re}\,\frac{L'}{L}(f,\sigma) + O(\log \mathfrak{q}(f)) \tag{5.37}$$

on ignoring zeros other than β_j's, by positivity since

$$\mathrm{Re}\,\frac{1}{\sigma-\rho} = \frac{\sigma-\beta}{|\sigma-\rho|^2} \geqslant 0$$

for any zero $\rho=\beta+i\gamma$ in (5.28). Since $\mathrm{Re}\,(\Lambda_f(n))\geqslant 0$ for $(n,q(f))=1$ we have

$$\mathrm{Re}\,\frac{L'_{nr}}{L_{nr}}(f,\sigma) \leqslant 0 \tag{5.38}$$

where $L_{nr}(f,s)$ is the Euler product of $L(f,s)$ restricted to unramified primes. We estimate the contribution of ramified primes by

$$\Big|\sum_{p|q(f)}\sum_{1\leqslant j\leqslant d}\frac{\alpha_j(p)p^{-\sigma}\log p}{1-\alpha_j(p)p^{-\sigma}}\Big|\leqslant d\sum_{p|q(f)}\log p\leqslant d\log q(f).$$

Hence by (5.37)

$$\sum_j\frac{1}{\sigma-\beta_j}<\frac{r}{\sigma-1}+O(d\log\mathfrak{q}(f))$$

where r is the order of the pole or zero of $L(f,s)$ at $s=1$. The above inequality shows that $\beta_j=1$ is not possible (in this case $r<0$) so we conclude that $L(f,1)\neq 0$, in other words $r\geqslant 0$. Suppose $\beta_j>1-c(d(r+1)\log\mathfrak{q}(f))^{-1}$ for $1\leqslant j\leqslant n$. We choose $\sigma=1+2c(d\log\mathfrak{q}(f))^{-1}$ and get

$$\frac{nd\log\mathfrak{q}(f)}{2c+c/(r+1)}<\Big(\frac{r}{2c}+O(1)\Big)d\log\mathfrak{q}(f)$$

which implies $n<r+r/2(r+1)+O(c)$ and $n\leqslant r$ if c is small enough. □

THEOREM 5.10. *Let $L(f,s)$ be an L-function of degree d such that the Rankin-Selberg convolutions $L(f\otimes f,s)$ and $L(f\otimes\bar f,s)$ exist, and the latter has a simple pole at $s=1$ while the former is entire if $f\neq\bar f$. Suppose that at the ramified primes $|\alpha_j(p)|^2\leqslant p/2$. There exists an absolute constant $c>0$ such that $L(f,s)$ has no zeros in the region*

$$\sigma\geqslant 1-\frac{c}{d^4\log(\mathfrak{q}(f)(|t|+3))}\tag{5.39}$$

except possibly for one simple real zero $\beta_f<1$, in which case f is self-dual.

PROOF. For $t\in\mathbb{R}$, let $L(g,s)$ be the L-function

$$L(g,s)=\zeta(s)L(f,s+it)^2L(\bar f,s-it)^2L(f\otimes f,s+2it)L(\bar f\otimes\bar f,s-2it)L(f\otimes\bar f,s)^2$$

of degree $(1+2d)^2$. Its analytic conductor satisfies

$$\mathfrak{q}(g)\leqslant\mathfrak{q}(f,it)^4\mathfrak{q}(f\otimes f,2it)^4\mathfrak{q}(f\otimes\bar f)^2\leqslant\mathfrak{q}(f)^{4+12d}(|t|+3)^{6d^2}\tag{5.40}$$

by (5.11).

The product is arranged so that its Dirichlet series coefficients are real, non-negative. It is sufficient for us to check this at unramified places. For an unramified prime p, the local Euler factor at p for $L(g,s)$ is of the form (5.1) with "roots" 1 with multiplicity one, α_jp^{it} and $\bar\alpha_jp^{-it}$ with multiplicity two, $\alpha_j\bar\alpha_k$ with multiplicity two, $\alpha_j\alpha_kp^{2it}$ and $\bar\alpha_j\bar\alpha_kp^{-2it}$ with multiplicity one, where α_j, α_k run over the roots (5.1) for $L(f,s)$. Therefore for any $k\geqslant 1$, the sum of the k-th powers of these roots is

$$\Big|1+\sum_j\alpha_j^kp^{kit}+\sum_j\bar\alpha_j^kp^{-kit}\Big|^2\geqslant 0$$

so that $\Lambda_g(n)\geqslant 0$ for any n coprime with $q(f)$.

Let $\rho = \beta + i\gamma$ be a zero of $L(f,s)$ with $\beta \geqslant \frac{1}{2}$ and $\gamma \neq 0$. Then we take $t = \gamma$ so $L(g,s)$ has a pole at $s = 1$ of order $\leqslant 3$ whereas β is a zero of $L(g,s)$ of order $\geqslant 4$. Hence by Lemma 5.9 we have

$$\beta < 1 - \frac{c}{d^2 \log \mathfrak{q}(g)} < 1 - \frac{c'}{d^4 \log(\mathfrak{q}(f)(|t|+3))} \tag{5.41}$$

for some absolute constants $c > 0$ and $c' > 0$.

Now consider real zeros of $L(f,s)$. For this purpose we take the product $L(g,s)$ with $t = 0$. Any real zero of $L(f,s)$ is a zero of $L(g,s)$ of multiplicity at least 4. On the other hand, the point $s = 1$ is a pole of $L(g,s)$ of order 3 if $f \neq \bar{f}$ and of order 5 if $f = \bar{f}$. Hence by Lemma 5.9 we conclude the proof, except that we must prove that the possible exceptional zero of a self-dual $L(f,s)$ satisfies $\beta_f < 1$.

Consider instead

$$L(h,s) = \zeta(s) L(f,s)^2 L(f \otimes f, s).$$

Assuming $L(f,1) = 0$, it follows that $L(h,s)$ is entire since the double zero compensates the poles of $\zeta(s)$ and $L(f \otimes f, s)$. It also has non-negative coefficients, precisely for an unramified prime p and $k \geqslant 1$ we have

$$\Lambda_h(p^k) = \Big(1 + \sum_j \alpha_j^k\Big)^2 \geqslant 0, \tag{5.42}$$

so the series

$$\log L(h,s) = \sum_p \sum_{k>0} \frac{1}{k} \Lambda_h(p^k) p^{-ks} \tag{5.43}$$

has non-negative coefficients. Let $\sigma_0 \leqslant 1$ be the largest real zero of $L(h,s)$, so the first singularity of $\log L(h,s)$ when s decreases. Note σ_0 exists since $\zeta(-2) = 0$. By Landau's Lemma (see Lemma 5.56) the series (5.43) converges for any real $\sigma > \sigma_0$, so by (5.42) we have $\log L(h,\sigma) \geqslant 0$ and $|L(h,\sigma)| \geqslant 1$. Letting $\sigma \to \sigma_0$, we get a contradiction. □

REMARK 1. If $L(f,s)$ is self-dual one may refine the real zero issue of Theorem 5.10 slightly, arguing with $L(h,s)$ more efficiently. Indeed (5.42) shows that $\Lambda_h(n) \geqslant 0$ for all n with $(n, \mathfrak{q}(f)) = 1$. Let l denote the order of the pole of $L(f \otimes f, s)$ at $s = 1$. Then $L(h,s)$ has a pole of order $l+1$ at $s = 1$ while any real zero of $L(f,s)$ is a zero of $L(h,s)$ of order $\geqslant 2$. Hence there are at most $\frac{l+1}{2}$ real zeros of $L(f,s)$ in the region (5.39).

REMARK 2. Note that if the Rankin-Selberg L-functions exist, but $L(f \otimes \bar{f}, s)$ has a pole of order > 1 at $s = 1$, it is often the case that $L(f,s)$ is a product of other L-functions. It is then more efficient to apply Theorem 5.10 to those than to $L(f,s)$ itself. For instance, consider the Dedekind zeta function of a quadratic field $\zeta_K(s) = \zeta(s) L(s,\chi)$. Then the Rankin-Selberg L-function is $\zeta(s)^2 L(s,\chi)^2$ which has a double pole at $s = 1$.

EXERCISE 4. Let $L(f,s)$ and $L(g,s)$ be two L-functions of degrees d and e respectively such that $g \neq f$ and $g \neq \bar{f}$. Assume that Rankin-Selberg L-functions $L(f \otimes g, s)$, $L(f \otimes \bar{g}, s)$, and hence their dual, exist, and that the local roots at ramified primes satisfy $|\alpha_i(p)|^2 < p/2$. Show that there exists an absolute constant $c > 0$ such that $L(f \otimes g, s)$ has no zero in the region

$$\sigma \geqslant 1 - \frac{c}{(d+e)^4 \log(\mathfrak{q}(f)\mathfrak{q}(g)(|t|+3))}.$$

What can you prove if $f = g$ or $f = \bar{g}$? [**Hint:** Use the auxiliary function

$$L(g,s) = L(f \otimes g, s + \tfrac{it}{2}) L(f \otimes \bar{g}, s) L(\bar{f} \otimes g, s) L(\bar{f} \otimes \bar{g}, s - \tfrac{it}{2})$$

and apply Lemma 5.9.]

In general nothing stronger is known than Theorem 5.10. For certain important L-functions, one can do better, individually, and also by considering families. Examples, including the most important one of Dirichlet characters, are discussed in Section 5.7 and 5.12.

5.5. Explicit formula.

A summation formula somewhat analogous to (5.12) can be derived by integrating the Dirichlet series for the logarithmic derivative of an L-function. For historical reasons, this type of formula is usually called an "explicit formula", emphasizing the link it expresses between sums over the zeros of an L-function with sums of coefficients at prime powers.

THEOREM 5.11. *Let $\varphi :]0, +\infty[\to \mathbb{C}$ be a C^∞ function with compact support, and*

$$\hat{\varphi}(s) = \int_0^{+\infty} \varphi(x) x^{s-1} dx,$$

be its Mellin transform. Put $\psi(x) = x^{-1}\varphi(x^{-1})$, which means $\hat{\psi}(s) = \hat{\varphi}(1-s)$. Then we have

$$(5.44)\quad \sum_n \left(\Lambda_f(n)\varphi(n) + \overline{\Lambda_f}(n)\psi(n)\right) = \varphi(1)\log q(f) + r\int_0^{+\infty} \varphi(x)dx$$
$$+ \frac{1}{2\pi i}\int_{(1/2)} \left(\frac{\gamma'}{\gamma}(f,s) + \frac{\gamma'}{\gamma}(\bar{f}, 1-s)\right)\hat{\varphi}(s)ds - \sum_\rho \hat{\varphi}(\rho)$$

where ρ runs over the zeros of $L(f,s)$ in the strip $0 \leqslant \sigma \leqslant 1$ (the non-trivial and the trivial ones) with relevant multiplicity.

PROOF. Start with the first term on the left side of (5.44) using Mellin inversion

$$\sum_n \Lambda_f(n)\varphi(n) = \frac{1}{2\pi i}\int_{(2)} -\frac{L'}{L}(f,s)\hat{\varphi}(s)ds.$$

Let c be a small positive number so that $L(f,s)$ has no zeros in $-c \leqslant \mathrm{Re}(s) < 0$. Move the line of integration to $\mathrm{Re}(s) = -c$, which is legitimate by (5.27). A simple pole is picked up at $s = 1$ with residue $r\int \varphi(x)dx$, and at the zeros $s = \rho$ with residues $m\hat{\varphi}(\rho)$, where m is the multiplicity. So those residues give the second and the fourth terms on the right side of (5.44).

The functional equation implies

$$-\frac{L'}{L}(f,s) = \log q(f) + \frac{\gamma'}{\gamma}(f,s) + \frac{\gamma'}{\gamma}(\bar{f},1-s) + \frac{L'}{L}(\bar{f},1-s)$$

for $\sigma = -c$. Integrating over $\sigma = -c$, the factor $\log q(f)$ gives the first term on the right side of (5.41) by Mellin inversion. Moreover, by absolute convergence

$$\frac{1}{2\pi i}\int\limits_{(-c)} \frac{L'}{L}(\bar{f},1-s)\hat{\varphi}(s)ds = \sum_{n\geqslant 1}\Lambda_{\bar{f}}(n)\frac{1}{2\pi i}\int\limits_{(-c)} \hat{\varphi}(s)n^{s-1}ds = -\sum_{n\geqslant 1}\Lambda_{\bar{f}}(n)\psi(n).$$

And finally one can move the line of integration for the gamma factors to $\sigma = \frac{1}{2}$ getting the integral of gamma factors in (5.44) without residues (the poles of $\frac{\gamma'}{\gamma}(f,s)$ cancel out with those of $\frac{\gamma'}{\gamma}(\bar{f},1-s)$). □

REMARK. Because of the appearance of the conductor $q(f)$ on the right side of (5.41), this formula turns out to provide a good analytic tool for its study. See Theorem 5.32 below for an example. However, it is quite useful for other things. First of all the explicit formula is seen as expressing a sum over primes by the sum over the zeros and vice versa. Certainly there are other explicit formulas connecting primes with zeros.

EXERCISE 5. Let φ be smooth and compactly supported on $[1,+\infty[$. Show that

$$\sum_{n\geqslant 1}\Lambda(n)\varphi(n) = \int_1^{+\infty}\Big(1-\frac{1}{(x-1)x(x+1)}\Big)\varphi(x)dx - \sum_{\rho}\hat{\varphi}(\rho),$$

where ρ runs over the non-trivial zeros of $\zeta(s)$. [**Hint:** Start as above, but move the line of integration far to the left instead of using the functional equation; the first term comes from the pole at $s=1$ and the trivial zeros of $\zeta(s)$ at $s=-2k$, $k\geqslant 1$ an integer.]

Sometimes it is more convenient to work with the explicit formula in a Fourier form rather than the Mellin form (5.44). This can be derived from (5.44) by simple substitutions $\varphi(x) = x^{-1/2}g(\log x)$ and $x = e^y$ getting

THEOREM 5.12. *Let $g(y)$ be a function of Schwartz class on $\mathbb{R}$ which is even. Let $h(t)$ be the Fourier transform of $g(y)$. We have*

$$\sum_n (\Lambda_f(n) + \overline{\Lambda_f}(n))\frac{g(\log n)}{\sqrt{n}} = g(0)\log q(f) + rh\Big(\frac{i}{4\pi}\Big) \tag{5.45}$$
$$+\frac{1}{2\pi}\int_{-\infty}^{+\infty}\Big(\frac{\gamma'}{\gamma}(f,\tfrac{1}{2}+it) + \frac{\gamma'}{\gamma}(\bar{f},\tfrac{1}{2}-it)\Big)h\Big(\frac{t}{2\pi}\Big)dt - \sum_\rho h\Big(\frac{\gamma}{2\pi}\Big)$$

where $\rho = \frac{1}{2}+i\gamma$ runs over the zeros of $L(f,s)$ in the strip $0\leqslant\sigma\leqslant 1$ (the non-trivial and the trivial ones) with relevant multiplicity.

5.6. The prime number theorem.

By Prime Number Theorem for an L-function $L(f,s)$, we mean first the asymptotic behavior of the sum

$$\psi(f,x)=\sum_{n\leqslant x}\Lambda_f(n) \tag{5.46}$$

which is essentially the sum of $\lambda_f(p)\log p$ over primes. When $L(f,s)=\zeta(s)$, this amounts indeed to counting primes $p\leqslant x$. Much of the arithmetic interest lies in the dependency of the error term in asymptotic approximations on extra parameters, notably on the conductor when f varies in a family. Therefore we are seeking results which are explicit in every possible parameter, although not the strongest ones.

The strength of results depends on the depth of the zero-free region for $L(f,s)$. The best we can hope for today is that $L(f,s)$ has no zeros in the region (5.39) with at most one exception of a simple zero β_f in case of f self-dual. By our definition (only for the purpose in this section) the exceptional zero is in the segment

$$1-\frac{c}{d^4\log(3\mathfrak{q}(f))}\leqslant\beta_f<1 \tag{5.47}$$

where c is the absolute positive constant from (5.39). This definition may cause some discomfort for new enthusiasts of analytic number theory, because it is loose with respect to the constant c. However, the practice shows that one benefits from such a flexible concept. After all, c can be improved by new tools in the future, not to mention that we don't believe in the existence of the exceptional zero of any L-function.

In full generality we do not yet have a strong bound for the individual coefficients of $L(f,s)$. But a crude bound for $\Lambda_f(n)$ on average will suffice. Specifically, we postulate the following:

$$\sum_{n\leqslant x}|\Lambda_f(n)|^2\ll xd^2\log^2(x\mathfrak{q}(f)) \tag{5.48}$$

for $x\geqslant 1$, where the implied constant is absolute.

EXERCISE 6. Derive from (5.48) that

$$\psi(f,x)=\sum_{p\leqslant x}\lambda_f(p)\log p+O(\sqrt{x}d^2\log^2(x\mathfrak{q}(f))) \tag{5.49}$$

where the implied constant is absolute.

Recall that the postulated zero-free region (5.39) was established in Theorem 5.10 under the assumption that the Rankin-Selberg convolution $L(f\otimes\bar{f},s)$ exists and has a simple pole at $s=1$. We shall see that the same assumption is also sufficient for the second postulate (5.48). To this end we apply Proposition 5.7 for $L(f\otimes\bar{f},s)$ rather than for $L(f,s)$. Combining (5.28) and (5.27) (see also (5.11)) we get

$$-\frac{L'}{L}(f\otimes\bar{f},\sigma)\ll\frac{d^2}{\sigma-1}\log\mathfrak{q}(f)$$

for $1<\sigma\leqslant 2$, where the implied constant is absolute. This with $\sigma=1+(\log 3x)^{-1}$ yields (5.49) by the non-negativity of coefficients.

From (5.49) we derive using Cauchy's inequality that

$$\sum_{x<n\leqslant x+y} |\Lambda_f(n)| \ll d\sqrt{xy}\log(x\mathfrak{q}(f)) \tag{5.50}$$

for $x \geqslant y \geqslant 1$ where the implied constant is absolute.

Now we are ready to state and to prove the main Prime Number Theorem.

THEOREM 5.13. *Let $L(f,s)$ be an L-function for which (5.39) is a zero-free region with at most one exceptional real zero β_f in the segment (5.47), where c is a positive absolute constant. Suppose (5.48) holds with an implied constant absolute. Then we have*

$$\psi(f,x) = rx - \frac{x^{\beta_f}}{\beta_f} + O\Big(x\exp\Big(\frac{-cd^{-4}\log x}{\sqrt{\log x}+3\log\mathfrak{q}(f)}\Big)(d\log x\mathfrak{q}(f))^4\Big), \tag{5.51}$$

for $x \geqslant 1$, the implied constant being absolute. By convention the term $-x^{\beta_f}/\beta_f$ is not in (5.51) if the exceptional zero does not exist.

REMARKS. The approximate formula (5.51) has meaning when the error term is smaller than the main term, and this is the case for

$$x \geqslant q^{4c^{-1}d^4\log(d\log q)}$$

where $q = \mathfrak{q}(f)$ is the conductor, d is the degree and c is the relevant absolute positive constant. One can simplify the error term by compromising slightly the strength of the result. For example, we have

$$\psi(f,x) = rx - \frac{x^{\beta_f}}{\beta_f} + O\Big(\sqrt{\mathfrak{q}(f)}x\exp\Big(-\frac{c}{2d^4}\sqrt{\log x}\Big)\Big) \tag{5.52}$$

where the implied constant is absolute. Indeed this holds true, because when the error term here is smaller than that in (5.51) the result is trivial. The conditions of Theorem 5.12 are satisfied if $L(f\otimes\bar{f},s)$ exists and has a simple pole at $s=1$, but we do not restrict (5.51) to this case only. Indeed we shall establish the zero-free region (5.39) and we shall see the crude bound (5.48) directly without appealing to the Rankin-Selberg convolution L-function in many classical cases.

PROOF OF THEOREM 5.12. First we smooth things out by writing

$$\psi(f,x) = \sum_n \Lambda_f(n)\phi(n) + O(d\sqrt{xy}\log(x\mathfrak{q}(f)))$$

where $\phi(z)$ is a function support on $[0,x+y]$, such that $\phi(z)=1$ if $1\leqslant z\leqslant x$ and $|\phi(z)|\leqslant 1$ elsewhere. The parameter y will be chosen later subject to $1\leqslant y\leqslant x$. For example, we take

$$\phi(z) = \min\Big(\frac{z}{y}, 1, 1+\frac{x-z}{y}\Big)$$

for $0\leqslant z\leqslant x+y$ and $\phi(z)=0$ if $z>x+y$. The Mellin transform of $\phi(z)$ satisfies

$$\hat{\phi}(s) = \int_0^{x+y} \phi(z)z^{s-1}dz \ll \frac{x^\sigma}{|s|}\min\Big(1,\frac{x}{|s|y}\Big)$$

for $s=\sigma+it$, $\frac{1}{2}\leqslant\sigma\leqslant 2$.

Now we can start with

$$\sum_n \Lambda_f(n)\phi(n) = \frac{1}{2\pi i}\int_{(2)} -\frac{L'}{L}(f,s)\hat{\phi}(s)ds,$$

the integral converging absolutely. We shall be operating in the region

$$\{s=\sigma+it \mid \sigma \geqslant 1-c_1/d^4\log(\mathfrak{q}(f)(|t|+3))\}$$

where $c_1 = 2c/3$, or $c_1 = c/3$, depending on whether the exceptional zero is in the right half of the segment (5.47), or elsewhere (with possibility of non-existence). Let $\mathcal{Z}$ denote the boundary of this region. The point is that all the zeros of $L(f,s)$ are distanced from $\mathcal{Z}$ by at least $c/6d^4\log(\mathfrak{q}(f)(|t|+3))$. Hence it follows by (5.27) and (5.28) that for $s\in\mathcal{Z}$,

$$-\frac{L'}{L}(f,s) \ll d^4\log^2(\mathfrak{q}(f)(|t|+3)).$$

Moving the integration from the line $\mathrm{Re}\,(s)=2$ to $\mathcal{Z}$ we get by Cauchy's theorem

$$\sum_n \Lambda_f(n)\phi(n) = r\hat{\phi}(1) - \hat{\phi}(\beta_f) + \frac{1}{2\pi i}\int_{\mathcal{Z}} -\frac{L'}{L}(f,s)\hat{\phi}(s)ds.$$

Here the presence of the exceptional zero term depends on whether we passed the point $s=\beta_f$ or not. For $s=1$, or $s=\beta_f$ we have

$$\hat{\phi}(s) = \int_0^x z^{s-1}dz + O(y) = \frac{x^s}{s} + O(y)$$

while the contour integral over $\mathcal{Z}$ is estimated by

$$d^4\int_{\mathcal{Z}}\frac{x^\sigma}{|s|}\min\Bigl(1,\frac{x}{|s|y}\Bigr)\log^2(\mathfrak{q}(f)(|t|+3))|ds| \ll d^4x^{\sigma(T)}\log^3(\mathfrak{q}(f)T)$$

where $T=x/y$ and $\sigma(T) = 1-c_1/d^4\log(\mathfrak{q}(f)T)$. Gathering the above results we arrive at

$$\psi(f,x) = rx - \frac{x_f^\beta}{\beta_f} + O\Bigl(d^4(xT^{-\frac{1}{2}} + x^{\sigma(T)})\log^3(x\mathfrak{q}(f))\Bigr)$$

where T is at our disposal subject to $1\leqslant T\leqslant x$. We choose $T=\exp(\frac{1}{3}\sqrt{\log x})$ getting (5.51). We still have to address the question of the exceptional zero term. An explanation about its contribution to the obtained formula is needed if the exceptional zero does exist and it lies in the left half of the segment (5.47), but in this case the term $-x^{\beta_f}/\beta_f$ is consumed by the existing error term, so it can be ignored. This completes the proof of Theorem 5.12. □

If one is willing to employ more zeros of $L(f,s)$, then one can derive along the above lines stronger approximations to $\psi(f,x)$.

EXERCISE 7. Assume that $L(f,s)$ satisfies the Ramanujan-Petersson conjecture. Derive (using Perron's formula (5.111)) the following approximate expansion

$$\psi(f,x) = rx - \sum_{|\gamma|\leqslant T}\frac{x^\rho-1}{\rho} + O\Bigl(\frac{x}{T}(\log x)\log(x^d\mathfrak{q}(f))\Bigr) \tag{5.53}$$

where $\rho = \beta + i\gamma$ runs over the zeros of $L(f, s)$ in the critical strip of height up to T, with any $1 \leqslant T \leqslant x$, and the implied constant is absolute.

EXERCISE 8. Assume that $L(f, s)$ satisfies the Ramanujan-Petersson conjecture and Theorem 5.13 holds for f. Prove that $r \leqslant d$.

5.7. The Grand Riemann Hypothesis.

The Grand Riemann Hypothesis (GRH for short) refers to the following conjectural statement about zeros of L-functions:

GRAND RIEMANN HYPOTHESIS. *Let $L(f, s)$ be an L-function. Then all zeros of $L(f, s)$ in the critical strip* $0 < \mathrm{Re}\,(s) < 1$ *are on the critical line* $\mathrm{Re}\,(s) = \frac{1}{2}$.

REMARKS. Needless to say we believe that every L-function (subject to our definition in Section 5.1) satisfies the Grand Riemann Hypothesis. Yet, proving this even for one L-function would be an achievement on a historical scale for human beings. Note that an L-function may have zeros on the line $\mathrm{Re}\,(s) = 0$ (certainly some trivial zeros, probably not the genuine ones), but as we already proved not on $\mathrm{Re}\,(s) = 1$. The GRH also ensures that the trivial zeros in the open critical strip are all on the critical line as well; do they exist is not clear, most likely they do not. It is very important to consider only L-functions with Euler products. Indeed it is known that Dirichlet series without an Euler product (still civilized ones) have many zeros even in the half-plane of absolute convergence. Arithmetic geometers should not be concerned, because their L-functions originate from an Euler product of some kind. Of course, this easy start makes it harder to investigate L-functions further in analytic aspects. For example, it was only recently that analytic continuation of the Hasse-Weil L-functions of elliptic curves was established by way of modularity. One should not underestimate the analytic continuation of an Euler product beyond the region of absolute convergence, for one often implements deep arithmetic resources in one way or another. As envisioned by Riemann, this venture into the critical strip illuminates prime numbers and relevant arithmetic objects built on them by revealing their dual companions – the zeros of L-functions. Mysterious as they are today, these zeros (most likely complex transcendental numbers) will eventually be cracked for further analysis. However, for the time being, as far as the GRH goes, we are only sure they are fine tuned to yield surprisingly strong estimates. Yes, the uniformity of estimates in terms of involved parameters which can be derived from the GRH are astonishing. On a negative note this superb uniformity could be a reminder how slim prospects are to prove the GRH in our lifetime. No tools of current analytic number theory are capable of treating the questions for which the GRH provides easy answers. In this section we present a few traditional analytic consequences of the GRH, paying attention to the uniformity in terms of the conductor.

First we state a few equivalent facts which help to grasp the analytic meaning of the Riemann hypothesis.

PROPOSITION 5.14. *Let $\frac{1}{2} \leqslant \alpha < 1$. The following statements are equivalent for an L-function:*

(1) There are neither zeros nor poles of $(s-1)^r L(f, s)$ in $\sigma > \alpha$, where r is a non-negative integer.

(2) The inverse, the logarithmic derivative and the logarithm of $(s-1)^r L(f, s)$, the latter normalized so that $\log L(f, s) \to 0$ as $\sigma \to +\infty$, are holomorphic in $\sigma > \alpha$.

(3) Let $\mu_f(n)$ denote the coefficients in the Dirichlet series expansion for the inverse $L(f,s)^{-1}$. Then

$$M(f,x) = \sum_{n\leqslant x} \mu_f(n) \ll x^{\alpha+\varepsilon}, \tag{5.54}$$

the implied constant depending on f and $\varepsilon > 0$.

(4) Let $r \geqslant 0$ be the order of the pole of $L(f,s)$ at $s=1$. Then

$$\psi(f,x) = rx + O(x^{\alpha+\varepsilon}), \tag{5.55}$$

the implied constant depending on f and $\varepsilon > 0$.

The Riemann hypothesis asserts that the above statements are true with $\alpha = \frac{1}{2}$. What is interesting (perhaps somewhat surprising for a beginner) is that the estimate (5.55) is self-improving by passage through zeros. Indeed, first (5.55) implies that the zeros are on the line $\mathrm{Re}\,(s) = \frac{1}{2}$, whence one derives by the expansion (5.53) with $T = x$ using the crude bound (5.26) the following

THEOREM 5.15. *Assuming the Riemann Hypothesis for $L(f,s)$ and the Ramanujan-Petersson Conjecture for $L(f,s)$ we have*

$$\psi(f,x) = rx + O(x^{\frac{1}{2}}(\log x)\log(x^d\mathfrak{q}(f))), \tag{5.56}$$

for $x \geqslant 1$, the implied constant being absolute.

The Riemann hypothesis shows that $L(f,s)$, the inverse $L(f,s)^{-1}$, the logarithmic derivative $\frac{L'}{L}(f,s)$ and the logarithm $\log L(f,s)$ can be well approximated by extremely short partial sums of the corresponding Dirichlet series uniformly in $\mathrm{Re}\,s = \sigma \geqslant \alpha > \frac{1}{2}$. Clearly this is not possible for s on the critical line, where the zeros are. We begin by considering

$$-\frac{L'}{L}(f,s) = \sum_n \frac{\Lambda_f(n)}{n^s}.$$

Let $\phi(y)$ be a continuous function on $[0,\infty[$ whose Mellin transform $\hat{\phi}(w)$ satisfies

$$w(w+1)\hat{\phi}(w) \ll 1, \text{ for } w \text{ such that } -\tfrac{1}{2} \leqslant \mathrm{Re}\,(w) \leqslant \tfrac{1}{2}. \tag{5.57}$$

Suppose $\hat{\phi}(w)$ has a simple pole at $w=0$ with residue 1 (normalization). For example, $\phi(y) = e^{-y}$ with $\hat{\phi}(w) = \Gamma(w)$, or $\phi(y) = \max(1-y,0)$ with $\hat{\phi}(w) = w^{-1}(w+1)^{-1}$.

PROPOSITION 5.16. *For $\frac{1}{2} < \mathrm{Re}\,(s) \leqslant \frac{5}{4}$ we have*

$$-\frac{L'}{L}(f,s) = \sum_n \frac{\Lambda_f(n)}{n^s}\phi\Big(\frac{n}{X}\Big) - r\hat{\phi}(1-s)X^{1-s} + \sum_\rho \hat{\phi}(\rho-s)X^{\rho-s} + O\Big(\frac{\log\mathfrak{q}(f,s)}{(2\sigma-1)\sqrt{X}}\Big) \tag{5.58}$$

for any $X \geqslant 1$, where the implied constant depends only on that in (5.57).

PROOF. By now it should be a standard argument for a reader to derive the formula (5.58) by contour integration with the error term being equal exactly

$$\frac{1}{2\pi i}\int_{(-\frac{1}{2})} -\frac{L'}{L}(f,s+w)\hat{\phi}(w)X^w dw.$$

Indeed this is obtained by starting with the above integral on the line $\mathrm{Re}\,(w) = \frac{1}{2}$ and moving it to the line $\mathrm{Re}\,(w) = -\frac{1}{2}$. The simple poles at $w = 0, 1-s, \rho - s$ produce the corresponding terms in (5.58). Now on the line $\mathrm{Re}\,(w) = -\frac{1}{2}$ we have $\mathrm{Re}\,(s+w) = \sigma - \frac{1}{2}$, and we deduce by (5.26) and (5.27) that

$$-\frac{L'}{L}(f,s+w) \ll (2\sigma-1)^{-1}\log \mathfrak{q}(f,s+w).$$

Integrating this against the bound from (5.57), we get the error term in (5.58). □

The sum over the zeros in (5.58) is estimated directly (without cancellation) by means of (5.26), (5.27) and (5.57) giving

$$(5.59)\qquad -\frac{L'}{L}(f,s) = \sum_n \frac{\Lambda_f(n)}{n^s}\phi\Big(\frac{n}{X}\Big) - r\hat{\phi}(1-s)X^{1-s} + O\Big(\frac{\log \mathfrak{q}(f,s)}{2\sigma-1}X^{\frac{1}{2}-\sigma}\Big)$$

for any $X \geqslant 1$, where the implied constant depends only on that in (5.57).

Next we are going to estimate the sum of $\Lambda_f(n)$. Essentially any crude bound, but independent of the conductor, like $|\Lambda_f(n)| \leqslant n$, would suffice for our purpose. Nevertheless, since we already assumed the GRH, there is no point in ignoring the Ramanujan-Petersson conjecture for the local roots, which yields $|\Lambda_f(n)| \leqslant d\Lambda(n)$. From this we get

$$\sum_n \frac{\Lambda_f(n)}{n^s}\phi\Big(\frac{n}{X}\Big) \ll dX^{1-\sigma} + d\log X.$$

Moreover, the polar term in (5.59) is

$$-r\hat{\phi}(1-s)X^{1-s} = r(s-1)^{-1} + O(rX^{1-\sigma} + r\log X).$$

Hence (5.59) simplifies to

$$-\frac{L'}{L}(f,s) = \frac{r}{s-1} + O\Big(\frac{\log \mathfrak{q}(f,s)}{2\sigma-1}X^{\frac{1}{2}-\sigma} + dX^{1-\sigma} + d\log X\Big).$$

Finally choosing $X = \log^2 \mathfrak{q}(f,s)$ we obtain

THEOREM 5.17. *Assume the Riemann Hypothesis and the Ramanujan-Petersson conjecture for $L(f,s)$. We have*

$$-\frac{L'}{L}(f,s) = \frac{r}{s-1} + O\Big(\frac{d}{2\sigma-1}(\log \mathfrak{q}(f,s))^{2-2\sigma} + d\log\log \mathfrak{q}(f,s)\Big)$$

for any s with $\frac{1}{2} < \sigma \leqslant \frac{5}{4}$, the implied constant being absolute.

COROLLARY 5.18. *For* $\sigma \geqslant \frac{1}{2} + (\log\log\log \mathfrak{q})/(\log\log \mathfrak{q})$ *we have*

$$-\frac{L'}{L}(f,s) = \frac{r}{s-1} + O\Big(\frac{\log \mathfrak{q}}{\log\log \mathfrak{q}}\Big)$$

where $\mathfrak{q} = \mathfrak{q}(f,s)$ *is the analytic conductor and the implied constant is absolute.*

To get an estimate for the logarithm of the L-function we integrate the logarithmic derivative along the horizontal line

$$\log L(f,s) = \log L(f, \tfrac{5}{4} + it) - \int_\sigma^{5/4} \frac{L'}{L}(f, \alpha + it) d\alpha.$$

Applying (5.59) we deduce the following estimates

THEOREM 5.19. *Assume the Riemann Hypothesis and the Ramanujan-Petersson conjecture for* $L(f,s)$. *We have*

$$\log\big(p_r(s)L(f,s)\big) \ll \frac{d(\log \mathfrak{q}(f,s))^{2-2\sigma}}{(2\sigma-1)\log\log \mathfrak{q}(f,s)} + d\log\log \mathfrak{q}(f,s)$$

for any s *with* $\frac{1}{2} < \sigma \leqslant \frac{5}{4}$, *the implied constant being absolute (recall* $p_r(s) = (s-1)^r(s+1)^{-r}$.

As immediate consequences of Theorem 5.19 we have lower and upper bounds for $L(f,s)$ in the half-plane $\sigma \geqslant \frac{1}{2} + \varepsilon$:

$$\mathfrak{q}(f,s)^{-\varepsilon} \ll p_r(s)L(f,s) \ll \mathfrak{q}(f,s)^{\varepsilon}$$

with implied constants depending only on ε. The upper bound can be extended up to the critical line by the convexity principle. In particular, we obtain

COROLLARY 5.20. *For* s *with* $\mathrm{Re}\,(s) = \frac{1}{2}$ *and any* $\varepsilon > 0$ *we have*

$$L(f,s) \ll \mathfrak{q}(f,s)^{\varepsilon} \tag{5.60}$$

where the implied constant depends only on ε.

The last estimate received considerable attention (it is known as the Lindelöf Hypothesis), because it is sufficient for solving many fundamental problems of arithmetical nature due to its uniformity in the conductor. We just proved that it follows from the Riemann hypothesis. One may think that the Lindelöf hypothesis should be easier to establish, but there is a popular opinion that the Riemann hypothesis must be solved first, the point being that the Riemann hypothesis is linked to more natural mathematical structures. Clearly the Lindelöf conjecture implies the following estimate for sums of coefficients of the L-function

$$\sum_{n\leqslant x} \lambda_f(n) = xP_f(\log x) + O(x^{\frac{1}{2}}(x\mathfrak{q}(f))^{\varepsilon}) \tag{5.61}$$

where $P_f(X)$ is the polynomial of degree $r-1$ given by

$$P_f(\log x) = \operatorname*{res}_{s=1} L(f,s)x^{s-1}.$$

Some unconditional results, weaker yet useful, are given in Chapter 4 (see Corollary 4.9 for $\zeta(s)L(s,\chi_4)$ and Exercise 7 of Chapter 4 for $\zeta(s)^2$ for instance).

5.8. Simple consequences of GRH.

We illustrate some of the simple arithmetic consequences of GRH, besides the Lindelöf Hypothesis, which has important applications of its own.

The first problem is to estimate the order of vanishing of an L-function at a given point on the critical line, in terms of the conductor of f. We consider the case of the central point $s = 1/2$. Note that (5.27) implies

$$\operatorname*{ord}_{s=1/2} L(f,s) \leqslant m(1,f) \ll \log \mathfrak{q}(f)$$

unconditionally, which we can improve slightly.

PROPOSITION 5.21. *Let $L(f,s)$ be an entire L-function satisfying GRH. We have*

$$\operatorname*{ord}_{s=1/2} L(f,s) \ll \frac{\log \mathfrak{q}(f)}{\log(\frac{3}{d}\log \mathfrak{q}(f))} \tag{5.62}$$

with an absolute implied constant (recall that $\log \mathfrak{q}(f) > d$).

PROOF. We apply the explicit formula (5.45) to a test function $g(y)$ supported on $[-2Y, 2Y]$ with $0 \leqslant g(y) \leqslant 2$ and $g(y) \geqslant 1$ if $|y| \leqslant Y$ such that its Fourier transform $h(t)$ is non-negative on $\mathbb{R}$. By positivity we can drop all the zeros $\rho = \frac{1}{2} + i\gamma \neq \frac{1}{2}$ in (5.45) getting

$$mh(0) \leqslant -2\sum_n \operatorname{Re}(\Lambda_f(n))n^{-\frac{1}{2}}g(\log n) + O(\log \mathfrak{q}(f)),$$

where m is the multiplicity of the zero $\rho = \frac{1}{2}$. On the left side we have $h(0) \asymp Y$. Since we have $|\Lambda_f(n)| \leqslant d\Lambda(n)n$ by (5.2), the sum over prime powers n on the right side is bounded by

$$2d \sum_{\log n \leqslant 2Y} \Lambda(n)n^{\frac{1}{2}} \ll de^Y,$$

where the implied constant is absolute. Choosing $2Y = \log(\frac{3}{d}\log \mathfrak{q}(f))$ we get

$$m \ll \frac{\log \mathfrak{q}(f)}{\log(\frac{3}{d}\log \mathfrak{q}(f))}$$

with an absolute implied constant. □

Proposition 5.21 is optimal although the estimate of the sum over primes is quite crude, which might suggest some improvement is possible. The issue is that this sum is extremely short with the choice of Y, in fact $n \leqslant e^{2Y} \leqslant \frac{3}{d}\log \mathfrak{q}(f)$ and it is conceivable that $\Lambda_f(n)$ does not change sign in this range. If we take $L(f,s)^k$ for $k \to +\infty$ instead of $L(f,s)$, we see that (5.62) cannot hold without the factor $\frac{3}{d}$. However, in a number of cases, it is possible to do better; see Proposition 5.34 and Exercise 13 below.

The problem of detecting sign changes of $\Lambda_f(n)$ for very small n is related to another classical application of the Grand Riemann Hypothesis: how many coefficients (in terms of the conductor $q(f)$) of $L(f,s)$ at primes are required to distinguish $L(f,s)$ among L-functions of a certain class?

PROPOSITION 5.22. *Let $L(f,s)$ and $L(g,s)$ be two L-functions with the same degree d and gamma factor. Assume that $L(f\otimes\bar{f},s)$ and $L(f\otimes\bar{g},s)$ exist and the latter is entire, that GRH holds for both and that the local roots at ramified primes for $L(f\otimes\bar{f},s)$ and $L(f\otimes\bar{g},s)$ are of modulus $\leqslant 1$. There exists a prime $p\leqslant C(d\log\mathfrak{q}(f)\mathfrak{q}(g))^2$ unramified for f and g such that the local roots of $L(f,s)$ and $L(g,s)$ at p are different. Here C is some absolute positive constant.*

PROOF. The point is that by assumption the L-function $L(f\otimes\bar{f},s)$ has a pole at $s=1$ whereas we assume that $L(f\otimes\bar{g},s)$ is entire. Assume the local roots of f and g coincide for all primes $p\leqslant 2X$ which are unramified for f and g. Then $\Lambda_{f\otimes\bar{g}}(n)=\Lambda_{f\otimes\bar{f}}(n)$ for $n\leqslant 2X$ such that $(n,q(f)q(g))=1$. By the explicit formula (5.44) applied to a function of the type $\phi(n/X)$ where $\phi\geqslant 0$ is smooth, compactly supported on $[1,2]$, and $\phi\neq 0$, we have

$$\sum_n \Lambda_{f\otimes\bar{g}}(n)\phi(n/X)\ll\sqrt{X}\log\mathfrak{q}(f\otimes g),$$

$$\sum_n \Lambda_{f\otimes\bar{f}}(n)\phi(n/X)=\hat{\phi}(0)X+O(\sqrt{X}\log\mathfrak{q}(f\otimes\bar{f}))$$

where $\hat{\phi}(0)=\int\varphi(t)dt>0$. These estimates are smooth versions of the prime number theorem for the Rankin-Selberg convolution L-functions. We used GRH to estimate the sum over zeros. On the other hand, the two left sides coincide by assumption up to the contribution from ramified primes. This contribution is small, precisely it is

$$\ll\sum_{n\leqslant 2X,(n,q(f)q(g))\neq 1}|\Lambda_{f\otimes\bar{f}}(n)|+|\Lambda_{f\otimes\bar{g}}(n)|\ll d^2(\log q(f)q(g))\log X.$$

Therefore $\hat{\phi}(0)X\ll\sqrt{X}\log\mathfrak{q}(f\otimes\bar{f})\mathfrak{q}(f\otimes\bar{g})+d^2(\log q(f)q(g))\log X$, or $X\ll(d\log\mathfrak{q}(f)\mathfrak{q}(g))^2$ where the implied constant is absolute (recall that the definition of the Rankin-Selberg convolution implies the bound $\mathfrak{q}(f\otimes g)\leqslant(\mathfrak{q}(f)\mathfrak{q}(g))^d$). □

REMARK. The best general unconditional result is much weaker. Suppose $L(f,s)$ has degree d with roots $|\alpha_i(p)|<p^{1/4}$ and that it admits $L(f\otimes\bar{f},s)$. Then for every $\varepsilon>0$ there exists $C>0$, depending on ε and on f with the following property: If $L(g,s)$ has the same gamma factor as $L(f,s)$ (so the same degree), if $L(f\otimes\bar{g},s)$ is entire and if $\lambda_g(p)=\lambda_f(p)$ for all $p\leqslant C\mathfrak{q}(g)^{d/2+\varepsilon}$, then $g=f$.

To prove this, notice that if $X\geqslant 1$ is the largest integer for which $\lambda_f(p)=\lambda_g(p)$ for $p\leqslant X$, then by multiplicativity we have $\lambda_f(n)=\lambda_g(n)$ for $n\mid N(X)$ where $N(X)$ is the product of primes $\leqslant X$. For $h=f$, or $h=g$, one can factor

$$L(f\otimes\bar{h},s)=L^\flat(f\otimes\bar{h},s)H(f,h,s)$$

where $L^\flat(f\otimes\bar{h},s)$ is the Dirichlet series restricted to squarefree integers and $H(f,h,s)$ converges absolutely for $\sigma>\frac{1}{2}$. In particular, $L^\flat(f\otimes\bar{h},s)$ is entire if $h=g$ and has a pole at $s=1$ for $h=f$, and satisfies the same convexity bound as the Rankin-Selberg L-function. We get by complex integration on the line $\sigma=\frac{1}{2}+\delta$, with $\delta>0$ small enough

$$\sum_{n\geqslant 1}\lambda_{f\otimes\bar{h}}\phi(n/X)=\delta(f,g)XP(\log X)+O(X^{\frac{1}{2}+\varepsilon}\mathfrak{q}(f\otimes g)^{\frac{1}{4}+\varepsilon})$$

where r is the order of the pole of $L(f\otimes\bar f,s)$ at $s=1$ and $P\neq 0$ is the polynomial of degree $r-1$ given by

$$P(\log X)=\operatorname*{res}_{s=1} L^\flat(f\otimes\bar h,s)X^{s-1}.$$

Comparing the formula for $h=f$ and that for $h=g$ yields the result.

5.9. The Riemann zeta function and Dirichlet L-functions.

The Riemann zeta function $\zeta(s)$ is a self-dual L-function, with conductor 1, gamma factor $\gamma(s)=\pi^{-s/2}\Gamma(s/2)$ and root number 1. Its Rankin-Selberg square is also $\zeta(s)$.

More generally, let χ be a primitive Dirichlet character modulo q. The Dirichlet L-function $L(s,\chi)$ is an L-function of degree 1 with conductor q, gamma factor

$$\gamma(s)=\pi^{-s/2}\Gamma\Bigl(\frac{s+\delta}{2}\Bigr),$$

where $\delta=0$ if $\chi(-1)=1$ and $\delta=1$ if $\chi(-1)=-1$, and its root number is $\varepsilon(\chi)=\tau(\chi)/\sqrt{q}$, where

$$\tau(\chi)=\sum_{x \pmod q}\chi(x)e\Bigl(\frac{x}{q}\Bigr)$$

is the Gauss sum associated to χ (see Section 3.4). If χ is non-trivial, then $L(s,\chi)$ is entire, otherwise it has a simple pole at $s=1$ with residue 1.

Any $L(s,\chi)$ satisfies trivially the Ramanujan-Petersson conjecture. The analytic conductor is given by $\mathfrak{q}(\chi,s)=q(|t+\delta|+3)\asymp q(|t|+3)$ and $\mathfrak{q}(\chi)=(2+\delta)q\asymp q$; see Theorem 4.15.

A Dirichlet L-function $L(s,\chi)$ is self-dual if and only if χ is quadratic. In this case $\varepsilon(\chi)=1$ by Theorem 3.3.

Any two Dirichlet characters χ_1 and χ_2 admit a Rankin-Selberg convolution given simply by $L(\chi_1\otimes\chi_2,s)=L(\chi_3,s)$, where χ_3 is the primitive character that induces the product $\chi_1\chi_2$. We have $\chi_3=\chi_1\chi_2$ if, for instance, the conductors q_i of χ_i are coprime, but not in general. In particular, the Rankin-Selberg square of $L(s,\chi)$ is $\zeta(s)$ and the finite product (5.10) is

$$\prod_{p|q}(1-p^{-s})^{-1}.$$

Fixing a modulus q and considering all primitive characters modulo q gives an example of a "family" of L-functions. Fixing $X\geqslant 2$ and considering all primitive quadratic characters of conductor $q\leqslant X$ gives another family.

By (5.22) we deduce the convexity bound for Dirichlet L-functions.

THEOREM 5.23. *Let χ modulo q be a primitive Dirichlet character. We have*

$$L(s,\chi)\ll (q|s|)^{\frac14} \tag{5.63}$$

for $\operatorname{Re} s=\frac12$, *the implied constant being absolute.*

The following exercise provides very simple estimates which are quite handy anyway.

EXERCISE 9. Let χ modulo q be a non-trivial Dirichlet character. Show that for $0 \leqslant \sigma \leqslant 1$ we have

$$L^{(k)}(\sigma, \chi) \ll q^{1-\sigma}(\log q)^{k+1} \tag{5.64}$$

for $k \geqslant 1$, the implied constant depending only on k. [**Hint:** Use the bound $|\sum_{n \leqslant x} \chi(n)| \leqslant \min(x, q)$.]

Theorem 5.8 becomes:

THEOREM 5.24. *Let χ modulo q be a primitive Dirichlet character, $N(T,\chi)$ the number of zeros $\rho = \beta + i\gamma$ of $L(s,\chi)$ in the critical strip $0 \leqslant \beta \leqslant 1$ with $|\gamma| \leqslant T$. We have*

$$N(T,\chi) = \frac{T}{\pi} \log \frac{qT}{2\pi e} + O(q(T+3)),$$

with an absolute implied constant.

In addition to the general explicit formula of Theorem 5.11, it is sometimes more practical to work with the following approximate formula for

$$\psi(x,\chi) = \sum_{n \leqslant x} \Lambda(n)\chi(n).$$

PROPOSITION 5.25. *For any character we have*

$$\psi(x,\chi) = \delta_\chi x - \sum_{\substack{L(\rho,\chi)=0 \\ |\mathrm{Im}\,(\rho)| \leqslant T}} \frac{x^\rho - 1}{\rho} + O\Big(\frac{x}{T}(\log xq)^2\Big) \tag{5.65}$$

where $\delta_\chi = 1$ if $\chi = \chi_0$ and $\delta_\chi = 0$ otherwise, $1 \leqslant T \leqslant x$ and the implied constant is absolute.

PROOF. This is a special case of (5.53) for primitive characters. However, (5.65) also holds as it is for χ not primitive. Indeed, suppose $\chi^*(\text{mod } q^*)$ with $q^* \mid q$ is the primitive character which induces $\chi(\text{mod } q)$. Then

$$|\psi(x,\chi) - \psi(x,\chi^*)| \leqslant \sum_{p^\alpha \leqslant x, p \mid q} \log p \ll (\log xq)^2.$$

Moreover, the zeros of $L(s,\chi)$ agree with these of $L(s,\chi^*)$ except for the ones coming from the product

$$\prod_{p \mid q, p \nmid q^*} (1 - p^{-s})$$

which are $\rho = 2\pi i l / \log p$. Hence these zeros contribute at most

$$\sum_{p \mid q} \sum_{\frac{2\pi |l|}{\log p} \leqslant T} \Big|\frac{x^\rho - 1}{\rho}\Big| \ll (\log xq)^2.$$

The above corrections are absorbed by the error term in (5.65) completing the proof. $\square$

By (5.65) we derive a strong approximation to

$$\psi(x,q,a)=\sum_{\substack{n\leqslant x\\ n\equiv a \pmod q}}\Lambda(n)$$

in terms of zeros of L-functions. First by the orthogonality of characters we have

$$\psi(x,q,a)=\frac{1}{\varphi(q)}\sum_{\chi \pmod q}\bar{\chi}(a)\psi(x,\chi)$$

Applying (5.65) we obtain

$$\psi(x;q,a)=\frac{x}{\varphi(q)}-\frac{1}{\varphi(q)}\sum_{\chi \pmod q}\bar{\chi}(a)\sum_{\substack{L(\rho,\chi)=0\\ |\mathrm{Im}\,(\rho)|\leqslant T}}\frac{x^\rho-1}{\rho}+O\Big(\frac{x}{T}(\log xq)^2\Big) \tag{5.66}$$

for $q\geqslant 1$, $(a,q)=1$ and $1\leqslant T\leqslant x$, with an absolute implied constant.

Theorem 5.10 gives the zero-free region for Dirichlet L-functions, except for the possible exceptional zero β_χ for a real primitive character χ. Because of its importance we include here the slightly different traditional proof.

THEOREM 5.26. *There exists an absolute constant $c>0$ such that for any primitive Dirichlet character χ modulo q, $L(s,\chi)$ has at most one zero in the region*

$$\sigma\geqslant 1-\frac{c}{\log q(|t|+3)}. \tag{5.67}$$

The exceptional zero may occur only if χ is real, and it is then a simple real zero, say β_χ, with

$$1-\frac{c}{\log 3q}\leqslant\beta_\chi<1. \tag{5.68}$$

PROOF. Assume χ is non-trivial (the case of $\zeta(s)$ is similar, but it needs a change of vocabulary because our definition of L-functions does not include $\zeta(s+it)$ with $t\neq 0$ for the reason that it has a pole at $s=1-it\neq 1$). Then consider the L-function of degree 8

$$L(f,s)=\zeta(s)^3L(s+it,\chi)^4L(s+2it,\chi_2)$$

for some $t\in\mathbb{R}$ where χ_2 is the primitive character modulo $q_2\mid q$ which induces χ^2. By the trigonometric inequality of de la Vallée Poussin

$$3|z|+4\mathrm{Re}\,(z)+\mathrm{Re}\,(z^2)=|z|(3+4\cos\theta+\cos 2\theta)=2|z|(1+\cos(\theta))^2\geqslant 0 \tag{5.69}$$

for $z=|z|\exp(i\theta)$, it follows that $\mathrm{Re}\,(\Lambda_f(n))\geqslant 0$ for all $n\geqslant 1$ such that $(n,q)=1$. In fact, a simple computation using the definition of χ_2 shows that this remains valid for all $n\geqslant 1$.

Let $\rho=\beta+i\gamma$ be a zero of $L(f,s)$ with $\beta\geqslant\frac{1}{2}$. If χ is not real, or if $\gamma\neq 0$, then taking $t=\gamma$ it follows that $L(f,s)$ has a pole of order 3 at $s=1$ while β is a real zero of order 4 of $L(f,s)$. Hence by Lemma 5.9 we have

$$\beta<1-\frac{c_2}{\log\mathfrak{q}(f,\rho)}<1-\frac{c}{\log(q(|\gamma|+3))}$$

with an absolute positive constant c. If $\chi^2 = 1$, $\rho = \beta$ is real we take $t = 0$, then $L(f,s)$ has a pole of order 4 at $s = 1$. By Lemma 5.9 again, $L(f,s)$ can have at most four real zeros with multiplicity satisfying the above inequality, which means that β must be a simple zero of $L(s,\chi)$. □

REMARK. One can also recover that $\beta < 1$ by Theorem 2.1.

As an application of Theorem 5.13, one derives the following prime number theorem .

THEOREM 5.27. *Let χ modulo q be a primitive Dirichlet character. We have*

$$\sum_{n\leqslant x}\chi(n)\Lambda(n) = \delta_\chi x - \frac{x^{\beta_\chi}}{\beta_\chi} + O\Big(x\exp\Big(\frac{-c\log x}{\sqrt{\log x}+\log q}\Big)(\log q)^4\Big) \tag{5.70}$$

for any $x \geqslant 1$, with some absolute effective constant $c > 0$, the implied constant being absolute.

For any $q \geqslant 1$ and $(a,q) = 1$ we have

$$\sum_{\substack{n\leqslant x\\ n\equiv a \ (\mathrm{mod}\ q)}}\Lambda(n) = \frac{x}{\varphi(q)} - \frac{\bar\chi(a)}{\varphi(q)}\frac{x^{\beta_\chi}}{\beta_\chi} + O\Big(x\exp\Big(\frac{-c\log x}{\sqrt{\log x}+\log q}\Big)(\log q)^4\Big) \tag{5.71}$$

for $x \geqslant 1$, where χ (mod q) is the possible exceptional real character modulo q and β_χ the corresponding exceptional zero.

Have in mind that there could be one or two real primitive characters modulo q, for example, if $q = 8r$ with r odd squarefree we have two such characters $\chi_8\chi_r$ and $\chi_4\chi_8\chi_r$. Nevertheless, at most one of these may give an exceptional zero. In fact, the exceptional characters are very rare, and the exceptional zero itself can be somewhat controlled.

THEOREM 5.28. *(1) (Landau) Let χ_1 and χ_2 be distinct real primitive characters modulo q_1 and q_2. Assume χ_1 and χ_2 have real zeros β_1 and β_2 respectively. There exists an absolute constant $c > 0$ such that*

$$\min(\beta_1,\beta_2) \leqslant 1 - \frac{c}{\log q_1q_2}. \tag{5.72}$$

(2) (Siegel) For any primitive real Dirichlet character χ modulo q, we have

$$\beta_\chi \leqslant 1 - \frac{c(\varepsilon)}{q^\varepsilon} \tag{5.73}$$

for any $\varepsilon > 0$, with a constant $c(\varepsilon) > 0$ depending only on ε. If $\varepsilon < \frac{1}{2}$, this constant is ineffective, meaning it is not numerically computable.

PROOF. (1) Consider the L-function of degree 4

$$L(f,s) = \zeta(s)L(s,\chi_1)L(s,\chi_2)L(s,\chi_1\chi_2) \tag{5.74}$$

which has conductor at most $(q_1q_2)^2$. Since the χ_1,χ_2 are non-trivial, and distinct, $L(f,s)$ has a pole of order 1 at $s = 1$. Moreover,

$$-\frac{L'}{L}(f,s) = \sum_{n\geqslant 1}(1+\chi_1(n))(1+\chi_2(n))\Lambda(n)n^{-s}$$

has non-negative coefficients. By Lemma 5.9, there exists $c > 0$ such that $L(f, s)$ has at most one real zero with $\beta \geqslant 1 - c/\log \mathfrak{q}(f)$. Since $\mathfrak{q}(f) \leqslant 9(q_1 q_2)^2$, this gives (5.72).

(2) For the proof of (5.73), we follow the nice argument of Goldfeld [**Go1**]. Let χ_1, χ_2 be (for the moment) arbitrary primitive real characters modulo q_1 and q_2 and consider again the L-function (5.74). We have $\lambda_f(n) \geqslant 0$. Let $\eta(x)$ be a function on $[0, +\infty[$ such that $0 \leqslant \eta \leqslant 1$, $\eta(x) = 1$ for $0 \leqslant x \leqslant 1$ and $\eta(x) = 0$ for $x \geqslant 2$. Consider the sums

$$Z(\beta, x) = \sum_n \frac{\lambda_f(n)}{n^\beta} \eta\Big(\frac{n}{x}\Big) \geqslant 1 \quad (\text{since } \lambda_f(1) = 1)$$

for $\frac{3}{4} \leqslant \beta \leqslant 1$ and $x \geqslant 1$. By contour integration we have

$$Z(\beta, x) = \frac{1}{2\pi i} \int_{(2)} L(f, s+\beta) x^s \hat{\eta}(s) ds = L(f, \beta)$$

$$+ L(1, \chi_1) L(1, \chi_2) L(1, \chi_1\chi_2) \hat{\eta}(1-\beta) x^{1-\beta} + \frac{1}{2\pi i} \int_{(\frac{1}{2}-\beta)} L(f, s+\beta) \hat{\eta}(s) x^s ds.$$

By (5.63), the last integral is $\ll q_1 q_2 x^{\frac{1}{2}-\beta}$. Thus we have

$$L(f, \beta) + L(1, \chi_1) L(1, \chi_2) L(1, \chi_1\chi_2) \hat{\eta}(1-\beta) x^{1-\beta} \geqslant 1 + O(q_1 q_2 x^{\frac{1}{2}-\beta}).$$

Now assume that $\beta_1 \geqslant \frac{3}{4}$ is a zero of $L(s, \chi_1)$. Then it follows

$$L(1, \chi_1) L(1, \chi_2) L(1, \chi_1\chi_2) \hat{\eta}(1-\beta_1) x^{1-\beta_1} \geqslant 1 + O(q_1 q_2 x^{\frac{1}{2}-\beta_1}).$$

We take $x = c(q_1 q_2)^4$ with c sufficiently large, so the term on the left is $> \frac{1}{2}$. From (5.64) we have $L(1, \chi_1) \ll \log q_1$ and $L(1, \chi_1\chi_2) \ll \log q_1 q_2$. Moreover, $\hat{\eta}(1-\beta_1) \ll (1-\beta_1)^{-1}$ since $\hat{\eta}(s)$ has a simple pole at $s = 0$ with residue 1. Gathering these estimates we get our basic lower bound

$$L(1, \chi_2) \gg (1-\beta_1)(q_1 q_2)^{-4(1-\beta_1)} (\log q_1 q_2)^{-2}. \tag{5.75}$$

We will use (5.75) to prove (5.73), using the hypothetical zero β_1 of $L(s, \chi_1)$ as a tool to control other characters. Let $0 < \varepsilon \leqslant \frac{1}{4}$. If all $L(s, \chi)$ for χ real modulo q do not vanish on the real segment $s \geqslant 1 - \varepsilon$, then (5.73) is obvious (just take $c(\varepsilon) = \varepsilon$). Suppose otherwise that for some χ_1 modulo q_1, the L-function $L(s, \chi_1)$ has a zero $s = \beta_1 \geqslant 1 - \varepsilon$. Then for any other real primitive character χ_2 modulo q_2, we can write (5.75) as

$$L(1, \chi_2) \gg q_2^{-4\varepsilon} (\log q_2)^{-2}$$

where the implied constant depends only on ε, since q_1, χ_1 and β_1 are fixed once ε is chosen. On the other hand, if β_2 is a real zero of $L(s, \chi_2)$ we have by the mean-value theorem

$$L(1, \chi_2) = (1-\beta_2) L'(\sigma, \chi_2) \ll (1-\beta_2) q_2^{1-\beta_2} (\log q_2)^2$$

for some σ with $\beta_2 \leqslant \sigma \leqslant 1$ by (5.64). Combining those two bounds gives $1 - \beta_2 \gg q_2^{-4\varepsilon}$, hence (5.73) after re-defining ε. □

REMARK. Essentially the bound (5.73) was first proved with $\varepsilon = \frac{1}{8}$ by Landau [**La2**] (also not effective), and with any $\varepsilon > 0$ shortly afterwards by Siegel [**Sie1**]. The ineffectiveness of the Siegel bound (5.73) is apparent in applications. For instance, it is not possible to use it to solve the class number one problem for imaginary quadratic fields. Yes, by Siegel's lower bound $L(1,\chi) \gg \Delta^{\frac{1}{2}-\varepsilon}$ and by the Dirichlet Class Number Formula (2.31), or (22.59), we get

$$h(\mathbb{Q}(\sqrt{-\Delta})) = \frac{w\sqrt{\Delta}}{2\pi} L(1,\chi) \gg \Delta^{\frac{1}{2}-\varepsilon} \tag{5.76}$$

for $-\Delta$ a negative fundamental discriminant, χ being the corresponding Kronecker symbol. Hence h tends to infinity as does $|\Delta|$. But when computing all Δ with $h(\mathbb{Q}(\sqrt{-\Delta})) = 1$ it is not possible to know when to stop. In Chapters 22 and 23, we use much more sophisticated ideas of Goldfeld involving L-functions of elliptic curves to prove an effective bound, although much weaker (see Theorem 23.2).

Siegel's estimate implies the Siegel-Walfisz theorem about primes in arithmetic progressions.

COROLLARY 5.29. *Let $q \geqslant 1$ and $A > 0$. We have*

$$\pi(x;q,a) = \frac{\mathrm{Li}(x)}{\varphi(q)} + O\Big(\frac{x}{(\log x)^A}\Big), \tag{5.77}$$

$$\psi(x;q,a) = \frac{x}{\varphi(q)} + O\Big(\frac{x}{(\log x)^A}\Big), \tag{5.78}$$

for any $x \geqslant 2$. Moreover, for any primitive Dirichlet character χ modulo $q > 2$, we have

$$\sum_{p \leqslant x} \chi(p) \ll \sqrt{q}x(\log x)^{-A}, \tag{5.79}$$

$$\sum_{n \leqslant x} \chi(n)\mu(n) \ll \sqrt{q}x(\log x)^{-A} \tag{5.80}$$

for any $A > 0$. The implied constants depend only on A, but are ineffective.

PROOF. If $q \leqslant (\log x)^{A+1}$ one gets (5.78) from the prime number formula (5.71) by employing the bound (5.73). For larger q there is nothing to prove, because the formula (5.78) is trivial. Then (5.77) follows from (5.78) by partial summation. Similarly one derives (5.79) from (5.52) by employing the bound (5.73)

The case of (5.80) is a bit harder. One could follow the proof of (5.51) using L^{-1} instead of L'/L, but there are serious complications which occur near multiple zeros or clusters of zeros (so far we cannot rule out this scenario). Therefore we are going to derive (5.80) directly from (5.79). First we write

$$\sum_{n \leqslant x} \chi(n)\mu(n) = 1 - \sum_{m \leqslant x} \chi(m)\mu(m) \sum_{p_m < p \leqslant \frac{x}{m}} \chi(p)$$

where p_m stands for the largest prime divisor of m. Let $r \geqslant 0$ be the number of prime divisors of m. Then the summation conditions imply $m^{1/r} \leqslant p_m < x/m$, so $m < x^{r/(r+1)}$. Hence, using (5.79) the inner sum over p satisfies

$$\sum_p \chi(p) \ll \frac{\sqrt{q}x}{m}\Big(\log\frac{x}{m}\Big)^{-A} \leqslant \frac{\sqrt{q}x}{m}\Big(\frac{r+1}{\log x}\Big)^A \ll \frac{\tau(m)\sqrt{q}x}{m(\log x)^A},$$

because $(r+1)^A \ll 2^r = \tau(m)$. Finally, summing over m trivially we obtain (5.80) with extra factor $(\log x)^2$. This completes the proof by resetting A. □

For Dirichlet characters, the problem discussed in Proposition 5.22 becomes that of finding the smallest prime p with $\chi(p) \neq 1$ for a non-trivial character χ modulo $q > 2$, say $p_{\min}(\chi)$. One derives directly from various estimates for short character sums the following bounds:

$$p_{\min}(\chi) \ll \sqrt{q}\log q,\ \ q^{\frac{1}{4}+\varepsilon},\ \ q^{\varepsilon},\ \ (\log q)^2$$

using respectively the estimates of Polyá-Vinogradov, Burgess, Lindelöf hypothesis, Riemann hypothesis. Actually we only get $(\log q)^4$ by (5.56), but with a little extra effort one can squeeze from the GRH the bound $(\log q)^2$. One can also do better unconditionally using Burgess's estimate enhanced by combinatorial arguments (see e.g. [**H1**]). Linnik proved (by the large sieve inequality) that the smallest quadratic non-residue for a primitive real character modulo $q \leqslant N$, q prime, is $\leqslant N^{\varepsilon}$ with at most finitely many exceptions, the number of exceptions depending only on $\varepsilon > 0$. See Section 7.4 for the proof and an analogue for elliptic curves.

The Riemann zeta function is further discussed in Chapters 10, 24, 25, and Dirichlet characters in Chapters 17, 18.

5.10. L-functions of number fields.

Let $K/\mathbb{Q}$ be a number field of degree d. The Dedekind zeta function

$$\zeta_K(s) = \sum_{\mathfrak{a}\neq 0} (N\mathfrak{a})^{-s} = \prod_{\mathfrak{p}} (1-(N\mathfrak{p})^{-s})^{-1}$$

defines a self-dual L-function of degree $d = [K:\mathbb{Q}]$ with conductor $D = |d_K|$, the absolute value of the discriminant of K, and root number $+1$. The gamma factor is

$$\gamma(s) = \pi^{-ds/2}\Gamma\Big(\frac{s}{2}\Big)^{r_1+r_2}\Gamma\Big(\frac{s+1}{2}\Big)^{r_2} \tag{5.81}$$

where r_1 is the number of real embeddings of K and r_2 the number of pairs of complex embeddings so that $d = r_1 + 2r_2$. The zeta function has simple pole at $s = 1$ with residue

$$\operatorname*{res}_{s=1} \zeta_K(s) = \frac{2^{r_1}(2\pi)^{r_2} hR}{w\sqrt{D}},$$

where h is the class number of K, R the regulator and w the number of roots of unity. For a proof, see e.g. [**La**].

The Ramanujan-Petersson conjecture is obviously satisfied for $\zeta_K(s)$. The analytic conductor is

$$\mathfrak{q}(\zeta_K, s) = D(|t|+3)^{r_1+r_2}(|t+1|+3)^{r_2} \leqslant D(|t|+4)^d$$

and $\mathfrak{q}(\zeta_K) = 2^{r_1+r_2}3^{r_2}D \leqslant 3^d D$. There exists an absolute constant $c > 1$ such that $D > c^{d-1}$ (see Exercise 10 below) so we have $\log \mathfrak{q}(\zeta_K) \asymp \log D$.

REMARK. Dedekind zeta functions factor as products of Artin L-functions (see Section 5.13). Let $E/\mathbb{Q}$ be the Galois closure of K, $G = \mathrm{Gal}(E/\mathbb{Q})$ and $H = \mathrm{Gal}(E/K)$. Let $r_{G/H}$ be the permutation representation of G on the coset space G/H. This is the representation induced by the trivial representation of H, hence by the invariance of Artin L-functions under induction we have $L(r_{G/H}, s) = \zeta_K(s)$. One can decompose $r_{G/H}$ in terms of irreducible representations of G, $r_{G/H} = \oplus_\rho n_\rho \rho$, getting the factorization

$$\zeta_K(s) = \prod_\rho L(s,\rho)^{n_\rho}. \tag{5.82}$$

In case of a failure of the Artin conjecture (a possibility we cannot yet rule out) this is not a factorization in terms of L-functions in the sense of Section 5.1.

If $K/\mathbb{Q}$ is abelian, then $E = K$, $H = 1$ and $r_{G/H} = \oplus\chi$ where χ runs over a group of Dirichlet characters so that

$$\zeta_K(s) = \prod_\chi L(s,\chi)$$

is a product of L-functions for distinct primitive Dirichlet characters of conductors whose product is $D = |d_K|$ and exactly one character is trivial.

We deduce from the above and the results of the previous sections:

THEOREM 5.30. *Let $K/\mathbb{Q}$ be a number field of degree d. We have*

$$(s-1)\zeta_K(s) \ll |Ds^d|^{(1-\sigma)/2+\varepsilon}$$

for $0 \leqslant \sigma \leqslant 1$, the implied constant depending only on d and ε.

THEOREM 5.31. *Let $K/\mathbb{Q}$ be a number field of degree d. Let $N_K(T)$ be the number of zeros $\rho = \beta + i\gamma$ of $\zeta_K(s)$ in the critical strip $0 \leqslant \beta \leqslant 1$ with $|\gamma| \leqslant T$. We have*

$$N_K(T) = \frac{T}{\pi}\log\frac{DT^d}{(2\pi e)^d} + O(\log DT^d)$$

for $T \geqslant 2$, the implied constant being absolute.

One can use the analytic properties of L-functions to deduce information on the discriminant of number fields. For instance, the explicit formula yields quite good lower bounds for $D = |d_K|$. This powerful method was implemented by Odlyzko [**Od**], Poitou [**Poi**], Serre [**Se7**] and others. Here is a simple result.

THEOREM 5.32. *Assume GRH for the Dedekind zeta functions of number fields. Then we have*

$$\liminf \frac{1}{d}\{\log D + (d-r_2)\frac{\Gamma'}{\Gamma}(\tfrac14) + r_2\frac{\Gamma'}{\Gamma}(\tfrac34)\} \geqslant \log\pi, \tag{5.83}$$

as $d = [K:\mathbb{Q}] = r_1 + 2r_2 \to +\infty$.

PROOF. Apply the explicit formula (5.44) to $\zeta_K(s)$ with

$$\phi(x) = x^{-\frac{1}{2}} e^{-a(\log x)^2},$$

for some $a > 0$. Although this test function is not of compact support, it decays very fast at infinity, so (5.44) holds. The Mellin transform is

$$\hat{\phi}(\tfrac{1}{2} + it) = \sqrt{\frac{2\pi}{a}} e^{-\frac{t^2}{4a}}$$

for $t \in \mathbb{R}$. Note that this is positive (compare with Proposition 5.55) and that $\phi(x) = x^{-1}\phi(x^{-1})$. Since $\zeta_K(s)$ is self-dual we get

$$\sum_n \Lambda_K(n)\phi(n) + \frac{1}{2}\sum_\rho \hat{\phi}(\rho) = \frac{1}{2}\log D + \frac{1}{2}\int_0^{+\infty} \phi(x)dx + \frac{1}{2\pi i}\int_{(\frac{1}{2})} \frac{\gamma'}{\gamma}(s)\hat{\phi}(s)ds.$$

By positivity, we derive

$$\frac{1}{2}\log D + \frac{1}{2}\int_0^{+\infty} \phi(x)dx + \frac{1}{2\pi i}\int_{(\frac{1}{2})} \frac{\gamma'}{\gamma}(s)\hat{\phi}(s)ds \geqslant 0. \tag{5.84}$$

By (5.81) and the Fourier inversion formula we have

$$\frac{1}{2\pi i}\int_{(\frac{1}{2})} \frac{\gamma'}{\gamma}(s)\hat{\phi}(s)ds = -\frac{d}{2}\log \pi + \frac{(d-r_2)}{2\pi}\int_{\mathbb{R}} \frac{\Gamma'}{\Gamma}(\tfrac{1}{4} + \tfrac{it}{2})\hat{\phi}(\tfrac{1}{2} + it)dt \tag{5.85}$$
$$+ \frac{r_2}{2\pi}\int_{\mathbb{R}} \frac{\Gamma'}{\Gamma}(\tfrac{3}{4} + \tfrac{it}{2})\hat{\phi}(\tfrac{1}{2} + it)dt.$$

When $a \to 0$, the function $(2\pi)^{-1}\hat{\phi}(\frac{1}{2} + it)$ converges to the Dirac delta at 0, hence

$$\frac{1}{2\pi}\int_{\mathbb{R}} \frac{\Gamma'}{\Gamma}(\tfrac{1}{4} + \tfrac{it}{2})\hat{\phi}(\tfrac{1}{2} + it)dt \to \frac{\Gamma'}{\Gamma}(\tfrac{1}{4})$$
$$\frac{1}{2\pi}\int_{\mathbb{R}} \frac{\Gamma'}{\Gamma}(\tfrac{3}{4} + \tfrac{it}{2})\hat{\phi}(\tfrac{1}{2} + it)dt \to \frac{\Gamma'}{\Gamma}(\tfrac{3}{4})$$

Therefore, dividing (5.84) by d, then letting $d \to +\infty$, and then letting $a \to 0$, we derive (5.83). □

Similar arguments can be applied unconditionally with slightly weaker result using test functions as constructed in Proposition 5.55.

EXERCISE 10. Show that there exists an absolute constant $c > 1$ such that $D = |d_K| > c^d$ for any number field $K/\mathbb{Q}$ of degree $d \geqslant 2$, and, in particular, $D \geqslant 3$ if $K \neq \mathbb{Q}$ so K has at least one ramified prime. [**Hint:** Proceed as in Theorem 5.51 below.]

One cannot apply directly Theorem 5.10 because the Rankin-Selberg square of $\zeta_K(s)$ involves convolutions of Artin L-functions of K which are not known to be entire, although they are conjectured to be. However, one can adapt the traditional arguments (as in the proof of Theorem 5.26) using $\zeta_K(s)^3\zeta_K(s+it)^4\zeta_K(s+2it)$

and derive the zero-free region, then the prime number theorem for K. We phrase it in terms of

$$\Lambda_K(\mathfrak{a}) = \begin{cases} \log N\mathfrak{p} & \text{if } \mathfrak{a} = \mathfrak{p}^k \text{ for some } k \geqslant 1, \\ 0 & \text{otherwise.} \end{cases}$$

Thus these are coefficients in the Dirichlet series

$$-\frac{\zeta_K'}{\zeta_K}(s) = \sum_{\mathfrak{a}} \Lambda_K(\mathfrak{a})(N\mathfrak{a})^{-s}.$$

THEOREM 5.33. *Let $K/\mathbb{Q}$ be a number field of degree d. There exists an absolute constant $c > 0$ such that $\zeta_K(s)$ has no zero in the region*

$$\sigma \geqslant 1 - \frac{c}{d^2 \log D(|t|+3)^d}$$

except possibly a simple real zero $\beta < 1$.

We have

$$\sum_{N\mathfrak{a} \leqslant x} \Lambda_K(\mathfrak{a}) = x - \frac{x^\beta}{\beta} + O(\sqrt{D}x \exp(-cd^{-2}\sqrt{\log x}))$$

for $x \geqslant 2$, where $c > 0$ is an absolute constant, and the term $-x^\beta/\beta$ should be removed if there is no exceptional zero.

Be aware that the zero-free region of Theorem 5.33 is not necessarily the best accessible. Indeed by (5.82), one can factorize $\zeta_K(s)$ and if one knows that the factors are L-functions, then zero-free regions for them produce one for ζ_K which can be much better because of a smaller conductor. For instance, if $K = \mathbb{Q}(\mu_p)$ is the field of p-th roots of unity, we have

$$\zeta_K(s) = \zeta(s) \prod_{\chi \neq 1} L(s, \chi)$$

where the product is over non-trivial Dirichlet characters modulo p. Thus by Theorem 5.26 and Theorem 5.28 there exists an absolute constant $c > 0$ such that $\zeta_K(s)$ has at most one simple real zero β in the region

$$\sigma \geqslant 1 - \frac{c}{\log p(|t|+3)}$$

whereas $d = p - 1$ and $D = |d_K| = p^{p-2}$ so Theorem 5.33 is much weaker.

Confirming that the zeros of real Dirichlet characters are the most difficult to deal with, Stark [**St3**] has shown that if $K/\mathbb{Q}$ does not contain a quadratic subfield, then there is no exceptional zero of $\zeta_K(s)$.

For Dedekind zeta functions, Proposition 5.21 can be improved.

PROPOSITION 5.34. *Let $K/\mathbb{Q}$ be a number field. Assume GRH holds for $\zeta_K(s)$. We have*

$$\operatorname*{ord}_{s=\frac{1}{2}} \zeta_K(s) \ll \frac{\log 3D}{\log\log 3D}$$

where $D = |d_K|$ and the implied constant is absolute.

PROOF. Let m be the order of vanishing at $\frac{1}{2}$. We argue as in the proof of (5.62). In applying the explicit formula (5.45) an additional term arises from the simple pole of $\zeta_K(s)$ at $s = 1$, hence we get

$$mh(0) + \sum_{\rho\neq\frac{1}{2}} h\Big(\frac{\gamma}{2\pi}\Big) = \int_{-\infty}^{+\infty} g(y)e^{y/2}dy$$

$$- 2\sum_{\mathfrak{a}} \frac{\Lambda_K(\mathfrak{a})}{\sqrt{N\mathfrak{a}}} g(\log N\mathfrak{a}) + O(\log 3D).$$

Both sums over the zeros and the prime ideals can be dropped by positivity. Since $h(0) \asymp Y$ and the integral is $O(e^Y)$ we infer that $mY \ll e^Y + \log 3D$. Taking $Y = \log\log 3D$ yields the result. □

Generalizing Dirichlet characters to number fields are Hecke characters (Grossencharakters). Those are defined thoroughly in Section 3.8 for imaginary quadratic fields, but here we use the basic terminology in the general case (for characters of weight 0). See e.g. [**La**] for details.

Let $K/\mathbb{Q}$ be a number field and ξ a primitive Hecke Grossencharakter modulo $(\mathfrak{m}, \Omega)$ where $\mathfrak{m}$ is a non-zero integral ideal in K and Ω is a set of real infinite places where ξ is ramified. The Hecke L-function is defined by

$$L(\xi, s) = \prod_{\mathfrak{p}} (1 - \xi(\mathfrak{p})N\mathfrak{p}^{-s})^{-1}$$

for $\mathrm{Re}\,(s) > 1$. Hecke showed that $L(\xi, s)$ is an L-function of degree $d = [K : \mathbb{Q}]$ which is entire if $\xi \neq 1$ and which coincides with $\zeta_K(s)$ for $\xi = 1$. The conductor is $\Delta = |d_K|N_{K/\mathbb{Q}}\mathfrak{m}$, where $N_{K/Q}$ is the norm and the gamma factor is given by

$$\gamma(\xi, s) = \pi^{-ds/2}\Gamma\Big(\frac{s}{2}\Big)^{r_1+r_2-|\Omega|}\Gamma\Big(\frac{s+1}{2}\Big)^{r_2+|\Omega|},$$

so $\mathfrak{q}(\xi) \leqslant 4^d|d_K|N_{K/\mathbb{Q}}\mathfrak{m} = 4^d\Delta$. The root number can be explicitly written as a normalized Gauss sum for ξ.

One gets the following generalization of Theorem 5.26.

THEOREM 5.35. *Let $K/\mathbb{Q}$ be a number field, ξ a Hecke Grossencharakter modulo $(\mathfrak{m}, \Omega)$. There exists an absolute effective constant $c > 0$ such that $L(\xi, s)$ has at most a simple real zero in the region*

$$\sigma > 1 - \frac{c}{d\log\Delta(|t| + 3)}.$$

The exceptional zero can occur only for a real character and it is < 1.

REMARK. This section highlights the fact that L-functions of various types are naturally defined over any number field K, using Dirichlet series over non-zero integral ideals, and Euler products over primes ideals. Thus there are Artin L-functions over K, automorphic L-functions over K, L-functions of varieties over K, whereas in a sense of analytic number theory we like to see these as L-functions over $\mathbb{Q}$. This view is possible because any L-function of degree d over K, where $[K : \mathbb{Q}] = f$, is an L-function over $\mathbb{Q}$, as defined in Section 5.1, of degree df. For instance, $\zeta_K(s)$ is of degree 1 over K. Thus it is easy for the reader to see what results there are for L-functions over number fields, with uniformity in the number field. Sometimes, however, the proof has to be adapted, as for Theorem 5.33, because the Rankin-Selberg L-functions have a different form depending on whether the L-functions are seen over K or over $\mathbb{Q}$. The reader should have no difficulty supplying new proofs if necessary. It is worth pointing out that for all known L-functions, the conductor of an L-function of degree d over K takes the form $q = |d_K|^d N_{K/\mathbb{Q}}\mathfrak{f}$ for some non-zero integral ideal $\mathfrak{f}$ of K. Thus the conductor and the analytic conductor contain the dependency in the discriminant of the base field K.

Analytic number theory flourished beautifully in the last two centuries in the garden of L-functions over the rationals. The extension to number fields is fruitful as well, but it needs to be more explored. To encourage the newcomers we end this section with a few cute applications of the theory over the Gaussian domain $\mathbb{Z}[i]$.

There are four units in $\mathbb{Z}[i]$, namely $1, -1, i, -i$. Every ideal of $\mathbb{Z}[i]$ is principal. Consider the "angle" characters

$$\xi_k(\mathfrak{a}) = \left(\frac{\alpha}{|\alpha|}\right)^{ik} = e^{ik\arg\alpha}$$

if $\mathfrak{a} = (\alpha) \neq 0$ for any $k \equiv 0 \pmod 4$. This is a primitive Hecke character of conductor (1). The associated L-function

$$L(s, \xi_k) = \sum_{\alpha} \left(\frac{\alpha}{|\alpha|}\right)^{ik} |\alpha|^{-s}$$

has conductor $D = 4$ and it satisfies the self-dual functional equation

$$\Lambda(s, \xi_k) = \pi^{-s}\Gamma(s + \tfrac{|k|}{2})L(s, \xi_k) = \Lambda(1 - s, \xi_k).$$

All characters are complex except the trivial character for $k = 0$ giving the Dedekind zeta function $L(s, \xi_0) = \zeta(s)L(s, \chi_4)$. Therefore there is no exceptional zero, and the Prime Number Theorem (5.52) gives

$$\sum_{|\pi| \leqslant x} \left(\frac{\pi}{|\pi|}\right)^{ik} = 2\delta_k \mathrm{Li}(x) + O\Big(|k| x e^{-c\sqrt{\log x}}\Big)$$

where $c > 0$ and the implied constant are absolute.

As an exercise we propose to derive from the above formula the following equidistribution property of Gaussian primes in sectors.

THEOREM 5.36. *Let $0 < \beta - \alpha \leqslant \frac{\pi}{2}$ and $x \geqslant 2$. Then*

$$|\{\pi \in \mathbb{Z}[i] \mid |\pi| \leqslant x, \ \alpha < \arg\pi \leqslant \beta\}| = \frac{\beta - \alpha}{\pi}\mathrm{Li}(x) + O\Big(x e^{-c\sqrt{\log x}}\Big)$$

where $c > 0$ and the implied constant are absolute.

[**Hint:** Approximate the characteristic function of $[\alpha, \beta]$ modulo 2π by a smooth function, periodic of period 2π, and symmetric with respect to $x \to \pm x \pm \pi$. Expand this function into Fourier series and apply for $x = \arg\pi$. By the symmetry the non-zero Fourier terms are on frequencies $k \equiv 0 \pmod 4$. At each frequency apply the Prime Number Theorem.]

See Section 11.8 for an application of this result to show that the Weil bound for the number of points on a curve over a finite field is optimal in horizontal sense.

EXERCISE 11. Playing with $\arg\alpha$ and $|\alpha|$ derive the equidistribution of Gaussian primes in regular domains of $\mathbb{C}$.

EXERCISE 12. Prove that there are infinitely many pairs of primes p, q such that $pq = a^2 + b^2$ with $0 < b < (3\log a)^2$.

We recommend W. Duke [**Du4**] for a powerful treatment of a variety of problems by means of Hecke characters.

5.11. Classical automorphic L-functions.

By classical automorphic forms we mean the forms on $GL(2)$, either holomorphic or eigenfunctions of the Laplace operator (the Maass forms).

Let f be a primitive holomorphic cusp form, of weight $k \geqslant 1$, level q, with nebentypus χ (see Chapter 14 for definitions). Let

$$f(z) = \sum_{n\geqslant 1} \lambda_f(n) n^{(k-1)/2} e(nz)$$

be its normalized Fourier expansion at the cusp ∞. Then

$$L(f,s) = \sum_n \lambda_f(n) n^{-s} = \prod_p (1 - \lambda_f(p)p^{-s} + \chi(p)p^{-2s})^{-1}$$

is an L-function of degree 2 with conductor q and gamma factor given by

$$\gamma(f,s) = \pi^{-s}\Gamma\Big(\frac{s+(k-1)/2}{2}\Big)\Gamma\Big(\frac{s+(k+1)/2}{2}\Big) = c_k(2\pi)^{-s}\Gamma\Big(s+\frac{k-1}{2}\Big) \tag{5.86}$$

with $c_k = 2^{(3-k)/2}\sqrt{\pi}$ by the Legendre duplication formula. Therefore the analytic conductor is

$$\begin{aligned}\mathfrak{q}(f,s) &= q(|s+\tfrac{k-1}{2}|+3)(|s+\tfrac{k+1}{2}|+3) \leqslant q(|s|+|k|+3)^2,\\ \mathfrak{q}(f) &= q(\tfrac{k-1}{2}+3)(\tfrac{k+1}{2}+3) \asymp qk^2.\end{aligned}$$

The root number is $i^k\bar{\eta}(f)$ where $\eta(f)$ satisfies $Wf = \eta\bar{f}$. It is known that $L(f,s)$ satisfies the Ramanujan-Petersson conjecture by work of Deligne [**De1**] for $k \geqslant 2$ and Deligne – Serre [**DeSe**] for $k = 1$. The dual form $\bar{f}$ satisfies

$$\overline{\lambda_f}(n) = \bar{\chi}(n)\lambda_f(n), \quad \text{if} \quad (n,q) = 1.$$

Have in mind then f can be self-dual for non-trivial nebentypus. For example, if ξ is a class group character of an imaginary quadratic field $K = \mathbb{Q}(\sqrt{D})$, then the form f with coefficients given by

$$\lambda_f(n) = \sum_{N\mathfrak{a}=n} \xi(n)$$

is self-dual of weight $k = 1$, of level $q = |D|$ and nebentypus $\chi = \chi_D$, the Kronecker symbol. For these facts, see Section 14.7 and Proposition 14.13, and the references there.

Similarly, let φ be a primitive Maass form of level q with nebentypus χ which is an eigenfunction of the Laplace operator with eigenvalue $\lambda = \frac{1}{4} + r^2$, where $r \in \mathbb{R}$ or $ir \in [0, \frac{1}{2}[$. Writing its Fourier expansion at infinity in the form

$$\varphi(z) = \sqrt{y} \sum_{n \neq 0} \rho(n) K_{ir}(2\pi |n| y) e(nx),$$

we associate with φ the L-function

$$L(\varphi, s) = \sum_{n \geqslant 1} \rho(n) n^{-s} = \prod_p (1 - \rho(p) p^{-s} + \chi(p) p^{-2s})^{-1}$$

of conductor q, with gamma factor

$$\gamma(\varphi, s) = \pi^{-s} \Gamma\Big(\frac{s + \delta + ir}{2}\Big) \Gamma\Big(\frac{s + \delta - ir}{2}\Big)$$

where $\delta = 0$ if φ is even and $\delta = 1$ otherwise. Thus the analytic conductor is

$$\mathfrak{q}(\varphi, s) = q(|s + ir| + 3)(|s - ir| + 3) \leqslant q(|s| + |r| + 3)^2,$$
$$\mathfrak{q}(\varphi) = q(|r| + 3)^2 \asymp \lambda q.$$

We have $\bar{\rho}(n) = \bar{\chi}(n)\rho(n)$ if $(n, q) = 1$. See Chapter 15 for the definition of Maass forms. Although L-functions and Hecke operators for Maass forms are not fully presented in this book, the theory is very similar to that of holomorphic forms; see for instance [**BG**], [**Bu**] and [**DFI3**].

In contrast with holomorphic forms, it is not known that the L-functions of primitive Maass cusp forms satisfy the Ramanujan-Petersson conjecture, although there is no doubt because the conjecture fits correctly in the general Langlands functoriality program. The current best known estimate is

$$|\alpha_p|, \ |\beta_p| \leqslant p^{7/64} \tag{5.87}$$

where $1 - \rho(p)p^{-s} + \chi(p)p^{-2s} = (1 - \alpha_p p^{-s})(1 - \beta_p p^{-s})$, and correspondingly $|\mathrm{Re}\, ir| \leqslant \frac{7}{64}$, hence

$$\lambda = \frac{1}{4} + r^2 \geqslant \frac{975}{4096} = 0.238\ldots \tag{5.88}$$

This is due to Kim and Sarnak [**KiS**]. For earlier results see e.g. [**S2**], [**Se8**], [**DuI2**], [**BDHI**], [**LRS**].

The original theory of Rankin [**Ra3**] and Selberg [**S5**], extended by W. Li [**L**], and its adélic version [**JPS**] show that any two cusp forms, holomorphic or non-holomorphic, admit a Rankin-Selberg convolution. The convolution $L(f \otimes g, s)$ has a simple pole at $s = 1$ if $g = \bar{f}$ and is entire otherwise. Assume f of weight k and g of weight $k \leqslant \ell$ (by symmetry). The gamma factor of $f \otimes g$ is given by

$$\gamma(f \otimes g, s) = \pi^{-2s} \Gamma\Big(\frac{s + \frac{\ell - k}{2}}{2}\Big) \Gamma\Big(\frac{s + \frac{\ell + k}{2}}{2}\Big) \Gamma\Big(\frac{s + \frac{\ell - k}{2} + 1}{2}\Big) \Gamma\Big(\frac{s + \frac{\ell + k}{2} - 1}{2}\Big).$$

Let now f have eigenvalue $\frac{1}{4}+r^2$, and parity $\delta=0$ or 1, and g have eigenvalue $\frac{1}{4}+u^2$ and parity $\eta=0$ or 1. Then

$$\gamma(f\otimes g,s)=\pi^{-2s}\Gamma\Big(\frac{s+i(r+u)+\nu}{2}\Big)\Gamma\Big(\frac{s+i(r-u)+\nu}{2}\Big)\\ \Gamma\Big(\frac{s-i(r-u)+\nu}{2}\Big)\Gamma\Big(\frac{s-i(r+u)+\nu}{2}\Big)$$

where $\nu=0,1$ according to whether $\delta=\eta$ or not. Finally assume f is holomorphic of weight k and g is a Maass form with eigenvalue $\frac{1}{4}+r^2$ and parity δ. Then

$$\gamma(f\otimes g,s)=\pi^{-2s}\Gamma\Big(\frac{s+ir+\frac{k-1}{2}}{2}\Big)\Gamma\Big(\frac{s+ir+\frac{k+1}{2}}{2}\Big)\\ \Gamma\Big(\frac{s-ir+\frac{k-1}{2}}{2}\Big)\Gamma\Big(\frac{s-ir+\frac{k+1}{2}}{2}\Big).$$

These factors can be derived by pleasant computations from the facts conveniently gathered in the Appendix to [**RS**].

Although the local factors of $L(f\otimes g,s)$ at primes $p\mid q(f\otimes g)$ cannot be expressed in a simple way from the Hecke eigenvalues $\lambda_f(p)$, $\lambda_g(p)$ or $\rho_f(p)$, $\rho_g(p)$, this is still the case if p divides exactly $[q(f),q(g)]$. In this case one can show that the factor over the ramified places in (5.10), or rather $1/H_p(p^{-s})$, is equal to

$$(1-\alpha_f(p)\alpha_g(p)p^{-s})(1-\alpha_f(p)\beta_g(p)p^{-s})(1-\beta_f(p)\alpha_g(p)p^{-s})(1-\beta_f(p)\beta_g(p)p^{-s}).$$

In other words, the local factor at all primes is given by (5.9). We emphasize again that this is not true in all cases. Note that since $p\mid q(f)q(g)$, one or both of $\alpha_f(p)$, $\beta_f(p)$ (resp. $\alpha_g(p)$, $\beta_g(p)$) is zero .

In terms of Dirichlet series, this implies that

$$L(f\otimes g,s)=L(2s,\chi_f\chi_g)\sum_{n\geqslant 1}\lambda_f(n)\lambda_g(n)n^{-s}$$

if $[q(f),q(g)]$ is squarefree, where f and g can be either holomorphic or Maass forms and χ_f, χ_g are the respective nebentypus. Note that $L(s,\chi_f\chi_g)$ is the Dirichlet series of the character $\chi_f\chi_g$ even if it is non-primitive, so, for instance, if $\chi_f=\chi_g=1$ then $L(2s,\chi_f\chi_g)=\zeta_{q(f)q(g)}(2s)$ is the Riemann zeta function with the Euler factors at $p\mid q(f)q(g)$ removed. In the general case, the local factors at ramified primes can be described using the local Langlands conjecture for $GL(2)$, but this is not very explicit. See Section 5.12 for the relation between $L(f\otimes f,s)$ and the symmetric square L-function.

REMARK. (1) The convolution of a Maass form and an holomorphic cusp form of weight 4 is important in the Phillips-Sarnak theory of spectral deformation [**PS**].

(2) One can also consider convolutions of $GL(1)$ with $GL(2)$. Those are the twists defined in Section 14.8, and the theory is rather elementary. Note, however, that the twisted modular form

$$f\otimes\psi=\sum_n\psi(n)\lambda_f(n)n^{(k-1)/2}e(nz)$$

(in case of holomorphic f) of level qm^2, where ψ is modulo m, may not be primitive, although it is always an eigenfunction of the unramified Hecke operators. In this case the Rankin-Selberg L-function $L(f\otimes\psi,s)$ is the L-function of the primitive form with conductor dividing qm^2 which has the same Hecke eigenvalues as $f \otimes \psi$ at almost all primes (see Exercise 5 of Section 14.7).

As an example, let f be a theta function for a class group character (see Section 14.3) for an imaginary quadratic field $K = \mathbb{Q}(\sqrt{-D})$, and χ_D the Kronecker symbol for this field, so that $f \in S_1(D, \chi_D)$. Then we have $L(f \otimes \chi_D, s) = L(f, s)$.

More generally, one can express the dual L-function $L(\bar{f}, s)$ of a modular form f having nebentypus χ as a twist

$$L(\bar{f}, s) = L(f \otimes \bar{\chi}, s) \tag{5.89}$$

(see (14.48)).

It is quite instructive to fix a modular form and consider its family of twists $f \otimes \chi$ by characters of conductor q, or by real characters of conductor $\leqslant Q$.

For ease of reading we restate some of the general results from previous sections now in the specific context of classical automorphic L-function.

THEOREM 5.37. *Let f be a holomorphic primitive cusp form of level q and weight $k \geqslant 1$ as above. We have*

$$L(f, s) \ll (\sqrt{q}(|s| + k))^{1-\sigma+\varepsilon}, \tag{5.90}$$

for $\frac{1}{2} \leqslant \sigma \leqslant 1$.

Let φ be a primitive Maass cusp form of level q with the Laplace eigenvalue $\lambda = \frac{1}{4} + r^2$. We have

$$L(\varphi, s) \ll (\sqrt{q}(|s| + |r|))^{1-\sigma+\varepsilon} \tag{5.91}$$

for $\frac{1}{2} \leqslant \sigma \leqslant 1$. In both cases, the implied constant depends only on ε.

PROOF. The inequality (5.90) is merely the re-formulation of (5.20) with the claimed uniformity, because $L(f, s)$ satisfies the Ramanujan-Petersson conjecture. To prove (5.91), one uses instead the estimate

$$\sum_{n \leqslant x} |\rho(n)|^2 \ll x(x + q + |r|)^{\varepsilon} \tag{5.92}$$

proved by Iwaniec [**I7**], hence again the implied constant in (5.91) depends only on ε, not on φ. □

Next (5.33) gives

THEOREM 5.38. *Let f be a holomorphic primitive cusp form of level q and weight $k \geqslant 1$, $N(T, f)$ the number of zeros $\rho = \beta + i\gamma$ of $L(f, s)$ in the critical strip $0 \leqslant \beta \leqslant 1$ with $|\gamma| \leqslant T$. We have*

$$N(T, f) = \frac{T}{\pi} \log \frac{qT^2}{(2\pi e)^2} + O(\log q(|T| + k)),$$

for $T \geqslant 2$ with an absolute implied constant.

Let φ be a primitive Maass form level q with eigenvalue $\lambda = \frac{1}{4} + r^2$, $N(T, \varphi)$ the number of zeros $\rho = \beta + i\gamma$ of $L(f, s)$ in the critical strip $0 \leqslant \beta \leqslant 1$ with $|\gamma| \leqslant T$. We have

$$N(T, \varphi) = \frac{T}{\pi} \log \frac{qT^2}{(2\pi e)^2} + O(\log q(T + |r|))$$

for $T \geqslant 2$ with an absolute implied constant.

Now we re-formulate our zero-free region theorems. The first result for modular forms beyond the classical context of Dirichlet characters (and their extensions to number fields) is due to C. Moreno [**Mor1**]. We can apply Theorem 5.10 for both holomorphic and non-holomorphic forms, getting the following zero-free region:

THEOREM 5.39. *Let f be a holomorphic primitive cusp form of level q and weight $k \geqslant 1$. There exists an absolute constant $c > 0$ such that $L(f, s)$ has no zero in the region*

$$\sigma \geqslant 1 - \frac{c}{\log q(|t| + k + 3)}$$

except possibly a one simple real zero $\beta < 1$ in which case f is self-dual.

Let φ be a primitive Maass form of level q and eigenvalue $\lambda = \frac{1}{4} + r^2$. There exists an absolute constant $c > 0$ such that $L(\varphi, s)$ has no zero in the region

$$\sigma \geqslant 1 - \frac{c}{\log q(|t| + |r| + 3)},$$

except possibly a one simple real zero $\beta < 1$ in which case f is self-dual.

Finally we state the prime number theorem for classical cusp forms.

THEOREM 5.40. *Let f be a holomorphic primitive cusp form of level q and weight $k \geqslant 1$. We have*

$$\sum_{p \leqslant x} \lambda_f(p) \log p = -\frac{x^\beta}{\beta} + O(\sqrt{q}x \exp(-c\sqrt{\log x}))$$

for $x \geqslant 2$, where $c > 0$ is some absolute constant and β is the exceptional zero of $L(f, s)$ if it exists and this term is omitted otherwise.

Let φ be a primitive Maass cusp form of level q and eigenvalue $\lambda = \frac{1}{4} + r^2$. We have

$$\sum_{p \leqslant x} \rho(p) \log p = -\frac{x^\beta}{\beta} + O(\sqrt{q}x \exp(-c\sqrt{\log x}))$$

for $x \geqslant 2$, where $c > 0$ is some absolute constant and β is the exceptional zero of $L(\varphi, s)$ if it exists and this term is omitted otherwise.

PROOF. Apply Theorem 5.13 (see the remark following) and (5.92) to verify (5.48). □

Classical cusp forms give examples of families of L-functions where many results known otherwise only under GRH can be proved unconditionally. For instance, the upper bound of Proposition 5.21 can be improved significantly.

Let q be a prime number and $S_2(q)^*$ the set of primitive weight 2 cusp forms of level q (see Chapter 14). We have $|S_2(q)^*| \sim \frac{q}{12}$. Define

$$L(J_0(q), s) = \prod_{f \in S_2(q)^*} L(f, s). \tag{5.93}$$

This is an L-function of degree $d = |S_2(q)^*|$ with $\mathfrak{q}(f) = (12q)^{|S_2(q)^*|}$, hence $d \sim \frac{q}{6}$ and $\log \mathfrak{q}(f) \sim \frac{q}{12} \log q$. Therefore it follows from Proposition 5.21 and GRH that the order of vanishing of $L(J_0(q), s)$ at $s = \frac{1}{2}$ is $\ll q(\log q)(\log\log q)^{-1}$. However, applying the Petersson formula (14.60) one can easily show (still using GRH) that

$$\operatorname*{ord}_{s=1/2} L(J_0(q), s) \ll q. \tag{5.94}$$

In Chapter 26, we show that this bound is sharp. Actually the bound (5.94) is established unconditionally in [**KM2**] (without recourse to the Grand Riemann Hypothesis). See Section 5.13 for links with Hasse-Weil zeta functions of abelian varieties.

5.12. General automorphic L-functions.

In this section we talk briefly of L-functions associated with cusp forms of higher rank, because they are about to make a permanent habitat in analytic number theory in years to come. Let f be a cusp form on $GL(m)/\mathbb{Q}$ with $m \geqslant 1$. Then Godement – Jacquet [**GJ**] have defined an L-function $L(f, s)$ of degree m associated to f. For $m = 1$, f corresponds uniquely to a primitive Dirichlet character χ modulo q and $L(f, s) = L(s, \chi)$. For $m = 2$, f corresponds either to a primitive cusp form, or to a primitive Maass cusp form. Hence the automorphic L-functions $L(f, s)$ generalize both Section 5.9 and Section 5.11. Since f is cuspidal, $L(f, s)$ is entire except for $L(f, s) = \zeta(s)$ with $m = 1$. In addition L-functions of non-cuspidal automorphic representations are also defined, but they factor as products of cuspidal ones. We recommend Cogdell's Chapter 9 of [**BG**] as a survey of L-functions on $GL(m)$.

It is expected that any automorphic L-function satisfies the Ramanujan-Petersson conjecture. Jacquet and Shalika proved that the local components satisfy $|\alpha_i(p)| < \sqrt{p}$ and $\mathrm{Re}\,(\kappa_j) > -\frac{1}{2}$. This has been improved by Luo, Rudnick and Sarnak [**LRS**], namely $|\alpha_i(p)| < p^c$ and $\mathrm{Re}\,(\kappa_j) > -c$ with

$$c = \frac{1}{2} - \frac{1}{m^2+1}. \tag{5.95}$$

The absolute convergence of $L(f, s)$ for $\mathrm{Re}\,(s) > 1$, due to Jacquet-Shalika and Shahidi [**JS**], follows from the existence of Rankin-Selberg L-functions (see the proof of (5.48)).

A useful feature of the association of L-functions with automorphic theory is that the root number $\varepsilon(f)$ is expressed as a product of local root numbers

$$\varepsilon(f) = \prod_p \varepsilon_p(f), \text{ with } \varepsilon_p(f) = 1 \text{ if } p \text{ is unramified.}$$

In fact, $\varepsilon_p(f)$ depends on the choice of a non-trivial additive character ψ_p of $\mathbb{Q}_p$, but the product is independent of such choices.

For any two cusp forms f and g on $GL(d)$ and $GL(e)$, the L-function $L(f \otimes g, s)$ exists in our sense by the far-reaching generalization of the classical Rankin-Selberg method due to Jacquet, Piatetski-Shapiro and Shalika [**JPS**]. Moeglin and Waldspurger [**MW**] proved that $L(f \otimes g, s)$ is entire if $g \neq \bar{f}$, and has a simple pole at $s = 1$ if $g = \bar{f}$. The bound $q(f \otimes g) \mid q(f)^e q(g)^d$ is due to Bushnell and Henniart [**BH**].

According to conjectures belonging to the Langlands Program, the "most general" L-function should in fact be a product of functions belonging to the class of L-functions of cusp forms on some $GL(m)/\mathbb{Q}$. Other parts of the Langlands conjectures imply that the Ramanujan-Petersson conjecture should hold for any automorphic L-function. Moreover, the conjectures imply the existence (and determination, in principle) of an equidistribution law for the coefficients $\lambda_f(p)$ (see the discussion of the Sato-Tate Conjecture in Chapter 21).

The convexity bound for automorphic L-functions can be made uniform.

THEOREM 5.41. *Let $L(f,s)$ be an automorphic L-function of a cusp form of degree d other than $\zeta(s)$. Then we have*

$$L(f,s) \ll \mathfrak{q}(f,s)^{\alpha+\varepsilon} \tag{5.96}$$

for $\sigma \geqslant \frac{1}{2}$, where $\alpha = \max(\frac{1}{2}(1-\sigma),0)$, the implied constant depending only on ε and d.

PROOF. This follows from (5.12) with $X=1$, Proposition 5.4 and the estimate

$$\sum_{n\leqslant x} |\lambda_f(n)|^2 \ll x(x\mathfrak{q}(f))^\varepsilon$$

for $x \geqslant 1$ with an implied constant depending only on ε and d. This is the analogue of (5.92) proved by Molteni [**Mol**]. □

The most commonly used L-functions of degree four in analytic number theory are the Rankin-Selberg convolutions $L(f\otimes g,s)$ of two classical cusp forms (described in Section 5.11). Close relatives to this are the symmetric square and adjoint square L-functions of classical cusp forms. Let f have nebentypus χ. The symmetric square of f (f is either holomorphic or real-analytic) is defined by

$$L(\mathrm{Sym}^2 f,s) = L(f\otimes f,s)L(s,\chi)^{-1} \tag{5.97}$$

whereas the adjoint square is defined by

$$L(\mathrm{Ad}^2 f,s) = L(f\otimes \bar{f},s)\zeta(s)^{-1}. \tag{5.98}$$

Thus both are L-functions of degree three, with conductors $q(\mathrm{Sym}^2 f) = q(f\otimes f)q(\chi)^{-1} \mid q(f)^2$ and $q(\mathrm{Ad}^2 f) = q(f\otimes\bar{f}) \mid q(f)^2$. If f has real coefficient, then the symmetric and adjoint squares coincide. One shows that the root number for the adjoint square is $\varepsilon(\mathrm{Ad}^2 f) = \varepsilon(f\otimes\bar{f}) = 1$. Both have gamma factor equal to $\gamma(f\otimes\bar{f},s)/\Gamma(s/2)$, hence

$$\gamma(\mathrm{Ad}^2 f,s) = \gamma(\mathrm{Sym}^2 f,s) = \pi^{-3s/2}\Gamma\Big(\frac{s+1}{2}\Big)\Gamma\Big(\frac{s+k-1}{2}\Big)\Gamma\Big(\frac{s+k}{2}\Big) \tag{5.99}$$

if f is holomorphic of weight k and

$$\gamma(\mathrm{Ad}^2 f,s) = \gamma(\mathrm{Sym}^2 f,s) = \pi^{-3s/2}\Gamma\Big(\frac{s}{2}\Big)\Gamma\Big(\frac{s}{2}+ir\Big)\Gamma\Big(\frac{s}{2}+ir\Big) \tag{5.100}$$

if f is real-analytic Maass form with eigenvalue $\lambda = \frac{1}{4}+r^2$ (note the parity does not matter). The local Euler factors for $L(\mathrm{Sym}^2 f,s)$ and $L(\mathrm{Ad}^2 f,s)$ are given by

$$(1-\alpha(p)^2p^{-s})^{-1}(1-\chi(p)p^{-s})^{-1}(1-\beta(p)^2p^{-s})^{-1},$$
$$(1-\tfrac{\alpha(p)}{\beta(p)}p^{-s})^{-1}(1-p^{-s})^{-1}(1-\tfrac{\beta(p)}{\alpha(p)}p^{-s})^{-1}$$

respectively if $p\nmid q(f)$, where $\alpha(p)$ and $\beta(p)$ are the local roots at p for $L(f,s)$. Note that from the twist formula (5.89) or the relation $\alpha(p)\beta(p)=\chi(p)$ for p unramified, one can also relate the adjoint and symmetric squares as twists (convolution of $GL(3)$ by $GL(1)$):

$$L_p(\mathrm{Ad}^2 f, s) = L_p(\mathrm{Sym}^2 f \otimes \bar{\chi}, s).$$

If p divides exactly $q(f)$, then this formula is still valid. In particular, for $q(f)$ squarefree a simple computation shows that

$$L(\mathrm{Sym}^2 f, s) = L(2s, \chi^2) \sum_{n \geqslant 1} \lambda_f(n^2) n^{-s},$$

$$L(\mathrm{Ad}^2 f, s) = L(2s, \bar{\chi}^2) \sum_{n \geqslant 1} \chi(n) \lambda_f(n^2) n^{-s}.$$

At the point $s = 1$ the adjoint square L-function takes a special value

$$L(\mathrm{Ad}^2 f, 1) = \operatorname*{res}_{s=1} L(f \otimes \bar{f}, s) = \frac{w \langle f, f \rangle}{\mathrm{Vol}\,(\Gamma_0(q) \backslash \mathbb{H})} \tag{5.101}$$

where $w = (4\pi)^k / \Gamma(k)$ or $w = \cosh(\pi r)/2\pi$ depending on the weight or eigenvalue of f respectively, while $\langle f, f \rangle$ is the square of the Petersson norm of f (see (14.11) for holomorphic cusp form, otherwise it is the standard L^2-norm in the Hilbert space of automorphic functions on $\Gamma_0(q) \backslash \mathbb{H}$; see Chapter 15 and also (26.47)). This formula shows, in particular, that $L(\mathrm{Ad}^2 f, 1) > 0$, and it is well exploited in the spectral analysis of cusp forms on $GL(2)$ (see for example [**Roy**]).

Theorem 5.10 gives a zero-free region for automorphic L-functions of any degree.

THEOREM 5.42. *Let $L(f, s)$ be an automorphic L-function of a cusp form of degree d. There exists an absolute constant $c > 0$ such that $L(f, s)$ has no zero in the region*

$$\sigma \geqslant 1 - \frac{c}{d^4 \log \mathfrak{q}(f)(|t| + 3)} \tag{5.102}$$

except possibly a one simple real zero $\beta_f < 1$. For this exceptional zero to exist, it is necessary that f be self-dual.

PROOF. By the properties of Rankin-Selberg convolutions of cusp forms recalled above, all assumptions of Theorem 5.10 are valid, which gives the result. □

REMARK. One can show that for an automorphic L-function $L(f, s)$, the auxiliary L-function $L(g, s)$ used in the proof of Theorem 5.10 has the property that $\Lambda_g(n) \geqslant 0$ for all n, not only unramified ones. The proof of this depends on the local Langlands conjecture for $GL(d)$ over p-adic fields, proved by Harris and Taylor [**HT**]. Roughly speaking, for any prime p, there exist continuous representations $\rho_p : W(\mathbb{Q}_p) \to GL(d, \mathbb{C})$, where W is the Weil-Deligne group of $\mathbb{Q}_p$, a variant of the Galois group, such that the p-factor of $L(f, s)$ is given by

$$L_p(f, s) = \det(1 - \rho_p(\mathrm{Fr}_p) p^{-s})^{-1}$$

and that of the Rankin-Selberg convolutions used in defining g are

$$L_p(f \otimes \bar{f}, s) = \det(1 - (\rho_p \otimes \tilde{\rho}_p)(\mathrm{Fr}_p) p^{-s})^{-1},$$

$$L_p(f \otimes f, s) = \det(1 - (\rho_p \otimes \rho_p)(\mathrm{Fr}_p) p^{-s})^{-1},$$

$$L_p(\bar{f} \otimes \bar{f}, s) = \det(1 - (\tilde{\rho}_p \otimes \tilde{\rho}_p)(\mathrm{Fr}_p) p^{-s})^{-1},$$

where $\tilde{\rho}_p$ is the contragredient representation of ρ_p, and Fr_p is the geometric Frobenius at p (see Tate [**Tat**] for more details); more precisely, Fr_p is meant to act on the invariants under inertia for all the above representations. This shows that, with the same convention, the p-factor of $L(g,s)$ is

$$\det(1-(\rho\otimes\tilde{\rho})(\mathrm{Fr}_p)p^{-s})^{-1}$$

where ρ is the representation

$$\rho = 1\oplus(\rho_p\otimes|\cdot|^{it})\oplus(\tilde{\rho}_p\otimes|\cdot|^{-it}).$$

But we have quite generally

$$\mathrm{Tr}\,(\rho\otimes\tilde{\rho})(g) = |\mathrm{Tr}\,\rho(g)|^2$$

for any g and ρ, and this gives $\Lambda_g(p^k)\geqslant 0$ for all $k\geqslant 1$.

This kind of very deep arithmetical argument depends crucially on the automorphic nature of $L(f,s)$ and is completely beyond reach when using only the bare data of the Dirichlet series or Euler product $L(f,s)$.

Statements like Proposition 5.22 for automorphic L-functions are usually called "multiplicity one theorems". The most basic result, called the strong multiplicity one principle, is due to Jacquet and Shalika [**JS**]:

PROPOSITION 5.43. *Let $L(f,s)$ and $L(g,s)$ be two automorphic L-functions of cusp forms of $GL(m)$. If the local components of f and g coincide at all but finitely many primes, then $f=g$.*

PROOF. Suppose the local components coincide for $p\notin S$, where S is finite (and contains the ramified primes for f or g). Since the local factors for the Rankin-Selberg convolution $L(f\otimes\bar{g},s)$ have no poles at $s=1$, the order of the pole of $L(f\otimes\bar{g},s)$ is the same as that for the product over primes $p\notin S$. But

$$\prod_{p\notin S} L_p(f\otimes\bar{g},s) = \prod_{p\notin S} L_p(f\otimes\bar{f},s)$$

by assumption. Therefore $L(f\otimes\bar{g},s)$ has a pole at $s=1$, which implies $f=g$. □

For classical modular forms, this is Theorem 14.12. If the local components satisfy $|\alpha_i(p)|<p^{1/4}$, then the remark after Proposition 5.22 gives an explicit version. Such a bound is known in full generality only for $n=2$, but using the logarithmic derivative of $L(f\otimes g,s)$ instead of the squarefree part of the convolution, Moreno [**Mor3**] gave a general explicit bound on the number of primes required.

Theorem 5.42 can be applied in particular to the symmetric and adjoint square L-functions. One can also apply it to the Rankin-Selberg L-functions $L(f\otimes g,s)$ for any two classical modular forms since Ramakrishnan [**Ra**] has shown that those L-functions of degree 4 are automorphic. Exercise 4 shows that one can in fact prove non-vanishing results for Rankin-Selberg L-functions without modularity.

THEOREM 5.44. *(1) Let f and g be classical primitive modular forms, either holomorphic or real-analytic. There exists an absolute constant $c>0$ such that $L(f\otimes g,s)$ has no zero in the region*

$$\sigma\geqslant 1-\frac{c}{\log\mathfrak{q}(f\otimes g)(|t|+3)}$$

except possibly a one simple real zero $\beta < 1$. For this exceptional zero to exist, it is necessary that $f \otimes g$ be self-dual.

(2) Let f be a classical primitive modular forms, either holomorphic or real-analytic. There exists an absolute constant $c > 0$ such that $L(Sym^2 f, s)$ and $L(Ad^2 f, s)$ have no zero in the region

$$\sigma \geqslant 1 - \frac{c}{\log \mathfrak{q}(f)(|t|+3)},$$

except possibly a simple real zero $\beta < 1$. If f is non-dihedral, then the exceptional zero does not exist.

PROOF. As mentioned, the first point follows directly from the fact that $L(f \otimes g, s)$ is an automorphic L-function.

For the second, $L(\mathrm{Sym}^2 f, s)$ is automorphic [**GJ**], so we need only prove that there is no exceptional zero if f is non-dihedral, which is due to Goldfeld, Hoffstein and Liehman [**GHL**] (since f is self-dual if there is an exceptional zero, $L(\mathrm{Ad}^2 f, s) = L(\mathrm{Sym}^2 f, s)$). To this end consider the L-function

$$\begin{aligned} L(g, s) &= \zeta(s) L(\mathrm{Sym}^2 f, s)^2 L(\mathrm{Sym}^2 f \otimes \mathrm{Sym}^2 f, s) \\ &= \zeta(s) L(\mathrm{Sym}^2 f, s)^3 L(\mathrm{Sym}^2 \mathrm{Sym}^2 f, s). \end{aligned}$$

The last L-function is a special case of the symmetric square of a cusp form on $GL(3)$ and has been shown by Bump and Ginzburg to have a simple pole at $s = 1$ if f is non-dihedral (this also follows by the first formula since $\mathrm{Sym}^2 f$ is a cusp form on $GL(3)$). Hence $L(g, s)$ has a pole of order 2 at $s = 1$, whereas any real zero of $L(\mathrm{Sym}^2 f, s)$ is a zero of $L(g, s)$ of order $\geqslant 3$. By easy local computations one checks that $\Lambda_g(n) \geqslant 0$ for $(n, q(g)) = 1$, hence the result follows from Lemma 5.9. □

REMARK. The formula (5.101) also proves that $L(\mathrm{Ad}^2 f, 1) \neq 0$.

The following is a useful corollary (compare with the lower bounds for class numbers).

COROLLARY 5.45. *Let f be a classical primitive modular form, holomorphic or real-analytic. We have*

$$\|f\|^2 \gg \mathrm{Vol}\,(\Gamma_0(q)\backslash\mathbb{H}) q^{-\varepsilon} \tag{5.103}$$

for any $\varepsilon > 0$. The implied constant depends only on ε and the weight k or the eigenvalue λ whatever is relevant.

If f is not dihedral, for instance, if f has trivial nebentypus and squarefree conductor q, then

$$\|f\|^2 \gg \frac{\mathrm{Vol}\,(\Gamma_0(q)\backslash\mathbb{H})}{\log 2q} \tag{5.104}$$

where the implied constant is effective and depends only on the weight k or the eigenvalue λ whatever is relevant.

PROOF. The first part is proved by Hoffstein and Lockhart [**HL**], and follows from (5.101) the bound $\beta \leqslant 1 - c(\varepsilon)q^{-\varepsilon}$ for the possible exceptional zero of $L(\mathrm{Ad}^2 f, s)$, which implies a lower bound for $L(\mathrm{Ad}^2 f, 1)$. The proof is similar to that of Siegel's bound (5.73), and the implied constant is non-effective.

If f is not dihedral, then there is no exceptional zero and one derives (5.104) by the same method from the zero-free region of Theorem 5.44. Finally, one can easily see that a dihedral form with trivial nebentypus has a non-squarefree level.□

Generally speaking the problem of exceptional zeros for the L-functions of cusp forms on $GL(m)$ with $m \geqslant 2$ turns out to be accessible, the hardest and stubborn case being with $m = 1$ for real Dirichlet characters. The above result of Goldfeld, Hoffstein and Liehman has been extended to the following theorem of Banks [**Ba**]:

PROPOSITION 5.46. *If $L(f, s)$ is the L-function of any cusp form on $GL(3)$, then the exceptional zero does not exist for $L(f, s)$.*

5.13. Artin L-functions.

In this Section we survey, mostly without proofs, some of the definitions and properties of Artin L-functions. In contrast with the previous sections, the fact that those are indeed L-functions as defined in Section 5.1 remains open. However, there is great confidence and evidence that the relevant conjectures are correct, and together with the analytic theory developed in this chapter, and possibly GRH, this provides heuristic evidence for results that sometimes are accessible by other methods. See e.g. [**CF**] for more about Artin L-functions. To avoid confusion in terminology, we will sometimes refer to the L-functions defined in Section 5.1 as analytic L-functions.

Artin L-functions are associated to a continuous finite dimensional Galois representation $\rho : \mathrm{Gal}(\bar{\mathbb{Q}}/K) \to GL(d, \mathbb{C})$ where $K/\mathbb{Q}$ is a number field. Define the Artin L-function of ρ by the following Euler product over prime ideals of K:

$$L(\rho, s) = \prod_{\mathfrak{p}} \det(1 - \rho(\mathrm{Fr}_{\mathfrak{p}})N\mathfrak{p}^{-s})^{-1},$$

where $\mathrm{Fr}_{\mathfrak{p}}$ is the Frobenius conjugacy class at $\mathfrak{p}$, acting on the invariants under the inertia group at $\mathfrak{p}$. Thus $L(\rho, s)$ is an Euler product of degree $\deg(\rho)$ over K, or $\deg(\rho)[K : \mathbb{Q}]$ over $\mathbb{Q}$.

Artin L-functions enjoy a formalism parallel to that of representations. For instance, the dual L-function is also the L-function of the contragredient representation $\tilde{\rho}$, and the Rankin-Selberg L-function of ρ_1 and ρ_2 is the L-function $L(\rho_1 \otimes \rho_2, s)$ of the tensor power $\rho_1 \otimes \rho_2$ (thus the notation is compatible). Moreover, $L(\rho_1 \oplus \rho_2, s) = L(\rho_1, s)L(\rho_2, s)$, and if ρ is induced from a representation ρ' from a finite index subgroup corresponding to a finite extension K'/K, then

$$L(\rho, s) = L(\rho', s). \tag{5.105}$$

In particular, taking $K = \mathbb{Q}$ and $K' = K$ any number field, it follows that any Artin L-function of degree d can be seen as one defined over $\mathbb{Q}$ of degree $\deg(\rho)[K : \mathbb{Q}]$.

For the trivial representation of $\mathrm{Gal}(\bar{\mathbb{Q}}/K)$, we get $L(1, s) = \zeta_K(s)$. More generally if $\deg(\rho) = 1$, class-field theory shows that to ρ is associated a unique Hecke character ξ such that $L(\rho, s) = L(\xi, s)$. So it is an analytic L-function in this case (see Section 5.10).

It is conjectured that for any ρ, $L(\rho, s)$ is an analytic L-function. Since $L(\rho_1 \oplus \rho_2, s) = L(\rho_1, s)L(\rho_2, s)$, one can restrict attention to irreducible representations ρ, and one expects that $L(\rho, s)$ is in fact automorphic.

It is known that $L(\rho, s)$ has meromorphic continuation to $\mathbb{C}$ and all of the conditions set in Section 5.1 are satisfied, except that some of these L-functions may have poles in the critical strip (because we still do not know if $L(\rho, s)$ is holomorphic in the strip $0 < \mathrm{Re}\,(s) < 1$). The gamma factor can be described as a product over infinite places v of K of local gamma factors $\gamma_v(\rho, s)$. Let σ_v be the Frobenius conjugacy class, which is of order 2 if v is a real place that extends to two complex places of L, and is $= 1$ otherwise. Then

$$\gamma_v(\rho, s) = \begin{cases} \pi^{-ds/2}\Gamma\Big(\dfrac{s}{2}\Big)^d \Gamma\Big(\dfrac{s+1}{2}\Big)^d & \text{if } v \text{ is a complex place,} \\ \pi^{-ds/2}\Gamma\Big(\dfrac{s}{2}\Big)^{d^+} \Gamma\Big(\dfrac{s+1}{2}\Big)^{d^-} & \text{if } v \text{ is a real place,} \end{cases}$$

where $d = \deg(\rho)$ is the dimension of ρ, and d^-, d^+ are the multiplicities, respectively, of the eigenvalue $+1, -1$ for $\rho(\sigma_v)$.

The conductor $q(\rho)$ is of the form

$$q(\rho) = |d_K|^{\deg(\rho)} N_{K/\mathbb{Q}}\mathfrak{f}(\rho)$$

for some integral ideal $\mathfrak{f}(\rho)$ of K, which is called the Artin conductor, defined using ramification groups (see Serre [**Se4**]). The root number $\varepsilon(\rho)$ can be in a sense defined from the functional equation, but there is also an a priori definition due to Dwork, Langlands and Deligne as a product of local factors. Thus the analytic conductor satisfies

$$\mathfrak{q}(\rho, s) \leqslant q(\rho)(|s| + 4)^{[K:\mathbb{Q}]\deg(\rho)} = (|d_K|(|s| + 4)^{[K:\mathbb{Q}]})^{\deg(\rho)} N\mathfrak{f}(\rho).$$

From the Brauer induction theorem (see e.g. [**Se5**]), we can write

$$\mathrm{Tr}\,\rho = \sum n_i \mathrm{Tr}\,\pi_i$$

where $n_i \in \mathbb{Z}$ and π_i is a representation induced by an abelian character ξ_i of a finite (in fact, cyclic) extension $K_i/\mathbb{Q}$. By the invariance under induction (5.105) we have

$$L(\rho, s) = \prod_i L(\pi_i, s)^{n_i} = \prod_i L(\xi_i, s)^{n_i} \tag{5.106}$$

and the case of Hecke characters (over number fields) implies that $L(\rho, s)$ is meromorphic with poles only in the critical strip, and satisfies a functional equation relating s and $1 - s$. That this functional equation is the right one involving the gamma factor above and the Artin conductor follow from the formalism of both, especially the behavior under direct sums and induction.

To show that $L(\rho, s)$ does not have extra poles, we need to prove the

ARTIN CONJECTURE. *Let ρ be an irreducible, non-trivial, Galois representation of a number field $K/\mathbb{Q}$. Then $L(\rho, s)$ is entire.*

Even the case $d = 2$ and $K = \mathbb{Q}$ is still not completely settled. More precisely, if an irreducible representation $\rho : \mathrm{Gal}(\bar{\mathbb{Q}}/\mathbb{Q}) \to GL(2, \mathbb{C})$ is given, then its image in $PGL(2, \mathbb{C})$ can be classified as either a dihedral group, the alternating group A_4, the symmetric group S_4 or the alternating group A_5. In all cases, except A_5, it is known that $L(\rho, s)$ is holomorphic (even automorphic), and there are partial results in the A_5 case. The dihedral case is classical, the A_4 case is due to Langlands, the S_4 case to Tunnell. The latter is crucial in Wiles's proof of Fermat's Great Theorem.

In a certain sense, the Artin conjecture is true on average: for instance, if $K/\mathbb{Q}$ is a fixed finite Galois extension, then

$$\zeta_K(s) = \zeta(s) \prod_{\rho \neq 1} L(\rho, s) \tag{5.107}$$

where ρ runs over the non-trivial irreducible representations of $\mathrm{Gal}(K/\mathbb{Q})$ (see (5.82)). Since both $\zeta(s)$ and $\zeta_K(s)$ are known to have only a simple pole at $s = 1$, the possible poles of $L(\rho, s)$ must be compensated by zeros for other representations.

The following is also useful.

COROLLARY 5.47. *Let ρ be a non-trivial irreducible Galois representation of $K/\mathbb{Q}$. Then $L(\rho, s)$ has neither poles nor zeros on the line* $\mathrm{Re}\,(s) = 1$.

PROOF. This follows by (5.106) because the L-functions of non-trivial Hecke Grossencharakters are entire and do not vanish on the line $\mathrm{Re}\,(s) = 1$ (Theorem 5.35). □

Assuming the Artin conjecture, one can deduce quite interesting arithmetic corollaries using the properties of analytic L-functions.

CHEBOTAREV DENSITY THEOREM. *Let L/K be a finite Galois extension of degree d of number fields. Let $C \subset \mathrm{Gal}(L/K)$ be a union of conjugacy classes and*

$$\psi(x, C) = \sum_{\substack{N\mathfrak{p} \leqslant x \\ \mathrm{Fr}_{\mathfrak{p}} \in C}} \log N\mathfrak{p}.$$

We have as $x \to +\infty$,

$$\psi(x, C) \sim \frac{|C| x}{|\mathrm{Gal}(L/K)|}. \tag{5.108}$$

If GRH holds for Artin L-functions, then for $x \geqslant 2$,

$$\psi(x, C) = \frac{|C| x}{|\mathrm{Gal}(L/K)|} + O\Big(\sqrt{x} (\log x) \sum_{\rho} |c_\rho| \log\big(x^{\deg(\rho)[K:\mathbb{Q}]} q(\rho)\big)\Big) \tag{5.109}$$

where ρ runs over the irreducible representations of $\mathrm{Gal}(L/K)$, *and*

$$c_\rho = \frac{1}{|\mathrm{Gal}(L/K)|} \sum_{x \in C} \overline{Tr\, \rho(x)}. \tag{5.110}$$

The implied constant is absolute.

PROOF. Although the proof below depends on the Artin conjecture, (5.108) is known unconditionally and (5.109) does not need the Artin conjecture (see e.g. [**Se6**]).

Let $G = \mathrm{Gal}(L/K)$. Functions of the form $\mathrm{Tr}\,\rho$, for ρ irreducible, form an orthonormal basis of the vector space of conjugacy-invariant functions on the finite group G with the scalar product

$$\langle f, g\rangle = \frac{1}{|G|}\sum_x f(x)\bar{g}(x).$$

Hence the characteristic function δ_C of $C \subset G$ is given by

$$\delta_C(x) = \sum_\rho c_\rho \mathrm{Tr}\,\rho(x)$$

where c_ρ is as in (5.110). For the trivial representation, we have $c_1 = |C|/|G|$, and thus (5.108) follows from the Artin Conjecture and the Prime Number Theorem 5.13 for $L(\rho, s)$. Similarly (5.109) comes from (5.56) and (5.110). □

The estimate (5.109) can be made explicit in various ways. Sometimes one can compute c_ρ explicitly. Otherwise, a simple general bound on GRH is given for instance by

$$\psi(x, C) = \frac{|C|x}{|\mathrm{Gal}(L/K)|} + O\Big(\sqrt{x}(\log x)(\sqrt{|C|}(\log x^{[K:\mathbb{Q}]}) + \log|d_L|)\Big).$$

To see this, notice that by orthogonality we have

$$\sum_\rho |c_\rho|^2 = \|\delta_C\|_2^2 = \frac{|C|}{|\mathrm{Gal}(L/K)|},$$

and it is well known that

$$\sum_\rho \deg(\rho)^2 = |\mathrm{Gal}(L/K)|.$$

Hence by Cauchy's inequality we derive

$$\begin{aligned}\sum_\rho |c_\rho| \log(x^{\deg(\rho)[K:\mathbb{Q}]}) &\leqslant [K:\mathbb{Q}](\log x)\Big(\frac{|C|}{|\mathrm{Gal}(L/K)|}\Big)^{\frac{1}{2}}|\mathrm{Gal}(L/K)|^{\frac{1}{2}}\\ &= [K:\mathbb{Q}](\log x)\sqrt{|C|}.\end{aligned}$$

On the other hand, since $|\mathrm{Tr}\,(\rho(x))| \leqslant \deg(\rho)$, we have $|c_\rho| \leqslant \deg(\rho)$, hence

$$\sum_\rho |c_\rho| \log q(\rho) \leqslant \log \prod_\rho q(\rho)^{\deg(\rho)} = \log|d_L|$$

by comparison of the conductors in the identity of L-functions

$$\zeta_L(s) = \prod_\rho L(\rho, s)^{\deg(\rho)}.$$

For other estimates and arrangements of the error term on GRH, and also for unconditional estimates with explicit error terms; see [**Se6**] and [**LO**].

Here is a sample application of the Chebotarev Density Theorem.

PROPOSITION 5.48. *Let $n \geqslant 2$ and $N \geqslant 1$. Let D_N be the set of monic polynomials $f \in \mathbb{Z}[X]$ of degree n whose coefficients have absolute value $\leqslant N$. Then we have*

$$\frac{1}{xN^n}\sum_{f\in D_N}\sum_{\substack{p\leqslant x\\ f \text{ has a root mod } p}} \log p = e_n + o(1)$$

as $x \to \infty$ and $N \to \infty$, where $e_n = 1 - \frac{1}{2} + \cdots + (-1)^{n-1}\frac{1}{n!}$ (note that $e_n \to 1 - \frac{1}{e}$ as $n \to \infty$).

PROOF. We have $|D_N| = N^n$. Gallagher ([**Ga4**], see Exercise 3 of Chapter 7) has shown that the set C_N of elements $f \in D_N$ such that either f is reducible or the Galois group of the equation $f(x) = 0$ is not the full symmetric group S_n satisfies

$$|C_N| \ll N^{n-\frac{1}{2}} \log N.$$

By trivial estimate, elements in C_N do not contribute to the limit above.

Let $f \notin C_N$. Choosing a root α of f in $\mathbb{C}$, let $L = \mathbb{Q}(\alpha)$ and K the Galois closure of L. We have $\mathrm{Gal}(K/\mathbb{Q}) = S_n$ and $H = \mathrm{Gal}(K/L) \simeq S_{n-1}$. By Galois theory, we see that

$$\{p \mid f \text{ has a root modulo } p\} = \{p \mid \mathrm{Fr}_p \in \bigcup_{\sigma\in S_n} \sigma^{-1}H\sigma\} = \{p \mid \mathrm{Fr}_p \in \bar{H}\}$$

where $\bar{H}$ is the set of elements in S_n fixing at least one element. Precisely, this holds up to finitely many primes at which the localizations of $\mathbb{Z}[X]/(f)$ and the ring of integers of L differ. By the Chebotarev Density Theorem we have

$$\sum_{\substack{p\leqslant x\\ f \text{ has a root mod } p}} \log p = \frac{|\bar{H}|}{n!}x + o(x)$$

By exclusion-inclusion, it follows that $|\bar{H}|/n! = e_n$, hence the result. □

See e.g. [**Se6**] for other applications of the Chebotarev Density Theorem.

5.14. L-functions of varieties.

The situation with L-functions of algebraic varieties over $\mathbb{Q}$ is even less understood than that for Galois representations. Let us call them "arithmetic geometry" L-functions, or just "geometric" L-functions. We will not discuss things in general, but mention the important cases of algebraic curves and abelian varieties. The only satisfactory theory exists for the intersection of these two cases, namely the abelian varieties of dimension one (that is the elliptic curves).

Let $E/\mathbb{Q}$ be an elliptic curve. Modularity of E, proved by Wiles, Taylor-Wiles and Breuil-Conrad-Diamond-Taylor [**W**], [**TW**], [**BCDT**], means that the Hasse-Weil zeta function of E, normalized to have critical strip $0 \leqslant \mathrm{Re}\,(s) \leqslant 1$, corresponds to a modular form of weight 2 and level equal to the conductor of E, hence is an L-function according to Section 5.11. See Section 14.4 for a more detailed statement, as well as a description of the root number in most cases.

Conjecturally, the "modularity" extends to the Hasse-Weil zeta functions of arbitrary smooth projective curves over number fields. Let $C/\mathbb{Q}$ be such a curve. Then for all primes $p \nmid N$, where $N \geqslant 1$ is some integer, the curve can be reduced modulo p to a smooth projective curve $C_p/\mathbb{F}_p$. As described in Exercise 2 of Chapter 11, the local zeta function

$$Z(C_p, T) = \exp\Big(\sum_{k\geqslant 1} \frac{|C_p(\mathbb{F}_{p^k})|T^k}{k}\Big)$$

is a rational function of the form

$$Z(C_p, T) = \frac{P_p(T)}{(1-T)(1-pT)}$$

for some monic polynomial P_p with $P_p(0) = 1$, of degree $2g$ where $g \geqslant 0$ is the genus of C. Write

$$P_p(T) = \prod_{1\leqslant j\leqslant 2g} (1 - \alpha_j(p)p^{\frac{1}{2}}T).$$

Then the Riemann Hypothesis for curves over finite fields (see Chapter 11) shows that $|\alpha_j(p)| = 1$. The Hasse-Weil zeta function of C outside N, analytically normalized, is defined by the Euler product

$$L_N(C, s) = \prod_{p\nmid N} (1 - \alpha_1(p)p^{-s})^{-1} \cdots (1 - \alpha_{2g}(p)p^{-s})^{-1}$$

It is conjectured that after multiplying by appropriately defined Euler factors at $p \mid N$, the complete L-function is entire, it is necessarily self-dual and the gamma factor should be

$$\gamma(C, s) = \pi^{-gs}\Gamma\big(\frac{s}{2} + \frac{1}{4}\big)^g\Gamma\big(\frac{s}{2} + \frac{3}{4}\big)^g = c_g(2\pi)^{-gs}\Gamma(s + \tfrac{1}{2})^g$$

for some constant $c_g > 0$ (see (5.86)). Note that the Riemann Hypothesis for the curves C_p over finite fields means that this L-function satisfies the Ramanujan-Petersson conjecture (up to the ramified places, but it is known that those still satisfy the required condition). In general for any $g > 1$, this is not yet known except for very special cases.

REMARK. In arithmetic geometry it is more practical to leave the local roots $\alpha_j(p)$ as they occur naturally without normalization, i.e. to write

$$P_p(T) = \prod_{1\leqslant j\leqslant 2g} (1 - \beta_j(p)T)$$

with $|\beta_j(p)| = \sqrt{p}$. Thus

$$L(C, s) = \prod_{p\nmid N} P_p(p^{-s+1/2})^{-1}.$$

In this notation the critical line is $\mathrm{Re}\,(s) = 1$ and the functional equation relates the values at s and $2 - s$. The same remark refers to abelian varieties.

Abelian varieties are higher-dimensional generalizations of elliptic curves, seen as algebraic groups. An abelian variety $A/\mathbb{Q}$ is a smooth projective variety with a given rational point $0 \in A(\mathbb{Q})$ and addition and inverse morphisms $p : A \times A \to A$, $i : A \to A$ which are maps of algebraic varieties over $\mathbb{Q}$ and satisfy the axioms of a commutative group (with i giving the opposite). If the dimension of A is one, this corresponds exactly to elliptic curves $E/\mathbb{Q}$ with the "usual" chord-and-tangent description of the group law (see [**Sil**] and Section 11.8).

There exists an integer $N \geqslant 1$ such that for $p \nmid N$, the reduced variety $A_p/\mathbb{F}_p$ is smooth. Weil showed that the local zeta function

$$Z(A_p, T) = \exp\Big(\sum_{k\geqslant 1} \frac{|A_p(\mathbb{F}_{p^k})|T^k}{k}\Big)$$

is a rational function of the form

$$Z(A_p, T) = \frac{P_{p,1}(T)\cdots P_{p,2g-1}(T)}{P_{p,0}(T)\cdots P_{p,2g}(T)}$$

where $P_{p,j}$ are polynomials. Weil proved that $P_{p,1}$ is of degree $2g$ and of the form

$$P_{p,1}(T) = \prod_{1\leqslant i\leqslant 2g} (1 - \alpha_i(p)p^{\frac{1}{2}}T)$$

with $|\alpha_i(p)| = 1$, which is the local Riemann Hypothesis. In addition one can arrange the $\alpha_i(p)$ such that $\alpha_i(p)\alpha_{i+g}(p) = 1$ for $1 \leqslant i \leqslant g$. The first polynomial $P_{p,1}$ determines $P_{p,j}$ as the j-th exterior power

$$P_{p,j}(T) = \prod_{i_1<\ldots<i_j} (1 - \alpha_{i_1}(p)\cdots\alpha_{i_j}(p)p^{\frac{j}{2}}T).$$

In particular, $P_{p,0} = 1 - T$, $P_{p,2g} = 1 - p^g T$. The global zeta function of $A/\mathbb{Q}$ outside N is

$$L_N(A, s) = \prod_{p\nmid N} (1 - \alpha_1(p)p^{-s})^{-1}\cdots(1 - \alpha_{2g}(p)p^{-s})^{-1}.$$

Again it is conjectured that after multiplying by appropriately defined Euler factors at $p \mid N$ the complete product $L(A, s)$ is an L-function, which is entire and self-dual and has gamma factor

$$\gamma(A, s) = \pi^{-gs}\Gamma(\frac{s}{2} + \frac{1}{4})^g\Gamma(\frac{s}{2} + \frac{3}{4})^g = c_g(2\pi)^{-gs}\Gamma(s + \tfrac{1}{2})^g$$

for some constant $c_g > 0$ (see (5.86)). It satisfies the Ramanujan-Petersson conjecture, since the correctly defined ramified factors also satisfy the required condition.

For $g = 1$, this conjecture is true by the modularity of elliptic curves over $\mathbb{Q}$. For any $g > 1$, except for some cases involving so-called CM varieties, the analytic continuation and functional equation remain open problems. The conductor $q(A)$ (which divides N) of this L-function is predicted as part of the conjecture and is a close relative of the Artin conductor for Artin L-functions.

Note the resemblance between the shape and definition of the L-functions of curves and abelian varieties. It is not a coincidence. The theory of the jacobian variety of curves associates to every smooth projective algebraic curve C of genus g over a field k an abelian variety $J(C)/k$ of dimension g, called its Jacobian, in such a way that for a curve $C/\mathbb{Q}$ we have $L(C, s) = L(J(C), s)$. For instance, the L-function $L(J_0(q), s)$ of (5.93) is the L-function of the Jacobian $J_0(q)$ of the modular curve $X_0(q) = \Gamma_0(q)\backslash\bar{\mathbb{H}}$, by work of Eichler and Shimura.

REMARK. An alternative definition of the local factors of $L(A,s)$ (or of $L(C,s)$ for a curve) is given in terms of the ℓ-adic Tate module of A: let

$$T_\ell(A) = (\lim_{\substack{\leftarrow \\ n}} A[\ell^n]) \otimes \mathbb{Q}_\ell,$$

then the Galois group of $\mathbb{Q}$ acts on $T_\ell(A)$ and for any prime $p \neq \ell$ where A has good reduction, we have

$$P_{p,1}(T) = \det(1 - T\mathrm{Fr}_p \mid T_\ell(A))^{-1},$$

denoting by Fr_p a Frobenius conjugacy class at p. Dually, $P_{p,1}$ is given in terms of the étale cohomology group $H^1(A,\mathbb{Q}_\ell)$ (see Section 11.11) by

$$P_{p,1}(T) = \det(1 - T\sigma_p \mid T_\ell(A))^{-1}$$

where σ_p is the inverse of Fr_p, because $T_\ell(A)$ is the dual of $H^1(A,\mathbb{Q}_\ell)$.

The Mordell-Weil theorem states that if $A/\mathbb{Q}$ is an abelian variety, then the group of rational points $A(\mathbb{Q})$ is finitely generated. The computation of its rank is one of the great problems of arithmetic geometry. We have the famous

BIRCH AND SWINNERTON-DYER CONJECTURE. *Let $A/\mathbb{Q}$ be an abelian variety. Then*

$$\operatorname{rank} A(\mathbb{Q}) = \operatorname*{ord}_{s=\frac{1}{2}} L(A,s).$$

Of course this conjecture assumes that $L(A,s)$ is an L-function, so $L(A,\frac{1}{2})$ is defined. If the Birch and Swinnerton-Dyer conjecture holds, one can get information on the rank of abelian varieties using analytic methods. Often this suggests very interesting problems. For instance, by Proposition 5.21 we derive:

COROLLARY 5.49. *Let $A/\mathbb{Q}$ be an abelian variety of dimension $g \geqslant 1$ with conductor q. Assume that its Hasse-Weil zeta function is an L-function and that the Birch and Swinnerton-Dyer conjecture holds for $A(\mathbb{Q})$. Then we have*

$$\operatorname{rank} A(\mathbb{Q}) \ll \frac{\log q}{\log(\frac{3}{g}\log q)},$$

with an absolute implied constant. Note that $q > 3^g$ by Theorem 5.51 below.

This can be turned over to give a lower bound for the conductor of an abelian variety with a given rank, for instance for elliptic curves.

COROLLARY 5.50. *Assume that the Birch and Swinnerton-Dyer conjecture holds for elliptic curves over $\mathbb{Q}$. Then if $E/\mathbb{Q}$ is an elliptic curve of rank $r \geqslant 1$ and conductor q, we have*

$$q \gg r^{cr}$$

where $c > 0$ is some absolute constant and the implied constant is also absolute.

It would be very interesting to prove any unconditional estimate comparable to those two corollaries using algebraic methods. In Chapter 26 we study the order of vanishing of the zeta function (often called the analytic rank) for the abelian varieties $J_0(q)/\mathbb{Q}$, obtaining quite precise results.

As for the discriminants of number fields in Section 5.10, one can use L-functions to investigate, conditionally, the conductor of elliptic curves, or of abelian varieties over $\mathbb{Q}$.

THEOREM 5.51. *Let $A/\mathbb{Q}$ be an abelian variety over $\mathbb{Q}$ of dimension $g \geqslant 1$ such that $L(A,s)$ is an analytic L-function of conductor $q = q(A)$. Then we have $q \geqslant e^{1.2g}$, and in particular $q > 3$.*

See [**Mes**] for even stronger bounds. The bound $q > 1$ is known unconditionally, by the celebrated theorem of Fontaine [**Fon**] according to which there is no abelian variety over $\mathbb{Q}$ everywhere unramified. For a variety of dimension $g = 1$, this means there is no elliptic curve defined over $\mathbb{Q}$ which is unramified everywhere, i.e. its discriminant is divisible by at least one prime. In this case it can also be proved without appealing to L-functions or modularity by showing that the discriminant of the Weierstrass equation (14.18) (with integral coefficients) cannot be equal to ± 1 (see [**Sil**], ex. 8.15). In the higher dimensional case, no such explicit reformulation is accessible, and Fontaine's unconditional proof is much more involved.

PROOF OF THEOREM 5.51. Apply (5.24) to $L(A,s)$ at $s = \sigma > 1$, then take the real part. Because $L(A,s)$ is self-dual, the constant b disappears by using (5.29). Moreover, by the positivity

$$\operatorname{Re}\frac{1}{s-\rho} = \frac{\sigma-\beta}{|\sigma-\rho|^2} \geqslant 0,$$

we derive the inequality

$$\frac{1}{2}\log q - g\log 2\pi + g\frac{\Gamma'}{\Gamma}(\sigma+\tfrac{1}{2}) + \frac{L'}{L}(A,\sigma) \geqslant 0.$$

Next by the Ramanujan-Petersson bound we have

$$\Big|\frac{L'}{L}(A,\sigma)\Big| \leqslant g\frac{\zeta'}{\zeta}(\sigma).$$

Hence we get

$$\frac{1}{2g}\log q \geqslant \log 2\pi - \frac{\Gamma'}{\Gamma}(\sigma+\tfrac{1}{2}) - \frac{\zeta'}{\zeta}(\sigma)$$

for any $\sigma > 1$. Taking $\sigma = 2.3$, one checks that the right side is 0.62... > 0.6, hence $q \geqslant e^{1.2g}$ and $q > 3$. □

REMARK. Since $e^{2.4} > 11$, and it is known (see (14.41)) that the elliptic curve with smallest conductor is

$$E : y^2 + y = x^3 - x^2$$

with conductor 11, the argument implies that 11 is in fact the smallest conductor of an abelian variety over $\mathbb{Q}$, assuming their zeta functions are L-functions.

5.A. Appendix: complex analysis.

In this section we present in concise form the concepts and results of complex analysis that are used in this chapter. We omit most proofs but refer to [**T1**], [**Ru**] or [**Ahl**] (among many references).

A.1. Functions of finite order.

DEFINITION A.1. Let $f : \mathbb{C} \to \mathbb{C}$ be an entire function. Then f is said to be of finite order if there exists $\beta > 0$ such that

$$|f(s)| \ll \exp(|s|^\beta)$$

for $s \in \mathbb{C}$. If one can take any $\beta > 1$, f is said to be of order $\leqslant 1$, and if no $\beta < 1$ is possible, then f is said to be of order 1.

Let f be a meromorphic function on $\mathbb{C}$. Then f is of order $\leqslant 1$ if there exists entire functions g and h of order $\leqslant 1$ such that $f = g/h$.

THEOREM 5.52. *Let f be an entire function of order* 1.

(1) We have

$$f(s) = s^r \prod_{\rho \neq 0} \Big(1 - \frac{s}{\rho}\Big) e^{s/\rho}$$

uniformly on all compact subsets of $\mathbb{C}$, where r is the order of the zero of f at $s = 0$ and ρ runs over zeros of f different from 0.

(2) For any $\varepsilon > 0$, the series

$$\sum_{\rho \neq 0} |\rho|^{-1-\varepsilon} < \infty.$$

A.2. The Phragmen–Lindelöf principle for a strip.

The Phragmen–Lindelöf "principle" is a generic term for theorems which extend the maximum modulus principle to various infinite regions of the complex plane. Only the case of strips $a \leqslant \sigma \leqslant b$ interests us.

THEOREM 5.53. *Let f be a function holomorphic on an open neighborhood of a strip $a \leqslant \sigma \leqslant b$, for some real numbers $a < b$, such that $|f(s)| \ll \exp(|s|^A)$ for some $A \geqslant 0$ and $a \leqslant \sigma \leqslant b$.*

(1) Assume that $|f(s)| \leqslant M$ for all s on the boundary of the strip, i.e. for $\sigma = a$ or $\sigma = b$. Then we have $|f(s)| \leqslant M$ for all s in the strip.

(2) Assume that

$$|f(a + it)| \leqslant M_a(1 + |t|)^\alpha,$$
$$|f(b + it)| \leqslant M_b(1 + |t|)^\beta$$

for $t \in \mathbb{R}$. Then

$$|f(\sigma + it)| \leqslant M_a^{\ell(\sigma)} M_b^{1-\ell(\sigma)} (1 + |t|)^{\alpha\ell(\sigma)+\beta(1-\ell(\sigma))}$$

for all s in the strip, where ℓ is the linear function such that $\ell(a) = 1$, $\ell(b) = 0$.

A.3. The Perron formula.

Let $h(x)$ be the function given by

$$h(x) = \begin{cases} 1 & \text{if } x > 1, \\ \frac{1}{2} & \text{if } x = 1, \\ 0 & \text{if } x < 1. \end{cases}$$

PROPOSITION 5.54. *We have*

$$\frac{1}{2\pi i}\int_{c-iT}^{c+iT} x^s\frac{ds}{s} = h(x) + O\Big(\frac{x^c}{T|\log x|}\Big) \tag{5.111}$$

for any $x > 0, x \neq 1$, $T > 0$ and $0 < c \leqslant 2$, with an absolute implied constant. If $x = 1$, the factor $|\log x|$ is omitted.

A.4. The Stirling formula.

The Stirling asymptotic formula

$$\Gamma(s) = \Big(\frac{2\pi}{s}\Big)^{\frac{1}{2}}\Big(\frac{s}{e}\Big)^s\Big(1 + O\Big(\frac{1}{|s|}\Big)\Big) \tag{5.112}$$

is valid in the angle $|\arg s| \leqslant \pi - \varepsilon$ with the implied constant depending on ε. Hence for $s = \sigma + it$, $t \neq 0$, σ fixed

$$\Gamma(\sigma + it) = \sqrt{2\pi}(it)^{\sigma-\frac{1}{2}}e^{-\frac{\pi}{2}|t|}\Big(\frac{|t|}{e}\Big)^{it}\Big\{1 + O\Big(\frac{1}{|t|}\Big)\Big\}. \tag{5.113}$$

Let

$$\gamma(s) = \prod_{j=1}^{d}\Gamma\Big(\frac{s+\kappa_j}{2}\Big)$$

be a gamma factor as in (5.3) with $\mathrm{Re}(\kappa_j) > -1$. It has no poles in $\mathrm{Re}(s) \geqslant 1$. Recall the corresponding conductor at infinity is

$$\mathfrak{q}_\infty(s) = \prod_{j=1}^{d}(|t+\kappa_j| + 3).$$

Let $-\frac{1}{2} \leqslant \mathrm{Re}(s) \leqslant 2$. From (5.112) we deduce

$$|\gamma(s)|\prod_j|s+\kappa_j| = \mathfrak{q}_\infty(s)^{\frac{1}{2}(k+\sigma+1)}\exp\Big(-\frac{\pi}{4}\sum_j|t + \mathrm{Im}(\kappa_j)| + O(d)\Big) \tag{5.114}$$

where $k = \mathrm{Re}\sum\kappa_j$. Hence we have

$$\frac{\gamma(1-s)}{\gamma(s)} \asymp \mathfrak{q}_\infty(s)^{(\frac{1}{2}-\sigma)}\prod_j\frac{s+\kappa_j}{1-s+\kappa_j}. \tag{5.115}$$

Moreover, using the recurrence $\Gamma(s+1) = s\Gamma(s)$ we derive

$$\frac{\gamma'}{\gamma}(s) - \sum_{|s+\kappa_j|<1}\frac{1}{s+\kappa_j} \ll \log\mathfrak{q}_\infty(s). \tag{5.116}$$

The implied constants in (5.114) and (5.116) are absolute.

A.5. Existence of test functions.

For applications of the explicit formula, especially in the absence of the GRH, it is useful to know that there exist test functions with certain positivity properties.

PROPOSITION 5.55. *There exist positive, non-zero, C^∞ functions η on $[0,+\infty[$ with compact support in $[e^{-1},e]$ such that $\eta(1)=1$, $\hat{\eta}(0)>0$, $\eta(x)=\eta(x^{-1})$ for all $x>0$ and moreover either $\hat{\eta}(it)\geqslant 0$ for $t\in\mathbb{R}$ or $\mathrm{Re}\,(\hat{\eta}(s))\geqslant 0$ for all $s\in\mathbb{C}$ with $|\sigma|\leqslant 1$.*

SKETCH OF PROOF. Define $f(y)=\eta(e^y)$, which is an even, compactly supported, C^∞ function on $\mathbb{R}$, and

$$\hat{\eta}(s)=\int_{\mathbb{R}} f(y)e^{sy}dy$$

for all $s\in\mathbb{C}$, hence

$$\hat{\eta}(it)=\int_{\mathbb{R}} f(y)e^{ity}dy$$

$$\mathrm{Re}\,(\hat{\eta}(s))=\int_{\mathbb{R}} f(y)e^{\sigma y}\cos(ty)dy=\int_{\mathbb{R}} f(y)\cosh(\sigma y)\cos(ty)dy.$$

By the first formula to have $\hat{\eta}(it)\geqslant 0$ it suffices to choose $f\geqslant 0$ such that its Fourier transform is positive. The Fourier pair (4.83), suitably smoothed, does the job.

From the second formula and the maximum modulus principle for harmonic functions, the inequality $\mathrm{Re}\,(\hat{\eta}(s))\geqslant 0$ will hold for $|\sigma|\leqslant 1$ if and only if the Fourier transform of the even function $g(y)=f(y)\cosh(y)$ is non-negative. Conversely, if a function g (even, smooth and compactly supported) with this property is given, a suitable test function η is easily obtained by reversing this procedure. Now if g_0 is any smooth compactly supported positive function on $\mathbb{R}$, then the convolution square $g=g_0\star g_0$ will work, since $\hat{g}=\hat{g}_0^2$.

The support of η and its value at 1 can be easily adjusted by homogeneity. □

Functions of this type were constructed by Poitou and others for the purpose of obtaining lower-bounds for the discriminant of number fields [**Poi**]. Note in a way they are simply better behaved versions of the basic function $f(s)=(\sigma-s)^{-1}$ for $\sigma>1$ (see (5.28) and the proof of Theorem 5.51). See also [**PP**] for another construction of a test function with useful properties in applications.

A.6. Landau's lemma.

LEMMA 5.56. *Let $\lambda_n\geqslant 0$ for all $n\geqslant 1$. Suppose the series*

$$D(s)=\sum_{n\geqslant 1}\lambda_n n^{-s}$$

converges absolutely for all s with $\mathrm{Re}\,(s)>\sigma_0$, but not for any s with $\mathrm{Re}\,(s)<\sigma_0$; in other words, σ_0 is the abscissa of absolute convergence of $D(s)$. Then $s=\sigma_0$ is a singularity of $D(s)$, i.e. $D(s)$ cannot be continued analytically to a neighborhood of σ_0.

CHAPTER 6

ELEMENTARY SIEVE METHODS

The sieve methods originated from works of Viggo Brun on the Goldbach and Twin Prime Conjectures [**Br1**], [**Br2**]. Modern sieve methods apply to a great variety of problems, and when combined with techniques of analytic number theory they are very advanced indeed (cf. [**FI1**]). In this chapter we provide only basic results from sieve theory. First we establish by refining the combinatorial arguments of Brun the so-called Fundamental Lemma. Then we develop an upper bound sieve of Selberg and illustrate its power by a few popular applications. More complete treatments of sieve methods can be found in the books of Halberstam and Richert [**HaRi**] and Greaves [**Gr**].

6.1. Sieve problems.

Let $\mathcal{A} = (a_n)$ be a sequence of non-negative numbers. The ultimate question is how often these numbers are supported on primes? For example, if $\mathcal{A}$ is the characteristic function of the shifted primes

$$a_n = \begin{cases} 1 \text{ if } n = p+2, \\ 0 \text{ otherwise,} \end{cases} \tag{6.1}$$

then we are asking how often is $p+2$ prime? In this case, the heuristic arguments described in Section 13.1 yield the following asymptotic formula

$$\pi_2(x) = |\{p \leqslant x; \quad p+2 \text{ is prime}\}| \sim Cx(\log x)^{-2} \tag{6.2}$$

where C is a positive constant given by

$$C = 2\prod_p \Big(1 - \frac{1}{(p-1)^2}\Big) = 1.32032\ldots \tag{6.3}$$

Hence there "are" infinitely many twin primes, but we do not have rigorous proof of this old conjecture by any method.

The sieve methods alone (in their classical format) are incapable of selecting primes, however they yield satisfactory results for slightly modified problems. Let $\mathcal{P}$ be a set of primes, $z \geqslant 2$ and

$$P(z) = \prod_{\substack{p<z \\ p\in\mathcal{P}}} p. \tag{6.4}$$

Now we seek estimates for the sifted sum

$$S(x,z) = \sum_{\substack{n\leqslant x \\ (n,P(z))=1}} a_n. \tag{6.5}$$

When $\mathcal{P}$ is the set of all primes and $z = \sqrt{x}$, then $S(x, \sqrt{x})$ agrees with

$$S(x) = \sum_{p \leqslant x} a_p \tag{6.6}$$

up to a few terms a_n with $n < \sqrt{x}$. The sieve methods produce upper and lower bounds for $S(x, z)$ of correct order of magnitude as far as $z \leqslant x^{\alpha}$, where $\alpha > 0$ is a small positive constant. Note that if $n \leqslant x$ has no prime divisors less than x^{α}, then n has at most $[1/\alpha]$ prime divisors, so one can say n is almost prime.

The remarkable universality of sieve is that it applies to quite general sequences $\mathcal{A} = (a_n)$. All we need are asymptotics for the partial sums

$$A_d(x) = \sum_{\substack{n \leqslant x \\ n \equiv 0 \pmod{d}}} a_n \tag{6.7}$$

with $d \mid P(z)$, $d < y$. We assume that

$$A_d(x) = g(d)X + r_d(x) \tag{6.8}$$

where $g(d)X$ is an expected main term and $r_d(x)$ is an error term which we think of as being relatively small if $d < y$. In the main term X is approximately equal to

$$A(x) = \sum_{n \leqslant x} a_n \tag{6.9}$$

so $g(d)$ stands for the density of the masses a_n attached to $n \equiv 0 \pmod{d}$. Thinking of the divisibility by distinct primes as independent events, it is not too surprising that we assume that $g(d)$ is a multiplicative function such that

$$0 \leqslant g(p) < 1 \text{ if } p \in \mathcal{P}. \tag{6.10}$$

Of course, the approximation of type (6.8) is not unique, however to get good results in practice there is not much room to choose $g(d)$ and X from. For some prime moduli q one may have a good approximation (6.8) with $g(q) = 1$, which means that almost all mass comes from the a_n with $n \equiv 0 \pmod{q}$. In this case the chance of finding terms a_n supported on primes is slim; that is why we exclude such prime moduli from the set $\mathcal{P}$. Since we use (6.8) only for $d \mid P(z)$, we are free to modify the multiplicative function $g(d)$ arbitrarily at d prime to $P(z)$ and indeed for notational simplicity we set

$$g(p) = 0 \text{ if } p \notin \mathcal{P}. \tag{6.11}$$

6.2. Exclusion-inclusion scheme.

Detecting the condition $(n, P(z)) = 1$ by means of the Möbius inversion formula

$$\delta(m) = \sum_{d \mid m} \mu(d) = \begin{cases} 1 & \text{if } m = 1, \\ 0 & \text{if } m \neq 1, \end{cases} \tag{6.12}$$

the sum (6.5) unfolds into

$$S(x, z) = \sum_{n \leqslant x} \Big(\sum_{\substack{d \mid n \\ d \mid P(z)}} \mu(d) \Big) a_n.$$

Changing the order of summation we obtain the Legendre formula

$$S(x, z) = \sum_{d|P(z)} \mu(d)A_d(x). \tag{6.13}$$

Next introducing (6.8) we get

$$S(x, z) = XV(z) + R(x, z) \tag{6.14}$$

where

$$V(z) = \prod_{p|P(z)} (1 - g(p)) \tag{6.15}$$

and

$$R(x, z) = \sum_{d|P(z)} r_d(x). \tag{6.16}$$

On probabilistic grounds one may expect that the product $V(z)$ represents the true density of the mass coming from a_n with $n \leqslant x$ having no prime divisors $p < z$ in $\mathcal{P}$. However, this is almost never correct except for z small relatively to x in the logarithmic scale. In other words, the divisibility by distinct primes are not completely independent events when these primes are relatively large. Technically speaking the problem is that, if z is large, the remainder sum $R(x, z)$ has many terms, so it exceeds the expected main term $XV(z)$.

In order to reduce the number of terms in the Legendre formula (6.13), and consequently in the remainder term (6.16), Brun truncated the Möbius function $\mu(d)$ to two sets, say $\mathcal{D}^+$ and $\mathcal{D}^-$, and considered the incomplete convolutions

$$\delta^+(n) = \sum_{\substack{d|n \\ d\in\mathcal{D}^+}} \mu(d) \tag{6.17}$$

$$\delta^-(n) = \sum_{\substack{d|n \\ d\in\mathcal{D}^-}} \mu(d) \tag{6.18}$$

in place of $\delta(n)$. He lost the exact property (6.12) but maintained the inequalities

$$\delta^-(n) \leqslant \delta(n) \leqslant \delta^+(n) \tag{6.19}$$

for all $n \geqslant 1$. Hence one obtains the lower and upper bounds

$$XV^-(z) + R^-(x, z) \leqslant S(x, z) \leqslant XV^+(z) + R^+(x, z) \tag{6.20}$$

where

$$V^\pm(z) = \sum_{d|P(z)} \lambda_d^\pm g(d), \tag{6.21}$$

$$R^\pm(x, z) = \sum_{d|P(z)} \lambda_d^\pm r_d(x) \tag{6.22}$$

and λ_d^+, λ_d^- denote the Möbius function on the sets $\mathcal{D}^+$, $\mathcal{D}^-$ respectively. Assuming that

$$\mathcal{D}^+, \mathcal{D}^- \subset [1, y) \tag{6.23}$$

we can estimate the remainder sums $R^{\pm}(x,z)$ by

$$R(x,y) = \sum_{\substack{d<y \\ d|P(z)}} |r_d(x)|. \tag{6.24}$$

Sometimes we can exploit the intrinsic structure of the sieve weights $\lambda_d^{\pm}$ to get estimates for $R^{\pm}(x,z)$ which are better than that for $R(x,y)$. This is due to a possible cancellation of the error terms $r_d(x)$ in $R^{\pm}(x,z)$, but we do not venture into this sophisticated area in this book (see e.g. [**I9**], [**I10**]).

For notational simplicity we are writing squarefree numbers as the product of distinct primes enumerated in decreasing order

$$d = p_1 \cdots p_r \quad \text{with} \quad p_1 > \cdots > p_r. \tag{6.25}$$

Put

$$\mathcal{D}^+ = \{d = p_1 \cdots p_r; \quad p_m < y_m \quad \text{for all } m \text{ odd}\} \tag{6.26}$$

$$\mathcal{D}^- = \{d = p_1 \cdots p_r; \quad p_m < y_m \quad \text{for all } m \text{ even}\} \tag{6.27}$$

where y_m are suitable parameters. By convention both sets $\mathcal{D}^+$ and $\mathcal{D}^-$ contain $d = 1$. We start from the recurrence formula

$$V(z) = 1 - \sum_{p<z} g(p)V(p). \tag{6.28}$$

Iterating this we derive by the well-known exclusion-inclusion procedure the following identities:

$$V(z) = V^+(z) - \sum_{n \text{ odd}} V_n(z), \tag{6.29}$$

$$V(z) = V^-(z) + \sum_{n \text{ even}} V_n(z) \tag{6.30}$$

where

$$V_n(z) = \sum_{\substack{y_n \leqslant p_n < \cdots < p_1 < z \\ p_m < y_m,\, m<n,\, m \equiv n \pmod 2}} \cdots \sum g(p_1 \cdots p_n)V(p_n). \tag{6.31}$$

Dropping the non-negative terms $V_n(z)$ in (6.29) and (6.30) it follows that

$$V^-(z) \leqslant V(z) \leqslant V^+(z). \tag{6.32}$$

Now we can easily verify the lower and the upper bound sieve conditions (6.19); indeed (6.32) becomes (6.19) on taking $P(z) = n$ and $g(d) = 1$.

From now on the combinatorial sieves $\Lambda^+ = (\lambda_d^+)$ and $\Lambda^- = (\lambda_d^-)$ depend on the truncation parameters y_m alone. Brun tried various parameters, his best choice was $y_m = y^{\alpha\beta^{-m}}$ for suitable constants $0 < \alpha < 1 < \beta$. We choose the parameters y_m depending on the primes $p_1, \ldots p_m$ which occurred prior to the m-th step of the iteration, precisely

$$y_m = (y/p_1 \cdots p_m)^{\frac{1}{\beta}} \tag{6.33}$$

where β is a fixed number, $\beta > 1$. Clearly every $d \in \mathcal{D}^+ \cup \mathcal{D}^-$ satisfies $d < y$ except possibly for $d = p \in \mathcal{D}^-$. However assuming $z \leqslant y$ in this case we also have $d = p < z \leqslant y$. Our choice (6.33) is inspired by heuristic arguments exploiting the concept of "sieving limit", however, as we are not going for optimal results, we are free to choose any $\beta > 1$.

6.3. Estimations of $V^+(z)$, $V^-(z)$.

We need an upper bound for $V^+(z)$ and a lower bound for $V^-(z)$. By virtue of (6.29) and (6.30) in both cases the problems reduce to getting upper bounds for $V_n(z)$. Since $V(p_n)$ in (6.31) are non-negative some of the summation conditions can be relaxed by positivity. We get

$$V_n(z) \leqslant \sum_{\substack{y_n \leqslant p_n < \cdots p_1 < z \\ p_1 \cdots p_{m-1} p_m^\beta < y}} \cdots \sum g(p_1 \cdots p_n) V(p_n)$$

where the conditions hold for all $1 \leqslant m < n$ regardless of parity. These conditions imply that

$$p_1 \cdots p_m < y^{1 - (\frac{\beta - 1}{\beta})^m} \quad \text{if} \quad m < n,$$

by induction on m. In particular, for $m = n - 1$ this yields the following lower bound for the least prime divisors of $d = p_1 \cdots p_n$ in $V_n(z)$,

$$p_n \geqslant (y/p_1 \cdots p_{n-1})^{\frac{1}{\beta+1}} \geqslant y^{\frac{1}{\beta+1}(\frac{\beta-1}{\beta})^{n-1}} \geqslant y^{\frac{1}{\beta}(\frac{\beta-1}{\beta})^n} \geqslant z_n,$$

say, where $z_n = z^{(\frac{\beta-1}{\beta})^n}$ with

$$z = y^s, \quad s \geqslant \beta.$$

Having established the lower bound $p_n \geqslant z_n$ we drop the other conditions and estimate as follows

$$\begin{aligned} V_n(z) &\leqslant \sum_{z_n \leqslant p_n < \cdots < p_1 < z} \cdots \sum g(p_1 \cdots p_n) V(p_n) \\ &\leqslant \frac{V(z_n)}{n!} \Big(\sum_{z_n \leqslant p < z} g(p) \Big)^n \leqslant \frac{V(z_n)}{n!} \Big(\log \frac{V(z_n)}{V(z)} \Big)^n. \end{aligned}$$

To proceed further we make the following assumption about the multiplicative function $g(d)$: for any w with $w < z$, we have

$$\prod_{w \leqslant p < z} (1 - g(p))^{-1} \leqslant K \Big(\frac{\log z}{\log w} \Big)^\kappa \tag{6.34}$$

where $\kappa > 0$ and $K > 1$ do not depend on w.

REMARKS. The condition (6.34) is relatively easy to verify in practice. We call the exponent κ the sieve dimension. Notice that κ is not uniquely defined; any larger number is also qualified. The constant K can be disturbingly large if the densities $g(p)$ are near one for many small p. Nevertheless the results can be refined in this respect. Indeed, in practice one can establish the following estimate

$$\prod_{w \leqslant p < z} (1 - g(p))^{-1} \leqslant \Big(\frac{\log z}{\log w} \Big)^{\kappa'} \Big(1 + \frac{K'}{\log w} \Big) \tag{6.35}$$

for any w with $2 \leqslant w < z$, where κ' and K' are positive numbers independent of w. This estimate implies (6.34) with $\kappa = 1 + \kappa'$ and

$$K = 1 + \frac{K'}{\log z}. \tag{6.36}$$

Therefore K can be reduced to a number near 1 at the expense of increasing the dimension κ by 1.

By (6.34) we derive using the inequality $1 + x \leqslant e^x$ that

$$V(z_n) \leqslant KV(z)\Big(\frac{\beta}{\beta-1}\Big)^{\kappa n} < KV(z)e^{n/b}$$

where $\beta = \kappa b + 1$. Hence

$$V_n(z) < \frac{K}{n!}\Big(\frac{n}{b} + \log K\Big)^n e^{n/b}V(z) \leqslant \frac{K}{n!}\Big(\frac{n}{b}e^{1/b}\Big)^n K^b V(z).$$

Inserting $n! \geqslant e(n/e)^n$ we obtain

$$V_n(z) < e^{-1}a^n K^{b+1}V(z) \tag{6.37}$$

where $a = b^{-1}e^{1+b^{-1}}$. We choose b large so that $a < 1$.

The condition $p_n \geqslant y_n$ in (6.31) implies $p_1^{n+\beta} \geqslant y$, and this shows that the sum (6.31) is void is $n \leqslant s - \beta$. Therefore (6.37) yields

$$\sum_{n>0} V_n(z) = \sum_{n>s-\beta} V_n(z) < \frac{a^{s-\beta}}{e(1-a)}K^{b+1}V(z). \tag{6.38}$$

We choose $\beta = 9\kappa + 1$, getting $b = 9$ and $a < e^{-1}$. Hence we conclude

THEOREM 6.1. *Let $\Lambda^+ = (\lambda_d^+)$, $\Lambda^- = (\lambda_d^-)$ be the combinatorial sieves of level $y > 1$ and the parameter $\beta = 9\kappa + 1$. Then for any multiplicative function $g(d)$ satisfying (6.10), (6.11) and (6.34) and any $z \leqslant y^{1/\beta}$ we have*

$$V^+(z) < (1 + e^{\beta-s}K^{10})V(z), \tag{6.39}$$

$$V^-(z) > (1 - e^{\beta-s}K^{10})V(z) \tag{6.40}$$

where $s = \log y/\log z$.

COROLLARY 6.2. *Let the conditions be as above. Then*

$$S(x,z) < (1 + e^{\beta-s}K^{10})V(z)X + R(x, z^s) \tag{6.41}$$

$$S(x,z) > (1 - e^{\beta-s}K^{10})V(z)X - R(x, z^s) \tag{6.42}$$

where $\beta = 9\kappa + 1$, s is any number $\geqslant \beta$, and $R(x,y)$ denotes the remainder sum (6.24).

6.4. Fundamental lemma of sieve theory.

Combining the inequalities (6.39), (6.40) with (6.32) we conclude the following

FUNDAMENTAL LEMMA 6.3. *Let $\kappa > 0$ and $y > 1$. There exist two sets of real numbers $\Lambda^+ = (\lambda_d^+)$ and $\Lambda^- = (\lambda_d^-)$ depending only on κ and y with the following properties:*

$$\lambda_1^\pm = 1, \tag{6.43}$$

$$|\lambda_d^\pm| \leqslant 1 \quad \text{if } 1 < d < y, \tag{6.44}$$

$$\lambda_d^\pm = 0 \quad \text{if } d \geqslant y, \tag{6.45}$$

and for any integer $n > 1$,

$$\sum_{d|n} \lambda_d^- \leqslant 0 \leqslant \sum_{d|n} \lambda_d^+. \tag{6.46}$$

Moreover, for any multiplicative function $g(d)$ with $0 \leqslant g(p) < 1$ and satisfying the dimension condition

$$\prod_{w \leqslant p < z} (1 - g(p))^{-1} \leqslant \Big(\frac{\log z}{\log w}\Big)^\kappa \Big(1 + \frac{K}{\log w}\Big) \tag{6.47}$$

for all $2 \leqslant w < z \leqslant y$ we have

$$\sum_{d|P(z)} \lambda_d^\pm g(d) = \Big(1 + O\Big(e^{-s}\Big(1 + \frac{K}{\log z}\Big)^{10}\Big)\Big) \prod_{p<z} (1 - g(p)) \tag{6.48}$$

where $P(z)$ denotes the product of all primes $p < z$ and $s = \log y / \log z$, the implied constants depending only on κ.

REMARK. The Fundamental Lemma plays an assisting role in sieve theory; it usually serves for preliminary elimination of the elements of a sequence which have small prime divisors, that is primes $p < z = y^{1/s}$ with $s \to +\infty$. In such applications the results are asymptotically precise, so one loses nothing while gaining a practical assumption that the sequence in question is supported on almost primes. Quite often one only needs an upper-bound of the right order of magnitude, that is

$$\sum_{d|P(z)} \lambda_d^\pm g(d) \ll K^{10} \prod_{p<z} (1 - g(p)). \tag{6.49}$$

This holds under the condition (6.34)(which is weaker than (6.47)) with the implied constant depending only on κ.

Although the sequence $\mathcal{A} = (a_n)$ accompanied us when constructing the sieves $\Lambda^+ = (\lambda_d^+)$ and $\Lambda^- = (\lambda_d^-)$, one should realize that Λ^+, Λ^- have actually nothing in common with $\mathcal{A}$. One can legally apply the particular sieves Λ^+, Λ^- to any sequence of non-negative numbers, the only risk is that the results may not be ideal if the parameter κ (the so-called dimension) and y (the level) of Λ^+, Λ^- do not match perfectly the characteristics of $\mathcal{A}$.

Of course one does not need to push s to infinity. The lower bound (6.42) is already interesting if $s > \beta$. Assuming that the remainder term $R(x, y)$ is negligible for a suitable level y, one infers from (6.42) that

$$\sum_{\substack{n \leqslant x \\ (n, P(z))=1}} a_n \gg X (\log z)^{-\kappa} \tag{6.50}$$

where $z = y^{1/(\beta+\varepsilon)}$ for any $\varepsilon > 0$ provided y is sufficiently large in terms of ε. Suppose we can control $R(x, y)$ for $y = x^{\alpha-\varepsilon}$ with $0 < \alpha \leqslant 1$. Then (6.50) holds for $z = x^{(\alpha-\varepsilon)/(\beta+\varepsilon)}$, while the condition $(n, P(z)) = 1$ together with $n \leqslant x$ imply that n has at most α/β prime divisors.

In general to establish an upper bound for $S(x, z)$ is much easier, because $S(x, z)$ is decreasing in z so one can reduce z at will.

EXERCISE 1. Derive from (6.41) the upper bound

$$\pi_2(x) \ll x(\log x)^{-2}. \tag{6.51}$$

EXERCISE 2. Show that the sequence (6.1) satisfies the sieve conditions for dimension $\kappa = 1$ and level $\alpha = 1/2$ (see Theorem 17.1). Then derive from (6.42) that

$$\pi_2(x, x^{1/20}) \gg x(\log x)^{-2} \tag{6.52}$$

where $\pi_2(x, z)$ denotes the number of primes $p \leqslant x$ for which $p + 2$ has no prime divisors $< z$. Therefore there are infinitely many primes p such that $p + 2$ has at most twenty prime divisors. Recent developments of sieve methods in this direction show (among other things) that

$$|\{p \leqslant x; p + 2 = p_1 \text{ or } p_1 p_2 \text{ with } p_1,\ p_2 > x^{3/11}\}| \gg x(\log x)^{-2}$$

for all sufficiently large x. Hence $p + 2$ has at most two prime divisors infinitely often. This last result is due to J.R. Chen [**Ch**]. We still do not know which case $p + 2 = p_1$ or $p + 2 = p_1 p_2$ occurs infinitely often? Probably both!

6.5. The Λ^2-sieve.

A powerful and elegant method of an upper bound sieve came from Atle Selberg [**S1**]. Selberg's method yields results of great generality, it is simpler than combinatorial sieves at the start, though equally complex in its advanced forms. We modify the notation of the previous section slightly to display clearly the economy of conditions which are required for performing Selberg's sieve.

Recall that an upper bound sieve of level D is a sequence of real numbers λ_d for $d < D$ with $\lambda_1 = 1$ such that

$$\sum_{d|m} \lambda_d \geqslant 0 \qquad \text{for all } m > 1. \tag{6.53}$$

Hereafter the superscript $+$ is omitted for notation simplicity. This positivity condition was quite difficult to get a hold on in combinatorial sieve. Selberg made it very easy by choosing λ_d such that

$$\sum_{d|m} \lambda_d = \Big(\sum_{d|m} \rho_d\Big)^2 \tag{6.54}$$

where $\{\rho_d\}$ is another sequence of real numbers with

$$\rho_1 = 1. \tag{6.55}$$

Since the squares are non-negative, such a choice guarantees (6.53) no matter what ρ_d are!

Selberg's choice amounts to

$$\lambda_d = \sum_{[d_1,d_2]=d} \rho_{d_1}\rho_{d_2}. \tag{6.56}$$

In order to control the level we assume ρ_d are supported on integers $< \sqrt{D}$,

$$\rho_d = 0 \qquad \text{if } d \geqslant \sqrt{D}. \tag{6.57}$$

Hence the resulting sieve $\{\lambda_d\}$ has level of support D. Following Selberg we call it the Λ^2-sieve of level D.

Throughout, $\mathcal{A} = (a_n)$ is a sequence of non-negative numbers, P is a squarefree number and for any $d \mid P$ we set

$$|\mathcal{A}_d| = \sum_{n\equiv 0 \pmod{d}} a_n = g(d)X + r_d(\mathcal{A}). \tag{6.58}$$

Here as before, $g(d)$ is a multiplicative function satisfying (6.10).

Applying the Λ^2-sieve to the sequence $\mathcal{A} = (a_n)$ in the sifting range P we get

$$\begin{aligned} S(\mathcal{A}, P) = \sum_{(n,P)=1} a_n &\leqslant \sum_n a_n \Big(\sum_{d|(n,P)} \rho_d \Big)^2 \\ &= \sum\sum_{d_1,d_2|P} \rho_{d_1}\rho_{d_2} |\mathcal{A}_{[d_1,d_2]}| = XG + R(\mathcal{A}, P) \end{aligned}$$

where

$$G = \sum\sum_{d_1,d_2|P} g([d_1,d_2])\rho_{d_1}\rho_{d_2} \tag{6.59}$$

and

$$R(\mathcal{A}, P) = \sum\sum_{d_1,d_2|P} \rho_{d_1}\rho_{d_2} r_{[d_1,d_2]}(\mathcal{A}). \tag{6.60}$$

The task before us is to make this general inequality optimal. Forgetting for a moment about the remainder term $R(\mathcal{A}, P, \Lambda^2)$ we wish to minimize G with respect to the unknown numbers ρ_d subject to (6.55) and (6.57). The ensuing numbers will satisfy

$$|\rho_d| \leqslant 1, \tag{6.61}$$

hence the remainder term is automatically under control.

The expression (6.59) is a quadratic form in ρ_d. In order to find the minimum of G it helps to diagonalize. In the presentation below it goes without saying that ρ_d is supported on, and the relevant variables of summation run over the divisors of P (thus over squarefree numbers). Furthermore we can assume that

$$\begin{aligned} 0 < g(p) < 1 \qquad & \text{if } p|P, \\ g(p) = 0 \qquad & \text{if } p \nmid P. \end{aligned} \tag{6.62}$$

Let $h(d)$ be the multiplicative function defined by

$$h(p) = \frac{g(p)}{1 - g(p)}. \tag{6.63}$$

We obtain

$$\begin{aligned} G &= \sum_{abc|P} g(abc)\rho_{ac}\rho_{bc} \\ &= \sum_{c} g(c) \sum\sum_{(a,b)=1} g(a)g(b)\rho_{ac}\rho_{bc} \\ &= \sum_{c} g(c) \sum_{d} \mu(d)g(d)^2 \Big(\sum_{m} g(m)\rho_{cdm}\Big)^2 \\ &= \sum_{d|P} h(d)^{-1} \Big(\sum_{m\equiv 0(\mathrm{mod}\ d)} g(m)\rho_m\Big)^2. \end{aligned}$$

Hence by the linear change of variables

$$\xi_d = \mu(d) \sum_{m\equiv 0(\mathrm{mod}\ d)} g(m)\rho_m \tag{6.64}$$

we obtain the diagonal form

$$G = \sum_{d|P} h(d)^{-1}\xi_d^2. \tag{6.65}$$

We still have to reinterpret the condition (6.55) in terms of the new variables ξ_d. To this end we use the Möbius inversion to convert (6.64) into

$$\rho_\ell = \frac{\mu(\ell)}{g(\ell)} \sum_{d\equiv 0(\mathrm{mod}\ \ell)} \xi_d. \tag{6.66}$$

In particular, for $\ell = 1$ this gives the linear equation

$$\sum_{d|P} \xi_d = 1. \tag{6.67}$$

Moreover, one observes by (6.64) and (6.66) that the support conditions (6.57) for ρ_d are equivalent to these for ξ_d,

$$\xi_d = 0 \qquad \text{if } d \geqslant \sqrt{D}. \tag{6.68}$$

Now our target is to minimize (6.65) on the hyperplane (6.67). Applying Cauchy's inequality to (6.67) we derive $GH \geqslant 1$ where

$$H = \sum_{d<\sqrt{D}, d|P} h(d) \tag{6.69}$$

so G cannot be smaller than H^{-1}. The equality

$$GH = 1 \tag{6.70}$$

holds for

$$\xi_d = h(d)H^{-1} \qquad \text{if } d < \sqrt{D}. \tag{6.71}$$

Note that

$$H \leqslant \sum_{d|P} h(d) = \prod_{p|P}(1+h(p)) = \prod_{p|P}(1-g(p))^{-1}, \tag{6.72}$$

i.e.,

$$\frac{1}{H} \geqslant \prod_{p|P}(1-g(p)). \tag{6.73}$$

Next we compute ρ_ℓ by inserting (6.71) into (6.66) getting

$$\mu(\ell)g(\ell)\rho_\ell H = \sum_{\substack{m<\sqrt{D} \\ m\equiv 0 (\mathrm{mod}\ \ell)}} h(m),$$

that is

$$\rho_\ell = \frac{\mu(\ell)h(\ell)}{g(\ell)H} \sum_{\substack{d<\sqrt{D}/\ell \\ (d,\ell)=1}} h(d). \tag{6.74}$$

Now we show (6.61). To this end we group the terms in (6.69) according to the greatest common divisor of d and ℓ getting

$$\begin{aligned} H &= \sum_{k|\ell} \sum_{\substack{d<\sqrt{D} \\ (d,\ell)=k}} h(d) = \sum_{k|\ell} h(k) \sum_{\substack{m<\sqrt{D}/k \\ (m,\ell)=1}} h(m) \\ &\geqslant \Big(\sum_{k|\ell} h(k)\Big) \sum_{\substack{m<\sqrt{D}/\ell \\ (m,\ell)=1}} h(m) = \mu(\ell)\rho_\ell H \end{aligned}$$

and this proves (6.61) (this neat estimate is due to J.H. van Lint and H.-E. Richert [**LR**]). From this one gets directly by (6.56)

$$|\lambda_d| \leqslant \tau_3(d). \tag{6.75}$$

From the above results we conclude the following

THEOREM 6.4. *Let $\mathcal{A} = (a_n)$ be a finite sequence of non-negative numbers and P be a finite product of distinct primes. For every $d|P$ we write*

$$|\mathcal{A}_d| = \sum_{n\equiv 0 (\mathrm{mod}\ d)} a_n = g(d)X + r_d(\mathcal{A}) \tag{6.76}$$

where $X > 0$ and $g(d)$ is a multiplicative function with $0 < g(p) < 1$ for $p|P$. Let $h(d)$ be the multiplicative function given by $h(p) = g(p)(1-g(p))^{-1}$ and

$$H = \sum_{d<\sqrt{D}, d|P} h(d) \tag{6.77}$$

for some $D > 1$. Then we have

$$S(\mathcal{A}, P) = \sum_{(n,P)=1} a_n \leqslant XH^{-1} + R(\mathcal{A}, P) \tag{6.78}$$

where

$$R(\mathcal{A}, P) = \sum_{d|P} \lambda_d r_d(\mathcal{A}) \tag{6.79}$$

with λ_d given by (6.56) and (6.74).

Using (6.75) one estimates the remainder term crudely by

$$|R(\mathcal{A}, P)| \leqslant \sum_{d<D, d|P} \tau_3(d)|r_d(\mathcal{A})|. \tag{6.80}$$

The Λ^2-sieve is very general indeed. Its upper bound does not require any regularity in the distribution of the density function $g(p)$ over primes, therefore the dimension of a sifting problem does not play explicitly a role. If $g(p) = \omega(p)p^{-1}$, where $\omega(p)$ is the number of residue classes (mod p) which one wants to exclude, then $h(p) = \omega(p)(p - \omega(p))^{-1}$ is the ratio of the numbers of rejected to admitted classes, and H incorporates these ratios. Of course, the larger $\omega(p)$ is, the smaller the upper bound (6.78). However, a few local deviations of $\omega(p)$ from its average have insignificant global effect. To the contrary our estimates derived by the combinatorial sieve are quite sensitive in this respect because the hypothesis (6.35) is assumed to hold for all segments of primes $w \leqslant p < z$, no matter how short.

The Λ^2-sieve yields fantastic results for sifting problems when the number of residue classes to be excluded is large, it can compete with the large sieve method of Linnik (see Section 7.4).

REMARKS. Note that the numbers ρ_d depend on the multiplicative function $g(d)$ so (indirectly) on the sifting sequence $\mathcal{A}$. However, assuming some regularity of $g(d)$ (as in the κ dimensional sieve, see (6.34)) one can establish fairly good approximation (for small d at any rate)

$$\rho_d = \mu(d)\Big(\frac{\log \sqrt{D}/d}{\log \sqrt{D}}\Big)^\kappa \Big\{1 + O\Big(\frac{1}{\log \sqrt{D}/d}\Big)\Big\}. \tag{6.81}$$

6.6. Estimate for the main term of the Λ^2-sieve.

To bring the Selberg upper bound (6.78) to a practical form we need a clear lower bound for

$$H = H(D) = \sum_{d<\sqrt{D}, d|P} h(d).$$

The upper bound (6.72) holds in general but a good lower bound requires some restrictions on the density function g, so on h, and the sifting range P. From now on P is the product of all primes $p < \sqrt{D}$, so we have

$$H(D) = \sum_{d<\sqrt{D}}^{\flat} h(d)$$

where the superscript $\flat$ restricts the summation to squarefree numbers.

We can estimate $H(D)$ strongly, elementarily and nicely for $g(d) = d^{-1}$ (this occurs when one is sifting numbers in an interval). In this case $h(d) = \varphi(d)^{-1}$ and

$$H(D) = \sum_{d<\sqrt{D}}^{\flat} d^{-1} \prod_{p|d} \left(\tfrac{1}{p} + \tfrac{1}{p^2} + \cdots\right) \geqslant \sum_{m<\sqrt{D}} m^{-1} > \log \sqrt{D}.$$

If $g(d)$ agrees with d^{-1} only for $(d,q) = 1$ (as for example in the case of sifting an arithmetic progression), then this lower bound holds with the following modification:

$$H(D) > \prod_{p|q} (1 - g(p))(\log \sqrt{D}). \tag{6.82}$$

The above example is rather special. Now suppose $g(p)p$ is κ on average, say

$$\sum_{p\leqslant x} g(p) \log p = \kappa \log x + O(1) \tag{6.83}$$

for all $x \geqslant 2$ where κ is a positive number. Hence $g(p) \log p \ll 1$. Suppose also that

$$\sum_p g(p)^2 \log p < \infty. \tag{6.84}$$

Since $h(p) = g(p) + O(g(p)^2)$, it follows that (6.83) holds for h as well (with a different implied constant). Applying Theorem 1.1 for the multiplicative function h in place of f we get

$$H(D) = c(\log \sqrt{D})^{\kappa} \left\{1 + O((\log D)^{-1})\right\}$$

where

$$c = \tfrac{1}{\Gamma(\kappa+1)} \prod_p (1 - g(p))^{-1} (1 - \tfrac{1}{p})^{\kappa}$$

(the required conditions (1.89) and (1.90) are satisfied by (6.83), (6.84)) and the implied constant depends on that in (6.83). If D is large in terms of this constant we can invert this approximation getting

$$H(D)^{-1} = 2^{\kappa}\Gamma(\kappa+1) H_g (\log D)^{-\kappa} \{1 + O((\log D)^{-1})\} \tag{6.85}$$

where

$$H_g = \prod_p (1 - g(p))(1 - \tfrac{1}{p})^{-\kappa}. \tag{6.86}$$

6.7. Estimates for the remainder term in the Λ^2-sieve.

Once again we consider a class of sifting problems in which the individual error terms satisfy

$$|r_d(\mathcal{A})| \leqslant g(d)d. \tag{6.87}$$

Naturally with this property one makes the condition

$$g(d)d \geqslant 1 \quad \text{if } d|P. \tag{6.88}$$

This implies $g([d_1,d_2])[d_1,d_2] \leqslant g(d_1)g(d_2)d_1d_2$, therefore

$$|R(\mathcal{A},P,\Lambda^2)| \leqslant \Big(\sum_{d<\sqrt{D}} |\rho_d| g(d)d\Big)^2 \leqslant \Big(\frac{1}{H}\sum_{m<\sqrt{D}} h(m)\sigma(m)\Big)^2 \tag{6.89}$$

by (6.60) and (6.72), where $\sigma(m)$ denotes the sum of divisors of m.

Next assume that the density function $g(p)$ satisfies the following crude bound

$$\sum_{y\leqslant p\leqslant x} g(p)\log p \ll \log(2x/y). \tag{6.90}$$

Hence we infer the same property for $h(p)\sigma(p)p^{-1}$. Then we apply (1.85) getting

$$\sum_{m<\sqrt{D}} h(m)\sigma(m) \ll \frac{\sqrt{D}}{\log D}\sum_{m<\sqrt{D}} h(m)\sigma(m)m^{-1}.$$

Here we have

$$\sum_{m<\sqrt{D}} h(m)\sigma(m)m^{-1} \leqslant H\sum_m h(m)m^{-1} \ll H.$$

Hence we conclude that

$$R(\mathcal{A},P) \ll D(\log D)^{-2}. \tag{6.91}$$

Combining with (6.78) we get

THEOREM 6.5. *Suppose the conditions of Theorem 6.4 hold. Moreover, assume (6.87), (6.88) and (6.90). Then we have*

$$S(\mathcal{A},P) \leqslant \frac{X}{H} + O\Big(\frac{D}{\log^2 D}\Big) \tag{6.92}$$

where $H = H(D)$ is given by (6.77), $D > 1$ is arbitrary, and the implied constant depends only on that in (6.90).

6.8. Selected applications of Λ^2-sieve.

Consider the sequence $\mathcal{A} = (a_n)$ which is the characteristic function of an arithmetic progression in a short interval

$$n \equiv a (\mathrm{mod}\ q), \qquad x < n \leqslant x+y \tag{6.93}$$

where $(a,q)=1$ and $1 \leqslant q < y$. Let P consist of primes $p \leqslant \sqrt{y}$ with $p \nmid q$. Then

$$\pi(x+y;q,a) - \pi(x;q,a) \leqslant S(\mathcal{A},P) + \sqrt{y}\,q^{-1}. \tag{6.94}$$

On the other hand, the conditions of Theorem 6.5 are satisfied with $X = yq^{-1}$ and $g(d) = d^{-1}$ if $(d,q)=1$. By (6.82) we have $H(D) > \frac{\varphi(q)}{2q}\log D$. Combining (6.92) with (6.94) we get

$$\pi(x+y;q,a) - \pi(x;q,a) < \frac{2y}{\varphi(q)\log D} + O\Big(\frac{D}{\log^2 D} + \frac{\sqrt{y}}{q}\Big).$$

Choosing $D = yq^{-1}$ we conclude

THEOREM 6.6. *For $(a,q)=1$ and $1\leqslant q<y$ we have*

$$\pi(x+y;q,a)-\pi(x;q,a)<\frac{2y}{\varphi(q)\log(y/q)}+O\left(\frac{y}{q\log^2(y/q)}\right) \tag{6.95}$$

where the implied constant is absolute.

This is called the Brun-Titchmarsh inequality. Using the large sieve method H.L. Montgomery and R.C. Vaughan [**MV1**] have shown that the error term in (6.95) can be deleted.

Next we take $\mathcal{A}=(a_n)$ the characteristic function of the polynomial values

$$n=(m-\alpha_1)\cdots(m-\alpha_k)$$

with $1\leqslant m\leqslant x$, where all α_j are distinct. In this case $g(p)=\nu(p)p^{-1}$ where $\nu(p)$ is the number of roots modulo p. If p is sufficiently large, $\nu(p)=k$ so we have a k-dimensional sieve problem. By (6.85) and (6.92) we deduce

THEOREM 6.7. *Let $\mathfrak{a}=(\alpha_1,\dots,\alpha_k)$ be distinct integers which do not cover all residue classes to any prime modulus. Then the number of integers $1\leqslant m\leqslant x$ for which $m-\alpha_1,\dots,m-\alpha_k$ are all primes satisfies*

$$\pi(x;\mathfrak{a})\leqslant 2^k k!Bx(\log x)^{-k}\Big(1+O\Big(\frac{\log\log x}{\log x}\Big)\Big) \tag{6.96}$$

where

$$B=\prod_p\Big(1-\frac{\nu(p)}{p}\Big)\Big(1-\frac{1}{p}\Big)^{-k}. \tag{6.97}$$

REMARKS. The upper bound (6.96) is larger by factor $2^k\,k!$ than the conjectured asymptotic

$$\pi(x;\mathfrak{a})\sim Bx(\log x)^{-k}.$$

CHAPTER 7

BILINEAR FORMS AND THE LARGE SIEVE

7.1. General principles of estimating double sums.

Certainly the most versatile instruments of analytic number theory are double sums

$$\Psi(\alpha,\beta)=\sum_m\sum_n \alpha_m\beta_n\phi(m,n). \tag{7.1}$$

Here $\alpha=(\alpha_m)$, $\beta=(\beta_n)$ are finite vectors of complex numbers and

$$\Phi=(\phi(m,n)) \tag{7.2}$$

is a finite matrix with complex entries. In matrix notation $\Psi(\alpha,\beta)=\alpha\Phi\beta^t$. Given Φ the goal is to give a sharp estimate for $\Psi(\alpha,\beta)$ which is valid for arbitrary vectors α, β. Yes, it is essential to assume nothing about the numbers α_m, β_n, because in practice they are quite complex objects. Very often an attempt to utilize specific structure of these vectors brings back a mirror image of the original situation. Hence a question, what is the benefit from arranging the original sums into bilinear forms? The key feature is that the sequences α, β do not see each other! Therefore a cancellation in $\Psi(\alpha,\beta)$ is possible, subject only to properties of the matrix coefficients $\phi(m,n)$. Indeed we have

$$|\Psi(\alpha,\beta)|^2\leqslant\Delta\|\alpha\|^2\|\beta\|^2 \tag{7.3}$$

where $\Delta=\Delta(\Phi)$ is the norm of the corresponding linear operator and

$$\|\alpha\|^2=\sum|\alpha_m|^2,\quad \|\beta\|^2=\sum|\beta_n|^2. \tag{7.4}$$

A more transparent way of estimating $\Psi(\alpha,\beta)$ goes by Cauchy's inequality. To this end choose first m to be in the outer sum and n in the inner sum getting

$$|\Psi(\alpha,\beta)|^2\leqslant\|\alpha\|^2\sum_m\Big|\sum_n\beta_n\phi(m,n)\Big|^2. \tag{7.5}$$

Note that the unknown coefficients α_m in the outer sum are gone. At this point one could be more flexible by introducing new coefficients $f(m)$ in place of α_m to smooth the next step and to optimize the final bound.

Smoothing is one among many optional devices which are put into practice with remarkable effects in analytic number theory. This is not just a technical device, it can be viewed as a kind of spectral completion.

Next squaring out the inner sum over n followed by interchanging the order of summation we get

$$\sum_m\Big|\sum_n\beta_n\phi(m,n)\Big|^2=\sum_{n_1}\sum_{n_2}\beta_{n_1}\bar\beta_{n_2}\sum_m\phi(m,n_1)\bar\phi(m,n_2). \tag{7.6}$$

For $n_1 = n_2$, the so-called diagonal terms, one cannot get cancellation in the inner sum

$$\sum_m |\phi(m,n)|^2, \tag{7.7}$$

so the saving factor comes solely from the fact that the diagonal is smaller than the full square! However, for most of n_1, n_2 off the diagonal one may find a considerable cancellation in the sum

$$\sum_m \phi(m,n_1)\bar{\phi}(m,n_2) \tag{7.8}$$

because of independent variations in the sign of $\phi(m,n_1)$, $\phi(m,n_2)$. How one executes the summation in (7.8) is a separate question depending on particular characteristics of Φ. At this point the sums (7.8) can be either estimated directly finishing the job, or it can be transformed to another sum (so to speak a dual sum) by a kind of spectral formula. In the latter scenario even a trivial estimation of the dual sum often produces a non-trivial bound for the original sum. Moreover, one can also exploit the extra summation over n_1, n_2. Our point in these remarks is to say that the general estimate (7.3) obstructs the visibility for improvements. We advise to carry out the arguments in practice always by Cauchy's inequality because it allows you to control your position.

Before launching Cauchy's inequality to the double sum $\Psi(\alpha,\beta)$, make sure to choose the outer and the inner sums variables conveniently as the results will be different. This option of interchanging m, n in $\Psi(\alpha,\beta)$ is the essence of the following duality principle.

DUALITY PRINCIPLE. *Suppose for any complex numbers β_n,*

$$\sum_m \Big|\sum_n \beta_n \phi(m,n)\Big|^2 \leqslant \Delta \|\beta\|^2. \tag{7.9}$$

Then for any complex numbers α_m

$$\sum_n \Big|\sum_m \alpha_m \phi(m,n)\Big|^2 \leqslant \Delta \|\alpha\|^2, \tag{7.10}$$

where Δ is the same in both inequalities.

PROOF. The left side of (7.10) is equal to (7.1) with

$$\bar{\beta}_n = \sum_m \alpha_m \phi(m,n). \tag{7.11}$$

By (7.5) and (7.9) we get $\Psi(\alpha,\beta)^2 \leqslant \Delta\|\alpha\|^2\|\beta\|^2$. But $\|\beta\|^2 = \Psi(\alpha,\beta)$ so (7.10) is derived from (7.9). □

Because of the application of Cauchy's inequality, it is inevitable that much attention concentrates on estimating sums of type (7.9) or (7.10). These equivalent statements for general coefficients α_m, β_n are called Large Sieve Inequalities for historical reasons although they do not resemble a sieve of any kind. It is better to regard (7.9), (7.10) as a near-orthogonality property of Φ.

We end this introduction to general bilinear forms with a few comments about obvious cases for which the ideas described definitely cannot work. Suppose $\phi(m,n)$

factors, say $\phi(m,n)=\chi(m)\psi(n)$, or $\phi(m,n)$ is a short linear combination of such products. In other words, the variables m, n of $\phi(m,n)$ are separated at no, or low, cost. Then we may choose the vectors $\alpha=(\alpha_m)$, $\beta=(\beta_n)$ with $\alpha_m=\bar{\chi}(m)$, $\beta_n=\bar{\psi}(n)$. For this biased choice of vectors the bilinear form is $\Psi(\alpha,\beta)=\|\alpha\|^2\|\beta\|^2$ showing no cancellation of terms.

The principles of estimating bilinear forms in analytic number theory have been articulated very slowly. Perhaps because it is not the simplicity of the ideas, but rather the additional devices introduced by researchers in particular cases which make this technology so effective today. First applications dealt with bilinear forms in which $\phi(m,n)$ depended only on the product mn, say

$$\phi(m,n)=g(mn), \tag{7.12}$$

where g is an arithmetic function in one variable. Of course, in this case g cannot be multiplicative. I.M. Vinogradov should be mentioned for the first impressive results. Using an exclusion-inclusion procedure (a sieve method) he managed to express exponential sums over primes by double sums and estimated the latter by elementary means (cancellation in geometric series). Then utilizing the result in the circle method Vinogradov gave a stunning solution of the ternary Goldbach problem. His ideas are still alive today (see Chapters 13 and 19); however, there are more instances which stimulated the whole business of bilinear forms. Some of these will be explicitly described in forthcoming chapters, and many are embedded in arguments without recognition. Linnik [**Li2**] has elaborated the idea of estimating double sums into the dispersion method, which is missing in this book.

The notation chosen for this introduction serves as a model. Obviously we do not need to consider $\phi(m,n)$ as a function in two integral variables. Actually in some works it is nicer to index the two variables by more appropriate parameters (rational fractions, characters, eigenvalues, etc.).

In this chapter we give a selection from the great variety of bilinear forms and large sieve type inequalities. A few basic ones will be given detailed proofs, a few more will be discussed in a broad context, and many will be just stated with references to original publications.

7.2. Bilinear forms with exponentials.

The bilinear forms of type (7.1) with

$$\phi(m,n)=e(x_m y_n) \tag{7.13}$$

appear quite often in classical analytic number theory. Here x_m, y_n are real numbers and $e(z)=\exp(2\pi iz)$ as usual.

LEMMA 7.1. *For any complex numbers α_m and real numbers x_m we have*

$$\int_{-Y}^{Y}\Big|\sum_m \alpha_m e(x_m y)\Big|^2 dy \leqslant 5Y \sum\sum_{2Y|x_{m_1}-x_{m_2}|<1} |\alpha_{m_1}\alpha_{m_2}|. \tag{7.14}$$

PROOF. Let $g(y)$ be a non-negative function on $\mathbb{R}$ with $g(y) \geqslant 1$ if $|y| \leqslant \frac{1}{2}$ and $\operatorname{supp}(\hat{g}) \subset [-1,1]$. Then the left side of (7.14) is bounded by

$$\int g\Big(\frac{y}{2Y}\Big)\Big|\sum_m \alpha_m e(x_m y)\Big|^2 dy = 2Y \sum\sum_{m_1,m_2} \alpha_{m_1}\bar{\alpha}_{m_2}\hat{g}(2Y(x_{m_1}-x_{m_2}))$$

$$\leqslant 2Y\hat{g}(0) \sum\sum_{2Y|x_{m_1}-x_{m_2}|<1} |\alpha_{m_1}\alpha_{m_2}|.$$

The Fourier pair $g(y) = \left(\frac{\sin \pi y}{2y}\right)^2$, $\hat{g}(v) = \frac{\pi^2}{4}\max(1-|v|,0)$ satisfies the above conditions, giving the result. □

Our basic inequality is

THEOREM 7.2. *Let x_m, y_m be real numbers with $|x_m| \leqslant X$, $|y_m| \leqslant Y$. Then for any complex numbers α_m, β_n*

$$\Big|\sum_m\sum_n \alpha_m\beta_n e(x_m y_n)\Big| \leqslant 5(XY+1)^{\frac{1}{2}}\Big(\sum\sum_{|x_{m_1}-x_{m_2}|Y<1}|\alpha_{m_1}\alpha_{m_2}|\Big)^{\frac{1}{2}}$$

$$\Big(\sum\sum_{|y_{n_1}-y_{n_2}|X<1}|\beta_{n_1}\beta_{n_2}|\Big)^{\frac{1}{2}} \tag{7.15}$$

PROOF. Put $\varepsilon = (4X)^{-1}$ and

$$\gamma_m = \frac{\pi x_m \alpha_m}{\sin(2\pi\varepsilon x_m)},$$

so $|\gamma_m| \leqslant \pi X|\alpha_m|$. Writing

$$e(xy) = \frac{\pi x}{\sin 2\pi\varepsilon x}\int_{y-\varepsilon}^{y+\varepsilon} e(xt)dt$$

we find that the left side of (7.15) is bounded by

$$\Big|\sum_n \beta_n \int_{y_n-\varepsilon}^{y_n+\varepsilon}\sum_m \gamma_m e(x_m y)dy\Big| \leqslant \int_{-Y-\varepsilon}^{Y+\varepsilon}\sum_{|y_n-y|<\varepsilon}|\beta_n|\Big|\sum_m \gamma_m e(x_m y)\Big|dy.$$

Hence by Cauchy's inequality and Lemma 7.1 we find that the square of the left side of (7.15) is bounded by

$$\Big(\int_{-Y-\varepsilon}^{Y+\varepsilon}\Big(\sum_{|y_n-y|<\varepsilon}|\beta_n|\Big)^2 dy\Big)\Big(\int_{-Y-\varepsilon}^{Y+\varepsilon}\Big|\sum_m \gamma_m e(x_m y)\Big|^2 dy\Big)$$

$$\leqslant 2\varepsilon\Big(\sum\sum_{|y_{n_1}-y_{n_2}|<\varepsilon}|\beta_{n_1}\beta_{n_2}|\Big)5(Y+\varepsilon)(\pi X)^2\Big(\sum\sum_{|x_{m_1}-x_{m_2}|Y<1}|\alpha_{m_1}\alpha_{m_2}|\Big).$$

Here $10\varepsilon(Y+\varepsilon)(\pi X)^2 = \frac{4\pi^2}{2}(XY+\frac{1}{4}) < 25(XY+1)$ completing the proof. □

If the points x_m and y_n are well-spaced, so that only the diagonal terms appear on the right side of (7.15), we get

COROLLARY 7.3. *Suppose* $|x_m| \leqslant X$, $|y_n| \leqslant Y$, *and for* $m_1 \neq m_2$, $n_1 \neq n_2$ *we have* $|x_{m_1} - x_{m_2}| \geqslant A$, $|y_{n_1} - y_{n_2}| \geqslant B$. *Then*

$$\Big|\sum_m \sum_n \alpha_m \beta_n e(x_m y_n)\Big| \leqslant 5(1+XY)^{\frac{1}{2}}\Big(1+\frac{1}{AY}\Big)^{\frac{1}{2}}\Big(1+\frac{1}{BX}\Big)^{\frac{1}{2}}\|\alpha\|\|\beta\|.$$

PROOF. Apply (7.15) with $\max(X, B^{-1})$ and $\max(Y, A^{-1})$ in place of X and Y respectively. □

The spacing problem can be easily solved for points given by smooth functions. For example we derive by Corollary 7.3 the following

COROLLARY 7.4. *Let* $f(m)$, $g(n)$ *be real functions on* $[M, 2M]$, $[N, 2N]$ *respectively, such that* $f \ll F$, $g \ll G$ *and* $|f'| \geqslant FM^{-1}$, $|g'| \geqslant GN^{-1}$. *Then for any complex numbers* α_m, β_n *we have*

$$\sum_m \sum_n \alpha_m \beta_n e(f(m)g(n)) \ll (FG)^{-\frac{1}{2}}(FG+M)^{\frac{1}{2}}(FG+N)^{\frac{1}{2}}\|\alpha\|\|\beta\|.$$

PROOF. We have $|f(m_1) - f(m_2)| \geqslant |m_1 - m_2| FM^{-1}$ and $|g(n_1) - g(n_2)| \geqslant |n_1 - n_2| GN^{-1}$ hence the result follows from Corollary 7.3 for $A = FM^{-1}$, $B = GN^{-1}$, $X = F$ and $Y = G$. □

Corollary 7.3 yields also interesting results for exponential sums with monomials, because the rational points are well-spaced.

LEMMA 7.5. *Let* $\alpha\beta \neq 0$, $\Delta > 0$, $K \geqslant 1$, $L \geqslant 1$. *The number of integral quadruples* (k, k', ℓ, ℓ') *with* $K \leqslant k, k' \leqslant 2K$, $L \leqslant \ell, \ell' \leqslant 2L$ *such that*

$$\Big|\Big(\frac{k'}{k}\Big)^{\alpha} - \Big(\frac{\ell'}{\ell}\Big)^{\beta}\Big| \leqslant \Delta \tag{7.16}$$

is bounded by $O(KL(\Delta KL + \log 2KL))$ *where the implied constant depends only on* α, β.

EXERCISE 1. Prove Lemma 7.5.

From Lemma 7.5 we derive

COROLLARY 7.6. *Let* $\alpha\beta\gamma\delta \neq 0$, $X > 0$, K, L, M, $N \geqslant 1$. *Let* $a_{k,\ell}$ *and* $b_{m,n}$ *be complex numbers with* $|a_{k,\ell}| \leqslant 1$ *and* $|b_{m,n}| \leqslant 1$ *for* $K \leqslant k \leqslant 2K$, $L \leqslant \ell \leqslant 2L$, $M \leqslant m \leqslant 2M$, $N \leqslant n \leqslant 2N$. *Then*

$$\begin{aligned}\sum_k \sum_\ell \sum_m \sum_n a_{k,\ell} b_{m,n} e\Big(X \frac{k^\alpha \ell^\beta m^\gamma n^\delta}{K^\alpha L^\beta M^\gamma N^\delta}\Big) \\ \ll \Big(1+\frac{KL}{X}\Big)^{\frac{1}{2}}\Big(1+\frac{MN}{X}\Big)^{\frac{1}{2}}(XKLMN)^{\frac{1}{2}}\log(2KLMN)\end{aligned} \tag{7.17}$$

where the implied constant depends only on α, β, γ, δ.

7.3. Introduction to the large sieve.

The large sieve type inequalities are ubiquitous in modern analytic number theory. The name is somewhat misleading however, as sieves do not enter the general picture: it was derived from Linnik's original idea (see [**Li1**]), which was subsequently transformed beyond recognition into general L^2-type estimates, from which sieve results can be derived in some cases.

First we explain in general terms the underlying philosophy, which is expected to have much larger applicability. Let $\mathcal{X}$ be a finite set of "harmonics", well-suited maybe to solve analytically some interesting equation. To each $x \in \mathcal{X}$, we assume that a sequence $(x(n))$ is associated, so to speak the "Fourier coefficients".

The large sieve problem for $\mathcal{X}$ is to find $C = C(\mathcal{X}, N) \geqslant 0$ such that the following "large sieve inequality" holds for the L^2-mean of a linear form in the $x(n)$:

$$\sum_{x\in\mathcal{X}} \Big|\sum_{n\leqslant N} a_n x(n)\Big|^2 \leqslant C(\mathcal{X}, N)\|a\|^2 \tag{7.18}$$

for any complex numbers a_n, where $\|a\|^2 = \sum |a_n|^2$. By Cauchy's inequality one gets (7.18) with $C(\mathcal{X}, N) = N|\mathcal{X}|$, but this is not useful in applications.

The idea is that the "harmonics" are orthonormal, or nearly so, hence if one expands the square

$$\sum_{x\in\mathcal{X}} \Big|\sum_{n\leqslant N} a_n x(n)\Big|^2 = \sum_{n_1,n_2} a_{n_1}\overline{a_{n_2}} \sum_{x\in\mathcal{X}} x(n_1)\overline{x(n_2)},$$

the diagonal terms corresponding to $n_1 = n_2$ should be dominating, contributing roughly $|\mathcal{X}|\|a\|^2$ because $x(n)$ should be of size 1 on average. In any case it is hard to imagine that $C(\mathcal{X}, N)$ would be much smaller than the number of elements in $\mathcal{X}$. Moreover, because of the duality principle for bilinear forms, which means that (7.18) is equivalent with

$$\sum_{n\leqslant N} \Big|\sum_{x\in\mathcal{X}} b_x x(n)\Big|^2 \leqslant C(\mathcal{X}, N) \sum_{x\in\mathcal{X}} |b_x|^2 \tag{7.19}$$

with diagonal contribution $N\sum |b_x|^2$, it is not possible to make $C(\mathcal{X}, N)$ smaller than the length of the vector (a_n). Therefore in view of these diagonal limitations the best estimate which one can hope for $C(\mathcal{X}, N)$ is roughly

$$C(\mathcal{X}, N) \simeq |\mathcal{X}| + N. \tag{7.20}$$

This holds indeed in a number of crucial cases.

A sharp large sieve inequality (7.20) says that the linear forms

$$\sum_{n\leqslant N} a_n x(n)$$

of length $N \leqslant |\mathcal{X}|$ with $|a_n| \leqslant 1$ are, on average over $x \in \mathcal{X}$, of size $N^{1/2}$. This is best possible, and although in many cases it is expected that the individual sums are of this size, very few cases are known. Especially interesting is the case $a_n = \mu(n)$

and $x(n) = \lambda_f(n)$ are coefficients of some L-function (see Chapter 5). Then the individual bound

$$\sum_{n\leqslant N} \mu(n)\lambda_f(n) \ll_\varepsilon N^{\frac{1}{2}+\varepsilon}$$

(for any $\varepsilon > 0$) is equivalent with the Riemann Hypothesis for $L(f,s)$ (see (5.52)). Hence, a sharp large sieve type inequality for a family of L-functions is as powerful on average as the Grand Riemann Hypothesis would give.

7.4. Additive large sieve inequalities.

In the additive large sieve inequality the harmonics considered are additive characters of $\mathbb{Z}$ with $x(n) = e(\alpha n)$ for some $\alpha \in \mathbb{R}$ (of course only α mod 1 matters). So the linear forms to be estimated are just the trigonometric polynomials

$$S(\alpha) = \sum_n a_n e(\alpha n) \tag{7.21}$$

where the complex coefficients are supported in $M < n \leqslant M+N$. Note that if α_r are well-spaced modulo 1, i.e.,

$$\|\alpha_r - \alpha_s\| \geqslant \delta \quad \text{if } r \neq s,$$

for some $\delta > 0$ (where $\|x\|$ is the distance to the nearest integer), then the number of distinct points α_r, i.e. of harmonics, is $\leqslant 1 + \delta^{-1}$.

THEOREM 7.7. *For any set of δ-spaced points $\alpha_r \in \mathbb{R}/\mathbb{Z}$ and any complex numbers a_n with $M < n \leqslant M+N$, where $0 < \delta \leqslant \frac{1}{2}$ and $N \geqslant 1$ is an integer, we have*

$$\sum_r \Big| \sum_{M<n\leqslant M+N} a_n e(\alpha_r n)\Big|^2 \leqslant (\delta^{-1} + N - 1)\|a\|^2. \tag{7.22}$$

This estimate is best possible; it was proved in the form stated independently by Selberg [**S3**] and Montgomery and Vaughan [**MV2**], and is of the strength of (7.20). We give the proof of Montgomery and Vaughan which is based on the following generalization of the Hilbert inequality.

LEMMA 7.8. *Suppose λ_r are distinct real numbers with $|\lambda_r - \lambda_s| \geqslant \delta$ if $r \neq s$. Then for any complex numbers z_r we have*

$$\Big|\sum\sum_{r\neq s} \frac{z_r \overline{z_s}}{\lambda_r - \lambda_s}\Big| \leqslant \frac{\pi}{\delta}\sum_r |z_r|^2. \tag{7.23}$$

PROOF. By Cauchy's inequality, it suffices to show that

$$\sum_r \Big|\sum_{s\neq r} \frac{\overline{z_s}}{\lambda_r - \lambda_s}\Big|^2 \leqslant \frac{\pi^2}{\delta^2}\sum_r |z_r|^2. \tag{7.24}$$

Squaring out we arrange the left side as follows

$$\begin{aligned} L &= \sum\sum_{s,t} \overline{z_s} z_t \sum_{r\neq s,t} \frac{1}{(\lambda_r - \lambda_s)(\lambda_r - \lambda_t)} \\ &= \sum_s |z_s|^2 \sum_{r\neq s} \frac{1}{(\lambda_r - \lambda_s)^2} + \sum\sum_{s\neq t} \frac{\overline{z_s} z_t}{\lambda_s - \lambda_t} \sum_{r\neq s,t} \Big(\frac{1}{\lambda_r - \lambda_s} - \frac{1}{\lambda_r - \lambda_t}\Big). \end{aligned}$$

For the last sum we have

$$\sum_{r\neq s,t}\Big(\frac{1}{\lambda_r-\lambda_s}-\frac{1}{\lambda_r-\lambda_t}\Big)=\sum_{r\neq s}\frac{1}{\lambda_r-\lambda_s}-\sum_{r\neq t}\frac{1}{\lambda_r-\lambda_t}+\frac{2}{\lambda_s-\lambda_t},$$

hence

$$\begin{aligned}L=\sum_s|z_s|^2\sum_{r\neq s}\frac{1}{(\lambda_r-\lambda_s)^2}&+2\sum\sum_{s\neq t}\frac{\overline{z_s}z_t}{(\lambda_s-\lambda_t)^2}\\&+\sum\sum_{s\neq t}\frac{\overline{z_s}z_t}{\lambda_s-\lambda_t}\Big\{\sum_{r\neq s}\frac{1}{\lambda_r-\lambda_s}-\sum_{r\neq t}\frac{1}{\lambda_r-\lambda_t}\Big\}.\end{aligned}$$

Note now that the estimate (7.24) amounts to finding the norm of the matrix $(\mu_{r,s})$ with $\mu_{r,s}=(\lambda_r-\lambda_s)^{-1}$ if $r\neq s$ and $\mu_{r,r}=0$, thus we may assume that the vector (z_r) is extremal. Since the matrix is skew-hermitian, the extremal vector is an eigenvector, i.e.

$$\sum_{r\neq s}\frac{z_r}{\lambda_r-\lambda_s}=\nu z_s$$

for some purely imaginary $\nu\in\mathbb{C}$. This shows that the last two sums in L cancel out. Therefore for the extremal vector we have

$$L=\sum_s|z_s|^2\sum_{r\neq s}\frac{1}{(\lambda_r-\lambda_s)^2}+2\sum\sum_{s\neq t}\frac{\overline{z_s}z_t}{(\lambda_s-\lambda_t)^2}.$$

Applying $2|z_sz_t|\leqslant|z_s|^2+|z_t|^2$, we obtain

$$L\leqslant 3\sum_s|z_s|^2\sum_r\frac{1}{(\lambda_r-\lambda_s)^2}.$$

Since we have $|\lambda_r-\lambda_s|\geqslant\delta|r-s|$ the innermost sum is bounded by $2\delta^{-2}\zeta(2)=\pi^2/3\delta^2$. This gives (7.24) completing the proof of (7.23). □

COROLLARY 7.9. *For any set of δ-spaced points $\alpha_r\in\mathbb{R}/\mathbb{Z}$ and any complex numbers z_r we have*

$$\Big|\sum\sum_{r\neq s}\frac{z_r\overline{z_s}}{\sin\pi(\alpha_r-\alpha_s)}\Big|\leqslant\delta^{-1}\sum_r|z_r|^2.\tag{7.25}$$

PROOF. We apply (7.23) to the doubly-indexed set of numbers $z_{m,r}=(-1)^mz_r$ and $\lambda_{m,r}=m+\alpha_r$ with $1\leqslant m\leqslant K$, getting

$$\Big|\sum\sum_{(r,m)\neq(s,n)}(-1)^{m-n}\frac{z_r\overline{z_s}}{m-n+\alpha_r-\alpha_s}\Big|\leqslant\frac{\pi K}{\delta}\sum_r|z_r|^2.$$

Here we can replace the summation condition $(r,m)\neq(s,n)$ by $r\neq s$ because for $r=s$ the remaining terms cancel out pairwise (m,n) against (n,m). If we put $k=m-n$ and divide by K we derive

$$\Big|\sum\sum_{r\neq s}z_r\overline{z_s}\sum_{k=-K}^{K}\Big(1-\frac{|k|}{K}\Big)\frac{(-1)^k}{k+\alpha_r-\alpha_s}\Big|\leqslant\frac{\pi}{\delta}\sum_r|z_r|^2.$$

This yields (7.25) by letting $K \to +\infty$ because for $\alpha \notin \mathbb{Z}$

$$\sum_{k\in\mathbb{Z}} \frac{(-1)^k}{k+\alpha} = \frac{\pi}{\sin \pi\alpha}$$

where the series is summed symmetrically. □

COROLLARY 7.10. *For any real x we have*

$$\Big|\sum_{r\neq s}\sum z_r\overline{z_s}\frac{\sin 2\pi x(\alpha_r-\alpha_s)}{\sin \pi(\alpha_r-\alpha_s)}\Big| \leqslant \delta^{-1}\sum_r |z_r|^2. \tag{7.26}$$

PROOF. This follows by applying Corollary 7.9 twice with z_r twisted by $e(x\alpha_r)$ and $e(-x\alpha_r)$. □

Now we are ready to prove (7.22), first with $\delta^{-1}+N$ instead of $\delta^{-1}+N-1$. By duality, it is equivalent to show that

$$\sum_{M<n\leqslant M+N}\Big|\sum_r z_r e(n\alpha_r)\Big|^2 \leqslant (N+\delta^{-1})\|z\|^2 \tag{7.27}$$

for any complex numbers z_r. Squaring out, we see that the diagonal term $r=s$ contributes $N\|z\|^2$, while the off-diagonal terms yield

$$\sum_{r\neq s}\sum z_r\overline{z_s}\sum_n e(n(\alpha_r-\alpha_s)) = \sum_{r\neq s}\sum z_r\overline{z_s}e(K(\alpha_r-\alpha_s))\frac{\sin \pi N(\alpha_r-\alpha_s)}{\sin\pi(\alpha_r-\alpha_s)}$$

with $K = M+\frac{1}{2}(N+1)$. This quantity is smaller than the left side of (7.26), hence is $\leqslant \delta^{-1}\|z\|^2$. Adding both contributions gives (7.27).

To save the -1, we apply Theorem 7.7 to the points $(\alpha_r+k)K^{-1}$ with $1\leqslant k\leqslant K$, for the trigonometric polynomial $T(\alpha)=S(\alpha K)$. We obtain

$$K\sum_r |S(\alpha_r)|^2 = \sum_k\sum_r \Big|T\Big(\frac{\alpha_r+k}{K}\Big)\Big|^2 \leqslant (\delta^{-1}K+NK-K+1)\|a\|^2$$

because the points $(\alpha_r+k)K^{-1}$ are spaced by δK^{-1}, and the sum $T(\alpha)$ ranges over $m=nK$ with $(MK+K-1)<m\leqslant(MK+K-1)+(NK+K-1)$. Hence dividing by K and letting K tend to infinity we obtain (7.22) as stated (this trick is due to Paul Cohen).

For many applications, a weaker estimate with the factor $\delta^{-1}+N-1$ replaced by $C(\delta^{-1}+N)$ with an absolute constant C is sufficient. This kind of large sieve inequality was first stated by H. Davenport and H. Halberstam [**DH1**]. We are going to give a quick proof of one such estimate. By duality we need to show that

$$\sum_{|n|<N}\Big|\sum_r \gamma_r e(\alpha_r n)\Big|^2 \leqslant C(\delta^{-1}+N)\|\gamma\|^2$$

for any complex numbers γ_r. Let $f(x)$ be a non-negative function with $f(x)\geqslant 1$ if $|x|\leqslant 1$. The left side above is estimated by the same expression with extra smoothing factor $f(n/N)$, and the latter equals

$$\sum_r\sum_s \gamma_r\overline{\gamma_s}\sigma(r,s)$$

where

$$\sigma(r,s) = \sum_n f\Big(\frac{n}{N}\Big) e((\alpha_r - \alpha_s)n).$$

By Poisson's formula

$$\sigma(r,s) = N \sum_h \hat{f}((\alpha_r - \alpha_s + h)N).$$

We choose $f(x)$ such that its Fourier transform satisfies $\hat{f}(y) \ll (1+y^2)^{-1}$ giving $\sigma(r,s) \ll N(1 + \|\alpha_r - \alpha_s\|^2 N^2)^{-1}$. Finally applying $2|\gamma_r\gamma_s| \leqslant |\gamma_r|^2 + |\gamma_s|^2$ and

$$\sum_s |\sigma(r,s)| \ll N \sum_{k=0}^{\infty} (1 + (\delta k N)^2)^{-1} \ll \delta^{-1} + N$$

we complete the proof.

The above arguments would yield the best possible constant $C = 1$ if the test function $f(x)$ was chosen appropriately. Selberg did just that, his optimal function being quite complicated.

Now specialize the large sieve inequality (7.4) by taking for the points α_r the rationals a/q with $1 \leqslant q \leqslant Q$ and $(a,q) = 1$. These points are spaced by $\delta = Q^{-2}$, indeed if $a/q \neq a'/q'$, then

$$\|a/q - a'/q'\| = \|(aq' - a'q)/qq'\| \geqslant (qq')^{-1} \geqslant Q^{-2}.$$

Therefore Theorem 7.7 yields

THEOREM 7.11. *For any complex numbers a_n with $M < n \leqslant M + N$, where N is a positive integer, we have*

$$\sum_{q \leqslant Q} \sideset{}{^\star}\sum_{a \pmod q} \Big| \sum_{M < n \leqslant M+N} a_n e\Big(\frac{an}{q}\Big)\Big|^2 \leqslant (Q^2 + N - 1)\|a\|^2. \tag{7.28}$$

Notice that if $\mathcal{A} = (a_n)$ is supported on an arithmetic progression $n \equiv \ell \pmod k$, and $(k,q) = 1$, we can change variables to derive

COROLLARY 7.12. *For any complex numbers a_n with $M < n \leqslant M + N$ we have*

$$\sum_{\substack{q \leqslant Q \\ (k,q)=1}} \sideset{}{^\star}\sum_{a \pmod a} \Big| \sum_{n \equiv \ell \pmod q} a_n e\Big(\frac{an}{q}\Big)\Big|^2 \leqslant (Q^2 + k^{-1}N) \sum_{n \equiv \ell \pmod k} |a_n|^2. \tag{7.29}$$

EXERCISE 2. Prove the following large sieve inequality in several variables: let $d \geqslant 1$ and $\delta > 0$ and let $\alpha_r = (\alpha_{r,1}, \dots, \alpha_{r,d})$ be δ-spaced points in $\mathbb{R}^d/\mathbb{Z}^d$, i.e. if $r \neq s$, we have $\max \|\alpha_{r,i} - \alpha_{s,i}\| \geqslant \delta$. Then

$$\sum_r \Big|\sum_n a_n e(n \cdot \alpha_r)\Big|^2 \ll (\delta^{-d} + N^d)\|a\|^2 \tag{7.30}$$

where $x \cdot y$ is the standard scalar product in $\mathbb{R}^d$, and a_n are arbitrary complex numbers for $n = (n_1, \dots, n_d)$ with $1 \leqslant n_i \leqslant N$, the implied constant depending only on the dimension d.

7.5. Multiplicative large sieve inequality.

Instead of additive characters, we now take as harmonics the Dirichlet characters. If one considers $\mathcal{X} = \{\chi \mid \chi \text{ is a character modulo } q\}$, then by expanding the square and applying the orthogonality relation one finds immediately

$$\sum_{\chi \pmod q} \Big|\sum_{n\leqslant N} a_n\chi(n)\Big|^2 \leqslant (q+N)\|a\|^2$$

which is comparable to (7.20).

Much more significant, however, is the case where the harmonics are all primitive Dirichlet characters to modulus $q \leqslant Q$. Here we have the basic result of Bombieri and Davenport [**BD**] (their original result is slightly weaker):

THEOREM 7.13. *For any complex numbers a_n with $M < n \leqslant M+N$, where N is a positive integer, we have*

$$\sum_{q\leqslant Q} \frac{q}{\varphi(q)} \sideset{}{^\star}\sum_{\chi \pmod q} \Big|\sum_{M<n\leqslant M+N} a_n\chi(n)\Big|^2 \leqslant (Q^2+N-1)\|a\|^2. \tag{7.31}$$

This is again as strong as (7.20). Note that without the restriction to primitive χ, this inequality would be false by a large factor (take $a_n = 1$ for all n, then the trivial characters modulo $q \leqslant Q$ contribute already about N^2Q). An interesting feature of (7.31) is that the bound depends only on the length of the interval, but not on its position. In this aspect the large sieve inequality for Dirichlet characters has some advantage over the GRH.

PROOF. We can in fact give a slightly more precise estimate, which is sometimes useful in applications.

The primitive multiplicative characters $\chi \pmod q$ can be expanded into additive characters by means of Gauss sums (see (3.12))

$$\tau(\chi)\bar{\chi}(n) = \sum_{a \pmod q} \chi(a)e\Big(\frac{an}{q}\Big)$$

for any n. If χ is not primitive but induced by a primitive χ_s modulo s with $q = rs$, then if $(r,s) = 1$, we have

$$\begin{aligned}\sum_{a \pmod q} \chi(a)e\Big(\frac{an}{q}\Big) &= \sideset{}{^\star}\sum_{b \pmod r}\ \sideset{}{^\star}\sum_{c \pmod s} \chi(bs+cr)e\Big(\frac{nb}{r}+\frac{cn}{s}\Big)\\ &= \bar{\chi}(n)\chi_s(r)c_r(n)\tau(\chi_s)\end{aligned}$$

where $c_r(n)$ is a Ramanujan sum (3.1). We derive

$$\sum_n a_n\bar{\chi}(n)c_r(n) = \frac{\bar{\chi_s}(r)}{\tau(\chi_s)} \sum_{a \pmod q} \chi(a)S(a/q)$$

for any complex numbers a_n, where $S(\alpha)$ is the trigonometric polynomial (7.21).

By orthogonality of characters we then have

$$\sum_{\substack{rs\leqslant Q\\(r,s)=1}} \frac{s}{\varphi(rs)} \sum_{\chi \,(\mathrm{mod}\, s)}^{\star} \Big|\sum_n a_n\bar{\chi}(n)c_r(n)\Big|^2 \leqslant \sum_{q\leqslant Q}\frac{1}{\varphi(q)} \sum_{\chi\,(\mathrm{mod}\, q)} \Big|\sum_a \chi(a)S\Big(\frac{a}{q}\Big)\Big|^2$$

$$= \sum_{q\leqslant Q}\ \sideset{}{^\star}\sum_{a \bmod q} \Big|\sum_n a_n e\Big(\frac{an}{q}\Big)\Big|^2.$$

Applying Theorem 7.11, we conclude that

$$\sum_{\substack{rs\leqslant Q\\(r,s)=1}} \frac{s}{\varphi(rs)} \sum_{\chi\,(\mathrm{mod}\, s)}^{\star} \Big|\sum_{M<n\leqslant M+N} a_n\bar{\chi}(n)c_r(n)\Big|^2 \leqslant (Q^2+N-1)\|a\|^2. \tag{7.32}$$

This inequality contains (7.31) by positivity (ignore all terms with $r \neq 1$). □

REMARK. Bombieri and Davenport did succeed in getting interesting applications of (7.32) by exploiting the extra summation over r.

7.4. Applications of the large-sieve to sieving problems.

In this section we explain how the large sieve inequality for additive characters implies a sieve result in the "ordinary" sense (see also Chapter 6 for background on sieves), and give Linnik's first application of such a result for estimation of the least quadratic non-residue.

Consider a finite set $\mathcal{M}$ of integers and a finite set $\mathcal{P}$ of prime numbers. For each $p \in \mathcal{P}$, let $\Omega_p \subset \mathbb{Z}/p\mathbb{Z}$ be a set of residue classes "to sieve out". The data $(\mathcal{M}, \mathcal{P}, \Omega)$ defines a *sieving problem*; the corresponding *sifted set* is

$$\mathcal{S}(\mathcal{M},\mathcal{P},\Omega) = \{m \in \mathcal{M};\quad m \,(\mathrm{mod}\, p) \notin \Omega_p \quad \text{for all } p \in \mathcal{P}\}. \tag{7.33}$$

The goal is to estimate $S = |\mathcal{S}(\mathcal{M},\mathcal{P},\Omega)|$, the cardinality of $\mathcal{S}(\mathcal{M},\mathcal{P},\Omega)$. More generally we consider, for an arbitrary sequence of complex numbers $a = (a_n)$, the sum

$$Z = \sum_{n\in\mathcal{S}(\mathcal{M},\mathcal{P},\Omega)} a_n, \tag{7.34}$$

and we seek a bound for Z in terms of the ℓ_2-norm of a. Naturally in this context the name "large sieve" is appropriate, because in contrast with other methods (like combinatorial sieves, see Chapter 6), we are looking for strong estimates even if $|\Omega_p|$ is rather large when p is large.

We derive now from Theorem 7.11 a result somewhat stronger than Linnik's original version (see [**Rot**], [**Mo1**], [**Hu1**], [**Bo2**]).

THEOREM 7.14. *Suppose $\mathcal{M}$ is contained in an interval of length $N \geqslant 1$, and assume $\Omega_p \neq \mathbb{Z}/p\mathbb{Z}$ for every $p \in \mathcal{P}$, i.e., $\omega(p) = |\Omega_p| < p$. Then we have*

$$|Z|^2 \leqslant \frac{N+Q^2}{H}\|a\|^2 \tag{7.35}$$

for any $Q \geqslant 1$, where

$$H = \sideset{}{^\flat}\sum_{q\leqslant Q} h(q) \tag{7.36}$$

and $h(q)$ is the multiplicative function supported on squarefree integers with prime divisors in $\mathcal{P}$ such that

$$h(p) = \frac{\omega(p)}{p - \omega(p)}. \tag{7.37}$$

In particular, taking for a_n the characteristic function of $\mathcal{S}(\mathcal{M}, \mathcal{P}, \Omega)$, we have

$$S \leqslant \frac{N + Q^2}{H}. \tag{7.38}$$

Let $S(\alpha)$ be the trigonometric polynomial (7.21), so $S(0) = Z$. First we establish the following inequality for individual moduli.

LEMMA 7.15. *For any positive squarefree number q we have*

$$h(q)|S(0)|^2 \leqslant \sum_{a \,(\mathrm{mod}\, q)}^{\star} \Big|S\Big(\frac{a}{q}\Big)\Big|^2. \tag{7.39}$$

PROOF. Let $X(q, \nu)$, for $\nu \in \mathbb{Z}/q\mathbb{Z}$, denote

$$X(q, \nu) = \sum_{n \equiv \nu \,(\mathrm{mod}\, q)} a_n.$$

Using additive characters we have

$$X(q, \nu) = \frac{1}{q} \sum_{a \,(\mathrm{mod}\, q)} e\Big(-\frac{a\nu}{q}\Big) S\Big(\frac{a}{q}\Big),$$

so by orthogonality (Plancherel formula) it follows that

$$q \sum_{\nu \,(\mathrm{mod}\, q)} |X(q, \nu)|^2 = \sum_{a \,(\mathrm{mod}\, q)} \Big|S\Big(\frac{a}{q}\Big)\Big|^2.$$

If $q = p$ is prime, we have $X(p, \nu) = 0$ for $\nu \in \Omega_p$, so by Cauchy's inequality we get

$$\begin{aligned}|Z|^2 = |S(0)|^2 &= \Big|\sum_{\nu \ (\mathrm{mod}\ p)} X(p, \nu)\Big| \\ &\leqslant (p - \omega(p)) \sum_{\nu} |X(p, \nu)|^2 = \Big(1 - \frac{\omega(p)}{p}\Big) \sum_{a \ (\mathrm{mod}\ p)} \Big|S\Big(\frac{a}{p}\Big)\Big|^2\end{aligned}$$

which yields (7.39) by subtracting the term for $a \equiv 0 \pmod p$. Now if $q = q_1 q_2$ with $(q_1, q_2) = 1$, we have

$$\sum_{a \,(\mathrm{mod}\, q)}^{\star} \Big|S\Big(\frac{a}{q}\Big)\Big|^2 = \sum_{a_1 \ (\mathrm{mod}\ q_1)}^{\star} \ \sum_{a_2 \ (\mathrm{mod}\ q_2)}^{\star} \Big|S\Big(\frac{a_1}{q_1} + \frac{a_2}{q_2}\Big)\Big|^2.$$

Hence if (7.39) holds for q_1 and q_2, we derive by successive applications that

$$\sum_{a \,(\mathrm{mod}\, q)}^{\star} \Big|S\Big(\frac{a}{q}\Big)\Big|^2 \geqslant h(q_2) \sum_{a_1 \ (\mathrm{mod}\ q_1)}^{\star} \Big|S\Big(\frac{a_1}{q_1}\Big)\Big|^2 \geqslant h(q_1)h(q_2)|S(0)|^2,$$

so the proof of (7.39) is finished by induction on the number of prime factors of q.□

PROOF OF THEOREM 7.14. Sum (7.39) over $q \leqslant Q$ and then apply Theorem 7.11. $\square$

Linnik [**Li1**] established an inequality somewhat weaker than (7.38), yet powerful enough to give a spectacular application to the problem of the least quadratic non-residue modulo a prime p, i.e., the smallest positive integer $q(p)$ such that $(q(p)/p) = -1$. Note that $q(p)$ is prime. It is conjectured that $q(p) \ll_\varepsilon p^\varepsilon$ whereas the best known estimate is

$$q(p) \ll_\varepsilon p^{\theta+\varepsilon}, \quad \theta = 1/4\sqrt{e} = 0.1516\dots$$

From the Grand Riemann Hypothesis for Dirichlet L-functions one derives that $q(p) \ll (\log p)^2$ with an absolute implied constant.

THEOREM 7.16 (LINNIK). *Let $\varepsilon > 0$. The number of primes $p \leqslant N$ such that $q(p) > N^\varepsilon$ is bounded by a constant depending only on ε.*

PROOF. Let X_ε be the number of primes $\leqslant \sqrt{N}$ with $q(p) > N^\varepsilon$, which we want to show is bounded in terms of ε. To this end consider the sieving problem with

$$\begin{aligned} \mathcal{M} &= \{1, 2, \dots, N\}, \\ \mathcal{P} &= \{p \leqslant \sqrt{N} \mid \left(\frac{n}{p}\right) = 1 \text{ for all } n \leqslant N^\varepsilon\}, \\ \Omega_p &= \{\nu \pmod{p} \mid \left(\frac{\nu}{p}\right) = -1\}. \end{aligned}$$

Note that $\omega(p) = \frac{1}{2}(p-1)$ and $h(p) = (p-1)/(p+1)$. This is indeed a "large sieve" because $h(p) \geqslant \frac{1}{3}$.

The sifted set contains the set, say $\mathcal{Z}_\varepsilon$, of all $n \leqslant N$ which have no prime divisors larger than N^ε. Therefore by (7.8) with $Q = \sqrt{N}$ we get $Z_\varepsilon = |\mathcal{Z}_\varepsilon| \leqslant 2NH^{-1}$. Combining this inequality with

$$\tfrac{1}{3}X_\varepsilon \leqslant \sum_{\substack{p \leqslant \sqrt{N} \\ q(p) \geqslant N^\varepsilon}} h(p) \leqslant H$$

we get $X_\varepsilon Z_\varepsilon \leqslant 6N$. It remains to estimate Z_ε.

It is known that $Z_\varepsilon \sim \delta(\varepsilon)N$ for some $\delta(\varepsilon) > 0$ as $N \longrightarrow \infty$ (see e.g. [**Bo2**], pp. 8–9), so the result follows. Alternatively, a sufficient lower bound for Z_ε follows by counting in the set Z_ε numbers of special type $n = mp_1 \cdots p_k \leqslant N$ with $N^{\varepsilon-\varepsilon^2} < p_j < N^\varepsilon$ for $1 \leqslant j \leqslant k = \varepsilon^{-1}$. We get

$$Z_\varepsilon \geqslant \sum_{p_1, \dots, p_k} \left[\frac{N}{p_1 \dots p_k}\right] \gg N.$$

Hence $X_\varepsilon \ll 1$. Finally changing N to N^2 and ε to $\varepsilon/2$ we conclude Linnik's theorem exactly as stated. $\square$

See Proposition 7.30 for an analogue for elliptic curve using the large sieve type inequality for symmetric square L-functions.

EXERCISE 3. (1) Consider the following sieve in dimension d : for any prime $p \in \mathcal{P}$, let $\Omega(p) \subset (\mathbb{Z}/p\mathbb{Z})^d$ be a subset with $\omega(p) = |\Omega(p)| < p$. Let $\mathcal{M} \subset [-N, N]^d$. Put

$$\mathcal{S}(\mathcal{M}, \mathcal{P}, \Omega) = \{m = (m_1, \dots, m_d) \in \mathcal{M} \mid \text{ for } p \in \mathcal{P},\ m \,(\mathrm{mod}\ p) \notin \Omega_p\}.$$

Show that

$$S = |\mathcal{S}(\mathcal{M}, \mathcal{P}, \Omega)| \ll (N^d + Q^{2d})H^{-1}$$

where H is given by (7.36) and the implied constant depends on d [**Hint:** Use (7.30)].

(2) Fix $n \geqslant 1$. Let D_N be the set of monic polynomials $f \in \mathbb{Z}[X]$ of degree n with coefficients in absolute value $\leqslant N$. For $r = (r_1, \dots, r_k)$ with $r_1 + 2r_2 + \cdots = n$ let $C_r \subset D_N$ be the subset of polynomials f for which $f \pmod p$ does not factor as a product of r_1 linear factors, r_2 quadratic factors, ..., for any prime p. Show that

$$|C_r| \ll N^{n-\frac{1}{2}} \log N,$$

the implied constant depending only on n.

(3) Deduce that the set $C \subset D_N$ of polynomials f for which the Galois group of the splitting field of f is not S_n satisfies

$$|C| \ll N^{n-\frac{1}{2}} \log N$$

the implied constant depending only on n. [**Hint:** Use the fact that if f has Galois group H which is a proper subgroup of S_n, the union of the conjugates of H is distinct from S_n, hence there is a "splitting type" r as in (2) such that $f \in C_r$.] This is due to Gallagher [**Ga4**].

Another interesting application of two-dimensional large sieve is the proof by Duke [**Du5**] that almost all elliptic curves $E/\mathbb{Q}$ have their p-torsion field with Galois group $GL(2, \mathbb{Z}/p\mathbb{Z})$ for all primes p.

7.6. Panorama of the large sieve inequalities.

There is a great variety of estimates of the large sieve type which are used these days in analytic number theory. In this section we selected a few important ones (without proofs) to make an impression about the scope of the matter.

Some applications require large sieve inequalities in which the set of harmonics consists of characters twisted by exponentials of different kind. The first case of such a hybrid large sieve is due to P.X. Gallagher [**Ga1**].

THEOREM 7.17. *Let N, Q, $T \geqslant 1$. For any complex numbers a_n we have*

$$\sum_{q \leqslant Q} \sum_{\chi \,(\mathrm{mod}\ q)}^{\star} \int_{-T}^{T} \Big| \sum_{n \leqslant N} a_n \chi(n) n^{it} \Big|^2 dt \ll (Q^2 T + N) \|a\|^2.$$

A discrete version of this result, among further generalizations, can be found in [**Mo2**].

Of course, twisting the multiplicative characters by additive ones does not produce stronger results, but if the exponential component is not linear, then it is likely to be orthogonal to the characters and one can receive extra saving (in Theorem 7.17 the exponential component is $e(\frac{t}{2\pi} \log n)$). Here is another example of some importance for applications (see [**DuI3**]).

THEOREM 7.18. *Let $v(x)$ be a real, smooth function on $\mathbb{R}^+$ such that $0 < x|v'(x)| < 1$ and $|v'(x)|^2 < |v''(x)|$. Then for $X \geqslant Q \geqslant 1$ we have*

$$\sum_{q\leqslant Q} \sum_{\chi \,(\mathrm{mod}\, q)}^{\star} \Big|\sum_{n\leqslant N} a_n\chi(n)e\Big(X\frac{v(x)}{q}\Big)\Big|^2 \ll (N + Q^{\frac{3}{2}}X^{\frac{1}{2}}\log X)\|a\|^2.$$

where the implied constant is absolute.

If the linear forms $\sum a_n\chi(n)$ are lacunary (many coefficients a_n vanish), then the power of the large sieve reduces significantly. There is a great demand for results which would be sensitive in this respect (cf. P. Elliott [**El**]). O. Ramaré worked out many interesting estimates of such kind (unpublished).

PROBLEM 7.19. *Prove that for any complex numbers a_n supported on primes*

$$\sum_{q\leqslant Q} \sum_{\chi \,(\mathrm{mod}\, q)}^{\star} \Big|\sum_{n\leqslant N} a_n\chi(n)\Big|^2 \ll (Q^2 + N(\log N)^{-1})\|a\|^2$$

where the implied constant is absolute.

A more general problem is to fix $f \in \mathbb{Z}[X]$ and establish good large sieve inequalities for linear forms

$$\sum_{n\leqslant N} a_n e\Big(\frac{af(n)}{q}\Big), \quad \sum_{n\leqslant N} a_n\chi(f(n)).$$

One should realize how fortunate is the case of large sieve for coefficients of full density, because the duality principle works efficiently. For sequences of coefficients a_n which are very sparse, the duality arguments are poor and very little can be done directly while predictions for the best possible estimates are risky.

In the same note it is a challenging problem to establish the best large sieve inequality for harmonics which are primitive characters of a fixed order. Here we have a powerful result of D. R. Heath-Brown [**HB1**] for quadratic characters (see also the earlier result of M. Jutila [**Ju2**]).

THEOREM 7.20. *For any complex numbers a_n we have*

$$\sum_{m\leqslant M}^{\flat} \Big|\sum_{n\leqslant N}^{\flat} a_n\Big(\frac{n}{m}\Big)\Big|^2 \ll (MN)^{\varepsilon}(M+N)\|a\|^2.$$

Here the superscript $\flat$ restricts the summation to squarefree numbers, $(\frac{n}{m})$ stands for the Jacobi symbol, ε is any positive number and the implied constant depends only on ε.

REMARKS. The fact that we can view $(\frac{n}{m})$ as either a character in n of modulus m, or a character in m of modulus $4n$ (by way of the quadratic reciprocity law) is the key feature of the proof. Note that the problem is self-dual, i.e. interchanging m and n does not matter. For interesting applications of Theorem 7.20, see [**HB1**], [**Sou**], [**IM**], [**PP**].

In connection with Kloosterman sums one meets characters $(\frac{f(n)}{m})$ where f is a quadratic polynomial. The corresponding large sieve was considered in [**I1**] with the following result

THEOREM 7.21. *For any complex numbers a_n we have*

$$\sum_{m\leqslant M}\Big|\sum_{n\leqslant N}a_n\Big(\frac{n^2-4}{m}\Big)\Big|^2\ll(MN)^\varepsilon(M^{\frac{3}{2}}+M^{\frac{2}{3}}N)\|a\|^2$$

where ε is any positive number and the implied constant depends only on ε.

EXERCISE 4. Let $S(m,n;c)$ denote the classical Kloosterman sum. For any complex numbers α_m, β_n we have

$$\sum_{c\leqslant C}\Big|\sum_{m\leqslant M}\sum_{n\leqslant N}\alpha_m\beta_nS(m,n;c)\Big|\leqslant(C^2+M+N)\|\alpha\|\|\beta\|.$$

EXERCISE 5. For any complex numbers α_m, β_n we have

$$\sum_{q\leqslant Q}\ \sideset{}{^\star}\sum_{a\,(\mathrm{mod}\,q)}\Big|\sum\sum_{\substack{m\leqslant M,n\leqslant N\\(mn,q)=1}}\alpha_m\beta_ne\Big(\frac{am\bar{n}}{q}\Big)\Big|^2\leqslant(Q^2+MN)\|\alpha\|^2\|\beta\|^2.$$

The estimates in the above exercises can be easily derived from the large sieve inequalities (7.11), (7.13) respectively. Much harder is the following estimate for bilinear forms with Kloosterman fractions due to W. Duke, J. Friedlander and H. Iwaniec [**DFI4**].

THEOREM 7.22. *Let a be a positive integer. For any complex numbers α_m, β_n we have*

$$\sum\sum_{\substack{m\leqslant M,n\leqslant N\\(m,n)=1}}\alpha_m\beta_ne\Big(a\frac{\bar{m}}{n}\Big)\ll(MN)^\varepsilon\Big(\frac{1}{M}+\frac{1}{N}\Big)^{\frac{1}{58}}(a+MN)^{\frac{1}{2}}\|\alpha\|\|\beta\|$$

where the implied constant depends only on ε.

REMARKS. Without the factor $(M^{-1}+N^{-1})^{1/58}$, this bound would be just trivial. It is often crucial for applications to have a saving factor, less important how large it is. The proof of Theorem 7.22 uses the amplification method; the principle of this is explained at the end of Chapter 26.

Bilinear forms in complex domains are also desired for applications. We select from [**FI1**] two results for the Jacobi-Dirichlet symbol $(\frac{z}{w})$ and for the Jacobi-Kubota symbol $[wz]$ (for definitions, see Section 3.7).

THEOREM 7.23. *Let α_w, β_z be any complex numbers with $|\alpha_w|\leqslant1$, $|\beta_z|\leqslant1$ for w, z in the discs $|w|^2\leqslant M$, $|z|^2\leqslant N$. Then*

$$\sideset{}{^\star}\sum_w\sum_z\alpha_w\beta_z\Big(\frac{z}{w}\Big)\ll(M+N)^{\frac{1}{12}}(MN)^{\frac{11}{12}+\varepsilon}$$

where the $*$ restricts the summation to the primary primitive numbers w, ε is any positive number and the implied constant depends only on ε.

Using the twisted multiplicativity (3.71) of the Jacobi-Kubota symbol one can derive from Theorem 7.23 the following estimate (subject to some minor conditions)

$$\sideset{}{^\star}\sum_w\sideset{}{^\star}\sum_z\alpha_w\beta_z[wz]\ll(M+N)^{\frac{1}{12}}(MN)^{\frac{11}{12}+\varepsilon}$$

With similar results but in a different fashion one can treat bilinear forms in $\psi_f(mn)$ where $\psi_f(\ell)$ are the normalized Fourier coefficients of a fixed cusp form $f \in S_k(\Gamma_0(N), \nu)$ of weight $k =$ half an odd integer, and ν is the compatible theta multiplier (see (1.53) and (3.42)). These coefficients are not multiplicative so a cancellation in general bilinear forms is possible. The following estimate was proved in [**DuI1**],

$$\sum_{m\leqslant M}^{\flat}\sum_{n\leqslant N}^{\flat}\alpha_m\beta_n\psi_f(mn) \ll (MN)^{\varepsilon}(M^{\frac{1}{2}}+M^{\frac{1}{4}}N)\|\alpha\|\|\beta\|$$

where α_m, β_n are any complex numbers and ε is any positive number, the implied constant depending only on ε and the cusp form f. Probably this general estimate holds true with $M^{1/2}+M^{1/4}N$ replaced by $M+N$. Have in mind that the bound $\psi_f(\ell) \ll \ell^{\varepsilon}$ is expected to be true for ℓ squarefree but it is not yet proved; it is essentially equivalent with the Lindelöf Hypothesis for a suitable automorphic L-function twisted by the real character χ_ℓ in the ℓ-aspect (via Shimura correspondence and Waldspurger formula).

7.7. Large-sieve inequalities for cusp forms.

The Fourier coefficients of cusp forms, or better the eigenvalues of Hecke operators in the space of cusp forms, are analogues of Dirichlet characters. As tools they are powerful due to the orthogonality in the sense of the large sieve inequalities. There are two essential aspects of the matter, the spectral and the level aspects. Moreover, some hybrid aspects also appear in practice.

Let (u_j) be an orthonormal basis of Maass cusp forms for $\Gamma = \Gamma_0(q)$ with $q \geqslant 1$ (see Theorem 15.5) and let $\lambda_j = s_j(1-s_j)$ with $s_j = \frac{1}{2}+it_j$ be the corresponding eigenvalues of the Laplace operator. Let

$$u_j(z) = \sqrt{y}\sum_{n\neq 0}\rho_j(n)K_{it_j}(2\pi|n|y)e(nx)$$

be the Fourier expansion of u_j at the cusp ∞ (see Lemma 15.1). Consider the normalized coefficients

$$\nu_j(n) = \Big(\frac{|n|q}{\cosh \pi t_j}\Big)^{\frac{1}{2}}\rho_j(n), \text{ for } n\neq 0, \quad t_j>0. \tag{7.40}$$

These are expected to be essentially bounded on average with respect to n, q, t_j. Using the Kuznetsov formula (Theorem 16.3) one can show

THEOREM 7.24. *Let $q \geqslant 1$, $T \geqslant 1$ and $N \geqslant 1$. For any complex numbers a_n we have*

$$\sum_{t_j\leqslant T}\Big|\sum_{n\leqslant N}a_n\nu_j(n)\Big|^2 \ll (qT^2+N\log N)\|a\|^2 \tag{7.41}$$

where the implied constant is absolute.

Many results of this type are established in [**DI**], but with a slightly weaker term $N^{1+\varepsilon}$ in place of $N\log N$ (probably N would suffice). Note that qT^2 is approximately the number of cusp forms $u_j(z)$ for $\Gamma_0(q)$ with $0 < t_j \leqslant T$ by Weyl's Law.

Still more orthogonality is expected from the primitive cusp forms. The following estimate is an open problem

PROBLEM 7.25. *Let $Q \geqslant 1$, $T \geqslant 1$ and $N \geqslant 1$. Prove that for any complex numbers a_n*

$$\sum_{q\leqslant Q}\sideset{}{^\star}\sum_{t_j\leqslant T}\Big|\sum_{n\leqslant N}a_n\nu_j(n)\Big|^2 \ll (Q^2T^2+N)\|a\|^2 \tag{7.42}$$

where the implied constant is absolute and $\sum^\star$ restricts u_j to the primitive cusp forms, so $T_nu_j = \lambda_j(n)u_j$ and $\nu_j(n) = \lambda_j(n)\nu_j(1)$ for all $n \geqslant 1$.

Similar inequalities hold true for holomorphic cusp forms. In this case more precise results can be derived from the Petersson formula (Proposition 14.5). Let $\mathcal{F}$ be an orthonormal basis of $S_k(q)$. Let

$$f(z) = \sum_{n\geqslant 1} a_f(n)e(nz)$$

be the Fourier expansion of f at ∞. We consider the normalized Fourier coefficients

$$\psi_f(n) = \Big(\frac{q\Gamma(k-1)}{(4\pi n)^{k-1}}\Big)^{\frac{1}{2}} a_f(n). \tag{7.43}$$

Recall that if $k \geqslant 2$ we have

$$|\mathcal{F}| = \dim S_k(q) \asymp kq\prod_{p\mid q}(1+p^{-1}). \tag{7.44}$$

THEOREM 7.26. *Let $\mathcal{F}$ be any orthonormal basis of $S_k(q)$ with $k > 2$. Then for any complex numbers a_n we have*

$$\sum_{f\in\mathcal{F}}\Big|\sum_{n\leqslant N}a_n\psi_f(n)\Big|^2 \ll (q+N)\|a\|^2 \tag{7.45}$$

where the implied constant is absolute.

PROOF. By (7.43) and Proposition 14.5, the Petersson formula implies that the left side of (7.45), say $L(q)$, equals

$$q\|a\|^2 + 2\pi q i^{-k}\sum_{c\equiv 0\,(\mathrm{mod}\ q)} c^{-1}\sum\sum_{m,n\leqslant N}\bar{a}_m a_n S(m,n;c)J_{k-1}\Big(\frac{4\pi\sqrt{mn}}{c}\Big).$$

Opening the Kloosterman sum we derive by orthogonality of additive characters and by Cauchy's inequality that

$$\Big|\sum\sum_{m,n\leqslant N}\bar{a}_m a_n S(m,n;c)\Big| \leqslant (c+N)\|a\|^2.$$

Next (to separate the variables) we derive via the power series expansion

$$J_{k-1}(2x) = \sum_{\ell\geqslant 0}\frac{(-1)^\ell x^{k-1+2\ell}}{\ell!\Gamma(k+\ell)}$$

that

$$\Big|\sum\sum_{m,n\leqslant N}\bar{a}_m a_n S(m,n;c)J_{k-1}\Big(\frac{4\pi\sqrt{mn}}{c}\Big)\Big|\leqslant I_{k-1}\Big(\frac{4\pi\sqrt{mn}}{c}\Big)(c+N)\|a\|^2$$

where

$$I_{k-1}(2x)=\sum_{\ell\geqslant 0}\frac{x^{k-1+2\ell}}{\ell!\Gamma(k+\ell)}\leqslant x^2,\ \text{if } x\leqslant 1.$$

The condition $x=2\pi\sqrt{mn}c^{-1}\leqslant 2\pi Nc^{-1}\leqslant 1$ holds for any $c\equiv 0\,(\mathrm{mod}\ q)$ if $q\geqslant 2\pi N$, giving

$$L(q)\leqslant q\|a\|^2+2\pi q\sum_{c\equiv 0\,(\mathrm{mod}\ q)}c^{-1}\Big(\frac{2\pi N}{c}\Big)^2(c+N)\|a\|^2,$$

hence (7.45) follows in this case.

To remove the condition $q\geqslant 2\pi N$ we observe that $S_k(q)\subset S_k(pq)$ and the index is $[\Gamma_0(pq):\Gamma_0(q)]\leqslant p+1$. Using this embedding with appropriate renormalization we find that $L(q)\leqslant(1+p^{-1})L(pq)$. If $q\leqslant 2\pi N$, we can choose p with $2\pi N\leqslant pq\leqslant 4\pi N$ and apply the result for $L(pq)$ getting (7.45). □

With a slight modification the above arguments work also for $S_k(q,\chi)$ with any character χ (mod q) and for $k=2$. However, the case $k=1$ is very different because the space $S_1(q,\chi)$ is small (probably $\dim S_1(q,\chi)\ll\sqrt{q}\log q$ if χ is a real character). Technically speaking this intrinsic distinction manifests itself by the lack of convergence of the series of Kloosterman sums in the Petersson formula. What lies behind the scene is the huge spectrum of Maass cusp forms near the bottom of which only a small part (the holomorphic forms at the bottom) is detected by the Petersson formula (the lack of spectral completeness causes the problem!). In this scenario one can proceed by the duality principle (no spectral completeness is required then), but a new obstacle emerges with the degree of the automorphic harmonics ($GL(2)\times GL(2)$ is essentially $GL(4)$; to the contrary for Dirichlet characters $GL(1)\times GL(1)$ is still $GL(1)$) which amplies the conductor. Nevertheless some non-trivial orthogonality can be established for sufficiently long vectors relative to the level. For example W. Duke succeeded in proving along such lines the following large sieve type inequality (not perfect but quite useful)

THEOREM 7.27. *Let χ be an odd primitive quadratic character modulo q. Let $\mathcal{F}$ be an orthonormal basis of $S_1(q,\chi)$ and $\psi_g(n)$ be the corresponding normalized coefficients. Then for any complex numbers a_n we have*

$$\sum_{g\in\mathcal{F}}\Big|\sum_{n\leqslant N}a_n\psi_g(n)\Big|^2\ll(q+N)\|a\|^2 \tag{7.46}$$

where the implied constant is absolute.

Duke [**Du1**] gave an important application of (7.46) for estimating the dimension of $S_1(q,\chi)$. Specifically he proved

$$\dim S_1(q,\chi)\ll q^{11/12+\varepsilon}$$

for q prime and χ a real primitive character modulo q, the first improvement over the bound $\dim S_1(q,\chi)\ll q/\log q$ which comes from the trace formula.

Using the existence of Rankin-Selberg L-functions for arbitrary degree automorphic forms, Duke and Kowalski [**DK**] have established by the duality principle a very general type of large sieve inequality which has applications especially to the study of automorphic L-functions close to the line $\mathrm{Re}(s) = 1$. We mention the special case of symmetric square coefficients (see Section 5.12) and give a sketch of proof to illustrate the wealth of ingredients.

THEOREM 7.28. *For q squarefree, let $S_2(q)^*$ be the basis of primitive forms of level q and weight 2. Let $\lambda_f(n)$ be the Hecke eigenvalues for $f \in S_2(q)^*$. For any complex numbers a_n we have*

$$\sum_{q\leqslant Q}\sum_{f\in S_2(q)^*}\Big|\sum_{n\leqslant N} a_n\lambda_f(n^2)\Big|^2 \ll (N(\log N)^{15} + N^{\frac{1}{2}+\varepsilon}Q^{\frac{7}{2}})\|a\|^2 \tag{7.47}$$

if $N \geqslant Q \geqslant 1$, with an implied constant depending only on ε.

SKETCH OF PROOF. One uses duality and one can attach a smooth test function to the sum over n, reducing the problem to estimating

$$H(f,g) = \sum_n \lambda_f(n^2)\lambda_g(n^2)\varphi(n/N)$$

where $f,\, g \in S_2(q)^*$, for some fixed smooth function φ with compact support such that $\varphi(x) = 1$ for $0 \leqslant x \leqslant 1$. (The Hecke eigenvalues $\lambda_f(n)$ for $f \in S_k^*(q)$ are real). By Mellin inversion we have

$$H(f,g) = \frac{1}{2\pi i}\int_{(2)} G(s)\hat{\varphi}(s)N^s ds$$

where

$$G(s) = \sum_n \lambda_f(n^2)\lambda_g(n^2)n^{-s}.$$

Using the fact that $\lambda_f(n^2)$ are, up to small perturbation, coefficients of the symmetric square L-function (see Section 5.12) and the description of Rankin-Selberg L-functions of cusp forms on $GL(3)$, it follows that

$$G(s) = L(\mathrm{Sym}^2 f \otimes \mathrm{Sym}^2 g, s)G_1(s)$$

where $G_1(s)$ is given by an Euler product absolutely convergent for $\mathrm{Re}(s) > \frac{1}{2}$. We have Deligne's bound $|\lambda_f(n)| \leqslant \tau(n)$ (see (14.54)), which implies

$$G_1(s) \ll |\zeta(\sigma + \tfrac{1}{2})|^A \ll (\sigma - \tfrac{1}{2})^{-A}$$

for $\mathrm{Re}(s) = \sigma > \frac{1}{2}$, for some absolute constant $A > 0$, with an absolute implied constant.

It is known (by properties of Rankin-Selberg convolutions) that $L(\mathrm{Sym}^2 f \otimes \mathrm{Sym}^2 g, s)$ is entire except if $\mathrm{Sym}^2 f = \mathrm{Sym}^2 g$, in which case there is a simple pole at $s = 1$. Moreover, in the Appendix to [**DK**], Ramakrishnan shows that this exception is equivalent with $f = g$ in the case of squarefree levels (in the general case, a quadratic twist can occur). So moving the contour to the line $\sigma = \frac{1}{2} + \delta$ for $\delta > 0$, we get

$$H(f,g) = \frac{1}{2\pi i} \int\limits_{(\frac{1}{2}+\delta)} G_1(s) L(\mathrm{Sym}^2 f \otimes \mathrm{Sym}^2 g, s) \hat{\varphi}(s) N^s ds$$

for $f \neq g$. For $\sigma = \frac{1}{2} + \delta$, we use the convexity bound (see Exercise 3 of Chapter 5)

$$L(\mathrm{Sym}^2 f \otimes \mathrm{Sym}^2 g, \sigma + it) \ll \mathfrak{q}(\mathrm{Sym}^2 f \otimes \mathrm{Sym}^2 g, \sigma + it)^{1/4-\delta/2+\varepsilon}$$

for this L-function of degree 9, where $\mathfrak{q}(\cdot, s)$ is the analytic conductor defined in (5.7). We have

$$\mathfrak{q}(\mathrm{Sym}^2 f \otimes \mathrm{Sym}^2 g, \sigma + it) \ll Q^6(|t|+1)^9$$

by the bound (5.11) of Bushnell and Henniart for the conductor of a Rankin-Selberg convolution, with an absolute implied constant. By the rapid decay of $\hat{\varphi}(s)$, we get for $f \neq g$,

$$H(f,g) \ll N^{\frac{1}{2}+\delta} Q^{3/2-3\delta+\varepsilon} \delta^{-A}$$

for any $\delta > 0$. We take $\delta = (\log N)^{-1}$ so

$$H(f,g) \ll N^{\frac{1}{2}+\varepsilon} Q^{3/2}$$

for $N \geqslant Q$ and $f \neq g$.

If $f = g$, by Deligne's bound we have $|\lambda_f(n^2)\lambda_g(n^2)| \leqslant \tau(n^2)^2 \leqslant \tau(n)^4$ hence by direct estimation

$$H(f,f) \ll N(\log N)^{15}.$$

Since there are $\ll Q^2$ forms on the left side of (7.47), it follows that for $N \geqslant Q$ we have (7.47). □

In this connection we suggest the following problem as a first case of a higher-degree large sieve inequality:

PROBLEM 7.29. *Prove that*

$$\sum_{f \in H_k(q)} \Big| \sum_{n \leqslant N} a_n \lambda_f(n^2) \Big|^2 \ll (qN)^\varepsilon (q+N) \|a\|^2$$

where $H_k(q)$ is a Hecke basis of eigenforms on $S_k(q)$, with q squarefree, and the implied constant depends only on ε and k.

We can deduce from Theorem 7.28 the analogue of Linnik's result (Theorem 7.16) for elliptic curves.

PROPOSITION 7.30. *Let $A > 0$ and $Q > 2$ be fixed. For $E/\mathbb{Q}$ a semistable elliptic curve of conductor $\leqslant Q$, let $M(E)$ be the number of semistable elliptic curves $F/\mathbb{Q}$ with conductor $\leqslant Q$ such that $a_F(p) = a_E(p)$ for all $p \leqslant (\log Q)^A$. Then we have*

$$M(E) \ll Q^{9/A}$$

where the implied constant depends only on A.

It is conjectured that the number of semistable elliptic curves of conductor $\leqslant Q$ is about $Q^{5/6}$; the lower bound is easy by constructing explicit families, but the best known upper bound is $Q^{1+\varepsilon}$ for any $\varepsilon > 0$ (see [**DK**]). Therefore our result is non-trivial for any $A > 11$.

PROOF. For a given weight 2 primitive form f of conductor $q \leqslant Q$, q squarefree, let $M'(f)$ be the number of primitive modular forms g, of weight 2 with squarefree conductor $\leqslant Q$ such that $\lambda_f(p) = \lambda_g(p)$ for $p \leqslant (\log Q)^A$. By modularity of (semistable) elliptic curves over $\mathbb{Q}$, we have $M(E) \leqslant M'(f_E)$ where f_E is the weight 2 form associated to E. Note that $\lambda_f(n) = \lambda_g(n)$ if all prime divisors $p \mid n$ are $\leqslant (\log Q)^A$.

As in Linnik's proof, the idea is to find coefficients a_n supported on prime numbers that make the linear form

$$L_f = \sum_{n \leqslant N} a_n \lambda_f(n)$$

large, hence $L_g = L_f$ is large and repeated for all g counted in $M'(f)$. On the other hand, the large sieve estimate (7.47) will show that this cannot happen very often producing a bound for $M'(f)$. Because the coefficients $\lambda_f(p)$ can be quite small (in contrast with Dirichlet characters), $a_n = \overline{\lambda_f(n)}$ may not work. Hence one is tempted to use Proposition 14.22, but N there would be too small (limited to $N \leqslant (\log Q)^A$). Instead, notice that by the Prime Number Theorem and the formula

$$\lambda_f(p)^2 - \lambda_f(p^2) = 1 \text{ for } (p, q(f)) = 1,$$

one of the sets T_1 or T_2 given by

$$T_i = \{p \leqslant (\log Q)^A \mid |\lambda_f(p^i)| \geqslant \tfrac{1}{2} \text{ and } (p, q(f)) = 1\}$$

satisfies $|T_i| \gg (\log Q)^A (A \log\log Q)^{-1}$ with an absolute implied constant. Assume it is T_2 (the case of T_1 is similar, but simpler).

We have $|\lambda_f(n)| \geqslant 2^{-\omega(n)}$ if n is squarefree with all its prime factors $p \in T_2$. Fix $m \geqslant 1$ (to be chosen later). Let $a_n = \overline{\lambda_f(n)}$ for such n with $\omega(n) = m$, and $a_n = 0$ otherwise. It follows that

$$L_f = \sum_{n \leqslant N} |a_n|^2 = \sum_{n \leqslant N} |\lambda_f(n)|^2 \geqslant 2^{-2m} U$$

where U is the number of values of $n \leqslant N$ satisfying the desired condition. By Theorem 7.28, we have

$$M'(f) L_f^2 \ll N (\log N)^{15} \|a\|^2 = N (\log N)^{15} L_f$$

for any $N \geqslant Q^8$. Hence for $N = Q^8$ we get

$$M'(f) \ll Q^8 (\log Q)^{15} 2^{2m} U^{-1}.$$

Choosing $m = [8(\log Q)(A \log\log Q)^{-1}]$ we get $2^{2m} \ll Q^{\varepsilon}$ and

$$U \geqslant \binom{|T_2|}{m} \gg Q^{8(1-A^{-1})-\varepsilon}$$

showing that

$$M'(f) \ll Q^{8/A+\varepsilon} \ll Q^{9/A}$$

for ε small enough. □

7.8. Orthogonality of elliptic curves.

A straightforward approach to the large sieve inequalities requires near-orthogonality and almost completeness of the outer harmonics. However, some spaces contain smaller subspaces which preserve completeness with respect to their own structure. Then proving a large sieve type inequality for the subspace becomes a harder yet feasible problem of independent interest. A clear example is offered by the family of elliptic curves

$$E : y^2 = x^3 + ax + b \tag{7.48}$$

with $a, b \in \mathbb{Z}$, $1 \leqslant a \leqslant A$, $1 \leqslant b \leqslant B$. The corresponding modular forms f_{ab} are of weight 2 and level $q \leqslant Q = 16(4A^3 + 27B^2)$. We have about AB such curves while the total number of all primitive cusp forms of weight two and level $q \leqslant Q$ is about Q^2. Therefore we are dealing with a very small subset indeed.

In this section we present a large sieve type inequality for the harmonics associated with the family of elliptic curves. If m is squarefree the Hecke eigenvalue for the cusp form associated to (7.48) is given by the character sum

$$\lambda_{ab}(m) = \mu(m) \sum_{x \,(\mathrm{mod}\, m)} \Big(\frac{x^3 + ax + b}{m}\Big) \tag{7.49}$$

(see Section 14.4). However, it is more interesting to consider the sum over the reduced classes $x \pmod m$, $(x, m) = 1$. Put

$$\lambda^*_{ab}(m) = \mu(m) \sideset{}{^\star}\sum_{x \,(\mathrm{mod}\, m)} \Big(\frac{x^3 + ax + b}{m}\Big). \tag{7.50}$$

Note that

$$\lambda_{ab}(m) = \sum_{d|m} \mu(d) \Big(\frac{b}{d}\Big) \lambda^*_{ab}\Big(\frac{m}{d}\Big).$$

THEOREM 7.31. *For any complex numbers α_a, β_b we have*

$$\sideset{}{^\flat}\sum_{1 \leqslant m \leqslant M} \Big|\sum\sum_{\substack{1\leqslant a\leqslant A\\ 1\leqslant b\leqslant B}} \alpha_a \beta_b \lambda^*_{ab}(m)\Big|^2 \ll \|\alpha\| \|\beta\| (M + \sqrt{A})(M + \sqrt{B}) M^\varepsilon \tag{7.51}$$

where ε is any positive number and the implied constant depends only on ε.

PROOF. We can assume that m is odd squarefree. By Gauss sums (see Theorem 3.2)

$$\begin{aligned}\lambda^*_{ab}(m) &= \mu(m) \frac{\bar{\varepsilon}_m}{\sqrt{m}} \sum_{z \,(\mathrm{mod}\, m)} \Big(\frac{z}{m}\Big) \sideset{}{^\star}\sum_{x \,(\mathrm{mod}\, m)} e_m(z(x^3 + ax + b)) \\ &= \mu(m) \frac{\bar{\varepsilon}_m}{\sqrt{m}} \sideset{}{^\star}\sum_{x \,(\mathrm{mod}\, m)} e_m(ax) \sum_{z \,(\mathrm{mod}\, m)} \Big(\frac{z}{m}\Big) e_m(\bar{z}^2 x^3 + zb).\end{aligned}$$

Hence the inner sum in (7.51) equals

$$\mu(m) \frac{\bar{\varepsilon}_m}{\sqrt{m}} \sideset{}{^\star}\sum_{x \,(\mathrm{mod}\, m)} \Big(\sum_a \alpha_a e_m(ax)\Big) \Big(\sum_b \beta_b \sideset{}{^\star}\sum_{z \,(\mathrm{mod}\, m)} e_m(\bar{z}^2 x^3 + zb)\Big)$$

By Cauchy's inequality the left side of (7.51) is bounded by $(\mathcal{A}\mathcal{B})^{1/2}$ where

$$\mathcal{A} = \sum_m \sideset{}{^\star}\sum_{x \,(\mathrm{mod}\, m)} \Big| \sum_a \alpha_a e_m(ax) \Big|^2 \leqslant \|\alpha\|^2 (A + M^2)$$

and

$$\begin{aligned}
\mathcal{B} &= \sum_m \frac{1}{m} \sideset{}{^\star}\sum_{x \,(\mathrm{mod}\, m)} \Big| \sum_b \beta_b \sum_{z \,(\mathrm{mod}\, m)} e_m(\bar{z}^2 x^3 + zb) \Big|^2 \\
&\leqslant \sum_m \frac{\tau_3(m)}{m} \sum_{x \,(\mathrm{mod}\, m)} \Big| \sum_b \beta_b \sum_{z \,(\mathrm{mod}\, m)} e_m(\bar{z}^2 x^3 + zb) \Big|^2 \\
&= \sum_m \frac{\tau_3(m)}{m} \sum_{z_1^2 \equiv z_2^2 \,(\mathrm{mod}\, m)} \Big(\frac{z_1 z_2}{m} \Big) \sum_{b_1, b_2}\sum \beta_{b_1} \bar{\beta}_{b_2} e_m(z_1 b_1 - z_2 b_2) \\
&\leqslant \sum_m \tau_3(m) \tau(m) \sideset{}{^\star}\sum_{z \,(\mathrm{mod}\, m)} \Big| \beta_b \, e_m(zb) \Big|^2 \ll \|\beta\|^2 (B + M^2) M^\varepsilon.
\end{aligned}$$

This completes the proof of (7.51). □

REMARKS. From the Hasse bound $\lambda_{ab}^*(m) \ll m^{1/2}\tau(m)$ (see Theorem 11.25) it follows that the left side of (7.51) is $\ll \|\alpha\|\|\beta\|(AB)^{1/2}M^{3/2}\log 2M$, which we regard as a trivial bound, because no cancellation is used. In applications one often wishes to have a saving slightly more than $M^{1/2}$. Our theorem guarantees this saving relative to the trivial bound if AB is larger than M^2. Choosing $A = X^{1/3}$ and $B = X^{1/2}$ we have the elliptic curves with discriminant $\Delta_{ab} = -16(4a^3+27b^2) \ll X$ and we obtained the desired saving if $M \ll X^{5/12}$. This is quite a large range for M, but not yet completely satisfactory. Indeed, for applications to L-functions one needs to pass the barrier $M = X^{1/2}$. The following conjecture would provide the passage

CONJECTURE 7.32. *For any complex numbers (γ_m) supported on squarefree numbers we have*

$$\sum_{1 \leqslant a \leqslant A} \Big| \sum_{m \leqslant M} \gamma_m \lambda_{ab}(m) \Big|^2 \ll (A + M) M \sum_{m \leqslant M} |\gamma_m \tau(m)|^2$$

where the implied constant may depend on b slightly.

When b is a square we are having elliptic curves of positive rank. Similar estimates are expected to be true for special families of elliptic curves of a given rank.

EXERCISE 6*. Let m be a positive, odd, squarefree number. Prove that for any complex numbers α_a, β_b,

$$\Big| \sum_{\substack{A \leqslant a \leqslant 2A \\ B \leqslant b \leqslant 2B}}\sum \alpha_a \beta_b \lambda_{ab}(m) \Big| \leqslant \tau_4(m) (A + m)^{\frac{1}{2}} (B + m)^{\frac{1}{2}} \|\alpha\| \|\beta\|.$$

7.9. Power-moments of L-functions.

We have remarked that a sharp large sieve type inequality (7.18), (7.20) for coefficients of a family of L-functions, has strength comparable to the Grand Riemann Hypothesis on average. So it is not surprising that the large sieve can also be employed to derive estimates for averages of L-functions which are as strong as the Lindelöf Hypothesis would give (see Corollary 5.20). There is a vast literature of relevant results. For classical L-functions we recommend the book by A. Ivic [**Iv**]. In this section we only grasp the industry by showing a few representative results.

THEOREM 7.33. *For $T \geqslant 2$ we have*

$$\int_{-T}^{T} |\zeta(\tfrac{1}{2}+it)|^2 dt \ll T \log T. \tag{7.52}$$

THEOREM 7.34. *We have*

$$\sum_{q \leqslant Q} \sideset{}{^\star}\sum_{\chi \,(\mathrm{mod}\, q)} |L(\tfrac{1}{2}+it,\chi)|^8 \ll Q^2(t^2+1)(\log Q(|t|+2))^{17} \tag{7.53}$$

for any $t \in \mathbb{R}$, where the implied constant is absolute.

THEOREM 7.35. *Let $k \geqslant 2$ and $\mathcal{F}$ be a Hecke orthogonal basis of $S_k(q)$. We have*

$$\sum_{f \in \mathcal{F}} |L(f,\tfrac{1}{2}+it)|^4 \ll q(t^2+1)(\log q(|t|+2))^{17} \tag{7.54}$$

for any $t \in \mathbb{R}$, where the implied constant depends only on k.

REMARKS. In all cases, slightly more precise estimates are known (in some cases asymptotic formulas). See [**T2**] for the Riemann zeta function, and [**KMV1**] for cusp forms. Also (7.53) and (7.54) are only comparable to the Lindelöf hypothesis in q-aspect. In t-aspect they are as good as the convexity bound, extra averaging over t would improve the situation.

In Chapter 26, as a by-product of a deeper study of special values of L-functions, we prove asymptotic formulas for the first and second moments of derivatives $L'(f,\frac{1}{2})$ with respect to the odd weight 2 cusp forms (see (26.35) and (26.36)). In fact, the modern methods used in the analytic study of special values of L-functions (and their zeros) rely heavily on various types of averaging, some of which are quite sophisticated indeed (they require orthogonality of a capacity beyond the diagonal, so to speak). See the introduction to Chapter 26 for some references, to which can be added [**CI1**] (among many others). See also Section 9.2.

SKETCH OF PROOF OF THEOREMS 7.33, 7.34, 7.35. The arguments are quite familiar so we only indicate the essential steps. The method starts with suitable short approximations of L-functions by Dirichlet polynomials (see Chapter 5), and then applies a large sieve inequality to the resulting linear forms. Without compromising estimates one can take a moment of order $2k$ as long as the length of the approximation for the k-th power is bounded by the total number of "harmonics". For instance, in Theorem 7.34, there are about Q^2 characters involved, and for each one the approximation to $L(\frac{1}{2}+it,\chi)$ is of length about $(q(|t|+1))^{1/2}$, so we can apply Theorem 7.13 for the mean square of $L(\frac{1}{2}+it,\chi)^4$, and get a bound for the eighth power-moment which is nearly best possible.

We sketch Theorem 7.34 this way, leaving the other two theorems as exercises (use Theorem 7.17 with $Q = 1$ and Theorem 9.1 in the first case, or Theorem 7.26 in the last case).

Let χ be primitive modulo q, $2 \leqslant q \leqslant Q$. We have

$$L(s,\chi)^4 = \sum_{n \geqslant 1} \chi(n)\tau_4(n)n^{-s}$$

for $\operatorname{Re}(s) > 1$. By Theorem 5.3 for $L(s,\chi)^4$ we see that $L(\frac{1}{2}+it,\chi)^4$ is equal to

$$\sum_{n\geqslant 1} \frac{\chi(n)\tau_4(n)}{n^{1/2+it}} V\Big(\frac{n}{q^2}\Big) + \varepsilon_\chi^4 q^{-4it}\frac{\gamma(\frac{1}{2}-it)}{\gamma(\frac{1}{2}+it)} \sum_{m\geqslant 1} \frac{\bar{\chi}(m)\tau_4(m)}{m^{1/2-it}} W\Big(\frac{m}{q^2}\Big)$$

where $V(y) = V_s(y)$ and $W(y) = W_s(y)$ are given by (5.13) and (5.14) with the gamma factor $\gamma(s) = \pi^{-2s}\Gamma(\frac{1}{2}(s+\kappa))^4$ for $\kappa = \frac{1}{2}(1-\chi(-1))$ and some suitable choice of auxiliary function $G(u)$. The functions $V(y)$ and $W(y)$ decay rapidly for $y \gg t^2+1$ (see Proposition 5.4) so practically we are left with sums of length $N \asymp q^2(t^2+1)$. By Theorem 7.13 we get

$$\sum_{q\leqslant Q} \sideset{}{^\star}\sum_{\chi \,(\mathrm{mod}\, q)} \Big|\sum_{n\leqslant N} \frac{\chi(n)\tau_4(n)}{n^{1/2+it}}\Big|^2 \ll (Q^2+N)(\log N)^{17}.$$

Using this estimate essentially for $N \asymp Q^2(t^2+1)$ one derives (7.53). □

CHAPTER 8

EXPONENTIAL SUMS

8.1. Introduction.

As usual we denote the additive character on $\mathbb{R}$ by $e(x) = e^{2\pi ix}$. Exponential sums which we consider in this chapter are of type

$$S_f(N) = \sum_{1\leqslant n\leqslant N} e(f(n)) \tag{8.1}$$

where f is a smooth, real valued function on the interval $[1, N]$ which is called the amplitude function and N is a positive integer called the length of the sum. We could begin the summation at any point and consider

$$S_f(M, N) = \sum_{M<n\leqslant M+N} e(f(n)) \tag{8.2}$$

where M is any integer and f is a smooth function on $[M + 1, M + N]$. Of course, the latter can be transformed to the former by shifting the argument,

$$S_f(M, N) = \sum_{1\leqslant n\leqslant N} e(f(n + M)).$$

We have the trivial bound $|S_f(M, N)| \leqslant N$ and our goal will be to improve on this bound as much as we can. However, in many cases even a slight improvement of the trivial bound is sufficient for applications.

Exponential sums $S_f(N)$ for suitable f are encountered in various areas of analytic number theory. For example, they can be used to estimate the Riemann zeta function $\zeta(s)$ on vertical lines. Indeed by the simple approximation

$$\zeta(s) = \sum_{1\leqslant n\leqslant N} n^{-s} + \frac{N^{1-s}}{s-1} + O(N^{-\sigma}) \tag{8.3}$$

for $s = \sigma + it$ with $\sigma \geqslant \frac{1}{2}$ and $1 \leqslant t \leqslant N$ the problem reduces to estimating sums

$$\sum_{1\leqslant n\leqslant N} n^{-it}$$

which are of type (8.2) with $f(x) = \frac{-t}{2\pi} \log x$. Another example is the problem of counting lattice points inside a planar domain. This reduces to estimating integral points under a curve $y = g(x)$, where g is a smooth, positive decreasing function on $[1, N]$. The number of points in question equals

$$\sum_{1\leqslant n\leqslant N} [g(n)] = \sum_{1\leqslant n\leqslant N} g(n) - \sum_{1\leqslant n\leqslant N} \psi(g(n)) - \frac{N}{2}.$$

On the right side the first sum is well approximated by an integral (see (1.67)) while the second sum is equal to

$$- \sum_{1\leqslant|h|\leqslant H} (2\pi i h)^{-1} \sum_{1\leqslant n\leqslant N} e(hg(n)) + O\Big(\sum_{1\leqslant n\leqslant N} \frac{1}{1+H\|g(n)\|}\Big)$$

by the truncated Fourier series (4.18) for $\psi(x)$. Hence the problem reduces to estimation of sums (8.1) with $f(x) = hg(x)$. In particular, in the Gauss circle problem (lattice points in the circle $x^2 + y^2 = R^2$) we encounter sums $S_f(R)$ with $f(x) = h\sqrt{R^2 - x^2}$, and in the Dirichlet divisor problem (lattice points under the hyperbola $xy = N$) we encounter sums $S_f(N)$ with $f(x) = hNx^{-1}$.

In this chapter we present classical methods for estimating general exponential sums $S_f(M, N)$ due to H. Weyl, J. G. Van der Corput and I. M. Vinogradov. Our results do not depend on particular properties of f but only on estimates for derivatives. It is always understood that the functions involved are of class C^k, where k is the largest order of derivation appearing in the statements.

Note that if f is well approximated by g, then $S_f(M, N)$ can be estimated by $S_g(M, N')$ for some $N' \leqslant N$. Precisely, if $f = g + h$, then by partial integration

$$S_f(M,N) = S_g(M,N) - \int_M^{M+N} S_g(M, x - M)de(h(x)). \tag{8.4}$$

Hence we get

$$|S_f(M,N)| \leqslant C_h \max_{N'\leqslant N} |S_g(M,N')| \tag{8.5}$$

where

$$C_h = 1 + 2\pi \int_M^{M+N} |h'(x)|dx.$$

If h is monotonic and bounded (so C_h is bounded), then the original sum S_f and the modified sum S_g are practically equal. Sometimes in applications we find that a small alteration of f is necessary to meet the conditions of our results, in which case (8.5) can be used.

8.2. Weyl's method.

The first applications of exponential sums in number theory were given in 1916 by H. Weyl [**W1**] to the problem of equidistribution of sequences modulo one. In the second paper of 1921 Weyl [**W2**] develops a general method which is particularly good for $S_f(N)$ where f is a polynomial

$$f(x) = \alpha x^k + \beta x^{k-1} + \cdots \in \mathbb{R}[x]$$

with $\alpha > 0$. In this case we call $S_f(N)$ a Weyl sum of degree k.

The Weyl sum for a linear polynomial $f(x) = \alpha x$ is a sum over geometric progression

$$S_f(N) = \sum_{1\leqslant n\leqslant N} e(\alpha n) = \frac{\sin \pi\alpha N}{\sin \pi\alpha} e\Big(\frac{\alpha}{2}(N+1)\Big).$$

Since $|\sin \pi\alpha| \geqslant 2\|\alpha\|$, where $\|\alpha\|$ denotes the distance of α to the nearest integer, we obtain for $\alpha \notin \mathbb{Z}$,

$$\Big|\sum_{1\leqslant n\leqslant N} e(\alpha n)\Big| \leqslant \min\left(N, \frac{1}{2\|\alpha\|}\right). \tag{8.6}$$

The Weyl sum for a quadratic polynomial $f(x) = \alpha x^2 + \beta x$ is also called a Gauss sum. In the special case where a, b are integers with $(2a, N) = 1$ we have the exact formula (3.38) from which it follows by completing the square that

$$\Big|\sum_{1\leqslant n\leqslant N} e\left(\frac{an^2+bn}{N}\right)\Big| = \sqrt{N}. \tag{8.7}$$

Though there is no simple expression for a general Gauss sum

$$S_f(N) = \sum_{1\leqslant n\leqslant N} e(\alpha n^2 + \beta n),$$

we can estimate this quite well. First we arrange $|S_f(N)|^2$ as follows

$$|S_f(N)|^2 = \sum_{|\ell|<N} e(\alpha\ell^2 + \beta\ell) \sum_{1\leqslant n, n+\ell\leqslant N} e(2\alpha\ell n).$$

Then applying (8.6) we get

$$|S_f(N)|^2 \leqslant N + \sum_{1\leqslant \ell<N} \min(2N, \|2\alpha\ell\|^{-1}). \tag{8.8}$$

Hence one can deduce that

$$|S_f(N)| \leqslant 2\sqrt{\alpha}N + \frac{1}{\sqrt{\alpha}}\log\frac{1}{\alpha} \tag{8.9}$$

if $0 < \alpha \leqslant \frac{1}{2}$, which restriction can always be arranged. However, a more precise estimate depends on the diophantine nature of the leading coefficient α. By Dirichlet's approximation theorem there exists a rational approximation to 2α of type

$$\left|2\alpha - \frac{a}{q}\right| \leqslant \frac{1}{2Nq} \tag{8.10}$$

with $(a, q) = 1$ and $1 \leqslant q \leqslant 2N$. Hence $\|2\alpha\ell\| \geqslant \frac{1}{2}\|a\ell/q\|$ for any $1 \leqslant \ell < N, \ell \not\equiv 0 (\mathrm{mod}\ q)$ and

$$\begin{aligned}\sum_{\substack{1\leqslant \ell<N\\ \ell\not\equiv 0(\mathrm{mod}\ q)}} \|2\alpha\ell\|^{-1} &\leqslant 2\left(\frac{N}{q}+1\right) \sum_{\substack{\ell(\mathrm{mod}\ q)\\ \ell\not\equiv 0(\mathrm{mod}\ q)}} \|\ell/q\|^{-1}\\ &= 2(N+q)\Big(\sum_{1\leqslant\ell\leqslant\frac{q}{2}} \ell^{-1} + \sum_{1\leqslant\ell<\frac{q}{2}} \ell^{-1}\Big) \leqslant 4(N+q)\log q.\end{aligned}$$

Then estimating the partial sum of (8.8) over $1 \leqslant \ell < N, \ell \equiv 0(\mathrm{mod}\ q)$ trivially by $2N^2q^{-1}$ we arrive at

$$|S_f(N)|^2 \leqslant N + 2N^2q^{-1} + 4(N+q)\log q.$$

This together with the trivial bound $|S_f(N)| \leqslant N$ implies

THEOREM 8.1. *If $f(x) = \alpha x^2 + \beta x$ with 2α satisfying (8.10), then*

$$|S_f(N)| \leqslant 2Nq^{-\frac{1}{2}} + q^{\frac{1}{2}} \log q. \tag{8.11}$$

Theorem 8.1 is essentially best possible. Slightly better results hold true for almost all coefficients α, β (mod 1) with respect to the Lebesque measure. Indeed we have the following estimate for the second, the fourth and the sixth power mean values

$$\int_0^1 \Big| \sum_{1 \leqslant n \leqslant N} e(\alpha n^2 + \beta n) \Big|^2 d\alpha = N, \tag{8.12}$$

$$\int_0^1 \Big| \sum_{1 \leqslant n \leqslant N} e(\alpha n^2 + \beta n) \Big|^4 d\alpha \ll (N \log 2N)^2, \tag{8.13}$$

$$\int_0^1 \int_0^1 \Big| \sum_{1 \leqslant n \leqslant N} e(\alpha n^2 + \beta n) \Big|^6 d\alpha d\beta \ll (N \log 2N)^3. \tag{8.14}$$

PROOF. The first formula is the Parseval identity. The second integral is bounded by the number of solutions to $n_1^2 + n_2^2 = n_3^2 + n_4^2$ in positive integers $n_1, n_2, n_3, n_4 \leqslant N$, and the latter is bounded by

$$\sum_{\ell \leqslant 2N^2} r^2(\ell) \ll (N \log 2N)^2.$$

The third integral is equal to the number of solutions to the system

$$\begin{cases} n_1 + n_2 + n_3 = n_4 + n_5 + n_6, \\ n_1^2 + n_2^2 + n_3^2 = n_4^2 + n_5^2 + n_6^2 \end{cases}$$

in positive integers $n_1, n_2, n_3, n_4, n_5, n_6 \leqslant N$. Putting $k_\nu = n_\nu - n_{\nu+3}$ and $\ell_\nu = n_\nu + n_{\nu+3}$ for $1 \leqslant \nu \leqslant 3$ we get the system $k_1 + k_2 + k_3 = 0$ and $k_1\ell_1 + k_2\ell_2 + k_3\ell_3 = 0$. This reduces to one equation $k_1(\ell_1 - \ell_3) + k_2(\ell_2 - \ell_3) = 0$ and k_3 is determined uniquely by k_1, k_2. If $k_1(\ell_1 - \ell_3) = 0$, then $k_2(\ell_2 - \ell_3) = 0$, therefore there are at most $16N^3$ such solutions. It remains to count the solutions with $k_1(\ell_1 - \ell_3) \neq 0$. Given m with $1 \leqslant m \leqslant 4N^2$ and ℓ_3 with $1 < \ell_3 \leqslant 2N$ there are at most $\tau^2(m)$ solutions to $m = k_1(\ell_1 - \ell_3) = -k_2(\ell_2 - \ell_3)$ in k_1, k_2, ℓ_1, ℓ_2. Thus the total number of solutions in question does not exceed

$$4N \sum_{1 \leqslant m \leqslant 4N^2} \tau^2(m) \ll (N \log 2N)^3.$$

This completes the proof of (8.14). □

Now we estimate Weyl's sum $S_f(N)$ for a polynomial of any degree $k \geqslant 2$. We begin by

$$\begin{aligned} |S_f(N)|^2 &= \sum_{0<m,\, n \leqslant N} \sum e(f(m) - f(n)) \\ &= \sum_{|\ell| < N} \sum_{\substack{0 < n \leqslant N \\ 0 < \ell + n \leqslant n}} e(f(\ell + n) - f(n)). \end{aligned}$$

If f is a polynomial of degree k, then $g(x) = f(\ell + x) - f(x)$ is of degree $k-1$ (for any $\ell \neq 0$). By repeated application of the above "differencing process" we ultimately arrive at a Weyl sum of degree one for which we can use the non-trivial bound (8.6). Precisely we derive by induction

PROPOSITION 8.2. *If* $f(x) = \alpha x^k + \cdots$ *with* $k \geqslant 1$, *then*

$$|S_f(N)| \leqslant 2N\Big\{N^{-k} \sum_{-N<\ell_1,\dots,\ell_{k-1}<N} \!\!\cdots\sum \min(N, \|\alpha k!\ell_1\cdots\ell_{k-1}\|^{-1})\Big\}^{2^{1-k}}.$$

PROOF. For $k=1$ this bound is interpreted as that in (8.6). Moreover, we have already proved this result for $k=2$ in (8.8). Suppose the result is true for $k \geqslant 2$. Let $f(x) = \alpha x^{k+1} + \dots$, then $g(x) = f(\ell + x) - f(x) = \alpha(k+1)\ell x^k + \dots$, therefore

$$|S_f(N)|^2 \leqslant \sum_{|\ell|<N} \Big| \sum_{\substack{0<n\leqslant N\\ 0<\ell+n\leqslant N}} e(\alpha(k+1)n^k + \cdots)\Big|$$

$$\leqslant 2N \sum_{|\ell|<N} \Big\{N^{-k} \sum_{-N<\ell_1,\dots,\ell_{k-1}<N}\!\!\cdots\sum \min(N, \|\alpha(k+1)!\ell_1\cdots\ell_{k-1}\ell\|^{-1}\Big\}^{2^{1-k}}$$

by the induction hypothesis. Hence by Hölder's inequality we get

$$|S_f(N)|^2 \leqslant 4N^2\Big\{N^{-k-1} \sum_{-N<\ell_1,\dots,\ell_k<N}\!\!\cdots\sum \min(N, \|\alpha(k+1)!\ell_1\cdots\ell_k\|^{-1})\Big\}^{2^{1-k}}.$$

This yields the result for $k+1$. □

Next we use Proposition 8.2 to estimate sums $S_f(M,N)$ for f which can be well approximated by polynomials. Let $k \geqslant 2$. Suppose

$$\frac{x^k}{k!}|f^{(k)}(x)| \leqslant F \tag{8.15}$$

in $M \leqslant x \leqslant M+N$. Then we have $f(x+M) = p(x) + r(x)$, where

$$p(x) = f(M) + xf'(M) + \cdots + \frac{x^{k-1}}{(k-1)!}f^{(k-1)}(M)$$

and the derivative of $r(x)$ satisfies

$$|r'(x)| = \frac{x^{k-1}}{(k-1)!}|f^{(k)}(y)| \leqslant kx^{k-1}M^{-k}F$$

for some $M \leqslant y \leqslant M+N$. Therefore by (8.5) we obtain

$$|S_f(M,N)| \leqslant (1 + 2\pi FN^kM^{-k})|S_p(N')|$$

for some $N' \leqslant N$. This estimate is reasonably good for short sums $S_f(M,N)$, i.e., if N is much smaller than M, because we can make the factor $1 + 2\pi FN^kM^{-k}$ to be bounded, by choosing k which is not very large. Applying Proposition 8.2 for the Weyl sum $S_p(N')$ we get

COROLLARY 8.3. *Let $f(x)$ be a smooth function in the interval $[M, M+N]$ in which it satisfies (8.15). Then*

$$|S_f(M,N)| \leqslant 2N(1+2\pi F N^k M^{-k}) V^{2^{2-k}}$$

where

$$V = N^{1-k} \sum_{-N<\ell_1,\dots,\ell_{k-2}<N} \cdots \sum \min(N, \|f^{(k-1)}(M)\ell_1 \cdots \ell_{k-2}\|^{-1}).$$

We shall use the above results to estimate $S_f(M, M')$ for any M' with $1 \leqslant M' \leqslant M$. To this end we assume that (8.15) holds in the whole interval $M \leqslant x \leqslant 2M$. We choose $1 \leqslant N \leqslant M$ and split $S_f(M, M')$ into shorter sums of length N getting

$$|S_f(M,M')| \leqslant \sum_{0\leqslant j<J} |S_f(M+jN, N)| + 2N$$

where $J = [M/N]$. Hence by Corollary 8.3 and Hölder's inequality we obtain

$$\begin{aligned}|S_f(M,M')| &\leqslant 2N(1+2\pi F N^k M^{-k}) \sum_{0\leqslant j<J} V_j^{2^{2-k}} + 2N \\ &\leqslant 2M(1+2\pi F N^k M^{-k}) W^{2^{2-k}} + 2N\end{aligned}$$

where V_j are the corresponding sums associated with the points $M + jM$ and W is the mean-value of these sums

$$\begin{aligned}W &= \frac{1}{J} \sum_{0\leqslant j<J} V_j \\ &= J^{-1} N^{1-k} \sum_{0\leqslant j<J} \sum_{-N<\ell_1\dots,\ell_{k-2}<N} \!\!\cdots\!\! \sum \min(N, \|f^{(k-1)}(M+jN)\ell_1 \dots \ell_{k-2}\|^{-1}).\end{aligned}$$

Gathering terms according to the product $\ell_1 \dots \ell_{k-2} = r$ and for $r = 0$ estimating trivially we arrive at

$$W \leqslant k2^k N^{-1} + 2^k N^{1-k} \sum_{1\leqslant r<R} c_r J^{-1} \sum_{0\leqslant j<J} \min(N, \|r f^{(k-1)}(M+jN)\|^{-1})$$

where $R = N^{k-2}$ and c_r denotes the number of representations $r = \ell_1 \dots \ell_{k-1}$ with $1 \leqslant \ell_1, \dots, \ell_{k-1} < N$. For each r we consider separately the inner sum

$$U = \frac{1}{J} \sum_{0\leqslant j<J} \min(N, \|y_j\|^{-1})$$

where $y_j = r f^{(k-1)}(M+jN)$. Note that $|y_j| \leqslant rk! M^{1-k} F = Y$, say. We want the points y_j to be well spaced, and for this reason we require a lower bound for the derivative of f of order k throughout the whole segment $[M, 2M]$. Precisely from now on we assume that

$$\frac{F}{A} \leqslant \frac{x^k}{k!} |f^{(k)}(x)| \leqslant F \tag{8.16}$$

with $A \geqslant 1$ for $M \leqslant x \leqslant 2M$. Then by the mean-value theorem we deduce that $|y_{j'} - y_j| \geqslant \Delta|j' - j|$ where $\Delta = rk!(2M)^{-k}NFA^{-1}$. Now it is clear that

$$\begin{aligned} U &\leqslant \frac{1}{J}\sum_{|u|\leqslant Y}\sum_{0\leqslant j<J}\min(N, |y_j - u|^{-1}) \\ &\leqslant \frac{2Y+1}{J}\Big(N + \sum_{0<j<J}\frac{2}{j\Delta}\Big) \leqslant (2Y + J^{-1})(N + 2\Delta^{-1}\log 3M). \end{aligned}$$

This yields

$$W \leqslant 4k!(4\log 3M)^k\Big(\frac{FN}{M^k} + \frac{A}{N} + \frac{N}{M} + \frac{AM^{k-1}}{FN^{k-1}}\Big).$$

Finally

$$S_f(M,M') \ll \Big(1 + F\frac{N^k}{M^k}\Big)\Big(\frac{FN}{M^k} + \frac{A}{N} + \frac{N}{M} + \frac{AM^{k-1}}{FN^{k-1}}\Big)^{2^{2-k}} M\log 3M$$

where the implied constant is absolute. This holds for any N with $1 \leqslant N \leqslant M$. Suppose $M^k \geqslant F \geqslant 1$, so we may take $N = [MF^{-1/k}]$ giving

THEOREM 8.4. *Let $k \geqslant 2$. Suppose f satisfies (8.16) in the segment $[M, 2M]$. Then for $1 \leqslant M' \leqslant M$ we have*

$$S_f(M,M') \ll A^{4/2^k}(FM^{-k} + F^{-1})^{4/k2^k} M\log 3M \tag{8.17}$$

where the implied constant is absolute.

REMARK. At the end of the proof of (8.17) we assumed that $M^k \geqslant F \geqslant 1$, however this condition is not necessary because otherwise the result is trivial.

COROLLARY 8.5. *Suppose $f(x)$ satisfies (8.16) for $k = 2$ and $k = 3$ in the segment $[M, 2M]$. Then for $1 \leqslant M' \leqslant M \leqslant F$ we have*

$$S_f(M,M') \ll AF^{\frac{1}{6}}M^{\frac{1}{2}}\log 3M. \tag{8.18}$$

where the implied constant is absolute.

PROOF. This follows by taking the minimum of the two bounds (8.17) for $k = 2$ and $k = 3$. □

Note that (8.18) is trivial if $M \leqslant F^{\frac{1}{3}}$. However, for any M which is comparable with F in the logarithmic scale Theorem 8.4 yields a non-trivial bound by choosing k appropriately. For example, choosing $k = [2\log F/\log M]$ we derive

COROLLARY 8.6. *Suppose $f(x)$ is smooth in $[M, 2M]$ and all its derivatives satisfy*

$$A^{-2^k}F \leqslant \frac{x^k}{k!}|f^{(k)}(x)| \leqslant A^{2^k}F \tag{8.19}$$

with some $A \geqslant 1$. Then for $1 \leqslant M' \leqslant M \leqslant F$ we have

$$S_f(M,M') \ll A^8 M^{1-4^{-\gamma}}\log 3M \tag{8.20}$$

where $\gamma = \log F/\log M$ and the implied constant is absolute.

As applications of the above results we derive estimates for the Riemann zeta function on the lines $\sigma = \frac{1}{2}$ and $\sigma = 1$. For $t \geqslant 3$ we have

$$\zeta(\sigma + it) = \sum_{n \leqslant t} n^{-\sigma - it} + O(t^{-\sigma}). \tag{8.21}$$

Hence we derive by (8.18) the following subconvexity bound

$$\zeta(\tfrac{1}{2} + it) \ll t^{\frac{1}{6}} (\log t)^2. \tag{8.22}$$

Moreover, we derive by (8.20) the following approximation:

$$\zeta(1 + it) = \sum_{n \leqslant y} n^{-1-it} + O(e^{-\sqrt{\log t}}) \tag{8.23}$$

with $y = t^{3/\log\log t}$, which yields by trivial estimation

$$\zeta(1 + it) \ll \frac{\log t}{\log\log t}. \tag{8.24}$$

8.3. Van der Corput method.

This method emerged from two papers of 1921 and 1922 by J. G. van der Corput [**Cor1**], [**Cor2**]. Briefly speaking the novelty consists in replacing the sum by integrals and estimating the latter. This is executed by an application of the Poisson summation formula followed by an asymptotic evaluation of the resulting Fourier integrals (by using various techniques such as stationary phase). One arrives at an approximate equation for two sums of different length; this transformation is called the B-process. The amplitude functions in both sums may change shape but not size, and they have comparable derivatives. Next these amplitude functions are reduced by a differencing process which is a refinement of that used in Weyl's method; this routine is called the A-process. The final estimate for the exponential sum $S_f(M, N)$ is obtained by repeated application of the two processes.

Now we proceed to the first step of the van der Corput method, which is a truncated version of the Poisson formula

$$\sum_n F(n) = \sum_m \hat{F}(m). \tag{8.25}$$

PROPOSITION 8.7. *Let $f(x)$ be a real function with $f''(x) > 0$ on the interval $[a, b]$. We then have*

$$\sum_{a<n<b} e(f(n)) = \sum_{\alpha-\varepsilon<m<\beta+\varepsilon} \int_a^b e(f(x) - mx)dx + O(\varepsilon^{-1} + \log(\beta - \alpha + 2)) \tag{8.26}$$

where $\alpha, \beta, \varepsilon$ are any numbers with $\alpha \leqslant f'(a) \leqslant f'(b) \leqslant \beta$ and $0 < \varepsilon \leqslant 1$, the implied constant being absolute.

PROOF. Let $\mathcal{M}$ stand for the interval $[\alpha-\varepsilon, \beta+\varepsilon]$. First we give the trivial bound

$$\begin{aligned}\int_a^b \Big|\sum_{m\in\mathcal{M}} e(-mx)\Big| dx &\leqslant 2(b-a+1)\int_0^{\frac12}\Big|\sum_{m\in\mathcal{M}} e(-mx)\Big| dx \\ &\leqslant 2(b-a+1)\int_0^{\frac12}\min(\beta-\alpha+2, x^{-1})dx \ll (b-a+1)\log(\beta-\alpha+2).\end{aligned}\tag{8.27}$$

This proves (8.26) if $0<b-a<2$. Therefore from now on we assume that $b-a\geqslant 2$. We apply the Poisson formula to the tempered sum

$$\sum_{a<n<b} g(n)e(f(n)) = \sum_m \int_a^b g(x)e(f(x)-mx)dx,$$

where $g(x)$ is a function which planes the edges, precisely $g(x)=\min(x-a,1,b-x)$. This function is introduced temporarily for convergence of the sum of the Fourier integrals and will be removed later. For $m\notin\mathcal{M}$ we write by partial integration

$$\begin{aligned}2\pi i\int_a^b g(x)e(f(x)-mx)dx &= -\int_a^b\Big(\frac{g(x)}{f'(x)-m}\Big)' e(f(x)-mx)dx \\ &= \Big(\int_{b-1}^b - \int_a^{a+1}\Big)\frac{e(f(x)-mx)}{f'(x)-m}dx + \int_a^b \frac{g(x)f''(x)}{(f'(x)-m)^2}e(f(x)-mx)dx \\ &= \mathcal{T}_b(m)-\mathcal{T}_a(m)+\mathcal{T}_{ab}(m),\end{aligned}$$

say. For $\mathcal{T}_{ab}(m)$ we have

$$|\mathcal{T}_{ab}(m)| \leqslant \int_a^b f''(x)(f'(x)-m)^{-2}dx = (\alpha-m)^{-1}-(\beta-m)^{-1}.$$

For $\mathcal{T}_a(m)$ we have

$$|\mathcal{T}_a(m)| \leqslant \int_a^{a+1}|f'(x)-m|^{-1}dx \leqslant |\alpha-m|^{-1}+|\beta-m|^{-1}.$$

Integrating by parts once more we obtain another bound

$$\begin{aligned}|\mathcal{T}_a(m)| &\leqslant (f'(a+1)-m)^{-2}+(f'(a)-m)^{-2}+\Big|\int_a^{a+1} d(f'(x)-m)^{-2}\Big| \\ &= 2\max\{(f'(a+1)-m)^{-2}, (f'(a)-m)^{-2}\} \leqslant 2(\alpha-m)^{-2}+2(\beta-m)^{-2}.\end{aligned}$$

The same bounds hold for $\mathcal{T}_b(m)$. Combining these bounds we derive

$$|\mathcal{T}_a(m)|+|\mathcal{T}_b(m)|+|\mathcal{T}_{ab}(m)| \ll \frac{\beta-\alpha}{(\beta-m)(\alpha-m)}$$

for any $m\notin\mathcal{M}$. Hence we conclude that

$$\sum_{a<n<b} g(n)e(f(n)) = \sum_{m\in\mathcal{M}}\int_a^b g(x)e(f(x)-mx)dx + O(\varepsilon^{-1}+\log(\beta-\alpha+2)).$$

It remains to remove the weights $g(n)$ and $g(x)$. The correction on the left side is $O(1)$ and on the right side it is $O(\log(\beta-\alpha+2))$ by (8.27) for $b=a+1$. This completes the proof of (8.26). □

EXERCISE 1. Derive (8.26) from the Euler-Maclaurin formula (4.7) using (4.18) or directly from (4.21).

It is easy to generalize (8.26) for a weighted sum as follows:

$$\sum_{a<n<b} g(n)e(f(n)) = \sum_{\alpha-\varepsilon<m<\beta+\varepsilon} \int_a^b g(x)e(f(x)-mx)dx + O\left(G(\varepsilon^{-1}+\log(\beta-\alpha+2))\right) \tag{8.28}$$

where g is any smooth function on $[a,b]$ and

$$G = |g(b)| + \int_a^b |g'(y)|dy. \tag{8.29}$$

For the proof put

$$\delta(y) = \sum_{a<n\leqslant y} e(f(n)) - \sum_{\alpha-\varepsilon<m<\beta+\varepsilon} \int_a^y e(f(x)-mx)dx.$$

Thus $\delta(y) \ll \varepsilon^{-1}+\log(\beta-\alpha+2)$ by virtue of (8.26). Now applying partial summation we obtain (8.28) with the error term

$$\int_a^b g(y)d\delta(y) = \delta(b)g(b) - \int_a^b \delta(y)g'(y)dy$$

which is bounded by $G(\varepsilon^{-1}+\log(\beta-\alpha+2))$ as claimed.

In a special case (8.28) simplifies to

LEMMA 8.8. *Let $f(x)$ be a real function with $|f'(x)|\leqslant 1-\theta$ and $f''(x)\neq 0$ on $[a,b]$. We then have*

$$\sum_{a<n<b} g(n)e(f(n)) = \int_a^b g(x)e(f(x))dx + O(G\theta^{-1}) \tag{8.30}$$

where G is given by (8.29) and the implied constant is absolute.

Our next task is to evaluate the exponential integrals in (8.26). We begin by proving simple estimates for

$$I_f(a,b) = \int_a^b e(f(x))dx \tag{8.31}$$

where $f(x)$ is a real, smooth function on $[a,b]$. Notice that the trivial bound $|I_f(a,b)|\leqslant b-a$ cannot be improved substantially if $f^{(k)} \ll (b-a)^{-k}$ for all $k\geqslant 1$, because such a function is more or less constant. However, if one of the derivatives is somewhat larger, then our result will be non-trivial.

LEMMA 8.9. *Suppose $f'(x)f''(x) \neq 0$ on $[a, b]$. Then*

$$|I_f(a,b)| \leqslant |f'(a)|^{-1} + |f'(b)|^{-1}. \tag{8.32}$$

PROOF. We can assume without loss of generality that $f''(x) > 0$. Then by partial integration we get

$$2\pi i I_f(a,b) = \frac{e(f(b))}{f'(b)} - \frac{e(f(a))}{f'(a)} - \int_a^b e(f(x))d\Big(\frac{1}{f'(x)}\Big).$$

Hence

$$2\pi |I_f(a,b)| \leqslant \frac{1}{|f'(b)|} + \frac{1}{|f'(a)|} + \frac{1}{f'(a)} - \frac{1}{f'(b)}$$

which is better than what is claimed. □

LEMMA 8.10. *Suppose that for some $k \geqslant 1$ we have*

$$|f^{(k)}(x)| \geqslant \Lambda \tag{8.33}$$

for any x on $[a, b]$ with $\Lambda > 0$. Then

$$|I_f(a,b)| \leqslant k2^k \Lambda^{-1/k}. \tag{8.34}$$

PROOF. We apply induction with respect to k. For $k = 1$ the result follows from (8.32). Suppose that $f^{(k+1)} \geqslant \Lambda > 0$ on the whole interval $[a, b]$. Thus $f^{(k)}$ is increasing, so it may have at most one zero in $[a, b]$, say c if it exists. If $f^{(k)} \neq 0$ on $[a, b]$, we still define c by putting $c = a$ or $c = b$ according to whether $f^{(k)} > 0$ or $f^{(k)} < 0$. Let δ be a positive number to be chosen later. We put $a_1 = \max(a, c - \delta)$ and $b_1 = \min(c + \delta, b)$ and we split $I_f(a, b) = I_f(a, a_1) + I_f(a_1, b_1) + I_f(b_1, b)$. For the integral over the middle segment we use the trivial bound

$$|I_f(a_1, b_1)| \leqslant b_1 - a_1 \leqslant 2\delta.$$

To estimate the integral over the left segment $[a, a_1]$ we may assume that $a_1 > a$, that is $a_1 = c - \delta > a$, or else we get nothing. Now we verify that for $a \leqslant x \leqslant a_1$,

$$\begin{aligned} -f^{(k)}(x) &= \int_x^c f^{(k+1)}(y)dy - f^{(k)}(c) \\ &\geqslant \int_x^c f^{(k+1)}(y)dy \geqslant (c - x)\Lambda \geqslant \delta\Lambda > 0. \end{aligned}$$

Hence by the induction hypothesis

$$|I_f(a, a_1)| \leqslant k2^k(\delta\Lambda)^{-1/k}.$$

The same bound holds true for the integral over the right segment $[b_1, b]$. Adding these three bounds we get

$$|I_f(a,b)| \leqslant 2\delta + 2k2^k(\delta\Lambda)^{-1/k}.$$

This is true for any $\delta > 0$. On putting $\delta = \Lambda^{-1/(k+1)}$ we complete the proof of (8.34) for $k + 1$. □

Combining Lemmas 8.8 and 8.9 we obtain

COROLLARY 8.11. *Let $f(x)$ be a real function with $\theta \leqslant |f'(x)| \leqslant 1-\theta$ and $f''(x) \neq 0$ on $[a,b]$. Then*

$$\sum_{a<n<b} g(n)e(f(n)) \ll G\theta^{-1} \tag{8.35}$$

where g is given by (8.29) and the implied constant is absolute.

EXERCISE 2. Prove that for any real numbers f_n satisfying $\theta \leqslant f_n - f_{n-1} \leqslant f_{n+1} - f_n \leqslant 1-\theta$ we have

$$\Big|\sum_n e(f_n)\Big| \leqslant \operatorname{cotan}\frac{\pi\theta}{2}. \tag{8.36}$$

Combining Proposition 8.7 with Lemma 8.10 for $k=2$ we derive

COROLLARY 8.12. *Let $b-a \geqslant 1$. Let $f(x)$ be a real function on $[a,b]$ with $f''(x) \geqslant \Lambda > 0$ on $[a,b]$. Then*

$$\sum_{a<n<b} e(f(n)) \ll (f'(b) - f'(a) + 1)\Lambda^{-\frac{1}{2}} \tag{8.37}$$

where the implied constant is absolute.

PROOF. By Lemma 8.10 the integrals in (8.26) are bounded by $8\Lambda^{-1/2}$ and the number of these integrals is less than $f'(b) - f'(a) + 1$ (take $\alpha = f'(a), \beta = f'(b)$ and $\varepsilon = \frac{1}{2}$), hence (8.37) follows. □

Note that $f'(b) - f'(a) = (b-a)f''(y)$ for some y with $a \leqslant y \leqslant b$, thus (8.37) implies

COROLLARY 8.13. *Let $b-a \geqslant 1$. Let $f(x)$ be a real function on $[a,b]$ such that $\Lambda \leqslant f''(x) \leqslant \eta\Lambda$ with $\Lambda > 0$, $\eta \geqslant 1$. Then*

$$\sum_{a<n<b} e(f(n)) \ll \eta\Lambda^{\frac{1}{2}}(b-a) + \Lambda^{-\frac{1}{2}} \tag{8.38}$$

where the implied constant is absolute.

Now we are going to establish an approximate formula for the exponential integrals (8.31) which is more precise than the estimate (8.34). We begin by a special case

LEMMA 8.14. *Let $h(x)$ be a real function on $[0,X]$ such that*

$$h(0) = 1,\ h(x) \gg 1,\ (xh(x))' \gg 1, \tag{8.39}$$

$$h'(x) \ll X^{-1},\ h''(x) \ll X^{-2}. \tag{8.40}$$

Then for $\alpha > 0$ we have

$$\int_0^X e(\alpha x^2 h(x))dx = e\Big(\frac{1}{8}\Big)\frac{1}{\sqrt{8\alpha}} + O\Big(\frac{1}{\alpha X}\Big) \tag{8.41}$$

where the constant implied in O depends only on the constant implied in $\gg$ of (8.39) and that in $\ll$ of (8.40).

PROOF. For notational simplicity we set $h(x) = g^2(x)$. Clearly $g(x)$ satisfies all the conditions of $h(x)$. We change the variable $x^2g^2(x) = t$ to obtain

$$\int_0^X e(\alpha x^2 g^2(x))dx = \int_0^T e(\alpha t)dt^{\frac{1}{2}} - \int_0^T e(\alpha t)f(x)dt$$

where $T = (Xg(X))^2$ and

$$f(x) = ((xg(x))' - 1)/2xg(x)(xg(x))'.$$

Using the conditions (8.39), (8.40) for $g(x)$ and the Taylor expansions

$$g(x) = 1 + xg'(0) + O(x^2X^{-2}), \qquad g'(x) = g'(0) + O(xX^{-2})$$

we deduce that $f(x) \ll X^{-1}$ and $f'(x) \ll X^{-2}$. Hence we obtain by partial integration

$$\begin{aligned}
2\pi i\alpha \int_0^T e(\alpha t)f(x)dt &= \int_0^T f(x)de(\alpha t)\\
&= e(\alpha T)f(X) - f(0) - \int_0^T e(\alpha t)f'(x)dx(t)\\
&\ll X^{-1} + \int_0^X |f'(x)|dx \ll X^{-1}
\end{aligned}$$

because $x'(t)$ is positive. This bound is absorbed by the error term in (8.41). Next we extend the first integral to

$$\int_0^\infty e(\alpha t)dt^{\frac{1}{2}} = e\Big(\frac{1}{8}\Big)\frac{1}{\sqrt{8\alpha}},$$

and we estimate the excess by partial integration as follows:

$$\begin{aligned}
2\pi i\alpha \int_T^\infty e(\alpha t)dt^{\frac{1}{2}} &= -T^{-\frac{1}{2}}e(\alpha T) - \int_T^\infty e(\alpha t)dt^{-\frac{1}{2}}\\
&\ll T^{-\frac{1}{2}} = (Xg(X))^{-1} \ll X^{-1}.
\end{aligned}$$

This bound is also absorbed by the error term in (8.41) completing the proof. □

COROLLARY 8.15. *Let $f(x)$ be a real function on $[a, b]$ such that*

$$f''(c) \geqslant \Lambda, \tag{8.42}$$

$$|f^{(3)}(x)| \leqslant \Lambda X^{-1},\ |f^{(4)}(x)| \leqslant \Lambda X^{-2} \tag{8.43}$$

for some $\Lambda > 0$ and $X > 0$. Suppose $f'(c) = 0$ at some point c in (a, b). Then we have

$$I_f(a, b) = e\big(f(c) + \tfrac{1}{8}\big)f''(c)^{-\frac{1}{2}} + O\Big(\frac{1}{\Lambda}\Big(\frac{1}{b-c} + \frac{1}{c-a} + \frac{1}{X}\Big)\Big) \tag{8.44}$$

where the implied constant is absolute.

PROOF. We write

$$I_f(a,b)=\int_a^b e(f(x))dx=\int_0^{b-c} e(f(c+x))dx+\int_0^{c-a} e(f(c-x))dx.$$

By Taylor's expansion we have $f(c+x)=f(c)+\alpha x^2h(x)$, say, where $\alpha=\frac{1}{2}f''(c)$ and $h(x)$ is a function which satisfies

$$h(x)\geqslant 1-\frac{x}{3X},\ (xh(x))'\geqslant 1-\frac{2x}{3X}-\frac{x^2}{12X^2}$$

and $h'(x)\ll X^{-1}, h''(x)\ll X^{-2}$. These estimates follow from (8.42), (8.43). Hence the conditions of Lemma 8.14 are satisfied in the interval $0\leqslant x\leqslant Y$ with $Y=\min(b-c,X)$ getting

$$\int_0^Y e(f(c+x))dx=e\big(f(c)+\tfrac{1}{8}\big)(4f''(c))^{-\frac{1}{2}}+O\Big(\frac{1}{\Lambda Y}\Big).$$

For the remaining part of the integral we apply Lemma 8.9 getting

$$\int_Y^{b-c} e(f(c+x))dx\ll|f'(c+Y)|^{-1}+|f'(b)|^{-1}\ll(\Lambda Y)^{-1}.$$

From both estimates we obtain

$$\int_c^b e(f(x))dx=e\big(f(c)+\tfrac{1}{8}\big)(4f''(c))^{-\frac{1}{2}}+O\Big(\frac{1}{\Lambda}\Big(\frac{1}{b-c}+\frac{1}{X}\Big)\Big).$$

Similarly we evaluate the integral over the segment $[a,c]$ completing the proof of (8.44). □

Now we are ready to show the main result of the van der Corput method

THEOREM 8.16. *Let $f(x)$ be a real function on $[a,b]$ whose derivatives satisfy the following conditions: $\Lambda\leqslant f''\leqslant\eta\Lambda$, $|f^{(3)}|\leqslant\eta\Lambda(b-a)^{-1}$, $|f^{(4)}|\leqslant\eta\Lambda(b-a)^{-2}$ for some $\Lambda>0$ and $\eta\geqslant 1$. Then we have*

$$\sum_{a<n<b} e(f(n))=\sum_{\alpha<m<\beta} e(f(x_m)-mx_m+\tfrac{1}{8})f''(x_m)^{-\frac{1}{2}}+R_f(a,b) \tag{8.45}$$

where $\alpha=f'(a),\beta=f'(b)$ and x_m is the unique solution to $f'(x)=m$. Here $R_f(a,b)$ is considered as an error term; it satisfies

$$R_f(a,b)\ll\Lambda^{-\frac{1}{2}}+\eta^2\log(\beta-\alpha+1) \tag{8.46}$$

where the implied constant is absolute.

PROOF. We apply (8.25) with $\varepsilon=\frac{1}{2}$, $\alpha=f'(a)$ and $\beta=f'(b)$. For m with $\alpha+\varepsilon<m<\beta-\varepsilon$ (note this range is void if $\beta-\alpha\leqslant 1$) we evaluate the involved exponential integrals by (8.44) getting

$$\int_a^b e(f(x)-mx)dx=e\big(f(x_m)-mx_m+\tfrac{1}{8}\big)f''(x_m)^{-\frac{1}{2}}$$
$$+O\Big(\frac{\eta}{\Lambda}\Big(\frac{1}{b-x_m}+\frac{1}{x_m-a}\Big)\Big).$$

Here we have $\eta\Lambda(b - x_m) \geqslant f'(b) - f'(x_m) = \beta - m$ and $\eta\Lambda(x_m - a) \geqslant f'(x_m) - f'(a) = m - \alpha$, therefore the error terms contribute at most

$$\eta^2 \sum_{\alpha+\varepsilon<m<\beta-\varepsilon} [(\beta - m)^{-1} + (m - \alpha)^{-1}] \ll \eta^2 \log(\beta - \alpha + 1).$$

For the two missing integrals with $\alpha - \varepsilon < m \leqslant \alpha + \varepsilon$ and $\beta - \varepsilon \leqslant m < \beta + \varepsilon$ we use the bound $8\Lambda^{-1/2}$ which follows from (8.34) with $k = 2$ completing the proof of (8.46). □

REMARKS. In general the bound (8.46) cannot be significantly improved because it is almost as small as the individual terms in both sums (8.45).

EXERCISE 3. Using partial summation derive from (8.45) the following formula for a weighted sum

(8.47)
$$\sum_{a<n<b} g(n)e(f(n)) = \sum_{\alpha<m<\beta} g(x_m)f''(x_m)^{-\frac{1}{2}}e(f(x_m) - mx_m + \tfrac{1}{8}) + O(G\Lambda^{-\frac{1}{2}} + G\eta^2 \log(\beta - \alpha + 1))$$

where g is any smooth function on $[a, b]$ and G is given by (8.29).

EXAMPLE. Let $X > 0, N > 0$ and $\alpha > 1, \nu > 1$. We then have

(8.48)
$$\sum_{N<n<\nu N} \Big(\frac{\alpha}{n}\Big)^{\frac{1}{2}} e\Big(\frac{X}{\alpha}\Big(\frac{n}{N}\Big)^{\alpha}\Big) = \sum_{M<m<\mu M} \Big(\frac{\beta}{m}\Big)^{\frac{1}{2}} e\Big(\frac{1}{8} - \frac{X}{\beta}\Big(\frac{m}{M}\Big)^{\beta}\Big) + O(N^{-\frac{1}{2}} \log(N + 2) + M^{-\frac{1}{2}} \log(M + 2))$$

where $\frac{1}{\alpha} + \frac{1}{\beta} = 1, \mu^{\beta} = \nu^{\alpha}$ and $MN = X$, the implied constant depending only on α, ν.

The approximate formula (8.47) (which constitutes the B-process of van der Corput) transforms an exponential sum of amplitude $f(x)$ to an exponential sum of amplitude $h(y)$ related by

(8.49)
$$h(y) = f(x) - xy, \quad y = f'(x).$$

Thus f and h have essentially the same size. However, the length of the sums changes from $b - a$ to $\beta - \alpha = f'(b) - f'(a)$. Note that the operation (8.47) is involutary, so it would be useless to apply it two times in a row. It is recommended to apply the B-process in the direction which reduces the length of the exponential sum. Then on the shorter side one can estimate trivially (getting for example (8.24)), but it is not necessary to terminate in this way. A new possibility is offered by a differencing process which reduces the amplitude function, but does not change the length of summation. This is called the A-process of Weyl-van der Corput. We begin by the following general inequality

LEMMA 8.17. *For any complex numbers z_n we have*

$$\Big|\sum_{a<n<b} z_n\Big|^2 \leqslant \Big(1+\frac{b-a}{Q}\Big)\sum_{|q|<Q}\Big(1-\frac{|q|}{Q}\Big)\sum_{a<n,n+q<b} z_{n+q}\bar{z}_n \tag{8.50}$$

where Q is any positive integer.

PROOF. Setting $z_n = 0$ if $n \notin (a,b)$ we obtain

$$S = \sum_{a<n<b} z_n = \sum_n z_n = \sum_n z_{n+q}$$

for any $q \in \mathbb{Z}$. Sum up this over q with $0 \leqslant q < Q$ getting

$$QS = \sum_{a-Q+1<n<b}\ \sum_{0\leqslant q<Q} z_{n+q}.$$

Hence by Cauchy's inequality

$$\begin{aligned} Q^2|S|^2 &\leqslant (b-a+Q)\sum_n \Big|\sum_{0\leqslant q<Q} z_{n+q}\Big|^2 \\ &= (b-a+Q)\sum_{0\leqslant q_1,q_2<Q}\ \sum_n z_{n+q_1}\bar{z}_{n+q_2} \\ &= (b-a+Q)\sum_{|q|<Q}\nu(q)\sum_n z_{n+q}\bar{z}_n \end{aligned}$$

where $\nu(q)$ is the number of integers $0 \leqslant q_1, q_2 < Q$ with $q_1 - q_2 = q$, i.e., $\nu(q) = Q - |q|$. This completes the proof of (8.50). □

Taking $z_n = e(f(n))$ we obtain the differencing process

PROPOSITION 8.18. *Let $f(x)$ be real function in (a,b) and Q a positive integer. We have*

$$\Big|\sum_{a<n<b} e(f(n))\Big|^2 \leqslant \Big(1+\frac{b-a}{Q}\Big)\sum_{|q|<Q}\Big(1-\frac{|q|}{Q}\Big)\sum_{a(q)<n<b(q)} e(f(n+q)-f(n)) \tag{8.51}$$

where $a(q) = \max(a, a-q)$ and $b(q) = \min(b, b-q)$.

REMARKS. Note that $q = 0$ is always there, so the resulting bound for the original sum $\sum e(f(n))$ can never be better than $(b-a)Q^{-1/2}$, i.e., one cannot save more than $Q^{1/2}$ from the trivial bound $b-a$. In practice Q is chosen to be much smaller than $b-a$, so the new amplitude function $h(x) = f(x+q) - f(x)$ has reduced derivatives, while $a(q)$ and $b(q)$ do not change much. Here the flexibility in choosing Q is another innovation introduced by van der Corput to the original differencing method of Weyl.

EXERCISE 4. Prove that for $1 \leqslant Q \leqslant (b-a)^{\frac{1}{2}}$ we have

$$\sum_{a<n<b} e(f(n)) \ll (b-a)Q^{-\frac{1}{2}} + (b-a)^{\frac{1}{2}}\Big|\sum_{a+Q<n<b-Q} e(h(n))\Big|^{\frac{1}{2}} \tag{8.52}$$

where $h(x) = f(x+q) - f(x-q)$ for some $0 < q < Q$ and the implied constant is absolute [**Hint:** Use only even integers for Weyl shifts.]

By successive application of Proposition 8.18 and Corollary 8.13 we derive

COROLLARY 8.19 (VAN DER CORPUT). *Let $b-a\geqslant 1$. Let $f(x)$ be a real function on (a,b) such that $\Lambda\leqslant f^{(3)}(x)\leqslant\eta\Lambda$ where $b-a\geqslant 1,\Lambda>0$ and $\eta\geqslant 1$. Then*

$$\sum_{a<n<b} e(f(n))\ll\eta^{\frac{1}{2}}\Lambda^{\frac{1}{6}}(b-a)+\Lambda^{-\frac{1}{6}}(b-a)^{\frac{1}{2}} \tag{8.53}$$

where the implied constant is absolute.

PROOF. For $0<q<Q$ and $a<x<b-q$ we have $h''(x)=f''(x+q)-f''(x)=qf^{(3)}(y)$ for some $y\in(a,b)$. Hence $\Lambda q\leqslant h''(x)\leqslant\eta\Lambda q$ and by (8.38)

$$\sum_{a<n<b-q} e(f(n+q)-f(n))\ll\eta(\Lambda q)^{\frac{1}{2}}(b-a)+(\Lambda q)^{-\frac{1}{2}}.$$

A similar result is true for the sums with negative q. For $q=0$ we have the trivial bound $b-a+1$. Collecting these results we deduce by (8.51) that

$$\begin{aligned}\Big|\sum_{a<n<b} e(f(n))\Big|^2&\ll\Big(1+\frac{b-a}{Q}\Big)\Big\{b-a+\sum_{0<q<Q}\Big(\eta\Lambda^{\frac{1}{2}}q^{-\frac{1}{2}}(b-a)+(\Lambda q)^{-\frac{1}{2}}\Big)\Big\}\\&\ll\Big(1+\frac{b-a}{Q}\Big)\Big(b-a+\eta\Lambda^{\frac{1}{2}}Q^{\frac{3}{2}}(b-a)+\Lambda^{-\frac{1}{2}}Q^{\frac{1}{2}}\Big)\end{aligned}$$

for any positive integer Q. However, the above estimate remains valid for any positive real number Q for obvious reasons. Taking $Q=\Lambda^{-1/3}$ we obtain (8.53) provided $\Lambda\geqslant(b-a)^{-3}$, and otherwise the result is trivial. □

The Weyl-van der Corput inequality (8.51) can be used repeatedly to reduce further the amplitude function until (8.38) becomes useful. This leads to the following generalization of (8.38) and (8.53).

THEOREM 8.20 (VAN DER CORPUT). *Let $b-a\geqslant 1$. Let $f(x)$ be a real function on (a,b) and $k\geqslant 2$ such that $\Lambda\leqslant f^{(k)}(x)\leqslant\eta\Lambda$, where $\Lambda>0$ and $\eta\geqslant 1$. Then*

$$\sum_{a<n<b} e(f(n))\ll\eta^{2^{2-k}}\Lambda^{\kappa}(b-a)+\Lambda^{-\kappa}(b-a)^{1-2^{2-k}} \tag{8.54}$$

where $\kappa=(2^k-2)^{-1}$ and the implied constant is absolute.

For large k the bounds (8.17) of Weyl and (8.54) of van der Corput are comparable.

8.4. Discussion of exponent pairs.

Throughout $S_f(N)$ denotes an exponential sum of type

$$S_f(N)=\sum_{N<n\leqslant N'} e(f(n))$$

with $N\leqslant N'\leqslant 2N$ and f is a smooth function on $[N,2N]$. We have established numerous estimates for $S_f(N)$ which depend essentially on the length N and on the size of the amplitude function f. In this section we describe these results in a unified form and discuss possible extensions. For the sake of simplicity we drop some minor, yet necessary conditions, so our assertions are not strictly correct in this section. Rigorous assertions can be found in the book by S. W. Graham and G. Kolesnik [**GK**] as well as in the original papers by van der Corput [**Cor1**], [**Cor2**], E. Phillips [**Ph**] and R. A. Rankin [**Ra1**].

Suppose f behaves like a monomial, precisely we assume that the derivatives satisfy

$$|f^{(j)}(x)| \asymp FN^{-j} \tag{8.55}$$

for $N \leqslant x \leqslant 2N$ and any $j \geqslant 0$, where the implied constant depends on j. In particular, $|f(x)| \asymp F$ and $|f'(x)| \asymp FN^{-1}$. If N is larger than F by a sufficiently large factor, then Corollary 8.11 yields

$$S_f(N) \ll NF^{-1}.$$

This bound is the best possible, because in this case the sum $S_f(N)$ is well approximated by the corresponding integral (see Lemma 8.8). In what follows we stay out of this special case by assuming that

$$\Lambda = FN^{-1} \geqslant 1. \tag{8.56}$$

Inspired by many examples we postulate the existence of universal numbers

$$0 \leqslant p, q \leqslant \tfrac{1}{2} \tag{8.57}$$

having the property that

$$S_f(N) \ll \Lambda^p N^{q+\frac{1}{2}} F^{\varepsilon} \tag{8.58}$$

for any exponential sum of length N and the frequency function f satisfying (8.55) and (8.56). Here ε is any positive number and the implied constant depends only on ε and on the sequence of constants implied in (8.55). We call (p, q) an exponent pair (notice that our definition of q differs by $\frac{1}{2}$ from that in the literature [**GK**]).

Clearly if (p, q) is an exponent pair, then any pair of larger numbers form an exponent pair. The trivial estimate for $S_f(N)$ corresponds to the exponent pair

$$(p, q) = \left(0, \frac{1}{2}\right). \tag{8.59}$$

The following statement is the most optimistic for this aspect of the theory of exponential sums, and is widely believed to be correct in its essentials (minor assumptions on f might be required to avoid pathologies):

EXPONENT PAIR HYPOTHESIS. *The pair $(p, q) = (0, 0)$ is an exponent pair. In other words, an exponential sum of type $S_f(N)$ with f, N satisfying (8.55) and (8.56) should satisfy*

$$S_f(N) \ll N^{\frac{1}{2}} F^{\varepsilon}. \tag{8.60}$$

This bound, if true for reasonable functions, would lead to solutions of many problems in analytic number theory. For example, it implies the Lindelöf Hypothesis

$$\zeta(\tfrac{1}{2} + it) \ll t^{\varepsilon}, \quad \text{if } t \geqslant 1.$$

Moreover, it would solve the Gauss circle problem and the Dirichlet divisor problem; namely it implies

$$\sum_{n\leqslant x} r(n) = \pi x + O(x^{\frac{1}{4}+\varepsilon}),$$

$$\sum_{n\leqslant x} \tau(n) = x\log x + (2\gamma - 1)x + O(x^{\frac{1}{4}+\varepsilon}),$$

where the exponents are best possible. Actually for these problems one only needs to know that $(p,q) = (0, \frac{1}{4})$ is an exponent pair. However, our current knowledge is very poor.

Corollary 8.13 tells us that

$$(p,q) = \Big(\frac{1}{2}, 0\Big) \tag{8.61}$$

is an exponent pair. Clearly a linear convex combination of exponent pairs is again an exponent pair. Therefore (8.59) and (8.61) imply that $(p,q) = (\frac{1}{4}, \frac{1}{4})$ is an exponent pair, however, we have already proved a stronger result (8.18) which can be now interpreted as saying that

$$(p,q) = \Big(\frac{1}{6}, \frac{1}{6}\Big) \tag{8.62}$$

is an exponent pair. Theorem 8.4 yields the exponent pairs

$$(p,q) = \Big(\frac{4}{k2^k}, \frac{1}{2} - \frac{4(k-1)}{k2^k}\Big) \tag{8.63}$$

and Theorem 8.20 yields

$$(p,q) = \Big(\frac{1}{2^k - 2}, \frac{1}{2} - \frac{k-1}{2^k - 2}\Big) \tag{8.64}$$

for any integer $k \geqslant 2$. Notice that as k tends to infinity in both sequences, $p = p(k)$ tends to zero slightly faster than $q = q(k)$ tends to $\frac{1}{2}$. This feature makes it possible to claim non-trivial bounds for the exponential sums $S_f(N)$ whose length N is comparable with the amplitude F in the logarithmic scale. In the next section we shall produce by Vinogradov's method a sequence of exponent pairs (p,q) in which p tends to zero much faster than q tends to $\frac{1}{2}$. So far we do not have an exponent pair with $p = 0$ and $q < \frac{1}{2}$; any such example would mark a major breakthrough in the theory of exponential sums.

A continuum of exponent pairs can be created by alternating the A and B processes. By the Weyl-van der Corput differencing inequality (8.51) one derives

A-PROCESS. *If (p,q) is an exponent pair, so is*

$$A(p,q) = \Big(\frac{p}{2p+2}, \frac{q+\frac{1}{2}}{2p+2}\Big). \tag{8.65}$$

By the van der Corput approximate functional equation (8.45) one derives

B-PROCESS. *If (p,q) is an exponent pair, so is*

$$B(p,q) = (q,p). \tag{8.66}$$

Here are some examples of exponent pairs generated from the trivial pair

$$B\Big(0,\frac{1}{2}\Big) = \Big(\frac{1}{2},0\Big),$$
$$AB\Big(0,\frac{1}{2}\Big) = \Big(\frac{1}{6},\frac{1}{6}\Big),$$
$$A^2B\Big(0,\frac{1}{2}\Big) = \Big(\frac{1}{14},\frac{2}{7}\Big),$$
$$A^3B\Big(0,\frac{1}{2}\Big) = \Big(\frac{1}{30},\frac{11}{30}\Big),$$
$$A^4B\Big(0,\frac{1}{2}\Big) = \Big(\frac{1}{62},\frac{13}{31}\Big),$$
$$ABABA^2B\Big(0,\frac{1}{2}\Big) = \Big(\frac{1}{11},\frac{1}{4}\Big).$$

For the purpose of estimating the Riemann zeta function on the critical line one needs an exponent pair (p,q) with $p+q$ as small as possible since

$$\zeta(\tfrac{1}{2}+it) \ll t^{\frac{1}{2}(p+q)+\varepsilon}, \quad \text{if } t \geqslant 1. \tag{8.67}$$

In 1955 Rankin [**Ra1**] computed the minimum of $\theta = \frac{1}{2}(p+q)$ over all exponent pairs generated by the A and B processes, namely $\theta_{\min} = 0.16451067....$ In other applications of exponent pairs one seeks the minimum of a fractional linear function $\theta(p,q) = (\alpha p + \beta q + \mu)/(\delta p + \gamma q + \nu)$. Having these applications in mind in 1985 Graham gave a truly elegant algorithm for the solution to this general problem; see [**GK**].

Today there are known exponent pairs which cannot be obtained by the van der Corput method. For example, Huxley and Watt [**HW**] showed that

$$(p,q) = \Big(\frac{9}{56},\frac{9}{56}\Big) \tag{8.68}$$

is an exponent pair by elaborating the work of Bombieri and Iwaniec [**BI**]. Further improvements along these lines are given in [**Hu4**] and a very nice exposition is contained in [**GK**].

The Weyl shift (from n to $n+q$) was introduced for the purpose of reducing the amplitude function (from $f(n)$ to $h(n) = f(n+q) - f(n)$). However, this shift also offers an additional variable of summation. One can take advantage of this by an appeal to the theory of two-dimensional exponent pairs, but the improvements are not impressive (in the context of the Riemann zeta function the results are still weaker than what follows from the Huxley-Watt exponent pair (8.68)).

8.5. Vinogradov's method.

In the late thirties I. M. Vinogradov [**Vi2**], [**Vi3**] invented a new method for estimating exponential sums of Weyl's type which is very strong for sums of large amplitude (relative to the length). He spent a considerable time of his life to massage the results, so we have today several elegant versions due to him and his followers.

Let $f(x)$ be a real smooth function in $[N, 2N]$. Our aim is to estimate exponential sums

$$S_f(a,b) = \sum_{a<n<b} e(f(n)) \tag{8.69}$$

for any a, b with $N \leqslant a < b \leqslant 2N$. We shall approach the problem in several distinct steps.

Step I (Small Shift). We begin, as in Weyl's method, by shifting the summation

$$S_f(a,b) = \sum_{a-q<n<b-q} e(f(n+q)).$$

We choose q to be relatively small so we can approximate $f(n+q)$ by a polynomial in q of reasonably small degree. Precisely we apply the Taylor expansion $f(n+q) = F_n(q) + R_n(q)$, where $F_n(q)$ is the polynomial

$$F_n(q) = \sum_{0\leqslant j\leqslant k} \alpha_j(n) q^j$$

with coefficients $\alpha_j(n) = f^{(j)}(n)/j!$ and $R_n(q)$ satisfies the bound

$$|R_n(q)| \leqslant q^{k+1} f^{(k+1)}/(k+1)!$$

with $f^{(k+1)} = \max |f^{(k+1)}(x)|$. Inserting this approximation we get

$$S_f(a,b) = \sum_{a<n<b} e(F_n(q)) + \theta(2q + q^{k+1} f^{(k+1)} N)$$

where $|\theta| \leqslant 1$. Here the first error term $2q$ represents the trivial bound for exponential sums over two non-overlapping intervals and the second error term comes from the inequality $|e(R_n(q)) - 1| \leqslant 2\pi |R_n(q)| \leqslant q^{k+1} f^{(k+1)}$. We shall require all the implied constants in the forthcoming estimates to be absolute. This is important, even with respect to k, because for some applications (first of all for bounding $\zeta(1+it)$) one uses k depending on the length of summation and the size of the amplitude function.

At this point Vinogradov's approach diverges from that of Weyl. The first new idea is that now we are going to average over q in a special subset, say $\mathcal{Q} \subset \{1, 2, \dots, Q\}$, rather than in the full set. We obtain

$$S_f(a,b) = \sum_{a<n<b} |\mathcal{Q}|^{-1} \sum_{q\in\mathcal{Q}} e(F_n(q)) + O(Q + Q^{k+1} f^{(k+1)} N) \tag{8.70}$$

where $|\mathcal{Q}|$ denotes the number of points in $\mathcal{Q}$. Though $F_n(q)$ is a polynomial in q the innermost sum

$$S_n = \sum_{q\in\mathcal{Q}} e(F_n(q))$$

cannot be considered as Weyl's sum of degree k, because of restrictions for $\mathcal{Q}$ (to be specified soon). Moreover, we are not going to apply the idea of differencing by way of squaring the sum $S_f(a,b)$, but rather we proceed directly to estimation of each inner sum S_n separately for $a < n < b$. In other words, we no longer need the variable n until the last step, and even there, it is not essential.

Since the variable n plays a minor role in what follows we simplify the notation by considering the exponential sum

$$S = \sum_{q \in \mathcal{Q}} e(F(q))$$

for arbitrary polynomial with real coefficients

$$F(q) = \sum_{0 \leqslant j \leqslant k} \alpha_j q^j .$$

Obviously S depends on the fractional part of the coefficients α_j. For this reason we shall exploit only the coefficients with $|\alpha_j| < 1$, because these can be easily controlled. On the other hand, the very small coefficients, specifically $|\alpha_j| < Q^{-j}$, do not make a variation in the argument of $e(F(q))$. Therefore only the middle terms of our polynomial $F(q)$ play a role, that is these with $Q^{-j} < |\alpha_j| < 1$, the other terms are wasted. In the final arguments we shall appeal to the original coefficients $\alpha_j = \alpha_j(n) = f^{(j)}(n)/j!$ whose particular shape is not important, we only need to estimate their size.

STEP II (creation of bilinear forms). We choose

$$\mathcal{Q} = \{xy : 1 \leqslant x, y \leqslant P\}$$

where the numbers $q = xy$ are counted with multiplicity of such representations, thus $|\mathcal{Q}| = Q = P^2$ and the exponential sum S becomes a bilinear form

$$S = \sum_{1 \leqslant x \leqslant P} \sum_{1 \leqslant y \leqslant P} e(F(xy)). \tag{8.71}$$

It is important that the variables x, y run independently over two sets (of some cardinality), yet it is of no significance what the points are. For example, we could restrict x, y to special numbers or consider a general bilinear form

$$\mathcal{B}(X, Y) = \sum_{\mathbf{x} \in \mathcal{X}} \sum_{\mathbf{y} \in \mathcal{Y}} a(\mathbf{x}) b(\mathbf{y}) e(\langle \mathbf{x}, \mathbf{y} \rangle)$$

with any complex coefficients $a(\mathbf{x}), b(\mathbf{y})$, where $\mathcal{X}, \mathcal{Y}$ are the sets of points of type $\mathbf{x} = [1, \dots, x^k]$, $\mathbf{y} = [1, y, \dots, y^k]$ with $x, y \in \mathbb{Z}, 1 \leqslant x \leqslant X, 1 \leqslant y \leqslant Y$ and $\langle \mathbf{x}, \mathbf{y} \rangle = F(xy)$. If $\mathcal{X}, \mathcal{Y}$ are sufficiently large and well-spaced sets then a non-trivial bound for $\mathcal{B}(X, Y)$ follows easily by Fourier technique (see Chapter 10). However, this is not the case under consideration here; our sets are quite sparse.

STEP III (expanding the variables). Our next goal is to create more points of summation and make many variables with no large gaps. To this end we raise the bilinear form (8.71) to a high power, apply Hölder's inequality and rearrange the resulting multiple sum as follows:

$$\begin{aligned}|S|^\ell &\leqslant P^{\ell-1}\sum_x\Big|\sum_y e(F(xy))\Big|^\ell\\ &= P^{\ell-1}\sum_x \zeta_x \sum_{y_1,\dots,y_\ell} e(F(xy_1)+\cdots+F(xy_\ell))\\ &= P^{\ell-1}\sum_{\lambda_1,\dots,\lambda_k}\nu(\lambda_1,\dots,\lambda_k)\sum_x \zeta_x e(\alpha_1\lambda_1 x+\cdots+\alpha_k\lambda_k x^k).\end{aligned}$$

Here ζ_x are complex numbers with $|\zeta_x| = 1$ and $\nu(\lambda_1,\dots,\lambda_k)$ is the number of solutions to the system of equations

$$\text{(8.72)}\qquad \begin{cases} y_1+\cdots+y_\ell=\lambda_1\\ y_1^2+\cdots+y_\ell^2=\lambda_2\\ \cdots\cdots\cdots\cdots\cdots\cdots\\ y_1^k+\cdots+y_\ell^k=\lambda_k\end{cases}$$

in integers $1\leqslant y_j\leqslant P$ for $j=1,2,\dots,\ell$. Thus $\ell\leqslant\lambda_j\leqslant \ell P^j$, or else there are no solutions. Since ℓ will be much larger than k (approximately $\ell\asymp k^2$) we expect that most vectors $(\lambda_1,\dots,\lambda_k)$ have good number of representations by the system (8.72). This number $\nu(\lambda_1,\dots,\lambda_k)$ may vary only slightly. In order to smooth the outer summation we raise $|S|^\ell$ again to power 2ℓ, getting

$$|S|^{2\ell^2}\leqslant P^{2\ell(\ell-1)}I_{\ell,k}(P)^{2(\ell-1)}J_{\ell,k}(P)Z_{\ell,k}(P)$$

where

$$\begin{aligned}I_{\ell,k}(P) &= \sum_{\lambda_1,\dots,\lambda_k}\nu(\lambda_1,\dots,\lambda_k),\\ J_{\ell,k}(P) &= \sum_{\lambda_1,\dots,\lambda_k}\nu^2(\lambda_1,\dots,\lambda_k),\\ Z_{\ell,k}(P) &= \sum_{\lambda_1,\dots,\lambda_k}\Big|\sum_x\zeta_x e(\alpha_1\lambda_1x+\cdots+\alpha_k\lambda_k x^k)\Big|^{2\ell}.\end{aligned}$$

For the average of $\nu(\lambda_1,\dots,\lambda_k)$ we have the exact value

$$I_{\ell,k}(P)=P^\ell,$$

but the estimation of the average of $\nu^2(\lambda_1,\dots,\lambda_k)$ is a much harder problem. Note that $J_{\ell,k}(P)$ is the number of solutions to the system of homogeneous equations

$$\text{(8.73)}\qquad \begin{cases} y_1+\cdots+y_\ell-y_{\ell+1}-\cdots-y_{2\ell}=0\\ y_1^2+\cdots+y_\ell^2-y_{\ell+1}^2-\cdots-y_{2\ell}^2=0\\ \cdots\cdots\cdots\cdots\cdots\cdots\cdots\cdots\cdots\cdots\cdots\\ y_1^k+\cdots+y_\ell^k-y_{\ell+1}^k-\cdots-y_{2\ell}^k=0\end{cases}$$

in integers $1\leqslant y_j\leqslant P$ for $j=1,2,\dots,2\ell$. We have

$$\text{(8.74)}\qquad J_{\ell,k}(P)=\int_0^1\cdots\int_0^1\Big|\sum_{1\leqslant x\leqslant P}e(\alpha_1x+\cdots+\alpha_kx^k)\Big|^{2\ell}d\alpha_1\dots d\alpha_k.$$

More generally, if the right-hand side of the system (8.73) is replaced by constants $\sigma_1, \ldots, \sigma_k$ then the number of solutions $J_{\ell,k}(P_j;\sigma_1,\ldots,\sigma_k)$ to the system of inhomogeneous equations is given by the integral

$$\int_0^1 \cdots \int_0^1 \Big| \sum_{1\leqslant x\leqslant P} e(\alpha_1 x+\cdots+\alpha_k x^k)\Big|^{2\ell} e(\alpha_1\sigma_1+\cdots+\alpha_k\sigma_k)d\alpha_1\ldots d\alpha_k.$$

Hence it is clear that

$$J_{\ell,k}(P_j;\sigma_1,\ldots,\sigma_k)\leqslant J_{\ell,k}(P_j;0,\ldots,0)=J_{\ell,k}(P). \tag{8.75}$$

STEP IV (tuning down to linear polynomials). Estimation of $J_{\ell,k}(P)$ lies in the heart of Vinogradov's method (see the next step), however, the vital saving comes from $Z_{\ell,k}(P)$ as it is the only sum left with non-trivial characters. We have

$$Z_{\ell,k}(P)\leqslant \sum_{x_1,\ldots,x_{2\ell}}\Big|\sum_{\lambda_1,\ldots,\lambda_k} e(\alpha_1\lambda_1\sigma_1+\cdots+\alpha_k\lambda_k\sigma_k)\Big|$$

where $\sigma_k=\sigma_k(x_1,\ldots,x_{2\ell})=\sum_1^\ell(x_j^h-x_{j+\ell}^h)$, so $|\sigma_k|\leqslant \ell P^k$. Hence by (8.75) we arrive at

$$Z_{\ell,k}(P)\leqslant J_{\ell,k}(P)\sum_{\sigma_1,\ldots,\sigma_k}\Big|\sum_{\lambda_1,\ldots,\lambda_k} e(\alpha_1\lambda_1\sigma_1+\cdots+\alpha_k\lambda_k\sigma_k)\Big|.$$

Here $\sigma_1,\ldots,\sigma_k,\lambda_1,\ldots,\lambda_k$ run independently over all integers with $|\sigma_h|\leqslant \ell P^h$ and $|\lambda_h|\leqslant \ell P^h$. Note that the multiple sum factors so we write the result as

$$Z_{\ell,k}(P)\leqslant J_{\ell,k}(P)\ell^{2k}P^{k(k+1)}\Delta,$$

where

$$\Delta=\prod_{1\leqslant h\leqslant k} D(\alpha_h,\ell P^h),$$

$$D(\alpha,X)=X^{-2}\sum_{|m|\leqslant X}\Big|\sum_{|n|\leqslant X} e(\alpha mn)\Big|.$$

Now, as in Weyl's method, the source of cancellation is in the exponential sums for linear polynomials. This is reached very differently here by Vinogradov's technique of bilinear forms than before by Weyl's differencing process. But above all, the new process is much faster. Indeed, Weyl's differencing requires applications of Cauchy's inequality $k-1$ times (each application reduced the degree by one), which amounts to raising the original sum to power 2^{k-1}, whereas Vinogradov's method gets the job done with the power $2\ell^2\asymp k^4$. Another new feature is that we produced linear polynomials out of every monomial in F, not just one from the highest degree; consequently the saving factors appear about k times.

Gathering the above estimates we obtain

$$|S|^{2\ell^2}\leqslant \ell^{2k}P^{4\ell(\ell-1)+k(k+1)}J_{\ell,k}^2(P)\Delta. \tag{8.76}$$

STEP V (estimation of $J_{\ell,k}(P)$). Let us give some thought about the inequality (8.76). Here S is the original sum

$$S(\alpha_1,\dots,\alpha_k)=\sum_{q\in\mathcal{Q}}e(\alpha_1 q+\dots+\alpha_k q^k)$$

while $J_{\ell,k}$ is the $L^{2\ell}$-norm of

$$P(\alpha_1,\dots,\alpha_k)=\sum_{1\leqslant x\leqslant \ell}e(\alpha_1 x+\dots+\alpha_k x^k)$$

with respect to the coefficients $\alpha_1,\dots,\alpha_k$ over the torus $(\mathbb{R}/\mathbb{Z})^k$. At first glance we may think that the problem of estimating $J_{\ell,k}(P)$ is not much different than that of the original sum S. Yes, this would be an even harder problem if we dealt with individual sums. But, due to complete integration over the coefficients, we are really having a diophantine problem (counting solutions to the system of homogeneous polynomial equations (8.73)).

By a heuristic argument one is lead to the bound

$$J_{\ell,k}(P)\ll P^{2\ell-\frac{1}{2}k(k+1)} \tag{8.77}$$

(the linear equation in (8.73) fixes one variable, the quadratic equation fixes two additional variables, etc.), but we cannot prove this. Vinogradov succeeded in proving a slightly weaker bound.

THEOREM 8.21. *Let k,m be positive integers and $\ell\geqslant k(k+m)$. We then have*

$$J_{\ell,k}(P)\leqslant 2^{4\ell m}P^{2\ell-\frac{1}{2}k(k+1)+\frac{1}{2}k(k+1)(1-1/k)^m} \tag{8.78}$$

for any $P\geqslant k^{k(1-1/k)^{-m}}$.

We begin by proving

LEMMA 8.22. *Let p be a prime number, $p>k$, and $\lambda_1,\dots,\lambda_k$ be integers. Then the number $T=T_p(\lambda_1,\dots,\lambda_k)$ of solutions to the system of congruences*

$$\begin{cases} x_1+\dots+x_k\equiv\lambda_1 \pmod{p}\\ \dots\dots\dots\dots\dots\dots\dots\dots\\ x_1^k+\dots+x_k^k\equiv\lambda_k \pmod{p^k}\end{cases} \tag{8.79}$$

in $x_1,\dots,x_k \pmod{p^k}$ such that $x_i\not\equiv x_j \pmod{p}$ is bounded by

$$T\leqslant k!\,p^{\frac{k(k-1)}{2}}. \tag{8.80}$$

PROOF. We write each x_r uniquely in the form

$$x_r=x_{r1}+x_{r2}p+\dots+x_{rk}p^{k-1}$$

with $0\leqslant x_{ri}<p$ for $1\leqslant r\leqslant k$ and $1\leqslant i\leqslant k$. The first coordinates x_{r1} satisfy the system (8.79) (mod p). From the theory of symmetric polynomials it follows that the j-th symmetric functions in $x_{11},x_{21},\dots,x_{k1}$, say s_j, are determined by $\lambda_1,\dots,\lambda_k$ and that x_{r1} are roots of $x^k-s_1x^{k-1}+\dots+(-1)^k s_k$. There are at most k roots, so there are at most $k!$ solutions $(x_{11},x_{21},\dots,x_{k1})$.

Now suppose $(x_{11}, x_{21}, \dots, x_{k1}), \dots, (x_{1h}, x_{2h}, \dots, x_{kh})$ with $1 \leqslant h < k$ are given. We shall find conditions for the coordinates of order $h+1$. Letting

$$x_{r1} + x_{r2}p + \cdots + x_{rh}p^{h-1} = y_{rh}$$

for $1 \leqslant r \leqslant k$, we have

$$x_r \equiv y_{rh} + x_{rh+1}p^h (\text{mod } p^{h+1}).$$

Consider the system (8.79) (mod p^{h+1}) reduced by omitting the first h congruences, i.e.,

$$\sum_{r=1}^{k} (y_{rh} + x_{rh+1}p^h)^m \equiv \lambda_m (\text{mod } p^{h+1})$$

for $h < m \leqslant k$. Hence

$$\sum_{r=1}^{k} x_{rh+1} y_{rh}^m \equiv \mu_m (\text{mod } p), \; h < m < k, \tag{8.81}$$

where μ_m are some numbers determined by the coordinates x_{rj} of order $1 \leqslant j \leqslant h$ and by the constants λ_m with $h < m < k$. The numbers $y_{rh}, 1 \leqslant r \leqslant k$, are different modulo p because so are x_r. In particular, at most one y_{rh} vanishes modulo p, say y_{1h}, if any. If we choose $x_{1h+1}, \dots, x_{hh+1}$ arbitrarily the remaining ones $x_{h+1h+1}, \dots x_{kh+1}$ are determined modulo p uniquely from the linear system (8.81) because its determinant (Vandermonde) is non-zero. Therefore, the number of coordinates of order $h+1$ does not exceed p^h. We conclude that

$$T \leqslant k! p p^2 \cdots p^{k-1} = k! p^{\frac{k(k-1)}{2}}.$$

□

COROLLARY 8.23. *Let p be a prime number, $p > k$, and let $U_{kp}(P)$ be the number of solutions to the system of congruences*

$$\sum_{1 \leqslant h \leqslant k} (x_h^m - x_{h+k}^m) \equiv 0 \; (\text{mod } p^m) \tag{8.82}$$

for $1 \leqslant m \leqslant k$, in $X < x_h \leqslant X + P$, $1 \leqslant h \leqslant 2k$, such that $x_{h_1} \not\equiv x_{h_2}$ (mod p) for $1 \leqslant h_1 \neq h_2 \leqslant k$. We have

$$U_{kp}(P) \leqslant k!(p^k + P)^k P^k p^{-\frac{k(k+1)}{2}}. \tag{8.83}$$

Our next aim is to establish a recurrence inequality for $J_{\ell k}(P)$. We assume that $\ell > k \geqslant 2$ and $P \leqslant pq$, where p is a prime number $> k$ to be chosen later and q is an integer. Clearly, we have $J_{\ell,k}(P) \leqslant J_{\ell,k}(pq)$. For estimating $J_{\ell,k}(pq)$ we divide the solutions of the system (8.73) into two classes. We say that $(y_1, \dots, y_{2\ell})$ is of the first class if both sequences $(y_1, \dots, y_\ell)$ and $(y_{\ell+1}), \dots, y_{2\ell})$ have at least k different numbers modulo p. The remaining solutions are of the second class.

Let us estimate the number J_1, say, of the first class solutions. Since k numbers can be put in ℓ places in $\ell(\ell-1)\dots(\ell-k+1)$ ways we obtain $J_1 \leqslant \ell^{2k} J_{11}$, where

J_{11} stands for the number of solutions $(y_1, \dots, y_{2\ell})$ such that all $(y_1, \dots, y_k)$ are different modulo p and all $(y_{\ell+1}, \dots, y_{\ell+k})$ are different modulo p. Letting

$$S(\alpha) = \sum_{1 \leqslant y \leqslant P} e(F_\alpha(y)),$$

where $F_\alpha(y) = \alpha_1 y + \cdots + \alpha_k y^k$, we split

$$S(\alpha) = \sum_{u (\text{mod } p)} \sum_{\substack{1 \leqslant y \leqslant P \\ y \equiv u (\text{mod } p)}} e(F_\alpha(y)) = \sum_{u (\text{mod } p)} S_u(\alpha),$$

say. By Hölder's inequality we obtain

$$\begin{aligned} J_{11} &= \int_0^1 \cdots \int_0^1 \Big| \sum_{u_1, \dots, u_k (\text{mod } p)}^{\star} S_{u_1}(\alpha) \cdots S_{u_k}(\alpha) \Big|^2 \Big| \sum_{u (\text{mod } p)} S_u(\alpha) \Big|^{2\ell - 2k} d\alpha \\ &\leqslant p^{2\ell - 2k - 1} \sum_{0 < u \leqslant p} J_{11}(u), \end{aligned}$$

say, where $\sum^\star$ means that the summation is restricted to the residue classes u_1, $\dots$, u_k modulo p that are all different, and

$$J_{11}(u) = \int_0^1 \cdots \int_0^1 \Big| \sum_{u_1, \dots, u_k}^{\star} S_{u_1}(\alpha) \cdots S_{u_k}(\alpha) \Big|^2 |S_u(\alpha)|^{2\ell - 2k} d\alpha.$$

This is equal to the number of solutions $(y_1, \dots, y_{2\ell})$ of the system (8.73) such that all $y_1, \dots, y_k$ are different modulo p, all $y_{\ell+1}, \dots, y_{\ell+1}, \dots, y_{\ell+k}$ are different modulo p and all the remaining ones are congruent to u modulo p. For these we set $y_h = u + p v_h$ with $0 \leqslant v_h < q$ if $k < h \leqslant \ell$ or $k + \ell < h \leqslant 2\ell$. We obtain

$$\sum_{0 < h \leqslant k} (y_h^m - y_{h+\ell}^m) + \sum_{k < h \leqslant \ell} [(u + p v_h)^m - (u + p v_{h+\ell})^m] = 0,$$

for $1 \leqslant m \leqslant k$. Now we apply an obvious fact that if $(y_1, \dots, y_{2\ell})$ satisfies the homogeneous system (8.73), then $(y_1 - u, \dots, y_{2\ell} - u)$ also does. Therefore we obtain

$$\sum_{0 < h \leqslant k} (x_h^m - x_{h+\ell}^m) + p^m \sum_{k < h \leqslant \ell} (v_h^m - v_{h+\ell}^m) = 0,$$

for $q \leqslant m \leqslant k$, where $x_h = y_h - u$ for $0 < h \leqslant k$ and $\ell < h \leqslant \ell + k$. In particular, we observe that x_h satisfy the system of congruences (8.79). Given a solution to (8.79) we see that v_h satisfy the inhomogeneous system of equations

$$\sum_{k < h \leqslant \ell} (v_h^m - v_{h+\ell}^m) = \lambda_m, \quad 1 \leqslant m \leqslant k,$$

with some constant λ_m independent of v_h. Thus the number of the v_h's is bounded by $J_{\ell-k,k}(q)$; see (8.75). Combining with (8.83) we conclude that

$$J_{11}(u) \leqslant k!(p^k + P)^k P^k p^{\frac{-k(k+1)}{2}} J_{\ell-k,k}(q)$$

and

$$J_1 \leqslant \ell^{2k} k! p^{2\ell - 2k - \frac{k(k+1)}{2}} P^k (p^k + P)^k J_{\ell-k,k}(q). \tag{8.84}$$

Now we estimate J_2, the number of solutions of the second class of (8.73). Among the numbers $(y_1, \dots, y_\ell)$ there are at most $k-1$ different residue classes modulo p, or among the number $(y_{\ell+1}, \dots, y_{2\ell})$ there are at most $k-1$ different residue classes modulo p. Therefore

$$J_2 = \int_0^1 \cdots \int_0^1 \sum_{u \in U} S_{u_1}(\alpha) \cdots S_{u_\ell}(\alpha) \overline{S}_{u_{\ell+1}}(\alpha) \dots \overline{S}_{u_{2\ell}}(\alpha) d\alpha$$

where U is the sequence of vectors $u = (u_1, \dots, u_{2\ell}) \pmod p$ with the relevant property, so $|U| \leqslant 2k^\ell p^{\ell+k-1}$.

By the inequality $(x_1 \cdots x_n)^{\frac{1}{n}} \leqslant \frac{1}{n}(x_1 + \dots + x_n)$ we obtain

$$J_2 \leqslant \sum_{(u_1, \dots, u_{2\ell}) \in U} (2\ell)^{-1} \sum_{h=1}^{2\ell} J_2(u_h),$$

say, where

$$J_2(u) = \int_0^1 \cdots \int_0^1 |S_u(\alpha)|^{2\ell} d\alpha \leqslant J_{\ell,k}(q) \leqslant q^{2k} J_{\ell-k,k}(q).$$

Hence

$$J_2 \leqslant 2k^\ell p^{\ell+k-1} q^{2k} J_{\ell-k,k}(q). \tag{8.85}$$

By (8.84) and (8.85) we conclude that

$$J_{\ell,k}(P) \leqslant \Big[\ell^{2k} k! p^{2\ell-2k-\frac{k(k+1)}{2}} P^k (p^k + P)^k + 2k^\ell p^{\ell+k-1} q^{2k}\Big] J_{\ell-k,k}(q).$$

Finally assuming that $P \geqslant k^k$ we may take a prime number p with $2 + P^{1/k} \leqslant p \leqslant 2P^{1/k}$ and the integer $q = [p^{-1}P] + 1$ getting

LEMMA 8.24. *Let $k \geqslant 2, , \ell \geqslant \frac{1}{2}k(k+3) - 1$ and $P \geqslant k^k$. We then have*

$$J_{\ell,k}(P) \leqslant 2^{4\ell} P^{\frac{2\ell}{k} + \frac{3k-5}{2}} J_{\ell-k,k}(P^{1-\frac{1}{k}}). \tag{8.86}$$

Now Theorem 8.21 follows easily from Lemma 8.14 by induction in m. Combining (8.76) with (8.78) for $\ell = k(k+m)$ we obtain

$$|S| \leqslant 4P^2 (\Delta P^{k(k+1)(1-\frac{1}{k})^m})^{\frac{1}{2k^2(k+m)^2}} \tag{8.87}$$

where m is any positive integer and $P > k^{k(1-\frac{1}{k})^{-m}}$.

STEP VI (estimation of Δ). It remains to estimate Δ. We have

$$\begin{aligned} X^2 D(\alpha, X) &\leqslant \sum_{|m| \leqslant X} \min\Big(X, \frac{1}{2\|\alpha m\|}\Big) \\ &\leqslant \sum_{|r| \leqslant \alpha X + \frac{1}{2}} \sum_{|\alpha m - r| < \frac{1}{2}} \min\Big(X, \frac{1}{2|\alpha m - r|}\Big) \\ &\leqslant 2(\alpha X + 1)(X + \sum_{0 < u < 1/2\alpha} \min\Big(2X, \frac{1}{\alpha u}\Big)). \end{aligned}$$

The last sum is bounded by an integral which equals $\alpha^{-1}\log eX$, therefore

$$D(\alpha,X)\leqslant 2X^{-2}(\alpha X+1)(X+\alpha^{-1}\log eX)\leqslant 4(\alpha+\alpha^{-1}X^{-2})\log 3X$$

for $\alpha>0$ and $X\geqslant 3$. We also have trivial bound $D(\alpha,X)\leqslant 4$. From both estimates we conclude that for any $I\subset\{1,2,\dots,k\}$,

$$\Delta\leqslant\prod_{j\in I}(|\alpha_j|+|\alpha_j|^{-1}P^{-2j})\,(4k\log 3\ell P)^k. \tag{8.88}$$

STEP VII (estimation of exponential sums).

We now assume that $f(x)$ is a real smooth function on $[N,2N]$ such that for all $j\geqslant 1$

$$\alpha^{-j^3}F\leqslant\frac{x^j}{j!}|f^{(j)}(x)|\leqslant\alpha^{j^3}F \tag{8.89}$$

on $[N,2N]$, where $F\geqslant N\geqslant 2$ and $\alpha\geqslant 1$. Hence for $\alpha_j(n)=f^{(j)}(n)/j!$ we get

$$\alpha^{-j^3}(2N)^{-j}F\leqslant|\alpha_j(n)|\leqslant\alpha^{j^3}N^{-j}F$$

and

$$\Delta\leqslant\prod_{j\in I}(FN^{-j}+F^{-1}N^jP^{-2j})\alpha^{k^4}(8k\log 3P)^k.$$

We assume that $P^4\leqslant N\leqslant F^{\frac{1}{4}}$ and $k\geqslant 2\log F/\log N$. We choose

$$I=\left\{j\,:\,\frac{\log F}{\log N}<j\leqslant\frac{\log F}{\log(N/P)}\right\}$$

getting

$$\Delta\leqslant\prod_{j\in I}(FN^{-j})(2ke^k)^{k^2}(\log 3N)^k\leqslant N^{-\frac{1}{2}J(J-1)}(2ke^k)^{k^2}(\log 3N)^k$$

where

$$J=|I|\geqslant\frac{\log F}{\log(N/P)}-\frac{\log F}{\log N}-1\geqslant\frac{(\log F)(\log P)}{2(\log N)^2}+1.$$

Finally

$$\Delta\leqslant\exp(-(\log F)^2(\log P)^2(2\log N)^{-3})a^{k^4}(k\log 3N)^k. \tag{8.90}$$

Now setting $P=N^{\frac{1}{4}}$, $k=[4\log F/\log N]$ and $m=8k$ we infer from (8.70), (8.87) and (8.90) the following

THEOREM 8.25. *Let $f(x)$ be a smooth function on $[N,2N]$ which satisfies (8.89) with $F\geqslant N^4$ and $\alpha\geqslant 1$. We then have*

$$S_f(a,b)\ll\alpha N\exp(-2^{-18}(\log N)^3(\log F)^{-2}), \tag{8.91}$$

where the constant implied in $\ll$ is absolute.

APPLICATIONS: We conclude the presentation of Vinogradov's method by drawing from (8.91) a few basic applications.

Corollary 8.26. *For $t \geqslant N \geqslant 2$ we have*

$$\sum_{1 \leqslant n \leqslant N} n^{it} \ll N \exp(-\beta(\log N)^3 (\log t)^{-2}), \tag{8.92}$$

where β and the constant implied in $\ll$ are absolute.

Theorem 8.27. *There exists an absolute constant $\alpha > 0$ such that for $s = \sigma + it$ with $t \geqslant 2$ and $\frac{1}{2} \leqslant \sigma \leqslant 1$,*

$$\zeta(s) \ll t^{\alpha(1-\sigma)^{\frac{3}{2}}} (\log t)^{\frac{2}{3}}. \tag{8.93}$$

The implied constant is absolute.

Proof. By (8.21) using partial summation and (8.92) we have

$$\begin{aligned}\zeta(s) = \sum_{1 \leqslant n \leqslant t} n^{-s} + O(1) &\ll \int_1^t x^{-\sigma} \exp(-\beta(\log x)^3 (\log t)^{-2}) dx \\ &= (\log t) \int_0^1 t^{(1-\sigma)u - \beta u^3} du \leqslant (\log t) \int_0^1 t^{f(v) - \beta(u-v)^3} du\end{aligned}$$

where $f(u) = (1-\sigma)u - \beta u^3$ and $v = \sqrt{(1-\sigma)/3\beta}$ so $f(v) = 2(1-\sigma)^{\frac{3}{2}}/3\sqrt{3\beta}$. This gives (8.93) with $\alpha = 2/3\sqrt{3\beta}$. □

Corollary 8.28 (Vinogradov-Korobov, 1957). *There exist absolute constants $\gamma > 0, \delta > 0$ such that*

$$|\zeta(\sigma + it)| \leqslant \gamma (\log t)^{\frac{2}{3}} \tag{8.94}$$

in the region

$$t \geqslant 2, \quad \sigma \geqslant 1 - \delta(\log t)^{-\frac{2}{3}}. \tag{8.95}$$

There are several interesting devices by means of which one can transform an upper bound for $\zeta(s)$ into a zero-free region and derive estimates for $1/\zeta(s)$, $\zeta'(s)/\zeta(s)$ in this region. The original arguments of Hadamard and de la Vallée Poussin (presented in Chapter 5) produce only a zero-free region of the type $\sigma \geqslant 1 - c(\log t)^{-1}$, regardless of how good the estimate for $\zeta(s)$ that one has. However combining these arguments with a method of Landau one can establish deeper relationships (see Theorem 3.10 and Theorem 3.11 of [**T2**]).

Lemma. *Let $\phi(t)$, $\psi(t)$ be positive increasing functions for $t \geqslant 0$ such that $\phi(t)\psi(t) = o(\exp(\phi(t)))$ as $t \to +\infty$. If*

$$\zeta(s) \ll \exp(\phi(t)) \quad \textit{for} \quad \sigma \geqslant 1 - \psi(t)^{-1},$$

then

$$\zeta(s) \neq 0 \quad \textit{for} \quad \sigma \geqslant 1 - c\psi(2t+3)^{-1}\phi(2t+3)^{-1}.$$

Moreover in this region

$$1/\zeta(s), \ \zeta'(s)/\zeta(s) \ \ll \ \phi(2t+3)\psi(2t+3).$$

Hence using Corollary 8.28 we obtain

THEOREM 8.29. *We have* $\zeta(s) \neq 0$ *and*

$$\frac{1}{\zeta(s)} \ll (\log t)^{\frac{2}{3}}(\log\log t)^{\frac{1}{3}},$$
$$\frac{\zeta'(s)}{\zeta(s)} \ll (\log t)^{\frac{2}{3}}(\log\log t)^{\frac{1}{3}}$$

for $\sigma \geqslant 1 - c(\log t)^{-2/3}(\log\log t)^{-1/3}$, $t \geqslant 3$, *where* $c > 0$ *is an absolute constant.*

Now the standard contour integral argument (see Chapter 5) yields the following strongest known Prime Number Theorem.

COROLLARY 8.30. *We have*

$$\psi(x) = x + O(x\exp(-c(\log x)^{\frac{3}{5}}(\log\log x)^{-\frac{1}{5}}))$$

for $x \geqslant 3$, *where* $c > 0$ *is some absolute constant.*

Let $\tau_k(n)$ stand for the number of ways of factoring n into k positive integers, so the Dirichlet generating series for $\tau_k(n)$ is $\zeta^k(s)$.

THEOREM 8.31. *For any* $\varepsilon > 0$ *and* $x \geqslant 2$ *we have*

$$D_k(x) = \sum_{n\leqslant x} \tau_k(n) = xP_k(\log x) + O(x^{\delta_k+\varepsilon}), \tag{8.96}$$

where P_k *is a polynomial of degree* $k-1$, $\delta_k = 1 - \gamma k^{-2/3}$, γ *is a positive absolute constant and the constant implied in* O *depends on* ε *and* k *only.*

PROOF. By Perron's formula (5.111) we obtain

$$D_k(x) = \frac{1}{2\pi i}\int_{1+\varepsilon-iT}^{1+\varepsilon+iT} \zeta^k(s)\frac{x^s}{s}ds + O(x^{1+2\varepsilon}T^{-1}),$$

where T is any number with $2 \leqslant T \leqslant x$. Move the integration to $\mathrm{Re}\,(s) = \sigma$, $\frac{1}{2} < \sigma < 1$, getting by (8.93)

$$D_k(x) = \operatorname*{res}_{s=1} \zeta^k(s)\frac{x^s}{s} + O((x^{\sigma}T^{\alpha k(1-\sigma)^{\frac{3}{2}}} + xT^{-1})x^{2\varepsilon}).$$

The residue is equal to $xP_k(\log x)$ and the error term becomes $O(x^{\delta_k+2\varepsilon})$ by setting $T = x^{k^{-2/3}}$ and $\sigma = 1 - (2\alpha k^{\frac{1}{3}})^{-2}$. □

Recall that for small k one gets stronger results using van der Corput method. It is conjectured that the best exponent in (8.96) is $\delta_k = \frac{k-1}{2k}$ for any k.

CHAPTER 9

THE DIRICHLET POLYNOMIALS

9.1. Introduction.

A Dirichlet polynomial is a finite Dirichlet series

$$D(s) = \sum_{1\leqslant n\leqslant N} a_n n^{-s} \tag{9.1}$$

with complex coefficients a_n. This is a special case of sums of type $\sum a_n e^{-\lambda(n)s}$ where $\lambda(n)$ are distinct real numbers called "frequencies." Our special case $\lambda(n) = \log n$ is distinguished by the following features:

- $\lambda(n)$ is smooth and slowly increasing,
- $\lambda(n)$ is additive.

The derivative of $D(s)$ is also a Dirichlet polynomial of length N,

$$D'(s) = \sum_{1\leqslant n\leqslant N} a'_n n^{-s},$$

its coefficients $a'_n = -a_n \log n$ are changed only slightly. The additive property of our frequencies implies that the product of two Dirichlet polynomials is a Dirichlet polynomial, namely we have

$$\Big(\sum_{n\leqslant N} a_n n^{-s}\Big)\Big(\sum_{m\leqslant M} b_m m^{-s}\Big) = \sum_{\ell\leqslant L} c_\ell \ell^{-s}$$

with $L = MN$ and the coefficients of the product are

$$c_\ell = \sum_{\substack{nm=\ell\\ n\leqslant N, m\leqslant M}} a_n b_m. \tag{9.2}$$

The main objective of the theory of Dirichlet polynomials is to estimate $D(s)$ at special points. Since the set of special points is not constructible in practice (think of the zeros $\rho = \beta+i\gamma$ of $\zeta(s)$ with $\beta > \frac{1}{2}$) it is necessary to deal with arbitrary points. These can be elected by combinatorial arguments into well-spaced sets. Moreover, the coefficients a_n are quite complicated in applications, therefore one has to treat them as arbitrary complex numbers. Though in such generality it is impossible to give a non-trivial bound for $D(s)$ at any fixed s, it is true that $|D(s)|$ cannot take large values for many well-spaced points. Hence our fundamental question will be how often $|D(s)|$ is larger than the mean value over the relevant set of points.

If $s = \sigma + it$ has fixed σ, then by changing the coefficients a_n into $a_n n^{-\sigma}$ we can assume without loss of generality that all the points in question are on the imaginary line $s = it$. After establishing results for points on the imaginary line we shall extend them for sets of well-spaced points in vertical strips. Actually we can also modify the Dirichlet polynomial $D(s)$ by twisting its terms by various functions $f_s(n)$ depending on s which change only slightly in n (have small derivatives in n), that is to say, the theory extends to sums of type

$$\sum_n a_n f_s(n) n^{-s}. \tag{9.3}$$

Such a twist can be handled using any standard Fourier analysis (separation of variables in $f_s(n)$, see Proposition 9.11).

9.2. The integral mean-value estimates.

By Cauchy's inequality we get

$$\Big| \sum_{1 \leqslant n \leqslant N} a_n n^{it} \Big|^2 \leqslant GN \tag{9.4}$$

where

$$G = \sum_{1 \leqslant n \leqslant N} |a_n|^2. \tag{9.5}$$

Of course, this bound is best possible, however, one can do better on average.

THEOREM 9.1. *For any complex numbers a_n we have*

$$\int_0^T \Big| \sum_{1 \leqslant n \leqslant N} a_n n^{it} \Big|^2 dt = (T + O(N))G \tag{9.6}$$

where the implied constant is absolute.

PROOF. Let $f(t)$ be the following continuous piecewise linear function

$$f(t) = \begin{cases} 0 & \text{if } t \leqslant -N, \\ 1 + \frac{t}{N} & \text{if } -N < t \leqslant 0, \\ 1 & \text{if } 0 < t \leqslant T, \\ 1 - \frac{t-T}{N} & \text{if } T < t \leqslant T + N, \\ 0 & \text{if } t > T + N. \end{cases}$$

Then our integral is majorized by

$$\int f(t) \Big| \sum_{1 \leqslant n \leqslant N} a_n n^{it} \Big|^2 dt = \sum_{1 \leqslant m,n \leqslant N} a_m \bar{a}_n F\Big(\frac{m}{n}\Big),$$

with

$$F(x) = \int f(t) x^{it} dt.$$

We have $F(1) = T + N$ and $F(x) \ll N^{-1}(\log x)^{-2}$ if $x \neq 1$. Hence for $m \neq n$ we have

$$F\Big(\frac{m}{n}\Big) \ll N^{-1}\Big(\log \frac{m}{n}\Big)^{-2} \ll \frac{1}{N}\Big(\frac{m+n}{m-n}\Big)^2 \ll \frac{N}{(m-n)^2}.$$

Therefore our integral is bounded from above by

$$(T+N)G+O\Big(N\sum_{1\leqslant m\neq n\leqslant N}|a_ma_n|(m-n)^{-2}\Big)=(T+O(N))G$$

by using the inequality $2|a_ma_n|\leqslant|a_m|^2+|a_n|^2$. Similarly we derive the same lower bound. Combining both estimates we complete the proof of (9.6). □

This result should be compared with the discussion of bilinear forms and large sieve in Chapter 7.

Theorem 9.1 implies that $G^{\frac{1}{2}}$ is the mean-value of $|D(it)|$ in the following asymptotic sense

$$\lim_{T\to\infty}\frac{1}{T}\int_0^T|D(t)|^2dt=G. \tag{9.7}$$

Note that the mean-value of the product of Dirichlet polynomials is essentially bounded by the product of mean-values. Precisely we derive for the coefficients (9.2) the following inequalities:

$$\begin{aligned}\sum_\ell|c_\ell|^2&\leqslant\sum\sum_{nm=n_1m_1}|a_nb_ma_{n_1}b_{m_1}|\\&\leqslant\sum_n\sum_m|a_nb_m|^2\tau(nm)\\&\leqslant\Big(\sum_n|a_n|^2\tau(n)\Big)\Big(\sum_m|b_m|^2\tau(m)\Big).\end{aligned}$$

Hence using the bound $\tau(n)\ll n^\varepsilon$ we obtain

$$\sum_\ell|c_\ell|\ll(MN)^\varepsilon\Big(\sum_n|a_n|^2\Big)\Big(\sum_m|b_m|^2\Big). \tag{9.8}$$

For the product of k polynomials we have

$$\Big(\sum_{n\leqslant N_1}a_n^{(1)}n^{-s}\Big)\cdots\Big(\sum_{n\leqslant N_k}a_n^{(k)}n^{-s}\Big)=\sum_{n\leqslant N}a_nn^{-s}$$

with $N=N_1\dots N_k$ and the coefficients are

$$a_n=\sum\cdots\sum_{n_1\cdots n_k=n}a_{n_1}^{(1)}\cdots a_{n_k}^{(k)}.$$

The mean-value of these coefficients satisfies

$$\sum_n|a_n|^2\leqslant\sum_{n_1}\cdots\sum_{n_k}|a_{n_1}^{(1)}\cdots a_{n_k}^{(k)}|\tau_k(n_1\cdots n_k)\leqslant G_k^{(1)}\cdots G_k^{(k)}$$

where for a given sequence $\mathcal{A}=(a_n)$ we put

$$G_k=\sum_n|a_n|^2\tau_k(n). \tag{9.9}$$

This follows by the inequality $\tau_k(n_1\cdots n_k)\leqslant\tau_k(n_1)\cdots\tau_k(n_k)$. Since $\tau_k(n)$ is multiplicative, it suffices to check this inequality for $n_j=p^{\alpha_j}$. In this case we have $\tau_k(p^\alpha)=(1+\alpha)(1+\alpha/2)\cdots(1+\alpha/(k-1))$, so it suffices to check that

$1 + (\alpha_1 + \cdots + \alpha_k)/\ell \leqslant (1 + \alpha_1/\ell)\cdots(1 + \alpha_k/\ell)$ for $1 \leqslant \ell < k$, which is obvious. Note that

$$G_k \ll GN^{\varepsilon}$$

by $\tau_k(n) \ll n^{\varepsilon}$, where the implied constant depends on ε and k; however, we shall often in practice have a better estimate $G_k \ll G(\log 2N)^K$ with some $K \geqslant k$.

The approximate formula (9.6) is the best possible of its kind in general. It is best used for polynomials of length N which is near T in the logarithmic scale. If the polynomial $D(s)$ is quite short we can better apply (9.6) to a suitable power $D(s)^k$ getting

COROLLARY 9.2. *For any integer $k \geqslant 1$ we have*

$$\int_0^T \Big| \sum_{1\leqslant n\leqslant N} a_n n^{it}\Big|^{2k} dt \ll (T + N^k)G_k^k \tag{9.10}$$

where G_k is given by (9.9) and the implied constant is absolute.

Still we may not be able to match T and N^k for k an integer, so (9.10) is not perfect in some ranges. In this connection H. Montgomery [**Mo2**] made the following

CONJECTURE M_k. *Suppose $|a_n| \leqslant 1$. Then for any real k with $1 \leqslant k \leqslant 2$ we have*

$$\int_0^T \Big| \sum_{1\leqslant n\leqslant N} a_n n^{it}\Big|^{2k} dt \ll (T + N^k)N^{k+\varepsilon} \tag{9.11}$$

where the implied constant depends only on ε.

It is not difficult to prove (9.11) for some special polynomials, for example, when $a_n = 1$ for all n. However, the bound (9.10) for general complex numbers a_n, with $N^{k+\varepsilon}$ replaced by G^kN^{ε}, was shown by J. Bourgain [**Bou**] to be false if k is not an integer. His counter-example is given by a Dirichlet polynomial supported in a short interval. But such cases are not common in applications, so the Conjecture M_k, if true, is still important. It yields, among other things, the density conjecture for zeros of the Riemann zeta function (see Chapter 10).

9.3. The discrete mean-value estimates.

Next we establish discrete versions of Theorem 9.1 and some of their consequences.

DEFINITION. A set $\mathcal{T}$ of real numbers t_r is said to be well-spaced if $|t_{r_1} - t_{r_2}| \geqslant 1$ for any $r_1 \neq r_2$.

LEMMA 9.3 (GALLAGHER). *Let $\mathcal{T}$ be a set of points t_r with $\frac{1}{2} \leqslant t_r \leqslant T - \frac{1}{2}$ which is well-spaced. Let $F(t)$ be a smooth function on $[0, T]$. Then we have*

$$\sum_{t_r\in\mathcal{T}} |F(t_r)|^2 \leqslant \int_0^T (|F(t)|^2 + |F(t)F'(t)|)dt. \tag{9.12}$$

PROOF. For a smooth function $f(x)$ on $[0,1]$ we get the identity

$$f(x)=\int_0^1 f(t)dt+\int_0^x tf'(t)dt+\int_x^1 (t-1)f'(t)dt$$

by partial integration. Hence

$$|f(\tfrac{1}{2})|\leqslant\int_0^1 (|f(t)|+\tfrac{1}{2}|f'(t)|)dt.$$

Applying this for $f^2(t)$ we infer

$$|f(\tfrac{1}{2})|^2\leqslant\int_0^1 (|f(t)|^2+|f(t)f'(t)|_dt.$$

This gives

$$|F(t_r)|^2\leqslant\int_{t_r-\frac{1}{2}}^{t_r+\frac{1}{2}} (|F(t)|^2+|F(t)F'(t)|)dt,$$

and summing over t_r in $\mathcal{T}$ we deduce (9.12) because the intervals $(t_r-\frac{1}{2},t_r+\frac{1}{2})$ do not overlap. □

Notice that by the Cauchy-Schwarz inequality we have

$$\int_0^T |F(t)F'(t)|dt\leqslant\Big(\int_0^T |F(t)|^2dt\Big)^{\frac{1}{2}}\Big(\int_0^T |F'(t)|^2dt\Big)^{\frac{1}{2}}. \tag{9.13}$$

Applying (9.12) together with (9.13) for $F(t)=D(it)$ one gets

THEOREM 9.4. *Let $\mathcal{T}$ be a set of well-spaced points in the segments $[0,T]$ with $T\geqslant 1$, and let a_n be any complex numbers. Then we have*

$$\sum_{t_r\in\mathcal{T}}\Big|\sum_{1\leqslant n\leqslant N} a_n n^{it_r}\Big|^2\ll(T+N)G\log 2N \tag{9.14}$$

where G is given by (9.5) and the implied constant is absolute.

COROLLARY 9.5. *Let the conditions be as in Theorem 9.4. For any integer $k\geqslant 1$ we have*

$$\sum_{t_r\in\mathcal{T}}\Big|\sum_{1\leqslant n\leqslant N} a_n n^{it_r}\Big|^{2k}\ll(T+N^k)G_k^k\log 2N^k \tag{9.15}$$

where G_k is given by (9.9) and the implied constant is absolute.

REMARKS. Gallagher's lemma yields a transition from an individual value $D(it)$ to an integral. It employs the derivative $D'(it)$ which is not acceptable in some applications. However, we can make a transition without differentiation of $D(s)$. To this end take a smooth function $f(x)\geqslant 0$ supported on $[\frac{1}{2},2N]$ with $f(x)=1$ on $[1,N]$ such that $x^af^{(a)}(x)\ll 1$ for $0\leqslant a\leqslant 2$. By Mellin inversion

$$f(x)=\frac{1}{2\pi}\int_{-\infty}^{\infty} F(t)x^{it}dt$$

where

$$F(t)=\int f(x)x^{it-1}dx\ll(1+|t|)^{-2}\log 2N$$

by partial integration. We multiply a_n by $f(n)$, which is a redundant factor if $1 \leqslant n \leqslant N$, and apply the above integral representation of $f(n)$ getting

$$D(s) = \frac{1}{2\pi} \int_{-\infty}^{\infty} D(s+it)F(t)dt.$$

Hence

$$D(it_r) \ll (\log 2N) \int_{-\infty}^{\infty} |D(it)|(1+|t-t_r|)^{-2}dt \tag{9.16}$$

where the implied constant is absolute. The same arguments apply to $D(s)^k$ with k a positive integer giving

$$D(it_r)^k \ll (\log 2N^k) \int_{-\infty}^{\infty} |D(it)|^k(1+|t-t_r)^{-2}dt \tag{9.17}$$

where the implied constant is absolute. Hence the estimates (9.14) and (9.15) follow from (9.6) and (9.10) respectively.

The bound (9.14) has deficiency of not depending on the cardinality of the set $\mathcal{T}$; it is relatively weaker for smaller sets. Montgomery [**Mon**] succeeded in improving (9.14) if $R = |\mathcal{T}|$ is smaller than $T^{\frac{1}{2}}$ (note that any well-spaced set $\mathcal{T} \in [0,T]$ has no more than $T+1$ points). The idea is based on the duality principle for bilinear forms:

Let $\mathcal{X} = (x_{mn})$ be a complex matrix. Then the following statements about $\mathcal{X}$ and a number D are equivalent:

(A) For any complex numbers a_n,

$$\sum_m \Big|\sum_n a_n x_{mn}\Big|^2 \leqslant D \sum_n |a_n|^2.$$

(B) For any complex numbers b_m,

$$\sum_n \Big|\sum_m b_m x_{mn}\Big|^2 \leqslant D \sum_m |b_m|^2.$$

For the proof, see Chapter 7.

THEOREM 9.6 (MONTGOMERY). *Let the condition be as in Theorem 9.4. Then we have*

$$\sum_{t_r \in \mathcal{T}} \Big| \sum_{1 \leqslant n \leqslant N} a_n n^{it_r} \Big|^2 \ll G(N + RT^{\frac{1}{2}}) \log 2T \tag{9.18}$$

where the implied constant is absolute.

PROOF. By the duality principle the problem is equivalent to showing that for any complex numbers c_r one has

$$\sum_{1 \leqslant n \leqslant N} \Big| \sum_{t_r \in \mathcal{T}} c_r n^{it_r} \Big|^2 \ll \Big(\sum_{t_r \in \mathcal{T}} |c_r|^2 \Big)(N + RT^{\frac{1}{2}}) \log 2T. \tag{9.19}$$

Squaring out and changing the order of summation on the left side of (9.19) we get

$$\sum_{r_1}\sum_{r_2} c_{r_1}\bar{c}_{r_2} Z(t_{r_1} - t_{r_2})$$

where

$$Z(t) = \sum_{1\leqslant n\leqslant N} n^{it}.$$

By (8.20), (8.26) and (8.34) we derive that

$$Z(t) \ll N|t|^{-1} + |t|^{\frac{1}{2}} \log |t|, \text{ if } |t| \geqslant 1. \tag{9.20}$$

This yields (9.18). □

By the Lindelöf Hypothesis $\zeta(\frac{1}{2}+it) \ll |t|^{\varepsilon}$ for $|t| \geqslant 1$, one should expect a better bound for $Z(t)$ than (9.20), namely that

$$Z(t) \ll N|t|^{-1} + N^{\frac{1}{2}}|t|^{\varepsilon}, \text{ for } |t| \geqslant 1. \tag{9.21}$$

Hence we get

$$\sum_{t_r\in\mathcal{T}} \Big| \sum_{1\leqslant n\leqslant N} a_n n^{it_r}\Big|^2 \ll G(N + RN^{\frac{1}{2}})T^{\varepsilon} \tag{9.22}$$

in place of (9.18). But an even stronger bound is predicted by Montgomery in the following

CONJECTURE M. *Suppose $|a_n| \leqslant 1$. Let $\mathcal{T}$ be a set of R well-spaced points t_r in the segment $[0,T]$ with $T \geqslant 1$. Then*

$$\sum_{t_r\in\mathcal{T}} \Big| \sum_{1\leqslant n\leqslant N} a_n n^{it_r}\Big|^2 \ll N(N + R)T^{\varepsilon}. \tag{9.23}$$

EXERCISE. Show that Conjecture M implies Conjecture M_k for any real $1 \leqslant k \leqslant 2$ (with extra factor T^{ε}).

9.4. Large values of Dirichlet polynomials.

The discrete mean-value estimates offer some answers to the question how often a Dirichlet polynomial assumes large values at a given set of well-spaced points. Thus (9.14) yields

THEOREM 9.7. *Let $D(s)$ be a Dirichlet polynomial of length $N \geqslant 1$ and let $\mathcal{T}$ be a set of well-spaced points t_r in the segment $[0,T]$ with $T \geqslant 1$ such that*

$$|D(it_r)| \geqslant V \tag{9.24}$$

for some $V > 0$. Then the cardinality of $\mathcal{T}$, say $R = |\mathcal{T}|$, satisfies

$$R \ll (T + N)GV^{-2} \log 2N \tag{9.25}$$

where the implied constant is absolute.

Notice that the bound (9.25) is non-trivial if $V \gg G^{\frac{1}{2}} \log 2N$. For V much larger than this Montgomery got a better result by applying (9.18).

THEOREM 9.8. *Let the conditions be as in Theorem 9.7 with*

$$V \geqslant G^{\frac{1}{2}} T^{\frac{1}{4}} \log 2T. \tag{9.26}$$

Then

$$R \ll GNV^{-2} \log 2T \tag{9.27}$$

where the implied constant is absolute.

PROOF. We can assume that T is larger than N by a large constant factor, because otherwise (9.27) follows by (9.25). Inserting (9.24) into (9.18) we get

$$RV^2 \ll G(N + RT^{\frac{1}{2}}) \log 2T.$$

This implies (9.27) under the restriction (9.26). □

The restriction (9.26) can be removed at a cost of introducing an extra term in the bound (9.27).

COROLLARY 9.9 (HUXLEY). *Let the conditions be as in Theorem 9.7. Then*

$$R \ll \{GNV^{-2} + G^3 NTV^{-6}\}(\log 2T)^6 \tag{9.28}$$

where the implied constant is absolute.

PROOF. Huxley's idea (the subdivision method) is a simple trick of dividing the set $\mathcal{T}$ to meet the condition (9.26). First we can assume that

$$V \geqslant G^{\frac{1}{2}} \log 2T \tag{9.29}$$

because otherwise the assertion (9.28) is trivial. Then we have $V \geqslant G^{\frac{1}{2}} T_0^{\frac{1}{4}} \log 2T_0$ with $T_0 = \min\{T, G^{-2}V^4(\log 2T)^{-4}\}$. The condition (9.29) ensures us that $1 \leqslant T_0 \leqslant T$. Let $\mathcal{T}_\ell$ be the subset of points of $\mathcal{T}$ in the interval $[\ell T, (\ell+1)T_0]$ with $0 \leqslant \ell \leqslant TT_0^{-1}$. For each subset $\mathcal{T}_\ell$ Theorem 9.8 is applicable giving $R_\ell = |\mathcal{T}_\ell| \ll GNV^{-2} \log 2T$. Hence

$$R = \sum_\ell R_\ell \ll TT_0^{-1} GNV^{-2} \log 2T$$

which yields (9.28). □

Applying the above results for the polynomial $D(it)^k$ with k a positive integer one gets

$$R \ll (T + N^k)(G_k V^{-2})^k \log 2N^k, \tag{9.30}$$

$$R \ll \{(G_k NV^{-2})^k + T(G_k^3 NV^{-6})^k\}(\log 2T)^6 \tag{9.31}$$

where the implied constant is absolute (see (9.25) and (9.28) respectively).

In 1975, M. Jutila came up with several innovations to improve the above estimates in some ranges. His best result (which we state here without proof, see [**Ju3**]) is

THEOREM 9.10 (JUTILA). *For any positive integer k we have*

$$R \ll \Big\{\frac{GN}{V^2} + \Big(\frac{GN}{V^2}\Big)^{-\frac{1}{k}}\frac{G^3NT}{V^6} + \Big(\frac{GN}{V^2}\Big)^{4k}\frac{T}{N^{2k}}\Big\}(NT)^{\varepsilon} \tag{9.32}$$

for any $\varepsilon > 0$, the implied constant depending only on ε and k.

REMARKS. Notice that if

$$V \geqslant G^{\frac{1}{2}}N^{\frac{1}{4}}(NT)^{\varepsilon} \tag{9.33}$$

then Jutila's bound (9.32) becomes essentially Huxley's bound (9.31) by letting k be sufficiently large in terms of ε.

The best (conditional) estimates for the number of large values of a Dirichlet polynomial with bounded coefficients comes from Montgomery's conjectures. By Conjecture M_k one derives

$$R \ll (T + N^k)N^kV^{-2k}(NT)^{\varepsilon} \tag{9.34}$$

for any real $k \geqslant 1$ and $\varepsilon > 0$, the implied constant depending on ε and k. In applications (9.34) is best used for k with $N^k = T$. By Conjecture M one derives

$$R \ll N^2V^{-2}T^{\varepsilon}, \qquad \text{if} \qquad V \geqslant N^{\frac{1}{2}}T^{\varepsilon}. \tag{9.35}$$

Actually the above estimates for R are equivalent to the Montgomery conjectures.

All the discrete mean-value estimates with respect to a well-spaced set $\mathcal{T} = \{t_1, \dots, t_R\} \subset [0, T]$ can be generalized slightly to sums of type

$$\sum_{1\leqslant n\leqslant N} a_nf_r(n)n^{it_r} \tag{9.36}$$

where $f_r(n)$ is a nice function which does not vary in n too much.

PROPOSITION 9.11. *Suppose that each $f_r(x)$ satisfies*

$$x^a|f_r^{(a)}(x)| \leqslant 2 \tag{9.37}$$

for $1 \leqslant x \leqslant N$ and $0 \leqslant a \leqslant 2$. Suppose that

$$\Big|\sum_{1\leqslant n\leqslant N} a_nf_r(n)n^{it_r}\Big| \geqslant V \tag{9.38}$$

for $1 \leqslant r \leqslant R$. Then the bounds (9.30) and (9.31) hold true with G_k replaced by $G_k(\log 2N)^2$. In particular, $f_r(n) = n^{-\sigma_r}$ with $0 \leqslant \sigma_r \leqslant 1$ qualifies.

PROOF. We may assume that $f_r(x)$ is supported in $[\frac{1}{2}, 2N]$ where it satisfies (9.37). Then we write

$$f_r(n) = \frac{1}{2\pi}\int_{-\infty}^{\infty} \hat{f}_r(it)n^{-it}dt$$

where $\hat{f}_r(s)$ is the Mellin transform of $f_r(x)$,

$$\hat{f}_r(s) = \int_0^{\infty} f_r(x)x^{s-1}dx.$$

This satisfies $\hat{f}_r(it) \ll (|t|+1)^{-2}\log 2N$ by partial integration two times. Hence we get

$$\sum_{1\leqslant n\leqslant N} a_n f_r(n) n^{it_r} \ll (\log 2N)\int_{-\infty}^{\infty} |D(it_r - it)|(|t|+1)^{-2}dt$$
$$\ll (\log 2N)\sum_m |D(it_r(m))|(|m|+1)^{-2}$$

where $t_r(m) = t_r - t$ is the point at which $|D(it_r - it)|$ is maximal with $m \leqslant t \leqslant m+1$. Notice that $|t_r(m)| \leqslant T + |m| + 1$. By the hypothesis (9.38) it follows that

$$|D(it_r(m))| \gg (|m|+1)^{\frac{3}{4}} V(\log 2N)^{-1}$$

for some $m \in \mathbb{Z}$. Let $\mathcal{T}_m$ be the set of such points $t_r(m)$ and R_m its cardinality, so we have $R \leqslant \sum_m R_m$. The set $\mathcal{T}_m$ can be divided into two well-spaced subsets. Applying (9.30) and (9.31) for each of these subsets we obtain corresponding bounds for R_m, and summing over m we get the asserted bounds for R. □

9.5. Dirichlet polynomials with characters.

A large part of the theory of Dirichlet polynomials (9.1) extends to sums of type

$$D(s,\chi) = \sum_{1\leqslant n\leqslant N} a_n \chi(n) n^{-s} \tag{9.39}$$

where χ is some sort of arithmetic harmonic, for example, $\chi(n)$ can be a Dirichlet character, or the Fourier coefficient of a modular form (see Chapter 14). However, in the latter example one is faced with technical difficulties caused by the lack of complete multiplicativity (the generating L-function has Euler product of degree two) while the orthogonality property is only approximate.

In this section we deal with the Dirichlet characters to various moduli. Let $k \geqslant 1$ and $Q \geqslant 1$. We consider the set $\mathcal{H}(k,Q)$ of characters $\chi(\text{mod } kq)$ such that $\chi = \xi\psi$ where ξ is any character to modulus k and ψ is any primitive character to modulus q with $1 \leqslant q \leqslant Q$, $(q,k) = 1$. Throughout we assume that $T \geqslant 3, N \geqslant 1$, and we denote

$$H = kQ^2T, \qquad \mathcal{L} = \log HN. \tag{9.40}$$

First we establish the following extension of Theorem 9.1.

THEOREM 9.12. *For any complex numbers a_n we have*

$$\sum_{\chi\in\mathcal{H}(k,Q)} \int_0^T \Big|\sum_{1\leqslant n\leqslant N} a_n \chi(n) n^{it}\Big|^2 dt \ll G(N+H)\mathcal{L}^3 \tag{9.41}$$

where the implied constant is absolute.

REMARKS. Our estimate (9.41) is not as strong as it could possibly be, namely the logarithmic factor $\mathcal{L}^3$ could be removed by taking a traditional approach based on the large sieve inequality for additive characters. The transition from multiplicative to additive characters being performed by Gauss sums (see Chapter 3), this approach has the extra advantage of working well for sequences (a_n) supported in short segments $M < n \leqslant M + N$. Nevertheless, we have chosen a direct method simply to show new ideas. Here the quantity $H = kQ^2T$ is about the number of harmonics $\chi(n)n^{it}$ being employed (think of t as a discrete variable ranging over a well-spaced set of points with $|t| \leqslant T$).

PROOF. For the proof of (9.41) we first assume that (a_n) is supported in a dyadic segment $X \leqslant n \leqslant 2X$ with $X \geqslant 1$. By the duality principle the problem reduces to the estimation of

$$\sum_{X\leqslant n\leqslant 2X}\Big|\sum_{\chi\in\mathcal{H}(k,Q)}\int_0^T c_\chi(t)\chi(n)n^{it}dt\Big|^2 \tag{9.42}$$

for any complex numbers $c_\chi(t)$. We enlarge the outer summation by introducing a smoothing function $f(n) \geqslant 0$ with $f(n) \geqslant 1$ for $X \leqslant n \leqslant 2X$. Then squaring out we see that the dual form (9.42) is bounded by

$$\sum_{\chi_1}\sum_{\chi_2}\int_0^T\int_0^T\Big|c_{\chi_1}(t_1)c_{\chi_2}(t_2)B(t_1-t_2,\chi_1\bar{\chi}_2)\Big|dt_1dt_2 \tag{9.43}$$

where

$$B(t,\chi) = \sum_n f(n)\chi(n)n^{it}. \tag{9.44}$$

Here $\chi = \chi_1\bar{\chi}_2$ is a character of modulus $\ell = k[q_1, q_2] \leqslant kQ^2$ and $|t| = |t_1 - t_2| \leqslant T$. By Poisson's summation

$$B(t,\chi) = \frac{1}{\ell}\sum_h G\Big(\frac{h}{\ell}\Big)F\Big(\frac{h}{\ell}\Big)$$

where G is the Gauss sum

$$G\Big(\frac{h}{\ell}\Big) = \sum_{a(\mathrm{mod}\ \ell)}\chi(a)e\Big(\frac{ah}{\ell}\Big)$$

and F is the Fourier transform of $f(x)x^{it}$,

$$F(y) = \int_0^\infty f(x)x^{it}e(xy)dx.$$

Have in mind that $F(y)$ is also a function of t. We choose $f(x)$ such that

$$F(y) \ll X\exp\Big(\frac{|t|}{T} - \Big(\frac{yX}{T}\Big)^{\frac{1}{2}}\Big), \text{ if } y > 0. \tag{9.45}$$

For this, there are many good choices, for example,

$$f(x) = \exp\Big(\frac{5}{2} - \frac{x}{X} - \frac{X}{x}\Big) \tag{9.46}$$

satisfies all the requirements. For the proof of (9.45), we move the path of integration in the Fourier integral of $f(x)x^{it}$ from $\mathbb{R}^+$ to $e^{i\theta}\mathbb{R}^+$ with $0 < \theta < \frac{\pi}{4}$ getting

$$|F(y)| \leqslant X \int_0^\infty \exp\Big(\frac{5}{2} + \theta|t| - (x + x^{-1})\cos\theta - 2\pi xyX\sin\theta\Big)dx.$$

Since $x^{-1}\cos\theta + 2\pi xyX\sin\theta \geqslant (\pi yX\sin 2\theta)^{\frac{1}{2}}$, this yields

$$|F(y)| \leqslant X\exp\Big(\frac{5}{2} + \theta|t| - (\pi yX\sin 2\theta)^{\frac{1}{2}}\Big)(\cos\theta)^{-1}.$$

We choose $\theta = T^{-1}$ getting (9.45).

By (9.45) we derive that (by the trivial bound $|G(h/\ell)| \leqslant \ell$)

$$B(t,\chi) = \delta_\chi \frac{\varphi(\ell)}{\ell} F(0) + O\Big(\ell T\exp\Big(\frac{|t|}{T} - \Big(\frac{X}{\ell T}\Big)^{\frac{1}{2}}\Big)\Big) \tag{9.47}$$

where

$$F(0) = \int_0^\infty f(x)x^{it}dx \ll X(|t|+1)^{-2}, \tag{9.48}$$

$\delta_\chi = 1$ if χ is principal, and $\delta\chi = 0$ otherwise. Note that $\chi = \chi_1\bar{\chi}_2$ is principal if and only if $\chi_1 = \chi_2$ (this is the place where our hypothesis that the characters $\psi(\text{mod } q)$ are primitive is essential). Inserting (9.47) and (9.48) into (9.43) we estimate (9.42) by

$$\{X + H^2\exp(-(X/H)^{\frac{1}{2}})\} \sum_{\chi\in\mathcal{H}(k,Q)} \int_0^T |c_\chi(t)|^2 dt.$$

Since this holds for any complex numbers $c_\chi(t)$, it follows by duality that for any complex numbers a_n,

$$\sum_{\chi\in\mathcal{H}(k,Q)} \int_0^T \Big|\sum_{X<n\leqslant 2X} a_n\chi(n)n^{it}\Big|^2 dt \ll \{X + H^2\exp(-(X/H)^{\frac{1}{2}})\} \sum_{X<n\leqslant 2X} |a_n|^2. \tag{9.49}$$

This result is as good as (9.41) provided $X \geqslant H(\log H)^2$, but it is very poor for smaller X. Note that we established (9.49) only for sums over n in dyadic segments. To derive a result for short sums we shall increase X artificially, while preserving the dyadic shape of the range of the character sum (this idea is due to E. Bombieri). We begin with the inequality

$$\sum_{\substack{P<p\leqslant 2P\\ p\nmid kq}} \log p \geqslant \sum_{P<p\leqslant 2P} \log p - \log kq \geqslant \tfrac{1}{2}P$$

which holds for $P \geqslant 9\log kq$. Hence the left side of (9.49) is bounded by

$$\frac{2}{P}\sum_{P<p\leqslant 2P}(\log p) \sum_{\chi\in\mathcal{H}(k,Q)} \int_0^T \Big|\sum_{X<n\leqslant 2X} a_n\chi(pn)(pn)^{it}\Big|^2 dt.$$

Here we used the multiplicativity of our harmonics $\chi(n)n^{it}$ to replace n by $m = pn$. Now m ranges over the dyadic segment $pX < m \leqslant 2pX$, so applying (9.49) with X replaced by pX we get a new estimate

$$\{PX + H^2 \exp(-(PX/H)^{\frac{1}{2}})\} \sum_{X<n\leqslant 2X} |a_n|^2$$

where P is any number $\geqslant 9 \log kQ$. We choose $P = (9 + HX^{-1})(\log H)^2$ showing that

$$\sum_{\chi \in \mathcal{H}(k,Q)} \int_0^T \Big| \sum_{X<n\leqslant 2X} a_n \chi(n) n^{it} \Big|^2 dt \ll (X + H)(\log H)^2 \sum_{X<n\leqslant 2X} |a_n|^2.$$

Finally we replace the dyadic segment $X < n \leqslant 2X$ by the whole interval $1 \leqslant n \leqslant N$ by subdividing the latter and applying Cauchy-Schwarz inequality. This produces an extra factor $\log 2N$, and it completes the proof of Theorem 9.12. □

To every $\chi \in \mathcal{H}(k, Q)$ we associate several points (possibly none)

$$s_r(\chi) = \sigma_r(\chi) + it_r(\chi) \quad \text{with } 0 \leqslant \sigma_r(\chi) \leqslant 1, \ |t_r(\chi)| \leqslant T. \tag{9.50}$$

Let $\mathcal{S}(k, Q, T)$ be the set of all such points counted with relevant multiplicity. We say that $\mathcal{S}(k, Q, T)$ is well-spaced if for any $s_{r_1}(\chi_1), s_{r_2}(\chi_2)$ in $\mathcal{S}(k, Q, T)$ with $(r_1, \chi_1) \neq (r_2, \chi_2)$ we have either $\chi_1 \neq \chi_2$, or $\chi_1 = \chi_2 = \chi$ and $|t_{r_1}(\chi) - t_{r_2}(\chi)| \geqslant 1$. Clearly, if $\mathcal{S}(k, Q, T)$ is well-spaced, then its cardinality satisfies

$$R = |\mathcal{S}(k, Q, T)| \leqslant 3kQ^2T = 3H. \tag{9.51}$$

Nowhere in the proof of Theorem 9.12 did we use essentially the continuous measure in t, therefore the same estimates hold true if one replaced the integration by a summation over any set of well-spaced points $t_r(\chi)$. Then one can change $n^{it_r(\chi)}$ into $n^{-s_r(\chi)}$ at the cost of an extra factor $(\log 2N)^2$ in (9.41) by the arguments from the proof of Proposition 9.11. This way Theorem 9.12 becomes

THEOREM 9.13. *Let $\mathcal{S} = \mathcal{S}(k, Q, T)$ be a well-spaced set of points (9.50). For any complex numbers a_n we have*

$$\sum_{s_r(\chi)\in\mathcal{S}} \Big| \sum_{1\leqslant n\leqslant N} a_n \chi(n) n^{-s_r(\chi)} \Big|^2 \ll G(N + H)\mathcal{L}^5 \tag{9.52}$$

where the implied constant is absolute.

Theorem 9.13 is an extension of Theorem 9.4. Next we establish an extension of Theorem 9.6, namely

THEOREM 9.14. *Let $\mathcal{S} = \mathcal{S}(k, Q, T)$ be a well-spaced set of points (9.50) of cardinality R. For any complex numbers a_n we have*

$$\sum_{s_r(\chi)\in\mathcal{S}} \Big| \sum_{1\leqslant n\leqslant N} a_n \chi(n) n^{-s_r(\chi)} \Big|^2 \ll G(N + RH^{\frac{1}{2}})\mathcal{L}^4 \tag{9.53}$$

where the implied constant is absolute.

PROOF. This goes by modifications of our arguments from the proof of Theorem 9.6 and Theorem 9.12. First we assume that $\sigma_r(\chi) = 0$, i.e., $s_r(\chi) = it_r(\chi)$. Then by duality we reduce the problem to the estimation of

$$\mathcal{D} = \sum_n f(n) \Big| \sum_\chi \sum_r c_\chi(t_r)\chi(n)n^{it_r} \Big|^2 \tag{9.54}$$

for any complex numbers $c_\chi(t_r)$, where $f(n)$ is any non-negative function on $\mathbb{R}^+$ such that $f(n) \geqslant 1$ if $1 \leqslant n \leqslant N$ (in what follows we assume $N \geqslant 2$, otherwise (9.53) is trivial). The dual form satisfies

$$\mathcal{D} \leqslant \sum_{\chi_1} \sum_{\chi_2} \sum_{r_1} \sum_{r_2} \Big| c_{\chi_1}(t_{r_1}) c_{\chi_2}(t_{r_2}) B(t_{r_1} - t_{r_2}, \chi_1\bar{\chi}_2) \Big| \tag{9.55}$$

where $B(t, \chi)$ is given by (9.44). At this point we are going to refine the estimation (9.47). We write

$$B(t, \chi) = \frac{1}{2\pi i} \int_{(2)} L(s - it, \chi) \hat{f}(s) ds \tag{9.56}$$

where $L(s, \chi)$ is the Dirichlet L-function and $\hat{f}(s)$ is the Mellin transform of f

$$\hat{f}(s) = \int_0^\infty f(x) x^{s-1} dx.$$

Moving the integration to the imaginary line we derive

$$B(t, \chi) = \delta_\chi \frac{\varphi(\ell)}{\ell} \hat{f}(1 + it) + O(\ell^{\frac{1}{2}}(|t| + 1)^{\frac{1}{2}}(\log N\ell(|t| + 1))^2). \tag{9.57}$$

Here the leading term comes from the pole of $L(s - it, \chi)$ at $s = 1 + it$ if χ is principal (of course, it agrees with that in (9.47)) and the error term is obtained from two estimates

$$L(s, \chi) \ll \ell^{\frac{1}{2}}(|s| + 1)^{\frac{1}{2}} \log \ell(|s| + 2) \tag{9.58}$$

(see Exercise 9 in Section 5.9) and

$$\hat{f}(s) \ll e^{-|s|} \log N \tag{9.59}$$

on $\operatorname{Re} s = 0$. To see the latter take

$$f(x) = 5(e^{-x/N} - e^{-x}) \tag{9.60}$$

so $\hat{f}(s) = 5\Gamma(s)(N^s - 1)$. Hence we also get $\hat{f}(1 + it) \ll Ne^{-|t|}$. Now (9.57) shows that the dual form satisfies

$$\mathcal{D} \ll (N + RH^{\frac{1}{2}}\mathcal{L}^2) \sum_\chi \sum_r |c_\chi(t_r)|^2,$$

and this proves (9.53) with the logarithmic factor $\mathcal{L}^2$ in place of $\mathcal{L}^4$, but only if all the points $s_r(\chi)$ are imaginary. Then we change n^{it} into $n^{-\sigma-it}$ at the cost of two extra logarithms completing the proof. □

There are two essential ingredients in our proof of Theorem 9.14, namely the reduction to the dual form (9.54) followed by the estimation (9.57) of the character sum $B(\chi, t)$. However, this estimation is not the best possible. Assuming the Lindelöf hypothesis

$$L(s,\chi) \ll (\ell|s|)^{\varepsilon} \tag{9.61}$$

for Re $s = \frac{1}{2}$ and $\chi(\mathrm{mod}\ \ell)$ one deduces (by moving the integration to the critical line rather than to the imaginary one) that

$$B(t,\chi) = \delta_\chi \frac{\varphi(\ell)}{\ell} \hat{f}(1+it) + O(N^{\frac{1}{2}}\ell^{\varepsilon}(|t|+1)^{\varepsilon}). \tag{9.62}$$

This yields

$$\sum_{s_r(\chi)\in\mathcal{S}} \Big| \sum_{1\leqslant n\leqslant N} a_n\chi(n)n^{-s_r(\chi)}\Big|^2 \ll G(N+RN^{\frac{1}{2}}H^{\varepsilon})\mathcal{L}^4 \tag{9.63}$$

in place of (9.53). The results (9.53) and (9.63) are due to H. Montgomery [**Mo2**].

9.6. The reflection method.

The Lindelöf hypothesis may be an open problem for a long time. Nevertheless, M. N. Huxley [**Hu2**] has succeeded to establish unconditionally a result which is almost as good as (9.63) in some important ranges. First he did it for polynomials $D(s)$ in Corollary 9.9 by the subdivision trick. However, this trick does not make sense for polynomials $D(s,\chi)$ twisted by characters. For these polynomials Huxley invented a completely different brilliant trick which he calls the reflection method for reasons soon to be clear. Using Huxley's ideas we are going to prove

THEOREM 9.15. *Let the conditions of Theorem 9.14 hold true. Then*

$$\sum_{s_r(\chi)\in\mathcal{S}} \Big| \sum_{1\leqslant n\leqslant N} a_n\chi(n)n^{-s_r(\chi)}\Big|^2 \ll G(N+R^{\frac{2}{3}}H^{\frac{1}{3}}N^{\frac{1}{3}})\mathcal{L}^6 \tag{9.64}$$

where the implied constant is absolute.

For the proof we can assume that the coefficients a_n are supported in a dyadic segment $N \leqslant n \leqslant 2N$, and as before, we can also assume that all the points $s_r(\chi)$ are on the imaginary line. The latter simplification costs a loss of factor $(\log 2N)^2$, so we must show (9.64) with $\mathcal{L}^4$ in place of $\mathcal{L}^6$.

The problem reduces to the estimation of the dual form $\mathcal{D}$. In the proof of Theorem 9.14 we have estimated $\mathcal{D}$ for any complex coefficients $c_\chi(t_r)$, however, it is sufficient to deal with

$$c_\chi(t_r) = D(s_r(\chi),\chi) = \sum_n a_n\chi(n)n^{-s_r(\chi)}. \tag{9.65}$$

Precisely, letting $\mathcal{C}$ be the left side of (9.64), we have

$$\mathcal{C}^2 \leqslant \mathcal{D}G \tag{9.66}$$

(see the derivation of the duality principle in Chapter 7). Now we are going to take advantage of these special coefficients (9.65). By (9.55) we are again faced with the character sum $B(\chi, t)$. We shall not improve the error term in (9.57) but rather replace it by another character sum; in other words, we use a kind of summation formula (see Chapter 4, in particular (4.26)). We derive such a formula from a functional equation for the corresponding L-function.

Without loss of generality, we can assume that all the characters in $\mathcal{C}$ have the same parity (by dividing $\mathcal{C}$ into two sums if necessary), so the character $\chi = \chi_1\bar{\chi}_2$ in $B(\chi, t)$ is always even. If $\chi(\text{mod } \ell)$ is primitive, then we have the functional equation (4.73)

$$L(s,\chi) = \varepsilon_\chi \gamma(s) \ell^{\frac{1}{2}-s} L(1-s,\bar{\chi}) \tag{9.67}$$

where $|\varepsilon_\chi| = 1$ and

$$\gamma(s) = \pi^{s-\frac{1}{2}} \Gamma\Big(\frac{1-s}{2}\Big) / \Gamma\Big(\frac{s}{2}\Big). \tag{9.68}$$

Suppose $\chi(\text{mod } \ell)$ is induced by the primitive character $\chi^\star(\text{mod } \ell^\star)$ with $\ell^\star|\ell$. Then

$$L(s,\chi) = L(s,\chi^\star) \prod_{p|\ell} (1 - \chi^\star(p)p^{-s}).$$

Hence the functional equation (9.67) for $L(s,\chi^\star)$ yields the following functional equation for $L(s,\chi)$,

$$L(s,\chi) = \varepsilon_\chi \gamma(s) P(s) L(1-s,\bar{\chi}) \tag{9.69}$$

where $\varepsilon_\chi = \varepsilon_{\chi^\star}$ and

$$P(s) = (\ell^\star)^{\frac{1}{2}-s} \prod_{p|\ell} (1 - \chi^\star(p)p^{-s})(1 - \bar{\chi}^\star(p)p^{s-1})^{-1}. \tag{9.70}$$

Notice that $\gamma(s), P(s)$ are holomorphic in $\text{Re } s < 1$, that $|P(s)| = 1$ on $\text{Re } s = \frac{1}{2}$, whence by the convexity bound or Phragmen-Lindelöf principle

$$|P(s)| \leqslant \ell^{\frac{1}{2}-\sigma} \quad \text{if } \sigma = \text{Re}\,(s) \leqslant \tfrac{1}{2}. \tag{9.71}$$

Moreover, $L(s,\chi)$ is holomorphic on $\mathbb{C}$ except for a simple pole at $s = 1$ with residue $\varphi(\ell)/\ell$ if $\chi(\text{mod } \ell)$ is principal. Therefore by moving the integration in (9.56) to the line $\text{Re } s = -\varepsilon$ and applying the functional equation (9.69) we obtain

$$B(t,\chi) = \delta_\chi \frac{\varphi(\ell)}{\ell} \hat{f}(1+it) + \varepsilon_\chi B^\star(t,\chi) \tag{9.72}$$

where

$$B^\star(t,\chi) = \frac{1}{2\pi i} \int_{(-\varepsilon)} \gamma(s-it) P(s-it) L(1-s+it,\bar{\chi}) \hat{f}(s) ds. \tag{9.73}$$

Expanding $L(1-s+it,\bar{\chi})$ into the Dirichlet series and integrating termwise we get

$$B^\star(t,\chi) = \sum_1^\infty g(m)\bar{\chi}(m) m^{-\frac{1}{2}-it} \tag{9.74}$$

where

$$g(y) = \frac{1}{2\pi i} \int_{(\frac{1}{2})} \gamma(s-it) P(s-it) \hat{f}(s) y^{s-\frac{1}{2}} ds. \tag{9.75}$$

Notice we moved the integration from $\text{Re } s = -\varepsilon$ in (9.73) back to $\text{Re } s = \frac{1}{2}$ in (9.75) as we can because $\gamma(s)$ and $P(s)$ have no poles in $\text{Re } s < 1$.

The formula (9.72) is the one we were talking about, but it is not yet ready for applications because the dual function $g(y)$ is not evaluated explicitly enough. To see its properties we take again the majorant function $f(x)$ given by (9.46) with $X = N$, so its Mellin transform is

$$\hat{f}(s) = \int_0^\infty \exp\Big(\frac{5}{2} - \frac{x}{N} - \frac{N}{x}\Big) x^{s-1} dx = 2e^{5/2} K_s(2) N^s \tag{9.76}$$

where $K_s(y)$ stands for the Bessel-Macdonald function. Hence $\hat{f}(s)$ has exponential decay on any vertical line $s = \sigma + it$. Using analytic properties of $\hat{f}(s)$, $\gamma(s)$ and $P(s)$ one shows, by moving the integration in (9.75) sufficiently far to the left (as in (10.64)), that

$$g(y) \ll N^{\frac{1}{2}} \exp\Big(-\frac{1}{5}\Big(\frac{yN}{\ell(|t|+1)}\Big)^{\frac{1}{2}}\Big) \tag{9.77}$$

where the implied constant is absolute (compare this estimate with (9.45)). Hence $g(m)$ is very small if $mN \geqslant \ell(|t|+1)(\log N\ell(|t|+1))^3$.

In our case $|t| \leqslant T$ and $\ell \leqslant kQ^2$, so the terms of $B^\star(t,\chi)$ with $mN \geqslant H\mathcal{L}^3$ contribute very little, precisely we get by (9.77)

$$B^\star(t,\chi) = \sum_{m \leqslant M} g(m)\bar{\chi}(m) m^{-\frac{1}{2}-it} + O(H^{-1}) \tag{9.78}$$

where

$$MN = H\mathcal{L}^3 \tag{9.79}$$

and the implied constant is absolute.

To make (9.78) ready for applications it is desired to separate m from χ and t involved in $g(m)$. We accomplish this goal quickly by introducing (9.75)

$$B^\star(t,\chi) = \frac{1}{2\pi i}\int_{(\frac{1}{2})} \gamma(s)P(s)\hat{f}(s+it)\Big(\sum_{m \leqslant M} \bar{\chi}(m) m^{s-1}\Big) ds + O(H^{-1}). \tag{9.80}$$

Since $|K_s(2)| \leqslant 8|\Gamma(s)| \ll e^{-|u|}$ for $s = \frac{1}{2} + iu$, we get

$$B^\star(t,\chi) \ll N^{\frac{1}{2}} \int_{-\infty}^{\infty} \Big|\sum_{m \leqslant M} \chi(m) m^{-\frac{1}{2}+i(t+u)}\Big| e^{-|u|} du + H^{-1} \tag{9.81}$$

where the implied constant is absolute.

This result is more precise than (9.57). Indeed, estimating trivially by (9.81) one gets

$$B^\star(t,\chi) \ll H^{\frac{1}{2}} \mathcal{L}^{\frac{3}{2}} \tag{9.82}$$

which recovers (9.57). But one can do better by putting the sum

$$\sum_{m \leqslant M} \chi_1\bar{\chi}_2(m) m^{-\frac{1}{2}+i(t_{r_1}-t_{r_2}+u)} \tag{9.83}$$

in the original polynomial

$$\sum_{N < n \leqslant 2N} a_n \chi_1(n) n^{it_{r_1}} \tag{9.84}$$

which appears as a coefficient in (9.55) (here and hereafter we use the abbreviated notation $t_r = t_r(\chi)$). By (9.55), (9.72) and (9.81) we derive

$$\begin{aligned}\mathcal{D} \ll N \sum_{\chi}\sum_{r_1}\sum_{r_2} |\mathcal{D}(it_{r_1},\chi)\mathcal{D}(it_{r_2},\chi)| \exp(-|t_{r_1}-t_{r_2}|)\\ + N^{\frac{1}{2}} \int_{-\infty}^{\infty} \sum_{\chi_1}\sum_{\chi_2}\sum_{r_1}\sum_{r_2} |\mathcal{D}(it_{r_2},\chi_2)\mathcal{P}(it_{r_1},\chi_1;it_{r_2},\chi_2;u)| e^{-|u|}du\\ + H^{-1}\sum_{\chi_1}\sum_{\chi_2}\sum_{r_1}\sum_{r_2} |\mathcal{D}(it_{r_1},\chi_1)\mathcal{D}(it_{r_2},\chi_2)|\end{aligned}$$

where $\mathcal{P}(it_{r_1},\chi_1,it_{r_2},\chi_2;u)$ denotes the product of the two sums (9.83) and (9.84). For fixed χ_2, t_{r_2}, u this product is a new polynomial of type (9.39). The length of $\mathcal{P}(it_{r_1},\chi_1) = \mathcal{P}(it_{r_1},\chi_1;it_{r_2},\chi_2;u)$ being $2MN = 2H\mathcal{L}^3$ is perfect for an application of Theorem 9.13. We obtain

$$\sum_{\chi_1}\sum_{r_1} |\mathcal{P}(it_{r_1},\chi_1)| \leqslant \Big(R\sum_{\chi_1}\sum_{r_1} |\mathcal{P}(it_{r_1},\chi_1)|^2\Big)^{\frac{1}{2}} \ll (RGH)^{\frac{1}{2}}\mathcal{L}^6.$$

After this bound, which does not depend on χ_2, t_2, u, we estimate

$$\sum_{\chi_2}\sum_{r_2} |\mathcal{D}(it_{r_2},\chi_2)| \leqslant \Big(R\sum_{\chi_2}\sum_{r_2} |\mathcal{D}(it_{r_2},\chi_2)|^2\Big)^{\frac{1}{2}} = (R\mathcal{C})^{\frac{1}{2}}.$$

Integrating in u, we find that the contribution of these polynomials to $\mathcal{D}$ is at most $O(R(\mathcal{C}GHN)^{\frac{1}{2}}\mathcal{L}^5)$. The polynomials from the diagonal $\chi_1 = \chi_2 = \chi$ contribute

$$N\sum_{\chi}\sum_{r_1}\sum_{r_2} |\mathcal{D}(it_{r_1},\chi)\mathcal{D}(it_{r_2},\chi)| \exp(-|t_{r_1}-t_{r_2}|) \ll N\mathcal{C},$$

and the remaining ones contribute

$$H^{-1}\sum_{\chi_1}\sum_{\chi_2}\sum_{r_1}\sum_{r_2} |\mathcal{D}(it_{r_1},\chi_1)\mathcal{D}(it_{r_2},\chi_2)| \ll H^{-1}R\mathcal{C} \ll \mathcal{C}.$$

Adding up these contributions we obtain $\mathcal{D} \ll N\mathcal{C} + R(\mathcal{C}GHN)^{\frac{1}{2}}\mathcal{L}^6$. Hence we derive by (9.66) that $\mathcal{C} \ll GN + GR^{\frac{2}{3}}(HN)^{\frac{1}{3}}\mathcal{L}^4$. This completes the proof of Theorem 9.15.

9.7. Large values of $D(s,\chi)$.

The notation from previous section holds here, in particular, we recall that $H = kQ^2T$ and $\mathcal{L} = \log HN$. In this section we deduce a few bounds for the number of large values of the polynomials $D(s,\chi)$. First, as in Theorem 9.7, we deduce from (9.52) the following

THEOREM 9.16. *Suppose for any $s_r(\chi)$ in a well-spaced set $\mathcal{S}(k,Q,T)$ we have*

$$|D(s_r(\chi),\chi)| \geqslant V. \tag{9.85}$$

Then the cardinality of this set, say $R = |\mathcal{S}(k,Q,T)|$, satisfies

$$R \ll (H+N)GV^{-2}\mathcal{L}^5 \tag{9.86}$$

where the implied constant is absolute.

If V is relatively large, then (9.53) yields a better bound (for details see the proof of Theorem 9.8), namely the following

THEOREM 9.17. *Let the conditions be as in Theorem 9.16 with*

$$V \geqslant G^{\frac{1}{2}} H^{\frac{1}{4}} \mathcal{L}^3. \tag{9.87}$$

Then

$$R \ll GNV^{-2}\mathcal{L}^4 \tag{9.88}$$

where the implied constant is absolute.

The estimate (9.88) is ideal for applications, unfortunately it is established only for very large V relative to the "conductor" H. One can remove the condition (9.87) at a cost of introducing an extra term in the bound (9.88). To this end we apply (9.64) and get the following result

THEOREM 9.18. *Let the conditions be as in Theorem 9.16. Then*

$$R \ll (GNV^{-2} + G^3NHV^{-6})\mathcal{L}^{18} \tag{9.89}$$

where the implied constant is absolute.

The last estimate (9.89) is the celebrated result of Huxley [**Hux**]. It is surprising to see that in the special case of the Riemann zeta function, his subdivision argument yields a bound (9.28) which agrees with (9.89) obtained by very different arguments (reflection method).

The assumption that the ordinates $t_r(\chi)$ of the points $s_r(\chi) = \sigma_r(\chi) + it_r(\chi)$ are well-spaced can be relaxed, it suffices that they do not cluster too much. Suppose that for any $s_r(\chi) \in \mathcal{S}(k, Q, T)$ we have

$$|\{r_1\, ;\, s_{r_1}(\chi) \in \mathcal{S}(k, Q, T), |t_{r_1} - t_r| \leqslant 1\}| \leqslant L. \tag{9.90}$$

Then clearly all Theorems 9.13 to 9.18 are true for the set $\mathcal{S}(k, Q, T)$ with the relevant upper bounds multiplied by L (to see this divide the set $\mathcal{S}(k, Q, T)$ into at most $2L$ subsets of well-spaced points and apply the results for each of these subsets separately).

In particular, we know that the number of zeroes $\rho = \beta + i\gamma$ of $L(s, \chi)$ for a given character $\chi(\text{mod } kq)$ with $|\gamma - t| \leqslant 1$ is bounded by $O(\log kq(|t| + 3))$ (see Chapter 5). Therefore, by virtue of the above remark, all Theorems 9.13 to 9.18 hold for any subset $\mathcal{S}(k, Q, T)$ of zeros of

$$\prod_{\substack{q \leqslant Q \\ (q,k)=1}} \; \prod_{\substack{\psi (\text{mod } q) \\ \psi \text{ primitive}}} \; \prod_{\xi (\text{mod } k)} L(s, \psi\xi) \tag{9.91}$$

in a rectangle $\sigma \geqslant \alpha$, $|t| \leqslant T$ (counted with multiplicity) with just one extra factor $\mathcal{L}$, and G replaced by

$$G(\alpha) = \sum_{n \leqslant N} |a_n|^2 n^{-2\alpha} \tag{9.92}$$

in the relevant upper bounds.

In the next chapter we apply these theorems for selected points $s_r(\chi)$ which are zeros of (9.91).

CHAPTER 10

ZERO-DENSITY ESTIMATES

10.1. Introduction.

In the absence of a proof of the Grand Riemann Hypothesis, it is natural to ask how many zeros of a given L-function can lie off the critical line. Since we know there are no zeros on the line $\mathrm{Re}\,(s) = 1$, a few such zeros would not normally influence applications, but a large number of these, especially when they are near this line, may distort substantially the results (an example being the distribution of primes in short intervals, see Section 10.5). Therefore, we should ask how many zeros can possibly be in a given region, and our estimates should reveal that the further the region is from the critical line, the smaller the number of zeros it contains. A quantitative statement of this kind is called a zero-density theorem.

The regions we need most in practice are the rectangles

$$R(\alpha, T) = \{s = \sigma + it; \quad \sigma \geqslant \alpha, \quad |t| \leqslant T\} \tag{10.1}$$

for $\frac{1}{2} \leqslant \alpha \leqslant 1$ and $T \geqslant 3$. Let $N(\alpha, T)$ denote the number of zeros of the Riemann zeta function in $R(\alpha, T)$ (counted with corresponding multiplicity), that is we count the zeros $\rho = \beta + i\gamma$ with

$$\beta \geqslant \alpha, \quad |\gamma| \leqslant T. \tag{10.2}$$

In this notation the Riemann hypothesis asserts that $N(\alpha, T) = 0$ for any $\alpha > \frac{1}{2}$ and $T \geqslant 3$, while the best known zero-free region (due to Vinogradov and Korobov) says that

$$N(\alpha, T) = 0 \quad \text{if} \quad \alpha > 1 - c(\log T)^{-2/3}(\log\log T)^{1/3} \tag{10.3}$$

where c is an absolute positive constant (see Chapter 8). Recall that the number of all zeros $\rho = \beta + i\gamma$ with $|\gamma| \leqslant T$ satisfies (see Chapter 5)

$$N(T) = \frac{T}{\pi} \log \frac{T}{2\pi e} + O(\log T). \tag{10.4}$$

By analogy to the convexity principle concerning the order of $\zeta(s)$ on vertical lines one may expect the following bound for $N(\alpha, T)$ to be true

Density Conjecture. *For $\frac{1}{2} \leqslant \alpha \leqslant 1$ and $T \geqslant 3$ we have*

$$N(\alpha, T) \ll T^{2(1-\alpha)} \log T \tag{10.5}$$

where the implied constant is absolute.

Of course, the convexity principle for zeros in rectangles does not hold for general analytic functions; nevertheless, the density conjecture for $\zeta(s)$ has a reliable backup – the Riemann Hypothesis. This conjecture is particularly attractive because it has a chance to be proved by the technology which is available today, while it replaces the Riemann hypothesis in various applications to the distribution of primes.

A great deal of effort was made to establish estimates of type

$$N(\alpha, T) \ll T^{c(\alpha)(1-\alpha)}(\log T)^A \tag{10.6}$$

with $c(\alpha)$ as small as possible (the logarithmic factor is of secondary concern here, although it will be important to eliminate it in Chapter 18). The first zero-density results were proved by Bohr and Landau [**BL**] in 1914. Their idea was basic in many works over three decades, notably in the works by F. Carlson [**Car**] and A. E. Ingham [**Ing**]. Two new devices were introduced by P. Turán in 1949, and in collaboration with G. Halász in 1958. Great refinements were made by H. Montgomery in 1969. The other significant contributors to this fascinating theory are M. N. Huxley, M. Jutila and D. R. Heath-Brown.

In this book we shall try to give a glimpse of the current state of knowledge. Our results will not be as strong as the best known ones, though nearly so. Moreover, we treat in considerable detail the zeros of Dirichlet L-functions. Let $N(\alpha, T, \chi)$ be the number of zeros $\rho_\chi = \beta_\chi + i\gamma_\chi$ of $L(s, \chi)$ with $\beta_\chi \geqslant \alpha$ and $|\gamma_\chi| \leqslant T$ (counted with multiplicity). The ultimate goal would be to prove the following

GRAND DENSITY CONJECTURE. *Let $k \geqslant 1, Q \geqslant 1, T \geqslant 3$ and $\frac{1}{2} \leqslant \alpha \leqslant 1$. Then*

$$\sum_{\substack{q \leqslant Q \\ (q,k)=1}} \sum_{\substack{\psi \pmod q \\ \psi \text{primitive}}} \sum_{\xi \pmod k} N(\alpha, T, \xi\psi) \ll H^{2(1-\alpha)}(\log H)^A \tag{10.7}$$

where $H = kQ^2T$ and A is an absolute constant, the implied constant being also absolute.

Our main result, the Grand Density Theorem 10.4, will give (10.7) in the range $\frac{5}{6} \leqslant \alpha \leqslant 1$. Moreover, it yields (10.7) in the whole segment $\frac{1}{2} < \alpha \leqslant 1$, but with the constant $c = 12/5$ in place of 2.

10.2. Zero-detecting polynomials.

First we outline the old and new strategies. Given a holomorphic function $L(s)$ in a region $D \subset \mathbb{C}$ there is a variety of tools for counting its zeros inside D. A typical one is the integral formula of Jensen

$$\int_0^R n(r)\frac{dr}{r} = \frac{1}{2\pi i}\int_{|s|=R} \log\left|\frac{L(s)}{L(0)}\right|\frac{ds}{s} \tag{10.8}$$

where $n(r)$ denotes the number of zeros of $L(s)$ in the disc $|s| \leqslant r$ (assuming $L(s) \neq 0$ for $s = 0$ and for s on $|s| = R$). There are similar formulas for other domains. Note that one needs both upper and lower bounds for $|L(s)|$ on the boundary to deduce an estimate for the number of zeros inside the region. Of the two, the lower bound is the hardest to find. In view of these requirements any integral formula is practical only for a region sufficiently wide so that its boundary is distanced from the area in which the zeros are concentrated (otherwise one cannot claim a reasonable lower bound for $|L(s)|$ on ∂D). But this is not the case for the narrow rectangles $R(\alpha, T)$.

To avoid the above barriers one gets the following idea: multiply $L(s)$ by another holomorphic function, say $M(s)$, which is suspected to be large when $L(s)$ is small, and count the zeros of the product $N(s) = L(s)M(s)$ rather than of $L(s)$ alone. Of course, the extra zeros of $M(s)$ will occur, but hopefully not so many as to change the order of magnitude of the original number (a positive proportion of the extra zeros is acceptable).

In reality the idea works only for counting "fictitious" zeros because $N(s)$ does vanish at the true zeros of $L(s)$. Thus the points at which we want the mollifier $M(s)$ to be large are only hypothetical. Having no possibility to inspect the distribution of these hypothetical points one may try $M(s)$ which mimics $1/L(s)$ at almost all points, so $N(s)$ is expected to be close to one almost everywhere in a given region. It would be tautological to take just $1/L(s)$ for the mollifier $M(s)$, or naive to deform $1/L(s)$ smoothly to get a useful holomorphic function. But suppose $1/L(s)$ is given by an absolutely convergent series in a certain region. Then the chances are that its partial sums do well approximate to $1/L(s)$ uniformly in a wider region which does not contain the true zeros of $L(s)$. Therefore, a finite but sufficiently long partial sum of such kind offers a good candidate for $M(s)$.

The above strategy serves well for counting fictitious zeros of Dirichlet series with multiplicative coefficients.

First to illustrate the ideas we give a quick treatment of $L(s) = \zeta(s)$. In this case

$$\frac{1}{\zeta(s)} = \prod_p \Big(1 - \frac{1}{p^s}\Big) = \sum_m \frac{\mu(m)}{m^s},$$

but only if Re $s > 1$. Nevertheless, one expects that the partial sum

$$M(s) = \sum_{m \leqslant M} \frac{\mu(m)}{m^s} \tag{10.9}$$

approximates to $1/\zeta(s)$ if Re $s = \sigma > \frac{1}{2}$. Indeed, the Riemann hypothesis ensures that

$$\frac{1}{\zeta(s)} = M(s) + O((|s|M)^{\varepsilon} M^{\frac{1}{2}-\sigma}). \tag{10.10}$$

REMARKS. The partial sum $M(s)$ was first used in the context of zero-density estimates by Carlson [**Car**] while Bohr and Landau [**BL**] were working earlier with the partial product

$$P(s) = \prod_{p \leqslant P} \Big(1 - \frac{1}{p^s}\Big). \tag{10.11}$$

Either device produces non-trivial results, yet $M(s)$ is more efficient than $P(s)$.

It helps to keep hold on the size of the points, so we restrict our analysis to the region

$$s = \sigma + it, \quad \sigma \geqslant \alpha, \quad T < |t| \leqslant 2T, \tag{10.12}$$

and we assume that T is large. In this region

$$\zeta(s) = \sum_{n \leqslant T} n^{-s} + O(T^{-\alpha}) \tag{10.13}$$

where the implied constant is absolute. Hence

$$\zeta(s)M(s) = \sum_{n \leqslant TM} a_n n^{-s} + O(T^{-\alpha}M^{1-\alpha}\log 2M) \tag{10.14}$$

where the coefficients are

$$a_n = \sum_{\substack{dm=n \\ m \leqslant M, d \leqslant T}} \mu(m), \qquad |a_n| \leqslant \tau(n), \tag{10.15}$$

and the error term is obtained by the trivial estimation

$$M(s) \ll M^{1-\alpha}\log 2M. \tag{10.16}$$

Assuming $1 \leqslant M \leqslant T$ we have $a_1 = 1$ and $a_n = 0$ if $1 < n \leqslant M$ or $n > MT$. We cover the interval $M < n \leqslant MT$ by dyadic segments $N < n \leqslant 2N$ with $N = 2^\ell M$, $0 \leqslant \ell < L = [\log T/\log 2]$. For each of these segments denote

$$D_\ell(s) = \sum_{N < n \leqslant 2N} a_n n^{-s}. \tag{10.17}$$

Thus we have

$$\zeta(s)M(s) = 1 + \sum_{0 \leqslant \ell < L} D_\ell(s) + E(s) \tag{10.18}$$

where the error term satisfies $|E(s)| \leqslant \frac{1}{2}$ for s in the region (10.12), provided $M^{1-\alpha} \leqslant T^\alpha(\log T)^{-2}$.

An important feature of the identity (10.18) is the absence of terms with $1 < n \leqslant M$ on the right side. The left side at a zero of $\zeta(s)$, say at $s = \rho$, vanishes so we must have

$$\Big|\sum_{0 \leqslant \ell < L} D_\ell(\rho)\Big| \geqslant \tfrac{1}{2}. \tag{10.19}$$

Hence

$$|D_\ell(\rho)| \geqslant (2L)^{-1} \tag{10.20}$$

for some $0 \leqslant \ell < L$. This lower bound is rather large by comparison to the mean-value of $|D_\ell(s)|^2$ on the line $\mathrm{Re}\, s = \alpha$. Indeed we have

$$G_\ell(\alpha) = \sum_{N < n \leqslant 2N} |a_n|^2 n^{-2\alpha} \ll N^{1-2\alpha}(\log 2N)^3. \tag{10.21}$$

For this reason we call $D_\ell(s)$ zero-detecting polynomials. What we showed is that every zero of $\zeta(s)$ in the region (10.12) is detected by at least one of a few fixed polynomials by way of producing unusually large value.

REMARKS. Of course, the lower bound (10.20) may not hold at points other than the zeros of $\zeta(s)$. In fact, one should expect that

$$D_\ell(s) \ll N^{\frac{1}{2}-\alpha+\varepsilon} \tag{10.22}$$

for any s in the region (10.12). Comparing (10.22) with (10.20) one could get a contradiction which proves the Riemann Hypothesis. Well, this argument is nothing but wishful thinking because one needs the Riemann Hypothesis to establish (10.22). These remarks reveal that the concept of a zero-detecting polynomial is somewhat superficial, it will loose its meaning when the GRH is fully established. In reality Dirichlet polynomials do not assume large values very often.

Now the theory of Dirichlet polynomials as developed in Chapter 9 tells us that (10.20) cannot happen too often, in particular, producing an interesting estimate for the number of zeros of the zeta function. Precisely, let R_ℓ be the number of zeros of $\zeta(s)$ in (10.12) which are detected by the polynomial $D_\ell(s)$. Then Theorem 9.7 (or Theorem 9.16) together with the closing remarks of Chapter 9 yield

$$R_\ell \ll (T+N)N^{1-2\alpha}(\log T)^{11} \tag{10.23}$$

by (10.20) and (10.21). Since $M \leqslant N < MT$, this gives

$$R_\ell \ll \{TM^{1-2\alpha} + (TM)^{2-2\alpha}\}(\log T)^{11} \tag{10.24}$$

for any $0 \leqslant \ell < L$. Choosing $M = T^{2\alpha-1}$ we get

$$R_\ell \ll T^{4\alpha(1-\alpha)}(\log T)^{11}. \tag{10.25}$$

Adding up the R_ℓ we get an upper bound for all zeros of $\zeta(s)$ in (10.12) (some of these may be detected by several polynomials $D_\ell(s)$), then we extend the result to the zeros in (10.2), and conclude

THEOREM 10.1. *For $\frac{1}{2} \leqslant \alpha \leqslant 1$ and $T > 2$ we have*

$$N(\alpha,T) \ll T^{4\alpha(1-\alpha)}(\log T)^{13}. \tag{10.26}$$

This estimate (apart from the logarithmic factor) was obtained by F. Carlson [**Car**] in 1920 (see also E. Landau [**La3**] of 1921). Twenty years later A. E. Ingham [**Ing**] proved

$$N(\alpha,T) \ll T^{3\frac{1-\alpha}{2-\alpha}}(\log T)^5. \tag{10.27}$$

Either (10.26) or (10.27) imply that the Riemann hypothesis is "almost true" in a statistical sense, namely that almost all zeros of $\zeta(s)$ lie arbitrarily close to the critical line. It follows by (10.27) that the number of zeros $\rho = \beta + i\gamma$ with $|\gamma| \leqslant T$ and

$$\beta > \tfrac{1}{2} + 3A\frac{\log\log}{\log T} \tag{10.28}$$

is bounded by $O(T(\log T)^{5-4A})$ for $A > 1$.

Concerning the zeros near the critical line in 1942 A. Selberg [**S4**] proved that

$$N(\alpha,T) \ll (\alpha - \tfrac{1}{2})^{-1}T, \qquad \text{if } \alpha > \tfrac{1}{2}, \tag{10.29}$$

where the implied constant is absolute. Hence, given a function $\Phi(T) \to \infty$ as $T \to \infty$, it follows that almost all zeros of the Riemann zeta function lie in the region

$$|\beta - \tfrac{1}{2}| < \frac{\Phi(T)}{\log T}. \tag{10.30}$$

10.3. Breaking the zero-density conjecture.

It is amazing how long the estimate (10.27) of Ingham resisted improvements. After sixty years it is still the best known result (apart from the logarithms) in the range $\frac{1}{2} < \alpha \leqslant \frac{3}{4}$. The new methods, which emerged from the work of Halász and Turán [**HaTu**] in the late sixties and flourished in the seventies, produced a new estimate for the number of large values of Dirichlet polynomials, namely (9.28) in place of (9.25). This new estimate turned out to be better than the old one for treating zeros near the line Re $s = 1$. First the estimate (10.27) was improved in the range $\frac{4}{5} < \alpha < 1$ by H. L. Montgomery [**Mo3**] in 1969, and shortly after in the range $\frac{3}{4} < \alpha < 1$ by M. N. Huxley [**Hu2**] in 1972 (see a more general Theorem 10.4).

For the zeros of $\zeta(s)$ near the line Re $s = 1$ one is also helped by having shorter partial sums which approximate to $\zeta(s)$. In this section we use a suitable short polynomial to establish a zero-density estimate which is even better than the zero-density conjecture for $\alpha > \frac{5}{6}$.

THEOREM 10.2 (HUXLEY). *For any $\alpha > \frac{5}{6}$ and $T \geqslant 2$ we have*

$$N(\alpha, T) \ll T^{\frac{3(1-\alpha)}{3\alpha-1}} (\log T)^A \tag{10.31}$$

where $A = 300(\alpha - \frac{5}{6})^{-2}$ and the implied constant depends only on α.

REMARKS. Actually Huxley proved (10.31) for any $\alpha \geqslant \frac{1}{2}$ with A and the implied constant being absolute. We could refine our arguments to get (10.31) in wider rectangles and with absolute constants, but it would obscure the presentation. Besides proving (10.31) our mission is also to introduce neatly some technical novelties due to M. Jutila [**Ju4**]. He realized that it was sufficient to take the mollifier $M(s)$ quite short. Then, to achieve best results, he makes an adjustment by raising the zero-detecting polynomials to appropriate high powers. This idea of Jutila is somewhat similar to the work of D. R. Heath-Brown on his identity for primes (13.37).

By Weyl's bound (8.18) (the van der Corput bound (8.53) would also do the job) we derive

$$\zeta(s) = \sum_{n \leqslant X} n^{-s} + O(X^{\frac{1}{2}-\sigma} T^{\frac{1}{6}} \log T) \tag{10.32}$$

for $s = \sigma + it$ with $\frac{1}{2} \leqslant \sigma \leqslant 1$, $T \leqslant |t| \leqslant 2T$, where $1 \leqslant X \leqslant T$ and the implied constant is absolute. Hence

$$\zeta(s)M(s) = 1 + \sum_{M < n \leqslant MX} a_n n^{-s} + E(s) \tag{10.33}$$

where $1 \leqslant M \leqslant X$, $|a_n| \leqslant \tau(n)$ (see (10.15) with T replaced by X) and

$$E(s) \ll M^{1-\sigma} X^{\frac{1}{2}-\sigma} T^{\frac{1}{6}} (\log T)^2. \tag{10.34}$$

We require

$$|E(s)| \leqslant \frac{1}{2} \tag{10.35}$$

for s in the rectangle (10.12), and this holds if

$$X^{\alpha-\frac{1}{2}} \geqslant M^{1-\alpha} T^{\frac{1}{6}} (\log T)^3 \tag{10.36}$$

because T is assumed to be large.

Now the situation is the same as in Section 10.2 except that we employ fewer zero-detecting polynomials (10.17); their length $N = 2^\ell M$ is restricted by $M \leqslant N < MX$ (i.e., $0 \leqslant \ell < L = [\log X/\log 2]$). Recall that R_ℓ denotes the number of zeros of $\zeta(s)$ in (10.12) satisfying (10.20). We shall estimate R_ℓ by an appeal to (9.31) with a suitable $k \geqslant 2$ rather than (9.30) which was used with $k = 1$ for deriving (10.23). In other words, we apply (9.28) for the polynomial $D_\ell(s)^k$. We choose $k \geqslant 2$ depending on N such that $P = N^k$ falls into the segment

$$Z < P \leqslant (MX)^2 + Z^{\frac{3}{2}}, \tag{10.37}$$

where Z is a given number to be specified later subject to $Z \geqslant MX$. To see that (10.37) is possible take $k \geqslant 2$ such that $Z^{1/k} < M \leqslant Z^{1/(k-1)}$, and in case $k = 2$ use also $N \leqslant MX$. The mean-value of $|D_\ell(s)|^{2k}$ on the line $\mathrm{Re}\, s = \alpha$ is bounded by

$$G_k^k = \Big(\sum_{N<n\leqslant 2N} |a_n|^2 n^{-2\alpha} \tau_k(n) \Big)^k \ll P^{1-2\alpha} (\log P)^{4k(k+1)} \tag{10.38}$$

where the implied constant depends only on k. Moreover, the lower bound $V = (2L)^{-1}$ of $|D_\ell(s)|$ at the zero is raised to the power $V^k = (2L)^{-k} \gg (\log T)^{-k}$. Therefore (9.31) yields

$$R_\ell \ll (P^{2(1-\alpha)} + TP^{4-6\alpha})(\log t)^{A_k} \tag{10.39}$$

where $A_k = 6(k+1)(12k+1)$ and the implied constant depends on k. Here it would be perfect to choose $P = T^{1/2(2\alpha-1)}$, producing the best bound

$$T^{(1-\alpha)/(2\alpha-1)} (\log T)^{A_k},$$

but it is not always possible because P is not precisely localized (since k had to be an integer!). However, we have reasonably good control of P given by the interval (10.37) where Z is at our disposal. Inserting the upper bound for P to the first term and the lower bound for P to the second term in the right side of (10.40) we get

$$R_\ell \ll \{(MX)^{4(1-\alpha)} + Z^{3(1-\alpha)} + TZ^{4-6\alpha}\} (\log T)^{A_k}. \tag{10.40}$$

Now it is clear that the best choice is $Z = T^{1/(3\alpha-1)}$. We also take $MX = Z^{\frac{3}{4}} (\log T)^9$ which yields

$$R_\ell \ll T^{\frac{3(1-\alpha)}{3\alpha-1}} (\log T)^{A_k+6}. \tag{10.41}$$

We still have some freedom to choose M and X subject to the condition (10.36). For the value $MX = X^{3/4(3\alpha-1)}(\log T)^9$ this condition is satisfied with

$$M = T^{(6\alpha-5)/12(3\alpha-1)} \tag{10.42}$$

$$X = T^{(7-3\alpha)/6(3\alpha-1)}(\log T)^9. \tag{10.43}$$

Finally, summing (10.41) we get (10.31) with $A = A_k + 8 = 6(K+1)(12K+1)+8$, where K is the maximal k which is needed to put $P = N^k$ into (10.37) for $M \leqslant N < MX$, i.e., $K = 1 + [\log Z/\log M] \leqslant 1 + 2(\alpha - \frac{5}{6})^{-1}$. Hence $A \leqslant 300(\alpha - \frac{5}{6})^{-2}$.

Having sufficiently strong estimates for exponential sums one can make a good approximation to $\zeta(s)$ by partial sums which are very short from which one can derive relatively sharp zero-density estimates. Thus, using the Vinogradov estimate (8.92) it was shown by Halász and Turán [**HaTu**] that for any α near one

$$N(\alpha, T) \ll T^{(1-\alpha)^{\frac{3}{2}}|\log(1-\alpha)|^3}. \tag{10.44}$$

EXERCISE 1. Prove along the above lines that the Lindelöf Hypothesis implies

$$N(\alpha, T) \ll T^{2(1-\alpha)+\varepsilon} \tag{10.45}$$

for $\frac{1}{2} \leqslant \alpha \leqslant 1$ and any $\varepsilon > 0$, the implied constant depending only on ε.

EXERCISE 2. Assuming the Lindelöf Hypothesis and the Montgomery Conjecture M (see (9.23)), prove along the above lines that

$$N(\alpha, T) \ll T^{\varepsilon} \tag{10.46}$$

if $\alpha > \frac{1}{2}$ for any $\varepsilon > 0$, the implied constant depending on α and ε.

10.4. Grand zero-density theorem.

Now we are going to exploit the ideas of the previous section and the results of Chapter 9 for polynomials with characters in their full force. We consider characters $\chi(\text{mod } kq)$ with $(k, q) = 1$ such that $\chi = \xi\psi$ with ξ any character to modulus k and ψ any primitive character of conductor q. Here k is given while q varies over the segment $1 \leqslant q \leqslant Q$. Moreover, we keep the points $s = \sigma + it$ in the rectangle $\sigma \geqslant \alpha, |t| \leqslant T$, where $T \geqslant 3$. Let $N(\alpha, k, Q, T)$ denote the total number of zeros of all $L(s, \chi)$ counted with multiplicity subject to the above restrictions, i.e.,

$$N(\alpha, k, Q, T) = \sum_{\substack{q \leqslant Q \\ (q,k)=1}} \sum_{\substack{\psi(\text{mod } q) \\ \psi \text{primitive}}} \sum_{\xi(\text{mod } k)} N(\alpha, T, \xi\psi).$$

Put $D = kQT$, $H = kQ^2T$ and $\mathcal{L} = 2\log D$. By the classical bound

$$N(\alpha, T, \chi) \ll T \log(kqT)$$

(see Theorem 5.24) it follows that $N(\alpha, k, Q, T) \ll H\mathcal{L}$, which we shall refer to as the trivial bound.

In this section we shall establish various non-trivial bounds for $N(\alpha, k, Q, T)$ which are generalizations rather than improvements of the results by Ingham and Huxley (compare (10.72) with (10.27) and (10.31)).

From the previous section it is clear that if one has shorter Dirichlet polynomials approximating $L(s,\chi)$, then a stronger zero-density theorem can be deduced. The best unconditional approximation is given by the approximate functional equation, which, roughly speaking, represents $L(s,\chi)$ as a sum of two polynomials of length X and Y respectively, with XY being the conductor. Classical formulas of such kind (see e.g. [**Lav**]) are quite messy, so we rather develop here a different expression for $L(s,\chi)$ of the same nature, but more friendly for applications in mind. See also Theorem 5.3 for a general version.

The quality of an approximate functional equation depends on the choice of a particular cut-off function. We have already experimented with such approximate equations when studying the reflection method in Section 9.6. Our present setting is close to that particular one. One may start from any smooth function f on $\mathbb{R}^+$ such that

$$f(0)=1 \text{ and } \lim_{x\to+\infty} f(x)=0 \tag{10.47}$$

with $f'(x)$ having rapid decay as x tends to 0 or ∞. Then its Mellin transform

$$\hat{f}(s)=\int_0^\infty f(x)x^{s-1}dx \tag{10.48}$$

has a simple pole at $s=0$ with residue 1 and $s\hat{f}(s)$ is an entire function. Indeed we have

$$s\hat{f}(s)=-\int_0^\infty f'(x)x^s dx \tag{10.49}$$

which shows what we claimed. Normally we would not recommend working with any specific cut-off function, but we do so here for clarity to keep track of uniformity in several variables side by side. A nice choice is

$$f(x)=\kappa\int_x^\infty \exp\Big(-y-\frac{1}{y}\Big)\frac{dy}{y}. \tag{10.50}$$

where κ is the normalizing constant such that $f(0)=1$. We shall see from computations below that $\kappa^{-1}=2K_0(2)=2\sum_1^\infty \Gamma'(k)\Gamma^{-3}(k)=0.2277\ldots$. This function satisfies

$$0<f(x)<\kappa e^{-x} \tag{10.51}$$

because $e^{-1/y}<y$, and

$$0<1-f(x)<\kappa e^{-1/x} \tag{10.52}$$

because

$$f(x)+f\Big(\frac{1}{x}\Big)=1. \tag{10.53}$$

By (10.49) for f given by (10.50) we get

$$s\hat{f}'(s)=\kappa\int_0^\infty \exp\Big(-y-\frac{1}{y}\Big)y^{s-1}dy=2\kappa K_s(2) \tag{10.54}$$

where $K_s(z)$ is the Bessel-Macdonald function. In particular, by the power series expansion of $K_s(z)$ at $z = 2$ we get

$$sf̂(s) = \frac{\pi\kappa}{\sin(\pi s)} \sum_1^\infty \frac{1}{\Gamma(k)} \left(\frac{1}{\Gamma(k-s)} - \frac{1}{\Gamma(k+s)} \right). \tag{10.55}$$

Hence we see that $\hat{f}(s)$ is odd

$$\hat{f}(s) = -\hat{f}(-s). \tag{10.56}$$

Moreover, we get the uniform bound for $s = \sigma + it \in \mathbb{C}$,

$$\hat{f}(s) \ll |s|^{|\sigma|-1} e^{-\frac{\pi}{2}|t|} \tag{10.57}$$

where the implied constant is absolute.

Now we are ready to derive the approximate formula for $L(s, \chi)$ in question. We assume that $s = \sigma + it$ with $\frac{1}{2} < \sigma < 1$ and χ is a character of modulus $\ell \geqslant 1$. To refresh your memory, we recommend reading again the arguments from (9.67) to (9.75) in the previous chapter. We begin by evaluating the sum

$$B(s, \chi) = \sum_1^\infty \chi(n) n^{-s} f\left(\frac{n}{X}\right). \tag{10.58}$$

By contour integration and the functional equation (9.69) we get

$$\begin{aligned} B(s, \chi) &= \frac{1}{2\pi i} \int_{(1)} L(s+u, \chi) X^u \hat{f}(u) du \\ &= \delta_\chi \frac{\varphi(\ell)}{\ell} X^{1-s} \hat{f}(1-s) + L(s, \chi) + \frac{1}{2\pi i} \int_{(-1)} L(s+u, \chi) X^u \hat{f}(u) du. \end{aligned}$$

Then by the functional equation (9.69) and by (10.56) we find that the integral on the line Re $u = -1$ is equal to $-\varepsilon_\chi B^\star(s, \chi)$ where

$$B^\star(s, \chi) = \frac{1}{2\pi i} \int_{(1)} \gamma(s-u) P(s-u) L(1-s+u, \bar{\chi}) X^{-u} \hat{f}(u) du.$$

Here we expand $L(1-s+u, \bar{\chi})$ into Dirichlet series and integrate each term getting

$$B^\star(s, \chi) = \sum_1^\infty \bar{\chi}(m) m^{s-1} g(mX) \tag{10.59}$$

where

$$g(y) = \frac{1}{2\pi i} \int_{(1)} \gamma(s-u) P(s-u) y^{-u} \hat{f}(u) du. \tag{10.60}$$

Gathering the above results we obtain the desired expression

$$L(s, \chi) = B(s, \chi) + \varepsilon_\chi B^\star(s, \chi) - \delta_\chi \frac{\varphi(\ell)}{\ell} \hat{f}(1-s) X^{1-s} \tag{10.61}$$

where $B(s, \chi)$ and $B^\star(s, \chi)$ are the reduced series given by (10.58) and (10.59) with arbitrary $X > 0$.

It is clear that $B(s,\chi)$ runs essentially over $n \ll X$, and we show that $B^\star(s,\chi)$ runs essentially over $m \ll Y$, where Y is somewhat larger than $\ell|s|X^{-1}$. To this end recall the estimates

$$|P(s)| \leqslant \ell^{\frac{1}{2}-\mathrm{Re}\ s}, \tag{10.62}$$

$$\gamma(s) \ll |s|^{\frac{1}{2}-\mathrm{Re}\ s} \tag{10.63}$$

which hold uniformly in the half-plane Re $s \leqslant \frac{1}{2}$, and the estimate (10.57) which holds for all s. Hence, moving the integration (10.60) sufficiently far to the right, say to the vertical line

$$\mathrm{Re}\ u = \max\Big(1, \tfrac{1}{5}\Big(\frac{y}{\ell|s|}\Big)^{\frac{1}{2}}\Big) \tag{10.64}$$

one derives

$$g(y) \ll \frac{\ell|s|}{y}\exp\Big(-\tfrac{1}{5}\Big(\frac{y}{\ell|s|}\Big)^{\frac{1}{2}}\Big) \tag{10.65}$$

where the implied constant is absolute. Hence $g(mX)$ is very small if mX is somewhat larger than $\ell|s|$. Precisely, we get by (10.65)

$$B^\star(s,\chi) = \sum_{m\leqslant Y} \bar{\chi}(m)m^{s-1}g(mX) + O\Big(\frac{1}{XY}\Big) \tag{10.66}$$

provided

$$XY \geqslant \ell|s|(\log \ell|s|)^3. \tag{10.67}$$

Next we write (10.60) as

$$g(y) = \frac{1}{2\pi i}\int_{(\eta)} \gamma(s-2\sigma+u)P(s-2\sigma+u)y^{u-2\sigma}\hat{f}(2\sigma-u)du \tag{10.68}$$

by changing u into $2\sigma - u$ and moving the integration to the line Re $u = \eta$, where $1 < \eta < 2\sigma$. Then we insert (10.68) into (10.66) getting

$$B^\star(s,\chi) = \frac{1}{2\pi i}\int_{(\eta)}\Big(\sum_{m\leqslant y}\bar{\chi}(m)m^{-\bar{s}+u-1}\Big)W(u)du + O(\tfrac{1}{XY}) \tag{10.69}$$

where

$$W(u) = \gamma(s-2\sigma+u)P(s-2\sigma+u)X^{u-2\sigma}\hat{f}(2\sigma-u).$$

By (10.63), (10.62), (10.57) we get for $u = \eta + iv$,

$$\begin{aligned} W(u) &\ll ((|s|+|v|)\ell)^{\frac{1}{2}+\sigma-\eta}X^{\eta-2\sigma}(2\sigma-\eta)^{-1}e^{-\frac{\pi}{2}|v|} \\ &\ll (2\sigma-\eta)^{-1}\Big(\frac{\ell|s|}{X^2}\Big)^{\frac{1}{2}+\sigma-\eta}X^{1-\eta}e^{-|v|}. \end{aligned}$$

We assume $X^2 \geqslant \ell|s|$ and $\eta = \alpha + \frac{1}{2}$, getting for $\sigma \geqslant \alpha$,

$$W(u) \ll (\alpha - \tfrac{1}{2})^{-1}X^{\frac{1}{2}-\alpha}e^{-|v|}. \tag{10.70}$$

Hence (10.69) yields

$$B^\star(s,\chi) \ll (\alpha-\tfrac{1}{2})^{-1}X^{\frac{1}{2}-\alpha}\int_{-\infty}^{\infty}\Big|\sum_{m\leqslant Y}\chi(m)m^{-s+\alpha-\frac{1}{2}+iv}\Big|e^{-|v|}dv + \frac{1}{XY}. \tag{10.71}$$

Before going further let us state what we have proved.

LEMMA 10.3. *Let χ be a character modulo ℓ and s a complex number with Re $s \geqslant \alpha > \frac{1}{2}$. Choose $X \geqslant (\ell|s|)^{\frac{1}{2}}$ and Y satisfying (10.67). Then we have (10.61) where $B(s,\chi)$ is given by (10.58) with f as in (10.50) and $B^\star(s,\chi)$ satisfies (10.71) with the implied constant being absolute.*

Now we proceed to the proof of our main result which is

GRAND DENSITY THEOREM 10.4. *Let $\frac{1}{2} < \alpha \leqslant 1$ and $\varepsilon > 0$. Then we have*

$$N(\alpha, k, Q, T) \ll (D^{(2+\varepsilon)(1-\alpha)} + H^{c(\alpha)(1-\alpha)})\mathcal{L}^A \tag{10.72}$$

where

$$c(\alpha) = \min\Big(\frac{3}{2-\alpha}, \frac{3}{3\alpha-1}\Big). \tag{10.73}$$

Here $D = kQT$, $H = kQ^2T$ and $\mathcal{L} = 2\log D$. The exponent A in the power of logarithm and the implied constant depend only on α and ε.

REMARK. One could remove ε and assert that both A and the implied constant are absolute, however, such a refinement is technically very complicated to make along our lines (try the original arrangements by H. L. Montgomery [**Mo2**]).

Note that (10.72) is essentially as strong as (10.7) in the range $\frac{5}{6} < \alpha < 1$, and considerably stronger with respect to Q. In particular, comparing with the number of terms in the sum, it shows that for T fixed and any $\alpha > \frac{1}{2}$, very few L-functions of primitive characters modulo $q \leqslant Q$ have a zero in the rectangle (10.1).

In what follows χ runs over characters of modulus $\ell = kq \leqslant Q$ and $s = \sigma + t$ is a complex variable restricted to the rectangle $\alpha \leqslant \sigma \leqslant 1, |t| \leqslant T$. Furthermore, if χ is principal we assume that $|t| \geqslant \mathcal{L}^2$, because otherwise the result is trivial. For the proof of Theorem 10.4 we use (10.61) with

$$X = D^{\frac{1}{2}}\mathcal{L}, \qquad Y = D^{\frac{1}{2}}\mathcal{L}^2. \tag{10.74}$$

The residual term in (10.61) is small

$$\delta_\chi \frac{\varphi(\ell)}{\ell}\hat{f}(1-s)X^{1-s} \ll D^{-1}$$

by virtue of (10.57), and the series (10.58) can be reduced to $n \leqslant Y$ up to a correction term $O(D^{-1})$ by virtue of (10.51). We obtain

$$\begin{aligned} L(s,\chi) &= \sum_{n\leqslant Y} \chi(n)n^{-s}f\Big(\frac{n}{X}\Big) \\ &+ O\Big(X^{\frac{1}{2}-\alpha}\int_{-\infty}^{\infty}\Big|\sum_{n\leqslant Y}\chi(n)n^{-s+\alpha-\frac{1}{2}+iv}\Big|e^{-|v|}dv + D^{-1}\Big). \end{aligned} \tag{10.75}$$

Multiply this by the polynomial

$$M(s,\chi) = \sum_{m\leqslant M} \mu(m)\chi(m)m^{-s} \tag{10.76}$$

where $1 \leqslant M \leqslant D^{\frac{1}{2}}$ getting

$$L(s,\chi)M(s,\chi) = \sum_{n \leqslant MY} a_n \chi(n) n^{-s} + O\Big(\mathcal{L}\int_{-\infty}^{\infty}\Big|\sum_{n \leqslant MY} a_n(v)\chi(n)n^{-s}\Big| e^{-|v|}dv + D^{-1}M^{\frac{1}{2}}\Big) \tag{10.77}$$

where the coefficients are given by

$$a_n = \sum_{\substack{dm=n \\ d \leqslant Y, m \leqslant M}} \mu(m) f\Big(\frac{d}{X}\Big), \tag{10.78}$$

$$a_n(v) = \sum_{\substack{dm=n \\ d \leqslant Y, m \leqslant M}} \mu(m) \Big(\frac{d}{Y}\Big)^{\alpha - \frac{1}{2} + iv}, \tag{10.79}$$

so $|a_n| \leqslant \tau(n)$ and $|a_n(v)| \leqslant \tau(n)$. For $n \leqslant M$ we have more precise estimates

$$a_n = \sum_{dm=n} \mu(m)\{1 + O(e^{-X/d})\} = \sum_{m|n} \mu(m) + O(D^{-2}),$$

$$a_n(v) = \Big(\frac{n}{Y}\Big)^{\alpha-\frac{1}{2}+iv} \prod_{p|n} (1 - p^{\frac{1}{2}-\alpha-iv}) \ll \tau(n)\Big(\frac{n}{Y}\Big)^{\alpha - \frac{1}{2}}.$$

The first yields $a_1 = 1 + O(D^{-2})$ and $a_n \ll D^{-2}$ if $1 < n \leqslant M$, while the latter gives

$$\Big|\sum_{n \leqslant M} a_n(v)\chi(n)n^{-s}\Big| \leqslant Y^{\frac{1}{2}-\alpha} \sum_{n \leqslant M} \tau(n) n^{-\frac{1}{2}} \ll Y^{\frac{1}{2}-\alpha} M^{\frac{1}{2}} \log 2M.$$

We want this to be bounded by $\mathcal{L}^{-2}$ which is the case by assuming

$$M \leqslant D^{\alpha - \frac{1}{2}} \mathcal{L}^{-6}. \tag{10.80}$$

By the above estimates we reduce (10.77) to

$$L(s,\chi)M(s,\chi) = 1 + \sum_{M < n \leqslant MY} a_n \chi(n) n^{-s} + O\Big(\mathcal{L}\int_{-\infty}^{\infty}\Big|\sum_{M < n \leqslant MY} a_n(v)\chi(n)n^{-s}\Big| e^{-|v|}dv + \mathcal{L}^{-1}\Big). \tag{10.81}$$

We want to treat the first sum and the integral of the second sum in one way. To this end we combine the sum and the integral into one integral with respect to the measure

$$d\mu = \tfrac{1}{3} e^{-|v|} dv + \tfrac{1}{3}\delta(v) \tag{10.82}$$

where dv is the Lebesque measure on $\mathbb{R}$ and $\delta(v)$ is the point measure at $v = 0$, the factor $\frac{1}{3}$ being introduced for the normalization

$$\int_{-\infty}^{\infty} d\mu = 1. \tag{10.83}$$

In this notation we write (10.81) as one inequality

$$L(s,\chi)M(s,\chi)-1\ll\mathcal{L}\int_{-\infty}^{\infty}\Big|\sum_{M<n\leqslant MY}a_n(v)\chi(n)n^{-s}\Big|d\mu(v)+\mathcal{L}^{-1}\tag{10.84}$$

where for $v=0$ we redefine $a_n(0)$ to be a_n. From now on we only need to know that $a_n(v)$ does not depend on s,χ and that $|a_n(v)|\leqslant\tau(n)$. For convenience we also set $a_n(v)=0$ if $n\leqslant M$ or $n>MY$.

Recall that (10.84) holds for s in the rectangle $R(\alpha,T)$ and a character χ modulo kq with $q\leqslant Q$, and in case χ is principal we require $|t|\geqslant\mathcal{L}^2$, where $\mathcal{L}=2\log D$ and $D=kQT$. In particular, if $s=\rho$ is a zero of $L(s,\chi)$ in this region, we get from (10.84)

$$\int_{-\infty}^{\infty}\Big|\sum_{M<n\leqslant MY}a_n(v)\chi(n)n^{-\rho}\Big|d\mu(v)\gg\mathcal{L}^{-1}$$

provided D is sufficiently large (which condition can be assumed, or else the goal is trivial). As in Section 10.3 we break the summation into dyadic segments $N<n\leqslant 2N$ with $N=2^{\ell}M$, $0\leqslant\ell<L=[\log Y/\log 2]$. Denote

$$D_\ell(s,\chi)=\int_{-\infty}^{\infty}\Big|\sum_{N<n\leqslant 2N}a_n(v)\chi(n)n^{-s}\Big|d\mu(v).\tag{10.85}$$

For every zero ρ being counted there exists ℓ such that

$$D_\ell(\rho,\chi)\geqslant L^{-3}.\tag{10.86}$$

Let R_ℓ denote the number of zeros satisfying (10.86). Then the total number of zeros is

$$R\leqslant\sum_{\ell}R_\ell\leqslant L\max_{\ell}R_\ell.$$

Raising $D_\ell(s,\chi)$ to a suitable power $2k$ with $k\geqslant 2$ (depending on N) we obtain

$$D_\ell(\chi)^{2k}\leqslant\int_{-\infty}^{\infty}\Big|\sum_{P<n\leqslant 2^kP}b_n(v)\chi(n)n^{-s}\Big|^2d\mu(v)$$

with $P=N^k$ which falls into the segment

$$Z\leqslant P\leqslant(MY)^2+Z^{\frac{3}{2}}\tag{10.87}$$

where Z is any fixed number to be chosen later, subject to $MY\leqslant Z\leqslant H$. At this point we require $M\geqslant D^{\varepsilon/4}$, so k is bounded in terms of ε. The coefficients $b_n(v)$ are bounded by a power of the divisor function, so we have

$$\sum_{P<n\leqslant 2^kP}|b_n(v)|^2n^{-2\alpha}\leqslant P^{1-2\alpha}\mathcal{L}^A$$

where A depends only on k. Adding up these inequalities for the points (ρ,χ) satisfying (10.86) we obtain

$$R_\ell\leqslant\mathcal{L}^{6k}\int_{-\infty}^{\infty}\sum_{(\rho,\chi)\in S_\ell}\Big|\sum_{P<n\leqslant 2^kP}b_n(v)\chi(n)n^{-\rho}\Big|^2d\mu(v).$$

Now we are ready to use two results from Chapter 9, namely Theorem 9.13 and Theorem 9.15 (see also the closing remarks in Chapter 9 concerning the spacing of points ρ). Applying Theorem 9.13 we get

$$R_\ell \ll P^{1-2\alpha}(P+H)\mathcal{L}^A, \tag{10.88}$$

and by Theorem 9.15

$$R_\ell \ll P^{1-2\alpha}(P+R_\ell^{\frac{2}{3}}H^{\frac{1}{3}}P^{\frac{1}{3}})\mathcal{L}^A,$$

the latter giving

$$R_\ell \ll P^{1-2\alpha}(P+H^{3-4\alpha})\mathcal{L}^A. \tag{10.89}$$

(here $H=kQ^2T$, and A is a constant depending on k which may be different in each appearance). A direct comparison of (10.88) and (10.89) shows that one is stronger than the other if $\frac{1}{2}<\alpha\leqslant\frac{3}{4}$ and $\frac{3}{4}\leqslant\alpha<1$ respectively.

Suppose $\frac{1}{2}+\varepsilon\leqslant\alpha<\frac{3}{4}$. Then introducing P from (10.87) into (10.88) one gets

$$R_\ell \ll ((MY)^{4(1-\alpha)}+Z^{3(1-\alpha)}+HZ^{1-2\alpha})\mathcal{L}^A.$$

Clearly the best choice is $Z=H^{1/(2-\alpha)}$. We also assume $MY\leqslant Z^{\frac{3}{4}}$ getting

$$R_\ell \ll H^{\frac{3(1-\alpha)}{2-\alpha}}\mathcal{L}^A. \tag{10.90}$$

It remains to verify whether the restriction $MY\leqslant Z^{\frac{3}{4}}$ does not contradict the conditions previously imposed. For $Y=D^{\frac{1}{2}}\mathcal{L}^2$, and since $D\leqslant H$, this restriction is verified with

$$M=H^{\frac{2\alpha-1}{4(2-\alpha)}}\mathcal{L}^{-2}.$$

Note that M satisfies (10.80), $M\geqslant D^{\varepsilon/4}$ and obviously $MY\leqslant Z$. The exponent A and the implied constant in (10.90) depend only on ε. Since ε is arbitrary, Theorem 10.4 is established in the range $\frac{1}{2}<\alpha\leqslant\frac{3}{4}$ (actually in this range (10.72) holds true without the first term because it is absorbed by the second one).

Suppose $\frac{3}{4}\leqslant\alpha\leqslant 1$. Then introducing P from (10.87) into (10.89) one gets

$$R_\ell \ll ((MY)^{4(1-\alpha)}+Z^{3(1-\alpha)}+HZ^{4-6\alpha})\mathcal{L}^A.$$

Now the best choice is $Z=H^{1/(3\alpha-1)}$ giving

$$R_\ell \ll ((DM^2)^{2(1-\alpha)}+H^{\frac{3(1-\alpha)}{3\alpha-1}})\mathcal{L}^A.$$

For $M=D^{\varepsilon/4}$ this establishes Theorem 10.4 in the range $\frac{3}{4}\leqslant\alpha\leqslant 1$.

EXERCISE 3. Show that for $\frac{1}{2}<\alpha\leqslant 1$ and $\varepsilon>0$,

$$N(\alpha,k,Q,T)\ll (D^{(3+\varepsilon)(1-\alpha)}+H^{b(\alpha)(1-\alpha)})\mathcal{L}^A, \tag{10.91}$$

where

$$b(\alpha)=\min\Big(\frac{8}{5-2\alpha},\frac{4}{5\alpha-2}\Big). \tag{10.92}$$

[**Hint:** Raise the polynomial $D_\ell(s,\chi)$ to a power $2k$ with $k\geqslant 3$ so that $Z\leqslant P\leqslant (MY)^3+Z^{4/3}$.]

EXERCISE 4. Show that for $\frac{1}{2} < \alpha \leqslant 1$ and $\varepsilon > 0$,

$$N(\alpha, k, Q, T) \ll (D^{(4+\varepsilon)(1-\alpha)} + H^{a(\alpha)(1-\alpha)})\mathcal{L}^A \tag{10.93}$$

where

$$a(\alpha) = \min\Big(\frac{5}{3-\alpha}, \frac{5}{7\alpha - 3}\Big). \tag{10.94}$$

[**Hint:** Raise the polynomial $D_\ell(s, \chi)$ to a power $2k$ with $k \geqslant 4$ so that $Z \leqslant P \leqslant (MY)^4 + Z^{5/4}$.]

REMARKS. For α near 1, the estimates (10.91) and (10.93) are stronger than (10.72) in terms of Q^2, but not in terms of kT. In particular, if $kT \leqslant Q^\varepsilon$, then (10.93) yields

$$N(\alpha, k, Q, T) \ll Q^{4(1-\alpha)+\varepsilon}$$

in the range $\frac{11}{14} \leqslant \alpha \leqslant 1$, which is essentially the density conjecture. By using his Theorem 9.10, M. Jutila [**Jut**] established the density conjecture for zeros of $\zeta(s)$ in the same range. Every exponent $a(\alpha), b(\alpha), c(\alpha)$ takes its maximum at $\alpha = \frac{3}{4}$, being $a(\frac{3}{4}) = \frac{20}{9}, b(\frac{3}{4}) = \frac{16}{7}$, $c(\frac{3}{4}) = \frac{12}{5}$. For $\alpha = 1$ we get $a(1) = \frac{5}{4}$, $b(a) = \frac{4}{3}$, $c(1) = \frac{3}{2}$, and for $\alpha = \frac{1}{2}$ all three exponents take value 2.

10.5. The gaps between primes.

This section is written for the purpose of giving an example of how the zero-density theorems can be applied. We have chosen a few questions about primes in short intervals, because they were inspirational for the development of density theorems in the first place, and for other statistical studies of zeros of $\zeta(s)$. By no means do we attempt to show the best results from the vast literature.

By the Prime Number Theorem,

$$\psi(x) = \sum_{n \leqslant x} \Lambda(n) = x + E(x),$$

where $E(x)$ is a suitable error term, it follows directly that

$$\psi(x + y) - \psi(x) \sim y \tag{10.95}$$

as $x \to \infty$, provided $y = y(x)$ is somewhat larger than $E(x)$. Hence there is a prime number in the short interval $(x, x + y]$ for all sufficiently large x.

Without improving the existing error term in the Prime Number Theorem, G. Hoheisel [**Ho**] succeeded, nevertheless, in showing in 1930 that (10.95) holds for $y = x^\theta$ with some absolute constant $\theta < 1$. This is an impressive achievement given the fact that the error term as good as $E(x) = O(x^\theta)$ seems to be far beyond the current technology. Recall that the latter bound translates into the non-vanishing of $\zeta(s)$ in Re $s > \theta$. However, Hoheisel required only two results which were available in his time. First is a zero-free region of $\zeta(\sigma + it)$ of the type

$$|t| \leqslant T, \qquad \sigma \geqslant 1 - B\frac{\log\log T}{\log T} \tag{10.96}$$

for some constant $B > 0$ and all T sufficiently large. The second ingredient is the zero-density estimate of type

$$N(\alpha, T) \ll T^{c(1-\alpha)}(\log T)^A \tag{10.97}$$

for all $\frac{1}{2} \leqslant \alpha \leqslant 1$ with some constants $c \geqslant 2$ and $A \geqslant 1$. From these results one derives

THEOREM 10.5. *Put* $\theta = 1 - (c + (A+1)/B)^{-1}$. *Then*

$$\psi(x+y) - \psi(x) = y + O\Big(\frac{y}{\log x}\Big) \tag{10.98}$$

for all y *with* $x^\theta (\log x)^3 \leqslant y \leqslant x$.

PROOF. One uses the approximate "explicit formula" (see Section 5.9)

$$\psi(x) = x - \sum_{|\gamma| \leqslant T} \frac{x^\rho}{\rho} + O\Big(\frac{x}{T}(\log x)^2\Big)$$

with $T = x^{1-\theta}$. Hence

$$\frac{\psi(x+y) - \psi(x)}{y} - 1 = \sum_{|\gamma| \leqslant T} \frac{(x+y)^\rho - x^\rho}{\rho y} + O\Big(\frac{1}{\log x}\Big).$$

Here the sum over the zeros is bounded by

$$\begin{aligned}
\sum_{|\gamma| \leqslant T} x^{\beta - 1} &\leqslant 2 \int_{\frac{1}{2}}^{1} x^{\alpha - 1} dN(\alpha, T) \\
&\leqslant 2x^{-\frac{1}{2}} N(\tfrac{1}{2}, T) + 2(\log x) \int_{\frac{1}{2}}^{1} x^{\alpha-1} N(\alpha, T) d\alpha \\
&\ll x^{-\frac{1}{2}} T \log T + (\log x)(\log T)^A \int_{\eta}^{\frac{1}{2}} (T^c/x)^\alpha d\alpha \\
&\ll x^{-\frac{1}{2}} T \log T + (T^c/x)^\eta (\log T)^A
\end{aligned}$$

where $\eta = B(\log\log T)/\log T$. Since

$$(T^c/x)^\eta = (\log T)^{(c - \frac{1}{1-\theta})B} = (\log T)^{-A-1},$$

the sum over zeros is $O(1/\log x)$ completing the proof of Theorem 10.5. □

Hoheisel used the zero-free region (10.96) with a positive but very small constant B due to J. E. Littlewood [**Lit**], and the zero-density estimate (10.97) with exponent $c = 4$ due to F. Carlson [**Car**] (see (10.26)). After Vinogradov's widening the region (10.96) (see Corollary 8.28) one can take B arbitrarily large so $\theta = 1 - c^{-1} + \varepsilon$ satisfies the conditions of Theorem 10.5. From now on the Hoheisel exponent θ depends only on c in the zero-density estimate. Note that one needs (10.97) to hold throughout the segment $\frac{1}{2} \leqslant \alpha \leqslant 1$ with the same c, so an improvement in a subrange does not help. In other words, all one gets from (10.6) is $c = \max c(\alpha)$. By the Grand Density Theorem 10.4 for $\zeta(s)$ one gets (10.97) with $c = 12/5$ (the maximum of $c(\alpha)$ is attained at $\alpha = \frac{3}{4}$), which yields (10.98) for $x^\theta \leqslant y \leqslant x$ with $\theta = \frac{7}{12} + \varepsilon$. This is the best result of its kind obtained so far (in 1972 by M. N. Huxley [**Hu3**]). The density conjecture $c = 2$ yields $\theta = \frac{1}{2} + \varepsilon$.

Various combinations of the above analytic arguments with sieve methods produced estimates

$$y \ll \psi(x+y) - \psi(x) \ll y \tag{10.99}$$

in place of the asymptotic formula (10.95), however, for shorter intervals. For example, R. Baker and G. Harman [**BH**] got (10.99) for $y = x^\theta$ with $\theta = 0.534$.

From (10.99) with $y = x^\theta$ it follows that the difference between consecutive primes satisfies

$$d_n = p_{n+1} - p_n \ll p_n^\theta. \tag{10.100}$$

Therefore we have (10.100) with $\theta = 0.534$ as a consequence of the Baker-Harman work. Recall that the Riemann hypothesis yields $d_n \ll p_n^{\frac{1}{2}} \log p_n$, but even on the Pair Correlation Conjecture (see Chapter 25), the best that has been achieved is (due to D. Goldston and D. R. Heath-Brown [**GHB**])

$$d_n \ll (p_n \log p_n)^{\frac{1}{2}}. \tag{10.101}$$

Creating a probabilistic model for primes in 1937 H. Cramer [**Cra**] was led to a conjecture that

$$d_n \ll (\log p_n)^2. \tag{10.102}$$

In the other direction R. Rankin [**Ra2**] showed by constructing special composite numbers that

$$d_n \geqslant (e^\gamma - \varepsilon)(\log p_n)(\log\log p_n)(\log\log\log\log p_n)(\log\log\log p_n)^{-2} \tag{10.103}$$

infinitely often, where γ is the Euler constant. Paul Erdös offered a price of \$10,000 to anyone who can replace e^γ in (10.103) by a function increasing to infinity (the largest price ever offered by Erdös for a solution to a mathematical problem).

It is conjectured that the normalized gaps between consecutive primes $d_n{}^* = (p_{n+1} - p_n)(\log p_n)^{-1}$ has a Poisson distribution, that is for any $t > 0$,

$$\lim_{x\to\infty} \frac{1}{x} |\{n \leqslant x;\ d_n{}^* \leqslant t\}| = 1 - e^{-t}. \tag{10.104}$$

Many estimates for $\psi(x+y) - \psi(x)$ were established on average with respect to x.

EXERCISE 5. Assuming (10.97) with $c \geqslant 2$ prove that

$$\int_X^{2X} (\psi(x+y) - \psi(x) - y)^2 dx \ll y^2 X (\log X)^{-A} \tag{10.105}$$

for $X^\theta \leqslant y \leqslant X$ with $\theta = 1 - 2c^{-1} + \varepsilon$, for any $\varepsilon > 0$ and any $A > 0$, the implied constant depending only on ε and A. In particular, by taking $c = 12/5$ deduce that (10.95) holds true with $y = x^\theta$ for almost all x, where $\theta > \frac{1}{6}$ is a fixed number.

We close this section by pointing out a formal similarity between the problems of primes in short intervals and primes in arithmetic progressions to large moduli.

EXERCISE 6. Assume the following estimates for the zeros of Dirichlet L-functions with characters $\chi(\text{mod } q)$:

(1) $L(s,\chi) \neq 0$ in the region $s = \sigma + it$ with $|t| \leqslant T$ and

$$\sigma \geqslant 1 - B\frac{\log\log qT}{\log qT} \tag{10.106}$$

where B is a positive constant, provided qT is sufficiently large,

(2) The number $N(\alpha, q, T)$ of zeros of all $L(s,\chi)$ with $\chi(\text{mod } q)$ in the rectangle $|t| \leqslant T, \sigma \geqslant \alpha$ satisfies

$$N(\alpha, q, T) \ll (qT)^{c(1-\alpha)}(\log qT)^A \tag{10.107}$$

for $\frac{1}{2} \leqslant \alpha \leqslant 1, T \geqslant 3$, where $c \geqslant 2$ and $A \geqslant 1$ are suitable constants.

Prove that

$$\psi(x; q, a) = \frac{x}{\varphi(q)}\Big(1 + O\Big(\frac{1}{\log x}\Big)\Big) \tag{10.108}$$

uniformly for $x \geqslant q^\theta(\log q)^3$ with $\theta = c + (A+1)/B$. See Chapter 17 for unconditional results.

CHAPTER 11

SUMS OVER FINITE FIELDS

11.1. Introduction.

In this chapter we consider a special type of exponential and character sums, called sometimes "complete sums", which can be seen as sums over the elements of a finite field. Although the methods of Chapter 8 can still be applied to the study of such sums, disregarding this special feature, the deepest understanding and the strongest results are obtained when the finite field aspect is taken into account and the powerful techniques of algebraic geometry are brought to bear.

We have already encountered in the previous chapters some examples of exponential sums which can be interpreted as sums over finite fields, for example, the quadratic Gauss sums

$$G_a(p) = \sum_{x \bmod p} \Big(\frac{x}{p}\Big) e\Big(\frac{ax}{p}\Big)$$

or the Kloosterman sums (1.56)

$$S(a,b;p) = \sideset{}{^\star}\sum_{x \bmod p} e\Big(\frac{ax+b\bar{x}}{p}\Big).$$

In this chapter we will study these sums in particular. The culminating point of our presentation is the elementary method of Stepanov which we apply for proving Weil's bound for Kloosterman sums

$$|S(a,b;p)| \leqslant 2\sqrt{p}$$

and Hasse's bound for the number of points of an elliptic curve over a finite field. Then we survey briefly, without proofs, the powerful formalism of ℓ-adic cohomology developed by Grothendieck, Deligne, Katz, Laumon and others, hoping to convey a flavor of the tools involved and to give the reader enough knowledge to make at least a preliminary analysis of any exponential sum he or she may encounter in analytic number theory.

11.2. Finite fields.

We first recall briefly some facts about finite fields, and establish the notations used in this chapter. For every prime p, the finite ring $\mathbb{Z}/p\mathbb{Z}$ of residue classes modulo p is a field, which we denote $\mathbb{F}_p$. The Galois theory of $\mathbb{F}_p$ is very easy to describe: for any $n \geqslant 1$, there exists a unique (up to isomorphism) field extension of $\mathbb{F}_p$ of degree n, written $\mathbb{F}_{p^n}$. Conversely, any finite field $\mathbb{F}$ with q elements is isomorphic (but not canonically) to a unique field $\mathbb{F}_{p^d}$, so $q = p^d$, and $\mathbb{F}$ admits also a unique finite extension of degree n for any $n \geqslant 1$, namely $\mathbb{F}_{p^{dn}}$.

Let now $\mathbb{F} = \mathbb{F}_q$ be a finite field with $q = p^d$ elements. In most of the chapter, p is fixed and we change notation slightly, denoting by $\overline{\mathbb{F}}$ an algebraic closure of $\mathbb{F}$

and by $\mathbb{F}_n \subset \bar{\mathbb{F}}$ the unique extension of degree n of $\mathbb{F}$ for $n \geqslant 1$. The context will always indicate clearly that the cardinality of $\mathbb{F}_n$ is q^n and not n.

The extension $\mathbb{F}_n/\mathbb{F}$ is a Galois extension, with Galois group G_n canonically isomorphic to $\mathbb{Z}/n\mathbb{Z}$, the isomorphism being the map $\mathbb{Z}/n\mathbb{Z} \to G_n$ defined by $1 \mapsto \sigma$, where σ is the Frobenius automorphism of $\mathbb{F}_n$ given by $\sigma(x) = x^q$.

Let $\bar{\mathbb{F}}$ be a given algebraic closure of $\mathbb{F}$, so by the above,

$$\bar{\mathbb{F}} = \bigcup_{n \geqslant 1} \mathbb{F}_n.$$

By Galois theory, for any $x \in \bar{\mathbb{F}}$, we have $x \in \mathbb{F} \iff \sigma(x) = x$ if and only if $x^q = x$ and more generally

$$x \in \mathbb{F}_n \iff \sigma^n(x) = x \iff x^{q^n} = x. \tag{11.1}$$

From this we can deduce that $\mathbb{F}_n$ is the splitting field of the polynomial $X^{q^n} - X \in \mathbb{F}[X]$. More precisely, one can state the following result of Gauss:

LEMMA 11.1. *For any integer $n \geqslant 1$, we have*

$$\prod_{d \mid n} \prod_{\deg(P) = d} P = X^{q^n} - X \tag{11.2}$$

where the product ranges over all irreducible monic polynomials P of degree d dividing n.

PROOF. This is an immediate consequence of the description of finite fields: the roots of the polynomial on the right side (in an algebraic closure) are exactly the elements $x \in \mathbb{F}_n$ with multiplicity one and, conversely, every such x has a minimal polynomial which must occur, exactly once, among the polynomials P on the left side. □

Associated to the extension $\mathbb{F}_n/\mathbb{F}$ are the trace map and the norm map. Because of the above description of the Galois group of $\mathbb{F}_n/\mathbb{F}$, the trace map $\mathrm{Tr} = \mathrm{Tr}_{\mathbb{F}_n/\mathbb{F}} : \mathbb{F}_n \to \mathbb{F}$ is given by

$$\mathrm{Tr}\,(x) = \sum_{0 \leqslant i \leqslant n-1} \sigma^i(x) = \sum_{0 \leqslant i \leqslant n-1} x^{q^i} \tag{11.3}$$

while the norm map $N = N_{\mathbb{F}_n/\mathbb{F}} : \mathbb{F}_n^* \to \mathbb{F}^*$ is similarly

$$N(x) = \prod_{0 \leqslant i \leqslant n-1} \sigma^i(x) = \prod_{0 \leqslant i \leqslant n-1} x^{q^i} = x^{\frac{q^n - 1}{q-1}}. \tag{11.4}$$

The equations $\mathrm{Tr}\,(x) = y$ and $N(x) = y$, for a fixed $y \in \mathbb{F}$ are very important. Because the extension $\mathbb{F}_n/\mathbb{F}$ is separable, the equation $\mathrm{Tr}\,(x) = y$ always has a solution. If x_0 is a given solution, then all solutions are in one-to-one correspondence with solutions of $\mathrm{Tr}\,(a) = 0$, by $x = x_0 + a$. Moreover, any solution of $\mathrm{Tr}\,(a) = 0$ is of the form $a = \sigma(b) - b = b^q - b$ for some $b \in \mathbb{F}_n$, unique up to addition of an element in $\mathbb{F}$.

Similarly, for any $y \in \mathbb{F}^*$, the equation $N(x) = y$ has a solution, and if x_0 is a given solution, the set of solutions is in one-to-one correspondence with solutions of $N(a) = 1$, which by Hilbert's Theorem 90 (or by direct proof) are all given by

$a = \sigma(b)b^{-1} = b^{q-1}$ for some $b \in \mathbb{F}_n^*$, unique up to multiplication by an element in $\mathbb{F}^*$.

As the additive group of $\mathbb{F}$ is finite, the general theory of characters of a finite abelian group (see Chapter 3) can be applied. Characters of $\mathbb{F}$ are called additive characters, and they are all of the form $x \mapsto \psi(ax)$ for some $a \in \mathbb{F}$, where ψ is some fixed non-trivial additive character. For instance, let $\mathrm{Tr} : \mathbb{F} \to \mathbb{Z}/p\mathbb{Z}$ be the trace map to the base-field, then

$$\psi(x) = e(\mathrm{Tr}\,(x)/p) \tag{11.5}$$

is a non-trivial additive character of $\mathbb{F}$. For a given additive character ψ and $a \in \mathbb{F}$, we denote by ψ_a the character $x \mapsto \psi(ax)$.

Applying the general theory of characters of finite abelian groups, we get the orthogonality relations

$$\sum_{\psi} \psi(x) = \begin{cases} q & \text{if } x = 1, \\ 0 & \text{otherwise} \end{cases}$$

(which is used to "solve" the equation $x = 0$ in $\mathbb{F}$) and

$$\sum_{x \in \mathbb{F}} \psi(x) = \begin{cases} q & \text{if } \psi = 1 \text{ is the trivial character,} \\ 0 & \text{otherwise.} \end{cases}$$

The description of characters of the multiplicative group $\mathbb{F}^*$ (also called multiplicative characters of $\mathbb{F}$) is not so explicit. The group structure of $\mathbb{F}^*$ is well-known (dating back to Gauss): it is a cyclic group of order $q-1$. Generators of $\mathbb{F}^*$ are called primitive roots, and there are $\varphi(q-1)$ of them, but no useful formula for a primitive root exists. Fixing one, say $z \in \mathbb{F}^*$, one has an isomorphism

$$\log : \begin{cases} \mathbb{F}^* \simeq \mathbb{Z}/(q-1)\mathbb{Z} \\ x \mapsto n \text{ such that } z^n = x \end{cases}$$

and all multiplicative characters of $\mathbb{F}$ are expressed as

$$\chi(x) = e\Big(\frac{a \log(x)}{q-1}\Big)$$

for some $a \in \mathbb{Z}/(q-1)\mathbb{Z}$, but such a description is usually of no use in analytic number theory.

As examples of multiplicative characters, suppose $\mathbb{F} = \mathbb{Z}/p\mathbb{Z}$ and $p \neq 2$. Then the Legendre symbol

$$x \mapsto \Big(\frac{x}{p}\Big)$$

is a non-trivial quadratic character. In general, if $\delta \mid (q-1)$, there is a cyclic group of order δ consisting of characters χ of $\mathbb{F}^*$ of order δ.

The orthogonality relations become

$$\sum_{\chi} \chi(x) = \begin{cases} q-1 & \text{if } x = 1, \\ 0 & \text{otherwise,} \end{cases}$$

(the sum over all multiplicative characters), and

$$\sum_{x\in\mathbb{F}^*}\chi(x)=\begin{cases} q-1 & \text{if } \chi=1 \text{ is the trivial character,}\\ 0 & \text{otherwise.}\end{cases}$$

It is usual to extend multiplicative characters to $\mathbb{F}$ by defining $\chi(0)=0$ if $\chi\neq 1$, and $\chi(0)=1$ if $\chi=1$. Notice then that for any $\delta\mid q-1$ the formula

$$\sum_{\chi^\delta=1}\chi(x)=|\{y\in\mathbb{F}\mid y^\delta=x\}| \tag{11.6}$$

(also a particular case of the orthogonality relations for the group $\mathbb{F}^*/(\mathbb{F}^*)^d$, as described in Chapter 3) is true for all $x\in\mathbb{F}$.

11.3. Exponential sums.

Let $\mathbb{F}=\mathbb{F}_q$ be a finite field with $q=p^m$ elements, p a prime. Exponential sums over $\mathbb{F}$ can be of various kinds. For the simplest case, consider a polynomial $P\in\mathbb{F}[X]$ and an additive character ψ, and define the sum

$$S(P)=\sum_{x\in\mathbb{F}}\psi(P(x)).$$

Slightly more generally, take a non-zero rational function $f=P/Q\in\mathbb{F}(X)$ and consider

$$S(f)=\sum_{\substack{x\in\mathbb{F}\\ Q(x)\neq 0}}\psi(f(x));$$

for instance, taking $q=p$ and $f(x)=ax+bx^{-1}$, we have $S(f)=S(a,b;p)$. Multiplicative characters can also be used, getting sums of the type

$$S_\chi(f)=\sum_{x\in\mathbb{F}}{}^{\star}\,\chi(f(x))$$

(where the star in $\sum^\star$ means here and henceforth that the summation extends to all x which are not poles of f). For $q=p$, $\chi=\left(\frac{\cdot}{p}\right)$ (the Legendre symbol) and $f(x)\in\mathbb{Z}[X]$ a cubic polynomial without multiple roots modulo p, we see that $-S_\chi(f)$ is the p-th coefficient a_p of the Hasse-Weil zeta function of the elliptic curve with equation $y^2=f(x)$ (see Section 14.4).

Still more generally, one can mix additive and multiplicative characters, and define sums such as

$$S_\chi(f,g)=\sum_{x\in\mathbb{F}}{}^{\star}\,\chi(f(x))\psi(g(x)), \tag{11.7}$$

an example of which is the Salié sum $T(a,b;p)$ defined by

$$T(a,b;p)=\sum_{x \bmod p}{}^{\star}\,\left(\frac{x}{p}\right)e\left(\frac{ax+b\bar{x}}{p}\right)$$

which occurs in the Fourier expansion of half-integral weight modular forms; see [**I6**] for instance. In contrast with the seemingly simpler Kloosterman sums $S(a,b;p)$, the Salié sums $T(a,b;p)$ can be explicitly computed (see Lemma 12.4, and Corollary 21.9 for the uniform distribution of the "angles" of the Salié sums).

To end this list, we mention that all these definitions can again be generalized to sums in more than one variable, and that the summation variables can be restricted to the rational points of an algebraic variety defined over $\mathbb{F}$: some examples will appear in the survey sections of this chapter.

The exponential sums which directly arise in analytic number theory are sums over the prime field $\mathbb{Z}/p\mathbb{Z}$. However, the deeper understanding naturally requires considering sums over the extension fields $\mathbb{F}_{p^n}$. Indeed, the very reason for the success of algebraic methods lies in the fact that an exponential sum over $\mathbb{F}_p$ doesn't really come alone, but has natural "companions" over all the extension fields $\mathbb{F}_{p^n}$, and it is really the whole family which is investigated and which is the natural object of study. Those companion sums are easily defined: take the most general sum $S = S_\chi(f, g)$ we have introduced, then for $n \geqslant 1$ let

$$S_n = \sum_{x \in \mathbb{F}_n}^{\star} \chi(N_{\mathbb{F}_n/\mathbb{F}}(f(x)))\psi(\operatorname{Tr}_{\mathbb{F}_n/\mathbb{F}}(g(x))) \tag{11.8}$$

where we use the multiplicative character $\chi \circ N$ and the additive character $\psi \circ \operatorname{Tr}$ of $\mathbb{F}_n$. All the sums S_n are incorporated into a single object, the zeta function of the exponential sum, which is the formal power series $Z = Z_\chi(f, g) \in \mathbb{C}[[T]]$ defined by the formula

$$Z = \exp\Bigl(\sum_{n \geqslant 1} \frac{S_n}{n} T^n\Bigr).$$

Justification for the introduction of the zeta function comes from the following rationality theorem, conjectured by Weil, and proved by Dwork.

THEOREM 11.2 (DWORK). *The zeta function Z is the power series expansion of a rational function; more precisely, there exist coprime polynomials P and Q in $\mathbb{C}[T]$, with $P(0) = Q(0) = 1$, such that $Z = \frac{P}{Q}$.*

As a corollary, denote by (α_i) (resp. (β_j)) the inverse of the roots (with multiplicity) of P (resp. Q), so

$$P = \prod_i (1 - \alpha_i T), \quad Q = \prod_j (1 - \beta_j T).$$

Then using the power-series expansion

$$\log \frac{1}{1 - T} = \sum_{n \geqslant 1} \frac{T^n}{n}$$

we find that the formula $Z = P/Q$ is equivalent to the formula

$$S_n = \sum_j \beta_j^n - \sum_i \alpha_i^n$$

for any $n \geqslant 1$, which shows how the various sums S_n are related. In particular, note that they satisfy a linear recurrence relation of order d equal to the number $\deg P + \deg Q$ of roots α_i, β_j.

COROLLARY 11.3. *We have for any $n \geqslant 1$ the upper bound*

$$|S_n| \leqslant \sum_j |\beta_j|^n + \sum_i |\alpha_i|^n.$$

In particular,

$$|S| \leqslant \sum_j |\beta_j| + \sum_i |\alpha_i|. \tag{11.9}$$

A common abuse of language is to speak of the α_i and β_j as the roots of the exponential sum S. We will describe in Section 11.11 a number of general facts about these roots. In the intervening sections, we will prove Dwork's Theorem and estimate the modulus of the roots in the important special cases of Gauss sums, Kloosterman sums and for the local zeta function of elliptic curves.

REMARKS. We have called the sums S_n, $n \geqslant 1$, "companions" of the original exponential sum S. However, one can consider other companions of S as well. If S involves an additive character, it can also be very useful sometimes to consider S as just one element of the family of sums obtained by varying the additive character, specifically if

$$S = \sum_{x \in \mathbb{F}}{}^{\star} \chi(f(x))\psi(g(x)),$$

we also introduce for $a \in \mathbb{F}$,

$$S_a = \sum_{x \in \mathbb{F}}{}^{\star} \chi(f(x))\psi_a(g(x)) = \sum_{x \in \mathbb{F}}{}^{\star} \chi(f(x))\psi(ag(x)).$$

Estimates on average over a for the first few power moments of S_a are often easily derived by elementary means, and they can be of great use in estimating S, even in addition to the methods of algebraic geometry. See the proof of Weil's bound for Kloosterman sums in Section 11.7 and the examples in Section 11.11. More general types of families have been (and still are) extensively studied by Katz; see for instance [**K1**].

11.4. The Hasse-Davenport relation.

We consider general Gauss sums over a finite field. Let $\mathbb{F} = \mathbb{F}_q$ be a finite field with $q = p^m$ elements, and let ψ be an additive character and χ a multiplicative character of $\mathbb{F}$. The Gauss sum $G(\chi, \psi)$ is

$$G(\chi, \psi) = \sum_{x \in \mathbb{F}} \chi(x)\psi(x). \tag{11.10}$$

(recall that χ is extended to $\mathbb{F}$ by $\chi(0) = 1, 0$ according to whether χ is trivial or not). When χ is the Legendre symbol, one recovers quadratic Gauss sums.

The associated sums over the extensions fields are

$$G_n(\chi, \psi) = \sum_{x \in \mathbb{F}_n} \chi(N_{\mathbb{F}_n/\mathbb{F}}(x))\psi(\mathrm{Tr}_{\mathbb{F}_n/\mathbb{F}}(x))$$

and the zeta function is

$$Z(\chi, \psi) = \exp\Bigl(\sum_{n \geqslant 1} \frac{G_n(\chi, \psi)}{n} T^n\Bigr). \tag{11.11}$$

In this case, Dwork's Theorem was proved by Hasse and Davenport and is known as the Hasse-Davenport Relation.

THEOREM 11.4 (HASSE-DAVENPORT). *Assume χ and ψ are non-trivial. Then we have for any $n \geqslant 1$,*

$$-G_n(\chi,\psi) = (-G(\chi,\psi))^n$$

or equivalently the zeta function is a linear polynomial

$$Z(\chi,\psi) = 1 + G(\chi,\psi)T.$$

Hence the only "root" for the Gauss sum is $G(\chi,\psi)$ itself. This can be estimated elementarily, as was done for the Gauss sums considered in Chapter 3.

PROPOSITION 11.5. *We have*

$$|G(\chi,\psi)| = \sqrt{q}$$

if neither χ nor ψ is trivial, while

$$|G(1,\psi)| = \begin{cases} 0 & \text{if } \psi \text{ non-trivial,} \\ q & \text{if } \psi = 1 \end{cases}$$

$$|G(\chi,1)| = \begin{cases} 0 & \text{if } \chi \text{ non-trivial,} \\ q & \text{if } \chi = 1. \end{cases}$$

PROOF. The last two statements are immediate, so assume neither χ nor ψ is trivial. We have

$$\begin{aligned} |G(\chi,\psi)|^2 &= \sum_{x,y\in\mathbb{F}^*} \chi(x)\bar{\chi}(y)\psi(x)\psi(-y) \\ &= \sum_{z\in\mathbb{F}} \chi(z) \sum_{y\in\mathbb{F}^*} \psi((z-1)y) \text{ (on writing } z = xy^{-1}) \\ &= q \text{ (by orthogonality, applied twice.)} \end{aligned}$$

□

We now turn to the proof of the Hasse-Davenport Relation. We consider the field $F = \mathbb{F}(X)$ of rational functions on $\mathbb{F}$ and the ring $R = \mathbb{F}[X]$ of polynomials. Recall that R is a principal ideal domain. For $h \in R$ of degree $d \geqslant 0$, we define the norm

$$N(h) = q^d.$$

The zeta function of F is the Dirichlet series (analogous to the Riemann zeta function)

$$\zeta_F(s) = \sum_{\substack{h\in R \\ h \text{ monic}}} N(h)^{-s}.$$

REMARK. This could also be written as a sum over the non-zero ideals $\mathfrak{a}$ in R,

$$\zeta_F(s) = \sum_{\mathfrak{a}} N(\mathfrak{a})^{-s}$$

where $N(\mathfrak{a}) = |R/\mathfrak{a}| = N(h)$ for any polynomial h such that $\mathfrak{a} = (h)$. But we will work with polynomials to emphasize the elementary spirit here.

The Dirichlet series $\zeta_F(s)$ converges absolutely for $\operatorname{Re}(s) > 1$. Indeed, putting

$$n(d) = \{h \in R \mid \deg(h) = d,\ h \text{ is monic}\} = q^d$$

we obtain immediately

$$\zeta_F(s) = \sum_{d \geqslant 0} n(d) q^{-ds} = \sum_{d \geqslant 0} q^{(1-s)d} = (1 - q^{1-s})^{-1}.$$

On the other hand, unique factorization into irreducible polynomials yields an expression of ζ_F as an Euler product

$$\zeta_F(s) = \prod_{\substack{P \in R \\ P \text{ monic irreducible}}} (1 - N(P)^{-s})^{-1}$$

which is convergent for $\operatorname{Re}(s) > 1$.

The first step in the proof of the Hasse-Davenport Relation consists of writing the zeta function of Gauss sums as an L-function for the field F. Let $H \subset F^*$ be the subgroup of rational functions which are quotients of monic polynomials, and $G \subset H$ a subgroup with the property

$$h_1 h_2 \in G \Rightarrow h_1,\ h_2 \in G.$$

Then if $\alpha : G \to \mathbb{C}^*$ is a character of the group G, it can be extended to a totally multiplicative function of the set of monic polynomials $h \in R$ by putting $\alpha(h) = 0$ if $h \notin G$. The corresponding L-function is defined analogously to the classical L-functions by the Dirichlet series

$$L(s, \alpha) = \sum_{\substack{h \in R \\ h \text{ monic}}} \alpha(h) N(h)^{-s} = \prod_P (1 - \alpha(P) N(P)^{-s})^{-1}$$

for $\operatorname{Re}(s) > 1$.

For dealing with Gauss sums we consider the subgroup $G \subset H$ of rational functions f defined and non-vanishing at 0. Define a character λ on G by

$$\lambda(h) = \chi(a_d)\psi(a_1)$$

for $h = X^d - a_1 X^{d-1} + \cdots + (-1)^d a_d \in R$. Clearly λ is multiplicative on monic polynomials, and extends to a character of G. In this case we get the following:

LEMMA 11.6. *We have* $L(s, \lambda) = 1 + G(\chi, \psi) q^{-s}$.

PROOF. We arrange the Dirichlet series for $L(s, \lambda)$ according to the degree of h:

$$L(s, \lambda) = \sum_{d \geqslant 0} \Big(\sum_{\deg(h) = d} \lambda(h) \Big) q^{-ds}$$

and evaluate each term in turn. For $d = 0$, the only monic polynomial occurring is $h = 1$, and $\lambda(1) = 1$. For $d = 1$, we have $h = X - a$ so that

$$\sum_{\deg(h) = 1} \lambda(h) = \sum_{a \in \mathbb{F}} \lambda(X - a) = \sum_{a \in \mathbb{F}} \chi(a)\psi(a) = G(\chi, \psi).$$

For any $d \geqslant 2$ we have

$$\sum_{\deg(h)=d} \lambda(h) = \sum_{a_1,\dots,a_d \in \mathbb{F}} \lambda(X^d - a_1 X^{d-1} + \cdots + (-1)^d a_d)$$
$$= q^{d-2} \sum_{a_1, a_d \in \mathbb{F}} \chi(a_d)\psi(a_1) = 0$$

by orthogonality, because at least one of the characters χ, ψ is non-trivial. □

On the other hand, appealing to the Euler product we will prove:

LEMMA 11.7. *We have $L(s, \lambda) = Z(q^{-s})$, where $Z = Z(\chi, \psi)$ is the zeta function (11.11) associated with Gauss sums.*

Theorem 11.4 follows from Lemmas 11.6 and 11.7.

PROOF OF LEMMA 11.7. Taking the logarithmic derivative of the Euler product, we get

$$-\frac{1}{\log q}\frac{L'(s,\lambda)}{L(s,\lambda)} = \sum_P \deg(P) \sum_{r \geqslant 1} \lambda(P)^r q^{-rds}$$
$$= \sum_{n \geqslant 1} \Big(\sum_{rd=n} d \sum_{\substack{P \\ \deg(P)=d}} d\lambda(P)^r \Big) q^{-ns}$$

while, on the other hand,

$$-\frac{1}{\log q}\frac{Z'(q^{-s})}{Z(q^{-s})} = \sum_{n \geqslant 1} G_n(\chi, \psi) q^{-ns}.$$

It therefore suffices to prove the formula

$$\sum_{\substack{P \\ d=\deg(P)|n}} d\lambda(P)^{n/d} = G_n(\chi, \psi) \tag{11.12}$$

for $n \geqslant 1$, the equality of the logarithmic derivatives being sufficient to imply Lemma 11.7 since both sides are Dirichlet series with leading coefficient 1.

To prove (11.12), let P be one of the irreducible polynomials appearing on the left side, of degree $d \mid n$. Its roots, say $x_1,\dots, x_d$, are in $\mathbb{F}_n$. Fix one root $x = x_j$ and write

$$P = X^d - a_1 X^{d-1} + \cdots + (-1)^d a_d.$$

We get

$$N(x) = (N_{\mathbb{F}_d/\mathbb{F}}(x))^{n/d} = a_d^{n/d},$$
$$\mathrm{Tr}\,(x) = \frac{n}{d}\mathrm{Tr}_{\mathbb{F}_d/\mathbb{F}}(x) = \frac{n}{d}a_1$$

hence

$$\lambda(P)^{n/d} = (\chi(a_d)\psi(a_1))^{n/d} = \chi(a_d^{n/d})\psi\Big(\frac{n}{d}a_1\Big) = \chi(N(x))\psi(\mathrm{Tr}\,(x)).$$

Summing over all roots of P we derive

$$d\lambda(P)^{n/d} = \sum_{i=1}^{d} \chi(N(x_i))\psi(\mathrm{Tr}\,(x_i)),$$

and summing over all P with $\deg(P) \mid n$, we get (11.12) by Lemma 11.1 since every element in $\mathbb{F}_n$ will appear exactly once as one of the roots x_i for some P. □

11.5. The zeta function for Kloosterman sums.

Next we consider Kloosterman sums. Let $\mathbb{F}$ be a finite field with $q = p^m$ elements and this time consider additive characters ψ and φ. We define the Kloosterman sum associated to ψ and φ by

$$S(\psi, \varphi) = -\sum_{x\in\mathbb{F}^*} \psi(x)\varphi(x^{-1}), \tag{11.13}$$

(the minus factor is only for cosmetic reasons). When $q = p$ is prime, and $\psi(x) = e(ax/p)$, $\varphi(x) = e(bx/p)$, we have therefore $S(\psi, \varphi) = -S(a, b; p)$.

The companion sums over the extension fields $\mathbb{F}_n$ are

$$S_n(\psi, \varphi) = -\sum_{x\in\mathbb{F}_n^*} \psi(\mathrm{Tr}\,(x))\varphi(\mathrm{Tr}\,(x^{-1}))$$

and the Kloosterman zeta function is

$$Z = Z(\psi, \varphi) = \exp\Big(\sum_{n\geqslant 1} \frac{S_n(\psi, \varphi)}{n} T^n\Big).$$

We will prove Dwork's Theorem in this case, which is due to Carlitz.

THEOREM 11.8. *Assume that ψ and φ are both non-trivial. Then*

$$Z(\psi, \varphi) = \frac{1}{1 - S(\psi, \varphi)T + qT^2}.$$

The proof is very similar to that of Theorem 11.4. We put $R = \mathbb{F}[X]$, $F = \mathbb{F}(X)$ as before, and consider the same group $G \subset F^*$ of quotients of monic polynomials defined and non-vanishing at 0. We define a character $\eta : G \to \mathbb{C}^*$ by putting

$$\eta(h) = \psi(a_1)\varphi(a_{d-1}/a_d)$$

for a monic polynomial $h \in G$, where we write (compare the previous section)

$$h = X^d + a_1X^{d-1} + \cdots + a_{d-1}X + a_d$$

(with $a_d \neq 0$ since $h \in G$). The following computation verifies that η is indeed a character of G: let $h' = X^e + b_1X^{e-1} + \cdots + b_{e-1}X + b_e$ with $b_e \neq 0$, then

$$hh' = X^{d+e} + (a_1 + b_1)X^{d+e-1} + \cdots + (a_{d-1}b_e + a_db_{e-1})X + a_db_e$$

and

$$\begin{aligned}\eta(hh') &= \psi(a_1 + b_1)\varphi\Big(\frac{a_{d-1}b_e + a_db_{e-1}}{a_db_e}\Big)\\ &= \psi(a_1)\varphi(a_{d-1}/a_d)\psi(b_1)\varphi(b_{e-1}/b_e)\\ &= \eta(h)\eta(h').\end{aligned}$$

Recall that we extend η to all $h \in R$ by putting $\eta(h) = 0$ for $h \notin G$.

LEMMA 11.9. *For ψ and ψ non-trivial, the L-function associated to η is given by*

$$L(s, \eta) = \sum_{h} \eta(h) N(h)^{-s} = 1 - S(\psi, \varphi) q^{-s} + q^{1-2s}.$$

PROOF. By arranging terms according to the degree of h, we write

$$L(s, \eta) = \sum_{d \geqslant 0} \Big(\sum_{\deg(h)=d} \eta(h) \Big) q^{-ds}$$

and evaluate the inner sums. For $d = 0$, we have only $h = 1$ and $\eta(1) = 1$. For $d = 1$, we have $h = X + a$ with $a \neq 0$, hence

$$\sum_{\deg(h)=1} \eta(h) = \sum_{a \in \mathbb{F}^*} \eta(X + a) = \sum_{a \in \mathbb{F}^*} \psi(a) \varphi(a^{-1}) = -S(\psi, \varphi).$$

For $d = 2$, we get

$$\begin{aligned}
\sum_{\deg(h)=2} \eta(h) &= \sum_{\substack{a \in \mathbb{F} \\ b \in \mathbb{F}^*}} \eta(X^2 + aX + b) = \sum_{\substack{a \in \mathbb{F} \\ b \in \mathbb{F}^*}} \psi(a) \varphi(ab^{-1}) \\
&= q - 1 + \Big(\sum_{a \in \mathbb{F}^*} \psi(a) \Big) \Big(\sum_{b \in \mathbb{F}^*} \varphi(b) \Big) = q
\end{aligned}$$

by applying twice the orthogonality of characters, since neither ψ nor φ is trivial.
Finally, for $d \geqslant 3$, we get

$$\begin{aligned}
\sum_{\deg(h)=d} \eta(h) &= \sum_{a \in \mathbb{F}^*} \sum_{a_1, \ldots, a_{d-1} \in \mathbb{F}} \eta(X^d + a_1 X^{d-1} + \cdots + a_{d-1} X + a) \\
&= q^{d-3} \sum_{\substack{a_1, a_{d-1} \in \mathbb{F} \\ a \in \mathbb{F}^*}} \psi(a_1) \varphi(a_{d-1} a^{-1}) = 0
\end{aligned}$$

since there is free summation over $a_1 \in \mathbb{F}$. □

LEMMA 11.10. *For ψ and φ non-trivial, we have the identity*

$$Z(\psi, \varphi)(q^{-s}) = L(s, \eta)^{-1} = \frac{1}{1 - S(\psi, \varphi) q^{-s} + q^{1-2s}}.$$

This lemma completes the proof of Theorem 11.8.

PROOF. The L-function has an Euler product

$$L(s, \eta) = \prod_{P} (1 - \eta(P) N(P)^{-s})^{-1}.$$

Taking the logarithmic derivative we get

$$\begin{aligned}
-\frac{1}{\log q} \frac{L'(s, \eta)}{L(s, \eta)} &= \sum_{P} \deg(P) \sum_{r \geqslant 1} \eta(P)^r q^{-r \deg(P) s} \\
&= \sum_{n \geqslant 1} \Big(\sum_{rd=n} d \sum_{\deg(P)=r} \eta(P)^r \Big) q^{-ns}
\end{aligned}$$

and as before it suffices to prove the formula

$$\sum_{d=\deg(P)\mid n} d\eta(P)^{n/d} = -S_n(\psi,\varphi) \tag{11.14}$$

for $n \geqslant 1$. Let

$$P = X^d + a_1X^{d-1} + \cdots + a_{d-1}X + a_d$$

be one of the irreducible polynomials on the left side of (11.14), of degree $d \mid n$, and $x_1,\ldots, x_d$ its roots, which lie in $\mathbb{F}_d$. We have for each i,

$$\mathrm{Tr}\,(x_i) = \frac{n}{d}\mathrm{Tr}_{\mathbb{F}_d/\mathbb{F}}(x_i) = -\frac{n}{d}a_1$$

(since $a_d^{-1}X^dP(X^{-1}) = X^d + \frac{a_{d-1}}{a_d}X^{d-1} + \cdots + a_d^{-1}$) and

$$\mathrm{Tr}\,(x_i^{-1}) = \frac{n}{d}\mathrm{Tr}_{\mathbb{F}_d/\mathbb{F}}(x_i^{-1}) = -\frac{n}{d}\frac{a_{d-1}}{a_d}.$$

Hence

$$\eta(P)^{n/d} = \psi\Big(\frac{n}{d}a_1\Big)\varphi\Big(\frac{n}{d}\frac{a_{d-1}}{a_d}\Big) = \psi(-x_i)\varphi(-x_i^{-1})$$

and summing over the roots x_i, then over the polynomials P of degree $d \mid n$, we obtain (11.14) by Gauss's Lemma again. □

Theorem 11.8 allows us to factor the Kloosterman zeta function

$$Z(\psi,\varphi) = (1 - S(\psi,\varphi)T + qT^2)^{-1} = (1-\alpha T)^{-1}(1-\beta T)^{-1}$$

where α and β are complex numbers, and of course $\alpha+\beta = S(\psi,\varphi)$, $\alpha\beta = q$. In sharp contrast to the case of Gauss sums, however, the roots α and β cannot be explicitly computed.

THEOREM 11.11 (WEIL). *Assume that ψ and φ are non-trivial and $p \neq 2$. Then the roots α and β for the Kloosterman sum $S(\psi,\varphi)$ satisfy $|\alpha| = |\beta| = \sqrt{q}$, and therefore we have*

$$|S(\psi,\varphi)| \leqslant 2\sqrt{q}. \tag{11.15}$$

We will prove Theorem 11.11 in the next two sections.

COROLLARY 11.12. *Let a, b, c be integers, c positive. We have*

$$|S(a,b;c)| \leqslant \tau(c)(a,b,c)^{1/2}c^{1/2}. \tag{11.16}$$

PROOF. By the twisted multiplicativity (1.59) for Kloosterman sums, it suffices to consider $c = p^\nu$ with p prime and $\nu \geqslant 1$. If $p \mid nm$, we have Ramanujan sums for which the result is easy (see (3.2), (3.3)). Otherwise, the case $\nu = 1$ follows from Theorem 11.11 for $p \geqslant 3$, and for $p = 2$ one checks immediately that the Kloosterman sums modulo 2 satisfy Theorem 11.11: we have $S(1,1;2) = 1$, and the associated zeta function is therefore $Z(T) = 1+T+2T^2$, with roots $(-1\pm i\sqrt{7})/4$ of modulus $1/\sqrt{2}$.

The case $p \nmid nm$ and $\beta \geqslant 2$ can be dealt with elementarily; see Exercise 1 of Chapter 12. □

EXERCISE 1. Consider a general Kloosterman-Salié sum

$$S(\chi;\psi,\varphi) = -\sum_{x\in\mathbb{F}} \chi(x)\psi(x)\varphi(x^{-1})$$

and its associated companions S_n and zeta function Z, where ψ and φ are additive characters of $\mathbb{F}$ and χ is multiplicative (so $\chi = 1$ is the case of Kloosterman sums). Show that

$$Z = (1 - S(\chi;\psi,\varphi)q^{-s} + \bar{\chi}(-a)\chi(b)q^{1-2s})^{-1}$$

where

$$\psi(x) = e\Big(\frac{\mathrm{Tr}\,(ax)}{p}\Big), \quad \varphi(x) = e\Big(\frac{\mathrm{Tr}\,(bx)}{p}\Big).$$

11.6. Stepanov's method for hyperelliptic curves.

We will prove Theorem 11.11 by deducing it from the Riemann Hypothesis for certain algebraic curves over finite fields. However, we use Stepanov's elementary method (see [**Ste**], [**Sch**], [**Bo3**]) instead of Weil's arguments.

Let $\mathbb{F}$ be a finite field with q element, of characteristic p. We will only consider algebraic curves C_f over $\mathbb{F}$ given by equations of the type

$$C_f \,:\, y^2 = f(x) \tag{11.17}$$

for some polynomial $f \in \mathbb{F}[X]$ of degree $m \geqslant 3$. We assume moreover the following condition

(11.18) The polynomial $Y^2 - f(X) \in \mathbb{F}[X,Y]$ is absolutely irreducible

(i.e. it is irreducible over the algebraic closure of $\mathbb{F}$). This is a minimal regularity assumption on the curve C_f. It is easily seen to be equivalent to the condition that f is not a square in $\bar{\mathbb{F}}[X]$, and we will use it in this form.

REMARK. Stepanov's method has been refined by Schmidt [**Sch**] and Bombieri [**Bo3**] and is capable of handling the general case of the Riemann Hypothesis for curves; the case of curves with equation of the type $y^d = f(x)$ is not much harder than the one treated here. We limit ourselves to the curves C_f for simplicity, and because it suffices for the application to Kloosterman sums and elliptic curves. Note that curves of the type $y^2 = f(x)$ are instances of so-called hyperelliptic curves, which are quite naturally distinguished among algebraic curves (but not all hyperelliptic curves are of this form; see for instance elliptic curves in characteristics 2 and 3).

The problem we consider is that of estimating the number $|C_f(\mathbb{F})|$ of $\mathbb{F}$-rational points of C_f, i.e., the number N of solutions $(x,y) \in \mathbb{F}^2$ to the equation $y^2 = f(x)$. We are especially interested in this question when q is large (typically, as with exponential sums, the polynomial $f \in \mathbb{F}[X]$ is fixed, and we consider the $\mathbb{F}_n$-rational points for all $n \geqslant 1$), although we will obtain completely explicit inequalities.

THEOREM 11.13. *Assume that $f \in \mathbb{F}[X]$ satisfies (11.18), and $m = \deg(f) \geqslant 3$. If $q > 4m^2$, then $N = |C_f(\mathbb{F})|$ satisfies*

$$|N - q| < 8m\sqrt{q}.$$

Clearly we can assume that $p > 2$, as otherwise the map $y \mapsto y^2$ is an automorphism of $\mathbb{F}$ and $N = q$.

Stepanov's idea, which was inspired by results of Thue [**Thu**] in diophantine approximation, is to construct an auxiliary polynomial of degree r, say, having zeros of high multiplicity (at least ℓ, say) at the x-coordinates of points of $C_f(\mathbb{F})$. Hence one gets easily the inequality

$$N \leqslant 2r\ell^{-1},$$

the factor two being the highest possible multiplicity of a given x-coordinate among points in $C_f(\mathbb{F})$. This inequality turns out to be so strong that it gives the upper-bound of the theorem (certainly a surprising fact!). A trick then deduces the lower-bound from this.

We first distinguish among the points (x, y) these with $y = 0$. Let N_0 be the number of distinct zeros of f in $\mathbb{F}$, which is also the number of points $(x, 0) \in C_f(\mathbb{F})$. If (x, y) is a point of C_f with $y \neq 0$, it follows that $f(x)$ is a square in $\mathbb{F}$, which is true if and only if $g(x) = 1$ where

$$g = f^c \quad \text{with} \quad c = \tfrac{1}{2}(q-1).$$

Conversely, given $x \in \mathbb{F}$ with $g(x) = 1$, there are exactly two elements $y \in \mathbb{F}^*$ with $y^2 = f(x)$. Hence, writing

$$N_1 = |\{x \in \mathbb{F} \mid g(x) = 1\}| \tag{11.19}$$

it follows that

$$N = N_0 + 2N_1. \tag{11.20}$$

We will estimate N_1 by following the strategy sketched above, but in order to handle the lower bound later, we generalize slightly and consider for any $a \in \mathbb{F}$ the set

$$\mathcal{S}_a = \{x \in \mathbb{F} \mid f(x) = 0 \text{ or } g(x) = a\}. \tag{11.21}$$

To produce polynomials vanishing to a large order, we wish to use derivatives to characterize when this occurs. In characteristic 0, a polynomial P has a zero of order ℓ at x_0 if and only if all the derivatives $P^{(i)}$ with $0 \leqslant i < \ell$ vanish at x_0. In characteristic $p > 0$, however, this is no longer true if $\ell > p$, as the example of the polynomial $P = X^p$ shows, since $P^{(k)} = 0$ for all $k \geqslant 1$, in particular, $P^{(p)}(0) = 0$. A satisfactory solution follows by considering other differential operators.

DEFINITION. Let K be any field. For any $k \geqslant 0$, the k-th Hasse derivative is the linear operator $E^k : K[X] \to K[X]$ defined by

$$E^k X^n = \binom{n}{k} X^{n-k}$$

for all $n \geqslant 0$, and extended to $K[X]$ by linearity. We also write $E = E^1$ (but beware that $E^k \neq E \circ E \circ \cdots \circ E$).

REMARK. From the binomial expansion

$$X^n = (X - a + a)^n = \sum_{k=0}^{n} \binom{n}{k} a^{n-k}(X-a)^k,$$

and by linearity, we see that the value of $E^k P$ at a point $a \in K$, for $P \in K[X]$, is simply the coefficient of $(X - a)^k$ in the Taylor expansion of P around a. This explains the properties of the Hasse derivatives, but we cannot take this as a definition, because the values of a polynomial over a finite field do not characterize the polynomial.

Note that for K of characteristic $p > 0$, we get $EX^p = E^2X^p = \cdots = E^{p-1}X^p = 0$, but $E^pX^p = 1 \neq 0$ and we see that the Hasse derivatives detect the zero of X^p of order exactly p at 0. This is a general fact, as Lemma 11.16 will show.

LEMMA 11.14. *The Hasse derivatives satisfy*

$$E^k(fg) = \sum_{j=0}^{k} (E^j f)(E^{k-j} g)$$

for all f, $g \in K[X]$, and more generally,

$$E^k(f_1 \cdots f_r) = \sum_{j_1+\cdots+j_r=k} (E^{j_1} f_1) \cdots (E^{j_r} f_r) \tag{11.22}$$

for f_1, ..., $f_r \in K[X]$.

PROOF. It suffices to consider $f = X^m$, $g = X^n$, and the first formula follows from the identity

$$\binom{n+m}{k} = \sum_{j=0}^{k} \binom{m}{j}\binom{n}{k-j}$$

which is obvious from the combinatorial interpretation of the binomial coefficients. Then the second formula follows by induction. $\square$

COROLLARY 11.15. *(1) For all k, $r \geqslant 0$, and all $a \in K$, we have*

$$E^k(X-a)^r = \binom{r}{k}(X-a)^{r-k}.$$

(2) For all k, $r \geqslant 0$ with $k \leqslant r$, and all f, $g \in K[X]$, we have

$$E^k(fg^r) = hg^{r-k}$$

for some polynomial h such that

$$\deg(h) \leqslant \deg(f) + k\deg(g) - k.$$

PROOF. For (1), we apply (11.22) to $f_1 = \cdots = f_r = X - a$, getting

$$E^k(X-a)^r = \sum_{j_1+\cdots+j_r=k} E^{j_1}(X-a) \cdots E^{j_r}(X-a)$$

and only terms with all $j_i \in \{0, 1\}$, $1 \leqslant i \leqslant r$, give non-zero contributions since $E^j(X-a) = 0$ for $j \geqslant 2$, from the definition. Hence (1) follows.

For (2), we observe that if $k \leqslant r$, we have $j_i = 0$ for at least $r - k$ indices in (11.22), which gives (2). $\square$

LEMMA 11.16. *Let $f \in K[X]$ and $a \in K$. Suppose that $(E^k f)(a) = 0$ for all $k < \ell$. Then f has a zero of order $\geqslant \ell$ at a, i.e., is divisible by $(X-a)^\ell$.*

PROOF. Let

$$f = \sum_{0 \leqslant i \leqslant d} \alpha_i (X-a)^i$$

be the Taylor expansion of f around a. By (1) of Corollary 11.15, we obtain

$$E^k f = \sum_{k \leqslant i \leqslant d} \alpha_i \binom{i}{k} (X-a)^{i-k}$$

and evaluating at a we get $\alpha_k = 0$ for all $k < \ell$, hence f is divisible by $(X-a)^\ell$ as claimed. □

We need another technical lemma.

LEMMA 11.17. *Let $K = \mathbb{F}$ be a finite field of characteristic p with q elements, and let $r = h(X, X^q) \in \mathbb{F}[X]$, where $h \in \mathbb{F}[X, Y]$. Then*

$$E^k r = (E_X^k h)(X, X^q)$$

for all $k < q$, where on the right side $E_X^k h$ denotes the Hasse derivative of h performed with respect to X.

PROOF. It suffices to consider $h = X^n Y^m$, so we must prove that $E^k X^{n+mq} = (E^k X^n) X^{mq}$. From Lemma 11.14, we get

$$E^k X^{n+mq} = \sum_{j=0}^{k} E^{k-j} X^n E^j X^{mq}$$

so it suffices to show that $E^j X^{mq} = 0$ for $0 < j < q$ to prove the lemma. But

$$\binom{mq}{j} = \frac{mq}{j} \binom{mq-1}{j-1} = 0$$

in characteristic p, and the result follows. □

We come to the heart of Stepanov's method, the construction of the auxiliary polynomial.

PROPOSITION 11.18. *Assume that $q > 8m$, and let ℓ be an integer satisfying $m < \ell \leqslant q/8$. Then there exists a polynomial $r \in \mathbb{F}[X]$ of degree*

$$\deg(r) < c\ell + 2m\ell(\ell - 1) + mq$$

which has a zero of order at least ℓ at all points $x \in \mathcal{S}_a$ (recall $c = \frac{1}{2}(q-1)$).

We will look, by the method of indeterminate coefficients, for a polynomial r of the special form

$$r = f^\ell \sum_{0 \leqslant j < J} (r_j + s_j g) X^{jq} \tag{11.23}$$

for some polynomials r_j, $s_j \in \mathbb{F}[X]$, to be constructed, each of which has degree bounded by $c - m$. Hence such a polynomial r has degree bounded by

$$\deg(r) \leqslant \ell m + c - m + cm + Jq \leqslant (J+m)q. \tag{11.24}$$

The next lemma is crucial to ensure that $r \neq 0$. This is where we need the assumption (11.18).

LEMMA 11.19. *We have $r = 0 \in \mathbb{F}[X]$ if and only if $r_j = s_j = 0 \in \mathbb{F}[X]$ for all j.*

PROOF. We can assume (by a shift $X \mapsto X + a$ if necessary) that $f(0) \neq 0$. Suppose that $r = 0$ but not all r_j, s_j are zero; let k be the smallest index for which one of r_k, s_k is non-zero. Dividing by $f^\ell X^{kq}$, we get from (11.22) the identity

$$\sum_{k \leqslant j < J} (r_k + s_k g) X^{(j-k)q} = 0$$

which we write in the form $h_0 + h_1 g = 0$, where

$$h_0 = \sum_{k \leqslant j < J} r_j X^{(j-k)q}, \qquad h_1 = \sum_{k \leqslant j < J} s_j X^{(j-k)q}.$$

We square this equation, then multiply both sides by f, getting

$$h_0^2 f = h_1^2 f^q.$$

Since $f \in \mathbb{F}[X]$, we have

$$f(X)^q = f(X^q) \equiv f(0) \pmod{X^q}$$

hence

$$r_k^2 f \equiv s_k^2 f(0) \pmod{X^q}.$$

However, the degree s of the polynomials in this congruence are bounded by $2\deg(r_k) + m \leqslant 2(c - m) + m < q$, and $2\deg(s_k) < 2(c - m) < q$, respectively. So there must be equality $r_k^2 f = s_k^2 f(0)$, which contradicts the assumption (11.18) that f is not a square in $\bar{\mathbb{F}}[X]$. □

We now evaluate the Hasse derivatives of r.

LEMMA 11.20. *Let $k \leqslant \ell$. Then there exist polynomials $r_j^{(k)}$, $s_j^{(k)}$ each one of degree $\leqslant c - m + k(m-1)$ such that*

$$E^k r = f^{\ell - k} \sum_{0 \leqslant j < J} (r_j^{(k)} + s_j^{(k)} g) X^{jq}.$$

PROOF. We can write $r = h(X, X^q)$ where $h \in \mathbb{F}[X, Y]$ is the polynomial

$$h = f^\ell \sum_{0 \leqslant j \leqslant J} (r_j + s_j f^c) Y^{jq}.$$

Hence by Lemma 11.17, we have

$$E^k r = (E_X^k h)(X, X^q) = \sum_{0 \leqslant j < J} (E^k(f^\ell r_j) + E^k(f^{\ell+c} s_j)) X^{jq}.$$

By (2) of Corollary 11.15 there exist polynomials $r_j^{(k)}$ and $s_j^{(k)}$ satisfying $E^k(f^\ell r_j) = f^{\ell-k} r_j^{(k)}$ and $E^k(f^{\ell+c} s_j) = f^{\ell-k+c} s_j^{(k)}$ with $\deg(r_j^{(k)}) \leqslant \deg(r_j) + k\deg(f) - k \leqslant c - m + k(m-1)$ and $\deg(s_j^{(k)}) \leqslant c - m + k(m-1)$. This is the desired result. □

Recall that we wish r to have zeros of order $\geqslant \ell$ at points in $\mathcal{S}_a$ (see (11.20)). If $f(x) = 0$, clearly this is the case. So let $x \in \mathcal{S}_a$, with $f(x) \neq 0$. Applying Lemma

11.21, we evaluate $E^k r$ at a point $x \in \mathcal{S}_a$, using $g(x) = a$, and most importantly $x^q = x$:

$$\begin{aligned} E^k r(x) &= f(x)^{\ell-k} \sum_{0\leqslant j<J} (r_j^{(k)}(x) + a s_j^{(k)}(x)) x^j \\ &= f(x)^{\ell-k} \sigma^{(k)}(x) \end{aligned}$$

where $\sigma^{(k)} \in \mathbb{F}[X]$ is the polynomial

$$\sigma^{(k)} = \sum_{0\leqslant j<J} (r_j^{(k)} + a s_j^{(k)}) X^j.$$

We can now prove Proposition 11.18: if $\sigma^{(k)} = 0$ for all $k < \ell$, Lemma 11.16 shows that r has a zero of order $\geqslant \ell$ at all points in $\mathcal{S}_a$. The system of equations

$$\sigma^{(k)} = 0, \text{ for all } k < \ell \tag{11.25}$$

is a homogeneous system of linear equations, the unknowns being the coefficients of the polynomials r_j, s_j, the equations corresponding to the coefficients of the $\sigma^{(k)}$. We observe that

$$\deg(\sigma^{(k)}) < c - m + k(m-1) + J,$$

so the number of equations does not exceed $B = \ell(c - m + J) + \frac{1}{2}\ell(\ell-1)(m-1)$ while, on the other hand, the number of coefficients of the r_j and s_j is at least $A = 2(c-m)J$. By choosing J large enough, we can make $A > B$. Then the system (11.25) has a non-trivial solution, and by Lemma 11.19 this produces $r \neq 0$ such that r has zeros of order $\geqslant \ell$ at all points $x \in \mathcal{S}_a$. Taking

$$J = \frac{\ell}{q}(c + 2m(\ell - 1))$$

one can check that $A > B$ (recall that $2c = q - 1$ and $8\ell \leqslant q$). The degree of r is bounded by (11.24), which gives Proposition 11.18.

We now prove Stepanov's Theorem 11.13. First, let a be arbitrary, and apply Proposition 11.18. Since the auxiliary polynomial r is non-zero and vanishes to order $\geqslant \ell$ at points in $\mathcal{S}_a$, we have $\ell|\mathcal{S}_a| \leqslant \deg(r) \leqslant c\ell + 2m\ell(\ell-1) + mq$ so $|\mathcal{S}_a| \leqslant c + 2m(\ell - 1) + mq\ell^{-1}$. We choose $\ell = 1 + [\sqrt{q}/2]$, which gives the bound

$$|\mathcal{S}_a| < c + 4m\sqrt{q}. \tag{11.26}$$

To prove Theorem 11.13, take first $a = 1$ getting

$$N_0 + N_1 = |\mathcal{S}_a| < \frac{q}{2} + 4m\sqrt{q}$$

hence the upper bound

$$N = N_0 + 2N_1 < 2(N_0 + N_1) < q + 8m\sqrt{q}. \tag{11.27}$$

To get a lower bound, by the factorization $X^q - X = X(X^c - 1)(X^c + 1)$ we have

$$f(x)(g(x) - 1)(g(x) + 1) = 0$$

for all $x \in \mathbb{F}$, hence $N_0 + N_1 + N_2 = q$ where $N_2 = |\{x \in \mathbb{F} \mid g(x) = -1\}|$. By (11.26) applied to $\mathcal{S}_{-1}$, we have

$$N_0 + N_2 = |\mathcal{S}_{-1}| < \frac{q}{2} + 4m\sqrt{q}$$

hence

$$N_1 = q - N_0 - N_1 > \frac{q}{2} - 4m\sqrt{q},$$

and finally,

$$N = N_0 + 2N_1 \geqslant 2N_1 > q - 8m\sqrt{q}. \tag{11.28}$$

Clearly (11.27) and (11.28) prove Theorem 11.13.

11.7. Proof of Weil's bound for Kloosterman sums.

Let $\mathbb{F}$ be a finite field with q elements, of characteristic $p \neq 2$. Let ψ be any fixed non-trivial additive character of $\mathbb{F}$. For any additive character φ there exists a unique $a \in \mathbb{F}$ such that $\varphi = \psi_a$, hence any Kloosterman sum $S(\psi, \varphi)$ is of the form

$$S(\psi_a, \psi_b) = -\sum_{x \in \mathbb{F}^*} \psi(ax + bx^{-1})$$

for some a, $b \in \mathbb{F}$. We consider a and b as fixed and write $g = aX + bX^{-1}$. We will prove Weil's bound (11.15) by relating the average of the Kloosterman sums $S(\psi_a, \psi_b)$ over ψ to the number of points on an hyperelliptic curve, where the contribution of the trivial character $\psi_0 = 1$ will be the main term.

LEMMA 11.21. *For any $n \geqslant 1$ and any $x \in \mathbb{F}_n$, we have*

$$|\{x \in \mathbb{F}_n \mid y^q - y = x\}| = \sum_{\psi} \psi(\mathrm{Tr}(x)) \tag{11.29}$$

where the sum ranges over all additive characters of $\mathbb{F}$ and Tr is the trace $\mathbb{F}_n \to \mathbb{F}$.

PROOF. If $\mathrm{Tr}\,(x) = 0$, then the equation $y^q - y = x$ has q solutions exactly, as recalled in Section 11.2, and in this case we have $\psi(\mathrm{Tr}\,(x)) = 1$ for all ψ, hence the right side of (11.29) is also equal to q. On the other hand, if $\mathrm{Tr}\,(x) \neq 0$, the equation $y^q - y = x$ has no solution, and the character sum is zero by orthogonality. □

From this lemma we deduce that

$$\begin{aligned} -\sum_{\psi} S_n(\psi_a, \psi_b) &= \sum_{\psi} \sum_{x \in \mathbb{F}_n^*} \psi(\mathrm{Tr}\, g(x)) \\ &= |\{(x, y) \in \mathbb{F}_n^* \times \mathbb{F}_n \mid y^q - y = g(x)\}| = N_n, \text{ say,} \end{aligned} \tag{11.30}$$

for $n \geqslant 1$. If $\psi = \psi_0$, the trivial character, we have

$$S(\psi_a, \psi_b) = S(\psi_0, \psi_0) = 1 - q^n.$$

For $\psi \neq \psi_0$, let α_ψ, β_ψ be the "roots" of the Kloosterman sum $S(\psi_a, \psi_b)$, so by Theorem 11.8 we have $\alpha_\psi \beta_\psi = q$ and

$$S_n(\psi_a, \psi_b) = \alpha_\psi^n + \beta_\psi^n,$$

for all $n \geqslant 1$.

We can therefore write

$$N_n = q^n - 1 - \sum_{\psi \neq \psi_0} (\alpha_\psi^n + \beta_\psi^n).$$

The equation $y^q - y = g(x)$ does not obviously describe a curve, since g is not a polynomial, but multiplying by x it is equivalent with

$$C_{a,b} : ax^2 - (y^q - y)x + b = 0$$

(note that $x = 0$ is not possible since $b \neq 0$). Because $p \neq 2$, the number of solutions is equal to the number of solutions of the discriminant equation of this quadratic equation

$$D_{a,b} : (y^q - y)^2 - 4ab = v^2,$$

i.e., $N_n = |D_{a,b}(\mathbb{F}_n)|$. This is of the form (11.17) with $\deg(f) = 2q$, and because $4ab \neq 0$ it satisfies (11.18). Hence by Theorem 11.13 we have

$$|N_n - q^n| < 16q^{1+n/2}$$

if n is large enough, so that $q^n > 16q$.

By (11.30) we get a sharp estimate for the roots α_ψ, β_ψ, on average

$$\frac{1}{q}\Big|\sum_{\psi\neq\psi_0}(\alpha_\psi^n + \beta_\psi^n)\Big| \leqslant 16q^{n/2} \tag{11.31}$$

for n large enough. The following simple lemma shows that the individual roots must be of modulus $\leqslant \sqrt{q}$:

LEMMA 11.22. *Let ω_1, ..., ω_r be complex numbers, A, B positive real numbers and assume that*

$$\Big|\sum_{j=1}^{r}\omega_j^n\Big| \leqslant AB^n$$

holds for all integers n large enough. Then $|\omega_j| \leqslant B$ for all j.

PROOF. One can do this by hand (using Dirichlet's box principle), but a nice trick gives the result immediately: consider the complex power series

$$f(z) = \sum_{n\geqslant 1}\Big(\sum_j \omega_j^n\Big)z^n = \sum_j \frac{1}{1-\omega_j z}.$$

The hypothesis implies that f converges absolutely in the disc $|z| < B^{-1}$, hence f is analytic in this region. In particular, it has no poles there, which means that we must have $|\omega_j|^{-1} \geqslant B^{-1}$ for all j. □

From this lemma applied with $A = 16q$, $B = \sqrt{q}$, we deduce the upper bounds $|\alpha_\psi| \leqslant \sqrt{q}$, $|\beta_\psi| \leqslant \sqrt{q}$ for all $\psi \neq \psi_0$. Since $\alpha_\psi\beta_\psi = q$, we have in fact $|\alpha_\psi| = |\beta_\psi| = \sqrt{q}$, and so Theorem 11.11 is proved.

REMARKS. (1) We see here twice how crucial the introduction of the companion sums K_n is: first because the curve $D_{a,b}$ has very high degree, so Stepanov's bound $|N - q| < 8m\sqrt{q}$ is trivial when applied to $\mathbb{F}$ itself, and secondly because only by the consideration of all extension fields can we determine the exact order of magnitude of the roots, and obtain Weil's bound $S(\psi, \varphi) \leqslant 2\sqrt{q}$ with the sharp constant 2.

(2) The constant 2 is optimal in Weil's bound for fixed a, b and q. Indeed we have

$$\sum_{n\geqslant 1} S_n(\psi_a, \psi_b)z^n = \frac{1}{1-\alpha_\psi z} + \frac{1}{1-\beta_\psi z}.$$

This is a non-zero rational function with poles on the circle $|z| = 1/\sqrt{q}$, hence this is its radius of convergence. Therefore

$$\limsup_{n\to+\infty} |S_n(\psi_a,\psi_b)|^{-1/n} = \frac{1}{\sqrt{q}}.$$

This means that for any $\varepsilon > 0$ there exist infinitely many n such that

$$|S_n(\psi_a,\psi_b)| \geqslant (2-\varepsilon)q^{n/2}.$$

It is conjectured (this follows from the Sato-Tate Conjecture for the angles of Kloosterman sums described in Chapter 21) that the Weil bound is also optimal when a, b are fixed, $n = 1$, and $q = p \to +\infty$. However, this remains very much open. See the remark at the end of the introduction to Section 11.8 for the case of elliptic curves.

(3) Using the extension of Stepanov's method to curves of the type $y^d = f(x)$ and an analysis of the corresponding zeta function, one can prove the following estimate for complete character sums:

THEOREM 11.23. *Let $\mathbb{F}$ be a finite field with q elements and let χ be a non-trivial multiplicative character of $\mathbb{F}^\star$ of order $d > 1$. Suppose $f \in \mathbb{F}[X]$ has m distinct roots and f is not a d-th power. Then for $n \geqslant 1$ we have*

$$\Big|\sum_{x\in\mathbb{F}_n} \chi(N(f(x)))\Big| \leqslant (m-1)q^{n/2}.$$

This is Theorem 2C', p. 43, of [**Sch**]. In particular, we get the following corollary which will be used in proving the Burgess bound for short character sums (Theorem 12.6).

COROLLARY 11.24. *Let χ (mod p) be a non-principal multiplicative character. If one of the classes b_v (mod p), $v = 1,\dots,2r$ is different from the remaining ones then*

$$\Big|\sum_{x(\mathrm{mod}\ p)} \chi((x+b_1)\dots(x+b_r))\bar{\chi}((x+b_{r+1})\dots(x+b_{2r}))\Big| \leqslant 2rp^{\frac{1}{2}}.$$

PROOF. Observe that

$$\chi((x+b_1)\dots(x+b_r))\bar{\chi}((x+b_{r+1})\dots(x+b_{2r}) = \chi(f(x))$$

with

$$f(x) = \prod_{1\leqslant j\leqslant r}(x+b_j)\prod_{r+1\leqslant j\leqslant 2r}(x+b_j)^{p-2}.$$

From the assumption, one of the b_i is a root of f of order either 1 or $p-2$, which is coprime with the order $d \mid (p-1)$ of χ, so we can apply Theorem 11.23. □

11.8. The Riemann Hypothesis for elliptic curves over finite fields.

A particularly important case of the Riemann Hypothesis is that of elliptic curves. Historically this was first established by Hasse using global methods. In the notation of Section 11.6, this means that we consider curves C_f with $\deg f = 3$, so the equation is of the form

$$C : y^2 = x^3 + a_2x^2 + a_4x + a_6 \tag{11.32}$$

(the numbering reflects the traditional notation for elliptic curves, see Section 14.4). In contrast with that section, we emphasize that we are considering the affine curve, without the point at infinity. The cubic polynomial $f(x) = x^3 + a_2x^2 + a_4x + a_6$ cannot be a square, so this curve satisfies the assumption (11.18). Moreover, we assume that f does not have a double root; this means that the curve C is smooth (see Section 11.9), and it is a necessary condition for what follows.

In this case, Theorem 11.13 implies that for $q > 36$ the number $N = |C(\mathbb{F})|$ satisfies

$$|N - q| < 24\sqrt{q}.$$

In Section 11.10 we will prove, as before for Kloosterman sums, the rationality and the functional equation of the corresponding zeta function, from which we will deduce:

THEOREM 11.25. *Let C be an elliptic curve over $\mathbb{F}$ given by*

$$C : y^2 = x^3 + a_2x^2 + a_4x + a_6$$

with $a_i \in \mathbb{F}$. Then for all $n \geqslant 1$ we have

$$\big||C(\mathbb{F}_n)| - q^n\big| \leqslant 2q^{n/2}. \tag{11.33}$$

REMARKS. Theorem 11.25 is optimal. Indeed letting $n \to +\infty$, this follows as for Kloosterman sums from Lemma 11.22. However, it is also true in the horizontal sense as the following example shows: let $E/\mathbb{Q}$ be the elliptic curve with equation

$$E : y^2 = x^3 - x$$

which has complex multiplication by $\mathbb{Z}[i]$. As before we consider the affine points, not the projective ones. The discriminant of E is 64 so E can be reduced modulo p to an elliptic curve over $\mathbb{Z}/p\mathbb{Z}$ for any odd prime p. One shows (for instance by relating E to the curve $y^2 = x^4 + 4$ by changing $(x, y) \mapsto (yx^{-1}, 2x - y^2x^{-2})$ for $(x, y) \neq (0, 0)$, see e.g. [**I4**] or [**IR**]) that $|E(\mathbb{Z}/p\mathbb{Z})| = p$ if $p \equiv 3 \pmod 4$ and $|E(\mathbb{Z}/p\mathbb{Z})| = p - 2a_p$ if $p \equiv 1 \pmod 4$, where

$$p = a_p^2 + b_p^2$$

with $\pi = a_p + ib_p \equiv 1 \pmod{2(1+i)}$ (this congruence determines π up to conjugation). For any $\varepsilon > 0$, Theorem 5.36 (generalized slightly to add the congruence condition) shows that there exist infinitely many Gaussian primes $\pi \equiv 1 \pmod{2(1+i)}$ such that $|\arg \pi| < \varepsilon$. Hence $|\mathrm{Im}\,(\pi)| \leqslant \varepsilon|\pi|$ and

$$|p - |E(\mathbb{Z}/p\mathbb{Z})|| = 2|a_p| \geqslant 2(1 - \varepsilon^2)\sqrt{p}$$

for infinitely many p.

Theorem 11.25 will be proved in Section 11.10 after some geometric and algebraic preliminaries. This goes a bit further away from the heart of analytic number theory, yet we include full details because the Hasse bound is also important as being the simplest case of the very important Deligne bound for Fourier coefficients of modular forms. The reader will also certainly appreciate the elegance and beauty of the geometry involved.

11.9. Geometry of elliptic curves.

In explaining the special geometric features of elliptic curves, we may as well consider a more general case. So let k be an arbitrary field, $\bar{k}$ an algebraic closure, and let C be the curve given by the equation

$$C : y^2 = f(x), \text{ with } f = X^3 + a_2X^2 + a_4X + a_6 \in k[X],$$

identified with the set of solutions $(x, y) \in \bar{k}^2$. We assume as before that f does not have a double root. If k'/k is any extension, we let $C(k')$ be the set of solutions in $(k')^2$.

The geometry of the elliptic curve becomes much clearer if we work with the projective version of the curve C, namely the curve E in the projective plane given, in homogeneous coordinates $(x : y : z)$, by the equation

$$E : y^2z = x^3 + a_2x^2z + a_4xz^2 + a_6z^3.$$

Putting $z = 1$ gives back C; on the other hand, "at infinity", we are only adding one point: taking $z = 0$ yields $x = 0$, and all elements $(0 : y : 0)$ (with $y \neq 0$) correspond to a single point $\infty = (0 : 1 : 0)$ in the projective plane. Notice that this point ∞ is rational over the base field k, so that for all extensions k'/k we have

$$E(k') = C(k') \cup \{\infty\}.$$

The main property of the curve E that we will use is the beautiful fact that its points form an abelian group, with identity element ∞. Throughout, p denotes points on E, not the characteristic of the field k. The group law (denoted by $+$) is described by the geometric condition that for any three (distinct) points p_1, p_2 and p_3 in E, we have $p_1 + p_2 + p_3 = 0$ if and only if the three points are collinear (in the projective plane), and the opposite of a point $(x : y : z)$ is the point $(x : -y : z)$ (symmetry with respect to the x-axis). This way one can construct the sum of any two distinct points, by computing the equation of the line joining them, and taking the opposite (in the sense above) of the third intersection point with the curve. That there are exactly three intersection points follows immediately from the fact that the polynomial f is of degree 3. In addition, to compute the double $p + p$ of a point p, the same construction is done with the tangent line at p; the condition that f has no double root ensures that this tangent line always exists.

Also, because $f \in k[X]$, it follows easily that the k'-rational points $E(k')$, for any extension k'/k, form a subgroup of $E(\bar{k})$.

We do not prove those facts here; completely elementary proofs, by computing explicitly the coordinates of the sum $p_1 + p_2$ of two points according to the recipe above and checking the abelian group axioms (associativity is the only difficulty), are fairly straightforward (see for instance [**IR**], ch. 18, 19).

We now introduce some further geometric objects related to E or, more generally, to any smooth, projective, algebraic curve[1]. Thus consider again a more general case: let $\bar{k}$ be an algebraically closed field, and let E be a plane algebraic curve over $\bar{k}$, i.e. given by an equation

$$f(x, y, z) = 0$$

for some homogeneous $f \in \bar{k}[X, Y, Z]$. We identify E with the set of points in the projective plane. We assume that E is smooth, which means here that for any point $p = (x : y : z)$ of E, not all partial derivatives $\partial f/\partial X(p)$, $\partial f/\partial Y(p)$, $\partial f/\partial Z(p)$, are zero. In this case the line with equation

$$\frac{\partial f}{\partial X}(p)(X - x) + \frac{\partial f}{\partial Y}(p)(Y - y) + \frac{\partial f}{\partial Z}(p)(Z - z) = 0$$

is well-defined and is the tangent line to E at p. For elliptic curves $y^2 = f(x)$, this smoothness condition is equivalent to the fact that the polynomial f has no double roots, by a simple calculation.

Let C be the affine curve corresponding to E given by

$$C : f(x, y, 1) = 0$$

in $\bar{k}^2$. Let $g(X, Y) = f(X, Y, 1) \in \bar{k}[X, Y]$. We define $\bar{k}[C] = \bar{k}[X, Y]/(g)$. Elements of $\bar{k}[C]$ can be interpreted as functions on C. We assume that (g) is a prime ideal (this is easily checked in the case of elliptic curves) so that $\bar{k}[C]$ is an integral domain, and we let $\bar{k}(C)$ or $\bar{k}(E)$ be its quotient field, called the function field of C or of E. It is a finite extension of the field $\bar{k}(X)$ of rational functions over $\bar{k}$ (for elliptic curves $y^2 = f(x)$, it is a quadratic extension $\bar{k}(X)(\sqrt{f})$). We interpret elements of the function field as rational functions on E, so given a point $p \in E$ and an element $\varphi \in \bar{k}(E)$, either φ has a pole at p or $\varphi(p) \in \bar{k}$ is defined.

Now the important point is that because E is smooth it is possible to define the order of φ at p for every p in E and every non-zero rational function $\varphi \in \bar{k}(E)^{\star}$. As expected, this order behaves like its analogue for holomorphic functions theory or rational functions. Precisely, for every $p \in R$, there is a discrete valuation

$$\mathrm{ord}_p : \bar{k}(E)^{\times} \to \mathbb{Z},$$

which gives the order of the zero (if $\geqslant 0$) or pole (if < 0) of a rational function at p. As a discrete valuation, it satisfies

$$\begin{aligned} \mathrm{ord}_p(c) &= 0, \text{ for } c \in \bar{k}^{\star}, \\ \mathrm{ord}_p(\varphi\psi) &= \mathrm{ord}_p(\varphi) + \mathrm{ord}_p(\psi), \\ \mathrm{ord}_p(\varphi + \psi) &\geqslant \min(\mathrm{ord}_p(\varphi), \mathrm{ord}_p(\psi)). \end{aligned}$$

We sketch a proof (see also [**Sil**], Prop. II.1.1): consider the ring $\mathcal{O}_p = \{\varphi \in \bar{k}(E) \mid \varphi \text{ is defined at } p\}$. This is a noetherian local domain with maximal ideal $\mathfrak{m}_p = \{\varphi \in \mathcal{O}_p \mid \varphi(p) = 0\}$ and residue field $\mathcal{O}_p/\mathfrak{m}_p \simeq \bar{k}$ (by evaluation at p). Because E is smooth at p (there is a tangent line), the $\bar{k}$-vector space $\mathfrak{m}_p/\mathfrak{m}_p^2$ is of dimension 1 (because the curve is in the plane, it is of dimension $\leqslant 2$; the equation of the tangent line gives a relation, and it is easy to see that $\mathfrak{m}_p/\mathfrak{m}_p^2 \neq 0$). By Nakayama's Lemma (see for instance [**AM**], p. 21), it follows that $\mathfrak{m}_p$ is a principal

[1] The reader can without damage assume that we are just dealing with the elliptic curves described before, with $k = \bar{\mathbb{F}}$. In this case every incomplete assertion can be checked by hand.

ideal. Let π be a generator; then $\mathfrak{m}_p^d$ is generated by π^d for any $d \geqslant 1$. The order ord_p can be defined for φ in $\mathcal{O}_p$, $\varphi \neq 0$, by

$$\mathrm{ord}_p(\varphi) = \max\{d \geqslant 0 \mid \varphi \in \mathfrak{m}_p^d\}$$

and extended to a homomorphism $\bar{k}(E)^\star \to \mathbb{Z}$. The properties above are then quite easy to check.

For elliptic curves $y^2 = f(x)$, one can easily see that if $p \neq \infty$, and $p = (x, y)$ with $y \neq 0$, it is possible to take $\pi = X - x$. For $p = \infty$, one can take $\pi = X/Y$, and one finds that $\mathrm{ord}_\infty(x) = -2$, $\mathrm{ord}_\infty(y) = -3$.

Every non-zero element $\varphi \in \bar{k}(E)$ has finitely many zeros and poles, and to package them conveniently we define a divisor on E to be a formal finite linear combination with coefficients in $\mathbb{Z}$ of symbols $[p]$, one for each point $p \in E$. Divisors form a free abelian group $\mathrm{Div}(E)$. Two homomorphisms are important. One associates to a non-zero rational function φ the divisor (denoted (φ)) of its zeros and poles:

$$\begin{cases} \bar{k}(E)^\times \to \mathrm{Div}(E) \\ \varphi \mapsto (\varphi) = \displaystyle\sum_{p \in E} \mathrm{ord}_p(\varphi) \end{cases}$$

and the second gives the degree of a divisor:

$$\deg \begin{cases} \mathrm{Div}(E) \to \mathbb{Z}, \\ [p] \mapsto 1. \end{cases}$$

As suggested by the notation, divisors of the type (φ) are called principal divisors.

Ordinary rational functions have as many zeros as poles, with multiplicity, and the same holds for $\varphi \in \bar{k}(E)^\star$: this means that for all $\varphi \in \bar{k}(E)^\times$, we have $\deg((\varphi)) = 0$ (see for instance [**Sil**], II-3). For elliptic curves $y^2 = f(x)$, an easy proof can be derived by observing that $\bar{k}(E)$ is a quadratic extension of $\bar{k}(X)$. The non-trivial element in the Galois group is $\varphi \mapsto \bar{\varphi}$ defined by $\bar{\varphi}(p) = \varphi(-p)$. It is clear that $\mathrm{ord}_p(\varphi) = \mathrm{ord}_{-p}(\bar{\varphi})$, hence $\deg(\varphi) = \deg(\bar{\varphi})$. Now $\varphi\bar{\varphi}$ is in $\bar{k}(X)$. One can check the following compatibility: if $\psi \in \bar{k}(X)$, with divisor $(\psi)_1 = \sum n_i x_i$ (as an ordinary rational function), then its divisor as an element of $\bar{k}(E)$ is $(\psi) = \sum n_i([p_i] + [-p_i])$, where p_i is any point of E with x-coordinate x_i. In particular, $0 = \deg((\psi)_1) = 2\deg((\psi))$. Applying this to $\varphi\bar{\varphi}$ gives

$$0 = \deg(\varphi\bar{\varphi}) = \deg(\varphi) + \deg(\bar{\varphi}) = 2\deg(\varphi).$$

DEFINITION. 1. Two divisors D_1 and D_2 such that $D_1 - D_2 = (\varphi)$ for some $\varphi \in \bar{k}(E)^\times$ are called linearly equivalent. This is denoted $D_1 \sim D_2$, and is an equivalence relation on $\mathrm{Div}(E)$.

2. The group $\mathrm{Div}(E)$ carries a partial ordering, compatible with the group structure, defined by $D \geqslant 0$ if and only if all the coefficients in the formal sum giving D are $\geqslant 0$. Such divisors are called effective divisors.

3. For a divisor $D \in \mathrm{Div}(E)$, let

$$L(D) = \{0\} \cup \{\varphi \in \bar{k}(E) \mid (\varphi) + D \geqslant 0\};$$

this is a $\bar{k}$-vector space. Let $\ell(D) = \dim L(D)$, an integer or $+\infty$.

If $D = n_1[p_1] + \dots + n_k[p_k] - m_1[q_1] - \dots - m_j[q_j]$ with n_i, $m_i \geqslant 0$, then $\varphi \in L(D)$, $\varphi \neq 0$, means simply that φ has

(1) Poles of order at most n_i at p_i, $1 \leqslant i \leqslant k$;
(2) Zeros of order at least m_i at q_i, $1 \leqslant i \leqslant j$.

It follows immediately that if $D_2 \leqslant D_1$, then $L(D_2) \subset L(D_1)$. Also, if $D_1 \sim D_2$, then writing $D_1 = D_2 + (\psi)$, the map $\varphi \mapsto \psi\varphi$ induces an isomorphism $L(D_1) \to L(D_2)$, in particular, $\ell(D_1) = \ell(D_2)$ only depends on the linear equivalence class of the divisor.

The following interpretation is also clear: there is a bijection

$$\begin{cases} \mathbf{P}(L(D)) \to \{\text{Effective divisors linearly equivalent to } D\}, \\ \varphi \mapsto (\varphi) + D \end{cases} \tag{11.34}$$

between the projective space $\mathbf{P}(L(D))$ of $L(D)$ and the effective divisors linearly equivalent to D (by definition of $L(D)$, the map has image in the set of effective divisors). This also requires the important fact that

$$L(0) = \bar{k}, \quad \ell(0) = 1. \tag{11.35}$$

In other words, an everywhere defined rational function on E is constant: this is obvious for elliptic curves, since regularity on C forces such a φ to be a polynomial $g(X) + Yh(X)$, and regularity at ∞ then forces $h = 0$, $g \in \bar{k}$.

We first notice the following simple lemma:

LEMMA 11.26. *Let D be a divisor on E such that $\ell(D) > 0$. Then either $\deg(D) > 0$ or $D \sim 0$.*

PROOF. If $\ell(D) > 0$, there is a non-zero element $\varphi \in L(D)$, so that $(\varphi) + D \geqslant 0$. Taking the degree we find that $\deg(D) \geqslant 0$. So we need only show now that if $\deg(D) = 0$, then $D \sim 0$. But $(\varphi) + D$ is then an effective divisor of degree 0; clearly it must be $= 0$, so $D = -(\varphi) = (\varphi^{-1}) \sim 0$. □

To prove the rationality of the local zeta function of elliptic curves, we will need to know the value of $\ell(D)$ for effective divisors. This is computed by the Riemann-Roch Theorem.

THEOREM 11.27. *Let E be an elliptic curve over an algebraically closed field $\bar{k}$. For any divisor D on E, $\ell(D)$ is finite and we have the formula*

$$\ell(D) - \ell(-D) = \deg D. \tag{11.36}$$

Equivalently, by Lemma 11.26, we can compute $\ell(D)$ for any D by:
1. If $\deg(D) \geqslant 0$, and $D \not\sim 0$, then $\ell(D) = \deg(D)$.
2. If $D \sim 0$, then $\ell(D) = 1$.
3. If $\deg(D) < 0$, then $\ell(D) = 0$.

This is the Riemann-Roch theorem specialized for elliptic curves. See the remark below for the general case.

The simple proof for the Riemann-Roch theorem hinges on the remarkable interaction of the group structure on E with the divisor group. Indeed, consider the map

$$\sigma : \mathrm{Div}(E) \to E$$

defined by $\sigma(n_1[p_1] + \dots + n_k[p_k]) = n_1 p_1 + \dots + n_k p_k$, the $+$ on the right side corresponding to the group law on E.

PROPOSITION 11.28. *Let D be a divisor on E. Then we have*

$$D \sim [\sigma(D)] + (\deg(D) - 1)[\infty].$$

PROOF. In essence this "is" the group law itself: by induction, we need only consider $D = [p] + [q]$ and $D = [p] - [q]$ for some points p and q. If p or q is the origin ∞, the result is obvious.

So consider first the case of $D = [p] + [q]$ with p, $q \neq \infty$ and $p \neq q$. Then the equation $aX + bY + c = 0$ of the line joining p and q defines an element $\varphi = aX + bY + c \in \bar{k}(E)$, which by definition of the group law satisfies

$$(\varphi) = [p] + [q] + [r] - 3[\infty]$$

where $-r = p + q = \sigma(D)$. Since $(\varphi) \sim 0$ we get

$$D = [p] + [q] \sim (\varphi) - [\sigma(D)] + 3[\infty] \sim -[\sigma(D)] + 3[\infty].$$

But similarly for any point p in E, the equation of the line joining p and $-p$ gives a function with divisor $[p] + [-p] - 2[\infty]$ so that for any p

$$-[p] + [\infty] \sim [-p] - [\infty] \tag{11.37}$$

and hence $D \sim [\sigma(D)] + [\infty]$, as desired.

If $p = q$, the equation of the tangent line gives $2[p] + [2p] - 2[\infty] \sim 0$, and the result again follows. Finally if $D = [p] - [q]$, use (11.37) to reduce to the previous case:

$$[p] - [q] = [p] + (-[q] + [\infty]) - [\infty] \sim [p] + [-q] - 2[\infty] \sim [p + q] - [\infty].$$

□

PROOF OF THE RIEMANN-ROCH THEOREM. Notice that (11.36) for D or $-D$ are equivalent. By Proposition 11.28, we have

$$\ell(D) = \ell([\sigma(D)] + (\deg D - 1)[\infty]),$$
$$\ell(-D) = \ell([-\sigma(D)] + (1 - \deg D)[\infty]).$$

One of the two divisors on the right is effective, so we can assume that D is effective and of the form $D = [p] + n[\infty]$ with $P \in E$ and $n \geqslant 0$.

If $p = \infty$, then $D = m[\infty]$ with $m \geqslant 1$. We must prove $\ell(D) = m$. Any element φ of $L(m[\infty])$ has no poles on C, hence is a polynomial $\varphi \in \bar{k}[X, Y]$, which satisfies $\mathrm{ord}_\infty(\varphi) \geqslant -m$. Since $\mathrm{ord}_\infty(X) = -2$ and $\mathrm{ord}_\infty(Y) = -3$, we have

$$\mathrm{ord}_\infty(\varphi) = \max(-2\deg(g), -3 - 2\deg(h)) \text{ for } \varphi = g(X) + Yh(X).$$

Let $V = \{2\deg(g), 3 + 2\deg(h)\}$ where g and h are polynomials in X. One sees immediately that V is the set of all positive integers, except 1. Now using monomials X^a or YX^b as basis elements, we check that

$$\ell(m[\infty]) = |\{n \in V \mid n \leqslant m\}| = m.$$

Now we consider $D = [p] + n[\infty]$ with $n \geqslant 0$ and $p \neq \infty$. If $n = 0$, we can use the automorphism $q \mapsto q - p$ which sends p to ∞ to get an isomorphism $L([p]) \simeq L([\infty])$ which implies $\ell([p]) = 1$.

If $n \geqslant 1$, we have $D \geqslant n[\infty]$ hence $L(n[\infty]) \subset L(D)$. Thus $n \leqslant \ell(D)$. Moreover, because only a simple pole is allowed at p, we have $\ell(D) \leqslant n+1$: if π_p is a function with a simple zero at p, we have a $\bar{k}$-linear map

$$\begin{cases} L(D)/L(n[\infty]) \to \bar{k} \\ \varphi \mapsto (\pi_p\varphi)(p) \end{cases}$$

which is tautologically injective. Thus we must show that there is in $L(D)$ one more $\bar{k}$-linearly independent element not in $L(n[\infty])$.

Write $p = (x, y)$ in affine coordinates. We have the following divisors:

$$(X - x) = [p] + [-p] - 2[\infty],$$
$$(Y + y) = [-p] + [p'] + [p''] - 3[\infty]$$

for some p' and p''. Let $\varphi = (Y+y)/(X-x)$, then $(\varphi) = -[p]+[p']+[p'']-[\infty] \geqslant -D$. We claim that $p \neq p'$, p''. Indeed p' and p'', by definition, are of the form $(x_1, -y)$, $(x_2, -y)$. This can be equal to p only if $y = 0$, but for $y = 0$ by assumption there are three distinct roots of $f(x) = 0$. Hence φ has a simple pole at p, so $\varphi \notin L(n[\infty])$, and we obtain the required formula $\ell(D) = n + 1 = \deg(D)$. □

We must now consider some rationality questions. We assume that we have an elliptic curve E given by $y^2 = f(x)$ with $f \in k[X, Y]$. The preceding analysis applies to an algebraic closure $\bar{k}$ of k. There is a natural action of the Galois group G_k of k on E (on the coordinates), and on the divisors. A point or a divisor is called k-rational if it is G_k-fixed. Notice that for a divisor $D = n_1p_1 + \dots + n_jp_j$ this does not mean that the p_i are in $E(k)$. Also G_k acts on $\bar{k}(E)$ and the field fixed by G_k is $k(E)$, the fraction field of $k[X, Y]/(f)$. The divisor of a function $\varphi \in k(E)$ is obviously k-rational.

If D is defined over k, we define

$$L_k(D) = \{0\} \cup \{\varphi \in k(E) \mid (\varphi) + D \geqslant 0\}$$

and we let $\ell_k(D) = \dim_k L_k(D)$. It is clear that $\ell_k(D) \leqslant \ell(D)$.

THEOREM 11.29. *For any k-rational divisor D, we have*

$$\ell_k(D) = \ell(D).$$

In particular the Riemann-Roch formula holds with $\ell_k(D)$ instead of $\ell(D)$.

PROOF. This is a special case of the following theorem, which is a formulation of Hilbert's Theorem 90 for $GL(n)$: let k be a field, $\bar{k}$ an algebraic closure, V a $\bar{k}$-vector space with an action of G_k. Then there is a basis of V made of elements which are G_k-fixed (equivalently, let $V_k = V^{G_k}$, then $V = V_k \otimes \bar{k}$, or $\dim_k V_k = \dim_{\bar{k}} V$). For a proof, see for instance [**Sil**], Lemma II.5.8.1. □

REMARK. This theory adapts in the following way to more general algebraic curves (see for instance [**Ha**], IV): if E is smooth and projective over $\bar{k}$, one can define a certain divisor class K (called the canonical class, and related to differentials on E). It has degree $\deg(K) = 2g - 2$ for some integer $g \geqslant 0$, called the genus of E, and the Riemann-Roch Theorem takes the form

$$\ell(D) - \ell(K - D) = \deg(D) + 1 - g. \tag{11.38}$$

Elliptic curves correspond to $g = 1$; in this case the canonical class is trivial, and this reduces to Theorem 11.27. The case $g = 0$ corresponds to the projective line, and is also very easy. The proof of (11.38) is much more involved than the one for elliptic curves, since there is no group law on the curve which would help.

11.10. The local zeta function of elliptic curves.

Let C be an elliptic curve over $\mathbb{F}$ given by (11.32). It is more convenient to use here the corresponding projective curve E, as described in Section 11.9. The zeta function of E is defined as the formal power series

$$Z(E) = \exp\Big(\sum_{n\geqslant 1} \frac{|E(\mathbb{F}_n)|}{n} T^n\Big). \tag{11.39}$$

We first relate $Z(E)$ to points on the curve, by giving its Euler product (compare Lemma 11.7). To do this we introduce some terminology, which comes from the language of schemes.

DEFINITION. Let E be an elliptic curve over a finite field $\mathbb{F}$ with q elements. A closed point of E is the Galois orbit of a point $x_0 \in E(\bar{\mathbb{F}})$. The degree $\deg(x)$ of a closed point x is the cardinality (necessarily finite) of the orbit and its norm is $Nx = q^{\deg(x)}$. The set of closed points of E is denoted $|E|$.

This notion is analogue to that of an irreducible polynomial in $\mathbb{F}[X]$ used for the zeta functions of Gauss sums and Kloosterman sums. To every closed point $x \in |E|$ is associated an $\mathbb{F}$-rational divisor which is simply the formal sum of all the elements in the orbit. The degree of this divisor is the degree of x. Moreover, it is easy to see that the group of $\mathbb{F}$-rational divisors is the free abelian group generated by the divisors associated to closed points.

LEMMA 11.30. *We have the Euler product expansion*

$$Z(E) = \prod_{x\in|E|} (1 - T^{\deg(x)})^{-1}, \tag{11.40}$$

where the product is over all closed points of E.

PROOF. This is very close to Lemmas 11.7 and 11.9. First by decomposing the points in $E(\mathbb{F}_n)$ in Galois orbits we obtain

$$|E(\mathbb{F}_n)| = \sum_{d|n} d \sum_{\substack{x\in|E| \\ \deg(x)=d}} 1,$$

which is the analogue of Lemma 11.1. Then we have

$$-T\frac{Z'(E)}{Z(E)} = \sum_{n\geqslant 1} |E(\mathbb{F}_n)| T^n$$

and, on the other hand, this operator applied to the right side of (11.40) yields

$$\sum_{x\in|E|} \deg(x) \sum_{n\geqslant 1} T^{n\deg(x)} = \sum_{n\geqslant 1} T^n \Big(\sum_{d|n} d \sum_{\substack{x\in|E| \\ \deg(x)=d}} 1\Big)$$

hence the result. □

Using the Riemann-Roch Theorem, we now prove the rationality and functional equation of the zeta function.

THEOREM 11.31. *The zeta function $Z(E)$ of an elliptic curve is a rational function. More precisely, it is of the form*

$$Z(E) = \frac{1 - aT + qT^2}{(1-T)(1-qT)} \tag{11.41}$$

where $a \in \mathbb{Z}$ is defined by the relations $|E(\mathbb{F})| = q+1-a$ or, in terms of C, $|C(\mathbb{F})| = q-a$. The zeta function satisfies the functional equation $Z(E,(qT)^{-1}) = Z(E,T)$.

LEMMA 11.32. *Let $d \geqslant 0$ and let $h_d(C)$ be the set of linear equivalence classes of $\mathbb{F}$-rational divisors of degree d. Then $h_d(C)$ is finite and $|h_d(C)| = |h_0(C)| \leqslant |E(\mathbb{F})|$.*

PROOF. For any rational divisor D, we have by Proposition 11.28 the equivalence $D \sim [\sigma(D)] + (\deg D - 1)[\infty]$, thus the linear equivalence class of D only depends on $\sigma(D)$. If D is $\mathbb{F}$-rational, it follows that $\sigma(D) \in E(\mathbb{F})$, and therefore the inequality $|h_d(C)| \leqslant |E(\mathbb{F})|$ holds.

Moreover, it is clear that the map $D \mapsto D + d[\infty]$ with inverse $D \mapsto D - d[\infty]$ induces a bijection between $h_d(C)$ and $h_0(C)$. □

PROOF OF THEOREM 11.31. Since $\mathbb{F}$-rational divisors on E are simply combinations with integer coefficients of divisors associated to closed points, the Euler product (11.40) gives the formal power series expression

$$Z(E) = \sum_{D \geqslant 0} T^{\deg(D)}$$

where the sum is over all effective $\mathbb{F}$-rational divisors on E.

Split the sum according to the degree d of D; for $d = 0$, the only effective divisor is $D = 0$ so

$$Z(E) = 1 + \sum_{d \geqslant 1} T^d \sum_{\substack{D \geqslant 0 \\ \deg(D) = d}} 1.$$

For each d, split further the sum over divisors of degree d in linear equivalence classes. By Lemma 11.32, there are $h_0(C)$ equivalence classes for each d. For a given class (that of D say), the contribution is the number of effective ($\mathbb{F}$-rational) divisors linearly equivalent to D. By (11.34) and Theorem 11.29, this is equal to

$$|\mathbf{P}(L(D))| = \frac{q^{\ell_{\mathbb{F}}(D)} - 1}{q-1} = \frac{q^{\ell(D)} - 1}{q-1}.$$

Since $d \geqslant 1$, the Riemann-Roch theorem implies $\ell(D) = \deg(D) = d$, so the computation of $Z(E)$ is now straightforward

$$\begin{aligned} Z(E) &= 1 + \sum_{d \geqslant 1} T^d \sum_{\substack{D \geqslant 0 \\ \deg(D) = d}} 1 = 1 + \frac{h_0(C)}{q-1} \sum_{d \geqslant 1} (q^d - 1) T^d \\ &= 1 + \frac{h_0(C)}{q-1} \Big(\frac{T^d}{1-qT} - \frac{T}{1-T} \Big) = 1 + \frac{h_0(C) T}{(1-T)(1-qT)} \\ &= \frac{1 - bT + qT^2}{(1-T)(1-qT)} \end{aligned} \tag{11.42}$$

where b is defined by $h_0(C) = q + 1 - b$.

This proves the rationality, and gives the precise form, except that we need to prove that $a = b$, where $|E(\mathbb{F})| = q+1-a$, or equivalently $h_0(C) = |E(\mathbb{F})|$ (actually in Lemma 11.32, we have already shown $|h_0(C)| \leqslant |E(\mathbb{F})|$, but we do not need it any more). To obtain this equality, start from the original definition (11.39) of $Z(E)$, and compare with (11.42): the latter is seen to imply that $|E(\mathbb{F})| = q+1-b = h_0(C)$.

Finally the functional equation of $Z(E)$ is a formal consequence of (11.42). □

It is worth recording separately one of the last steps of the proof.

PROPOSITION 11.33. *Let E be an elliptic curve over a finite field $\mathbb{F}$, let D be a divisor on E. Then D is principal if and only if $\deg(D) = 0$ and $\sigma(D) = 0 \in E$. More precisely, the map $j : D \mapsto \sigma(D)$ is an isomorphism between the group of divisor classes of degree 0 and $E(\overline{\mathbb{F}})$.*

PROOF. A divisor D is $\mathbb{F}_n$-rational for some $n \geqslant 1$; looking at E over $\mathbb{F}_n$, it suffices to prove the isomorphism between classes of $\mathbb{F}$-rational divisors of degree 0 and $\mathbb{F}$-rational points. But j is a surjective ($j([x] - [\infty]) = x$) map between finite sets with the same cardinality ($|h_0(C)| = |E(\mathbb{F})|$). □

This is the special case of the so-called Abel-Jacobi Theorem, for an elliptic curve over a finite field. It actually holds over any field, and a generalization to all (smooth projective) curves is the content of the theory of jacobian varieties associated to curves.

To conclude the proof of Theorem 11.25, we proceed as in the case of Kloosterman sums: from (11.41), we derive

$$|E(\mathbb{F}_n)| - (q^n + 1) = \alpha^n + \beta^n$$

where $1 - aT + qT^2 = (1 - \alpha T)(1 - \beta T)$. Then Lemma 11.22, applied with the input from Stepanov's Theorem 11.13, shows that $|\alpha| \leqslant \sqrt{q}$, $|\beta| \leqslant \sqrt{q}$, and since $\alpha\beta = q$, this concludes the proof.

EXERCISE 2. Assuming the general Riemann-Roch formula (11.38), prove that for a smooth projective algebraic curve E of genus g over a finite field $\mathbb{F}$ with q elements, the zeta function

$$Z(E) = \exp\Bigl(\sum_{n\geqslant 1} \frac{|E(\mathbb{F}_n)|}{n} T^n\Bigr)$$

is a rational function of the form

$$Z(E) = \frac{P(T)}{(1-T)(1-qT)}$$

for some polynomial P with integral coefficients and degree $2g$.

[**Hint:** The question will arise whether there exist $\mathbb{F}$-rational divisor classes on E of degree 1 (which is obvious for elliptic curves since the point ∞ is $\mathbb{F}$-rational). The image of the degree map is $\delta\mathbb{Z}$ for some $\delta \mid (2g-2)$ (the degree of the canonical class). Using this fact, find a preliminary form of the zeta function and analyze the poles to show that actually $\delta = 1$ (see [**Mor2**], 3.3).]

11.11. Survey of further results: a cohomological primer.

The methods of Stepanov are very useful and, in certain circumstances they provide the best tools available today, especially when the genus of the curve is large compared to the cardinality of the finite field (see for instance the proof by Heath-Brown of non-trivial estimates for Heilbronn sums [**HB2**]).

However, the deepest understanding of exponential sums over finite fields and the greatest impact on classical problems of analytic number theory comes from the sophisticated concepts of algebraic number theory, especially the ℓ-adic cohomology theory as developed by Grothendieck and his collaborators, which give a very powerful and flexible framework for working with very general exponential sums.

The proof of the Riemann Hypothesis for varieties by Deligne [**De1**], and even more his far-reaching generalization [**De2**], are the basis for the extensive work of Katz, Laumon and others. It is beyond the scope of this book to discuss this theory in great detail. Let us direct the interested reader to the survey articles [**Lau**], [**K2**]. Study of the foundational basis of the ℓ-adic theory can be started in [**De3**] and continued together with applications in the books of Katz, for instance [**K3**], [**K4**].

We will limit this section to a short introduction of the basic vocabulary and we will state a few of the most fundamental results in this language. We then include examples to show that such knowledge can already be very useful even when one is not familiar with the details and background of algebraic geometry.

In Sections 11.4 and 11.5, we have shown that Gauss sums and Kloosterman sums can be related to analogues of Dirichlet characters over finite fields. The ℓ-adic cohomological formalism which we now discuss can be thought as relating exponential sums, dually, to objects which are Galois-theoretic in nature.

The exponential sums S_n defined by (11.8) can be interpreted as sums over the algebraic curve $U_{f,g}$ consisting of $\overline{\mathbb{F}}_p^*$ minus the poles of the rational functions f and g. More generally, one wishes to consider exponential sums not only over curves but over more general varieties. We will use some basic vocabulary of algebraic geometry to describe such situations, but will illustrate them in the simpler case of curves. Already the case of (11.8) and $U_{f,g}$ are quite interesting.

Let $\mathbb{F}$ be a finite field and $U/\mathbb{F}$ be a smooth algebraic variety of dimension $d \geqslant 0$ (technically, we assume as part of the smoothness assumption that U is geometrically connected, and as part of being a variety that U is quasi-projective). The simplest examples in dimension 1 are $U_{f,g}$, or smooth projective curves. In dimension $d > 1$, the most important examples are the affine d-space $\mathbb{A}^d$, with set of points $\mathbb{A}^d(\bar{\mathbb{F}}) = \bar{\mathbb{F}}^d$, and the projective d-space. The exponential sums over U will be of the type

$$S_n = \sum_{x \in U(\mathbb{F}_n)} \chi(N(f(x)))\psi(\mathrm{Tr}\,(g(x))) \tag{11.43}$$

where f and g are $\mathbb{F}$-rational functions defined on U.

To $U/\mathbb{F}$ is associated the so-called arithmetic étale fundamental group $\pi_1(U)$ which "classifies" étale coverings $V \to U$ of U, and is the analogue both of the Galois group of a field, or of the "ordinary" topological fundamental group. A morphism of algebraic varieties is étale if it is flat and unramified; if U is a curve, this means V is a curve, f is non-constant and unramified. For the simpler purposes of exponential sums, the fundamental group can be considered somewhat as a black box in what follows, but one should keep in mind that the elements of V in $\pi_1(U)$ act as automorphisms of any étale covering $\pi : V \to U$ (i.e. $\pi(\gamma x) = \pi(x)$ for any $\gamma \in \pi_1(U)$ and $x \in V$), and that it is a functor: any map $U \to V$ between varieties induces a continuous group homomorphism $\pi_1(U) \to \pi_1(V)$. (One should fix a base-point in defining $\pi_1(U)$, but a more or less canonical choice exists, the so-called "generic point" of the scheme U.)

EXAMPLES. (1) Let U be a single point $\{x\}$ defined over $\mathbb{F}$. Then $\pi_1(U)$ is the Galois group $\mathrm{Gal}(\bar{\mathbb{F}}/\mathbb{F})$.

(2) Let $U/\mathbb{F}$ be a smooth curve, not necessarily projective. There is a an associated smooth projective curve $C/\mathbb{F}$ such that $U \subset C$ with complement a finite set T of points. If $U = \overline{\mathbb{F}}^*$ for instance, then $C = \mathbb{P}^1$ is the projective line, and $T = \{0, \infty\}$.

The fundamental group can be described concretely as follows: let $K = \mathbb{F}(U) = \mathbb{F}(C)$ be the function field of U, i.e. the field of rational functions on U or C (if $C = \mathbb{P}^1$, then $K = \mathbb{F}(t)$ is the usual field of rational fractions). We have the Galois group $G_K = \mathrm{Gal}(\bar{K}/K)$ of K. For every closed point x of C, there is the corresponding discrete valuation ord_x of K. This extends to the separable closure $\bar{K}$ of K, and gives rise to a decomposition group $D_x < G_K$ and an inertia group $I_x < D_x$ as in classical algebraic number theory, with the property that $D_x/I_x \simeq \mathrm{Gal}(\bar{\mathbb{F}}_q/\mathbb{F}_q)$, where $\mathbb{F}_q$ is the residue field of x, a finite field with $q = Nx$ elements. Then $\pi_1(U)$ "is" the quotient of G_K by the smallest closed normal containing all inertia groups I_x for x a closed point of U.

Fix a prime number $\ell \neq p$. The objects used to interpret exponential sums over U are the so-called ℓ-adic sheaves on U. In the simpler cases, those will be "lisse", in which case there is a simpler alternate Galois-theoretic description which we take as definition.

DEFINITION. Let $U/\mathbb{F}$ be a smooth variety over a finite field. A lisse ℓ-adic sheaf on U is a continuous representation $\rho : \pi_1(U) \longrightarrow GL(V)$ where V is a finite dimensional $\bar{\mathbb{Q}}_\ell$-vector space. Continuity refers to the profinite topology on $\pi_1(U)$ and the ℓ-adic topology on V.

Note the similarity with the definition of Galois representations of number fields (see Section 5.13). Because of the original definition of a sheaf, one usually denotes ℓ-adic sheaves by curly letters $\mathcal{F}$, $\mathcal{G}$, etc. Notice that one can obviously speak of direct sums, tensor product, symmetric powers, etc., of lisse ℓ-adic sheaves by performing the corresponding operations on the representations. Also one can speak of irreducible sheaves, etc.

An important ℓ-adic sheaf, denoted $\bar{\mathbb{Q}}_\ell(1)$, is obtained by considering the natural action of $\pi_1(U)$ on ℓ-power roots of unity, which arises from the étale coverings where one simply extends the base field from $\mathbb{F}$ to its extension by roots of unity. This action is given by a certain character $\chi_\ell : \pi_1(U) \to \bar{\mathbb{Q}}_\ell^*$. Using this sheaf, one defines Tate twists: if $\mathcal{F}$ is a lisse ℓ-adic sheaf and $i \in \mathbb{Z}$, then one denotes $\mathcal{F}(i)$ ($\mathcal{F}$ twisted i times) the sheaf which corresponds to the action ρ' of $\pi_1(U)$ on the same vector space but with

$$\rho'(\gamma) = \chi_\ell^i(\gamma)\rho(\gamma);$$

in other words, $\mathcal{F}(1) = \mathcal{F} \otimes \bar{\mathbb{Q}}_\ell(1)$ for instance.

Exponential sums arise by looking at the action of the Frobenius elements at points of U. Let x be a closed point x of U, which can be seen as a Galois orbit of points in $U(\bar{\mathbb{F}})$. The fundamental group of the "point" x is the Galois group D_x of the residue field of U at x, isomorphic to $\mathbb{F}_n$ where n is the degree of x. By functoriality there is a map $D_x \to \pi_1(U)$. We have $D_x \simeq \mathrm{Gal}(\bar{\mathbb{F}}_n/\mathbb{F}_n)$ and the latter is generated (topologically) by the Frobenius morphism σ, so taking the image we get in $\pi_1(U)$ a well-defined conjugacy class, called the arithmetic Frobenius conjugacy class at x. In particular, for any ℓ-adic sheaf one can speak of the trace $\mathrm{Tr}\,\rho(\sigma_x)$ without ambiguity. However, it turns out that it is the inverse F of σ (the so-called geometric Frobenius) which appears naturally in the cohomological description of exponential sums. We denote by F_x the corresponding conjugacy class; it is called simply the Frobenius conjugacy class at x (omitting the adjective geometric).

THEOREM 11.34. *Let $U/\mathbb{F}$ be a smooth variety, let $f \neq 0$ and g be $\mathbb{F}$-rational functions on U, let ψ be an additive character and χ a multiplicative character of $\mathbb{F}$. Let $S_n = S_n(U, f, g, \chi, \psi)$ be the associated exponential sums over $U(\mathbb{F}_n)$ as in (11.43). Then there exists a lisse ℓ-adic sheaf $\mathcal{F}$ on U of degree 1 with the property that for all $n \geqslant 1$ we have*

$$S_n = \sum_{x \in U(\mathbb{F}_n)} Tr(F_x \mid \mathcal{F}) \tag{11.44}$$

where we denote $Tr(g \mid \mathcal{F}) = Tr(\rho(g) \mid V)$, $\mathcal{F}$ corresponding to the representation $\rho : \pi_1(U) \to GL(V)$.

To compare with the characters used to describe Gauss sums and Kloosterman sums, one should think of the latter as analogues of Dirichlet characters or Hecke characters, whereas the ℓ-adic sheaves given by this theorem are analogues of Galois characters. The correspondence between the two concepts is an instance of reciprocity or class-field theory.

We sketch the construction of $\mathcal{F}$ in the case where $\chi = 1$ and g is a non-zero rational function on $U = \mathbb{A}^1 - \{\text{poles of } g\}$, over $\mathbb{F}$, which makes it clear that this is very closely related to the argument in Section 11.7. Consider the curve

$$C : y^q - y = g(x) \tag{11.45}$$

and notice that there is a surjective map $\pi : (x, y) \mapsto x$ from C to U. For any $a \in \bar{\mathbb{F}}$, the equation $y^q - y - a = 0$ is separable, hence it has q distinct roots in $\bar{\mathbb{F}}$. In fact the additive group of $\mathbb{F}$ acts on the roots by translation: if y is a root and $z \in \mathbb{F}$, then $(y + z)^q - (y + z) = y^q - y = a$. Moreover, $\pi : C \to U$ is an étale covering (we've just seen it is everywhere unramified and surjective). In other words, π is an étale Galois covering with Galois group isomorphic to the additive group $\mathbb{F}$ (coverings given by such equations are called Artin-Shreier coverings).

The fundamental group $\pi_1(U)$ acts on C by automorphisms of the covering, which means as translations by elements of $\mathbb{F}$ as above. This defines a surjective map $\varphi : \pi_1(U) \to \mathbb{F}$ such that $\varphi(\gamma) = \gamma y - y$ for any $y \in C$. (This doesn't depend on the choice of y because the action of γ on C must be a morphism of curves.)

Consider the trivial ℓ-adic sheaf $\bar{\mathbb{Q}}_\ell$ on C, or equivalently the trivial representation of $\pi_1(C)$. By the above we can construct the induced representation ρ from

$\pi_1(C)$ to $\pi_1(U)$, which can be described as the space

$$V = \{f : \pi_1(U) \to \bar{\mathbb{Q}}_\ell \mid f(\tau\gamma) = f(\gamma) \text{ for any } \tau \in \pi_1(C)\}$$

(where $\tau \in \pi_1(C)$ is seen through the map $\pi_1(C) \to \pi_1(U)$ coming from π), on which $\pi_1(U)$ acts by translation on the right

$$\rho(\gamma)f(\tau) = f(\tau\gamma).$$

The elements $f \in V$ depend only on $\pi_1(C)\backslash\pi_1(U) \simeq \mathbb{F}$ (i.e. on the automorphisms of the covering $C \to U$), which implies that $V \simeq \bar{\mathbb{Q}}_\ell^q$ is an ℓ-adic sheaf on U of degree q. The representation space V can be decomposed over the additive characters ψ of $\mathbb{F}$,

$$V = \bigoplus_\psi \mathcal{L}_\psi$$

where $\mathcal{L}_\psi$ is the ψ-eigencomponent of V, namely

$$\mathcal{L}_\psi = \{f \in V \mid \rho(\gamma)f = \psi(\varphi(\gamma))f \text{ for all } \gamma \in \pi_1(U)\}.$$

It is easy to see that each $\mathcal{L}_\psi$ is an ℓ-adic sheaf on U, and because ρ is induced from the trivial representation, each $\mathcal{L}_\psi$ is of degree 1.

Then for every additive character ψ, the ℓ-adic sheaf on U corresponding to $\mathcal{L}_{\bar{\psi}}$ is the sheaf satisfying (11.44) for the exponential sums $S_n(U, g, \psi)$.

Indeed, if $x \in U(\mathbb{F}_n)$, and y satisfies $y^q - y = g(x)$, then the Frobenius of x acts on y by $y^{q^n} = y + \mathrm{Tr}_{\mathbb{F}_n/\mathbb{F}}(g(x))$ since

$$\begin{aligned} y^{q^n} - y &= y^{q^n} - y^{q^{n-1}} + y^{q^{n-1}} - \cdots + y^q - y \\ &= (y^q - y)^{q^{n-1}} + \cdots + y^q - y = \mathrm{Tr}\,(y^q - y) = \mathrm{Tr}\, g(x). \end{aligned}$$

Hence $\varphi(\sigma_x) = \mathrm{Tr}\, g(x)$ and by definition of $\mathcal{L}_\psi$ it follows that σ_x acts on $\mathcal{L}_\psi$ by multiplication by $\psi(\mathrm{Tr}\, g(x))$, hence $F_x = \sigma_x^{-1}$ acts by $\bar{\psi}(\mathrm{Tr}\, g(x))$, which gives (11.44).

In particular, note that taking the trace for $\mathbb{Q}_\ell$ on C we derive

$$|C(\mathbb{F}_n)| = \sum_\psi S_n(U, f, \psi),$$

as in (11.30).

EXERCISE 3. (1) Let S_n be the character sum (11.8) with $g = 0$ for some multiplicative character χ of $\mathbb{F}^\star$ and some non-zero rational function $f \in \mathbb{F}(x)$, on the variety $U = \mathbb{A}^1 - \{\text{zeros and poles of } f\}$. Describe as above the construction of the sheaf $\mathcal{L}$ satisfying (11.44) in this case. [**Hint:** Use the cover $y^d = f(x)$, where d is the order of the multiplicative character χ.]

(2) Let S_n be as in (11.8), $U \subset \mathbb{A}^1$ the complement of the zeros and poles of f and the poles of g. If $\mathcal{L}_\psi$ is the sheaf satisfying (11.44) for $f = 1$ and $\mathcal{L}_\chi$ is the sheaf satisfying (11.44) for $g = 0$, show that $\mathcal{L} = \mathcal{L}_\psi \otimes \mathcal{L}_\chi$ satisfies (11.44) for S_n.

EXAMPLES. (1) Even the case $\rho = 1$ is interesting when dealing with a general variety U. This "trivial" ℓ-adic sheaf is denoted $\bar{\mathbb{Q}}_\ell$, and one has $S_n = |U(\mathbb{F}_n)|$.

(2) For the sheaf $\bar{\mathbb{Q}}_\ell(1)$, notice that σ_x acts by $\xi \mapsto \xi^q$ for any root of unity if $Nx = q$. Therefore F_x acts by $\xi \mapsto \xi^{1/q}$ and in particular the only eigenvalue of F_x is q^{-1}.

Now in addition to $U/\mathbb{F}$ we consider its "extension of scalars" $\bar{U}/\bar{\mathbb{F}}$ over the algebraic closure of $\mathbb{F}$. There is a corresponding geometric fundamental group $\pi_1(\bar{U})$, which sits in an exact sequence

$$1 \to \pi_1(\bar{U}) \to \pi_1(U) \to \mathrm{Gal}(\bar{\mathbb{F}}/\mathbb{F}) \to 1. \tag{11.46}$$

To every ℓ-adic sheaf $\mathcal{F}$ on U are associated the ℓ-adic cohomology groups with compact support of $\bar{U}$ with coefficients in $\mathcal{F}$. Those are finite-dimensional $\bar{\mathbb{Q}}_\ell$-vector spaces, denoted, $H^i_c(\bar{U}, \mathcal{F})$ for $i \geqslant 0$. The key point is that the Galois group of $\mathbb{F}$ acts naturally on $H^i_c(\bar{U}, \mathcal{F})$, and in particular, so do the Frobenius σ and its inverse F, the geometric Frobenius. The key to the cohomological interpretation of exponential sums is the

GROTHENDIECK-LEFSCHETZ TRACE FORMULA. *Let $U/\mathbb{F}$ be a smooth variety of dimension $d \geqslant 0$, $\mathcal{F}$ an ℓ-adic sheaf on U. We have $H^i_c(\bar{U}, \mathcal{F}) = 0$ if $i > 2d$ and for any $n \geqslant 1$*

$$\sum_{x \in U(\mathbb{F}_n)} Tr(F_x \mid \mathcal{F}) = Tr(F^n \mid H^0_c(\bar{U}, \mathcal{F})) - Tr(F^n \mid H^1_c(\bar{U}, \mathcal{F})) + \cdots \tag{11.47}$$
$$- Tr(F^n \mid H^{2d-1}_c(\bar{U}, \mathcal{F})) + Tr(F^n \mid H^{2d}_c(\bar{U}, \mathcal{F})).$$

Therefore to evaluate the exponential sums (11.43) using the associated sheaf, we need to know the traces, or equivalently the eigenvalues, of F (equivalently, of $\sigma = F^{-1}$) acting on H^i_c for $0 \leqslant i \leqslant 2d$. It turns out that in most cases H^0_c and H^{2d}_c are easy to compute:

PROPOSITION 11.35. *Let $\mathcal{F}$ be a lisse ℓ-adic sheaf on a smooth variety $U/\mathbb{F}$, corresponding to the representation ρ of $\pi_1(U)$ on the $\bar{\mathbb{Q}}_\ell$-vector space V. We have*

$$H^0_c(\bar{U}, \mathcal{F}) \simeq \begin{cases} V^{\pi_1(\bar{U})} & \text{if } U \text{ is projective,} \\ 0 & \text{if } U \text{ is not projective,} \end{cases} \tag{11.48}$$

and

$$H^{2d}_c(\bar{U}, \mathcal{F}) \simeq V_{\pi_1(\bar{U})}(-2d) \tag{11.49}$$

where V^G denotes the space of vectors invariant under the action of a group G on an abelian group, and V_G denotes the space of co-invariants, the largest quotient of V on which G acts trivially. In both cases, the isomorphisms are canonical isomorphisms of vector spaces with an action of the Galois group of $\mathbb{F}$.

Since V is a representation of $\pi_1(U)$, the exact sequence (11.46) shows that $V^{\pi_1(\bar{U})}$ and $V_{\pi_1(\bar{U})}$ are acted on by $\mathrm{Gal}(\bar{\mathbb{F}}/\mathbb{F})$, "through" the given representation ρ.

This proposition shows that for a curve $U/\mathbb{F}$, the only "difficult" cohomology group is $H^1_c(\bar{U}, \mathcal{F})$.

EXAMPLE. Let $U = E/\mathbb{F}_p$ be an elliptic curve, $\mathcal{F} = \bar{\mathbb{Q}}_\ell$ the trivial sheaf. By the proposition one has

(1) $H^0_c(\bar{E}, \bar{\mathbb{Q}}_\ell) = \bar{\mathbb{Q}}_\ell$, with trivial action of F (since $\bar{\mathbb{Q}}_\ell$ is the trivial sheaf).

(2) $H^2_c(\bar{E}, \bar{\mathbb{Q}}_\ell) = \bar{\mathbb{Q}}_\ell(-1)$, so by definition of the twist, F acts by multiplication by p (on roots of unity, i.e. on $\bar{\mathbb{Q}}_\ell(1)$, σ acts by $\xi \mapsto \xi^p$, hence F by multiplication by p^{-1}).

The Lefschetz trace formula (11.47) gives

$$|E(\mathbb{F}_n)| = p^n + 1 - \operatorname{Tr}(F^n \mid H_c^1(\bar{E}, \bar{\mathbb{Q}}_\ell))$$

(compare Theorem 11.31).

More generally, one derives the rationality of the zeta function directly from the Trace Formula.

COROLLARY 11.36. *Let U, S_n and $\mathcal{F}$ be as in Theorem 11.34. For $0 \leqslant i \leqslant 2d$, let $b_i = \dim H_c^i(\bar{U}, \mathcal{F})$ and*

$$P_i(T) = \det(1 - FT \mid H_c^i(\bar{U}, \mathcal{F})) = \prod_{j=1}^{b_i} (1 - \alpha_{i,j} T).$$

We have

$$Z(\mathcal{F}) = \exp\Big(\sum_{n \geqslant 1} \frac{S_n}{n} T^n\Big) = \frac{P_1(T) \cdots P_{2d-1}(T)}{P_0(T) \cdots P_{2d}(T)} = \prod_{i=0}^{2d} \prod_j (1 - \alpha_{i,j} T)^{(-1)^{i+1}},$$

and for $n \geqslant 1$,

$$S_n = \sum_{0 \leqslant i \leqslant 2d} (-1)^i \alpha_{i,j}^n. \tag{11.50}$$

Theorems 11.4, 11.8 and the result of Exercise 1 are all special cases of this corollary, together with suitable computations of cohomology groups. The numbers b_i are called the ℓ-adic Betti numbers for $\mathcal{F}$.

Of much greater importance, however, is Deligne's vast generalization of the Riemann Hypothesis [**De2**]. One starts with the following "local" definition:

DEFINITION. Let $w \in \mathbb{Z}$ be an integer. A lisse ℓ-adic sheaf $\mathcal{F}$ on $U/\mathbb{F}$ is said to be pure of weight w if for any closed point x of U, all eigenvalues of F_x acting on the $\bar{\mathbb{Q}}_\ell$ vector space V associated to $\mathcal{F}$ are algebraic integers all conjugates of which have the same absolute value equal to $q^{w/2}$ where $q = Nx$ is the cardinality of the residue field.

For instance, the trivial sheaf $\bar{\mathbb{Q}}_\ell$ is pure of weight 0 (all eigenvalues 1). For any $i \in \mathbb{Z}$, $\bar{\mathbb{Q}}_\ell(i)$ is pure of weight $-2i$, and if $\mathcal{F}$ is pure of weight w, then $\mathcal{F}(i)$ is pure of weight $w - 2i$. For any exponential sum (11.43), the associated sheaf $\mathcal{F}$ is pure of weight 0 because the only eigenvalue at x is the root of unity $\chi(Nf(x))\psi(\operatorname{Tr} g(x))$.

THEOREM 11.37 (DELIGNE). *Let $U/\mathbb{F}$ be a smooth variety and $\mathcal{F}$ a lisse ℓ-adic sheaf on U, pure of weight w. Let $i \geqslant 0$ and let ξ be any eigenvalue of the geometric Frobenius F acting on $H_c^i(\bar{U}, \mathcal{F})$. Then ξ is an algebraic integer, and if $\alpha \in \mathbb{C}$ is a conjugate of ξ, we have*

$$|\alpha| \leqslant q^{(w+i)/2}. \tag{11.51}$$

The conclusion is also phrased as saying that $H_c^i(\bar{U}, \mathcal{F})$ is mixed of weights $\leqslant i + w$. If there is equality in (11.51), then $H_c^i(\bar{U}, \mathcal{F})$ is said to be pure (of weight $w + i$). In certain cases, one can apply duality theorems (for instance Poincaré duality) to deduce further that (11.51) is an equality.

REMARK. Although Deligne's proof is a monumental achievement of very deep algebraic geometry, it is an interesting fact that a crucial use is made of a generalization of the method of Hadamard and de la Vallée Poussin for proving non-vanishing of L-functions on the line $\mathrm{Re}\,(s) = 1$ (see Section 5.4). Similarly, in Deligne's first proof [**De1**], the ideas of the classical Rankin-Selberg method for modular forms are essential (specifically, Deligne acknowledges the influence of [**Ra3**]).

EXAMPLE. Let $C/\mathbb{F}$ be a smooth connected projective curve (for instance, an elliptic curve). By Proposition 11.35 as in the previous example, we have easily:

(1) $H_c^0(\bar{C}, \bar{\mathbb{Q}}_\ell) = \bar{\mathbb{Q}}_\ell$, with F acting trivially.

(2) $H_c^2(\bar{C}, \bar{\mathbb{Q}}_\ell) = \bar{\mathbb{Q}}_\ell(-1)$, with F acting by multiplication by p.

It is more difficult to show that

(3) $H_c^1(\bar{C}, \bar{\mathbb{Q}}_\ell) \simeq \bar{\mathbb{Q}}_\ell^{2g}$, as $\bar{\mathbb{Q}}_\ell$ vector spaces (not as Galois-modules!), where $g \geqslant 0$ is the genus of $\bar{C}$ (for an elliptic curve $g = 1$).

Moreover, there is a Galois-invariant perfect pairing

$$H_c^1(\bar{C}, \bar{\mathbb{Q}}_\ell) \times H_c^1(\bar{C}, \bar{\mathbb{Q}}_\ell) \longrightarrow \bar{\mathbb{Q}}_\ell(-1).$$

It follows that if α is one of (the complex conjugates of) the eigenvalues of F on $H_c^1(\bar{C}, \bar{\mathbb{Q}}_\ell)$, then p/α is one also. Hence from Theorem 11.37, since $\bar{\mathbb{Q}}_\ell$ is pure of weight 0, one deduces that $|\alpha| = \sqrt{p}$. Thus

$$|C(\mathbb{F}_n)| = p^n + 1 - \sum_{i=1}^{2g} \alpha_i^n$$

where the $\alpha_i \in \bar{\mathbb{Q}}$ are the eigenvalues of F on H_c^1. Estimating trivially now, we get

$$\Big| |C(\mathbb{F}_n)| - (p^n + 1) \Big| \leqslant 2gp^{n/2}$$

recovering the Riemann Hypothesis, and in particular, Theorem 11.25 for the case $g = 1$.

In the case of exponential sums (11.43), the sheaf $\mathcal{F}$ is pure of weight 0, hence denoting

$$d(\mathcal{F}) = \max\{i \mid H_c^i(\bar{U}, \mathcal{F}) \neq 0\},$$

we derive directly from (11.50) and (11.51) the bound

$$|S_n| \leqslant \sum_{0 \leqslant i \leqslant d(\mathcal{F})} b_i q^{ni/2}, \tag{11.52}$$

for $n \geqslant 1$ and, in particular,

$$|S_n| \leqslant q^{nd(\mathcal{F})/2} \Big(\sum_i b_i\Big). \tag{11.53}$$

As in the case of Kloosterman sums, the exponent $d(\mathcal{F})/2$ is best possible in this inequality. The bound $d(\mathcal{F}) \leqslant 2d$ gives a trivial estimate (because U is smooth of dimension d, it has about q^{nd} points, as proved by the Riemann Hypothesis for the trivial sheaf $\bar{\mathbb{Q}}_\ell$). Any improvement of this trivial bound is equivalent with $H_c^{2d}(\bar{U}, \mathcal{F}) = 0$, and the square root cancellation often expected from heuristic reasonings is equivalent with $H_c^i(\bar{U}, \mathcal{F}) = 0$ for $i > d$. Although not always true, this turns out to hold "generically", as the analytic intuition suggests (see for instance Theorem 11.43 below).

For the exponential sums (11.43) we have $d(\mathcal{F}) < 2d$, unless $\mathcal{F}$ is the trivial sheaf $\bar{\mathbb{Q}}_\ell$, so there is always a non-trivial bound. This follows from (11.49), since $\mathcal{F}$ is of degree 1 so the space of co-invariants is either the whole space (meaning the representation is trivial) or 0. However, this small gain is usually insufficient in applications.

Another surprising consequence of Deligne's result and the discreteness of integers is the following "self-improving" statement:

COROLLARY 11.38. *Let S_n be an exponential sum as in (11.43) and $\mathcal{F}$ the associated sheaf. Suppose $w \geqslant 0$ is an integer such that*

$$|S_n| \ll q^{w/2+\delta}$$

for some $\delta \in [0, \frac{1}{2}[$ and $n \geqslant 1$. Then we have $d(\mathcal{F}) \leqslant w$, hence $|S_n| \ll q^{w/2}$.

A second issue in applying the estimates (11.52) or (11.53) in the context of applications to analytic number theory is that we usually have $\mathbb{F} = \mathbb{Z}/p\mathbb{Z}$, with the prime number p varying. In this case, whereas the variety U can be defined over $\mathbb{Q}$ (or $\mathbb{Z}$) so that the sum is, for all p, over the $\mathbb{F}_p$-points of the reduction U_p of U modulo p, the sheaves $\mathcal{F}_p$ genuinely depend on p (see the equation (11.45)), i.e. there is no theory of sheaves over $U/\mathbb{Z}$ giving each $\mathcal{F}_p$ by "reduction modulo p". (Katz has asked a number of times for such a theory of "exponential sums over $\mathbb{Z}$"; see e.g. [**K2**], but it remains elusive.) Thus, the Betti numbers

$$b_i(p) = \dim H^i_c(\bar{U}_p, \mathcal{F}_p)$$

of the cohomology groups can depend on p, and the applicability of the results above would be ruined, even with the Riemann Hypothesis, if these dimensions were not bounded in a reasonable way in terms of p.

This is in fact the case. The first general result in this direction is due to Bombieri [**Bo4**] for additive character sums (11.43) where $f = 1$, and was generalized by Adolphson and Sperber [**AS1**], [**AS2**] for general sums (their methods are p-adic, based on Dwork's original ideas). In general, those results bound the Euler characteristic

$$\chi_c(\mathcal{F}) = \sum_{i=0}^{2d} (-1)^i \dim H^i_c(\bar{U}, \mathcal{F}) = \sum_{0\leqslant i\leqslant 2d} (-1)^i b_i,$$

of a sheaf $\mathcal{F}$ on $U/\mathbb{F}$, but further arguments of Katz [**K5**] show how to deduce bounds for

$$\sigma_c(\mathcal{F}) = \sum_{i=0}^{2d} \dim H^i_c(\bar{U}, \mathcal{F}) = \sum_{0\leqslant i\leqslant 2d} b_i,$$

(hence for $b_i \leqslant \sigma_c(\mathcal{F})$) from those for $\chi_c(\mathcal{F})$.

THEOREM 11.39. *Let $U/\mathbb{Q}$ be a smooth variety over $\mathbb{Q}$, f and g functions on U with f invertible. Let ℓ be a prime number and for all $p \neq \ell$ such that the reduction U_p of U modulo p is smooth, let χ and ψ be any multiplicative and additive characters of $\mathbb{F}_p$. Let $\mathcal{F}_p$ be an ℓ-adic sheaf on U_p such that*

$$\sum_{x\in U(\mathbb{F}_{p^n})} \chi(Nf(x))\psi(Tr g(x)) = \sum_{x\in U(\mathbb{F}_{p^n})} Tr(F_x \mid \mathcal{F}_p)$$

for $n \geqslant 1$. We have $\sigma_c(\mathcal{F}_p) \leqslant C$ where C is a constant depending only on U, f and g.

A simple explicit bound is given in [**AS3**] if $f(x) = 1$, so only the additive characters occur, and g is a Laurent polynomial on $U = (\bar{\mathbb{Q}} - \{0\})^d$. The sums over $\mathbb{Z}/p\mathbb{Z}$ in question are therefore sums in d variables of the type

$$S_{f,p} = \sum_{x_1,\dots,x_d \in (\mathbb{Z}/p\mathbb{Z})^*} \psi(f(x_1,\dots,x_d)) \tag{11.54}$$

where $f \in \mathbb{Q}[x_1, x_1^{-1}, \dots, x_d, x_d^{-1}]$ is a non-zero Laurent polynomial. Writing

$$f = \sum_{j \in J} a_j x^j$$

for some (finite) set $J \subset \mathbb{Z}^d$, the Newton polyhedron $W(f)$ of f is defined to be the convex hull in $\mathbb{R}^d$ of $J \cup \{0\}$.

PROPOSITION 11.40. *With the above assumptions, denoting by $\mathcal{F}_{f,p}$ the associated sheaf for the sums $S_{f,p}$, we have*

$$|\chi_c(\mathcal{F}_{f,p})| \leqslant d!\mathrm{Vol}\,(W(f)),$$
$$\sigma_c(\mathcal{F}_{f,p}) \leqslant 10^d d!\mathrm{Vol}\,(W(f))$$

for any p not dividing the denominator of any coefficient of f, where Vol $(W(f))$ *is the volume of the Newton polyhedron in the subspace spanned by $W(f)$ in $\mathbb{R}^d$, with respect to Lebesgue measure.*

Note that by using exclusion-inclusion and detecting polynomials equations by means of multiplicative characters, one can use combinations of sums of the type (11.54) to describe much more general ones. Also, in many cases, one can show that only all the odd (or even) cohomology groups vanish, in which case $|\chi_c(\mathcal{F})| = \sigma_c(\mathcal{F})$. See also Theorems 11 and 12 of [**K5**] for explicit estimates in quite general cases.

We now give examples of computations using these fundamental results. For exponential sums arising in analytic number theory, one often needs nothing more, if one uses skillfully some other simple tricks such as averaging over extra parameters to analyze the weight of the roots.

EXAMPLE 1. The Kloosterman sums $S(a, b; p)$ for $ab \neq 0$ can be treated using Proposition 11.40 with $d = 1$ and $f(x) = ax + bx^{-1}$. Then $W(f)$ is the interval $[-1, 1]$. By Proposition 11.35, we have $H_c^0 = H_c^2 = 0$ in this case since $U = \mathbb{P}^1 - \{0, \infty\}$ is not projective, so $\sigma_c = -\chi_c$. By Theorem 11.37, H_c^1 is mixed of weight $\leqslant 1$. Hence we recover the Weil bound:

$$|S(a, b; p)| \leqslant \sigma_c p^{1/2} \leqslant 2p^{1/2}.$$

(Of course, in fact we have $b_1 = 2$ and the last inequality is an equality).

EXAMPLE 2. The previous example generalizes to the multiple Kloosterman sums defined by

$$K_r(a, q) = \sum_{x_1 \cdots x_r = a} e\Big(\frac{\mathrm{Tr}\,(x_1 + \cdots + x_r)}{p}\Big) \tag{11.55}$$

for $r \geqslant 2$ and $a \neq 0$, so $K_2(a,p) = S(a,1;p)$ (see [**Bo4**], [**De1**]). Without appealing to the L-function, one can nevertheless get some information by averaging over a. We get

$$\sum_{a\neq 0} |K_r(a,q)|^2 = q^r - q^{r-1} - \cdots - q - 1. \tag{11.56}$$

Hence $|K_r(a,q)| \leqslant q^{r/2}$. To improve this elementary bound we appeal to the following Lemma

LEMMA 11.41. *Given a finite set of distinct angles θ_i modulo 2π and complex numbers α_i we have*

$$\sum_{n\leqslant N} \Big|\sum_i \alpha_i e(n\theta_i)\Big|^2 = N\|\alpha\|^2 + O(1)$$

where the implied constant does not depend on N. Hence

$$\limsup_{n\to+\infty} \Big|\sum_i \alpha_i e(n\theta_i)\Big| \geqslant \|\alpha\|. \tag{11.57}$$

PROOF. We have

$$\sum_{n\leqslant N} \Big|\sum_i \alpha_i e(n\theta_i)\Big|^2 = N\sum_i |\alpha_i|^2 + \sum_{i\neq j}\sum \alpha_i\alpha_j \sum_{n\leqslant N} e(n(\theta_i - \theta_j)).$$

The inner sum is bounded by a constant independent of N, so the first result follows and (11.57) is an obvious consequence. □

From (11.56) and (11.57) it follows that among the $K_r(a,p)$, there is at most one root of weight r, say for $K_r(a_0,p)$, and all other roots are of weight $\leqslant r-1$.

Notice that $K_r(a_0,p) \in \mathbb{Q}(\mu_p)$, the cyclotomic field of p-th roots of unity. Using the Galois action on $\mathbb{Q}(\mu_p)$, the conjugates of $K_r(a_0,p)$ are $K_r(a_0v^r,p)$ for $v \in \mathbb{F}_p^\star$. By the Riemann Hypothesis, this means that the conjugate of the root ξ of weight r is still a root of weight r for $K_r(a_0v^2,p)$. Hence $v^r = 1$ for all $v \in \mathbb{F}_p$, which is only possible if $p-1 \mid r$. In particular all roots are of weight $\leqslant r-1$ if $p > r+1$. One therefore gets by Proposition 11.40

$$K_r(a,q) \ll q^{(r-1)/2} \tag{11.58}$$

where the implied constant depends only on r.

In the case of Kloosterman sum the Newton polyhedron is the simplex with vertices $(1,0,\dots,0)$, $\dots$, $(0,\dots,0,1), (-1,\dots,-1)$ whose volume is $1/r!$. Moreover, it is known that the zeta function is a polynomial so $\chi_c = -\sigma_c$ and we get the precise estimate

$$|K_r(a,q)| \leqslant rq^{(r-1)/2}.$$

This was first proved by Deligne [**De3**], without any assumption on p and r.

EXAMPLE 3. Here is another higher-dimensional example. In [**CI1**], the following exponential sum over finite fields appears:

$$W(\chi,\psi;p)=\sum_{x,y \bmod p}\chi(xy(x+1)(y+1))\psi(xy-1) \tag{11.59}$$

where p is prime, χ is a non-trivial quadratic character modulo p and ψ is any multiplicative character modulo p.

THEOREM 11.42 (CONREY-IWANIEC). *There exists an absolute positive constant C such that*

$$|W(\chi,\psi;p)|\leqslant Cp \tag{11.60}$$

for all p and all ψ as above.

The first step of the proof is to apply Theorem 11.34 to say there exists an ℓ-adic sheaf $\mathcal{F}$ on the algebraic surface

$$U=\{(x,y)\mid xy(x+1)(y+1)\neq 0\},$$

pure of weight 0 for which we have, for $q=p^n$, $n\geqslant 1$, the formula

$$\begin{aligned}W(\chi,\psi;q)&=\sum_{x,y\in U(\mathbb{F}_q)}\chi(N(xy(x+1)(y+1)))\psi(N(xy-1))\\&=\sum_{x,y\in U(\mathbb{F}_q)}\mathrm{Tr}\,(F_{(x,y)}\mid\mathcal{F}).\end{aligned}$$

The Lefschetz trace formula (11.47) takes the form

$$W(\chi,\psi;q)=\sum_{i=0}^{4}(-1)^i\mathrm{Tr}\,(F^n\mid H^i_c(\bar{U},\mathcal{F})).$$

By Theorem 11.37, each $H^i_c(\bar{U},\mathcal{F})$ is mixed of weights $\leqslant i$. Let (α_ν,i_ν,w_ν) be the family of eigenvalues of F acting on the whole cohomology (with multiplicities), together with their index and weight. We have $|\alpha_\nu|=p^{w_\nu/2}$ and

$$W(\chi,\psi;q)=\sum_\nu(-1)^{i_\nu}\alpha_\nu^m.$$

By Theorem 11.39 in this case, the total number of roots α_ν is bounded by a constant independent of p. Thus, we gain on the trivial bound $W\ll p^2$ if $w_\nu\leqslant 3$ (instead of 4), and the statement of the theorem is that $w_\nu\leqslant 2$.

The second step is to show that there is at most one root of weight $\geqslant 3$ (actually, it must be $=3$) and, if it exists, then $\psi=\chi$ is the non-trivial quadratic character. This will be derived from the following average formula:

$$A=\frac{1}{q-1}\sum_\psi|W(\chi,\psi;q)|^2=q^2-2q-2. \tag{11.61}$$

To prove this formula, open the square and sum over ψ first getting by the orthogonality of characters

$$A = \sum\sum_{u_1v_1=u_2v_2} \chi(u_1v_1(u_1+1)(v_1+1))\bar{\chi}(u_2v_2(u_2+1)(v_2+1))$$
$$= \sum\sum_{u_1,v_1,u_2} \chi((u_1+1)(v_1+1))\bar{\chi}((u_2+1)(u_1v_1+u_2))\chi(u_2)$$

(where we shorten the notation from $\chi \circ N$ to χ). Next the summation over u_1 is performed giving

$$B(v_1,u_2) = \sum_{u_1\neq 0} \chi(u_1+1)\bar{\chi}(u_1v_1+u_2) = \begin{cases} \bar{\chi}(v_1)(q-2) & \text{if } u_2 = v_1, \\ -\bar{\chi}(v_1) - \bar{\chi}(u_2) & \text{if } u_2 \neq v_1. \end{cases}$$

Then the sum over u_2 is performed giving

$$\sum_{u_2\neq 0} \bar{\chi}(u_2+1)\chi(u_2)B(v_1,u_2) = q\bar{\chi}(v_1+1) + \bar{\chi}(v_1) + 1,$$

and finally the sum over v_1 gives

$$A = \sum_{v_1\neq 0} \chi(v_1+1)\Big(q\bar{\chi}(v_1+1) + \bar{\chi}(v_1) + 1\Big) = q(q-2) - 2.$$

From (11.61), using Lemma 11.41, it is clear that for all ψ and ν we have $w_\nu \leqslant 3$ and moreover, $w_\nu \leqslant 2$ except for at most one root, for one character ψ. If this case occurs for ψ, it happens for $\bar{\psi}$ too, so the only possibility is ψ being a real character. Since

$$W(\chi,1;q) = \Big(\sum_u \chi(u(u+1))\Big)^2 = 1,$$

we must have $\psi = \chi$.

The last step is to treat the case $\psi = \chi$ separately. Precisely, one can show that

$$|W(\chi,\chi;q)| \leqslant 4q$$

for any p and $q = p^n$. This is done in a purely elementary manner without appealing to the Riemann Hypothesis, and we refer to [**CI1**] for the details. In fact, W. Duke showed that $W(\chi,\chi;p) = 2\mathrm{Re}\,(J^2(\chi,\xi))$, where $J(\chi,\xi)$ is the Jacobi sum and ξ is a quartic character modulo $p \equiv 1 \pmod 4$. Also $W(\chi,\chi;p)$ is the p-th Fourier coefficient of the modular form $\eta(4z)^6$ of weight 3 and level 12.

When U is not a curve, numerous geometric subtleties can be involved in dealing with the non-trivial cohomology groups H_c^i with $i \neq 0$, $2d$. Here are two general bounds, among many: the first one is due to Deligne [**De1**], and the second is the recent version in [**FK**] of a general "stratification" theorem of Katz and Laumon.

THEOREM 11.43. *(1) Let $f \in \mathbb{Z}[X_1,\dots,X_m]$ be a non-zero polynomial of degree d such that the hypersurface H_f in $\mathbb{P}^{m-1}$ defined by the equation*

$$H_f \,:\, f_d(x_1,\dots,x_m) = 0,$$

where f_d is the homogeneous component of degree d of f, is non-singular. For any $p \nmid d$ such that the reduction of H_f modulo p is smooth, any non-trivial additive character ψ modulo p and any $n \geqslant 1$ we have

$$\Big|\sum\cdots\sum_{x_1,\dots,x_m\in\mathbb{F}_{p^n}} \psi(Tr(f(x_1,\dots,x_m)))\Big| \leqslant (d-1)^m q^{nm/2}. \tag{11.62}$$

(2) Let $d \geqslant 1$, $n \geqslant 1$ be integers, V a locally closed subscheme of $\mathbb{A}^n_{\mathbb{Z}}$ of dimension $\dim V(\mathbb{C}) \leqslant d$ and $f \in \mathbb{Z}[X_1,\dots,X_n]$ a polynomial.

Then there exists $C = C(n,d,V,f)$ and closed subschemes $X_j \subset \mathbb{A}^n_{\mathbb{Z}}$ of relative dimension $\leqslant n-j$ such that

$$X_n \subset \ldots \subset X_2 \subset X_1 \subset \mathbb{A}^n_{\mathbb{Z}}$$

and for any rational function g non-zero on V, any $h \in (\mathbb{Z}/p\mathbb{Z})^n - X_j(\mathbb{Z}/p\mathbb{Z})$, any prime p, any non-trivial additive character ψ and multiplicative character χ modulo p, we have

$$\sum_{x\in V(\mathbb{Z}/p\mathbb{Z})} \chi(g(x))\psi(f(x)+h_1x_1+\cdots h_nx_n) \leqslant Cp^{\frac{d}{2}+\frac{j-1}{2}}.$$

Note that in (1), subject to a geometric condition on H_f, we obtain square root cancellation in the exponential sum. In (2) the assumptions are much less stringent, but the conclusion is weaker: we have a family of exponential sums parameterized by $h \in \mathbb{A}^n$, and roughly speaking we have square root cancellation for "generic" sums (for h outside an exceptional subvariety X_1 of codimension $\geqslant 1$ in $\mathbb{A}^n$), while worse and worse bounds can occur only on smaller and smaller subvarieties. We will give an application of (2) in Chapter 21.

The ℓ-adic theory and formalism are much more developed than what we have surveyed here. It can also deal with sums over singular varieties, but the necessary algebraic notions become rather formidable, and we concede being unable to discuss the perverse sheaves that arise in more advanced situations.

We wish to emphasize, however, that this theory is also particularly well-suited to the study of *families* of exponential sums. The parameters defining those, say S_x, are most naturally themselves points on some algebraic variety $X/\mathbb{F}$. In favorable circumstances there exists a lisse ℓ-adic sheaf $\mathcal{F}$ on X (corresponding to an action of $\pi_1(X)$ on V) such that

$$S_{x,n} = \operatorname{Tr}(F_x^n \mid V) \tag{11.63}$$

for every value of the parameter $x \in X$ and any $n \geqslant 1$. For example a non-zero polynomial f of degree $\leqslant d$ over $\mathbb{F}$ can be described as an $\mathbb{F}$-rational point of the affine parameter space $X = \mathbb{A}^{d+1} - \{0\}$ by

$$f = \sum_{i=0}^{d} a_i X^i \mapsto (a_0,\dots,a_d).$$

There is an ℓ-adic sheaf $\mathcal{F}$ on X such that

$$\sum_{x\in\mathbb{F}_n} \psi(f(x)) = \operatorname{Tr}(F_x^n \mid V).$$

Purity of a sheaf satisfying (11.63) depends on a first application of the Riemann Hypothesis. If it holds, the application of the ℓ-adic theory typically results in equidistribution statements (following from [**De2**]) for the arguments of the exponential sums. This equidistribution is in some space of conjugacy classes of the "monodromy group" of the situation. We refer for instance to [**KS**] for a very lucid introduction to these profound aspects.

11.12. Comments.

In this closing section we give some impressions about how the exponential and character sums over finite fields interact with analytic number theory. There are more subtle issues between the two subjects than just applications of results concerning the first one for solving problems of the latter. We could be quite specific by covering completely a few representative examples, but it would be long and not transparent enough. Rather we decided to discuss principles, ideas and tricks in general terms and guide the reader to particular publications.

First of all some exponential sums appear when one uses Fourier analysis to get a hold on the sequence under investigation. There are no finite fields in the background, so the resulting sums are not immediately related to objects of algebraic geometry. However, one can complete these sums (by another use of Fourier analysis) and then factor them into sums of prime power moduli. Usually one can evaluate these local sums explicitly, or give strong estimates by elementary or ad hoc methods, except when the modulus is prime. But in the prime case one may naturally consider the sum as being over a finite field. This scheme allows us to appeal to the powerful results from algebraic geometry. However, the drawback of finishing by estimates for every complete sum individually is that one cannot exploit a possible cancellation from extra variables offered by the varying moduli (finite fields of different characteristics do not interact in algebraic geometry). Sometimes a kind of reciprocity formula can help turn the modulus into a variable (see for example [**I11**] or [**M3**]). Another scenario is that the sums over modulus appear in the spectral resolution of a differential operator, in which case the spectral theory produces estimates far stronger than those derived by algebraic geometry. For example, this is the case of sums of Kloosterman sums; see Chapter 16. One can also think this way about the real character sums with cubic polynomials; they are coefficients of a cusp form associated with an elliptic curves, so the modularity gives extra cancellation in summation over the modulus.

As a rule the exponential/character sum of a given modulus which comes out of analytic number theory is incomplete. This itself is not a problem because various completing techniques are available as mentioned above. Completing is a natural step to take, but is it useful or wasteful? At this point one should realize that a bound for a complete sum holds essentially the same for the original incomplete sum. This means that the result is relatively weaker for a shorter sum. Still it is non-trivial when the length of the original sum is larger than the square root of the modulus. Very short sums cannot be treated this way. We do not have an absolute recommendation when to complete or not a given incomplete sum. Our experience suggests executing the Fourier method as long as the resulting summation over the frequencies is shorter than the range of the original sum: at least one can feel that one is progressing. But sometimes it is worth acting otherwise, accepting a step backwards in this respect while opening a position for a stronger second move.

For example, imagine that the amplitude in the exponential sum is not a rational function, but nevertheless becomes one after an application of the stationary phase method to the relevant Fourier transform. In this case the losses from the range of summation can be recovered with extra savings by applications of algebraic arguments (see Section 8 of [**CI1**], where this game is played in several variables simultaneously).

Whatever the arguments which lead to complete sums may be, in the final step one cannot beat the square root saving factor. Therefore to receive a non-trivial result one must first produce somehow a sum with a number of terms larger than the square root of the modulus. There are several ways to get started, depending on the shape of the exponential/character sum. First, one can try to apply a Weyl shift with the effect of squaring the number of terms. Similarly one may just square the whole sum, or raise it to a higher power to produce even more points. Note that shifting the variable and squaring the sum are not the same things; the first requires some additive features of the variable while the latter nothing at all. These operations seem to be quite superficial at first glance (we are taking essentially replicas of the original sum), yet with ingenuity one can rearrange the points so the summation goes in a skewed direction, the consecutive terms repel violently and randomly producing a considerable cancellation. This is easy to say, much harder to execute. In fact one needs many other devices, such as gluing several variables with small multiplicity in order to arrive at a single variable over a range larger than the square root of the modulus. One also must smooth out this composite variable before applying algebraic arguments. Usually an application of Cauchy's inequality and enlarging the outer summation (due to positivity) does the job. A powerful example how this works is given by Burgess [**Bur1**]. In this paper a short character sum is estimated by an appeal to the Riemann Hypothesis for algebraic curves. An interesting point is that after all the tricks one comes to a complete character sum for a curve of a large genus, although the original sum is over a line segment.

Different arguments for building one extra large variable are applied in [**FI4**], consequently the final complete sum comes in three variables, or equivalently in four variables over a hypersurface. Here the Deligne theory applies (see the Appendix by Birch and Bombieri), although the related variety is singular. One should not be surprised and afraid of that singularity, because, after all, the process of creating more points of summation at the start is quite superficial. In this game one must be experienced when mixing the points to be sure that it is quite random. Another interesting case of creating and estimating exponential sums in three variables is given by N. Pitt [**Pi**].

Applications of the Riemann Hypothesis for curves over finite fields are by now customary. Much less successful are the use of genuinely higher-dimensional varieties. There are reasons for the difficulties involved. First of all when more variables appear, stronger restrictions are imposed on them which are harder to resolve (a kind of uncertainty principle). Just imagine having an abundance of points to work with, but which are not free because of some side conditions. For example how would you cope with a requirement that the determinants of a family of elliptic curves match the conductors?

Of course, there are also direct applications of the Riemann Hypothesis for varieties to traditional problems of solvability of diophantine equations by means of the circle method (see examples in Chapter 20). If the number of variables is sufficiently large, one needs nothing to manipulate, except for completing the sum by the standard Fourier method. Some ingenuity, however, is required to apply the circle method (a variant of Kloosterman) to treat diophantine equations with a relatively small number of variables, an excellent example being the work of Heath-Brown on cubic forms [**HB6**].

It is possible in some circumstances to beat the bound for exponential sums which is derived by the Riemann Hypothesis. This is because the angles of the roots of the L-function themselves vary so that additional cancellation may occur. Deligne and Katz have established such occurrences for families parameterized by points on curves or varieties. In other words, in their cases one is actually considering exponential sums in more variables. However, the cancellation of roots can also occur for families parameterized by points over small irregular sets. More important for analytic number theory is that these sets can be quite general, no structure of a subvariety is needed, but instead a kind of a bilinear form structure would suffice. In practice it is not clear how to work with the roots, so one returns to the corresponding exponential sums where manipulations with the parameters (grouping, gluing, etc.) can be performed properly according to the shape of the involved rational function which is seen with the naked eye. In this process one must not destroy the complete variables since in our mind the corresponding summations are already executed in terms of the roots. Therefore when applying Cauchy's inequality to smooth the one variable composed out of the parameters, we put all the complete variables to the inner summation, say n of them, together with some remaining parameters which where not used in the composition. These inner parameters are critical for enlarging the diagonal. After squaring out we get a complete exponential sum in $2n + 1$ variables which depends on the inner parameters. Except for a few configurations of those, the complete exponential sum satisfies the best possible bound derived from the Riemann Hypothesis, thus saving the factor of square root of the modulus per each variable. Since the number of variables is larger than twice the original, we win the game. The above recipe is somewhat oversimplified, yet it reveals the source of extra saving. One can see how it works in the particular case of [**CI1**]. Speaking of [**CI1**] we would like to add that here the exponential/character sums in several variables emerge after applications of harmonic analysis with respect to the hyperbolic Laplace operator rather than by the traditional Fourier analysis.

The profound theory of Deligne and other geometers is being used in analytic number theory with spectacular effects, yet more ideas need to be invented to fully exploit its potential. Perhaps one should go beyond borrowing estimates and penetrate deeply inside the theory. This is a great subject for future research. P. Michel [**M3**] made the first significant steps (see also [**FK**] and [**KS1**]).

CHAPTER 12

CHARACTER SUMS

12.1. Introduction.

In analytic number theory one often encounters sums of type

$$S = \sum_{x \in V} F(x) \tag{12.1}$$

where $V \subset \mathbb{Z}^n$ is a finite set and $F : V \to \mathbb{C}$ is a periodic function of period q. Because V does not match the periodicity, S is called an incomplete sum. Suppose

$$F(x) = \chi(f(x))\psi(g(x)) \tag{12.2}$$

where χ, ψ are multiplicative and additive characters to modulus q, and f, g are rational functions with integer coefficients. Precisely, we assume that

$$f = f_1/f_0, \quad g = g_1/g_0 \tag{12.3}$$

where $f_0, f_1, g_0, g_1 \in \mathbb{Z}[x]$ and

$$(f_0(x)g_0(x), q) = 1 \quad \text{if } x \in V. \tag{12.4}$$

Having fixed these polynomials (which are not unique) we define

$$\chi(f(x)) = \chi(f_1(x)\bar{f}_0(x)), \tag{12.5}$$

$$\psi(g(x)) = \psi(g_1(x)\bar{g}_0(x)) \tag{12.6}$$

where, as usual, $\bar{a}$ denotes the multiplicative inverse of a modulo q. Then the resulting sum

$$S = \sum_{x \in V}{}^{\star} \chi(f(x))\psi(g(x)) \tag{12.7}$$

is called an incomplete character sum. Here, and hereafter, the star restricts the summation to the points of V satisfying (12.4). The residue classes $x(\text{mod } q)$ which satisfy (12.4) will be called admissible. An important case, but by no means the only one, is V being a box. When the box has size exactly q we get

$$S(\chi, \psi) = \sum_{x (\text{mod } q)}{}^{\star} \chi(f(x))\psi(g(x)) \tag{12.8}$$

which is called a complete character sum.

In this chapter we give basic techniques of estimating incomplete character sums in one variable over an interval. As an exercise for the reader, we suggest generalizing some of the forthcoming results to several variables.

12.2. Completing methods.

The most common treatment of an incomplete sum

$$S(M;N) = \sideset{}{^\star}\sum_{M<n\leqslant M+N} F(n) \tag{12.9}$$

goes by expanding it into complete sums (this is called the completing technique),

$$S\Big(\frac{a}{q}\Big) = \sideset{}{^\star}\sum_{x(\mathrm{mod}\ q)} F(x)e\Big(-\frac{ax}{q}\Big) \tag{12.10}$$

and treating the latter by various arithmetic means. To do so we split the summation into the $\star$-admissible residue classes $n \equiv x(\mathrm{mod}\ q)$, and detect these classes by the orthogonality of additive characters getting

$$S(M;N) = \frac{1}{q}\sum_{a(\mathrm{mod}\ q)} \lambda\Big(\frac{a}{q}\Big)S\Big(\frac{a}{q}\Big) \tag{12.11}$$

where

$$\lambda\Big(\frac{a}{q}\Big) = \sum_{M<n\leqslant M+N} e\Big(\frac{an}{q}\Big). \tag{12.12}$$

Usually the main contribution to $S(M;N)$ comes from $a = 0$ in (12.11) for which $\lambda(0) = N$ (we assume that M and $N \geqslant 1$ are integers). For $0 < |a| \leqslant \frac{q}{2}$, we have $|\lambda(\frac{a}{q})| \leqslant q|a|^{-1}$. Hence

LEMMA 12.1. *Let $F(x)$ be a complex-valued function defined on the residue classes x (mod q) which are $\star$-admissible. Then the corresponding sums satisfy*

$$\Big|S(M;N) - \frac{N}{q}S(0)\Big| \leqslant \sum_{0<|a|\leqslant\frac{q}{2}} |a|^{-1}\Big|S\Big(\frac{a}{q}\Big)\Big|. \tag{12.13}$$

Suppose that the complete sums satisfy

$$\Big|S\Big(\frac{a}{q}\Big)\Big| \leqslant c(\theta)(a+b,q)^{\frac{1}{2}}q^\theta \tag{12.14}$$

for some $b \in \mathbb{Z}$ and $\theta > \frac{1}{2}$. This bound is often true with $\theta = \frac{1}{2} + \varepsilon$ for any $\varepsilon > 0$. Then (12.13) becomes

$$\Big|S(M;N) - \frac{N}{q}S(0)\Big| \leqslant c(\theta)\ell(b,q)q^\theta \tag{12.15}$$

where

$$\ell(b,q) = \sum_{0<|a|\leqslant\frac{q}{2}} |a|^{-1}(a+b,q)^{\frac{1}{2}}. \tag{12.16}$$

Since $\ell(b,q)$ is usually small (for example, we have $\ell(0,q) \leqslant 2\tau(q)\log q$), the inequality (12.15) shows that the incomplete sum $S(M;N)$ satisfies essentially the same bound as does the corresponding complete sums $S(\frac{a}{q})$, up to the main term $\frac{N}{q}S(0)$.

Clearly the above method of completing the sum of a periodic function $F(x)$ works for more general sums of type

$$S = \sum_n F(n)G(n) \tag{12.17}$$

where $G(x)$ is a nice function of analytic character which decays to zero rapidly as $|x| \to \infty$. Now we present another method. Suppose $G(x)$ is of Schwartz class on $\mathbb{R}$. Then one can apply the Poisson summation formula, giving

$$S = \frac{1}{q} \sum_{a \in \mathbb{Z}} S\Big(\frac{a}{q}\Big) \hat{G}\Big(\frac{a}{q}\Big) \tag{12.18}$$

where $\hat{G}(y)$ is the Fourier transform of $F(x)$. In principle both methods are equivalent, however the latter may be preferable in case $G(x)$ propagates some vibrations. If the vibrations are regular, $\hat{G}(\frac{a}{q})$ can be evaluated asymptotically by the stationary phase method (see for example Corollary 8.15) giving an asymptotic expansion for S in terms of $S(\frac{a}{q})$. On the other hand, if $G(x)$ is wild, one should have some reservation for completing S in terms of additive characters. One should try to use "harmonics" which are more adequate for the particular case. The Fourier coefficients of cusp forms, holomorphic or non-holomorphic, may serve well as the relevant harmonics in many cases.

12.3. Complete character sums.

Let χ_q and ψ_q be multiplicative and additive characters respectively to the modulus q. The complete character sum

$$S(\chi_q, \psi_q) = \sum_{x (\mathrm{mod}\ q)}^{\star} \chi_q(f(x)) \psi_q(g(x))$$

is multiplicative with respect to q. Indeed, let $q = rs$ with $(r, s) = 1$. Then χ_q factors uniquely into $\chi_r \chi_s$ where χ_r, χ_s are multiplicative characters to moduli r, s respectively. The additive character ψ_q is given by

$$\psi_q(x) = e\Big(\frac{ax}{q}\Big) \tag{12.19}$$

for some $a \in \mathbb{Z}$. By the "reciprocity" formula

$$\frac{\bar{s}}{r} + \frac{\bar{r}}{s} = \frac{1}{q} (\mathrm{mod}\ 1) \tag{12.20}$$

where $s\bar{s} \equiv 1 (\mathrm{mod}\ r)$ and $r\bar{r} \equiv 1 (\mathrm{mod}\ s)$, the additive character factors into $\psi_q = \psi_r^{\bar{s}} \psi_s^{\bar{r}}$. Hence it follows that

$$S(\chi_q, \psi_q) = S(\chi_r, \psi_r^{\bar{s}}) S(\chi_s, \psi_s^{\bar{r}}). \tag{12.21}$$

Therefore the problem of evaluating a complete character sum of modulus q reduces to that of prime power moduli.

The complete sum $S(\chi, \psi)$ of modulus $q = p^\beta$ shouldn't be confused with the character sum over the finite field $\mathbb{F}_q$ which was considered in Chapter 11, except for $q = p$ in which case $S(\chi, \psi)$ is indeed one of such sums. The case of prime modulus belongs to the theory of L-functions for curves over finite fields, as described in Chapter 11. The rationality of the relevant L-function together with the Riemann

Hypothesis (both proved by A. Weil [**We1**] in this case) yield algebraic numbers $g_1, \dots, g_r$ with $|g_\nu| = p, p^{\frac{1}{2}}, 1$ such that

$$S(\chi, \psi) = g_1 + \cdots + g_r, \tag{12.22}$$

The number r is bounded independently of the characteristic p. Moreover, assuming some non-singularity conditions for the rational functions f, g with respect to the characters χ, ψ, there are no roots g_ν with $|g_\nu| = p$, so (12.22) yields

$$|S(\chi, \psi)| \leqslant r p^{\frac{1}{2}}. \tag{12.23}$$

See Chapter 11 and in particular Section 11.11 for more complete discussion of this situation.

There is no need to algebraic geometry for the complete character sums $S(\chi, \psi)$ to modulus $q = p^\beta$ with $\beta \geqslant 2$, because in this case elementary arguments are available.

LEMMA 12.2. *Let $q = p^{2\alpha}$ with $\alpha \geqslant 1$. Then we have*

$$S(\chi, \psi) = p^\alpha \sideset{}{^\star}\sum_{\substack{y (\mathrm{mod}\ p^\alpha) \\ h(y) \equiv 0 (\mathrm{mod}\ p^\alpha)}} \chi(f(y)) \psi(g(y)) \tag{12.24}$$

where $h(y)$ is the rational function given by

$$h(y) = a g'(y) + b \frac{f'}{f}(y) \tag{12.25}$$

with the integers a, b depending on the characters ψ, χ which are determined by (12.19) and (12.27) below.

REMARK. Since the characters χ, ψ have modulus $p^{2\alpha}$ it is required for correctness of the summation (12.24) to know that $\chi(f(y))\psi(g(y))$ does not depend on the choice of the representative of $y (\mathrm{mod}\ p^\alpha)$ on the curve $h(y) \equiv 0 (\mathrm{mod}\ p^\alpha)$. This property follows in the course of the proof.

PROOF. Write $x = y + z p^\alpha$, where y and z run independently over any fixed systems of residue classes modulo p^α, and y is restricted by the condition $p \nmid f_0(y) g_0(y)$. We have

$$f(x) \equiv f(y) + f'(y) z p^\alpha \pmod{p^{2\alpha}} \tag{12.26}$$

Indeed, the congruence (12.26) is easily seen for monomials x^n by the binomial formula

$$(y + z p^\alpha)^n = y^n + n y^{n-1} z p^\alpha + \cdots.$$

Then it extends to arbitrary polynomials $f_1(x), f_0(x)$ with integral coefficients by linearity. Moreover, if $f_0(y) \equiv 0 (\mathrm{mod}\ p)$, then (12.26) is verified for $f(x)/f_0(x)$ as follows:

$$\begin{aligned} f_1(x)/f_0(x) &\equiv (f_1(y) + f_1'(y) z p^\alpha) \bar{f}_0(y)(1 - \bar{f}_0(y) f_0'(y) z p^\alpha) \\ &\equiv f_1(y)\bar{f}_0(y) + (f_1'(y)\bar{f}_0(y) - \bar{f}_0^2(y) f_0'(y) f_1(y)) z p^\alpha. \end{aligned}$$

By (12.26) we get

$$\chi(f(x)) = \chi(f(y)) \chi\Big(1 + \frac{f'}{f}(y) z p^\alpha\Big).$$

Clearly $\chi(1+zp^\alpha)$ is an additive character to modulus p^α, so there exists an integer b (uniquely determined modulo p^α) such that

$$\chi(1+zp^\alpha) = e\Big(\frac{bz}{p^\alpha}\Big). \tag{12.27}$$

Hence

$$\chi(f(x)) = \chi(f(y))e\Big(b\frac{f'}{f}(y)zp^{-\alpha}\Big). \tag{12.28}$$

Similarly we get (12.26) for the rational function $g(x)$, whence

$$\psi(g(x)) = \psi(g(y))e(ag'(y)zp^{-\alpha}). \tag{12.29}$$

Multiplying (12.28), (12.29) and summing over the residue classes y, z modulo p^α we get

$$S(\chi,\psi) = \sideset{}{^\star}\sum_{y(\mathrm{mod}\ p^\alpha)} \chi(f(y))\psi(g(y)) \sum_{z(\mathrm{mod}\ p^\alpha)} e(h(y)zp^{-\alpha}).$$

The inner sum vanishes unless $h(y) \equiv 0(\mathrm{mod}\ p^\alpha)$ in which case it equals p^α. This completes the proof of (12.24). □

LEMMA 12.3. *Let $q = p^{2\alpha+1}$ with $\alpha \geqslant 1$. Then we have*

$$S(\chi,\psi) = p^\alpha \sideset{}{^\star}\sum_{\substack{y(\mathrm{mod}\ p^\alpha)\\ h(y)\equiv 0(\mathrm{mod}\ p^\alpha)}} \chi(f(y))\psi(g(y))G_p(y) \tag{12.30}$$

where $G_p(y)$ is the Gauss sum

$$G_p(y) = \sum_{z(\mathrm{mod}\ p)} e_p(d(y)z^2 + h(y)p^{-\alpha}z). \tag{12.31}$$

Here $h(y)$ is the rational function (12.25) but with b given by (12.35) below, and

$$d(y) = \frac{a}{2}g''(y) + \frac{b}{2}\frac{f''}{f}(y) + (p-1)\frac{b}{2}\Big(\frac{f'}{f}(y)\Big)^2. \tag{12.32}$$

REMARK. As z runs mod p, so does $z^2/2p$, therefore the Gauss sum (12.31) is correctly defined.

PROOF. We write $x = y + zp^\alpha$ with y running modulo p^α subject to $p \nmid f_0(y)g_0(y)$, and z running modulo $p^{\alpha+1}$. As before we argue that

$$f(x) \equiv f(y) + f'(y)zp^\alpha + \frac{1}{2}f''(y)z^2p^{2\alpha}(\mathrm{mod}\ p^{2\alpha+1}). \tag{12.33}$$

Note that the rational function $\frac{1}{2}f''(y)$ has integral coefficients. This is clear for the monomial y^n because $n(n-1) \equiv 0(\mathrm{mod}\ 2)$, then for any integral polynomial by linearity, and finally for a rational function $f(y) = f_1(y)/f_0(y)$ by the identity

$$\tfrac{1}{2}f'' = \tfrac{1}{2}f_1''f_0^{-1} - f_1'f_0'f_0^{-2} - \tfrac{1}{2}f_1f_0''f_0^{-2} + f(f_0')^2f_0^{-3}.$$

By (12.33) we get

$$\chi(f(x)) = \chi(f(y))\chi\Big(1 + \Big(\frac{f'}{f}(y)z + \frac{1}{2}\frac{f''}{f}(y)z^2p^\alpha\Big)p^\alpha\Big). \tag{12.34}$$

Consider the function

$$\xi(1+zp^\alpha) = e\Big(\frac{z}{p^{\alpha+1}} + (p-1)\frac{z^2}{2p}\Big).$$

This is a character of the subgroup of residue classes $x(\bmod p^{2\alpha+1})$ with $x \equiv 1(\bmod p^\alpha)$. Indeed

$$\begin{aligned}\xi((1+zp^\alpha)(1+wp^\alpha)) &= \xi(1+(z+w+zwp^\alpha)p^\alpha)\\ &= e\Big(\frac{z+w}{p^{\alpha+1}} + (p-1)\frac{z^2+w^2}{2p}\Big)\\ &= \xi(1+zp^\alpha)\xi(1+wp^\alpha).\end{aligned}$$

Since the subgroup has order $p^{\alpha+1}$ and ξ^b are all different for distinct b modulo $p^{\alpha+1}$, it follows that there exists an integer b (uniquely determined modulo $p^{\alpha+1}$) such that

$$\chi(1+zp^\alpha) = e\Big(\frac{bz}{p^{\alpha+1}} + (p-1)\frac{bz^2}{2p}\Big). \tag{12.35}$$

Using (12.34), (12.35) and

$$\psi(g(x)) = \psi(g(y))e\Big(\frac{ag'(y)}{p^{\alpha+1}}z + \frac{a}{2}\frac{g''(y)}{p}z^2\Big) \tag{12.36}$$

we derive that

$$S(\chi,\psi) = \sum_{y(\bmod p^\alpha)} \chi(f(y)\psi(g(y)) \sum_{z(\bmod p^{\alpha+1})} e_p(d(y)z^2 + h(y)p^{-\alpha}z).$$

Here the innermost sum vanishes unless $h(y) \equiv 0 \pmod{p^\alpha}$ in which case it equals $p^\alpha G_p(y)$. This completes the proof of (12.30). □

The Gauss sums $G_p(y)$ were computed in Chapter 3. If $p \nmid 2d(y)$, we have

$$G_p(y) = \varepsilon_p p^{\frac12}\Big(\frac{d(y)}{p}\Big)e_p\Big(-\overline{4d(y)}\Big(\frac{h(y)}{p^\alpha}\Big)^2\Big). \tag{12.37}$$

The formulas (12.24) and (12.30) represent truly the final stage of computation. The only terms which are not determined on the right side are the roots of the congruence $h(y) \equiv 0(\bmod p^\alpha)$. However, in practice there are not many roots (essentially a bounded number) so one gets the estimate $|S(\chi,\psi)| \leqslant cq^{1/2}$ with c depending mildly on the coefficients of the rational functions f, g.

EXERCISE 1. Using Lemma 12.2 and Lemma 12.3 together with (12.37) evaluate the Kloosterman sum

$$S(m,n;q) = \sideset{}{^\star}\sum_{x(\bmod q)} e\Big(\frac{m\bar{x}+nx}{q}\Big) \tag{12.38}$$

for $q = p^\beta$ with $\beta \geqslant 2$. Suppose $p \nmid 2mn$. Show that $S(m,n;q)$ vanishes, unless $(\frac{m}{p}) = (\frac{n}{p})$ in which case

$$S(m,n;q) = 2\Big(\frac{\ell}{q}\Big)q^{\frac12}\,\mathrm{Re}\ \varepsilon_q e\Big(\frac{2\ell}{q}\Big) \tag{12.39}$$

where $\ell^2 \equiv mn (\bmod q)$.

The Kloosterman sums to modulus $q = p^\beta$ with $\beta \geqslant 2$ were computed first by H. Salié [**Sal**]. He also computed the so-called Salié sums

$$T(m,n;q) = \sum_{x (\bmod q)} \Big(\frac{x}{q}\Big) e\Big(\frac{m\bar{x}+nx}{q}\Big) \tag{12.40}$$

where $(\frac{x}{q})$ is the Jacobi-Legendre symbol, including the case of prime modulus. One can do it for composite moduli as well.

LEMMA 12.4. *Suppose* $(q, 2n) = 1$. *Then*

$$T(m,n;q) = \varepsilon_q q^{\frac{1}{2}} \Big(\frac{n}{q}\Big) \sum_{v^2 \equiv mn (\bmod q)} e\Big(\frac{2v}{q}\Big). \tag{12.41}$$

PROOF. Consider the function $F(u) = T(m, nu^2; q)$ defined for u (mod q). The Fourier transform of $F(u)$ is

$$\begin{aligned}\hat{F}(v) &= \sum_{u(\bmod q)} F(u) e\Big(-\frac{uv}{q}\Big) \\ &= \sum_{x(\bmod q)} \Big(\frac{x}{q}\Big) e\Big(\frac{m\bar{x}}{q}\Big) \sum_{u (\bmod q)} e\Big(\frac{nxu^2 - uv}{q}\Big) \\ &= \varepsilon_q q^{\frac{1}{2}} \Big(\frac{n}{q}\Big) \sideset{}{^\star}\sum_{x (\bmod q)} e\Big(\frac{m\bar{x} - \overline{4nx}v^2}{q}\Big)\end{aligned}$$

by the formula (3.21) for quadratic Gauss sums. Notice that the Jacobi-Legendre symbol canceled out, therefore the last sum is the Ramanujan sum

$$S(0, m - \overline{4n}v^2; q) = \sum_{d | (4mn - v^2, q)} d\mu(q/d).$$

Hence by Fourier inversion

$$F(u) = \frac{1}{q} \sum_{v (\bmod q)} \hat{F}(v) e\Big(\frac{uv}{q}\Big) = \varepsilon_q q^{-\frac{1}{2}} \Big(\frac{n}{q}\Big) \sum_{d|q} d\mu\Big(\frac{q}{d}\Big) \sum_{\substack{v (\bmod q) \\ v^2 \equiv 4mn (\bmod d)}} e\Big(\frac{uv}{q}\Big).$$

If $(u, q) = 1$, this simplifies to

$$F(u) = \varepsilon_q q^{\frac{1}{2}} \Big(\frac{n}{q}\Big) \sum_{v^2 \equiv mn (\bmod q)} e\Big(\frac{2uv}{q}\Big).$$

In particular, for $u = 1$ we get (12.41). □

Suppose $(q, 2mn) = 1$. Then $T(m, n; q)$ vanishes unless there exists ℓ with

$$\ell^2 \equiv mn (\bmod q). \tag{12.42}$$

Given ℓ, all the solutions to $v^2 \equiv mn \pmod q$ can be written explicitly as $v = (r\bar{r} - s\bar{s})\ell$, where r, s run over the factorizations $rs = q$ with $(r,s) = 1$. Hence the formula (12.41) can be written more explicitly as follows:

$$T(m,n;q) = \varepsilon_q q^{\frac{1}{2}} \Big(\frac{n}{q}\Big) \sum_{\substack{rs=q \\ (r,s)=1}} e\Big(2\ell\Big(\frac{\bar{r}}{s} - \frac{\bar{s}}{r}\Big)\Big). \tag{12.43}$$

The Kloosterman sums to prime modulus cannot be computed in elementary terms. This shouldn't be surprising in view of the results (and conjecture) concerning the distribution of angles of Kloosterman sums (see Section 21.2, in particular Theorem 21.7 and the Sato-Tate Conjecture).

12.4. Short character sums.

In this section we are dealing with character sums over a short interval

$$S_\chi(N) = \sum_{M<n\leqslant M+N} \chi(n), \tag{12.44}$$

where χ is a non-principal Dirichlet character of modulus q. By Lemma 12.1 we obtain

$$|S_\chi(N)| \leqslant 2 \sum_{0<a\leqslant \frac{q}{2}} a^{-1} |g_\chi(a)| \tag{12.45}$$

where

$$g_\chi(a) = \sum_{x \pmod q} \chi(x) e\Big(\frac{ax}{q}\Big) \tag{12.46}$$

are the classical Gauss sums. Let $\chi^\star \pmod{q^\star}$ be the primitive character which induces χ. We have

$$g_\chi(a) = g_{\chi^\star}(1) \sum_{d|(a,q/q^\star)} d\bar{\chi}^\star(a/d)\mu(q/dq^\star) \tag{12.47}$$

by Lemma 3.2. Hence

$$|g_\chi(a)| \leqslant \sigma((a, q/q^\star))\sqrt{q^\star}. \tag{12.48}$$

Inserting this bound into (12.45) we derive

$$|S_\chi(N)| \leqslant 3\tau(q/q^\star)\sqrt{q^\star}\log q. \tag{12.49}$$

Since $\tau(m) \leqslant 2\sqrt{m}$, this yields

THEOREM 12.5. *For any non-principal character* $\chi \pmod q$ *we have*

$$\Big|\sum_{M<n\leqslant M+N} \chi(n)\Big| \leqslant 6\sqrt{q}\log q. \tag{12.50}$$

This inequality was proved in 1918 independently by G. Pólya [**Pol**] and I. M. Vinogradov [**V4**] except for the constant 6. By the Grand Riemann Hypothesis for Dirichlet L-functions one can derive a slightly stronger estimate

$$S_\chi(N) \ll \sqrt{q}\log\log 3q,$$

however, nothing better than the Pólya-Vinogradov inequality is known in general except for the constant factor. For recent improvements of this constant see A. Hildebrand [**Hi2**].

Next we give another derivation of the Pólya-Vinogradov inequality without completing $S_\chi(N)$. We have

$$S_\chi(N) = \sum_n \chi(n) f(n)$$

where $f(x) = \min(x - M, 1, M + N + 1 - x)$ if $M \leqslant x \leqslant M + N + 1$, and $f(x) = 0$, otherwise. This redundant cut-off function will help to separate the variables in the forthcoming expressions. For any integer a we have

$$S_\chi(N) = \sum_n \chi(n + a) f(n + a).$$

Hence

$$AS_\chi(N) = \sum_{0<a\leqslant A} \sum_{M-A<n\leqslant M+N} \chi(n + a) f(n + a).$$

Now separate the variables in $f(n + a)$ by the Fourier transform

$$f(n + a) = \int_{-\infty}^{\infty} \hat{f}(t) e((n + a)t) dt$$

getting

$$|S_\chi(N)| \leqslant \int_{-\infty}^{\infty} |\hat{f}(t)| \mathcal{B}(t) dt$$

where $\mathcal{B}(t)$ is the sum in two independent variables

$$\mathcal{B}(t) = \frac{1}{A} \sum_{0<a\leqslant A} \Big| \sum_{M-A<n\leqslant M+N} \chi(n + a) e(nt) \Big|.$$

By Cauchy's inequality

$$\begin{aligned}\mathcal{B}(t)^2 &\leqslant \Big(\frac{1}{A} + \frac{1}{q}\Big) \sum_{x(\mathrm{mod}\ q)} \Big| \sum_n \chi(n + x) e(nt) \Big|^2 \\ &\leqslant \Big(\frac{1}{A} + \frac{1}{q}\Big) \sum_{n_1} \sum_{n_2} \Big| \sum_{x(\mathrm{mod}\ q)} \chi(n_1 + x) \bar{\chi}(n_2 + x) \Big|.\end{aligned}$$

Suppose $\chi(\mathrm{mod}\ q)$ is primitive. Then

$$\sum_{x(\mathrm{mod}\ q)} \chi(n_1 + x) \bar{\chi}(n_2 + x) = S(0, n_1 - n_2; q) \tag{12.51}$$

is the Ramanujan sum. Hence we derive that

$$\mathcal{B}(t)^2 \leqslant \Big(\frac{1}{A} + \frac{1}{q}\Big) \Big(\frac{A + N}{q} + 1\Big) (A + N) R(q)$$

where

$$R(q) = \sum_{y(\mathrm{mod}\ q)} |S(0, y; q)|. \tag{12.52}$$

Assuming that $N \leqslant \frac{q}{2}$ (as we can by virtue of periodicity), and choosing $A = N$ the above bound simplifies to $6R(q)$. Hence

$$|S_\chi(N)| \leqslant (6R(q))^{\frac{1}{2}} \int_{-\infty}^{\infty} |\hat{f}(t)|dt.$$

By $|\hat{f}(t)| = (\pi t)^{-2}|\sin(\pi t)\sin(\pi t N)|$ we infer that the integral is bounded by $3\log 2N$, and conclude that

$$|S_\chi(N)| \leqslant 8R(q)^{\frac{1}{2}} \log q.$$

Note that $R(q) = 2^{\omega(q)}\varphi(q) \leqslant \tau(q)q$, therefore

$$|S_\chi(N)| \leqslant 8(\tau(q)q)^{\frac{1}{2}} \log q. \tag{12.53}$$

For q prime this estimate gives the Pólya-Vinogradov inequality while for composite q it is only slightly weaker.

REMARK. The factor $\log q$ emerged in the process of separation of variables. Since this factor cannot be entirely ignored, we make a point that in general the losses from separation of variables are not exclusively of technical nature, but occur inevitably for arithmetic reasons as well. This comment sounds even stronger in the context of bilinear forms, where factorization by superficial separation of variables at no cost would be fatal (think of Corollary 7.3).

The Pólya-Vinogradov inequality (12.50) is trivial for character sums $S_\chi(N)$ of length $N \ll q^{\frac{1}{2}}$, while the true bound is expected to be

$$S_\chi(N) \ll N^{\frac{1}{2}}q^{\varepsilon}, \tag{12.54}$$

which is non-trivial if $N \gg q^{3\varepsilon}$. In a series of papers D. Burgess [**Bur1**], [**Bur2**] established several results for relatively short character sums, one of them being

THEOREM 12.6 (D. BURGESS). *Let χ be a primitive character of conductor $q > 1$. Then*

$$S_\chi(N) \ll N^{1-\frac{1}{r}} q^{\frac{r+1}{4r^2}+\varepsilon} \tag{12.55}$$

for $r = 2, 3$, and for any $r \geqslant 1$ if q is cubefree, the implied constant depending only on ε and r.

The hypothesis that q is cubefree can be removed at the cost of weakening the result. Let $q = k\ell$, where ℓ is the maximal cubefree factor. By factoring the character $\chi(\text{mod } q)$ accordingly and splitting the summation into classes modulo k one derives from Theorem 12.6,

$$S_\chi(N) \ll N^{1-\frac{1}{r}} k^{\frac{1}{r}} \ell^{\frac{r+1}{4r^2}+\varepsilon} \tag{12.56}$$

(if $\ell = 1$, then $k = q$ and (12.56) follows by the periodicity of χ).

Burgess's bound (12.55) for q cubefree is non-trivial if $N \gg q^{\frac{1}{4}+\frac{1}{4r}+\varepsilon}$, and by taking r sufficiently large, one derives

$$S_\chi(N) \ll N^{1-\delta}, \qquad \text{if } N \geqslant q^{\frac{1}{4}+\sqrt{\delta}} \tag{12.57}$$

for any $0 < \delta \leqslant \frac{1}{4}$, the implied constant depending only on δ.

We shall give a proof of a slightly stronger estimate

$$|S_\chi(N)| \leqslant cN^{1-\frac{1}{r}}p^{\frac{r+1}{4r^2}}(\log p)^{\frac{1}{r}} \tag{12.58}$$

but only for characters $\chi(\text{mod } p)$ to prime modulus. Here c is an absolute constant ($c = 30$ is fine). Our arguments go by induction with respect to N. First notice that the assertion (12.58) is either trivial or it follows from (12.50), unless

$$c^r p^{\frac{1}{4}+\frac{1}{4r}}\log p \leqslant N \leqslant p^{\frac{1}{2}+\frac{1}{4r}}\log p \tag{12.59}$$

which condition we henceforth assume. Applying a shift $n \mapsto n+h$ with $1 \leqslant h \leqslant H < N$ we obtain

$$S_\chi(N) = \sum_{M<m\leqslant M+N} \chi(n+h) + 2\theta E(H) \tag{12.60}$$

where $|\theta| \leqslant 1$ and

$$E(H) = cH^{1-\frac{1}{r}}p^{\frac{r+1}{4r^2}}(\log p)^{\frac{1}{r}} \tag{12.61}$$

by the induction hypothesis (12.58) for the two character sums of length h which do not overlap with the original segment. Let $H = AB$, where A, B are positive integers. We use the shifts of type

$$h = ab \quad \text{with } 1 \leqslant a \leqslant A, \ \ 1 \leqslant b \leqslant B. \tag{12.62}$$

Averaging (12.60) over a, b we get

$$S_\chi(N) = \frac{1}{H}\sideset{}{^\star}\sum_{\substack{1\leqslant a\leqslant A\\ 1\leqslant b\leqslant B}} \sum_{M<n\leqslant M+N} \chi(n+ab) + 2\theta E(H)$$

where $|\theta| \leqslant 1$. By $\chi(n+ab) = \chi(a)\chi(\bar{a}n+b)$, where $\bar{a}$ is the multiplicative inverse of a modulo p (here we employ two properties of characters, the periodicity and the multiplicativity) we deduce that

$$|S_\chi(N)| \leqslant H^{-1}V + 2E(H) \tag{12.63}$$

where

$$V = \sum_{x(\text{mod } p)} \nu(x)\Big|\sum_{1\leqslant b\leqslant B} \chi(x+b)\Big|$$

and $\nu(x)$ is the number of representations of x as $\bar{a}n(\text{ mod } p)$ with $1 \leqslant a \leqslant A$ and $M < n \leqslant M+N$. We shall estimate V without using the induction hypothesis, so the implied constant will be independent of c.

As a and n vary over short segments relative to q, many residue classes $x(\text{mod } q)$ are not represented by $\bar{a}n$; in other words, $\nu(x)$ is often zero. Moreover, $\nu(x)$ is essentially bounded, but it is impossible to analyze its random changes of size. Therefore we relax $\nu(x)$ in V by applying the Hölder inequality

$$V \leqslant V_1^{1-\frac{1}{4}}V_2^{\frac{1}{2r}}W^{\frac{1}{2r}},$$

where

$$V_1 = \sum_{x(\mathrm{mod}\ p)} \nu(x), \qquad V_2 = \sum_{x(\mathrm{mod}\ p)} \nu^2(x),$$

$$W = \sum_{x(\mathrm{mod}\ p)} \Big| \sum_{1\leqslant b\leqslant B} \chi(x+b)\Big|^{2r}.$$

We could have restricted the outer summation in W to the residue classes $x(\mathrm{mod}\ p)$ for which $\nu(x) \neq 0$, but we are not able to take advantage of such a condition. The extension to the complete sum is massive, nevertheless it is nominal relative to the length of the character sum raised to power $2r$. This explains our choice of Hölder's exponents.

Clearly, we have $V_1 = AN$. We shall show that V_2 is also of this magnitude (essentially).

LEMMA 12.7. *We have*

$$V_2 \leqslant 8AN(ANp^{-1} + \log 3A). \tag{12.64}$$

PROOF. V_2 is the number of quadruples (a_1, a_2, n_1, n_2) with $1 \leqslant a_1, a_2 \leqslant A$ and $M < n_1, n_2 \leqslant M + N$ such that $a_1 n_2 \equiv a_2 n_1 (\mathrm{mod}\ p)$. Fix a_1, a_2 and put $a_1 n_2 - a_2 n_1 = kp$. We have

$$\Big|k - (a_1 - a_2)\frac{M}{p}\Big| \leqslant \frac{AN}{p}$$

and $(a_1, a_2)|k$. Given a_1, a_2 and k as above we find that the number of pairs (n_1, n_2) satisfying the equation $a_1 n_2 - a_2 n_1 = kp$ is bounded by $2N(a_1, a_2)/\max(a_1, a_2)$. Therefore

$$V_2 \leqslant 2N \sideset{}{^\star}\sum_{1\leqslant a_1, a_2 \leqslant A} \frac{(a_1, a_2)}{\max(a_1, a_2)} \Big(\frac{2AN}{(a_1, a_2)p} + 1\Big).$$

Hence (12.64) follows after simple estimates. □

LEMMA 12.8. *We have*

$$W \leqslant (2rB)^r p + 2rB^{2r} p^{\frac{1}{2}}. \tag{12.65}$$

This estimate for W lies at the heart of Burgess' method. Assuming (12.65) we complete the proof of (12.58) as follows. We choose $B = [rp^{\frac{1}{2r}}]$ giving

$$W \leqslant (2r)^{2r} p^{\frac{3}{2}}.$$

Then we choose $A = [N/9rp^{\frac{1}{2r}}]$. Note that $A \geqslant 1$ by the left side of (12.59), and $AN \leqslant N^2/9rp^{\frac{1}{2r}} \leqslant p(\log p)^2$ by the right side of (12.59). Therefore Lemma 12.7 gives

$$V_2 \leqslant AN(4\log p)^2.$$

Recalling that $V_1 = AN$ we derive

$$V \leqslant 2r(AN)^{1-\frac{1}{2r}}(4p^{\frac{3}{4}}\log p)^{\frac{1}{r}} \leqslant N^{2-\frac{1}{r}}(p^{\frac{r+1}{4r}}\log p)^{\frac{1}{r}}.$$

Next note that $H = AB \leqslant N/9$ and $H \geqslant N/10$. Therefore we derive by (12.61) and (12.63) that

$$|S_\chi(N)| \leqslant (10 + \frac{2}{3}c)N^{1-\frac{1}{r}}p^{\frac{r+1}{4r^2}}(\log p)^{\frac{1}{4}}.$$

Taking $c = 30$ we obtain (12.58).

It remains to prove Lemma 12.8, for which we can assume $B < p$. As in Burgess' original proof we appeal to the Riemann Hypothesis for curves over finite fields. By Corollary 11.24, for any χ modulo p which is non-trivial, we have

$$\Big|\sum_{x(\mathrm{mod}\ p)} \chi((x+b_1)\cdots(x+b_r))\bar{\chi}((x+b_{r+1})\cdots(x+b_{2r}))\Big| \leqslant 2rp^{\frac{1}{2}} \tag{12.66}$$

if one of the classes $b_v(\mathrm{mod}\ p)$, $v = 1, \dots, 2r$ is different from any other one.

We have

$$W = \sum_{1\leqslant b_1,\dots,b_{2r}\leqslant B}\cdots\sum\ \sum_{x(\mathrm{mod}\ p)} \chi((x+b_1)\cdots(x+b_r))\bar{\chi}((x+b_{r+1})\cdots(x+b_{2r})).$$

The complete sum satisfies (12.66) except possibly for the $(b_1, \dots, b_{2r})$ which can be arranged in r equal pairs. Thus the number of exceptions is bounded by $r\binom{2r}{r}B^r \leqslant (2rB)^r$. Hence (12.65) follows.

REMARK. Choosing the involved parameters in the above proof more economically one can get (12.58) with the factor $c(\log p)^{\frac{1}{r}}$ replaced by $c'(\log p)^{\frac{1}{2r}}$ for some absolute constant $c' \geqslant 1$ (for other comments see [**Fri**]).

EXERCISE 2. Prove along the above lines the estimate (12.55) with $r = 2$ for any modulus $q > 1$.

The Burgess bound (12.55) with $r = 2$ implies a subconvexity bound for Dirichlet L-functions in conductor aspect (compare Section 5.9).

THEOREM 12.9. *Let χ be a primitive Dirichlet character modulo $q \geqslant 2$, and $s = \frac{1}{2} + it$. We have*

$$L(s,\chi) \ll |s|q^{\frac{3}{16}+\varepsilon} \tag{12.67}$$

for any $\varepsilon > 0$, the implied constant depending only on ε.

PROOF. By Theorem 5.3 we have

$$|L(s,\chi)| \leqslant 2\Big|\sum_n \frac{\chi(n)}{n^s}V_s\Big(\frac{n}{\sqrt{q}}\Big)\Big|$$

where $V_s(y)$ is a rapidly decaying function given by (5.13). Precisely it satisfies

$$V_s(y) \ll \Big(1 + \frac{y}{|s|}\Big)^{-1} \qquad \text{and} \qquad V_s'(y) \ll \frac{|s|}{y}\Big(1 + \frac{y}{|s|}\Big)^{-1}.$$

Hence

$$\Big(x^{-s}V_s(x/\sqrt{q})\Big)' \ll \frac{|s|}{x^{3/2}}\Big(1 + \frac{x}{q|s|}\Big)^{-1}.$$

By (12.55) with $r = 2$ we have

$$S(x) = \sum_{n \leqslant x} \chi(n) \ll \min(q, x^{\frac{1}{2}} q^{\frac{3}{16}+\varepsilon}).$$

Then by partial summation we get

$$|L(s, \chi)| \leqslant \int_1^\infty \left| S(x) \left(x^{-s} V_s(x/\sqrt{q}) \right)' \right| dx$$

and the result follows by applying the relevant estimates. □

12.5. Very short character sums to truly composite modulus.

It is natural that Burgess's method fails to produce a non-trivial estimate for the character sum $S_\chi(N)$ of length $N \ll q^{\frac{1}{4}}$, where q is the conductor of χ. On the other hand, we do have effective treatments of exponential sums of Weyl's type (see Chapter 8) which are very short relative to the size of the amplitude function. Here we apply the differencing process to reduce the amplitude function several times until it is small enough (however unbounded) so that a non-trivial estimate for the resulting sum in the final step is available from special sources. In this respect the case of the character sum $S_\chi(N)$ is quite different because one normally cannot reduce the conductor by shifting the argument; absolutely not, if the conductor is a prime number. There is, however, a possibility to imitate the Weyl-Van der Corput differencing ideas for character sums of modulus which factor into relatively small numbers; we say that such a modulus is truly composite. An important estimate for character sums of special composite moduli was given by S. W. Graham and J. Ringrose [**GRi**] (see also [**H-B**] and [**FI3**]). In this section we present our version of the Graham-Ringrose magnificent result.

We shall proceed by induction with respect to the number of factors (not necessarily primes) in the modulus of the character. For this reason, in order to meet the induction hypothesis, we treat more general sums than necessary for the final result.

Let $\xi \pmod{k}$ be a multiplicative character. We consider

$$S_\xi(N) = \sum_{M < m \leqslant M+N} \xi(f(n)) \tag{12.68}$$

where $f(x)$ is a rational function with all zeros and poles being integers. We write

$$f(x) = \prod_{\nu=1}^{m} (x - a_\nu)^{d_\nu} \tag{12.69}$$

where d_ν are non-zero integers and a_ν are any integers. We do not assume that a_1, ..., a_m are distinct, so the representation (12.69) does not define the exponents $d_1, \dots, d_m$ uniquely. Recall that for a given representation $f = f_1/f_0$ with $f_1, /f_0 \in \mathbb{Z}[x]$ we defined $\xi(f(x)) = \xi(f_1(x))\bar{\xi}(f_0(x))$, therefore the sum (12.68) is actually restricted by

$$((n - a_1) \cdots (n - a_m), k) = 1. \tag{12.70}$$

THEOREM 12.10. *Let $\xi = \chi_1 \cdots \chi_{r-1}\chi$, where χ_ℓ (mod q_ℓ) for $1 \leqslant \ell < r$ are any characters and χ (mod q) is a primitive character of conductor $q > 1$, q squarefree. Let f be given by (12.69) with*

$$(d_1, \dots, d_m, h) = 1 \tag{12.71}$$

where h is the order of χ. Put $N_0 = \max\{q_1, \dots, q_{r-1}, q^{\frac{1}{4}}\}q^{\frac{5}{4}}$, and

$$\Delta = \prod_{\nu_1 \neq \nu_2} \prod (a_{\nu_1} - a_{\nu_2}). \tag{12.72}$$

Then for any $N \geqslant N_0$ we have

$$|S_\xi(N)| \leqslant 4N((\Delta q_1 \cdots q_{r-1}, q)q^{-1}\tau(q)^{r^2}\tau_m(q)^{2r})^{2^{-r}}. \tag{12.73}$$

The power of proof stems from estimation of the complete character sum

$$S(\chi) = \sum_{x (\mathrm{mod}\ q)} \chi(f(x)). \tag{12.74}$$

If $q = p$ is prime we are dealing with a sum over the field $\mathbb{F}_p$. In this case we derive from [**Sch**] the following

PROPOSITION 12.11. *Let χ (mod p) be a non-principal character of order h which satisfies (12.71). Then*

$$|S(\chi)| \leqslant (m-1)(\Delta, p)^{\frac{1}{2}} p^{\frac{1}{2}}. \tag{12.75}$$

PROOF. First note that $p \neq 2$ because there is only the trivial character of modulus two. For $m = 1$ we have $f(x) = (x - a_1)^{d_1}$ with $(d_1, h) = 0$ and

$$S(\chi) = \sum_{x (\mathrm{mod}\ p)} \chi^{d_1}(x) = 0,$$

so (12.75) holds (in this case $\Delta = 1$). Let $m \geqslant 2$. If $\Delta \equiv 0 (\mathrm{mod}\ p)$, then (12.75) is trivial. Now suppose all the numbers $a_1, \dots, a_m$ are distinct modulo p. If f is a polynomial (i.e., all the exponents $d_1, \dots, d_m$ are positive), then the polynomial $F(X, Y) = Y^h - f(X)$ is absolutely irreducible (see Lemma 2C of [**Sch**]) and our assertion follows from Theorem 11.23. If $f(x) = f_1(x)/f_0(x)$ with $f_1(x), f_0(x) \in \mathbb{Z}[x]$ having all distinct zeros, then we consider the polynomial $g(x) = f_1(x)f_0(x)^{p-2}$ in place of the rational function $f(x)$. Note that $\chi(f(x)) = \chi(g(x))$ and $(p-2, h) = 1$, because $h|(p-1)$. Therefore the condition (12.71) holds for $g(x)$, so (12.75) is established. □

Proposition 12.11 extends by multiplicativity as follows

COROLLARY 12.12. *Let χ (mod q) be a primitive character of squarefree conductor q and order h which satisfies (12.71). Then*

$$|S(\chi)| \leqslant (m-1)^{\omega(q)}(\Delta, q)^{\frac{1}{2}} q^{\frac{1}{2}} \tag{12.76}$$

where $\omega(q)$ is the number of prime divisors of q.

PROOF. If $q = p_1 \cdots p_t$, then $\chi = \chi_1 \cdots \chi_t$, where $\chi_1, \dots, \chi_t$ are primitive characters of order $h_1, \dots, h_t$ respectively which are divisors of h. Therefore the condition (12.71) is satisfied for each character $\chi_s \pmod{p_s}$. Multiplying (12.75) for $\chi_1, \dots, \chi_t$ we get (12.76). □

Now we are ready to prove that for $N \geqslant N_0$,

$$|S_{\chi_1 \cdots \chi_{r-1}\chi}(N)| \leqslant c_r(m)((\Delta q_1 \cdots q_{r-1}, q)q^{-1})^{2^{-r}} N \tag{12.77}$$

by induction with respect to r. The factors $c_r(m) \geqslant 1$ will be determined in the induction process; they depend on q, but it is not necessary to display this because q will not change throughout the process. Moreover, we do not change the set of exponents $\{d_1, \dots, d_m\}$ in the representation (12.69), although every induction step will double the number of zeros and poles of the resulting rational functions.

For $r = 1$ we have $\xi = \chi$ and

$$S_\chi(N) = \sum_{M < m \leqslant M+N} \chi(f(n)) = \frac{N}{q} S(\chi) + \theta q$$

where $S(\chi)$ is the complete sum (12.74) and $|\theta| \leqslant 1$. Hence we obtain by (12.76) that (because $q^{\frac{3}{2}} \leqslant N$)

$$|S_\chi(N)| \leqslant m^{\omega(q)} \Big(\frac{(\Delta, q)}{q}\Big)^{\frac{1}{2}} N. \tag{12.78}$$

Now let $r \geqslant 2$. We apply the shift $n \mapsto n + hq_1$ and average over $1 \leqslant h \leqslant H$ getting

$$S_\xi(N) = \tfrac{1}{H} \sum_{M < n \leqslant M+N} \chi_1(f(n)) \sum_{1 \leqslant h \leqslant H} \tilde{\xi}(f(n + hq_1)) + 2\theta H q_1$$

where $\tilde{\xi} = \chi_2 \cdots \chi_{r-1}\chi$ and $|\theta| \leqslant 1$. Here the error term is obtained by trivial estimation of the short character sums which do not overlap with the original sum. Hence $|S_\xi(N)| \leqslant T + 2Hq_1$, where

$$T = \frac{1}{H} \sum_{M < n \leqslant M+N} \Big| \sum_{1 \leqslant h \leqslant H} \tilde{\xi}(f(n + hq_1)) \Big|.$$

By Cauchy's inequality

$$T^2 \leqslant \frac{N}{H^2} \sum_{1 \leqslant h_1, h_2 \leqslant H}^{\star} \Big| \sum_{M < n \leqslant M+N} \tilde{\xi}(f(n + h_1 q_1)/f(n + h_2 q_1)) \Big|.$$

Here the inner sum is of type (12.68) for the reduced character $\tilde{\xi}$ and the quotient function $\tilde{f}(x) = f(x + h_1 q_1)/f(x + h_2 q_1)$. This has $2m$ zeros and poles. For $f(x)$ written as (12.69) we introduce the polynomial

$$\Delta(x) = \prod_{\nu_1 \neq \nu_2} \prod (x + a_{\nu_1} - a_{\nu_2}), \quad \deg \Delta(x) = m(m-1),$$

so $\Delta = \Delta(0)$. Then the corresponding polynomial for $\tilde{f}(x)$ is

$$\tilde{\Delta}(x) = \Delta^2(x)\Delta^2(x + y)(x + y)^{2m}$$

where $y = (h_1 - h_2)q_1$. Hence $\tilde{\Delta} = \tilde{\Delta}(0) = \Delta^2\Delta^2(y)y^{2m}$ and

$$(\tilde{\Delta}q_2 \cdots q_{r-1}, q) \leqslant (\Delta q_1 \cdots q_{r-1}, q)(y\Delta(y), q/(q, q_1)).$$

By the induction hypothesis we get

$$\Big|\sum_n \tilde{\xi}(\tilde{f}(n))\Big| \leqslant c_{r-1}(2m)g(h_1 - h_2)((\Delta q_1 \cdots q_{r-1}, q)q^{-1})^{2^{1-r}}N$$

where $g(h) = (hq_1\Delta(hq_1), q/(q, q_1))^{2^{1-r}}$. We choose $H = q$ and estimate as follows

$$\begin{aligned}\frac{1}{H^2}\sideset{}{^\star}\sum_{1\leqslant h_1,h_2\leqslant H} g(h_1 - h_2) &= \frac{1}{q}\sum_{h(\text{mod } q)} (hq_1\Delta(hq_1), q/(q, q_1))^{2^{1-r}}\\ &\leqslant \frac{1}{q}\sum_{y(\text{mod } q)} (y\Delta(h), q)^{2^{1-r}}\\ &\leqslant \Big(\frac{1}{q}\sum_{y(\text{mod } q)} (y\Delta(y), q)\Big)^{2^{1-r}}.\end{aligned}$$

For the inner sum we have

$$\begin{aligned}\frac{1}{q}\sum_{y(\text{mod } q)} (y\Delta(h), q) &\leqslant \sum_{d|q} |\{y(\text{mod } d); y\Delta(y) \equiv 0(\text{mod } d)\}|\\ &\leqslant \sum_{d|q}(1 + \deg\Delta)^{\omega(d)} = (2 + \deg\Delta)^{\omega(q)}.\end{aligned}$$

We have $\deg\Delta = m(m-1)$, so $2 + \deg\Delta \leqslant 2m^2$ and $(2 + \deg\Delta)^{\omega(q)} \leqslant \tau(q)\tau_m^2(q)$. Gathering the above estimates we obtain

$$|S_\xi(N)| \leqslant c_{r-1}^{\frac{1}{2}}(2m)(\tau(q)\tau_m^2(q)(\Delta q_1 \cdots q_{r-1}, q)/q)^{2^{-r}}N + 2qq_1.$$

Since $q_1 q^{\frac{5}{4}} \leqslant N$, this yields (12.77) for $r \geqslant 2$, provided

$$c_r(m) \geqslant c_{r-1}^{\frac{1}{2}}(2m)(\tau(q)\tau_m^2(q))^{2^{-r}} + 2.$$

By (12.78) we get (12.77) for $r = 1$, provided $c_1(m) \geqslant \tau_m(q)$. One can check that the numbers

$$c_r(m) = 4(\tau(q)^{r^2}\tau_m(q)^{2r})^{2^{-r}}$$

satisfy the above (recurrence) inequality (use $\tau_{2m}(q) = \tau(q)\tau_m(q)$) giving (12.73).

Our main interest in Theorem 12.10 is for $f(x) = x$ (so $m = 1, \Delta = 1$) in which case it yields

THEOREM 12.13. *Let χ_ℓ (mod q_ℓ) for $1 \leqslant \ell < r$ be any characters and let χ (mod q) be a primitive character of conductor $q > 1$, q squarefree with $(q, q_1 \cdots q_{r-1}) = 1$. Then for $N \geqslant N_0 = \max\{q_1, \ldots, q_{r-1}, q^{\frac{1}{4}}\}q^{\frac{5}{4}}$ we have*

$$\Big|\sum_{M<m\leqslant M+N} \chi_1 \cdots \chi_{r-1}\chi(n)\Big| \leqslant 4N\Big(\frac{\tau(q)^{r^2}}{q}\Big)^{2^{-r}}. \tag{12.79}$$

COROLLARY 12.14. *Let χ be a primitive character of conductor $q > 1$, q squarefree. Suppose all prime factors of q are $\leqslant N^{\frac{1}{9}}$. Then*

$$\Big|\sum_{M<n\leqslant M+N}\chi(n)\Big| \leqslant 4N\tau(q)^{r/2^r}q^{-1/r2^r} \tag{12.80}$$

where r is any integer with $N^r \geqslant q^3$.

PROOF. Let p be the largest prime factor of q. We can write $q = q_1\cdots q_r$ with $q_\ell \leqslant pq^{\frac{1}{4}}$ for $1 \leqslant \ell \leqslant r$ (some of the factors can be one). We have $q_\ell \leqslant N^{\frac{4}{9}}$. Moreover, one of the factors, say q_r, satisfies $q_r \geqslant q^{\frac{1}{r}}$. We can also require that $\omega(q_r) \leqslant \omega(q)/r$. To meet these conditions define q_r as the smallest product of the largest prime factors of q such that $q_r \geqslant q^{\frac{1}{r}}$. Then define q_ℓ successively for $1 \leqslant \ell < r$ as the largest product of the largest prime factors of $q/q_r\cdots q_{\ell+1}$ such that $q_\ell \leqslant pq^{\frac{1}{r}}$ (if $q = q_r\cdots q_{\ell+1}$, then put $q_\ell = \cdots = q_1 = 1$). Clearly, every prime factor of q is used once in these r steps, so $q = q_1\cdots q_r$. Accordingly $\chi = \chi_1\cdots\chi_r$, where χ_ℓ are characters modulo q_ℓ and χ_r is primitive. Therefore (12.80) follows from (12.79). □

The divisor function in the estimates (12.79), (12.80) is a nuisance. In general we have $\tau(q) = (\alpha_1+1)\cdots(\alpha_n+1) \leqslant 2^{\alpha_1+\cdots+\alpha_n} = 2^{\Omega(q)}$ if $q = p_1^{\alpha_1}\cdots p_n^{\alpha_n}$, where $\Omega(q)$ is the number of prime factors of q counted with multiplicity. This can be quite large. We estimate $\Omega(q)$ as follows,

$$\Omega(q) = \sum_{d|q}\frac{\Lambda(d)}{\log d} \leqslant \sum_{d|q}\frac{\Lambda(d)}{\log z} + \sum_{d<z}\Big(\frac{\Lambda(d)}{\log d} - \frac{\Lambda(d)}{\log z}\Big) \leqslant \frac{\log q}{\log z} + \frac{cz}{\log^2 z}$$

where $z > 1$ and c is an absolute constant. Choosing $z = \log q$ for $q \geqslant 3$ we get

$$\Omega(q) \leqslant \frac{\log q}{\log\log q}\Big(1 + \frac{c}{\log\log q}\Big). \tag{12.81}$$

Hence

$$\tau(q) \leqslant 2^{(1+c/\log\log q)\log q/\log\log q}. \tag{12.82}$$

EXERCISE 3. Prove that $\tau(q) \leqslant q^{1/\log\log 3q}$ for all $q \geqslant 1$.

Using this crude estimate we derive from (12.80)

COROLLARY 12.15. *Let χ be a primitive character of conductor $q \geqslant 3$, q squarefree. Let p be the largest prime factor of q. Then, for*

$$N \geqslant q^{\varepsilon(q)} + p^9 \quad \textit{with } \varepsilon(q) = 4(\log\log q)^{-\frac{1}{2}} \tag{12.83}$$

we have

$$\sum_{M<n\leqslant M+N}\chi(n) \ll N\exp(-\sqrt{\log q}) \tag{12.84}$$

where the implied constant is absolute.

EXERCISE 4. Let the conditions be as in Corollary 12.15. Show that

$$L(1,\chi) \leqslant \varepsilon(q)\log q + 9\log p + b \tag{12.85}$$

where b is an absolute constant.

12.6. Characters to powerful modulus.

The truly composite moduli (note that the numbers having no large prime divisors, the so-called "smooth numbers", or "entiers friables" in French, are truly composite) appear in practice not as often as the prime moduli. Another extreme are the moduli which are high powers of a fixed prime. These are examples of powerful numbers. We say that q is powerful if its kernel

$$k = \prod_{p|q} p \tag{12.86}$$

is small relative to q in the logarithmic scale.

Let χ be a primitive character of conductor q which is a powerful number. In this case a large part of character values can be described as an exponential of a polynomial, so the character sum $S_\chi(N)$ can be reduced to Weyl sums. To this end consider the following polynomial:

$$L(x) = x - \frac{x^2}{2} + \cdots \pm \frac{x^d}{d}. \tag{12.87}$$

One can show that (see [**Ga2**], p. 192)

$$L(x+y+xy) = L(x) + L(y) + \sum_{d<a+b\leqslant 2d}^{\star} c(a,b)x^a y^b \tag{12.88}$$

with some rational coefficients $c(a,b)$ such that $c(a,b) \in \mathbb{Z}[a,b]$.

Suppose $4 \nmid q$ (for a technical reason). Let d be sufficiently large so that $q^2|k^d$. Let D be the product of all $m \leqslant d, (m,q) = 1$. Clearly $DL(kx) \in \mathbb{Z}[x]$, and it follows from (12.88) that

$$DL(kx + ky + k^2xy) \equiv DL(kx) + DL(ky) (\mathrm{mod}\ q).$$

Hence the function $\xi(1+kx) = e(DL(kx)/q)$ is a character of the multiplicative group of residue classes mod q congruent to 1 mod k. It is easy to see that the order of ξ is q/k, which is the order of the group. Therefore there exists an integer B such that

$$\chi(1+kx) = e(BDL(kx)/q). \tag{12.89}$$

Since q is the conductor of χ it follows that B is prime to q/k. On the other hand, b is determined only mod q/k so we can assume that B is prime to q.

By (12.89) we arrange the character sum (12.44) as follows:

$$S_\chi(N) = \sum_{0<a<k} \chi(a) \sum_n \chi(1+k\bar{a}n) = \sum_{0<a<k} \chi(a) \sum_n e(F(\bar{a}n)/q)$$

where n runs over the interval $(M-a)k^{-1} < n \leqslant (M+N-a)k^{-1}$, $a\bar{a} \equiv 1 (\mathrm{mod}\ q)$, and $F(x) = BDL(kx)$ is a polynomial of degree d with integer coefficients. The last sum is of Weyl's type which can be estimated by Vinogradov's method. This has been done in detail in [**I2**], getting

THEOREM 12.16. *Let* χ *be a primitive character of conductor* $q, 2 \nmid q$. *Let* k *be the product of prime factors of* q. *Then, for* $k^{100} < N < N' \leqslant 2N$ *we have*

$$\Big|\sum_{N<n\leqslant N'} \chi(n)\Big| \leqslant \gamma N^{1-\delta} \tag{12.90}$$

where

$$\gamma = \exp(200r\log^2(1200r)), \quad \delta = (1800r)^{-2}(\log 3600r)^{-1}, \quad r = \frac{\log 3q}{\log N}.$$

REMARKS. The first result of this type for $q = p^\ell$ was established by A. G. Postnikov [**Pos**]. Our generalization of the Postnikov formula (12.89) for the powerful modulus is based on the formula (12.88) which is due to P. X. Gallagher [**Ga2**]. There are applications of the results to bounding the $L(s,\chi)$, to widening the zero-free region, and to estimating the least prime $p \equiv a(\text{mod } q)$ (see [**Ga2**] and [**I2**]).

CHAPTER 13

SUMS OVER PRIMES

13.1. General principles.

Given $f : \mathbb{N} \to \mathbb{C}$ we are interested in estimation of the sum

$$V = \sum_p f(p) \tag{13.1}$$

where p runs over prime numbers, or of the closely related sum

$$S = \sum_n \Lambda(n) f(n) \tag{13.2}$$

where $\Lambda(n)$ is the von Mangoldt function. Note that S is obtained from V for $f(n)$ replaced by $f(n)\log n$ up to a small contribution of terms $n = p^\nu$ with $\nu \geqslant 2$. Of course, the problems of estimating S and V are equivalent by partial summation, but it is often simpler in practice to deal with S.

If f is a nice, smooth function with compact support on $\mathbb{R}^+$ one can express S by the zeros of $\zeta(s)$ by integrating $-\hat{f}(s)\zeta'(s)/\zeta(s)$, where $\hat{f}$ is the Mellin transform of f. One gets

$$S = \hat{f}(1) - \sum_{n>0} \hat{f}(-2n) - \sum_\rho \hat{f}(\rho). \tag{13.3}$$

However, even assuming the Riemann Hypothesis, this expression (the so-called explicit formula, see Exercise 4 in Chapter 5) is not always useful. For example, if $f : \mathbb{R}^+ \to \mathbb{C}$ is an oscillatory function, such as

$$f(x) = e(\alpha x) g(x) \tag{13.4,)}$$

$$f(x) = e(\alpha\sqrt{x}) g(x) \tag{13.5}$$

with $\alpha \in \mathbb{R}^\star$ and $g(x)$ a bump function, then the Mellin transform $\hat{f}(s)$ is spread over a large range of zeros so that the expansion (13.3) fails to produce good results. As with any summation type formula (see Chapter 4) the decision to apply it or not can be reasonably guided according to whether the original sum has more terms over primes than the resulting sum over the zeros. Recently, equipped with powerful computers, researchers do not hesitate to download from the internet[1] a list of numerical values of the zeros, and are not afraid of using them to test sums over primes indirectly via the explicit formula. Primes and zero are no longer different beasts, it is a matter of choice to work with one or the other sets.

[1] See Odlyzko's site: **http://www.dtc.umn.edu/~odlyzko/zeta_tables/**

Many interesting arithmetic functions $f : \mathbb{N} \to \mathbb{C}$ are not even defined on the real numbers, in which case the explicit formula (13.3) is not applicable (true, one can take smooth extension of f to $\mathbb{R}^+$, but it does not work efficiently if the set of values $f(n)$ for $n \in \mathbb{N}$ is erratic). In such a case we prefer to use a sequence notation $\mathcal{A} = (a_n)$ in place of the function $f(n) = a_n$. Since $\mathcal{A}$ is infinite, we consider the finite sums

$$S(\mathcal{A}, x) = \sum_{n \leqslant x} \Lambda(n) a_n \tag{13.6}$$

in place of (13.2) with the aim of estimating these for all sufficiently large x.

It is not difficult to predict the asymptotic behavior of $S(\mathcal{A}, x)$ by appealing to a randomness of the Möbius function:

MÖBIUS RANDOMNESS LAW. *The Möbius function $\mu(m)$ changes sign randomly so that for any "reasonable" sequence of complex numbers $\mathcal{A} = (a_m)$ the twisted sum*

$$M(\mathcal{A}, x) = \sum_{m \leqslant x} \mu(m) a_m \tag{13.7}$$

is relatively small due to the cancellation of its terms.

Of course a reasonable sequence means chosen with no bias. In practice it is often the case, except when playing with sieve weights which do conspire with the Möbius function.

To apply this law to prime numbers we write $\Lambda = \mu \star L$, or

$$\Lambda(n) = \sum_{d|n} \mu(d) \log \frac{n}{d} = -\sum_{d|n} \mu(d) \log d, \tag{13.8}$$

and insert the latter into (13.6) getting

$$S(\mathcal{A}, x) = -\sum_{d} \mu(d)(\log d) A_d(x), \tag{13.9}$$

where

$$A_d(x) = \sum_{\substack{n \leqslant x \\ n \equiv 0 (\mathrm{mod}\ d)}} a_n. \tag{13.10}$$

The quantities $A_d(x)$, so also $A_d(x) \log d$, are "reasonable" if the original arithmetic terms a_n are themselves "reasonable". Therefore, according to the principle, we believe that the contribution of large d's in (13.9) is negligible, so $S(\mathcal{A}, x)$ is well approximated by

$$S(\mathcal{A}, D, x) = -\sum_{d \leqslant D} \mu(d)(\log d) A_d(x) \tag{13.11}$$

for some D which is relatively small. Now for $d \leqslant D$ the quantities $A_d(x)$ may satisfy (due to large averaging) a simple, yet strong approximate formula

$$A_d(x) = g(d) A(x) + r_d(x) \tag{13.12}$$

where $g(d)$ is a nice multiplicative function, $A(x) = A_1(x)$, and $r_d(x)$ is a small error term. Ignoring the error term we arrive at $S(\mathcal{A}, D, x) \sim H(D)A(x)$, where

$$H(D) = -\sum_{d \leqslant D} \mu(d)(\log d)g(d).$$

Usually g is regularly distributed over primes so there is a limit of $H(D)$ as D tends to ∞, say $H(D) \sim H(\infty) = H$ with

$$H = -\sum_{d} \mu(d)(\log d)g(d) = \prod_{p} (1 - g(p))\Big(1 - \frac{1}{p}\Big)^{-1}. \tag{13.13}$$

Therefore, the above transformations lead us to the following asymptotic formula

$$S(\mathcal{A}, x) \sim HA(x), \quad \text{as } x \to \infty. \tag{13.14}$$

Of course, proving this rigorously is a different matter. What needs to be said, however, is that the heuristic formula (13.14) has never failed to hold in every natural case in which $S(\mathcal{A}, x)$ is evaluated rigorously by special means.

In 1934, I. M.Vinogradov (see [**V5**], [**V6**]) gave a very good estimate for the exponential sum

$$V(\alpha; x) = \sum_{p \leqslant x} e(\alpha p) \tag{13.15}$$

which was needed for solving the Goldbach problem with three primes (see Chapter 19). In the absence of the Riemann Hypothesis, this was a surprising accomplishment. His method applies to the sum $S(\mathcal{A}, x)$ for a great variety of oscillating sequences $\mathcal{A} = (a_n)$ which are not multiplicative, such as $a_n = e(\alpha n)$, $a_n = e(\alpha\sqrt{n})$, $a_n = \chi(n + 1)$, where χ is a multiplicative non-principal character.

Here is how one can explain the principles of Vinogradov's method in abstract settings. One starts by repeated applications of the exclusion-inclusion procedure to arrange $S(\mathcal{A}, x)$ into sums of two types

$$S_1 = \sum_{d \leqslant D} \sum_{dn \leqslant x} \alpha_d a_{dn}, \tag{13.16}$$

$$S_2 = \sum\sum_{\substack{mn \leqslant x \\ M < m \leqslant N}} \beta_m \gamma_n a_{mn}, \tag{13.17}$$

where $\alpha_d, \beta_m, \gamma_n$ are independent real numbers (essentially bounded) and D, M, N are suitable parameters (neither small nor large relative to x). The sum of type S_1 can be written as

$$S_1 = \sum_{d \leqslant D} \alpha_d A_d(x) \tag{13.18}$$

where $A_d(x)$ is given by (13.10). Here the terms a_n of $A_d(x)$ are cleared of complicated coefficients except for the condition $n \equiv 0 \pmod{d}$ which is easy to handle for small d. Assuming some regularity in the variation of the argument of a_n one shows that $A_d(x)$ is small for every $d \leqslant D$ due to the cancellation of terms.

The sum of type S_2 is considered as a bilinear form in the coefficients β_m, γ_n (see Chapter 7 for more general considerations). For estimation it is convenient to split S_2 into bilinear forms over dyadic segments

$$\mathcal{B}(M,N) = \sum_{M<m\leqslant 2M} \beta_m \sum_{N<n\leqslant 2N} \gamma_n a_{mn}. \tag{13.19}$$

Applying Cauchy's inequality one removes the coefficients β_m, γ_n as follows:

$$|\mathcal{B}(M,N)|^2 \leqslant \Big(\sum_m |\beta_m|^2\Big) \sum_{n_1}\sum_{n_2} |\gamma_{n_1}\gamma_{n_2}| \mathcal{C}(n_1,n_2) \tag{13.20}$$

where

$$\mathcal{C}(n_1,n_2) = \sum_m a_{mn_1}\bar{a}_{mn_2}. \tag{13.21}$$

If $n_1 = n_2$ we can do nothing better than the trivial bound. However, since N is quite large the number of such cases (the diagonal terms) is relatively small. On the other hand, for a majority of $n_1 \neq n_2$ the sum $\mathcal{C}(n_1,n_2)$ is small because of cancellation of terms (the variation of the argument of a_{mn_1} is not completely annihilated by that of $\bar{a}_{mn_2}$). In this way one arrives at a non-trivial estimate for $\mathcal{B}(M,N)$ and also for the sums of type S_2.

Combining the results for sums of type S_1 and S_2 one deduces a bound for $S(\mathcal{A},x)$ which is often remarkably strong. In the forthcoming sections we provide details of the Vinogradov method in particular contexts. We shall also present some other methods.

13.2. A variant of Vinogradov's method.

In this section we develop a formula for sums over primes using some modifications of the original ideas of Vinogradov. The arguments are of combinatorial nature, they have much in common with sieve methods.

For $1 < n \leqslant x$ we put $\beta_n(x) = (1+\omega(n,\sqrt{x}))^{-1}$, where $\omega(n,\sqrt{x})$ is the number of primes $p|n$ with $p \leqslant \sqrt{x}$. If n is squarefree, we have

$$\sum_{p|n,p\leqslant\sqrt{x}} \beta_{n/p}(x) = \begin{cases} 1 & \text{if } n \text{ has a prime factor} \leqslant \sqrt{x}, \\ 0 & \text{if } n \text{ is prime with } \sqrt{x} < n \leqslant x. \end{cases}$$

This nice formula is due to O. Ramaré. Hence for any sequence of complex numbers $\mathcal{A} = (a_n)$ we have

$$\sum_{\sqrt{x}<p\leqslant x} a_p = \sideset{}{^\flat}\sum_{1<n\leqslant x} a_n - \sideset{}{^\flat}{\sum\sum}_{\substack{p\leqslant\sqrt{x},\ mp\leqslant x\\(m,p)=1}} \beta_m(x) a_{mp}$$

where $\sum^\flat$ restricts the summation to squarefree numbers. Let $P(z)$ be the product of all primes $p \leqslant z$ with $2 \leqslant z \leqslant \sqrt{x}$. We apply the above identity to the subsequence of numbers a_n with $(n,P(z)) = 1$ getting

$$\sum_{\sqrt{x}<p\leqslant x} a_p = \sideset{}{^\flat}\sum_{\substack{1<n\leqslant x\\(n,P(z))=1}} a_n - \sideset{}{^\flat}{\sum\sum}_{\substack{z<p\leqslant\sqrt{x} \quad mp\leqslant x\\ p\nmid m,(m,P(z))=1}} \beta_m(x) a_{mp}.$$

Now we remove the restriction to squarefree numbers n, and the condition $p \nmid m$. The correction quantity does not exceed

$$\sum_{\substack{n \leqslant x \\ \forall p>z, p^2 | n}} |a_n| \leqslant \Big(\sum_{n \leqslant x} |a_n|^2\Big)^{\frac{1}{2}} \Big(\sum_{p>z} xp^{-2}\Big)^{\frac{1}{2}} \leqslant \|\mathcal{A}(x)\|\, x^{\frac{1}{2}} z^{-\frac{1}{2}}$$

where $\|\mathcal{A}(x)\|$ is defined by

$$\|\mathcal{A}(x)\| = \Big(\sum_{n \leqslant x} |a_n|^2\Big)^{\frac{1}{2}}. \tag{13.22}$$

To this we add the terms a_1 and a_p with $p \leqslant \sqrt{x}$ and estimate their contribution by

$$\sum_{n \leqslant \sqrt{x}} |a_n| \leqslant \|\mathcal{A}(x)\| x^{\frac{1}{4}}.$$

Next we remove the condition $(n, P(z)) = 1$ by the exact sieve (the Legendre formula)

$$\sum_{\substack{n \leqslant x \\ (n,P(z))=1}} a_n = \sum_{d | P(z)} \mu(d) A_d(x).$$

We keep the terms $A_d(x)$ with $d \leqslant \sqrt{x}$ and estimate the remaining ones. First by Cauchy's inequality we get

$$\begin{aligned} \sum_{\substack{d|P(z) \\ d>\sqrt{x}}} |A_d(x)| &\leqslant \sum_{\substack{dn \leqslant x \\ d|P(z), d>\sqrt{x}}}\sum |a_{dn}| \leqslant \|\mathcal{A}(x)\| \Big(\sum_{\substack{dn \leqslant x \\ d|P(z), d>\sqrt{x}}}\sum \tau(dn)\Big)^{\frac{1}{2}} \\ &\leqslant \|\mathcal{A}(x)\| (x \log 3x)^{\frac{1}{2}} \Big(\sum_{d|P(z), d>\sqrt{x}} \tau(d) d^{-1}\Big)^{\frac{1}{2}}. \end{aligned}$$

To the last sum we apply Rankin's trick as follows

$$\begin{aligned} \sum_{\substack{d|P(z) \\ d>\sqrt{x}}} \tau(d) d^{-1} &\leqslant x^{-\varepsilon} \sum_{d|P(z)} \tau(d) d^{2\varepsilon - 1} \leqslant x^{-\varepsilon} \prod_{p \leqslant z} (1 + 2p^{2\varepsilon-1}) \\ &\leqslant x^{-\varepsilon} \prod_{p} (1 + p^{-1-\varepsilon})^{2z^{3\varepsilon}} \leqslant x^{-\varepsilon} \zeta(1+\varepsilon)^{2z^{3\varepsilon}} \\ &\leqslant x^{-\varepsilon} \Big(1 + \frac{1}{\varepsilon}\Big)^{2z^{3\varepsilon}} = \exp\Big(-\frac{\log x}{\log z}\Big)(1 + \log z)^{2e^3} \end{aligned}$$

by choosing $\varepsilon = 1/\log z$. Collecting the above estimates we get

PROPOSITION 13.1. *Let $2 \leqslant z \leqslant \sqrt{x}$. For any sequence of complex numbers $\mathcal{A} = (a_n)$ we have*

$$\sum_{p \leqslant x} a_p = \sum_{\substack{d|P(z) \\ d \leqslant \sqrt{x}}} \mu(d) A_d(x) - \sideset{}{^\flat}\sum_{(m,P(z))=1} \beta_m(x) \sum_{\substack{mp \leqslant x \\ z<p \leqslant \sqrt{x}}} a_{mp} + \varepsilon(x,z)\sqrt{x}\|\mathcal{A}(x)\| \tag{13.23}$$

where $P(z)$ is the product of all primes $p \leqslant z$, $\sum^\flat$ restricts the summation to square-free numbers, $A_d(x)$ is given by (13.10), $\|\mathcal{A}(x)\|$ is given by (13.22), $|\beta_m(x)| \leqslant 1$ and $\varepsilon(x,z)$ (which also depends on $\mathcal{A}$) satisfies

$$|\varepsilon(x,z)| \leqslant \frac{2}{\sqrt{z}} + \exp\Big(-\frac{\log x}{2\log z}\Big)(\log 3x)^{21}. \tag{13.24}$$

REMARKS. More precisely we have (13.23) with $\beta_m(x) = (1+\omega(m,\sqrt{x}))^{-1}$, where $\omega(m,\sqrt{x})$ is the number of primes $p|m, p \leqslant \sqrt{x}$, but we didn't specify this to emphasize that one doesn't need anything explicit about these coefficients. The first sum on the right side of (13.23) is of type S_1 with $D = \sqrt{x}$ (see (13.16) and (13.18)). One can also split this into two sums $S_1 + S_2$ according to $d \leqslant z$ and $z < d \leqslant \sqrt{x}$. Then S_1 is of type (13.16) with $D = z$ while S_2 is of type (13.17) with $M = z$ and $N = \sqrt{x}$. The double sum on the right side of (13.23) is also of type (13.17) with $M = z$ and $N = \sqrt{x}$. The estimation (13.24) is fine for

$$z = \exp(\sqrt{\log x}). \tag{13.25}$$

For this choice of z we have

$$|\varepsilon(x,z)| \leqslant 3\exp(-\tfrac{1}{2}\sqrt{\log x})(\log 3x)^{21}. \tag{13.26}$$

If $\mathcal{A} = (a_n)$ is an oscillatory sequence (so the main term in (13.12) is expected to be small), then without great loss one can simplify (13.23) by writing the inequality

$$\Big|\sum_{p\leqslant x} a_p\Big| \leqslant \sum_{d\leqslant\sqrt{x}} |A_d(x)| + \sum_m \Big| \sum_{\substack{mp\leqslant x\\ z<p\leqslant\sqrt{x}}} a_{mp}\Big| + \varepsilon(x,z)\sqrt{x}\|\mathcal{A}(x)\|. \tag{13.27}$$

We shall return to the above constructions in Section 13.6.

13.3. Linnik's identity.

In 1960 Yu.V. Linnik [**Li2**] gave a marvelous expression for

$$\Lambda'(n) = \frac{\Lambda(n)}{\log n} = \begin{cases} a^{-1} & \text{if } n = p^a, a > 0,\\ 0 & \text{otherwise}\end{cases} \tag{13.28}$$

in terms of the strict divisor functions

$$\tau_k'(n) = |\{n_1, \dots, n_k \geqslant 2;\ n_1 \dots n_k = n\}|. \tag{13.29}$$

To this end he computes the logarithm of

$$\zeta(s) = \sum_1^\infty n^{-s} = \prod_p (1-p^{-s})^{-1}$$

in two ways using the power series expansion

$$\log(1-x) = -\sum_{k=1}^\infty \frac{x^k}{k}.$$

First by the Euler product for $\zeta(s)$ one gets

$$\log\zeta(s) = \sum_{n\geqslant 2} \Lambda'(n) n^{-s}. \tag{13.30}$$

On the other hand, by the Dirichlet series for $\zeta(s)$,

(13.31)
$$\log \zeta(s) = \log(1 - (1 - \zeta(s)))$$
$$= -\sum_{k=1}^{\infty} \frac{(-1)^k}{k} \Big(\sum_{n \geqslant 2} n^{-s}\Big)^k = -\sum_{k=1}^{\infty} \frac{(-1)^k}{k} \tau_k'(n) n^{-s}.$$

Comparing the coefficients in both expansions Linnik gets the identity

$$\Lambda'(n) = -\sum_{k} \frac{(-1)^k}{k} \tau_k'(n). \tag{13.32}$$

Note that $\tau_k'(n) = 0$ if $2^k > n$, so (13.32) is a finite sum over $1 \leqslant k \leqslant \log n / \log 2$. In applications Linnik uses (13.32) for numbers n having no small prime factors, say for $(n, P(z)) = 1$ with $z \geqslant 2$. Then (13.32) runs over $k \leqslant \log n / \log z$. This yields the following

PROPOSITION 13.2. *Let $2 \leqslant z \leqslant \sqrt{x}$. For any sequence of complex numbers $\mathcal{A} = (a_n)$ we have*

$$\sum_{\substack{p^\nu \leqslant x \\ p > z}} a_{p^\nu} = -\sum_{k \leqslant K} \frac{(-1)^k}{k} \sum_{\substack{n \leqslant x \\ (n, P(z)) = 1}} a_n \tau_k'(n) \tag{13.33}$$

where $K = \log x / \log z$ and $\tau_k'(n)$ is the strict divisor function (13.29).

If one prefers the exact divisor functions $\tau_\ell(n)$ one can replace the strict ones $\tau_k'(n)$ in the Linnik formula by

$$\tau_k'(n) = \sum_{0 \leqslant \ell \leqslant k} (-1)^{k-\ell} \binom{k}{\ell} \tau_\ell(n). \tag{13.34}$$

Moreover, the condition $(n, P(z)) = 1$ can be relaxed at small cost by applying the exact sieve in the same way as in the Vinogradov formula.

On the right-hand side of (13.33) the first term ($k = 1$) is a sum of type S_1, precisely

$$S_1 = \sum_{\substack{1 < n \leqslant x \\ (n, P(z)) = 1}} a_n, \tag{13.35}$$

while any other sum

$$S_k = \sum_{\substack{n \leqslant x \\ (n, P(z)) = 1}} a_n \tau_k'(n) \tag{13.36}$$

for $k \geqslant 2$ can be regarded as a bilinear form (a sum of type S_2). However, Linnik gave a special treatment for S_k for a few small k (in the solution of the Hardy-Littlewood equation $p + x^2 + y^2 = N$), and only for larger k he arranged S_k into a suitable bilinear form. By extending the class of special forms involving the divisor functions of degree $k > 1$ one gains more flexibility in the remaining parts of the identity (13.33) from which to arrange bilinear forms. Of course, not every form S_k can be treated by special means. Usually one treats S_1 by classical Fourier analysis, and one can often apply to S_2 the spectral theory of automorphic forms. One may hope that S_k with $k \geqslant 3$ can be treated by Fourier analysis in GL_k, but so far this has not been successful.

In 1981 D. R. Heath-Brown [**HB3**] developed a formula for $\Lambda(n)$ which is reminiscent of that of Linnik. Let

$$M(s) = \sum_{m \leqslant z} \mu(m) m^{-s}.$$

Then

$$\zeta(s) M(s) = 1 + \sum_{n > z} a_n(z) n^{-s}.$$

Consider the identity

$$\frac{\zeta'}{\zeta}(s)(1 - \zeta(s)M(s))^K = \frac{\zeta'}{\zeta}(s) + \sum_{1 \leqslant k \leqslant K} (-1)^k \binom{K}{k} \zeta'(s)\zeta^{k-1}(s) M^k(s).$$

Comparing the Dirichlet coefficients on both sides we obtain

PROPOSITION 13.3. *Let $K \geqslant 1, z \geqslant 1$. Then for any $n < 2z^K$ we have*

$$\Lambda(n) = - \sum_{1 \leqslant k \leqslant K} (-1)^k \binom{K}{k} \sum_{\substack{m_1 \dots m_k n_1 \dots n_k = n \\ m_1, \dots, m_k \leqslant z}} \cdots \sum \mu(m_1) \dots \mu(m_k) \log n_k. \tag{13.37}$$

This identity of Heath-Brown has a few advantages over that of Linnik. Most of all, the number of terms K can be chosen smaller by regulating with the parameter z more efficiently.

EXERCISE 1. Show that for $n \leqslant z^K$ we have

$$\mu(n) = - \sum_{1 \leqslant k \leqslant K} (-1)^k \binom{K}{k} \sum_{\substack{m_1 \dots m_k n_1 \dots n_{k-1} = n \\ m_1, \dots, m_k \leqslant z}} \cdots \sum \mu(m_1) \dots \mu(m_k). \tag{13.38}$$

13.4. Vaughan's identity.

The most popular formula for $\Lambda(n)$ is due to R. C. Vaughan [**Va**]. We derive it from

$$\Lambda(n) = \sum_{b|n} \mu(b) \log \frac{n}{b}.$$

Here we keep the terms with $b \leqslant y$ and transform the remaining sum as follows

$$\sum_{\substack{b|n \\ b>y}} \mu(b) \log \frac{n}{b} = \sum_{\substack{bc|n \\ b>y}} \sum \mu(b) \Lambda(c).$$

Next we keep the terms with $c > z$ and transform the remaining sum as follows

$$\sum_{\substack{bc|n \\ b>y, c \leqslant z}} \sum \mu(b)\Lambda(c) = \sum_{\substack{bc|n \\ c \leqslant z}} \sum \mu(b)\Lambda(c) - \sum_{\substack{bc|n \\ b \leqslant y, c \leqslant z}} \sum \mu(b)\Lambda(c).$$

Here the complete sum over all b dividing n/c vanishes unless $c = n$, which is not possible if $n > z$. Adding up the above expressions one obtains

PROPOSITION 13.4. *Let $y, z \geqslant 1$. Then for any $n > z$ we have*

$$\Lambda(n) = \sum_{\substack{b|n \\ b\leqslant y}} \mu(b)\log\frac{n}{b} - \sum\sum_{\substack{bc|n \\ b\leqslant y, c\leqslant z}} \mu(b)\Lambda(c) + \sum\sum_{\substack{bc|n \\ b>y, c>z}} \mu(b)\Lambda(c). \tag{13.39}$$

Similarly we derive a formula for the Möbius function.

PROPOSITION 13.5. *Let $y, z \geqslant 1$. Then for any $m > \max(y, z)$ we have*

$$\mu(m) = -\sum\sum_{\substack{bc|m \\ b\leqslant y, c\leqslant z}} \mu(b)\mu(c) + \sum\sum_{\substack{bc|m \\ b>y, c>z}} \mu(b)\mu(c). \tag{13.40}$$

PROOF. We start from

$$\mu(m) = \sum\sum_{bc|m} \mu(b)\mu(c)$$

and split the summation into four ranges according to $b \leqslant y, b > y, c \leqslant z, c > z$. Then we transform the two middle sums into

$$\sum\sum_{\substack{bc|m \\ b\leqslant y, c>z}} \mu(b)\mu(c) = -\sum\sum_{\substack{bc|m \\ b\leqslant y, c\leqslant z}} \mu(b)\mu(c),$$

$$\sum\sum_{\substack{bc|m \\ b>y, c\leqslant z}} \mu(b)\mu(c) = -\sum\sum_{\substack{bc|m \\ b\leqslant y, c\leqslant z}} \mu(b)\mu(c).$$

Adding up the above expressions one obtains (13.40). □

Vaughan's identity (13.39) is not as flexible for arranging bilinear forms as those of Linnik or Heath-Brown, but it proved to be sufficient for basic applications and it is the simplest of all (ready to use without re-grouping its terms).

EXERCISE. Derive a formula à la Vaughan for $\Lambda_k(n)$.

13.5. Exponential sums over primes.

As an example of numerous applications of the methods described in the previous sections we prove Vinogradov's estimate for the exponential sum (13.15). Assuming

$$\left|\alpha - \frac{a}{q}\right| \leqslant \frac{1}{q^2} \quad \text{with } (a, q) = 1, \tag{13.41}$$

Vinogradov [**V6**] showed that

$$V(\alpha; x) \ll q^{\frac{1}{2}} x^{\frac{1}{2}} + q^{-\frac{1}{2}} x + x\exp(-\tfrac{1}{2}\sqrt{\log x}). \tag{13.42}$$

Using Vaughan's identity (13.39) we derive practically the same bound for

$$S(\alpha; x) = \sum_{n\leqslant x} e(\alpha n)\Lambda(n). \tag{13.43}$$

THEOREM 13.6. *Suppose α satisfies (13.41). Then for $x \geqslant 2$ we have*

$$S(\alpha, x) \ll (q^{\frac{1}{2}}x^{\frac{1}{2}} + q^{-\frac{1}{2}}x + x^{\frac{4}{5}})(\log x)^3 \tag{13.44}$$

where the implied constant is absolute.

We begin the proof of (13.44) by estimating special exponential sums. We have

$$\Big|\sum_{1\leqslant n\leqslant N} e(\alpha n)\Big| \leqslant \min\Big(N, \frac{1}{2\|\alpha\|}\Big).$$

Hence for any numbers $x(m) > 0$ we get

$$\sum_{|m|\leqslant M}\Big|\sum_{\substack{1\leqslant n\leqslant N\\ n\leqslant x(m)}} e(\alpha mn)\Big| \leqslant \sum_{|m|\leqslant M} \min\Big(N, x(m), \frac{1}{2\|\alpha m\|}\Big).$$

As m varies over a segment of length $q/2$, the points $\|\alpha m\|$ are all distinct and spaced by $1/2q$ at least. By this observation we derive

$$\begin{aligned}\sum_{|m|\leqslant M} \min\Big(N, \frac{1}{2\|\alpha m\|}\Big) &\leqslant (1 + 4Mq^{-1})\Big(N + \sum_{1\leqslant \ell\leqslant q} q\ell^{-1}\Big)\\ &\ll (M + N + MNq^{-1} + q)\log 2q.\end{aligned}$$

Similarly for $x(m) = x/m$ we derive (consider separately the range $1 \leqslant m \leqslant q/2$)

$$\sum_{1\leqslant m\leqslant M} \min\Big(\frac{x}{m}, \frac{1}{2\|\alpha m\|}\Big) \ll (M + xq^{-1} + q)\log 2qx.$$

From these estimates we get

LEMMA 13.7. *For any numbers $x(m) > 0$ we have*

$$\sum_{|m|\leqslant M}\Big|\sum_{\substack{1\leqslant n\leqslant N\\ n\leqslant x(m)}} e(\alpha mn)\Big| \ll (M + N + MNq^{-1} + q)\log 2q, \tag{13.45}$$

$$\sum_{1\leqslant m\leqslant M}\Big|\sum_{mn\leqslant x} e(\alpha mn)\Big| \ll (M + xq^{-1} + q)\log 2qx. \tag{13.46}$$

Next we derive a bound for general bilinear forms

$$\mathcal{B}(x; N) = \sum_{N<n\leqslant 2N}\Big|\sum_{mn\leqslant x} \gamma_m e(\alpha mn)\Big| \tag{13.47}$$

where γ_m are complex numbers with $|\gamma_m| \leqslant 1$. Applying Cauchy's inequality and (13.45) we get

$$\begin{aligned}\mathcal{B}^2(x; N) &\leqslant 2N \sum\sum_{1\leqslant m_1\leqslant m_2\leqslant \frac{x}{N}}\Big|\sum_{\substack{N<n\leqslant 2N\\ nm_2\leqslant x}} e(\alpha(m_1 - m_2)n)\Big|\\ &\ll \Big(\frac{x}{N} + N + \frac{x}{q} + q\Big)x\log 2q.\end{aligned}$$

From this bound we deduce

LEMMA 13.8. *For any complex numbers α_m, β_n with $|\alpha_m| \leqslant 1, |\beta_n| \leqslant 1$ we have*

$$\sum\sum_{\substack{mn\leqslant x\\ m>M,n>N}} \alpha_m\beta_n e(\alpha mn) \ll \Big(\frac{x}{M}+\frac{x}{N}+\frac{x}{q}+q\Big)^{\frac12} x^{\frac12}(\log x)^2. \tag{13.48}$$

PROOF OF THEOREM 13.6. By the identity (13.39) we have

$$\begin{aligned} S(\alpha;x) = \sum\sum_{\substack{\ell m\leqslant x\\ m\leqslant M}} \mu(m)(\log \ell)e(\alpha\ell m) - \sum\sum\sum_{\substack{\ell mn\leqslant x\\ m\leqslant M,n\leqslant N}} \mu(m)\Lambda(n)e(\alpha\ell mn) \\ + \sum\sum\sum_{\substack{\ell mn\leqslant x\\ m\geqslant M,n\geqslant N}} \mu(m)\Lambda(n)e(\alpha\ell mn) + O(N). \end{aligned}$$

We choose $M = N = x^{\frac25}$. To the first and the second sums we apply (13.46) getting $O((x^{\frac45} + xq^{-1} + q)(\log x)^2$. In the last sum we consider $\ell n = k$ as one variable with coefficient

$$c(k) = \sum_{n|k,n\geqslant N} \Lambda(n) \leqslant \log k.$$

Then we apply (13.48) getting $O((x^{\frac35} + xq^{-1} + q)^{\frac12}x^{\frac12}(\log x)^3$. Adding up these estimates we get (13.44). □

By the same arguments, but using the identity (13.40) in place of (13.39) (also apply Cauchy's inequality along the way to remove the corresponding coefficients $c(k)$ which are bounded by $\tau(k)$, not by $\log k$ as above), one shows

THEOREM 13.9. *Suppose α satisfies (13.41). Then for $x \geqslant 2$ we have*

$$\sum_{m\leqslant x} \mu(m)e(\alpha m) \ll (q^{\frac12}x^{\frac12} + q^{-\frac12}x + x^{\frac45})^{\frac12}x^{\frac12}(\log x)^4. \tag{13.49}$$

If α is a rational number with small denominator, then the sum (13.49) can be well estimated by an appeal to the zero-free region of L-functions. Indeed, for any Dirichlet character $\chi(\operatorname{mod} k)$ we have (see (5.80))

$$\sum_{m\leqslant x} \mu(m)\chi(m) \ll k^{\frac12}x(\log x)^{-A} \tag{13.50}$$

for any $x \geqslant 2$ and $A \geqslant 0$, the implied constant depending only on A. Hence we derive by means of Gauss sums that

$$\sum_{m\leqslant x} \mu(m)e\Big(\frac{am}{q}\Big) \ll qx(\log x)^{-A}. \tag{13.51}$$

Now let α be any real number. Given $Q \geqslant 1$ there exists a rational number a/q with $(a,q) = 1, 1 \leqslant q \leqslant Q$ such that

$$\Big|\alpha - \frac{a}{q}\Big| \leqslant \frac{1}{qQ}. \tag{13.52}$$

Hence we derive by partial summation that

$$\Big|\sum_{m\leqslant x}\mu(m)e(\alpha m)\Big| \leqslant \Big(1+\frac{2\pi x}{qQ}\Big)\Big|\sum_{m\leqslant y}\mu(m)e\Big(\frac{am}{q}\Big)\Big|$$

for some $1 \leqslant y \leqslant x$. Using (13.51) one gets

$$\sum_{m\leqslant x}\mu(m)e(\alpha m) \ll \Big(q+\frac{x}{Q}\Big)x(\log x)^{-5A} \tag{13.53}$$

for any $x \geqslant 2$ and $A \geqslant 0$, the implied constant depending only on A. We apply (13.53) if $q \leqslant xQ^{-1}$ and (13.49) if $xQ^{-1} < q \leqslant Q$ getting

$$\sum_{m\leqslant x}\mu(m)e(\alpha m) \ll Q^{-1}x^2(\log x)^{-5A} + Q^{\frac{1}{4}}x^{\frac{3}{4}}(\log x)^4 + x^{\frac{9}{10}}(\log x)^4.$$

Finally, choosing $Q = x(\log x)^{-4A}$, we conclude the following

THEOREM 13.10. *For any real number α and $x \geqslant 2$ we have*

$$\sum_{m\leqslant x}\mu(m)e(\alpha m) \ll x(\log x)^{-A} \tag{13.54}$$

for any $A \geqslant 0$, where the implied constant depends only on A.

This estimate was first proved in 1937 by H. Davenport [**Da1**] by applying Vinogradov's method. The complete uniformity in α of this bound is its pleasant feature. We shall exploit (13.54) in our treatment of the Goldbach problems in Chapter 19.

EXERCISE 2. Using the identity (13.39) prove that

$$\sum_{n\leqslant x}e(\alpha\sqrt{n})\Lambda(n) \ll x^{\frac{5}{6}}(\log x)^4 \tag{13.55}$$

for any real number $\alpha \neq 0$, where the implied constant depends on α.

13.6. Back to the sieve.

We elaborate further the ideas of Section 13.2 to draw results which can be useful in more delicate situations than that in the last section. The goal is to produce the bound

$$\Big|\sum_{p\leqslant x}a_p\Big| \leqslant \varepsilon\pi(x) \tag{13.56}$$

for any $\varepsilon > 0$ and $x \geqslant x_0(\varepsilon)$ subject to hypotheses on sums of type (13.18) and (13.19) which can be verified by existing technology. One of such technology will be the spectral theory of automorphic forms which we present in Chapter 21 for the problem of equidistribution of quadratic roots to prime moduli.

Recall that $P(z)$ denotes the product of all primes $p < z$. Given x we consider the sums

$$Q(\mathcal{A}, z) = \sum_{\substack{n\leqslant x\\(n,P(z))=1}} a_n \tag{13.57}$$

as a function of z for the sequence $\mathcal{A} = (a_n)$ and its subsequences $\mathcal{A}_d = (a_{md})$. (See also Chapter 6 for an elementary introduction to sieve methods). Suppose that $|a_n| \leqslant \tau(n)$.

Let $x^{\frac{1}{3}} < z \leqslant x^{\frac{1}{2}}$. Then our sum over primes is approximated by $Q(\mathcal{A}, z)$ as follows:

$$\begin{aligned}\sum_{p\leqslant x} a_p &= Q(\mathcal{A}, z) - \sum_{\substack{pq\leqslant x\\ z\leqslant p\leqslant q}}\sum a_{pq} + \sum_{p<z} a_p + a_1\\ &= Q(\mathcal{A}, z) + O\left(\frac{x}{\log x}\log\left(\frac{\log x}{2\log z}\right) + \frac{x}{(\log x)^2}\right).\end{aligned}$$

One can reduce the "sieving level" z by applying the Buchstab identity, namely for any sequence $\mathcal{A}$ and any $w < z$ we have

$$Q(\mathcal{A}, z) = Q(\mathcal{A}, w) - \sum_{w\leqslant p<z} Q(\mathcal{A}_p, p). \tag{13.58}$$

Applying this identity twice we get

$$Q(\mathcal{A}, z) = Q(\mathcal{A}, w) - \sum_{w\leqslant p<z} Q(\mathcal{A}_p, w) + \sum_{w\leqslant q<p<z}\sum Q(\mathcal{A}_{pq}, q). \tag{13.59}$$

Opening $Q(\mathcal{A}, w)$ and $Q(\mathcal{A}_p, w)$ by the exact sieve of Legendre we get

$$Q(\mathcal{A}, w) - \sum_{w\leqslant p<z} Q(\mathcal{A}_p, w) = \sum_d{}' \mu(d)A_d(x)$$

where the dash restricts the summation to the divisors d of $P(z)$ which have at most one prime factor $p \geqslant w$. We fix $y \geqslant z$ and estimate the partial sum over $d > y$ by applying Rankin's trick as follows:

$$\Big|\sum_{d>y}{}' \mu(d)A_d(x)\Big| \leqslant \sum_{d>y}{}' \sum_{md\leqslant x} \tau(md) \ll x(\log x)\sum_{d>y}{}' \tau(d)d^{-1}$$

and

$$\begin{aligned}\sum_{d>y}{}' \tau(d)d^{-1} &\leqslant y^{-\varepsilon}\sum_d{}' \tau(d)d^{\varepsilon-1}\\ &\leqslant y^{-\varepsilon}\Big(1 + \sum_{w\leqslant p<z} \tau(p)p^{\varepsilon-1}\Big)\prod_{p<w}(1 + \tau(p)p^{\varepsilon-1}).\end{aligned}$$

Choosing $\varepsilon = (\log w)^{-1}$ we get

$$\sum_{d>y}{}' \tau(d)d^{-1} \leqslant \left(\frac{z}{y}\right)^{\varepsilon}\prod_{p<z}\left(1 + \frac{1}{p}\right)^{2e+1} \ll \left(\frac{z}{y}\right)^{1/\log w}(\log z)^7.$$

Hence

$$\sum_{d>y}{}' \mu(d)A_d(x) \ll \left(\frac{z}{y}\right)^{1/\log w} x(\log x)^8.$$

The remaining sum over $d \leqslant y$ is estimated by

$$R(x) = \sum_{d\leqslant y} |A_d(x)|. \tag{13.60}$$

Now we proceed to the double sum of $Q(\mathcal{A}_{pq},q)$ over primes p,q with $w \leqslant q < p < z$. Let $x^{\frac14} < v \leqslant x^{\frac13}$. We estimate the partial sum over $q \geqslant v$ trivially as follows:

$$\sum_{v\leqslant q\leqslant p<z} Q(\mathcal{A}_{pq},q) \ll \frac{x}{\log x}\sum_{v\leqslant q\leqslant x^{\frac13}} q^{-1} + \left(\frac{z}{\log z}\right)^2$$
$$\ll \frac{x}{\log x}\log\left(\frac{\log x}{3\log v}\right) + \frac{x}{(\log x)^2}.$$

It remains to evaluate the sum of $Q(\mathcal{A}_{pq},q)$ over primes p,q with $w \leqslant q < v$ and $q < p < z$. We write this in the form

$$\sum_{\substack{w\leqslant q<v\\ q<p<z}}\sum Q(\mathcal{A}_{pq},q) = \sum_{\substack{mq\leqslant x\\ w\leqslant q<v\\ q<p_m<z}}\sum \gamma_m a_{mp} \tag{13.61}$$

where p_m is the least prime factor of m and γ_m is the number of prime divisors of m which are $< z$, so $\gamma_m \leqslant \omega(m)$. We want to make it a bilinear form. This requires us to relax the relation $q < p_m$. To this end we use the following lemma which is convenient for separation of variables:

LEMMA 13.11. *There exists a function $h(t)$ depending only on z such that*

$$\int_{-\infty}^{\infty} |h(t)|dt < \log 6z \tag{13.62}$$

and for all integers $1 \leqslant a,b \leqslant z$,

$$\int_{-\infty}^{\infty} h(t)\left(\frac{a}{b}\right)^{it}dt = \begin{cases} 1 & \textit{if } a \leqslant b,\\ 0 & \textit{if } a > b.\end{cases} \tag{13.63}$$

PROOF. Put $g(u) = \min\{uz, 1, 1+(1-u)z\}$ for $0 \leqslant u \leqslant 1+z^{-1}$ and $g(u) = 0$ otherwise. Then for positive integers $a,b \leqslant z$ we have $g(a/b) = 1$ if $a \leqslant b$, and $g(a/b) = 0$ if $a > b$. We also have

$$g(\frac{a}{b}) = \frac{1}{2\pi i}\int_{(0)} \hat{f}(s)\left(\frac{a}{b}\right)^{-s}ds$$

where $\hat{f}(s)$ is the Mellin transform of $g(u)$, i.e.,

$$\hat{f}(s) = \int_0^\infty g(u)u^{s-1}du = \frac{(z+1)^{s+1} - z^{s+1} - 1}{s(s+1)z^s}.$$

For $s = it$ with $t > 0$ we have three estimates $|\hat{f}(s)| \leqslant \min\{\log 6z, 2t^{-1}, 4zt^{-2}\}$ from which it follows that the function $h(t) = \frac{1}{2\pi}\hat{f}(-it)$ has the asserted properties. □

Applying (13.63) to (13.61) and changing the notation q to p we obtain the integral of bilinear forms

$$\int_{-\infty}^{\infty} \sum_{\substack{mp\leqslant x\\ w\leqslant p<v\\ w\leqslant p_m<z}}\sum \gamma_m a_{mp}(p_m/p)^{it}h(t)dt$$

which we estimate by $B(x)(\log x)^2$ with

$$B(x) = \sum_{(m,P(w))=1} \Big| \sum_{\substack{mp\leqslant x\\ w\leqslant p<v}} a_{mp}p^{it} \Big| \tag{13.64}$$

for some real number t (use $\omega(m) < \log 2m \leqslant \log x$, (13.62) and $6z \leqslant x$).

Collecting the above estimates we obtain

$$\Big|\sum_{p\leqslant x} a_p\Big| \leqslant R(x) + B(x)(\log x)^2 + E(x) \tag{13.65}$$

where

$$E(x) \ll \frac{x}{\log x}\log\left(\frac{(\log x)^2}{6(\log v)(\log z)}\right) + \frac{x}{(\log x)^2} + \left(\frac{z}{y}\right)^{1/\log w} x(\log x)^8 \tag{13.66}$$

and the implied constant is absolute.

Let $2 \leqslant \Delta \leqslant x^{\frac{1}{12}}$. We choose $z = \Delta^{-2}x^{\frac{1}{2}}$, $y = \Delta^{-1}x^{\frac{1}{2}}$, $v = \Delta^{-1}x^{\frac{1}{3}}$, and $w = \Delta^{1/10\log\log x}$ getting $E(x) \ll x(\log x)^{-2}\log\Delta$. Hence we conclude (take $\Delta = x^{\varepsilon}$)

THEOREM 13.12. *Let $\mathcal{A} = (a_n)$ be a sequence of complex numbers with $|a_n| \leqslant \tau(n)$. Let $x \geqslant e^{12}$ and $(\log x)^{-1} \leqslant \varepsilon \leqslant \frac{1}{12}$. Suppose that*

$$\sum_{d\leqslant y}\Big|\sum_{dm\leqslant x} a_{dm}\Big| \leqslant \frac{x}{(\log x)^2}, \tag{13.67}$$

$$\sum_{(m,P(w))=1}\Big|\sum_{\substack{pm\leqslant x\\ w\leqslant p<v}} p^{it}a_{pm}\Big| \leqslant \frac{x}{(\log x)^4} \tag{13.68}$$

for any $t \in \mathbb{R}$, where the parameters y, w, v are given in terms of x and ε by

$$y = x^{\frac{1}{2}-\varepsilon}, \quad v = x^{\frac{1}{3}-\varepsilon}, \quad w = x^{\varepsilon/10\log\log x} \tag{13.69}$$

and $P(w)$ is the product of primes $< w$. Then

$$\sum_{p\leqslant x} a_p \ll \frac{\varepsilon x}{\log x} \tag{13.70}$$

where the implied constant is absolute.

REMARKS. The condition (13.68) is regarded as an estimate for general bilinear forms. The factor p^{it} (which comes on our way of separation of variables) contaminates the coefficients, but it has no effect in practice. One could remove this factor, however, a precise statement would involve another parameter, which compromises the clarity. Some of the restrictions in the bilinear form (13.68) can also be modified. For example, one can remove the coprimality condition in the outer sum, but our experience shows that it is convenient to have it (note that $(m, P(w)) = 1$ implies $(m,p) = 1$). A concept of "prime producing exponents" was discussed in [**DFI5**]; Theorem 13.12 asserts that the pair $(\frac{1}{2}, \frac{1}{3})$ is prime producing. The presence of ε in the parameters (13.69) is very important; without it the requirement (13.68) would be impossible to verify for most interesting sequences $\mathcal{A} = (a_n)$. A well balanced choice is $\varepsilon = \varepsilon(x) = (\log\log x)^{-1}$ giving a bound for the sum over primes which shows only little cancellation, nevertheless it is a satisfactory result, particularly in applications to the equidistribution problems of degree two objects related to the GL_2 theory (see Chapter 21).

CHAPTER 14

HOLOMORPHIC MODULAR FORMS

14.1. Quotients of the upper half-plane and modular forms.

Holomorphic modular forms were historically discovered and studied for purposes of complex analysis and algebraic geometry, in particular, in connection with elliptic functions. Then they were introduced in algebraic number theory, particularly for the development of class field theory, Galois representations and arithmetic of elliptic curves. More recently modular forms have entered the territory of analytic number theory. Besides providing new tools, they are also a source of deep problems with numerous connections to arithmetic geometry, notably through the theory of L-functions.

See also Chapter 15 for a survey of the theory of non-holomorphic forms, as well as Chapter 16 for applications to Kloosterman sums and Chapter 26 for some results about L-functions. There are many fine books on various aspects of this vast subject, for instance [**Bu**], [**BG**], [**Mi**], [**Sh3**], [**I4**], [**Se1**].

We start with $\mathbb{H}$, the Poincaré upper half-plane

$$\mathbb{H} = \{z \in \mathbb{C} \mid y = \mathrm{Im}\,(z) > 0\} \tag{14.1}$$

which is an open subset of $\mathbb{C}$ (in the next chapter, it will be seen as a riemannian manifold of constant negative curvature with the Poincaré metric).

There is an action of the group $G = SL(2, \mathbb{R})$ on $\mathbb{H}$ by linear fractional transformations

$$gz = \frac{az+b}{cz+d}, \text{ for } g = \begin{pmatrix} a & b \\ c & d \end{pmatrix} \tag{14.2}$$

because

$$\mathrm{Im}\,(gz) = \frac{\mathrm{Im}\,(z)}{|cz+d|^2} \tag{14.3}$$

which is positive if $\mathrm{Im}\,(z) > 0$.

The center of G, namely $\{\pm 1\}$, acts trivially, and one shows that the group of holomorphic automorphisms of $\mathbb{H}$ is the quotient $PSL(2, \mathbb{R}) = SL(2, \mathbb{R})/\{\pm 1\}$. Note that this action is also isometric for the hyperbolic Poincaré metric; see Chapter 15.

The action of this large group gives $\mathbb{H}$ a large amount of symmetry. A useful manifestation of this is the following property: define first the action of G on the extended upper half-plane $\mathbb{H}^* = \mathbb{H} \cup \mathbb{R} \cup \infty$, by the same formula (14.2) in addition to the rules $g\infty = a/c$ if $c \neq 0$ and $g\infty = \infty$ if $c = 0$, then for any three distinct points v, w and z in $\mathbb{R} \cup \infty$, there exists $g \in G$ such that

$$gv = 0, \quad gw = 1, \quad gz = \infty.$$

Associated to a Riemannian metric is a natural invariant measure $d\mu$, which in the case of $\mathbb{H}$ is the hyperbolic measure

$$d\mu(z) = y^{-2}dxdy \quad \text{if } z = x + iy \tag{14.4}$$

which is invariant by G, i.e. for any f integrable on $\mathbb{H}$ we have

$$\int_{\mathbb{H}} f(z)d\mu(z) = \int_{\mathbb{H}} f(gz)d\mu(z).$$

Arithmetic comes when we consider discrete subgroups of G acting on $\mathbb{H}$ and the corresponding quotient spaces. One should keep in mind the analogy of $\mathbb{Z} \subset \mathbb{R}$ (or $\mathbb{Z}[i] \subset \mathbb{C}$). This is a very rich theory; there are many "different" discrete subgroups of G, and to classify them is almost the same as classifying all Riemann surfaces. For arithmetic applications, the most important examples are the congruence subgroups, particularly those which contain one of the so-called Hecke congruence subgroups $\Gamma_0(q)$ defined by

$$\Gamma_0(q) = \left\{ \begin{pmatrix} a & b \\ c & d \end{pmatrix} \in SL(2,\mathbb{Z}) \mid c \equiv 0 \ (q) \right\}.$$

The groups $\Gamma_0(q)$ have finite index in $SL(2,\mathbb{Z})$, precisely

$$[\Gamma_0(1) : \Gamma_0(q)] = q \prod_{p|q} (1 + p^{-1}).$$

For q varying one obtains very interesting quotients $\Gamma_0(q)\backslash\mathbb{H}$ (their genus increases).

One prefers to deal with compact Riemann surfaces, however, quotients of $\mathbb{H}$ are not always compact. It is even possible to have $\mathrm{Vol}\,(\Gamma\backslash\mathbb{H}) = +\infty$, where the volume refers to the natural measure induced from (14.4). For example, the group of translations by integers

$$\Gamma_\infty = \left\{ \begin{pmatrix} 1 & b \\ 0 & 1 \end{pmatrix} \mid b \in \mathbb{Z} \right\}$$

has the vertical strip $0 < \mathrm{Re}\,(z) \leqslant 1$ "representing" the quotient $\Gamma_\infty\backslash\mathbb{H}$, and it has infinite area.

More generally, a fundamental domain for the action of a discrete subgroup $\Gamma \subset G$ on $\mathbb{H}$ is a nice subset $F \subset \mathbb{H}$ which "almost" contains one point per orbit. Precisely, F must satisfy the following conditions:

(1) F is an open subset in $\mathbb{H}$.

(2) The closure $\overline{F}$ of F in $\mathbb{C}$ intersects every orbit of Γ, i.e. for all $z \in \mathbb{H}$ there exists $\gamma \in \Gamma$ with $\gamma z \in \bar{F}$.

(3) No two points in F are Γ-equivalent.

A fundamental domain always exists, but it is certainly not unique. One can even choose F to be a polygon with (hyperbolic) geodesic sides.

For the congruence subgroups $\Gamma_0(q)$, one can first consider the usual fundamental domain for $SL(2,\mathbb{Z})$,

$$F_1 = \{z \in \mathbb{H} \mid |\mathrm{Re}\,(z)| < \tfrac{1}{2}, \quad |z| > 1\}$$

and then take the union of its images under coset representatives in $SL(2,\mathbb{Z})$:

$$F_q = \bigcup_{\gamma\in\Gamma_0(q)\backslash\Gamma_0(1)} \gamma F_1$$

is a fundamental domain for $\Gamma_0(q)$. (It is not necessarily connected; for other choices, see [**I5**], Section 2.2.)

On a picture of F_1, the non-compactness appears because F_1 touches the boundary $\mathbb{R}\cup\infty$ at infinity, so $(i, 2i, 3i, \dots)$ is a sequence without converging subsequence. However, by the Gauss-Bonnet formula or a simple direct computation, the volume of $SL(2,\mathbb{Z})\backslash\mathbb{H}$ (which is the volume of any fundamental domain) remains finite, namely $\mathrm{Vol}\,(SL(2,\mathbb{Z})\backslash\mathbb{H}) = \frac{\pi}{3}$, and the same is true for $\Gamma_0(q)$ with

$$\mathrm{Vol}\,(\Gamma_0(q)\backslash\mathbb{H}) = \frac{\pi}{3}[\Gamma_0(1) : \Gamma_0(q)].$$

For Γ such that $\mathrm{Vol}\,(\Gamma\backslash\mathbb{H}) < +\infty$, the points at which F touches the boundary are called cusps of Γ (this intuitive description will be made more precise in Section 15.7 when we discuss the classification of elements in G), and they are in finite number (because to each cusp one can attach non-overlapping small neighborhoods in $\mathbb{H}$, which have fixed positive volume). From the above, ∞ is the only cusp for $SL(2,\mathbb{Z})$. So for $\Gamma_0(q)$, they are easily classified by considering the action of $\Gamma_0(q)\backslash\Gamma_0(1)$ on ∞. One finds that every cusp has a unique representative $\frac{a}{c}$ with $(a,c) = 1$, $c \geqslant 1$, $c \mid q$, and a is modulo $(c, \frac{q}{c})$. Their number is therefore

$$h = \sum_{cd=q} \varphi((c,d)).$$

This is equal to $\tau(q)$ if q is squarefree.

For any cusp $\mathfrak{a}$, its stability group $\Gamma_{\mathfrak{a}} = \{\gamma\in\Gamma \mid \gamma\mathfrak{a} = \mathfrak{a}\}$ is infinite cyclic (up to ± 1), generated by $\gamma_{\mathfrak{a}}$ say, and there is a $\sigma_{\mathfrak{a}} \in G$, called a scaling matrix, such that $\sigma_{\mathfrak{a}}\infty = \mathfrak{a}$ and

$$\sigma_{\mathfrak{a}}^{-1}\gamma_{\mathfrak{a}}\sigma_{\mathfrak{a}} = \begin{pmatrix} 1 & 1 \\ 0 & 1 \end{pmatrix}. \tag{14.5}$$

Using $\sigma_{\mathfrak{a}}$, most computations and properties related to the various cusps of Γ can be studied for the cusp at infinity, with stability group generated by the translation $z \mapsto z+1$, for the conjugate group $\sigma_{\mathfrak{a}}^{-1}\Gamma\sigma_{\mathfrak{a}}$ (it is an advantage of the Poincaré model over other models of the hyperbolic plane, for instance the unit disc, that it has such a distinguished cusp).

We can now define modular forms as holomorphic functions on $\mathbb{H}$ transforming in a simple way under the action of a discrete subgroup. Let $k \geqslant 1$ be an integer. Then we have an action (the "weight k action") of $G = SL(2,\mathbb{R})$ on functions $f : \mathbb{H} \to \mathbb{C}$ by

$$(f \mid_k g)(z) = j(g,z)^{-k} f(gz), \quad \text{where } j(g,z) = cz+d \text{ for } g = \begin{pmatrix} a & b \\ c & d \end{pmatrix}, \tag{14.6}$$

Let $q \geqslant 1$ be an integer, and χ a Dirichlet character modulo q (not necessarily primitive). Clearly χ induces a character of $\Gamma_0(q)$ by $\chi(g) = \chi(d)$ for $g = \begin{pmatrix} a & b \\ c & d \end{pmatrix}$.

A modular form of weight k, level q, and nebentypus (or character) χ is a holomorphic function f on $\mathbb{H}$ which satisfies

$$f\,|_k\,\gamma = \chi(\gamma)f, \text{ for all } \gamma \in \Gamma_0(q) \tag{14.7}$$

and is holomorphic at all cusps of $\Gamma_0(q)$. The holomorphy at cusps is explained as follows. First, for a cusp $\mathfrak{a}$ of $\Gamma_0(q)$, the function $f_{\mathfrak{a}} = f\,|_k\,\sigma_{\mathfrak{a}}$ is seen by modularity to be periodic of period one, $f_{\mathfrak{a}}(z+1) = f_{\mathfrak{a}}(z)$. This implies now that $f_{\mathfrak{a}}$ is a function of the parameter $q = e(z)$, namely $f_{\mathfrak{a}}(z) = g_{\mathfrak{a}}(q)$, where $g_{\mathfrak{a}}$ is holomorphic in a punctured disc $\{z \in \mathbb{C} \mid 0 < |z| < r\}$. We then say that f is meromorphic at $\mathfrak{a}$ if $g_{\mathfrak{a}}$ is meromorphic at 0. Therefore $f_{\mathfrak{a}}$ has a Laurent series expansion

$$f_{\mathfrak{a}}(z) = \sum_{n \geqslant n_{\mathfrak{a}}} \hat{f}_{\mathfrak{a}}(n)e(nz)$$

for some integer $n_{\mathfrak{a}}$, and $\hat{f}_{\mathfrak{a}}(n_{\mathfrak{a}}) \neq 0$. We say that f is holomorphic at $\mathfrak{a}$ if it is meromorphic and $n_{\mathfrak{a}} \geqslant 0$. If, moreover, we have $n_{\mathfrak{a}} > 0$, we say that f vanishes at $\mathfrak{a}$. If f vanishes at all cusps, it is called a cusp form.

EXERCISE 1. Show from the definition that a modular form f is a cusp form if and only if the Γ-invariant function

$$g(z) = y^{k/2}|f(z)| \tag{14.8}$$

is bounded on $\mathbb{H}$ (this criterion is very convenient in practice).

EXERCISE 2. Show that if we take the same definition but $k = 0$, the corresponding space of modular forms is reduced to 0.

Clearly modular forms of fixed level, weight and character form a vector space, which is denoted $M_k(q, \chi)$, and the cusp forms a subspace denoted $S_k(q, \chi)$. Taking $\gamma = -1$, we see that f is identically zero unless $\chi(-1) = (-1)^k$, so we assume that χ satisfies this consistency condition.

The first fundamental result is that $M_k(q, \chi)$ is a finite dimensional vector space. This can be proved in general using the Riemann-Roch theorem for compact Riemann surfaces. In the case of $\Gamma_0(q)$, a simpler proof is possible. For $k \geqslant 2$, the dimension of $M_k(q, \chi)$ can be computed exactly (see e.g. [**Sh3**]), and then estimated very precisely (the case of forms of weight one is a tantalizing problem of deep arithmetical significance because of links with Artin L-functions, see [**Se2**], [**Du1**]). One shows for instance that

$$\dim M_k(q, \chi) = \frac{k-1}{12}\nu_q + O(\sqrt{qk})$$

where $\nu_q = [\Gamma_0(1) : \Gamma_0(q)]$.

For any modular form f we denote by $a_f(n)$ its Fourier coefficients at ∞, so that

$$f(z) = \sum_{n \geqslant 0} a_f(n)e(nz). \tag{14.9}$$

Those coefficients are very important objects (they are arithmetic harmonics), and their order of magnitude, in particular, is the subject of many inquiries. For cusp forms, the criterion that $y^{k/2}|f(z)|$ is bounded on $\mathbb{H}$ implies quickly the trivial bound

$$a_f(n) \ll n^{k/2} \tag{14.10}$$

which is good enough to start with (see below (14.54) for the much deeper Deligne bound).

One also defines the Petersson inner product on $M_k(q, \chi)$ by

$$\langle f, g\rangle = \int_F f(z)\overline{g(z)}y^k d\mu(z). \tag{14.11}$$

This is well-defined because the integrand is $\Gamma_0(q)$-invariant, as a simple computation reveals (compare (14.8)), and the integral is finite as soon as one of f or g is a cusp form. In particular, $S_k(q, \chi)$ is a (finite dimensional) Hilbert space with this inner product, and this analytic fact is very important in what follows (see the definition of "newforms" in Section 14.7).

Notice also that it is possible to multiply modular forms of the same level: if $f \in M_k(q, \chi)$, $g \in M_{k'}(q, \chi')$, we have $fg \in M_{k+k'}(q, \chi\chi')$. In particular, the direct sum of all $M_k(q)$, $k \geqslant 0$, is a graded $\mathbb{C}$-algebra. This important algebraic fact will not be of much use in this book, however.

It should be emphasized that in proving that a function is modular, the modularity formula (14.7) need only be proved for γ in a family of generators of $\Gamma_0(q)$. Such groups are always finitely generated so this is a finite set of conditions on f.

For instance $SL(2, \mathbb{Z})$ is generated by the two elements

$$\begin{pmatrix} 1 & 1 \\ 0 & 1 \end{pmatrix} \text{ and } \begin{pmatrix} 0 & -1 \\ 1 & 0 \end{pmatrix}$$

acting as the translation $z \mapsto z + 1$ and as inversion $z \mapsto -z^{-1}$ respectively, and the two conditions for f to be modular become

$$f(z) = f(z+1) \quad \text{and} \quad f\Big(\frac{-1}{z}\Big) = z^k f(z).$$

We now give examples of modular forms and in particular of cusp forms.

14.2. Eisenstein and Poincaré series.

The basic examples of modular forms are the Eisenstein and the Poincaré series. We only define those series associated to the cusp ∞ for simplicity, but similar constructions can be done at every cusp. Let $k > 2$ (to ensure absolute convergence) and put

$$E_k(z) = \sum_{\gamma \in \Gamma_\infty \backslash \Gamma_0(q)} \overline{\chi}(\gamma) j(\gamma, z)^{-k}.$$

Clearly, E_k is in $M_k(q, \chi)$, as it is constructed by the averaging technique; holomorphy at the cusps is checked by explicit computation of the Fourier expansion, which also shows that E_k is not a cusp form. It is called an Eisenstein series.

More generally let $m \geqslant 0$. The m-th Poincaré series is obtained by the averaging technique applied to $e(mz)$:

$$P_m(z) = \sum_{\gamma \in \Gamma_\infty \backslash \Gamma_0(q)} \overline{\chi}(\gamma) j(\gamma, z)^{-k} e(m\gamma z)$$

for $k \geqslant 2$ (the convergence for $k = 2$ is not absolute but can be seen for $m \geqslant 1$ from the Fourier expansion (14.12) below, using the Weil bound for Kloosterman sums (1.60)). We do not consider $m < 0$ as the series diverges then.

The important fact is that for $m \geqslant 1$, we get cusp forms.

PROPOSITION 14.1. (1) *If $m = 0$, then $P_0 = E_k$.*
(2) *If $m \geqslant 1$, then P_m is a cusp form in $S_k(q, \chi)$.*

We sketch part of the proof, namely the following lemma.

LEMMA 14.2. *The Poincaré series P_m has Fourier expansion*

$$P_m(z) = \delta(m, 0) + \sum_{n \geqslant 1} p(m, n) e(nz) \tag{14.12}$$

with coefficients

$$p(0, n) = \Big(\frac{2\pi}{i}\Big)^k \frac{n^{k-1}}{\Gamma(k)} \sum_{\substack{c>0 \\ c \equiv 0 \,(\mathrm{mod}\, q)}} c^{-k} S_\chi(0, n; c)$$

and

$$p(m, n) = \Big(\frac{m}{n}\Big)^{\frac{k-1}{2}} \Big(\delta(m, n) + 2\pi i^{-k} \sum_{\substack{c>0 \\ c \equiv 0 \,(\mathrm{mod}\, q)}} c^{-1} S_\chi(m, n; c) J_{k-1}\Big(\frac{4\pi\sqrt{mn}}{c}\Big)\Big)$$

if $m \geqslant 1$. Here $S_\chi(m, n; c)$ is the Kloosterman sum with character χ,

$$S_\chi(m, n; c) = \sideset{}{^\star}\sum_{d \,(\mathrm{mod}\, c)} \chi(d) e\Big(\frac{md + n\bar{d}}{c}\Big) \tag{14.13}$$

and J_{k-1} is the Bessel function of order $k - 1$.

PROOF. We first make a convenient parametrization of the cosets $\Gamma_\infty \backslash \Gamma_0(q)$, namely using the action of an element of Γ_∞ on the right and on the left, we see that $\Gamma_0(q)$ is the disjoint union

$$\Gamma_0(q) = \Gamma_\infty \bigcup \Big(\bigcup_{\substack{c>0 \\ c \equiv 0 \,(\mathrm{mod}\, q)}} \bigcup_{\substack{d \,(\mathrm{mod}\, c) \\ (c,d)=1}} \Gamma_\infty \begin{pmatrix} a & b \\ c & d \end{pmatrix} \Gamma_\infty \Big) \tag{14.14}$$

(where for given c and d with $(c, d) = 1$, a and b are any two integers such that $ad - bc = 1$).

Thus we have correspondingly

$$P_m(z) = e(mz) + \sum_{\substack{c>0 \\ c \equiv 0 \,(\mathrm{mod}\, q)}} \sideset{}{^\star}\sum_{d \,(\mathrm{mod}\, c)} \overline{\chi}(d) I(c, d; z)$$

where

$$I(c, d; z) = \sum_{n \in \mathbb{Z}} \big(c(z + n) + d\big)^{-k} e\Big(m\Big(\frac{a}{c} - \frac{1}{c(c(z + n) + d)}\Big)\Big).$$

Next, by the Poisson summation formula

$$I(c,d;z)=\sum_{n\in\mathbb{Z}}\int_{\mathbb{R}}(c(z+v)+d)^{-k}e\Big(\frac{am}{c}-\frac{m}{c(c(z+v)+d)}-nv\Big)dv$$

$$=\sum_{n\in\mathbb{Z}}e\Big(\frac{am}{c}+\frac{nd}{c}\Big)\Big\{\int_{-\infty+iy}^{+\infty+iy}(cv)^{-k}e\Big(\frac{-m}{c^2v}-nv\Big)dv\Big\}e(nz).$$

Here we observe that the inner integral vanishes for $n\leqslant 0$ (move the line of integration upwards; the case $n=0$ is the most important, so the reader is encouraged to check carefully). Then putting together the different terms, and recognizing the Bessel function in the integral (see [**GR, 8.315.1, 8.412.2**]), we obtain the lemma. □

EXERCISE 3. Prove the decomposition (14.14).

Notice that there are infinitely many Poincaré series P_m, but they all live in the finite dimensional space $S_k(q,\chi)$. This means that there must be many linear relations between the P_m's. On the other hand, such abundance tends to suggest that the Poincaré series should span the whole space. This is indeed the case.

LEMMA 14.3. *Let $f\in M_k(q,\chi)$ be a modular form with expansion (14.9). Then for any $m\geqslant 1$,*

$$\langle f,P_m\rangle=\frac{\Gamma(k-1)}{(4\pi m)^{k-1}}a_f(m).$$

COROLLARY 14.4. *The Poincaré series with $m\neq 0$ span $S_k(q,\chi)$.*

Indeed, any cusp form f orthogonal to every P_m must have all its Fourier coefficients at ∞ equal to zero, and hence must be zero. Because the space spanned by Poincaré series is closed (this uses crucially the finite-dimensionality of the space) the result follows.

PROOF OF LEMMA 14.3. By definition of P_m we have

$$\langle f,P_m\rangle=\int_F f(z)\Big\{\sum_{\gamma\in\Gamma_\infty\backslash\Gamma_0(q)}\chi(\gamma)\overline{j(\gamma,z)}^{-k}\,\overline{e(m\gamma z)}\Big\}y^k d\mu(z)$$

$$=\int_F\sum_{\gamma\in\Gamma_\infty\backslash\Gamma_0(q)}\mathrm{Im}\,(\gamma z)^k f(\gamma z)\overline{e(m\gamma z)}d\mu(z)$$

(by modularity of f and $\mathrm{Im}\,(\gamma z)=|j(g,z)|^{-2}\mathrm{Im}\,(z)$)

$$=\int_0^1\int_0^{+\infty}y^{k-2}f(z)e(-m\bar z)dxdy$$

$$=\sum_{n\geqslant 0}a_f(n)\int_0^{+\infty}y^{k-2}e^{-2\pi(n+m)y}dy\int_0^1 e((n-m)x)dx$$

$$=\frac{\Gamma(k-1)}{(4\pi m)^{k-1}}a_f(m)$$

□

Taken together, the two previous lemmas imply a formula which shows how the sequences of Fourier coefficients of a basis of $S_k(q,\chi)$ are almost orthogonal, thereby giving another decomposition in arithmetic harmonics of the diagonal symbol $\delta(m,n)$. Let $\mathcal{F}$ be any orthonormal basis of $S_k(q,\chi)$ (distinguished basis will be described in Section 14.7). Since P_m is in $S_k(q,\chi)$, we can expand it as a linear combination of the $f\in\mathcal{F}$, namely we have

$$P_m = \sum_{f\in\mathcal{F}} \langle f, P_m\rangle f$$

so that by taking the n-th Fourier coefficient of both sides (on the right with Lemma 14.3, on the left with Lemma 14.2) we obtain the Petersson formula

PROPOSITION 14.5 (PETERSSON). *For any $n\geqslant 1$ and $m\geqslant 1$,*

$$\frac{\Gamma(k-1)}{(4\pi\sqrt{mn})^{k-1}}\sum_{f\in\mathcal{F}} a_f(n)\overline{a_f(m)} = \delta(m,n) + 2\pi i^{-k}\sum_{\substack{c>0\\ c\equiv 0\,(\mathrm{mod}\, q)}} c^{-1}S_\chi(m,n;c)J_{k-1}\Big(\frac{4\pi\sqrt{mn}}{c}\Big). \tag{14.15}$$

EXAMPLE. The "first" cusp form occurs for level 1 and weight 12; it is the famous Ramanujan Δ function, which has the elegant product expansion

$$\Delta(z) = q\prod_{n\geqslant 1}(1-q^n)^{24}, \quad \text{with } q = e(z)$$

(see e.g. [**Se1**] for one of many direct proofs of the modularity).

The existence of a non-zero $f\in S_{12}(1)$ can be predicted by considering the function $f = E_4^3 - E_6^2$, where E_4 and E_6 are the Eisenstein series of level 1 and weight 4 and 6. These have the Fourier expansions (at the only cusp ∞)

$$E_4(z) = 1 + 240\sum_{n\geqslant 1}\sigma_3(n)e(nz)$$

$$E_6(z) = 1 - 504\sum_{n\geqslant 1}\sigma_5(n)e(nz)$$

so that a direct computation of the small order terms gives

$$E_4^3 - E_6^2 = 1728q + O(q^2)$$

whence f is non-zero and is a cusp form. Since $\dim S_{12}(1) = 1$, the equality $f = 1728\Delta$ must follow once we know that Δ is indeed modular. The coefficients in the expansion of Δ are denoted $\tau(n)$ (not to be mistaken with the divisor function). These are integers having many fascinating properties, for example,

$$\tau(n) \equiv \sum_{d\mid n} d^{11} \bmod 691.$$

This kind of congruences belongs properly to the theory of Galois representations (see e.g. [**Se3**]).

Evidently, Δ must also be proportional to the non-zero Poincaré series in $S_{12}(1)$, but it does not arise naturally in this way. Also, the construction of f exploits the fact that holomorphic modular forms can be multiplied. Actually, it is not very difficult to show that the graded ring $\oplus_{k\geqslant 0} M_k(1)$ is isomorphic to the polynomial ring $\mathbb{C}[E_4, E_6]$, and E_4 and E_6 are algebraically independent.

14.3. Theta functions.

The Eisenstein and the Poincaré series are constructed by averaging over the group cosets. Theta functions provide another important class of examples of modular forms which come out in quite a different way. The most basic theta series is Jacobi's θ function and equation (1.53) is indeed a modularity property of weight $\frac{1}{2}$. However, we have not defined half-integral weight forms (see for instance [**Sh1**], [**I4**] for this theory) so we square θ and obtain a weight 1 form

$$\theta^2(z) = \sum_{n\geqslant 0} r(n)e(nz).$$

More generally we consider first theta series attached to binary quadratic forms. Definite forms are related to holomorphic theta series, while indefinite ones would give non-holomorphic forms, which were first constructed by Maass [**Ma**].

Recall the notation of Section 3.8: $K = \mathbb{Q}(\sqrt{D})$ is an imaginary quadratic field, $D < 0$ being a fundamental discriminant, $\mathcal{O} = \mathbb{Z} \oplus \omega\mathbb{Z}$ is the ring of integers of K, $\omega = \frac{1}{2}(D + \sqrt{D})$, w is the number of units of $\mathcal{O}$ and $\mathcal{H}$ is the ideal class group.

Let $\mathcal{A} \in \mathcal{H}$ be an ideal class. The theta series attached to $\mathcal{A}$ is defined to be

$$\theta_{\mathcal{A}}(z) = \frac{1}{w} + \sum_{\mathfrak{a}\in\mathcal{A}} e(zN\mathfrak{a}) \tag{14.16}$$

where $\mathfrak{a}$ runs over the non-zero integral ideals belonging to the class $\mathcal{A}$. As shown in Section 22.2, there is a one-to-one correspondence between ideal classes and classes of definite binary quadratic forms of discriminant D. Let $\Phi_{\mathcal{A}} = [a, b, c]$ be the quadratic form associated to $\mathcal{A}$. If $\mathfrak{b} \in \mathcal{A}$, we have $\mathfrak{b} = (\alpha)\mathfrak{a}$ for some $\alpha \in \mathfrak{a}^{-1}$, $\alpha = m + n\overline{z_{\mathfrak{a}}}$. Then $N\mathfrak{b} = \Phi_{\mathcal{A}}(m, n)$ so taking the units into account we obtain the formula

$$\theta_{\mathcal{A}}(z) = \frac{1}{w} \sum_{m,n\in\mathbb{Z}} e\big(\Phi_{\mathcal{A}}(m, n)\big). \tag{14.17}$$

Poisson summation (in two variables) shows again that $\theta_{\mathcal{A}}$ is modular, precisely $\theta_{\mathcal{A}} \in M_1(|D|, \chi_D)$, where χ_D is the Kronecker symbol (the Dirichlet character associated to the field K). It also follows that

$$\theta_{\mathcal{A}}(z) = \frac{i}{z\sqrt{|D|}}\theta_{\mathcal{A}^{-1}}\Big(\frac{-1}{|D|z}\Big).$$

From (14.16) or (14.17) we see that $\theta_{\mathcal{A}}$ is not a cusp form. For a class group character $\chi \in \hat{\mathcal{H}}$ we define the theta series

$$f_\chi(z) = \sum_{\mathfrak{a}} \chi(\mathfrak{a})e(zN\mathfrak{a})$$

where the sum is over all integral ideals of $\mathcal{O}$ this time. Splitting into classes we derive

$$f_\chi(z) = \sum_{\mathcal{A}\in\mathcal{H}} \chi(\mathcal{A})\theta_{\mathcal{A}}(z)$$

for $\chi \neq 1$, so $f_\chi \in M_1(|D|, \chi_D)$ also. Computing the Fourier expansion at other cusps, one proves that f_χ is a cusp form if $\chi^2 \neq 1$. Recall that the characters of order 2 of $\mathcal{H}$ are called genus characters. They were determined (essentially) by Gauss, who showed that the subgroup of genus characters is isomorphic to $(\mathbb{Z}/2\mathbb{Z})^{t-1}$, where $t = \omega(|D|)$ is the number of distinct prime divisors of $|D|$.

For χ not a genus character, we have $f_\chi \in S_1(|D|, \chi_D)$, and it also satisfies

$$f_\chi\Big(\frac{-1}{|D|z}\Big) = -if_\chi(z).$$

All class group characters are particular cases of Hecke characters of K (see Section 3.8), so they have an L-function (over K) attached to them

$$L_K(s,\chi) = \sum_{\mathfrak{a}} \chi(\mathfrak{a})(N\mathfrak{a})^{-s}$$

and by Mellin inversion the completed L-function

$$\Lambda_K(s,\chi) = \Big(\frac{\sqrt{|D|}}{2\pi}\Big)^s \Gamma(s) L_K(s,\chi)$$

is expressed by the formula

$$\Lambda_K(s,\chi) = \int_1^{+\infty} (y^{s-1} + y^{-s}) f_\chi\Big(\frac{iy}{\sqrt{|D|}}\Big)dy.$$

A few other representations of $\Lambda_K(s,\chi)$ are given in Section 22.3. From the above integral representation we derive, as for the Riemann zeta function or the Dirichlet L-function, that $L_K(s,\chi)$ has meromorphic continuation to the whole complex plane and is entire if $\chi \neq 1$. Moreover, we have the functional equation

$$\Lambda_K(s,\chi) = \Lambda_K(1-s,\chi).$$

Much more general theta series are obtained by considering any integral positive definite quadratic form, and associated harmonic polynomials in n variables instead of only binary quadratic forms. However, there is no interpretation in terms of ideal classes of number fields in the general case (for 4 variables, of course, there are links with quaternion algebras).

Let $A = (a_{ij})$ be a real, symmetric, positive definite matrix of rank $r \equiv 0 \pmod 2$, and

$$A[x] = {}^t x A x = \sum_{i,j} a_{ij} x_i x_j$$

the corresponding quadratic form (the theory of forms in odd number of variables is somewhat harder because of a complicated multiplier system for the corresponding modular forms). A homogeneous polynomial $P \in \mathbb{C}[X]$ is called a harmonic polynomial if $\Delta_A P = 0$, where

$$\Delta_A = \sum_{i,j} a^*_{ij} \frac{\partial^2}{\partial x_i \partial x_j}$$

is the Laplace operator associated with the matrix $A^{-1} = (a^*_{ij})$. With this data we define the theta series

$$\theta(z) = \sum_{m\in\mathbb{Z}^r} P(m) e\Big(\frac{z}{2}A[m]\Big).$$

Suppose A and NA^{-1} have integral coefficients, and even diagonal, for some suitable integer $N \geqslant 1$. Then $\theta(z) \in M_k(N, \chi_D)$ where $k = \frac{r}{2} + \deg P$ and $\chi_D(d) = \left(\frac{D}{d}\right)$ is the Kronecker symbol with $D = (-1)^{\frac{r}{2}}|A|$. Moreover, if $\deg P > 0$, then $\theta(z) = \theta(z; A, P)$ is a cusp form which satisfies

$$\theta(z; A, P) = i^{\frac{r}{2}}|A|^{-\frac{1}{2}} z^{-k} \theta\Big(-\frac{1}{z}; A^{-1}, P^*\Big)$$

where $P^*(x) = P(A^{-1}x)$ is a harmonic polynomial associated with A^{-1} (see also Section 20.4 and, for instance, [**I4**], Chapters 9 and 10 for complete proofs).

14.4. Modular forms associated to elliptic curves.

Cusp forms associated to elliptic curves have both arithmetic and analytic flavors. We begin by introducing the basic invariants of elliptic curves over an arbitrary field (see [**Sil**] and also Section 11.9). Any elliptic curve E, defined over a field k, has a plane model given by a Weierstrass equation

$$E : y^2 + a_1xy + a_3y = x^3 + a_2x^2 + a_4x + a_6 \tag{14.18}$$

with coefficients $a_j \in k$. Associated with the a-coefficients is the set of coefficients b_2, b_4, b_6, b_8 given by $b_2 = a_1^2 + 4a_2$, $b_4 = a_1a_3 + 2a_4$, $b_6 = a_3^2 + 4a_6$, $b_8 = a_1^2a_6 - a_1a_3a_4 + 4a_2a_6 + a_2a_3^2 - a_4^2$. The b-coefficients are related by $4b_8 = b_2b_6 - b_4^2$. Having these quantities we express the discriminant of (14.18) by

$$\Delta = -b_2^2b_8 - 8b_4^3 - 27b_6^2 + 9b_2b_4b_6. \tag{14.19}$$

The cubic equation (14.18) defines an elliptic curve over k if and only if it is non-singular, which means that its discriminant Δ does not vanish. To simplify equations we associate with (14.18) another set of coefficients c_4, c_6 given by $c_4 = b_2^2 - 24b_4$, $c_6 = -b_2^3 + 36b_2b_4 - 216b_6$. In these terms we have

$$12^3\Delta = c_4^3 - c_6^2. \tag{14.20}$$

REMARKS. The first set of b-coefficients arises from completing the square. Assuming the characteristic of k is not 2, we can change the coordinates (x, y) in (14.18) into $(x, y + \frac{a_1}{2}x + \frac{a_3}{2})$ getting

$$E : y^2 = x^3 + \frac{b_2}{4}x^2 + \frac{b_4}{2}x + \frac{b_6}{4}. \tag{14.21}$$

The second set of c-coefficients arises from completing the cube. If further the characteristic is not 3, we can change the coordinates (x, y) in (14.21) into $(x - \frac{b_2}{12}, y)$ getting

$$E : y^2 = x^3 - \frac{c_4}{48}x - \frac{c_6}{864} = x^3 + ax + b, \tag{14.22}$$

say. This has the discriminant $\Delta = 12^{-3}(c_4^3 - c_6^2) = -16(4a^3 + 27b^2)$. In this equation we change further (x, y) into $(x, \frac{y}{2})$ and multiply through by 4 getting

$$E : y^2 = 4x^3 + Ax + B \tag{14.23}$$

with $A = -\frac{c_4}{12} = 4a$ and $B = -\frac{c_6}{216} = 4b$. Now the discriminant of (14.23) is simply given by

$$\Delta = -(A^3 + 27B^2) = (c_4^3 - c_6^2)/1728 \tag{14.24}$$

(notice that $1728 = 12^3$).

We return to the general equation (14.18) over any field k. We define the so-called j-invariant by

$$j = c_4^3 \Delta^{-1} = (12c_4)^3 (c_4^3 - c_6^2)^{-1}; \tag{14.25}$$

this makes sense because we assume $\Delta \neq 0$. The j-invariant plays an important role in classification of elliptic curves. First we single out two special cases.

CASE $j = 0$. This is equivalent to $c_4 = 0$, therefore if E is given by (14.23), then $A = 0$ and $\Delta = -27B^2$.

CASE $j = 1728$. This is equivalent to $c_6 = 0$, therefore if E is given by (14.23), then $B = 0$ and $\Delta = -A^3$.

EXAMPLE 1. Suppose $j \neq 0, 1728$. Then the equation

$$E : y^2 + xy = x^3 - \frac{36}{j - 1728} x - \frac{1}{j - 1728} \tag{14.26}$$

gives an elliptic curve over any field k which contains j; its j-invariant is equal to j, the c-coefficients are $c_4 = j(j-1728)^{-1}, c_6 = -j(j-1728)^{-1}$ and the discriminant is $\Delta = j^2(j-1728)^{-3}$. Also the cubic equation

$$E_j : y^2 + jxy = x^3 - \frac{j(j-1)}{4} x^2 - \frac{36j^2}{j - 1728} x - \frac{j^3}{j - 1728} \tag{14.27}$$

(which is the quadratic twist of (14.26) by j) gives an elliptic curve over the field k; its invariants are $c_4 = j^3(j-1728)^{-1}$, $c_6 = -j^4(j-1728)^{-1}$ and $\Delta = j^8(j-1728)^{-3}$.

EXAMPLE 2. Suppose $j \neq 0, 1728$ and char.$k \neq 2, 3$. Then the cubic equation

$$E : y^2 = 4x^3 - \frac{27j}{j - 1728} x - \frac{27j}{j - 1728} \tag{14.28}$$

with $j \in k$ gives an elliptic curve over k whose j-invariant is equal to j and the discriminant is $\Delta = 2^6 \cdot 3^{12} \cdot j^2 (j - 1728)^{-3}$.

From now on we consider elliptic curves E defined over $\mathbb{Q}$. We can assume that E is given by the Weierstrass equation (14.18) with integer coefficients a_j (if necessary, change the coordinates (x, y) into $(c^{-2}x, c^{-3}y)$ and multiply through by c^6 to kill the denominators of a_j by choosing a suitable $c \in \mathbb{N}$). Every elliptic curve $E/\mathbb{Q}$ has a minimal model, that is E is given by the equation (14.18) with integer coefficients such that $\mathrm{ord}_p \Delta$ is minimal for every p among all such equations defining E (this special equation is possible to arrange by a rational change of variables because $\mathbb{Z}$ has unique factorization property). The following condition

$$\min\{\mathrm{ord}_p \Delta,\ 3\mathrm{ord}_p c_4\} < 12 \tag{14.29}$$

for all p is necessary and sufficient for the model (14.18) with $a_j \in \mathbb{Z}$ to be minimal.

Given a minimal equation (14.18) for E we look at its reduction modulo p. If $p \nmid \Delta$, then the reduced cubic is nonsingular, so it defines an elliptic curve over the finite field $\mathbb{F}_p$. In this case it is said E has good reduction at p. If $p \mid \Delta$, then the reduced curve is singular, and it is said E has bad reduction at p. The bad reduction divides into two types:

(1) Multiplicative reduction (or semistable) if $p \mid \Delta$ and $p \nmid c_4$ (this means $E/\mathbb{F}_p$ has a node).

(2) Additive reduction (or unstable) if $p \mid \Delta$ and $p \mid c_4$ (this means $E/\mathbb{F}_p$ has a cusp).

The multiplicative reduction is said to be split or non-split if the slopes of the tangent lines at the node are rational or not, respectively.

For every p we define the conductor exponent f_p as follows:
(1) $f_p = 0$ if E has good reduction at p.
(2) $f_p = 1$ if E has multiplicative reduction at p.
(3) $f_p = 2$ if E has additive reduction at $p \neq 2, 3$.
When E has additive reduction at $p = 2$ or $p = 3$, then the definition of f_p is quite involved (see [**Sil**]), but we always have $f_p \geqslant 2$ and $f_p - 2$ is a certain "measure of wild ramification". In these cases one can show that $f_2 \leqslant 8$ and $f_3 \leqslant 5$. Having the exponents f_p for all p the conductor of $E/\mathbb{Q}$ is set to be

$$N = \prod_p p^{f_p}. \tag{14.30}$$

Note that N is squarefree if and only if E has no additive reduction at all places. In this case E is called semistable.

Hasse defined the L-function of E as an Euler product of local factors defined by looking at the reduction E_p of E modulo p for all primes p, which is a curve over the finite field $\mathbb{F}_p$. The local zeta function is defined by the formal power series

$$Z_p(E) = \exp\Big(\sum_{n \geqslant 0} |E(\mathbb{F}_{p^n})| \frac{X^n}{n}\Big)$$

where $|E(\mathbb{F}_{p^n})|$ is the number of points of E_p in a finite field of cardinality p^n (with the point at infinity included).

Let $a(p)$ be the integer defined by $|E(\mathbb{F}_p)| = p + 1 - a(p)$. Notice that if E is given by a simpler Weierstrass equation

$$y^2 = x^3 + ax + b$$

(to which one can reduce for $p \neq 2$ or 3), then $a(p)$ can be expressed by the quadratic character sum

$$a(p) = -\sum_{x \bmod p} \Big(\frac{x^3 + ax + b}{p}\Big).$$

For $p \mid \Delta$, the coefficient $a(p)$ is given by

$$a(p) = \begin{cases} 0 & \text{if } E \text{ has additive reduction,} \\ 1 & \text{if } E \text{ has split multiplicative reduction,} \\ -1 & \text{if } E \text{ has non-split multiplicative reduction.} \end{cases}$$

Hasse proved that $Z_p(E)$ is a rational function for every p, and for $p \nmid \Delta$ (the reduced curve is then smooth)

$$Z_p(E) = \frac{1 - a(p)X + pX^2}{(1 - X)(1 - pX)}. \tag{14.31}$$

Moreover, Hasse proved the Riemann Hypothesis for E_p in the form of the inequality

$$|a(p)| \leqslant 2\sqrt{p} \tag{14.32}$$

which implies that both roots of the numerator of Z_p have modulus equal to $1/\sqrt{p}$. See Section 11.8 for elementary proofs of (14.31) and (14.32).

The Hasse-Weil zeta function of E, already mentioned in Section 5.14, is defined by the Euler product

$$L(E,s) = \prod_{p|\Delta}(1-a(p)p^{-s})^{-1}\prod_{p\nmid\Delta}(1-a(p)p^{-s}+p^{1-2s})^{-1}. \tag{14.33}$$

Define $a(n)$ for all $n \geqslant 1$ by expanding (14.33) into the Dirichlet series

$$L(E,s) = \sum_1^\infty a(n)n^{-s}. \tag{14.34}$$

This L-function is holomorphic in the region $\mathrm{Re}\,(s) > \frac{3}{2}$ by the Riemann hypothesis (14.32). Hasse further conjectured that the L-function thus defined from local data could be extended to an entire function satisfying the functional equation

$$\Lambda(E,s) = w\Lambda(E,2-s)$$

where $w = \pm 1$ and

$$\Lambda(E,s) = \Big(\frac{\sqrt{N}}{2\pi}\Big)^s\Gamma(s)L(E,s),$$

N being the conductor of E.

This functional equation is the same as that satisfied by the Hecke L-functions of cusp forms of weight 2 and level N with trivial character (see Section 14.7).

Around 1950, Shimura and Taniyama made the conjecture that for any $E/\mathbb{Q}$, there is a cusp form whose L-function is equal to $L(E,s)$. Later Weil refined the conjecture by postulating that the level of this form should equal exactly the conductor of E. This refined conjecture can be tested computationally since the space of cusp forms of a given level, weight and character has finite dimension.

This deep modularity conjecture was proved in 1995 by Wiles [**W**], Wiles-Taylor [**TW**] for semistable curves (with Fermat's Great Theorem as a consequence, as had been shown before by Ribet), and is now known for all elliptic curves over $\mathbb{Q}$, the final work being due to Breuil, Conrad, Diamond and Taylor [**BDCT**].

THEOREM 14.6. *Let $E/\mathbb{Q}$ be any elliptic curve of conductor N. There exists a primitive cusp form*

$$f(z) = \sum_1^\infty \lambda(n)n^{\frac{1}{2}}e(nz) \in S_2(\Gamma_0(N)) \tag{14.35}$$

such that $a(n) = \lambda(n)n^{\frac{1}{2}}$, that is

$$L(E,s+\tfrac{1}{2}) = L(f,s) = \sum_1^\infty \lambda(n)n^{-s}. \tag{14.36}$$

Since the cusp form f associated to E is primitive of level N it is an eigenfunction of the Fricke involution $(Wf)(z) = N^{-1}z^{-2}f(-1/Nz)$, i.e.,

$$Wf = -wf, \qquad \text{with } w = \pm 1. \tag{14.37}$$

Hence the functional equation of the completed L-function is

$$\Lambda(f,s) = w\Lambda(f,1-s). \tag{14.38}$$

The sign $w = \pm 1$ in this equation is also called the root number of E. It is not difficult to compute w computers for any concrete curve, yet there is no simple formula for w in general. In the important special case of semistable curves, that is when N is squarefree, we obtain by the theory of Hecke operators

$$w = -\mu(N)a(N) = -\mu(N)\lambda(N)N^{\frac{1}{2}} = -(-1)^m \tag{14.39}$$

where m is the number of places with split multiplicative reduction.

The celebrated Birch and Swinnerton-Dyer Conjecture asserts that $L(E,s)$ (resp. $L(f,s)$) vanishes at $s=1$ (resp. $s=\frac{1}{2}$) to order exactly equal to the rank of the group $E(\mathbb{Q})$ of rational points on E. Though $r = \mathrm{rank}E(\mathbb{Q})$ is not easy to compute, its parity is conjectured to be determined by

$$w = (-1)^r, \tag{14.40}$$

and this has been proved by Nekovár [**Ne**] under the assumption that the Tate-Shafarevitch group of E is finite (which is known if the order of vanishing of $L(E,s)$ at $s=1$ is $\leqslant 1$).

Proofs of the modularity of special elliptic curves existed before, notably for the curves with complex multiplication, due to Deuring (for a self-contained proof in the case of the "congruent number" curves, see for instance [**I4**], Chapter 8). An early example (due to Shimura) of a modular curve which is not a CM-curve is the following elliptic curve

$$E : y^2 + y = x^3 - x^2$$

whose j-invariant is $j = -4096/11$ and discriminant is $\Delta = -11$.

It follows from the properties of the conductor that the level of the weight 2 cusp form f corresponding to this particular E must be 11. But it is easy to show that $S_2(11)$ is a 1-dimensional space, and actually it is not too hard to construct an element in this space from the knowledge of the Ramanujan Δ function, namely

$$f(z) = \big(\Delta(z)\Delta(11z)\big)^{1/12} = q\prod_{n\geqslant 1}(1-q^n)^2(1-q^{11n})^2$$

($q = e(z)$ as before). In other words, the modularity of E is equivalent with the identity

$$q\prod_{n\geqslant 1}(1-q^n)^2(1-q^{11n})^2 = \sum_{n\geqslant 1} a(n)q^n. \tag{14.41}$$

No elementary proof of this seems to be known. One proceeds in two steps. The first one consists of showing that $Y_0(11) = \Gamma_0(11)\backslash\mathbb{H}$ (which, suitably compactified, is a compact Riemann surface) is an algebraic curve isomorphic to the (affine) elliptic curve E; this is done by constructing meromorphic functions g_1 and g_2 on $X_0(11)$ (as quotients of modular forms of the same weight) which satisfy the equation of E, and goes back to Fricke and Klein. The second step, which is arithmetic and much less elementary, is to deduce from this the required identity of L-functions. This last argument is an example of the so-called Eichler-Shimura theory.

Another interesting example is the Gross-Zagier curve given by the twisted Weierstrass equation

$$-139y^2 = x^3 + 10x^2 - 20x + 8.$$

This curve has conductor $N = 37 \cdot 139^2$ and the sign of the functional equation is $\varepsilon = -1$. Moreover, the rank of the group of rational points on this curve is 3 while one can show using the Gross-Zagier formula that the Hasse-Weil L-function vanishes at $s = 1$ to order exactly 3 (see Appendix to Chapter 23 for a sketch). Hence this example confirms the conjecture of Birch and Swinnerton-Dyer. In combination with an earlier work of Goldfeld it yields a non-trivial effective lower-bound for the class number of imaginary quadratic fields, as will be explained in Chapter 23.

14.5. Hecke L-functions.

Hecke showed how to produce Dirichlet series out of modular forms satisfying many of the properties of Dirichlet L-functions. Consider $f \in M_k(q, \chi)$, so f has the Fourier expansion at ∞

$$f(z) = \sum_{n \geqslant 0} a_f(n) e(nz).$$

One defines its Hecke L-function to be the Dirichlet generating series for the Fourier coefficients

$$L(f, s) = \sum_{n \geqslant 1} a_f(n) n^{-s}. \tag{14.42}$$

This L-function and the completed L-function

$$\Lambda(s, f) = \Big(\frac{\sqrt{q}}{2\pi}\Big)^s \Gamma(s) L(f, s) \tag{14.43}$$

are holomorphic in the region $\mathrm{Re}\,(s) > 1 + \frac{k}{2}$ by the trivial bound (14.10).

We also introduce an operator W (sometimes called the Fricke involution) on $M_k(q, \chi)$ by

$$W f(z) = q^{-k/2} z^{-k} f\Big(\frac{-1}{qz}\Big) \tag{14.44}$$

which, because the element $H_q = \begin{pmatrix} 0 & -1 \\ q & 0 \end{pmatrix}$ normalizes $\Gamma_0(q)$, i.e. $H_q \Gamma_0(q) = \Gamma_0(q) H_q$, induces maps $W : M_k(q, \chi) \to M_k(q, \overline{\chi})$ and $W : S_k(q, \chi) \to S_k(q, \overline{\chi})$.

THEOREM 14.7 (HECKE). *With notation as above, the L-function of f has a meromorphic continuation to the whole complex plane and the completed L-function satisfies the functional equation*

$$\Lambda(f, s) = i^k \Lambda(W f, k - s). \tag{14.45}$$

Moreover, $L(f, s)$ is entire if f is a cusp form, and otherwise it has only a simple pole at $s = k$.

PROOF. The proof is very much like Riemann's (second) proof of the functional equation of $\zeta(s)$. By the definition of the gamma function we have for all $n \geqslant 1$,

$$\left(\frac{\sqrt{q}}{2\pi}\right)^s \Gamma(s) n^{-s} = \int_0^{+\infty} e^{-2\pi n y/\sqrt{q}} y^s \frac{dy}{y}$$

so in the region of absolute convergence we have the representation

$$\Lambda(s,f) = \int_0^{+\infty} \Big(f\Big(\frac{iy}{\sqrt{q}}\Big) - a_0(f)\Big) y^s \frac{dy}{y}.$$

Here we split the integral into the part from 1 to $+\infty$ and the part from 0 to 1 and we transform the latter as follows

$$\begin{aligned}\int_0^1 \Big(f\Big(\frac{iy}{\sqrt{q}}\Big) - a_0(f)\Big) y^s \frac{dy}{y} &= \int_1^{+\infty} \Big(f\Big(\frac{i}{y\sqrt{q}}\Big) - a_0(f)\Big) y^{-s} \frac{dy}{y} \\ &= i^k \int_1^{+\infty} \Big(Wf\Big(\frac{iy}{\sqrt{q}}\Big) - a_0(f)\Big) y^{k-s} \frac{dy}{y}\end{aligned}$$

by applying (14.44). Adding both parts we obtain the integral representation

$$\begin{aligned}\Lambda(f,s) = &\int_1^{+\infty} \Big(f\Big(\frac{iy}{\sqrt{q}}\Big) - a_0(f)\Big) y^s \frac{dy}{y} \\ &+ i^k \int_1^{+\infty} \Big(Wf\Big(\frac{iy}{\sqrt{q}}\Big) - a_0(f)\Big) y^{k-s} \frac{dy}{y}.\end{aligned}$$

Since $f - a_0(f)$ decays exponentially fast at infinity, the meromorphic continuation follows, and the function equation is then clear. □

If $f = E_k$ is an Eisenstein series, one can see from the Fourier expansion that $L(E_k, s)$ is a product of two Dirichlet L-functions, so we do not get any really new L-functions. But when f is a cusp form, the associated Hecke L-function is a genuine $GL(2)$ object.

REMARK. Looking back at the theta series in Section 14.3, one can observe that it provides an equality

$$L_K(s,\chi) = L(f_\chi, s)$$

where on the left side is the L-function attached to a class group character of the quadratic extension K, which has an Euler product

$$L_K(s,\chi) = \prod_{\mathfrak{p}} (1 - \chi(\mathfrak{p}) N\mathfrak{p}^{-s})^{-1}$$

of degree 1 over prime ideals in $\mathcal{O}$, whereas on the right side we have the Hecke L-function of the modular form $f_\chi \in M_1(|D|, \chi_D)$ which has an Euler product of degree 2 over rational primes. This is an early sign of the existence of an instance of Langlands functoriality. Similarly, L-functions defined over number fields are expected to be identical with particular higher-degree L-functions over $\mathbb{Q}$. Another case of this phenomenon appears in the theory of elliptic curves, where the Hasse-Weil zeta function of an elliptic curve $E/\mathbb{Q}$ with complex multiplication was also shown by Deuring to coincide with the L-function of a Hecke character of weight 2 of an imaginary quadratic field.

14.6. Hecke operators and automorphic L-functions.

We now sketch the theory of Hecke operators and the Atkin-Lehner theory of primitive forms, which establishes a link between Dirichlet characters and automorphic forms, showing clearly how the latter generalize the former.

The Hecke operators are certain linear operators acting on spaces of modular forms, whose existence and properties can be seen to be intimately related to the arithmeticity of the subgroups $\Gamma_0(q)$ of $SL(2,\mathbb{R})$, although this will not be apparent in this survey (see [**Mar**]).

Fix integers $k \geqslant 1$, $q \geqslant 1$ and a character χ modulo q satisfying the consistency property $\chi(-1) = (-1)^k$. The operator $T(n)$ is defined by the formula

$$T(n)f(z) = \frac{1}{n}\sum_{ad=n}\chi(a)a^k\sum_{0\leqslant b<d} f\Big(\frac{az+b}{d}\Big) \tag{14.46}$$

for $n \geqslant 1$ an integer (notice that $T(n)$ also depends on χ and so implicitly on q; this can be sometimes significant). The following proposition is fundamental.

Proposition 14.8. *For any $n \geqslant 1$, $T(n)$ acts on modular forms and on cusp forms: in other words, it induces linear maps*

$$\begin{aligned} T(n) &: M_k(q,\chi) \to M_k(q,\chi),\\ T(n) &: S_k(q,\chi) \to S_k(q,\chi). \end{aligned}$$

The proof of this proposition requires identifying the formula (14.46) as another instance of an averaging operator, this time over the (finite) orbit space $\Delta_n = \Gamma_0(q)\backslash\Gamma_n$, where Γ_n is the set of integral matrices of order 2 and determinant n, on which $SL(2,\mathbb{Z})$ acts naturally.

From the definition, we compute the action of $T(n)$ on the Fourier expansion (at ∞) of a modular form (or even of any 1-periodic function), namely

$$\begin{aligned} T(n)f(z) &= \frac{1}{n}\sum_{ad=n}\chi(a)a^k\sum_{0\leqslant b<d}\sum_m a_f(m)e\Big(m\frac{az+b}{d}\Big)\\ &= \frac{1}{n}\sum_m a_f(m)\sum_{ad=n}\chi(a)a^k e\Big(\frac{amz}{d}\Big)\sum_{0\leqslant b<d} e\Big(\frac{mb}{d}\Big)\\ &= \sum_m\Big(\sum_{\substack{ad=n\\ a\ell=m}}\chi(a)a^{k-1}a_f(d\ell)\Big)e(mz). \end{aligned}$$

Therefore

$$T(n)f(z) = \sum_m\Big(\sum_{d\mid(n,m)}\chi(d)d^{k-1}a_f\Big(\frac{nm}{d^2}\Big)\Big)e(mz). \tag{14.47}$$

This formula could also be taken as a definition of $T(n)$, but the modularity of $T(n)f$ would be harder to prove. However, since the Fourier expansion determines a modular form, it can be used to give quick proofs of many properties of the Hecke operators.

PROPOSITION 14.9. *The Hecke operators commute. More precisely, for any $n \geqslant 1$ and $m \geqslant 1$ we have*

$$T(m)T(n) = \sum_{d|(n,m)} \chi(d)d^{k-1}T\Big(\frac{mn}{d^2}\Big)$$

or equivalently

$$T(mn) = \sum_{d|(n,m)} \mu(d)\chi(d)d^{k-1}T\Big(\frac{m}{d}\Big)T\Big(\frac{n}{d}\Big).$$

In particular, the $T(n)$ are multiplicative: $T(mn) = T(m)T(n)$ if m and n are coprime. Therefore any $T(n)$ is a product of Hecke operators $T(p^\nu)$ for some primes p and integers ν. Those operators obey a second order linear recurrence relation (showing that they are determined by $T(p)$):

$$T(p^{\nu+1}) = T(p)T(p^\nu) - \chi(p)p^{k-1}T(p^{\nu-1})$$

which can be summarized by the identity of generating functions

$$\sum_{\nu\geqslant 0} T(p^\nu)X^\nu = (1 - T(p)X + \chi(p)p^{k-1}X^2)^{-1},$$

so the Dirichlet generating series of all the $T(n)$ has an Euler product (of degree 2)

$$\sum_{n\geqslant 1} T(n)n^{-s} = \prod_p (1 - T(p)p^{-s} + \chi(p)p^{k-1-2s})^{-1}.$$

For p prime, we have the formula

$$T(p)f(z) = \chi(p)p^{k-1}f(pz) + \frac{1}{p}\sum_{0\leqslant b<p} f\Big(\frac{z+b}{p}\Big).$$

If p divides q (the level), this simplifies to

$$T(p)f(z) = \frac{1}{p}\sum_{0\leqslant b<p} f\Big(\frac{z+b}{p}\Big).$$

The operators $T(p)$ for p not dividing q turn out to have very different properties than those with p dividing q; the latter are often referred to as the "bad" Hecke operators, and the corresponding p as the bad or ramified primes. The main difference between those two kinds of operators, from the analytic point of view, lies in their relation with the inner product. Indeed, the next lemma fails for $(n, q) > 1$.

LEMMA 14.10. *Let n be coprime with q. Then the operator $T(n)$ acting on the space of cusp forms $S_k(q, \chi)$ is normal with respect to the Petersson inner product; more precisely, its adjoint is*

$$T(n)^* = \overline{\chi}(n)T(n). \tag{14.48}$$

The lemma means that for f and g two cusp forms we have

$$\langle T(n)f, g\rangle = \chi(n)\langle f, T(n)g\rangle;$$

note that indeed this formula cannot hold for n not coprime with q, i.e. for n with $\chi(n) = 0$, because it would imply that $T(n) = 0$, which is certainly not always true.

The proof of this lemma (due to Hecke) is quite involved (see [**14**], pages 104-106 for instance).

From the normality of the $T(n)$ with $(n,q)=1$ and the fact that they commute, we deduce by standard linear algebra the following

PROPOSITION 14.11. *There is an orthonormal basis of the space $S_k(q,\chi)$ of cusp forms which consists of eigenfunctions of all the Hecke operators $T(n)$ for $(n,q)=1$.*

Any non-zero modular form $f \in M_k(q,\chi)$ for which there exist complex numbers $\lambda(n)$ such that

$$T(n)f = \lambda(n)f$$

for all n coprime with q is called a Hecke form. If f is a cuspidal Hecke form, the adjointness formula (14.48) shows that $\lambda(n) = \chi(n)\overline{\lambda(n)}$, for $(n,q)=1$.

What does it mean that f is a Hecke form for the Fourier coefficients $a_f(n)$ of f? From (14.47) applied to f we obtain by taking the first Fourier coefficient of both sides that

$$\lambda(n)a_f(1) = a_f(n), \quad \text{for all } n \text{ with } (n,q)=1 \tag{14.49}$$

so if $a_f(1) \neq 0$, the Hecke eigenvalues and the Fourier coefficients are the same up to a constant factor! This already proves that the Ramanujan τ function in Section 14.2 is multiplicative, as already conjectured by Ramanujan himself (and proved by Mordell). Indeed, since $S_{12}(1)$ is one-dimensional and spanned by the Δ function, perforce Δ is an eigenfunction of all Hecke operators (there are no ramified primes here), and since $\tau(1)=1$, the eigenvalue is $\tau(n)$ and the multiplicativity follows from Proposition 14.9.

14.7. Primitive forms and special basis.

Proposition 14.11 is not exactly what we wish to be true. It would be more interesting to have a basis of eigenfunctions of all the Hecke operators without exceptions. Then (14.49) holds for all n, and so $a_f(1)$ must be non-zero or otherwise $f=0$. After normalization, the Fourier coefficients will be multiplicative, and more importantly the Hecke L-function of f will have an Euler product

$$L(f,s) = \prod_p (1 - a_f(p)p^{-s} + \chi(p)p^{k-1-2s})^{-1}.$$

Such a wonderful world does not exist, however. Indeed, let χ^* modulo q^* be the primitive character which induces χ. Take any q' and d with $q^* \mid q'$, $dq' \mid q$ and let χ' be the character modulo q' induced by χ^*. Now if $f \in S_k(q',\chi')$, the function $f\mid_d(z) = f(dz)$ is in $S_k(q,\chi)$, as follows from the matrix identity

$$\begin{pmatrix} d & 0 \\ 0 & 1 \end{pmatrix}\begin{pmatrix} \alpha & \beta \\ \gamma & \delta \end{pmatrix} = \begin{pmatrix} \alpha & \beta d \\ \gamma/d & \delta \end{pmatrix}\begin{pmatrix} d & 0 \\ 0 & 1 \end{pmatrix}.$$

It is also easy to show that the Hecke operators $T(n)$ with $(n,q)=1$ commute with the operation $f \mapsto f\mid_d$, so f being a Hecke form implies that $f\mid_d$ is one. However, since

$$f\mid_d(z) = \sum_{n\geqslant 1} a_f(n)e(dnz),$$

the first Fourier coefficient of $f\mid_d$ vanishes if $d>1$.

The theory of primitive forms (or "newforms") remedies to this defect by considering that cusp forms such as $f\mid_d$ in $S_k(q,\chi)$ are not really of level q but come from lower levels, and that a better theory of Hecke operators should hold when such "oldforms" are put aside. This was developed by Atkin and Lehner [**AL**].

Therefore, let $S^\flat(q,\chi)$ be the subspace of $S_k(q,\chi)$ spanned by all cusp forms of the type $f\mid_d$ where $f\in S_k(q',\chi')$ with $q'<q$ and $dq'\mid q$. Then let $S^*(q,\chi)$ be the orthogonal complement of $S_k^\flat(q,\chi)$ with respect to the Petersson inner product, so

$$S_k(q,\chi)=S^\flat(q,\chi)\oplus S^*(q,\chi)$$

(note that if χ is primitive, we have $S_k^*(q,\chi)=S_k(q,\chi)$, and that if χ is trivial, q is prime and $k\leqslant 10$ or $k=14$, we also have $S_k^*(q)=S_k(q)$, because $S_k(1)=0$ for those k).

Since $(T(n)f)\mid_d=T(n)(f\mid_d)$ for $(n,q)=1$, it is immediate that $T(n)$ acts on $S_k^\flat(q,\chi)$ for $(n,q)=1$. Since those Hecke operators are normal, they act also on the orthogonal complement $S_k^*(q,\chi)$. In particular, both spaces still have a basis of Hecke forms. By definition, a Hecke form f which is in $S_k^*(q,\chi)$ is called a primitive form (of weight k, level q, and character χ). The main result is the following:

THEOREM 14.12 (MULTIPLICITY ONE PRINCIPLE). *Given a sequence $(\lambda(n))$ of complex numbers, the subspace of $S_k^*(q,\chi)$ spanned by Hecke forms with eigenvalues $\lambda(n)$ for all $(n,q)=1$ is at most one-dimensional. In other words, such a Hecke form f, if it exists, is unique up to multiplication by a scalar.*

EXERCISE 4. Prove that the multiplicity-one principle is equivalent to the assertion that any primitive form $f\in S_k^*(q,\chi)$ has non-zero first Fourier coefficient.

Suppose now that T is a linear operator on the space $S_k^*(q,\chi)$ which commutes with all $T(n)$ for $(n,q)=1$. Then it follows that T acts on the common eigenspaces of those $T(n)$. By the multiplicity-one principle, those are one-dimensional, each spanned by a primitive form f, and therefore f is also an eigenfunction of T. In particular applying this property to the "bad" $T(n)$, i.e. for $(n,q)\neq 1$, we obtain the desired result.

PROPOSITION 14.13. *Let $f\in S_k^*(q,\chi)$ be a primitive form. Then f is an eigenfunction of all Hecke operators,*

$$T(n)f=\lambda_f(n)f,\quad \textit{for all } n\geqslant 1$$

and for all $n\geqslant 1$, $a_f(n)=\lambda_f(n)a_f(1)$. Hence the first Fourier coefficient $a_f(1)$ is non-zero.

It is natural to normalize a primitive form, so that it becomes a distinguished basis of its common eigenspace; then the basis of $S_k^*(q,\chi)$ of normalized primitive forms is also unique. Two natural normalizations exist: we will say that f is Hecke-normalized if its first Fourier coefficient is $a_f(1)=1$, so that the Fourier coefficients are the same as the Hecke eigenvalues, and that f is Petersson-normalized if $\|f\|=1$ (such normalization is often more useful in the context of the Petersson formula). If f is Hecke-normalized, then of course $\|f\|^{-1}f$ is Petersson-normalized. The norm $\|f\|$ is related to the value of the adjoint square L-function of f at 1, and plays a role in some important applications. See Section 5.12, and in particular Corollary 5.45 for the comparison between the two normalizations.

When no other indication is present, "primitive" will always mean "Hecke-normalized primitive". In any case, once a normalization is chosen, the basis of primitive forms of $S_k^*(q,\chi)$ is unique. In particular, it is possible to speak of the set of primitive forms without ambiguity.

The Hecke L-function of a (Hecke-normalized) primitive form f, from the previous arguments, has an Euler product

$$L(f,s)=\prod_p (1-\lambda_f(p)+\chi(p)p^{k-1-2s})^{-1}.$$

This is equivalent with the formula

$$\lambda_f(mn)=\sum_{d|(m,n)}\chi(d)d^{k-1}\lambda_f\Big(\frac{mn}{d^2}\Big). \tag{14.50}$$

This can be also written by Möbius inversion as

$$\lambda_f(m)\lambda_f(n)=\sum_{d|(m,n)}\mu(d)\chi(d)d^{k-1}\lambda_f\Big(\frac{m}{d}\Big)\lambda_f\Big(\frac{n}{d}\Big). \tag{14.51}$$

Moreover, the functional equation of Theorem 14.7 is also simpler because the primitivity of f will show that Wf is expressed in terms of f. The linear operator W is an isometry, and it is easy to show that W commutes with the good Hecke operators:

$$WT^{\chi}(n)=\chi(n)T^{\overline{\chi}}(n)W,\quad \text{for}\quad (n,q)=1$$

(where the superscript χ indicates for which character the $T(n)$ is defined).

To come back from $S_k(q,\overline{\chi})$ to the original space $S_k(q,\chi)$, we use another operator, the K-operator defined by

$$Kf(z)=\overline{f(-\overline{z})}.$$

Note that K acts on Fourier expansions by conjugating the coefficients,

$$Kf(z)=\sum_{n\geqslant 0}\overline{a_f}(n)e(nz)=\overline{f}(z),\ \text{say}.$$

One must be careful because K is not $\mathbb{C}$-linear, namely $K(af)=\overline{a}Kf$. But from the definition it is clear that $K^2=Id$, that K is an isometry, and also that K commutes with all $T(n)$. Consider now the composition (also only $\mathbb{R}$-linear)

$$\overline{W}=KW\,:\,S_k(q,\chi)\to S_k(q,\chi)$$

which satisfies the commutation relation

$$T(n)\overline{W}=\chi(n)\overline{W}T(n),\quad \text{for } (n,q)=1.$$

Let f be a primitive form. Then for $(n,q)=1$ we have

$$T(n)\overline{W}f=\chi(n)\overline{W}T(n)f=\chi(n)\overline{W}(\lambda_f(n)f)=\chi(n)\overline{\lambda_f(n)}\overline{W}(f)=\lambda_f(n)\overline{W}(f)$$

by the formula for the adjoint of $T(n)$. By the multiplicity one principle, it follows that $\overline{W}f$ must be a multiple of f, hence

PROPOSITION 14.14. *If f is a primitive form, then there exists $\eta \in \mathbb{C}$, with $|\eta| = 1$, such that $\overline{W} f = \eta f$.*

That $|\eta| = 1$ follows from the fact that $\overline{W}$ is an involution, $\overline{W}^2 = Id$. The eigenvalue η is another very intricate invariant of f. It is determined by f but only in special cases do we have a formula for its value in terms of more accessible data.

PROPOSITION 14.15. *Let χ be a primitive character of conductor q. Then η is given by*

$$\eta = \tau(\overline{\chi})\lambda_f(q)q^{-k/2}.$$

PROPOSITION 14.16. *Let q be squarefree, and χ trivial. Then η is given by*

$$\eta = \mu(q)\lambda_f(q)q^{1-\frac{k}{2}}.$$

From Proposition 14.14 and Theorem 14.7 we deduce the main result about automorphic L-functions of primitive forms.

THEOREM 14.17. *Let f be a primitive form. Then the Hecke L-function of f has an Euler product expansion*

$$L(f, s) = \prod_p (1 - \lambda_f(p) + \chi(p)p^{k-1-2s})^{-1}$$

and has analytic continuation to an entire function. The completed L-function

$$\Lambda(f, s) = \left(\frac{\sqrt{q}}{2\pi}\right)^s \Gamma(s)L(f, s)$$

satisfies the functional equation

$$\Lambda(f, s) = i^k \overline{\eta} \Lambda(\overline{f}, k - s).$$

Notice that this theorem is in perfect analogy with the corresponding result for primitive Dirichlet characters. Even deeper analogies are revealed by the insight of the Langlands program, which uses representation theory to unify the whole theory of L-functions (cf. [**BG**]).

We conclude this section by stating a non-trivial generalization of the multiplicity-one principle which is very useful in applications.

THEOREM 14.18 (THE STRONG MULTIPLICITY ONE PRINCIPLE). *Given a sequence $(\lambda(n))$ of complex numbers and a fixed positive integer M, there exists at most one primitive form $f \in S_k^*(q, \chi)$ such that*

$$\lambda_f(n) = \lambda(n), \text{ for all } n \text{ prime to } M.$$

The main point is that M has nothing to do with the level q. In effect, this shows that a primitive form is known when one knows its eigenvalues for all primes, except finitely many. In fact one can show using Rankin-Selberg L-functions that finitely many primes suffice (depending on the weight and level); see Proposition 5.22 and the remark following.

EXERCISE 5. Show that if $f \in S_k(q,\chi)$ is a Hecke form, then there exists a unique primitive form $g \in S_k(q',\chi')$ with $q' \mid q$ such that $\chi'(n) = \chi(n)$ and $\lambda_g(n) = \lambda_f(n)$ for all n prime to q.

14.8. Twisting modular forms.

We have already mentioned that we intend to exploit the eigenvalues $\lambda_f(n)$ of primitive modular forms as arithmetic harmonics; however, before they can be used effectively for such a purpose, it is necessary to get information on the eigenvalues themselves. For this, following our strategy, it is natural to probe their behavior when twisted by other harmonics. The case of primitive Dirichlet characters is basic and the results are nice.

PROPOSITION 14.19. *Let $f \in M_k(q,\chi)$ be a modular form with Fourier coefficients $a_f(n)$, not necessarily a Hecke form, q^* the conductor of the Dirichlet character χ and let ψ be a primitive Dirichlet character modulo r. Let $f\otimes\psi$ be the function on $\mathbb{H}$ given by the Fourier expansion*

$$(f\otimes\psi)(z) = \sum_{n\geqslant 0} \psi(n)a_f(n)e(nz).$$

Then $f\otimes\psi$ is also a modular form, more precisely $f\otimes\psi \in M_k(N,\chi\psi^2)$, where the level N is the least common multiple of q, q^*r and r^2. If f is a cusp form, then so is $f\otimes\psi$.

PROOF. Since ψ is primitive, we have for any n the expression for $\psi(n)$ in terms of additive characters

$$\tau(\overline{\psi})\psi(n) = \sum_{u \,(\mathrm{mod}\, r)} \overline{\psi}(u)e\Big(\frac{un}{r}\Big)$$

where $\tau(\overline{\psi}) \neq 0$ is the Gauss sum, whence

$$(f\otimes\psi)(z) = \tau(\overline{\psi})^{-1} \sum_{u \,(\mathrm{mod}\, r)} f\Big(z+\frac{u}{r}\Big). \tag{14.52}$$

It is now a matter of writing, for $\gamma = \begin{pmatrix} a & b \\ c & d\end{pmatrix}$ in $\Gamma_0(N)$, the matrix identity in $SL(2,\mathbb{R})$,

$$\begin{pmatrix} 1 & \frac{u}{r} \\ 0 & 1\end{pmatrix}\begin{pmatrix} a & b \\ c & d\end{pmatrix} = \begin{pmatrix} a+\frac{uc}{r} & b-\frac{bcdu}{r}-\frac{cd^2u^2}{r^2} \\ c & d-\frac{cd^2u}{r}\end{pmatrix}\begin{pmatrix} 1 & 1+\frac{du^2}{r} \\ 0 & 1\end{pmatrix}$$

to obtain

$$\begin{aligned}((f\otimes\psi)\mid_k \gamma)(z) &= \chi(d)\tau(\overline{\psi})^{-1} \sum_{u \,(\mathrm{mod}\, r)} \overline{\psi}(u) f\Big(z+\frac{d^2u}{r}\Big) \\ &= \chi(d)\psi(d)^2(f\otimes\psi)(z)\end{aligned}$$

which shows the modularity. As for $f\otimes\psi$ being a cusp form when f is, the criterion of Exercise 1 applies immediately from (14.52). □

By multiplicativity of ψ and the formula (14.47), we see that the operation of twisting by ψ takes Hecke forms to Hecke forms. On the other hand, it might not take a primitive form to a primitive form, because the level N as described in Proposition 14.19 might not be optimal. This happens, for instance, when $f = f_\chi$ is one of the forms of weight 1 associated to an ideal-class character χ in Section 14.3. Indeed since the Fourier coefficients of f_χ are supported on integers which are norms of ideals of the field $K = \mathbb{Q}(\sqrt{D})$, so twisting by the field character $\psi = \chi_D$ has no effect at all, since it is identically 1 on norms of ideals. By Exercise 5, however, there is a unique primitive form g, with level dividing N, which is such that

$$\psi(n)\lambda_f(n) = \lambda_g(n), \quad \text{for all } n \text{ coprime with } q.$$

In the special case when q and r are coprime, however, there is no problem and $f \otimes \psi$ is primitive whenever f is (as might indeed be expected). In this case the functional equation of the twisted L-function takes the following form.

PROPOSITION 14.20. *Let $f \in S_k^*(q, \chi)$ be a primitive form and ψ be a primitive Dirichlet character modulo r with $(r, q) = 1$. Then $f \otimes \psi$ is primitive of level $N = qr^2$ and the Hecke L-function of $f \otimes \psi$ is entire and polynomially bounded in vertical strips. Moreover, the completed L-function (14.43) satisfies the functional equation*

$$\Lambda(s, f \otimes \psi) = i^k w \overline{\eta}_f \Lambda(k - s, \overline{f} \otimes \overline{\psi})$$

where η_f is the eigenvalue of f for the operator $\overline{W}$, and the root number w depends only on χ and ψ, namely

$$w = \chi(r)\psi(q)\frac{\tau(\psi)^2}{r}.$$

The fact that there is no pole for any of the twists of a cusp form shows that the sign of the Fourier coefficients are changing randomly and independently of those of any Dirichlet character. The lack of periodicity of the Fourier coefficients is a welcome feature in analytic number theory. See [**DuI2**], [**DuI3**], [**DuI4**] for some applications of twisting as a replacement of positivity.

Twists are important also as a simple case of Rankin-Selberg convolution $GL(2) \otimes GL(1)$ (see Chapter 5), and they occur in so-called converse theorems which characterize modular forms by means of analytic properties of L-functions. The first converse theorem using twists is due to Weil [**We3**]. Deeper variants are important in proving instances of Langlands functoriality, see the survey by Cogdell in [**BG**]. We quote Weil's result:

THEOREM 14.21. *Let*

$$L_1(s) = \sum_{n \geqslant 1} a(n)n^{-s}, \qquad L_2(s) = \sum_{n \geqslant 1} b(n)n^{-s}$$

be two Dirichlet series absolutely convergent for $\mathrm{Re}\,(s) > C$ *for some* $C > 0$.

Assume that there exists integers $k \geqslant 1$, $q \geqslant 1$ and $M > 0$ such that for any ψ primitive modulo m, with $(m, Mq) = 1$, the Dirichlet series

$$L(f \otimes \psi, s) = \sum_{n \geqslant 1} a(n)\psi(n)n^{-s}, \qquad L(g \otimes \psi, s) = \sum_{n \geqslant 1} b(n)\psi(n)n^{-s}$$

admit analytic continuation to entire functions bounded in vertical strips such that the functions

$$\Lambda(f \otimes \psi, s) = (2\pi)^{-s}\Gamma(s)L(f \otimes \psi, s), \quad \Lambda(g \otimes \psi, s) = (2\pi)^{-s}\Gamma(s)L(g \otimes \psi, s)$$

are entire and satisfy the functional equation

$$\Lambda(f \otimes \psi, s) = w_\psi (qm^2)^{k/2-s} \Lambda(g \otimes \overline{\psi}, k-s),$$

with

$$w_\psi = i^k \chi(m)\psi(q)\tau(\psi)^2 r^{-1}.$$

Then there exists a cusp form $f \in S_k(q,\chi)$ for some χ modulo q such that $L(f,s) = L_1(s)$ and $L(Wf,s) = L_2(s)$.

14.9. Estimates for the Fourier coefficients of cusp forms.

The question of the size of the Fourier coefficients (or Hecke eigenvalues) of cusp forms is one of the classical problems in the study of automorphic forms for themselves. Consider $f \in S_k(q,\chi)$. By the criterion for cusp forms of Exercise 1 and the Parseval formula we obtain, from the Fourier expansion of f,

$$\sum_{n\geqslant 1} |a_f(n)|^2 e^{-4\pi n y} = \int_0^1 |f(z)|^2 dx \ll y^{-k}$$

for any $y > 0$, so that for any $N \geqslant 1$,

$$\sum_{n\leqslant N} |a_f(n)|^2 \ll y^{-k} e^{4\pi N y}$$

and by choosing $y = N^{-1}$ we derive a bound for the second-moment of the Fourier coefficients

$$\sum_{n\leqslant N} |a_f(n)|^2 \ll N^k. \tag{14.53}$$

Hence the trivial bound (14.10), namely $a_f(n) \ll n^{k/2}$, follows by positivity. This last step is rather crude and it is therefore expected that the resulting bound for the individual coefficient $a_f(n)$ can be improved. On the other hand, the bound obtained on average turns out to have already the correct order of magnitude, as shown by the properties of the Rankin-Selberg L-functions. Thus, it shows that $a_f(n)$, on average, is of order $n^{\frac{k-1}{2}}$.

The statement that $a_f(n) \ll n^{\frac{k-1}{2}+\varepsilon}$ for any $\varepsilon > 0$ holds individually is known as the Ramanujan-Petersson Conjecture. For primitive cusp forms, this was proved by P. Deligne [**De1**] for $k \geqslant 2$ as a consequence of the Riemann Hypothesis for varieties over finite fields (the Weil conjectures), in the very sharp form

$$|\lambda_f(n)| \leqslant \tau(n) n^{\frac{k-1}{2}} \tag{14.54}$$

for any $n \geqslant 1$, where τ is the divisor function. The corresponding formula for $k = 1$ holds and is due to Deligne and Serre [**DeSe**]. This is an extremely powerful result, in particular, because it is completely uniform. If f is not primitive, by writing it as a linear combination of Hecke forms, we derive from (14.54)

$$a_f(n) \ll \tau(n) n^{\frac{k-1}{2}},$$

but the implied constant now depends on f quite badly.

It is often convenient to normalize the Fourier coefficients so as to extract the expected order of magnitude, writing

$$f(z) = \sum_{n \geqslant 0} a_f(n) n^{\frac{k-1}{2}} e(nz) \tag{14.55}$$

instead of (14.9), and similarly for primitive forms with $\lambda_f(n)$. This has the effect of putting the critical line of the L-function at $\mathrm{Re}\,(s) = \frac{1}{2}$, and the functional equation then relates $\Lambda(f, s)$ and $\Lambda(\overline{f}, 1-s)$. This normalization is most convenient in analytic number theory and will be in effect from now on, except where explicitly stated. It is also the normalization used in Chapter 5, see Section 5.11 for the necessary adjustments. In this context, Deligne's bound takes the form $|\lambda_f(n)| \leqslant \tau(n)$ for f primitive, and the Hecke L-function becomes

$$\sum_{n \geqslant 1} \lambda_f(n) n^{-s} = \prod_p (1 - \lambda_f(p) p^{-s} + \chi(p) p^{-2s})^{-1}.$$

In recent applications (for instance, of the "amplification method"), the need for lower bounds on the Fourier coefficients has also arisen. Since an individual lower bound is impossible, as there exist forms f (primitive) and n (coprime with the level) with $\lambda_f(n) = 0$, this can only be expected to hold on average. Indeed, from the Rankin-Selberg method we know that the upper-bound (14.53) can be sharpened to the asymptotic (remember that we now normalize $a_f(n)$ as in (14.55))

$$\sum_{n \leqslant N} |a_f(n)|^2 = c_f N + O(N^{\frac{3}{5}}), \tag{14.56}$$

with $c_f > 0$ the implied constant depending on f. In particular, for a fixed f, the Fourier coefficients are not too small too often. However, the dependence of the constant on f in this formula is not explicit enough and so the uniformity which is often required for applications is not present. In this context a very useful observation is that if f is a primitive form

$$\lambda_f(p^2) - \lambda_f(p)^2 = \chi(p).$$

From this simple formula we see that for p not dividing the level q it is not possible for $\lambda_f(p)$ and $\lambda_f(p^2)$ to be simultaneously very small! This is very useful because it is completely uniform in all parameters involved. Here is an example of the use of this formula.

PROPOSITION 14.22. *Let $f \in S_k^*(q, \chi)$ be a primitive form, $\lambda_f(n)$ its Hecke eigenvalues. There exists a sequence of complex numbers $c(n)$ such that*

$$\sum_{n \leqslant N} c(n) \lambda_f(n) \asymp \sqrt{N} (\log N)^{-1}, \tag{14.57}$$

$$\sum_{n \leqslant N} |c(n)|^2 \asymp \sqrt{N} (\log N)^{-1} \tag{14.58}$$

if $N \gg (\log 2q)^2$, where the implied constants are absolute.

PROOF. It is enough to take $c(n) = \overline{\chi}(p)$ if $n = p^2 \leqslant N$, $c(n) = -\overline{\chi}(p)\lambda_f(p)$ if $n = p \leqslant \sqrt{N}$, and $c(n) = 0$ otherwise. Then the left side of (14.57) is equal to

$$\sum_{p\leqslant\sqrt{N}} \overline{\chi}(p)\big(\lambda_f(p^2) - \lambda_f(p)^2\big) = \sum_{\substack{p\leqslant\sqrt{N}\\ p\nmid q}} 1 \asymp \sqrt{N}(\log N)^{-1}$$

by Tchebyshev's estimate for $\pi(\sqrt{N})$ and $\omega(q) \ll (\log 2q)(\log\log 3q)^{-1}$. On the other hand, $c(n)$ satisfies (14.58) by Deligne's bound (14.54). □

REMARK. From the proof one sees that the numbers $c(n)$ can be chosen to have support on primes or squares of primes and that $|c(n)| \leqslant 2$. These extra properties are not important but could be convenient for technical reasons.

14.10. Averages of Fourier coefficients.

The Petersson formula (Proposition 14.5) is very useful in applications; it permits averaging over natural families of automorphic forms, which are so to speak "complete". In doing so, it shows that the Fourier coefficients of automorphic forms behave as "harmonics", they are nearly orthogonal (see also Section 7.3 for the underlying philosophy). This is also very true with the similar effects of the Kuznetsov formula (see Chapter 16).

It is especially important to deal with coefficients of primitive forms, which benefit also from multiplicative properties (among other things). The issue arises that in general there is no basis of $S_k(q,\chi)$ of primitive forms. However, taking a basis of Hecke forms (see Proposition 14.11), say $H_k(q,\chi)$, and denoting

$$f(z) = \sum_{n\geqslant1} \lambda_f(n)n^{(k-1)/2}e(nz)$$

their Fourier expansion, Proposition 14.5 yields

COROLLARY 14.23. *Let $q \geqslant 1$, $k \geqslant 2$. Let χ be a character modulo q and $H_k(q,\chi)$ any Hecke basis of $S_k(q,\chi)$. For any m, $n \geqslant 1$, we have*
(14.59)
$$\sum_{f\in H_k(q,\chi)}^{h} \lambda_f(n)\overline{\lambda_f(m)} = \delta(m,n) + 2\pi i^{-k} \sum_{\substack{c>0\\ c\equiv0\,(\mathrm{mod}\,q)}} c^{-1}S_\chi(m,n;c)J_{k-1}\Big(\frac{4\pi\sqrt{mn}}{c}\Big).$$

where the superscript h means that f has been spectrally normalized to be of Petersson norm 1. Without normalization this means the terms of summation are appropriately weighted, namely

$$\sum_{f\in H_k(q,\chi)}^{h} \alpha_f = \frac{\Gamma(k-1)}{(4\pi)^{k-1}} \sum_{f\in H_k(q,\chi)} \frac{\alpha_f}{\|f\|^2}. \tag{14.60}$$

It is important to know how good the approximate orthogonality can really be, that is, to estimate the series of Kloosterman sums. A simple but effective treatment yields the following corollary (see Theorem 16.7 for a more refined estimate).

COROLLARY 14.24. *With notation as above, we have for any m, $n \geqslant 1$,*

$$(14.61)\qquad \sum_{f\in H_k(q,\chi)}^{h} \lambda_f(n)\overline{\lambda_f(m)} = \delta(m,n) + O\Big(\tau_3((m,n))(m,n,q)^{\frac{1}{2}}(mn)^{\frac{1}{4}}\frac{\tau(q)}{q\sqrt{k}}\log\Big(1+\frac{(mn)^{\frac{1}{4}}}{\sqrt{qk}}\Big)\Big)$$

where the implied constant is absolute.

PROOF. Use the estimate $J_{k-1}(x) \ll \min(1, x/k)$ and the Weil bound for Kloosterman sums (Corollary 11.12)

$$(14.62)\qquad |S(m,n;c)| \leqslant (m,n,c)^{1/2}c^{1/2}\tau(c).$$

For $c = qr$, say, this yields the bound $(m,n,q)^{\frac{1}{2}}(m,n,r)^{\frac{1}{2}}(qr)^{\frac{1}{2}}\tau(q)\tau(r)$. Hence the sum of Kloosterman sums in (14.59) is estimated by

$$(m,n,q)^{\frac{1}{2}}\frac{\tau(q)}{\sqrt{q}}\sum_r \frac{\tau(r)}{\sqrt{r}}(m,n,r)^{\frac{1}{2}}\min\Big(1,\frac{\sqrt{mn}}{kqr}\Big).$$

Hence the result follows by the elementary estimate

$$\sum_r \frac{\tau(r)}{\sqrt{r}}\min\Big(1,\frac{X}{r}\Big) \ll \sqrt{X}\log(1+\sqrt{X}),$$

which holds for any positive X. □

If the space of old forms vanishes, then the set of primitive forms $S_k(q,\chi)^*$ is a distinguished basis of $S_k(q,\chi)$. In particular, this is so if χ is primitive for instance.

However, we will discuss further the special case where q is prime, χ is trivial and $k < 12$, because this will be used in Chapter 26 for studying non-vanishing of L-functions. In that case $S_k^*(q) = S_k(q)$ because $S_k(1) = 0$ (the first cusp form for $SL(2,\mathbb{Z})$ is of weight 12). If $f \in S_k(q)^*$ is a primitive form it has real eigenvalues, and since q is squarefree the sign $\varepsilon_f = \pm 1$ of the functional equation is given by $\varepsilon_f = -\lambda_f(q)q^{1/2}$ (Proposition 14.16; note we have changed normalization of Fourier coefficients slightly). It is often desirable to be able to average over forms with a fixed sign, say $\varepsilon = \pm 1$. This can be done by inserting a factor $(1+\varepsilon\varepsilon_f)$, and applying the summation formula twice:

$$2\sum_{\varepsilon_f=\varepsilon}^{h}\lambda_f(m)\overline{\lambda_f(n)} = \sum_f^{h}(1+\varepsilon\varepsilon_f)\lambda_f(m)\overline{\lambda_f(n)} = \sum_f^{h}\lambda_f(m)\overline{\lambda_f(n)} - \varepsilon q^{1/2}\sum_f^{h}\lambda_f(q)\lambda_f(m)\overline{\lambda_f(n)}.$$

Because f is primitive and q is the level, $\lambda_f(q)\lambda_f(n) = \lambda_f(qn)$ for all n. We deduce

PROPOSITION 14.25. *Let q be prime, $1 < k < 12$, $\varepsilon = \pm 1$. For any m, $n \geqslant 1$, we have*

$$(14.63)\quad 2\sum_{\varepsilon_f=\varepsilon}^{h} \lambda_f(m)\overline{\lambda_f(n)} = \delta(m,n) - \varepsilon q^{1/2}\delta(m,nq)$$

$$+ 2\pi i^{-k} \sum_{\substack{c>0 \\ c\equiv 0 \,(\mathrm{mod}\, q)}} c^{-1}(S(m,n;c) - \varepsilon q^{1/2}S(m,nq;c))J_{k-1}\Big(\frac{4\pi\sqrt{mn}}{c}\Big).$$

Here also we can estimate some or all Kloosterman sums getting an approximation to the diagonal symbol in terms of the Hecke eigenvalues of primitive cusp forms.

COROLLARY 14.26. *Let q be prime, $1 < k < 12$ and $\varepsilon = \pm 1$. For any m, $n \geqslant 1$ with $(m,q) = 1$ we have*

$$(14.64)\quad 2\sum_{\varepsilon_f=\varepsilon}^{h} \lambda_f(m)\overline{\lambda_f(n)} = \delta(m,n)$$

$$- \frac{2\varepsilon\pi i^{-k}}{\sqrt{q}} \sum_{(r,q)=1} \frac{1}{r} S(m\bar{q},n;r)J_{k-1}\Big(\frac{4\pi}{r}\sqrt{\frac{mn}{q}}\Big)$$

$$+ O\Big(\tau_3((m,n))\frac{(mn)^{\frac{1}{4}}}{q\sqrt{k}}\log\Big(1 + \frac{(mn)^{\frac{1}{4}}}{\sqrt{qk}}\Big)\Big)$$

and

$$(14.65)\quad 2\sum_{\varepsilon_f=\varepsilon}^{h} \lambda_f(m)\overline{\lambda_f(n)} = \delta(m,n) + O\Big(\frac{\tau_3((m,n))}{\sqrt{q}}\Big(\frac{mn}{qk^2}\Big)^{\frac{1}{4}}\log\Big(1 + \Big(\frac{mn}{qk^2}\Big)^{\frac{1}{4}}\Big)\Big).$$

The implied constant in both estimates is absolute.

PROOF. Since q does not divide m we have $\delta(m,qn) = 0$. Moreover, the first term involving Kloosterman sums on the right side of (14.63) is already estimated in Corollary 14.24. For the second, we write $c = qr$ and estimate the contribution of the terms with $q \mid r$, by the same method getting a bound which is of smaller order of magnitude.

When $(r,q) = 1$, the Kloosterman sum factorizes

$$S(m,n;qr) = S(m\bar{q},n;r)S(0,m;q) = -S(m\bar{q},n;r).$$

This gives (14.64). The remaining sum of Kloosterman sums in (14.64) can also be estimated directly by using Weil's bound, and the result is given in the error term of (14.65) (this is worse than the error term in (14.64) in terms of q). □

The error term of (14.65) is larger than the error term in (14.64) and will not be sufficient (just) for applications in Chapter 26. Better estimates will be derived from (14.64) by exploiting a cancellation in sums of Kloosterman sums over the variables m and n.

CHAPTER 15

SPECTRAL THEORY OF AUTOMORPHIC FORMS

15.1. Motivation and geometric preliminaries.

Abelian harmonic analysis, as developed in the previous chapters (see especially Chapter 4), gives powerful means of transforming or estimating summations over integers, or lattice points in various geometrical domains. However, in analytic number theory, it is desired to have such tools for sums over other subsets of integers which are not defined by analytic conditions alone. Common among these are sums over $(a, b, c, d) \in \mathbb{Z}^4$ restricted by the determinant equation

$$\begin{vmatrix} a & b \\ c & d \end{vmatrix} = ad - bc = h$$

where h is fixed. For example, one encounters the determinant equation when investigating the power moments (with amplification) of L-functions on the critical line to handle off-diagonal contributions.

Although the classical Fourier analysis applies to the determinant equation quite well (especially when enhanced by estimates for Kloosterman sums) the spectral theory on the hyperbolic plane produces much stronger results. The key point is that for the analysis of the determinant equation the automorphic forms are more natural harmonics than the exponential functions or Dirichlet characters. They enter the modern analytic number theory through many other doors. For example, the old problems concerning the class group of quadratic fields (Lagrange, Gauss, Dirichlet) cannot be properly understood without speaking of automorphic theory. Another incentive for new analytic tools is to study automorphic L-functions.

References for this chapter are for instance [**Bu**], [**I5**]. The basic theory was constructed by Maass [**Ma**] and Selberg [**S6**].

As in the previous chapter we consider first the group $G = SL(2, \mathbb{R})$. Just as $\mathbb{Z}$ is a discrete subgroup of $\mathbb{R}$, from which the theory of periodic functions and Fourier series arises, $\Gamma = \Gamma_0(1) = SL(2, \mathbb{Z})$ is a discrete subgroup of G. Studying the determinant equation with $h = 1$ by harmonic analysis means taking a kernel (smooth, compactly supported) $k : G \to \mathbb{C}$ and finding a summation formula à la Poisson for the function $\mathcal{K} : G \to \mathbb{C}$ given by

$$\mathcal{K}(g) = \sum_{\gamma \in \Gamma} k(\gamma g)$$

which is Γ-periodic; $\mathcal{K}(\gamma g) = \mathcal{K}(g)$ for $\gamma \in \Gamma$.

However, a useful simplification is possible by performing first an ordinary Fourier expansion with respect to the compact abelian subgroup

$$K = \left\{ \begin{pmatrix} \cos\theta & -\sin\theta \\ \sin\theta & \cos\theta \end{pmatrix} \mid \theta \in [0, 2\pi] \right\} \subset SL(2, \mathbb{R})$$

(which is isomorphic to the circle $\mathbb{R}/2\pi\mathbb{Z}$). Any function k can be expanded as a series

$$k(g) = \sum_{m \in \mathbb{Z}} k_m(g)$$

where k_m satisfies a simple transformation rule $k_m(gr) = e(m\theta)k(g)$ for all $r = r_\theta \in K$. Therefore, instead of functions on G, we can look at functions on the quotient G/K. This quotient is diffeomorphic to the Poincaré upper half-plane by

$$\text{(15.1)} \qquad \begin{cases} G/K \to \mathbb{H} \\ \begin{pmatrix} a & b \\ c & d \end{pmatrix} \mapsto \dfrac{ai+b}{ci+d}. \end{cases}$$

Instead of doing this reduction explicitly, we work on $\mathbb{H}$ and have G act on it.

Recall from Chapter 14 that the Poincaré upper half-plane is $\mathbb{H} = \{z \in \mathbb{C} \mid \mathrm{Im}\,(z) > 0\}$. We will see it now as a model of the hyperbolic plane, a complete Riemannian manifold of dimension 2 with constant negative curvature -1 when equipped with the Poincaré metric

$$\text{(15.2)} \qquad ds^2 = y^{-2}(dx^2 + dy^2) = y^{-2}|dz|^2.$$

The geometry of $\mathbb{H}$ with this metric is accessible because $\mathbb{H}$ has a large isometry group. Indeed, the action (14.2) of $G = SL(2, \mathbb{R})$ on $\mathbb{H}$ by linear fractional transformations

$$\text{(15.3)} \qquad gz = \frac{az+b}{cz+d}, \quad \text{for } g = \begin{pmatrix} a & c \\ c & d \end{pmatrix}$$

is an action by isometries: this follows directly from the formula above and from

$$\text{(15.4)} \qquad d(gz) = (cz+d)^{-2}dz.$$

Hence $G \subset \mathrm{Isom}(\mathbb{H})$, the isometry group of $\mathbb{H}$, with the center of G acting trivially. Actually, $PSL(2, \mathbb{R}) = SL(2, \mathbb{R})/\{\pm 1\}$ is the full group of orientation-preserving isometries of $\mathbb{H}$, and $\mathrm{Isom}(\mathbb{H})$ is generated by G and the reflection $\sigma : z \mapsto -\bar{z}$.

Using this large isometry group, many geometric properties of $\mathbb{H}$ can be established fairly quickly:

(1) Angles in $\mathbb{H}$ are the same as euclidean angles (simply because the metric is conformal to the euclidean metric $|dz|^2$).

(2) The hyperbolic geodesics in $\mathbb{H}$ are the half-circles (in the euclidean sense) orthogonal to the "boundary" $\mathbb{R} \cup \infty$, and the vertical lines $\mathrm{Re}\,(z) =$ constant. In particular, such a circle or line is transformed into another by $g \in G$.

(3) A geodesic circle of center w, $\{z \in \mathbb{H} \mid d(z, w) = r\}$, is also a euclidean circle (with different center and radius).

(4) The distance (in the hyperbolic metric) is explicitly given by

$$d(z, w) = \log \frac{|z - \overline{w}| + |z - w|}{|z - \overline{w}| - |z - w|},$$

but it is actually often more convenient to work with the function $u(z, w)$ given by

$$u(z,w) = u(d(z,w)) = \frac{|z-w|^2}{4\mathrm{Im}\,(z)\mathrm{Im}\,(w)} \tag{15.5}$$

where $u : \mathbb{R}^+ \to \mathbb{R}^+$ is the function of the distance

$$u(d) = \frac{1}{2}(\cosh d - 1) = 2\Big(\sinh \frac{d}{2}\Big)^2. \tag{15.6}$$

REMARKS. (1) One has to be careful that the topology of $\mathbb{H}$ induced by the Riemannian metric is different from its (euclidean) topology as a subset of $\mathbb{C}$. For instance, $\mathbb{H}$ is complete with the Riemannian metric, but of course not with the euclidean one.

(2) Notice that the compact subgroup K is the stabilizer of $i \in \mathbb{H}$, hence the isomorphism $G/K \simeq \mathbb{H}$ described above is simply the map $g \mapsto gi$ on G.

Recall that the invariant measure associated to the Poincaré metric is

$$d\mu(z) = y^{-2}dxdy \tag{15.7}$$

(see (14.4)).

When we come to harmonic analysis on quotients of $\mathbb{H}$ by discrete subgroups, the congruence subgroup $\Gamma_0(q)\backslash\mathbb{H}$ in particular, see Section 14.1, thinking of the abelian case of $\mathbb{R}$ and its own discrete subgroups, many new phenomena occur. First, any discrete subgroup of $\mathbb{R}$ is necessarily of the form $a\mathbb{Z}$ with $a \in \mathbb{R}$, and they are therefore basically "all the same". This similarity allows us, for instance, to deduce quickly a Poisson summation formula for an arithmetic progression from the one for $\mathbb{Z}$ (see Section 4.3). On the other hand, there are very subtle questions concerning harmonic analysis on $\Gamma_0(q)\backslash\mathbb{H}$ (similar to their complex-analytic differences mentioned in Chapter 14).

15.2. The laplacian on $\mathbb{H}$.

Harmonic analysis on $\mathbb{H}$ is associated with the hyperbolic Laplace operator, which is given by

$$\Delta = -y^2\Big(\frac{\partial^2}{\partial x^2} + \frac{\partial^2}{\partial y^2}\Big). \tag{15.8}$$

This is a G-invariant differential operator, i.e. for any function f which is twice differentiable we have

$$\Delta(f \mid g) = \Delta f.$$

Here and hereafter the action on the right of G on functions is of course given by

$$(f \mid g)(z) = f(gz).$$

There are many ways to analyze Δ on $\mathbb{H}$, depending on which coordinates are chosen. Using rectangular coordinates $z = x + iy$ and working by separation of variables, one considers functions f on $\mathbb{H}$ which are periodic of period one in the variable x; then f has a Fourier expansion

$$f(z) = \sum_{n\in\mathbb{Z}} \hat{f}_n(y)e(nx) \tag{15.9}$$

with coefficients

$$\hat{f}_n(y) = \int_0^1 f(x+iy)e(-nx)dx. \tag{15.10}$$

Applying the Laplace operator one gets

$$\Delta f(z) = y^2 \sum_{n\in\mathbb{Z}} (4\pi^2 n^2 \hat{f}_n(y) - \hat{f}_n''(y))e(nx).$$

Of special concern are the eigenfunctions of Δ, the functions f such that $\Delta f = \lambda f$ for some $\lambda \in \mathbb{C}$. It is best to write the eigenvalue λ in terms of a parameter s such that $\lambda = s(1-s)$. Because Δ is an elliptic differential operator, any such f is automatically real-analytic on $\mathbb{H}$.

Continuing with a function f periodic in x, we see that it is an eigenfunction with eigenvalue λ if and only if $\hat{f}_n$ is a solution to the ordinary differential equation

$$f'' + \Big(\frac{\lambda}{y^2} - 4\pi^2 n^2\Big) f = 0.$$

For $n = 0$ all solutions are linear combinations of y^s and y^{1-s} (replace y^{1-s} by $y^{1/2}\log y$ if $s = 1/2$). Then for $n \geqslant 1$, two linearly independent solutions to the equation are given by standard Bessel functions

$$f_n(y) = 2y^{1/2} K_{s-1/2}(2\pi|n|y)$$

and

$$f_n(y) = 2y^{1/2} I_{s-1/2}(2\pi|n|y)$$

which are distinguished by their asymptotic behavior as $y \to +\infty$. This leads to

LEMMA 15.1. *Let $f : \mathbb{H} \to \mathbb{C}$ be an eigenfunction of Δ which is 1-periodic in x, with eigenvalue $\lambda = s(1-s)$, and suppose f satisfies the growth condition*

$$f(z) = o(e^{2\pi y}), \quad \textit{as } y \to +\infty.$$

Then there exist complex numbers a, b, and a_n for $n \neq 0$, such that

$$f(z) = ay^s + by^{1-s} + y^{1/2} \sum_{n\neq 0} a_n K_{s-1/2}(2\pi|n|y)e(nx)$$

(if $s = 1/2$, replace y^{1-s} by $y^{1/2}\log y$).

In particular, this explicit Fourier expansion implies that f is in fact of at most polynomial growth; precisely, we have

$$f(z) \ll y^\sigma + y^{1-\sigma}$$

for $y \to +\infty$, where $\sigma = \mathrm{Re}\,(s)$ as usual.

15.3. Automorphic functions and forms.

We now consider a discrete subgroup of G for which $\mathrm{Vol}\,(\Gamma\backslash\mathbb{H}) < +\infty$ (such subgroups are called Fuchsian groups of the first kind). Of particular importance are the Hecke groups $\Gamma_0(q)$ for $q \geqslant 1$.

We define different spaces of Γ-periodic functions on $\mathbb{H}$:

$$\mathcal{A}(\Gamma\backslash\mathbb{H}) = \{f : \mathbb{H} \to \mathbb{C} \mid f \mid \gamma = f, \quad \text{for all } \gamma \in \Gamma\}$$

(the space of automorphic functions),

$$L^2(\Gamma\backslash\mathbb{H}) = \{f \in \mathcal{A}(\Gamma\backslash\mathbb{H}) \mid \|f\| = \int_F |f(z)|^2 d\mu(z) < +\infty\}$$

(the space of square-integrable automorphic functions),

$$\mathcal{A}_s(\Gamma\backslash\mathbb{H}) = \{f \in \mathcal{A}(\Gamma\backslash\mathbb{H}) \mid \Delta f = s(1-s)f\}$$

(the space of automorphic forms, or Maass forms, with eigenvalue $\lambda = s(1-s)$).

Let $\mathfrak{a}$ be any cusp of Γ and $\sigma_{\mathfrak{a}}$ a scaling matrix as in (14.5). For any $f \in L^2(\Gamma\backslash\mathbb{H})$ the function $f_{\mathfrak{a}} = f \mid \sigma_{\mathfrak{a}}$ is thus 1-periodic in x, and has a Fourier expansion as in (15.9). We define further

$$L_0^2(\Gamma\backslash\mathbb{H}) = \{f \in L^2(\Gamma\backslash\mathbb{H}) \mid \hat{f}_{\mathfrak{a},0} = 0 \quad \text{for any cusp } \mathfrak{a}\}$$

(the space of cuspidal automorphic functions),

$$\mathcal{A}_s^0(\Gamma\backslash\mathbb{H}) = L_0^2(\Gamma\backslash\mathbb{H}) \cap \mathcal{A}_s(\Gamma\backslash\mathbb{H})$$

(the space of cusp forms with eigenvalue $s(1-s)$).

Note, in particular, that any automorphic form in $L^2(\Gamma\backslash\mathbb{H})$ (or with moderate growth) admits a Fourier expansion at the cusp ∞ as in Lemma 15.1.

Our first goal will be to describe how to decompose elements of $L^2(\Gamma\backslash\mathbb{H})$ in "spectral" terms, similar to the Fourier decomposition (series or integrals) on $\mathbb{R}/\mathbb{Z}$ or $\mathbb{R}$. This will be done in terms of the eigenvalues of the laplacian acting on automorphic functions.

We define the domain $\mathcal{D}$ of Δ to be

$$\mathcal{D} = \{f \in L^2(\Gamma\backslash\mathbb{H}) \mid f \text{ and } \Delta f \text{ are of class } C^\infty \text{ and bounded}\},$$

thereby turning Δ into an unbounded linear operator on $L^2(\Gamma\backslash\mathbb{H})$, defined on the dense domain $\mathcal{D}$. Using Stoke's formula, it is shown that Δ is symmetric and positive, and by general functional analysis, it therefore has a self-adjoint extension to all of $L^2(\Gamma\backslash\mathbb{H})$.

If the quotient $\Gamma\backslash\mathbb{H}$ were compact, then the general Hilbert space theory would imply that Δ has pure point spectrum, and that $L^2(\Gamma\backslash\mathbb{H})$ is spanned by eigenfunctions of Δ, as is the case with $\mathbb{R}/\mathbb{Z}$. Due to the lack of compactness, things are more complicated because a continuous spectrum exists but, on the other hand, the presence of the cusps gives rise to Fourier expansions, which are powerful tools.

15.4. The continuous spectrum.

We start the spectral decomposition by constructing a space of "obvious" automorphic functions, namely those constructed by the familiar averaging technique. More precisely, for a test function $\psi : \mathbb{R}^+ \to \mathbb{C}$ which is smooth with compact support and a cusp $\mathfrak{a}$, we define the incomplete Eisenstein series

$$E_{\mathfrak{a}}(z,\psi) = \sum_{\gamma \in \Gamma_{\mathfrak{a}}\backslash\Gamma} \psi(\operatorname{Im} \sigma_{\mathfrak{a}}^{-1}\gamma z) \tag{15.11}$$

which is obviously in $L^2(\Gamma\backslash\mathbb{H})$ since ψ has compact support. Of course, for the same reason, $E_{\mathfrak{a}}(\cdot,\psi)$ can never be an eigenfunction of Δ. However, using Mellin inversion, we find that

$$E_{\mathfrak{a}}(z,\psi)=\frac{1}{2\pi i}\int_{(\sigma)}E_{\mathfrak{a}}(z,s)\hat{\psi}(s)ds \tag{15.12}$$

where $\sigma>1$, $\hat{\psi}$ is the Mellin transform of ψ and $E_{\mathfrak{a}}(\cdot,s)$ is the Eisenstein series

$$E_{\mathfrak{a}}(z,s)=\sum_{\gamma\in\Gamma_{\mathfrak{a}}\backslash\Gamma}(\operatorname{Im}\sigma_{\mathfrak{a}}^{-1}\gamma z)^s$$

the series being absolutely convergent, uniformly on compact sets, for $\operatorname{Re}(s)>1$. Because $\Delta(\operatorname{Im}z)^s=s(1-s)\operatorname{Im}(z)$, the Eisenstein series $E_a(z,s)$ is an eigenfunction of Δ, but unfortunately it is not square integrable so it cannot be used directly for the spectral decomposition.

We define a new space $\mathcal{E}(\Gamma\backslash\mathbb{H})$ to be the closure in $L^2(\Gamma\backslash\mathbb{H})$ of the space spanned by the incomplete Eisenstein series. Clearly, Δ acts on $\mathcal{E}(\Gamma\backslash\mathbb{H})$. The solution to the spectral decomposition in this subspace lies in the meromorphic continuation of the Eisenstein series (with respect to the s-variable) to the whole complex plane, which was proved by Selberg. In the case of $\Gamma_0(q)$, the analytic continuation can be seen quite easily by computing explicitly its Fourier expansion at the cusp ∞.

Let $\mathfrak{a}$, $\mathfrak{b}$ be two (not necessarily distinct) cusps. Then $E_{\mathfrak{a}}(\cdot,s)$ has a Fourier expansion at the cusp $\mathfrak{b}$, which we write

$$E_{\mathfrak{a}}(\sigma_{\mathfrak{b}}z,s)=\delta_{\mathfrak{a},\mathfrak{b}}y^s+\varphi_{\mathfrak{a},\mathfrak{b}}y^{1-s}+y^{1/2}\sum_{n\neq 0}\varphi_{\mathfrak{a},\mathfrak{b}}(n,s)K_{s-1/2}(2\pi|n|y)e(nx).$$

The square matrix of constant terms $\Phi(s)=(\varphi_{\mathfrak{a},\mathfrak{b}}(s))_{\mathfrak{a},\mathfrak{b}}$ is called the scattering matrix of Γ.

In the case of $\Gamma=SL(2,\mathbb{Z})$, the Fourier coefficients are (we drop the subscripts $\mathfrak{a}$, $\mathfrak{b}$ since there is only one cusp)

$$\varphi(s)=\sqrt{\pi}\frac{\Gamma(s-1/2)}{\Gamma(s)}\frac{\zeta(2s-1)}{\zeta(2s)},$$

$$\varphi(n,s)=\pi^s\Gamma(s)^{-1}\zeta(s)^{-1}|n|^{-1/2}\sum_{ab=|n|}\left(\frac{a}{b}\right)^{s-1/2},$$

and the meromorphic continuation can indeed be seen from that of the Riemann zeta function $\zeta(s)$. In this case the full Fourier expansion of $E(z,s)$ is

$$E(z,s)=\theta(s)y^s+\theta(1-s)y^{1-s}+4\sqrt{y}\sum_{1}^{\infty}\tau_{s-1/2}(n)K_{s-1/2}(2\pi ny)\cos(2\pi nx). \tag{15.13}$$

For $\Gamma_0(q)$ in general, similar expressions hold in terms of Dirichlet L-functions.

Let $\operatorname{Re}(s)\geqslant 1/2$. It is easily seen that in this region $E(z,s)$ has only a simple pole at $s=1$ with constant residue V^{-1}, where $V=\operatorname{Vol}(\Gamma\backslash\mathbb{H})$. For general fuchsian groups of the first kind, there might be other poles in the half-plane $\operatorname{Re}(s)\geqslant 1/2$, although only a finite number, all in the interval $1/2<s\leqslant 1$, none on the line $\operatorname{Re}(s)=1/2$. All residues are in $L^2(\Gamma\backslash\mathbb{H})$ and one denotes $\mathcal{R}(\Gamma\backslash\mathbb{H})$ the subspace that they span; this is called the space of residual spectrum.

For $\Gamma_0(q)$, one further sees that $s = 1$ is the only pole of $E(z, s)$, with constant residue

$$V_q = \mathrm{Vol}\,(\Gamma_0(q)\backslash\mathbb{H}) = \frac{\pi}{3} q \prod_{p|q} (1 + p^{-1}).$$

Using the self-adjointness of Δ, one can prove the functional equation of the Eisenstein series

$$E(z, s) = \Phi(s)E(z, 1 - s)$$

where $E(z, s)$ is the column vector $(E_{\mathfrak{a}}(z, s))_{\mathfrak{a}}$; correspondingly, for the scattering matrix we have

$$\Phi(s)\Phi(1 - s) = \mathrm{Id}.$$

One shows that the orthogonal complement to $\mathcal{R}(\Gamma\backslash\mathbb{H})$ in $\mathcal{E}(\Gamma\backslash\mathbb{H})$ has continuous spectrum of multiplicity equal to the number of cusps, described by the eigenpacket of Eisenstein series for $\mathrm{Re}\,(s) = 1/2$. For $\Gamma_0(q)$, we therefore derive:

THEOREM 15.2. *Let $f \in \mathcal{E}(\Gamma_0(q)\backslash\mathbb{H})$. Then f has the expansion*

$$f(z) = \frac{1}{V_q} \int_F f(z) d\mu(z) + \sum_{\mathfrak{a}} \frac{1}{4\pi} \int_{\mathbb{R}} \langle f, E_{\mathfrak{a}}(\cdot, \tfrac{1}{2} + it)\rangle E_{\mathfrak{a}}(z, \tfrac{1}{2} + it) dt \tag{15.14}$$

where $V_q = \mathrm{Vol}\,(\Gamma_0(q)\backslash\mathbb{H})$. This formula is valid in L^2 sense, and the integrals are absolutely convergent point-wise and uniformly on compact sets if $f \in \mathcal{D}$.

The main point in the proof is that, while $E_{\mathfrak{a}}(\cdot, s)$ is not square-integrable (because of term $\delta_{\mathfrak{a},\mathfrak{b}} y^s + \varphi_{\mathfrak{a},\mathfrak{b}}(s) y^{1-s}$ in the Fourier expansion at the cusp $\mathfrak{b}$), it is almost so when $\mathrm{Re}\,(s) = 1/2$. Indeed, one can show that $E_{\mathfrak{a}}(\sigma_{\mathfrak{b}} z, s) \ll \sqrt{y}$ when $y \to \infty$, and that the extra integration in t in (15.14) saves a factor $\log y$, which is sufficient to turn the right-hand side into an L^2 function. On the other hand, one might be tempted to say that (15.12) on the line $\mathrm{Re}\,(s) = 1/2$ gives the spectral decomposition, but it is not true because the coefficient $\hat{\psi}(s)$ is not the inner product $\langle E_{\mathfrak{a}}(\cdot, \psi), E_{\mathfrak{a}}(\cdot, s)\rangle$ as it should be (since $\langle E_{\mathfrak{a}}(\cdot, s), E_{\mathfrak{a}}(\cdot, s)\rangle$ diverges, multiplying by the Eisenstein series and integrating is not permitted in (15.12). It is necessary to appeal to the functional equation to get the proper formula which (despite what (15.12) might suggest) shows that it is necessary to use the Eisenstein series at all cusps $\mathfrak{b}$ to get the expression of the incomplete Eisenstein series at the cusp $\mathfrak{a}$ (see [**I5**], 22.1-22.3).

In the case of a general subgroup Γ, the Eisenstein series remain in many respects rather mysterious objects, but they are comparatively less so for $\Gamma_0(q)$, where the explicit Fourier expansion yields crucial estimates in a straightforward way, with the consequence that, for the applications in this book, the continuous spectrum will be easily manageable.

15.5. The discrete spectrum.

We now turn our attention towards the complement of the space $\mathcal{E}(\Gamma\backslash\mathbb{H})$.

LEMMA 15.3. *The orthogonal complement of $\mathcal{E}(\Gamma\backslash\mathbb{H})$ is the space $L^2_0(\Gamma\backslash\mathbb{H})$ of cuspidal automorphic functions.*

This follows by a simple computation, which we write down completely to give a feeling of the whole theory: we simply compute the scalar product $\langle f, E_{\mathfrak{a}}(\cdot, \psi)\rangle$

of an L^2-automorphic function against an incomplete Eisenstein series

$$\begin{aligned}\langle f, E_{\mathfrak{a}}(\cdot,\psi)\rangle &= \int_F f(z)\overline{E}_{\mathfrak{a}}(z,\psi)d\mu(z)\\ &= \int_F f(z) \sum_{\gamma\in\Gamma_{\mathfrak{a}}\backslash\Gamma} \overline{\psi}(\operatorname{Im}\sigma_{\mathfrak{a}}^{-1}\gamma z)d\mu(z)\\ &= \sum_{\gamma\in\Gamma_{\mathfrak{a}}\backslash\Gamma}\int_F f(z)\overline{\psi}(\operatorname{Im}\sigma_{\mathfrak{a}}^{-1}\gamma z)d\mu(z)\end{aligned}$$

which by automorphy of f transforms to

$$\begin{aligned}\langle f, E_{\mathfrak{a}}(\cdot,\psi)\rangle &= \sum_{\gamma\in\Gamma_{\mathfrak{a}}\backslash\Gamma}\int_{\sigma_{\mathfrak{a}}^{-1}\gamma F} f(\sigma_{\mathfrak{a}} z)\overline{\psi}(z)d\mu(z)\\ &= \int_0^1\int_0^{+\infty} f(\sigma_{\mathfrak{a}} z)\overline{\psi}(z)y^{-2}dydx\end{aligned}$$

since the sets $\sigma_{\mathfrak{a}}^{-1}\gamma F$ are disjoint and make a partition of the strip $0<\operatorname{Re}(z)<1$ (see (14.5) or think of $\mathfrak{a}=\infty$).

This we can now write as

$$\langle f, E_{\mathfrak{a}}(\cdot,\psi)\rangle = \int_0^{+\infty} \hat{f}_{\mathfrak{a},0}(y)\overline{\psi}(y)y^{-2}dy \tag{15.15}$$

where $\hat{f}_{\mathfrak{a},0}(y)$ is the constant term in the Fourier expansion of f at $\mathfrak{a}$, and the lemma is now clear.

Again, Δ acts on the space $L_0^2(\Gamma\backslash\mathbb{H})$. But now one can show that the resolvent of the laplacian is a compact operator on $L_0^2(\Gamma\backslash\mathbb{H})$. From this it readily follows that Δ has pure point spectrum on $L_0^2(\Gamma\backslash\mathbb{H})$ and thus this space is spanned by eigenfunctions of Δ, which are called Maass cusp forms.

Because Δ is self-adjoint, the eigenvalues of Δ must also be non-negative, and indeed they are positive since the eigenvalue 0 corresponds to bounded harmonic functions on $\mathbb{H}$, which are necessarily constant. In terms of s this means that either $1/2<s\leqslant 1$, or $\operatorname{Re}(s)=1/2$. There can be only finitely many eigenvalues $\lambda=s(1-s)$ with $1/2<s\leqslant 1$, and they are called exceptional eigenvalues for Γ. Although it is fairly easy to construct fuchsian groups having arbitrarily many exceptional eigenvalues, Selberg conjectured that they do not exist for congruence (arithmetic) groups.

Selberg's Eigenvalue Conjecture. *For any $q\geqslant 1$, we have*

$$\lambda_1\geqslant\frac{1}{4}$$

where λ_1 is the smallest eigenvalue of Δ acting on $L_0^2(\Gamma_0(q)\backslash\mathbb{H})$.

Further, Selberg proved

Theorem 15.4. *For any $q\geqslant 1$, we have $\lambda_1\geqslant\frac{3}{16}$.*

This has been improved, as mentioned in Section 5.11, the best result to date being $\lambda_1\geqslant 975/4096$ by Kim and Sarnak [**KiS**].

As will appear clearly in Sections 15.8 and 15.9, exceptional eigenvalues are somewhat analogous to exceptional zeros of Dirichlet L-functions (see (5.51)), and too many of them can compromise analytic estimates. However, Theorem 15.4 is analogous to a zero-free strip and is thus much stronger than what is known in this case. The more fruitful analogy is with the Ramanujan-Petersson conjecture, and in representation theoretic terms there is a common formulation.

For general Γ, cusp forms are shrouded in mystery; indeed, it is by no means clear that they must exist (see [**PS**]), except in certain cases where this is ensured by symmetry (for instance, if the reflection $z \mapsto -\bar{z}$ descends to the quotient, odd functions must come from odd cusp forms since all Eisenstein series are even). For congruence subgroups, however, it was shown by Selberg using the trace formula that not only do cusp forms exist, but they are in a certain sense "dominating" over the Eisenstein series. Their presence is measured by the counting function

$$N_{\Gamma_0(q)}(T) = |\{j \mid |s_j| \leqslant T\}|$$

which is shown to satisfy the Weyl Law

$$N_{\Gamma_0(q)}(T) = \frac{\mathrm{Vol}\,(\Gamma_0(q)\backslash\mathbb{H})}{4\pi}T^2 + O(\sqrt{q}T\log qT)$$

for $T \geqslant 2$, where the constant is absolute. Hence there are many eigenvalues, on average about qT between T and $T+1$, but the proof is indirect and doesn't produce any of the corresponding cusp forms!

15.6. Spectral decomposition and automorphic kernels.

From the two previous sections we can deduce the full spectral decomposition theorem for the automorphic laplacian on $\Gamma_0(q)\backslash\mathbb{H}$.

THEOREM 15.5. *Let u_0, u_1, u_2, ... be an orthonormal basis of the residual and cuspidal spaces, i.e. u_0 = constant = $\mathrm{Vol}\,(\Gamma_0(q)\backslash\mathbb{H})^{-\frac{1}{2}} \in \mathcal{R}(\Gamma_0(q)\backslash\mathbb{H})$ with eigenvalue $\lambda_0 = 0$ and $u_j \in \mathcal{A}^0_{s_j}(\Gamma_0(q)\backslash\mathbb{H})$, with eigenvalue $\lambda_j = s_j(1-s_j)$ for $j = 1, 2, \ldots$. Then any $f \in L^2(\Gamma_0(q)\backslash\mathbb{H})$ has the spectral decomposition*

$$f(z) = \sum_{j\geqslant 0}\langle f, u_j\rangle u_j(z) + \sum_{\mathfrak{a}}\frac{1}{4\pi}\int_{\mathbb{R}}\langle f, E_{\mathfrak{a}}(\cdot, \tfrac{1}{2}+it)\rangle E_{\mathfrak{a}}(z, \tfrac{1}{2}+it)dt \tag{15.16}$$

valid in L^2 sense, and also converging absolutely and uniformly on compact sets if $f \in \mathcal{D}$. Moreover, the Parseval formula holds

$$\|f\|^2 = \sum_{j\geqslant 0}|\langle f, u_j\rangle|^2 + \sum_{\mathfrak{a}}\frac{1}{4\pi}\int_{\mathbb{R}}|\langle f, E_{\mathfrak{a}}(\cdot, \tfrac{1}{2}+it)\rangle|^2 dt. \tag{15.17}$$

We will now apply Theorem 15.5 to get a kind of Poisson summation formula for quotients of the hyperbolic plane. For this we consider a kernel function in two variables[1] $k : \mathbb{H}\times\mathbb{H} \to \mathbb{C}$, say continuous and compactly supported, which is a "point-pair invariant", i.e. $k(gz, gw) = k(z, w)$ for all $g \in G$. This means that $k(z, w)$ depends only on the distance $d(z, w)$. We express this property by saying that there exists another function in one variable $k : \mathbb{R}^+ \to \mathbb{C}$ such that $k(z, w) = k(u(z, w))$ where u is the function in (15.5).

[1] The second variable is useful because there is no privileged basepoint.

Associated to k is an invariant integral operator L acting on functions on $\mathbb{H}$ by

$$Lf(z) = \int_{\mathbb{H}} f(w)k(z,w)d\mu(w)$$

and, from the assumption that k depends only on the distance, it follows that L is G-invariant. In particular, L maps automorphic functions to automorphic functions, and indeed for $f \in \mathcal{A}(\Gamma_0(q)\backslash\mathbb{H})$ which is bounded we have another formula

$$Lf(z) = \int_F f(w)K(z,w)d\mu(w).$$

where the kernel K is defined by

$$K(z,w) = \sum_{\gamma\in\Gamma_0(q)/\{\pm 1\}} k(z,\gamma w). \tag{15.18}$$

The new kernel K is automorphic in both variables: indeed, in w by the averaging construction, and in z because k is symmetric.

Applying the spectral decomposition in the z-variable gives the expression

$$\begin{aligned} K(z,w) &= \sum_{j\geqslant 0} \langle K(\cdot,w), u_j\rangle u_j(z) \\ &+ \sum_{\mathfrak{a}} \frac{1}{4\pi}\int_{\mathbb{R}} \langle K(\cdot,w), E_{\mathfrak{a}}(\cdot,\tfrac{1}{2}+it)\rangle E_{\mathfrak{a}}(z,\tfrac{1}{2}+it)dt \end{aligned}$$

where the coefficients $\langle K(\cdot,w),u_j\rangle$ (respectively $\langle K(\cdot,w),E_{\mathfrak{a}}(\cdot,\frac{1}{2}+it)\rangle$) are again automorphic and can be also expanded similarly. However, the next lemma shows that they are simply proportional to the cusp forms u_j (respectively the Eisenstein series).

LEMMA 15.6. *Let $k : \mathbb{R}^+ \to \mathbb{C}$ be smooth and compactly supported, $k(z,w) = k(u(z,w))$ the associated kernel and L the integral operator as above. If $f : \mathbb{H} \to \mathbb{C}$ (not necessarily automorphic) is any eigenfunction of the laplacian with eigenvalue $\lambda = s(1-s)$, $s = 1/2+it$, $t \in \mathbb{C}$, then f is also an eigenfunction of the integral operator L, with an eigenvalue which depends on λ and L, but not on f. More precisely, we have*

$$Lf = \Lambda f$$

where $\Lambda = h(t)$, the transform $k \mapsto h$ being obtained by the following steps:

$$\begin{aligned} q(v) &= \int_v^{+\infty} k(u)(u-v)^{-1/2}du, \\ g(r) &= 2q\Big((\sinh\frac{r}{2})^2\Big), \\ h(t) &= \int_{\mathbb{R}} g(r)e^{irt}dt. \end{aligned}$$

For a proof, see e.g. [**I5, Th. 1.16**]. The transform above is called the Harish-Chandra/Selberg transform for $SL(2,\mathbb{R})$. It can be inverted by

$$\begin{aligned} g(r) &= \frac{1}{2\pi}\int_{\mathbb{R}} h(t)e^{-irt}dt, \\ q(v) &= \frac{1}{2}g\big(2\log(\sqrt{v}+\sqrt{v+1})\big), \\ k(u) &= -\frac{1}{\pi}\int_u^{+\infty}(v-u)^{-1/2}dq(v), \end{aligned}$$

sufficient conditions for h being

$$\begin{cases} h(t) = h(-t), \\ h \text{ is holomorphic in } |\mathrm{Im}\,(t)| \leqslant \frac{1}{2}+\delta, \\ h(t) \ll (|t|+1)^{-2-\delta} \end{cases} \tag{15.19}$$

for some $\delta > 0$. These conditions are in fact sufficient for the analysis of the automorphic kernel $K(z,w)$, i.e. the restriction that k has compact support can be relaxed.

We deduce from this the spectral decomposition which was our goal.

THEOREM 15.7. *Let h be a function satisfying the conditions (15.19), k the Harish-Chandra/Selberg transform of h and K the automorphic kernel on $\Gamma_0(q)$ associated to k. Then the following expansion holds*

$$\begin{aligned} K(z,w) = \sum_{\gamma\in\Gamma_0(q)/\{\pm 1\}} k(u(z,\gamma w)) &= \sum_{j\geqslant 0} h(t_j)u_j(z)\overline{u_j(w)} \\ &+ \sum_{\mathfrak{a}}\frac{1}{4\pi}\int_{\mathbb{R}} h(t)E_{\mathfrak{a}}(z,\tfrac{1}{2}+it)\overline{E_{\mathfrak{a}}(w,\tfrac{1}{2}+it)}dt \end{aligned}$$

where the convergence is absolute and uniform on compact sets.

It is important to have some control on the growth of the eigenvalues and eigenfunctions in this formula. The following is often sufficient for first estimations.

PROPOSITION 15.8. *Let $T \geqslant 1$ and $z \in \mathbb{H}$. We have*

$$\sum_{|t_j|<T}|u_j(z)|^2 + \sum_{\mathfrak{a}}\frac{1}{4\pi}\int_{-T}^{T}|E_{\mathfrak{a}}(z,\tfrac{1}{2}+it)|^2dt \ll T^2 + Ty(z)$$

where the implied constant depends on Γ alone and

$$y(z) = \max_{\mathfrak{a}}\max_{\gamma\in\Gamma}\mathrm{Im}\,(\sigma_{\mathfrak{a}}^{-1}\gamma z).$$

For the proof, see [**I5**], Section 7.2.

15.7. The Selberg trace formula.

In Section 15.6, we have obtained the spectral expansion of an automorphic kernel, and in doing so, we have seen that it is useful to consider the invariant integral operator L associated to it, namely

$$Lf(z) = \int_F f(w)K(z,w)d\mu(w)$$

where K is given by (15.18). Another formula, which is also a generalization of the Poisson summation formula, arises when one computes the trace of L, using the spectral expansion of K on the one hand, and its definition on the other hand: this is the celebrated Selberg trace formula. Of course, the correct definition of the trace of an operator on an infinite dimensional space is already a problem, but here we take a pragmatic approach and define the trace of L (or of K) to be the integral of the kernel on the diagonal (which is well-defined because our kernels are always at least continuous):

$$\operatorname{Tr} K = \int_F K(z,z)d\mu(z). \tag{15.20}$$

If there were no Eisenstein series in Theorem 15.7, the spectral expansion would yield immediately

$$\operatorname{Tr} K = \sum_{j\geqslant 0} h(t_j)$$

since (u_j) is an orthonormal basis, and we would obtain the pre-trace formula

$$\sum_j h(t_j) = \sum_{\gamma\in\Gamma}\int_F k(u(\gamma z,z))d\mu(z).$$

The right-hand side is still not very explicit. To go further, Selberg partitioned the sum over γ in such a way that the partial sums are computable. This is done by dividing the group into its conjugacy classes (that those are non-trivial is a feature of the non-commutativity of Γ). For one conjugacy class C, say of $\gamma\in\Gamma$, we have

$$C = \{\tau^{-1}\gamma\tau \mid \tau\in\Gamma\}$$

which is in bijection with $Z(\gamma)\backslash\Gamma$, where $Z(\gamma)$ is the centralizer of γ in Γ,

$$Z(\gamma) = \{\tau\in\gamma \mid \gamma\tau=\tau\gamma\}$$

so the contribution of the class C to the sum is equal to

$$\sum_{\tau\in Z(\gamma)\backslash\Gamma} k(u(\tau^{-1}\gamma\tau z,z)) = \sum_{\tau\in Z(\gamma)\backslash\Gamma} k(u(\gamma\tau z,\tau z))$$

which, by invariance, gives the contribution

$$\operatorname{Tr}_C K = \int_{Z(\gamma)\backslash\mathbb{H}} k(u(\gamma z,z))d\mu(z)$$

to the trace of K. The advantage we gain in performing this summation over $\gamma\in C$ is that the above expression for $\operatorname{Tr}_C K$ now depends on the conjugacy class of γ in $SL(2,\mathbb{R})$, not in Γ. Hence in further computations we can replace γ by its most standard representative $\tau^{-1}\gamma\tau$ for suitable τ in G. Such classes have geometric meaning, and the fact that γ is an element of Γ will translate this meaning into information about geometric data of the quotient space $\Gamma\backslash\mathbb{H}$.

We first describe the classification of the conjugacy classes in $G = SL(2,\mathbb{R})$ using the geometric action of G on $\mathbb{H}$, namely by considering the fixed points of the isometry of $\mathbb{H}$ induced by an element $g\in G$. This gives information about conjugacy classes, because if z is fixed by g, then τz is fixed by the conjugate $\tau^{-1}g\tau$ of g. Because the group is very symmetric, the fixed points will be enough to classify g up to conjugacy.

Let $g \in G$, $g = \begin{pmatrix} a & b \\ c & d \end{pmatrix}$, $g \neq \pm 1$, acting non-trivially on $\mathbb{H} \cup \mathbb{Q} \cup \infty$. The equation $gz = z$ is a quadratic equation $cz^2 + (d-a)z - b = 0$, with discriminant $\Delta = (d-a)^2 + 4bc = (\operatorname{Tr} g)^2 - 4$, and there are three cases.

• First case: $|\operatorname{Tr} g| > 2$ (g is called hyperbolic). Then $\Delta > 0$, and there are two distinct fixed points in $\mathbb{R}$ (the case $c = 0$, $a \neq 1$ also belongs to this case, but one of the fixed points is ∞). There exists $\tau \in G$ which sends one of the fixed points to 0 and the other to ∞; then by the observation above the conjugate $\tau^{-1} g \tau$ fixes 0 and ∞, and clearly (changing τ into τ^{-1} if necessary), we find that g is conjugate to a unique element of the form

$$a(p) = \pm \begin{pmatrix} p^{1/2} & 0 \\ 0 & p^{-1/2} \end{pmatrix} \tag{15.21}$$

for some $p > 1$, which acts geometrically as a dilation $z \mapsto pz$. We call p the norm of g and write $Ng = p$; the norm classifies exactly a hyperbolic conjugacy class. Notice that p can be recovered from g by solving the equation $z + z^{-1} = \operatorname{Tr} g$ and taking the square root of the solution which is larger than 1.

Associated to a hyperbolic motion is the geodesic γ_g in $\mathbb{H}$ which is the half-circle perpendicular to $\mathbb{R}$ at the two fixed points, or in the case $c = 0$, the vertical line from the real fixed point to ∞. This geodesic is preserved by g (this is obvious for $a(p)$, and follows by conjugation in the general case). Note that the norm of g can be recovered by $\log Ng = d(z, gz)$ for $z \in \gamma_g$ (g acts as a translation on the geodesic).

Conversely, to every geodesic γ in $\mathbb{H}$, one can associate the isotropy subgroup Γ_γ in G, containing elements which preserve γ. This is generated by ± 1, an element of order 2 which interchanges the end-points of γ in $\mathbb{R} \cup \{\infty\}$, and the subgroup of those elements fixing the end-points (which is isomorphic to the multiplicative group of positive real numbers, and so to $\mathbb{R}$ by the logarithm map). All these facts follow by conjugation to the case when the geodesic is the imaginary axis, and Γ_γ is generated by the involution $z \mapsto -z^{-1}$ and the group of elements $a(p)$ above.

Let again g be hyperbolic. The centralizer $Z(g)$ of g in G is a subgroup of Γ_{γ_g}; one checks that the involution is not in it, but the other subgroup is, so $Z(g)$ is isomorphic to $\mathbb{R}$ (times ± 1, of course).

• Second case: $|\operatorname{Tr} g| = 2$ (g is called a parabolic motion). Then $\Delta = 0$, so g has a unique fixed point $\mathfrak{a}$, $\mathfrak{a} \in \mathbb{R}$ if $c \neq 0$, and $\mathfrak{a} = \infty$ if $c = 0$. As before, some τ sends $\mathfrak{a}$ to ∞ and one sees this way that g is conjugate to a unique element

$$n(x) = \pm \begin{pmatrix} 1 & x \\ 0 & 1 \end{pmatrix} \tag{15.22}$$

for some $x \in \mathbb{R}$, which acts as a horizontal translation $z \mapsto z + x$. The parameter x classifies the parabolic conjugacy classes. The group $N = \{n(x) \mid x \in \mathbb{R}\}$ is the isotropy group of ∞ in G, and it is the centralizer in G of any of its elements. By conjugation, the centralizer of a parabolic element is thus isomorphic to $\mathbb{R}$.

• Third case: $|\operatorname{Tr} g| < 2$ (g is called an elliptic motion). Then $\Delta < 0$, so the quadratic equation has two conjugate complex roots, one of which, say z_g, is in $\mathbb{H}$. Moving z_g to i by an element $\tau \in G$, a short computation reveals that g is conjugate to a unique element

$$r(\theta) = \pm \begin{pmatrix} \cos\theta & \sin\theta \\ -\sin\theta & \cos\theta \end{pmatrix} \tag{15.23}$$

with $\theta \in \mathbb{R}/2\pi\mathbb{Z}$, which acts by hyperbolic rotation of angle θ around i. The group $K = \{r(\theta) \mid \theta \in \mathbb{R}/2\pi\mathbb{Z}\}$, isomorphic to $\mathbb{R}/2\pi\mathbb{Z}$, is the centralizer of any of its (non-trivial) elements, and the centralizer of any elliptic element is isomorphic to K.

Now consider a discrete subgroup Γ of G and an element $\gamma \in \Gamma$. Let $Z_\Gamma(\gamma)$ be the centralizer of γ in Γ; it is of course a subgroup of the centralizer $Z(\gamma)$ of γ in G, and it is a discrete subgroup. When γ is hyperbolic or parabolic and $Z(\gamma)$ is isomorphic to $\mathbb{R}$, this implies that $Z_\Gamma(\gamma)$ is infinite cyclic, and when γ is elliptic, this implies that $Z_\Gamma(\gamma)$ is finite cyclic. In all cases, we say that γ (or its conjugacy class in Γ) is primitive when γ is itself a generator of the centralizer $Z_\Gamma(\gamma)$. It is clear that any element γ is of the form $\pm\gamma_0^m$ for some integer $m \geqslant 0$ and a unique primitive element γ_0. By conjugating the generator γ_0 to one of the standard forms described above, we see that the action of $Z_\Gamma(\gamma)$ on $\mathbb{H}$ has the following simple fundamental domain, depending on the type of γ:

(1) If γ is hyperbolic, γ_0 is conjugate to (15.21) for $p = N\gamma_0$, acting by $z \mapsto pz$, and a fundamental domain is the horizontal strip

$$Z_\Gamma(\gamma)\backslash\mathbb{H} = \{z = x + iy \in \mathbb{H} \mid 1 < y < p\}.$$

(2) If γ is parabolic, γ_0 is conjugate to (15.22) for some x, and a fundamental domain is the vertical strip

$$Z_\Gamma(\gamma)\backslash\mathbb{H} = \{z = x + iy \mid 0 < x < 1\}.$$

(3) If γ is elliptic, γ_0 is conjugate to (15.23) for some θ. Let w be the fixed point of γ_0 (the center of the rotation); a fundamental domain for $Z_\Gamma(\gamma)$ is obtained by taking the domain between two geodesics going through w with an angle equal to θ (by discreteness, notice that θ must be of the form $2\pi/\ell$ for some integer $\ell \geqslant 1$).

In the special case of $\Gamma = SL(2,\mathbb{Z})$, the parabolic and elliptic conjugacy classes are very easy to describe. First, the fixed point of a parabolic element must be in $\mathbb{Q} \cup \{\infty\}$. But all such points are Γ-equivalent, by Bezout's theorem, and since $n_0 = n(1)$ clearly generates the isotropy group of ∞ in $SL(2,\mathbb{Z})$, we find that its conjugacy class is the only parabolic conjugacy class in $SL(2,\mathbb{Z})$. The centralizer $Z_\Gamma(n_0)$ is $\{n(x) \mid x \in \mathbb{Z}\}$, and a fundamental domain for $Z_\Gamma(n_0)$ is the vertical strip $\{z = x + iy \mid 0 < x < 1\}$.

By simple computations, one finds also that $SL(2,\mathbb{Z})$ contains only two primitive conjugacy classes of elliptic elements, with representatives $\left(\begin{smallmatrix} 0 & -1 \\ 1 & 0 \end{smallmatrix}\right)$, $\left(\begin{smallmatrix} 1 & 1 \\ -1 & 0 \end{smallmatrix}\right)$ of order 2 (as an isometry, but 4 as a matrix), fixing i and of order 3 (as an isometry, but 6 as a matrix), fixing j, respectively.

EXERCISE 1. Find the primitive parabolic conjugacy classes in $\Gamma_0(q)$, and show that they correspond to the set of cusps as described in Section 14.1.

We come back to the notation of the introduction to this section, with Γ a discrete subgroup such that the quotient $\Gamma\backslash\mathbb{H}$ has finite volume, and has cusps (finitely many). The lack of compactness creates some analytic difficulties, because it turns out that the kernel is not integrable on the diagonal, so the trace, as we have defined it, is infinite. This problem is solved by considering truncated fundamental domains $F(Y)$ which tend to F for $Y \to +\infty$, in the sense that $\int_{F(Y)} f(z)d\mu(z) \to \int_F f(z)d\mu(z)$ for any integrable f. The kernel K restricted to $F(Y)$ is now integrable on the diagonal, and one can compute the truncated trace

$$\operatorname{Tr}^Y K = \int_{F(Y)} K(z,z)d\mu(z)$$

using the spectral expansion for the restriction on the one hand, and the decomposition into conjugacy classes on the other hand. On the spectral side, this gives for $Y \to +\infty$ an expression of the form $\operatorname{Tr}^Y K = A\log Y + T_1 + o(1)$, and on the geometric side another expression $\operatorname{Tr}^Y K = A\log Y + T_2 + o(1)$, for the some $A > 0$. Hence we have the tautology $A = A$, and the non-trivial formula $T_1 = T_2$, which is the trace formula.

A small analytical miracle occurs when doing the exact computations: the complicated Harish-Chandra/Selberg transform disappears from the picture, and all that is left is the function h on one side, and its Fourier transform $\hat{h}$ on the other side.

We now give indications on the shape of the various parts, and state the complete formula in the special case $\Gamma = SL(2,\mathbb{Z})$. On the spectral side, the cusp forms and residual spectrum u_j contribute, as in the case of a compact quotient

$$\sum_j h(t_j) \tag{15.24}$$

and the Eisenstein series are shown to contribute

$$\frac{h(0)}{4}\operatorname{Tr}\Phi(\tfrac{1}{2}) - \frac{1}{4\pi}\int_{-\infty}^{+\infty} h(t)\frac{\varphi'}{\varphi}(\tfrac{1}{2}+it)dt \tag{15.25}$$

where $\Phi(s)$ is the scattering matrix formed by the first Fourier coefficients of the Eisenstein series and $\varphi(s) = \det\Phi(s)$ is its determinant (see Section 15.4).

The geometric computations are done independently for each type of (primitive) conjugacy classes. The class of the identity motion has a contribution given by

$$\frac{\operatorname{Vol}\Gamma\backslash\mathbb{H}}{4\pi}\int_{-\infty}^{+\infty} th(t)\tanh(\pi t)dt. \tag{15.26}$$

A hyperbolic conjugacy class P with associated primitive class P_0 contributes

$$\frac{1}{2\pi}\frac{\log NP_0}{NP^{1/2} - NP^{-1/2}}\hat{h}\Big(\frac{1}{2\pi}\log NP\Big). \tag{15.27}$$

A primitive elliptic class R of order m (as an isometry) contributes

$$\sum_{0<\ell<m}\Big(2m\sin\frac{\pi\ell}{m}\Big)^{-1}\int_{-\infty}^{+\infty}\frac{e^{-2\pi t\ell/m}}{1-e^{-2\pi t}}h(t)dt. \tag{15.28}$$

Finally, all primitive parabolic conjugacy classes contribute the same amount, which is

$$\frac{h(0)}{4} - \frac{1}{2\pi}\Big\{\hat{h}(0)\log 2 - \int_{-\infty}^{+\infty}\psi(1+it)h(t)dt\Big\} \tag{15.29}$$

where $\psi(s) = \frac{\Gamma'}{\Gamma}(s)$ is the logarithmic derivative of the Gamma function.

For $SL(2,\mathbb{Z})$, after some rearranging, and taking into account the description of the cusps and of the scattering matrix, we obtain

THEOREM 15.9. *Let h be a function satisfying the conditions (15.19), $\hat{h}$ its Fourier transform. Let $\varphi(s) = \theta(1-s)\theta(s)^{-1}$, where $\theta(s) = \pi^{-s}\Gamma(s)\zeta(2s)$, be the scattering matrix of $SL(2,\mathbb{Z})$. Then we have the formula*

$$
(15.30)\quad \sum_j h(t_j) + \frac{1}{4\pi}\int_{\mathbb{R}} \frac{-\varphi'}{\varphi}(\tfrac{1}{2}+it)h(t)dt =
$$

$$
\int_{\mathbb{R}} h(t)\Big(\frac{t}{12}\tanh(\pi t) - \frac{1}{2\pi}\log 2 - \frac{1}{2\pi}\psi(1+it)\Big)dt
$$

$$
+ \frac{1}{2\pi}\sum_{\substack{P\\ \text{hyperbolic}}} \frac{\log NP_0}{NP^{\frac{1}{2}} - NP^{-\frac{1}{2}}}\hat{h}\Big(\frac{1}{2\pi}\log NP\Big)
$$

$$
+ \sum_{\substack{R\\ \text{elliptic}}}\sum_{0<\ell<m} \Big(2m\sin\frac{\pi\ell}{m}\Big)^{-1}\int_{\mathbb{R}} \frac{e^{-2\pi t\ell/m}}{1-e^{-2\pi t}}h(t)dt.
$$

15.8. Hyperbolic lattice point problems.

One of the most direct applications of Theorem 15.7, similar in spirit to the examples given for the Poisson summation formula and the Voronoi formula in Chapter 4, is to the study of hyperbolic lattice point problems.

Let $D \subset \mathbb{H}$ be an open subset. The lattice point problem for D is the problem of counting the number of translates of a given $z_0 \in \mathbb{H}$ by elements of a discrete subgroup $\Gamma \subset G$ which are in D. The classical problems of Gauss and Dirichlet already mentioned in Chapter 4 are exactly analogue, taking the action of the lattice $\mathbb{Z}^2$ on $\mathbb{R}^2$ by translation and D the points in a circle or under an hyperbola. As in these cases, analytic number theory can have its say only if one considers a sequence of subsets getting larger, and sufficiently regular.

We take a geodesic ball of radius tending to infinity, in the more convenient form

$$
B(w,R) = \{z \in \mathbb{H} \mid 4u(w,z) + 2 \leqslant R\}
$$

and look for good asymptotics for the counting function

$$
N(R) = |\{\gamma \in \Gamma \mid \gamma w \in B(w,R)\}|.
$$

As in the euclidean case, the expected value of $N(R)$ is the area of the geodesic ball, but whereas Gauss's classical argument (packing with unit squares) establishes the corresponding result easily with a fairly good error term $R^{1/2}$ in the euclidean case, this simple argument is very inefficient in the context of hyperbolic geometry. Indeed, the error term is proportional to the length of the boundary of the disc, which is small in euclidean geometry, but is of the same order of magnitude as the area in the hyperbolic case. This fact follows from the isoperimetric inequality: if the boundary ∂D of D is smooth, $A = \mathrm{Vol}\,(D)$ and $L = \mathrm{length}(\partial A)$, then we have

$$
4\pi A + A^2 \leqslant L^2, \qquad \text{hence } L \geqslant A.
$$

One can at least prove elementary upper bounds which are quite useful (it is used in the proof of Proposition 15.8 and in Chapter 22).

PROPOSITION 15.10. *Let Γ be a discrete subgroup of $SL(2,\mathbb{R})$ such that the volume* $\mathrm{Vol}(\Gamma\backslash\mathbb{H})$ *is finite and let $\mathfrak{a}$ be a cusp of Γ.*

(1) *For any $z \in \mathbb{H}$ we have*

$$|\{\gamma \in \Gamma_{\mathfrak{a}}\backslash\Gamma \mid \mathrm{Im}\, \sigma_{\mathfrak{a}}^{-1}\gamma z > Y\}| < 1 + \frac{10}{c_{\mathfrak{a}}Y}.$$

(2) *For any z, $w \in \mathbb{H}$, and $\delta > 0$, we have*

$$|\{\gamma \in \sigma_{\mathfrak{a}}^{-1}\Gamma\sigma_{\mathfrak{a}} \mid u(\gamma z, w) < \delta\}| \ll \sqrt{\delta(\delta+1)}(\mathrm{Im}\,(w) + c_{\mathfrak{a}}^{-1}) + \frac{\delta+1}{c_{\mathfrak{a}}\mathrm{Im}\,(w)} + 1$$

with an absolute implied constant.

Here we let

$$c_{\mathfrak{a}} = \min\Big\{c > 0 \mid \begin{pmatrix} a & b \\ c & d \end{pmatrix} \in \sigma_{\mathfrak{a}}^{-1}\Gamma\sigma_{\mathfrak{a}} \textit{ for some } a,\ b,\ d\Big\}.$$

See [**I5**], Lemma 2.11 for the proof.

Applying Theorem 15.7 for suitable kernels, one can get an asymptotic formula.

THEOREM 15.11. *Let Γ be a discrete subgroup of $SL(2,\mathbb{R})$ with* $\mathrm{Vol}(\Gamma\backslash\mathbb{H}) < +\infty$. *Let $w \in \mathbb{H}$. We have*

$$N(R) = \frac{2\pi R}{\mathrm{Vol}\,(\Gamma\backslash\mathbb{H})} + 2\sqrt{\pi}\sum_{\frac{1}{2}<s_j<1} \frac{\Gamma(s_j - \frac{1}{2})}{\Gamma(s_j+1)}|u_j(w)|^2 R^{s_j} + O(R^{2/3})$$

with an implied constant depending on Γ and z only.

SKETCH OF PROOF. We consider only $\Gamma = \Gamma_0(q)$ as before. If $k(u) \geqslant 0$ is any function with $k(u) \geqslant 1$ for $u \leqslant (R-2)/4$, and if the corresponding automorphic kernel $K(z,w)$ satisfies the assumptions of Theorem 15.7, we have by positivity

$$N(R) \leqslant \sum_{\gamma\in\bar{\Gamma}} k(u(w,\gamma w)) = 2K(w,w). \tag{15.31}$$

By Theorem 15.7, we have

$$K(w,w) = \sum_{j\geqslant 0} h(t_j)|u_j(w)|^2 + \sum_{\mathfrak{a}} \frac{1}{4\pi}\int_{\mathbb{R}} h(t)|E_{\mathfrak{a}}(w, \tfrac{1}{2}+it)|^2 dt \tag{15.32}$$

with notation as in the theorem. We take $k(u) = 1$ for $u \leqslant (R-2)/4$ and $k(u) = 0$ for $u \geqslant (R+S-2)/4$, and k linear between.

We estimate $h(t)$ using Lemma 15.6, for a suitable function f on $\mathbb{H}$ which is eigenvalue of Δ with $\lambda = \frac{1}{4} + t^2$. For example, $f(z) = y^s$ yields

$$h(t) = \int_{\mathbb{H}} k(i,z) y^s d\mu(z).$$

One finds first that

$$h(t) = \sqrt{\pi}\frac{\Gamma(s-\frac{1}{2})}{\Gamma(s+1)} R^s + O(R^{1/2} + S)$$

for $\frac{1}{2} < s \leqslant 1$, where the implied constant depends on s. The constant eigenvalue for $j = 0$ yields the main term $\pi R(\mathrm{Vol}\,(\Gamma\backslash\mathbb{H}))^{-1}$, and the exceptional eigenvalues (if any) with $0 < \lambda_j < \frac{1}{4}$ yield the second term, up to the error term $O(R^{1/2} + S)$.

Let $T = RS^{-1}$. For $\mathrm{Re}\,(s) = \frac{1}{2}$, integrating by parts yields

$$h(t) \ll |s|^{-5/2} R^{1/2}(\min(|s|, T) + \log R),$$

with an absolute implied constant. Hence we estimate the Eisenstein series contribution to (15.32) at a given cusp $\mathfrak{a}$ by

$$\begin{aligned}\int_{\mathbb{R}} h(t)|E_{\mathfrak{a}}(w, \tfrac{1}{2} + it)|^2 dt &\ll R^{1/2} \int_{-\sqrt{T}}^{\sqrt{T}} ((1+t)^2 + \log R) \frac{|E_{\mathfrak{a}}(w, \frac{1}{2} + it)|^2}{(1+t)^5} dt \\ &\quad + R^{1/2} \int_{|t|>\sqrt{T}} (T + \log R) \frac{|E_{\mathfrak{a}}(w, \frac{1}{2} + it)|^2}{(1+t)^5} dt \\ &\ll (RT)^{1/2}\end{aligned}$$

by integration by parts using Proposition 15.8. The same bound holds for the contribution of the non-exceptional discrete spectrum by partial summation from Proposition 15.8. Using these estimates with $S = R^{2/3}$, (15.31) and (15.32) give

$$N(R) \leqslant \frac{2\pi R}{\mathrm{Vol}\,(\Gamma\backslash\mathbb{H})} + 2\sqrt{\pi} \sum_{\frac{1}{2} < s_j < 1} \frac{\Gamma(s_j - \frac{1}{2})}{\Gamma(s_j + 1)} |u_j(w)|^2 R^{s_j} + O(R^{2/3}).$$

Similarly with $k(u) = 1$ for $u \leqslant (R - S - 2)/4$ and $k(u) = 0$ for $u \geqslant (R-2)/4$, we get the lower bound and Theorem 15.11 is proved. □

REMARK. Note that for $\Gamma_0(q)$ we have $\lambda_1 \geqslant \frac{2}{9}$ so $s_j < \frac{2}{3}$ by (5.88), so the contribution of exceptional eigenvalues is smaller than the error term.

COROLLARY 15.12. *(1) Let $N_1(X)$ be the number of integer solutions to the equation $ad - bc = 1$ with $a^2 + b^2 + c^2 + d^2 \leqslant X$. We have*

$$N_1(X) = 6X + O(X^{2/3}).$$

(2) Let $r(n)$ be the number of representations of n as a sum of two squares. We have

$$\sum_{n \leqslant X} r(n)r(n+1) = 8X + O(X^{2/3}).$$

PROOF. Note first that for $w = i$, we have $4u(\gamma i, i) + 2 = a^2 + b^2 + c^2 + d^2$ for any $\gamma = \begin{pmatrix} a & b \\ c & d \end{pmatrix}$. So (1) follows immediately from Theorem 15.11 for $\Gamma = SL(2, \mathbb{Z})$ and $w = i$, since $\mathrm{Vol}\,(SL(2,\mathbb{Z})\backslash\mathbb{H}) = 3\pi^{-1}$ and $\lambda_1 = 91.14\ldots > \frac{1}{4}$ for the modular group.

For (2), consider the group

$$\Gamma = \left\{ \begin{pmatrix} a & b \\ c & d \end{pmatrix} \in SL(2, \mathbb{Z}) \mid \begin{matrix} a \equiv d \pmod 2 \\ c \equiv b \pmod 2 \end{matrix} \right\} = \begin{pmatrix} 0 & -1 \\ 1 & 1 \end{pmatrix}^{-1} \Gamma_0(2) \begin{pmatrix} 0 & -1 \\ 1 & 1 \end{pmatrix}.$$

We have $a + d = 2k$, $a - d = 2\ell$, $b + c = 2m$, $b - c = 2n$, with $ad - bc = k^2 - \ell^2 - m^2 + n^2 = 1$ and $a^2 + b^2 + c^2 + d^2 = 2(k^2 + \ell^2 + m^2 + n^2)$, so Theorem 15.11 gives

$$\sum_{m \leqslant X} r(m)r(m+1) = N(4X + 2) = 8X + O(X^{2/3}).$$

□

15.9. Distribution of length of closed geodesics and class numbers.

The simplest application of the Selberg Trace Formula is to the distribution of the lengths of closed geodesics on the quotient space $X = \Gamma\backslash\mathbb{H}$, also called the length spectrum. In fact, closed geodesics are in one-to-one correspondence with the hyperbolic conjugacy classes of Γ. Indeed, if $g \in \Gamma$ is an hyperbolic element, then g maps the geodesic γ (half-circle) joining its two fixed-points to itself. In particular, for any $z \in \gamma$, the geodesic segment from z to gz is mapped in $\Gamma\backslash\mathbb{H}$ to a closed geodesic. Taking the model situation where $g = a(p)$ with $p > 1$ as in (15.21) and $z = i$, observe that the length of this closed geodesic is $d(i, g(i)) = \log p = \log Ng$. Conjugates of p, which have the same fixed points, give rise to the same closed geodesic, and conversely any closed geodesic is obtained in this way.

THEOREM 15.13. *Let Γ be a subgroup of $SL(2,\mathbb{R})$ with* $\mathrm{Vol}\,(\Gamma\backslash\mathbb{H}) < +\infty$. *We have*

$$\sum_{NP\leqslant X} \log NP = X + \sum_{\frac{1}{2}<s_j<1} s_j^{-1}X^{s_j} + O(X^{3/4})$$

where the sum is over primitive hyperbolic conjugacy classes and the implied constant depends on Γ.

REMARKS. (1) For Γ with compact quotient, this was proved by Huber [**Hub**].

(2) The exponent can be improved, see e.g. Luo, Rudnick and Sarnak [**LRS**], where $3/4$ is replaced by $7/10 + \varepsilon$ for any $\varepsilon > 0$. Although the analogy with the Prime Number Theorem suggests that it should be $X^{1/2+\varepsilon}$, and the analogue of the Riemann Hypothesis holds, this is yet unknown. The main difficulty is that there are many more eigenvalues than zeros of the zeta function.

Also, as in Theorem 15.11, if $\Gamma = \Gamma_0(q)$, Selberg's inequality $\lambda_1 \geqslant \frac{3}{16}$ means that the exceptional eigenvalues have smaller contribution than the error term.

SKETCH OF PROOF. The argument is quite similar in principle to that of the previous theorem. We choose h such that the Fourier transform $\hat{h}$ localizes the sum over hyperbolic conjugacy classes to those with $NP \leqslant X + Y$ for some parameter $Y \geqslant 1$, say $\hat{h}(t) = 2\cosh(\pi t)q(2\pi t)$ where q is a smooth function, even, supported on $|t| \leqslant \log(X + Y)$ and with $0 \leqslant q \leqslant 1$ and $q = 1$ on $[-\log X, \log X]$. The contribution of the discrete and residual spectrum to the spectral trace is

$$\sum_j h(t_j) = X + \sum_{\frac{1}{2}<s_j<1} s_j^{-1}X^{s_j} + O(Y + X^{1/2}T)$$

with $T = XY^{-1}$, and the same estimate holds for Eisenstein series (to see this, estimate h by Fourier inversion on $\frac{1}{2} < s \leqslant 1$ and on $\mathrm{Re}\,(s) = 1$ separately). The identity, elliptic and parabolic motions have contributions $\ll X^{1/2}T$ by simple estimations, so one gets

$$\sum_P q(\log NP)(\log NP) = X + \sum_{\frac{1}{2}<s_j<1} s_j^{-1}X^{s_j} + O(Y + X^{1/2}T).$$

Subtracting this applied at $X + Y$ and X, by positivity, we see that

$$\sum_{X<NP<X+Y} (\log NP) \ll Y + X^{1/2}T$$

hence the result by choosing $Y = X^{3/4}$. □

In [**Sa2**], P. Sarnak deduces the following corollary.

THEOREM 15.14. *For any integer $d \geqslant 1$ such that $d \equiv 0,\ 1 \pmod 4$, let $h(d)$ be the class number of $\mathbb{Q}(\sqrt{d})$ and ε_d be the fundamental unit. We have*

$$\sum_{\varepsilon_d \leqslant X} h(d) = \mathrm{Li}(X^2) + O(X^{3/2}(\log X)^2)$$

for $X \geqslant 2$, with an absolute implied constant.

PROOF. The proof depends on the following map between indefinite quadratic forms and closed geodesics. Let $Q(x, y) = ax^2 + bxy + cy^2$ be a primitive integral indefinite quadratic form of discriminant $d > 0$, i.e. $(a, b, c) = 1$ and $d = b^2 - 4ac > 0$. By linear change of variable, $SL(2, \mathbb{Z})$ acts on such forms and the number of equivalence classes is $h(d)$.

The equations $Q(\theta, 1) = 0$ has two real roots, $\theta_1 = (-b + \sqrt{d})/2$ and $\theta_2 = (-b - \sqrt{d})/2$, to which is associated the geodesic in $\mathbb{H}$ joining θ_1 and θ_2. The stabilizer of Q under the action of $SL(2, \mathbb{Z})$ is equal to the stabilizer of θ_1 (or θ_2), and a generator is given explicitly by

$$g(Q) = \begin{pmatrix} (t_0 - bu_0)/2 & -cu_0 \\ au_0 & (t_0 + bu_0)/2 \end{pmatrix}$$

where $\varepsilon_d = (t_0 + u_0\sqrt{d})/2$ (in other words, (t_0, u_0) is the fundamental solution to the equation $t_0^2 - du_0^2 = 4$). The norm of this hyperbolic element is then ε_d^2.

The map $Q \mapsto g$ sends primitive integral quadratic forms to hyperbolic conjugacy classes of $SL(2, \mathbb{Z})$. We claim it induces a bijection between classes of forms and primitive hyperbolic conjugacy classes.

To see this, let g be any primitive hyperbolic element of $SL(2, \mathbb{Z})$. Let θ_1 and θ_2 be the fixed points of g and $K = \mathbb{Q}(\theta_1, \theta_2)$. Clearly K is a real quadratic field, say $K = \mathbb{Q}(\sqrt{d})$ for a unique $d > 0$, $d \equiv 0, 1 \pmod 4$. Then the stabilizer of θ_1 is generated by g because g is primitive. The form $Q(x, y) = (x - \theta_1 y)(x - \theta_2 y)$ is a primitive quadratic form with $g(Q) = g$. This implies surjectivity of the map and injectivity is easy to check.

Hence the norms of conjugacy classes of primitive hyperbolic elements are ε_d^2 with multiplicity $h(d)$. In this case Theorem 15.13 becomes

$$\sum_{\varepsilon_d \leqslant \sqrt{X}} 2h(d) \log \varepsilon_d = X + O(X^{3/4}), \tag{15.33}$$

hence the result after summation by parts. □

REMARK. Siegel [**Sie2**] has shown that

$$\sum_{d \leqslant X} h(d) \log \varepsilon_d = \frac{\pi^2 X^{3/2}}{18\zeta(3)} + O(X \log X) \tag{15.34}$$

for $X \geqslant 2$. Notice the difference between the ordering of the discriminants in (15.33) and (15.34). An asymptotic formula for $h(d)$ over $d \leqslant X$ is not known.

EXERCISE 2. Prove (15.34) using the Dirichlet Class Number Formula (2.31).

CHAPTER 16

SUMS OF KLOOSTERMAN SUMS

16.1. Introduction.

Individual Kloosterman sums $S(m,n;c)$ are quite well examined by algebraic methods (see Chapter 11). The result is Weil's bound

$$|S(m,n;c)| \leqslant (m,n,c)^{\frac{1}{2}} c^{\frac{1}{2}} \tau(c) \tag{16.1}$$

which is the best possible. This and earlier estimates have been used in analytic number theory indirectly in many ingenious ways (see for instance [**Li2**], [**At**], [**I10**], [**Wi3**]). Yu. V. Linnik [**Li3**] pointed out that for certain applications one actually needs estimates for sums of type

$$S_{mn}(X) = \sum_{\substack{c\leqslant X\\ c\equiv 0 (\mathrm{mod}\ q)}} c^{-1} S(m,n;c). \tag{16.2}$$

The upper bound (16.1) implies

$$S_{mn}(X) \ll \tau((m,n)q)(m,n,q)^{\frac{1}{2}} X^{\frac{1}{2}} \log X \tag{16.3}$$

for $X \geqslant 2$ where the implied constant is absolute. However, Linnik suggested that a smaller bound should be true, if X is sufficiently large, because of cancellation of Kloosterman sums, which is due to a regular change of sign. Linnik conjectured that

$$S_{mn}(X) \ll X^{\varepsilon}, \tag{16.4}$$

where the implied constant depends on m, n and ε, and he also provided some heuristics to support this bound.

In his investigations Linnik was driven by an analogy of the zeta function

$$Z_{mn}(s) = \sum_{c\equiv 0 (\mathrm{mod}\ q)} c^{-2s} S(m,n;c) \tag{16.5}$$

to the hypothesis of Hasse on the L-functions of elliptic curves (to be precise Linnik spoke only for $q=1$ and used a different notation). This analogy turns out to be close only in the analytic aspects of the series (16.5) (like analytic continuation, but not the Euler product). By independent analysis A. Selberg [**S7**] revealed that $Z_{mn}(s)$ admits a spectral decomposition with respect to automorphic forms (for the group $\Gamma_0(q)$). Selberg led the foundation for the spectral theory of Kloosterman sums yet he did not deliver the necessary technical tools to cultivate this fertile soil.

In this chapter we sketch proofs of central results of this profound theory. There are solid presentations of selected topics in the articles [**I3**], [**I13**], [**DI**] and in the books [**I5**], [**Sa3**], [**Mot1**], however, a conclusive exposition is yet to be written.

16.2. Fourier expansion of Poincaré series.

There are various methods to develop the spectral decompositions of sums of Kloosterman sums. Arguably the best structural treatments appeal directly to the Green function (see [**I5**]), or the resolvent operator (see [**GS**]). Nevertheless, we follow the traditional method which begins by dual treatments of the Poincaré series (for the group $\Gamma = \Gamma_0(q)$)

$$P_m(z) = \sum_{\gamma \in \Gamma_\infty \backslash \Gamma} F(m\gamma z), \qquad m = 1, 2, 3, \dots . \tag{16.6}$$

Here $F(z)$ is a function on $\mathbb{H}$ which is periodic in x of period one and it satisfies $F(z) \ll y^\sigma$ with some $\sigma > 1$ so the series (16.6) converges absolutely ($\sigma = 1$ will be attained by continuity). Our treatments of $P_m(z)$ are parallel to those already given for the holomorphic forms in Section 14.2, and our goal is the Kuznetsov formula (16.34) which is an analog of the Petersson formula (14.15). The extra care is needed to justify infinite summations over the spectrum of the Laplace operator. To this end one needs some crude estimates for the Fourier coefficients of cusp forms in the spectral parameters. We shall prove much stronger estimates than necessary in Section 16.5, but only after the final Kuznetsov formulas are established. However, one could derive the required auxiliary estimates from the early forms of Kuznetsov's formulas. We encourage the reader to do so and to organize our arguments into a correct logical order (we compromised the order, but not the completeness, for reducing the length of presentation).

Applying the double coset decomposition (14.14) and then the Poisson summation formula as in the proof of Lemma 14.2 we arrive at the Fourier series for $P_m(z)$,

$$P_m(z) = F(mz) + \sum_{q|c} \sum_{ad \equiv 1 (\text{mod } c)} \sum_n e(nz) \int_{\text{Im } (\xi) = y} F\Big(m \begin{pmatrix} a & \star \\ c & d \end{pmatrix} \xi\Big) e(-n\xi) d\xi.$$

Now assume that the generating function $F(z)$ satisfies the rule $F(z+t) = e(t)f(z)$ for $t \in \mathbb{R}$ so by the action of Γ on $\mathbb{H}$ we have

$$F\Big(m \begin{pmatrix} a & \star \\ c & d \end{pmatrix} \xi\Big) = e\Big(\frac{am}{c}\Big) F(-m/c(c\xi + d)).$$

From this action one gets the Kloosterman sums in the Fourier series

$$P_m(z) = F(mz) + \sum_n \sum_{c \equiv 0 (\text{mod } q)} S(m, n; c) F_c(m, n; y) e(nz) \tag{16.7}$$

where

$$F_c(m, n; y) = \int_{\text{Im } \xi = y} F(-m/c^2\xi) e(-n\xi) d\xi$$

by shifting ξ to $\xi - d/c$. We put

$$F(z) = p(4\pi y) e(z) \tag{16.8}$$

and call $p(y)$ the test function for the corresponding $P_m(z)$, getting

$$F_c(m,n;y)=\int_{\mathrm{Im}\ \xi=y} p\Big(\frac{4\pi my}{c^2|\xi|^2}\Big)e\Big(-\frac{m}{c^2\xi}-n\xi\Big)d\xi. \tag{16.9}$$

In this generality we cannot compute the integral (16.9). An interesting choice for the test function is

$$p(y)=y^s \tag{16.10}$$

where s is a complex parameter at our disposal with $\mathrm{Re}\ s>1$ (there are other eigenfunctions of Δ which lead also to interesting results). In this case the Poincaré series becomes

$$P_m(z,s)=\sum_{\gamma\in\Gamma_\infty\backslash\Gamma}(4\pi m\mathrm{Im}\ \gamma z)^s e(m\gamma z) \tag{16.11}$$

and its Fourier expansion becomes

$$P_m(z,s)=(4\pi m)^s\{e(mz)+\sum_n Z_{mn}(s;y)e(nz)\} \tag{16.12}$$

where

$$Z_{mn}(s;y)=\sum_{c\equiv 0(\mathrm{mod}\ q)} c^{-2s}S(m,n;c)I_s(m,n;y)$$

and

$$I_s(m,n;y)=\int_{\mathrm{Im}\xi=y}|\xi|^{-2s}e(-n\xi-m/\xi c^2)d\xi.$$

The general Poincaré series $P_m(z)$ can be represented by $P_m(z,s)$ by applying the Mellin transform to $p(y)$.

The particular Poincaré series $P_m(z,s)$ was introduced by A. Selberg [**S7**]. This is nearly a cusp form, but not exactly so, because it fails to be an eigenfunction of Δ. Precisely we have

$$(\Delta+s(1-s))P_m(z,s)=sP_m(z,s+1), \tag{16.13}$$

because $(\Delta+s(1-s))F(z)=4\pi syF(z)$ for $F(z)=y^s e(z)$. Hence we get the recurrence formula

$$P_m(z,s)=sR_\lambda P_m(z,s+1) \tag{16.14}$$

where $R_\lambda=(\Delta+\lambda)^{-1}$ is the resolvent of Δ at $\lambda=s(1-s)$. By the spectral theory R_λ is meromorphic, therefore $P_m(z,s)$ can be continued meromorphically to the whole complex s-plane. Moreover, in the half-plane $\mathrm{Re}\,(s)>\frac{1}{2}$ the poles of R_λ are at the points $\lambda_j=s_j(1-s_j)>0$ in the discrete spectrum of Δ, and they are simple, so are the poles of $P_m(z,s)$ at $s=s_j$. In fact it is conjectured by Selberg that $\lambda_j\geqslant\frac{1}{4}$, so $s_j=\frac{1}{2}+it_j$ with t_j real and $P_m(z,s)$ is holomorphic in $\mathrm{Re}\,(s)>\frac{1}{2}$.

In what follows we avoid the analytic continuation of the Selberg-Poincaré series and the resolvent operator altogether. Instead, we rely heavily on the spectral decomposition of automorphic functions.

16.3. The projection of Poincaré series on Maass forms.

We return to the Poincaré series $P_m(z)$ for a generating function of type (16.8). Let $f(z)$ be a Maass form of eigenvalue $\lambda = \frac{1}{4} + r^2$ given by the Fourier series

$$f(z) = ay^{\frac{1}{2}+ir} + by^{\frac{1}{2}-ir} + y^{\frac{1}{2}} \sum_{n \neq 0} a_n K_{ir}(2\pi|n|y)e(nx) \tag{16.15}$$

(see Lemma 15.1). By the unfolding method we derive

$$\langle f, P_m \rangle = \int_0^\infty \int_0^1 f(z)\bar{F}(mz) d\mu z.$$

Inserting (16.15) and (16.8) we pick up the term a_m by integration in $0 < x < 1$. Then, changing the variable $y \to y/2\pi m$ we obtain the formula

$$\langle f, P_m \rangle = (2\pi m)^{\frac{1}{2}} a_m \int_0^\infty e^{-y} K_{ir}(y) \bar{p}(2y) y^{-\frac{3}{2}} dy. \tag{16.16}$$

Hence observe that $P_m(z)$ is orthogonal to constant functions. In particular for $p(y) = y^s$ the above integral is a product of gamma functions giving

$$\langle f, P_m \rangle = 2\pi\Gamma(s - \tfrac{1}{2} + r)\Gamma(s - \tfrac{1}{2} - ir)\Gamma(s)^{-1} m^{\frac{1}{2}} a_m. \tag{16.17}$$

This is clearly an analog of the formula for the inner product of a holomorphic modular form with a holomorphic Poincaré series given in Lemma 14.3.

16.4. Kuznetsov's formulas.

Now we are ready to derive explicit relations between Kloosterman sums and Fourier coefficients of automorphic forms. To this end we compute in two ways the inner product $\langle P_m, Q_n \rangle$ of two Poincaré series $P_m(z)$ and $Q_n(z)$ generated by $F(z) = (4\pi y)^{\frac{1}{2}} p(4\pi y) e(z)$ and $G(z) = (4\pi y)^{\frac{1}{2}} q(4\pi y) e(z)$ respectively (note we modified the test functions $p(y), q(y)$ by the factor $y^{\frac{1}{2}}$).

First by the spectral decomposition (see Theorem 15.2) we get

(16.18)

$$\begin{aligned}\langle P_m, Q_n \rangle = &\sum_{j=1}^{\infty} \langle P_m, u_j \rangle \langle u_j, Q_n \rangle \\ &+ \sum_{\infty} \frac{1}{4\pi} \int_{-\infty}^{\infty} \langle P_m, E_\infty(\star, \tfrac{1}{2} + ir) \rangle \langle E_\infty(\star, \tfrac{1}{2} + ir), Q_n \rangle dr.\end{aligned}$$

where $(u_j(z))$ is an orthonormal basis of cusp forms and $(E_\infty(z, \frac{1}{2} + ir); r \in \mathbb{R})$ is the eigenpacket of Eisenstein series associated with a system of inequivalent cusps. Suppose $u_j(z)$ has the Fourier expansion

$$u_j(z) = \rho_j(0) y^{s_j} + y^{\frac{1}{2}} \sum_{n \neq 0} \rho_j(n) K_{it_j}(2\pi|n|y) e(nx) \tag{16.19}$$

where $s_j = \frac{1}{2} + it_j$ with t_j real, or $\frac{1}{2} < s_j < 1$ because $\lambda_j = s_j(1 - s_j) = \frac{1}{4} + t_j^2 > 0$. Then

$$\langle u_j, Q_n \rangle = (4\pi n)^{\frac{1}{2}} \rho_j(n) \bar{\omega}_q(t_j) \tag{16.20}$$

where

$$\omega_q(t) = \int_0^\infty e^{-y} K_{it}(y) q(2y) y^{-1} dy \tag{16.21}$$

by the formula (16.16). Similarly for the Eisenstein series

$$E_\mathfrak{a}(z, \tfrac{1}{2} + ir) = \delta_\mathfrak{a} y^{\frac{1}{2}+ir} + \varphi_\mathfrak{a}(\tfrac{1}{2} + ir) y^{\frac{1}{2}-ir} + y^{\frac{1}{2}} \sum_{n \neq 0} \tau_\mathfrak{a}(n, r) K_{ir}(2\pi|n|y) e(nx) \tag{16.22}$$

we have

$$\langle E_\mathfrak{a}(\star, \tfrac{1}{2} + ir), Q_n \rangle = (4\pi n)^{\frac{1}{2}} \tau_\mathfrak{a}(n, r) \omega_q(r). \tag{16.23}$$

Inserting (16.19) and (16.23) into (16.18) we obtain

$$\langle P_m, Q_n \rangle = 4\pi\sqrt{mn} \Big\{ \sum_{j=1}^\infty \bar\rho_j(m) \rho_j(n) \omega_p(t_j) \bar\omega_q(t_j) + \sum_\mathfrak{a} \frac{1}{4\pi} \int_{-\infty}^\infty \bar\tau_\mathfrak{a}(m, r) \tau_\mathfrak{a}(n, r) \omega_p(r) \bar\omega_q(r) dr \Big\}. \tag{16.24}$$

Next we compute $\langle P_m, Q_n \rangle$ by unfolding with respect to Q_n and applying the Fourier expansion (16.7) for P_m. We obtain

$$\begin{aligned} \langle P_m, Q_n \rangle &= \int_0^\infty \int_0^1 P_m(z) \bar G(nz) d\mu z \\ &= \int_0^\infty \Big(\delta(m, n) F(imy) + \sum_{q|c} S(m, n; c) F_c(m, n; y) e(iny) \Big) \bar G(iny) \frac{dy}{y^2}. \end{aligned}$$

Changing the variables we arrive at

$$\langle P_m, Q_n \rangle = 4\pi\sqrt{mn} \Big(\delta(m, n)(p, q) + \sum_{q|c} c^{-1} S(m, n; c) V_{pq}\Big(\frac{4\pi\sqrt{mn}}{c}\Big) \Big) \tag{16.25}$$

where

$$(p, q) = \int_0^\infty e^{-y} p(y) \bar q(y) y^{-1} dy \tag{16.26}$$

and

$$V_{pq}(x) = \int_{\mathrm{Im}\xi=1} \int_0^\infty p\Big(\frac{x}{y|\xi|}\Big) \bar q\Big(\frac{xy}{|\xi|}\Big) e\Big(\frac{-x|\xi|}{4\pi\xi}\Big(\frac{1}{y} + y\Big)\Big) \frac{dy d\xi}{y|\xi|}. \tag{16.27}$$

We simplify $V_{pq}(x)$ by changing $\xi = 2i(\zeta^2 + 1)^{-1}$, where ζ runs over the semi-circle $|\zeta| = 1$, $\mathrm{Re}(\zeta) > 0$ clockwise. Letting $\mathrm{Re}(\zeta) = \eta$ we have $\xi = i(\zeta\eta)^{-1}, |\xi| = \eta^{-1}$ and $|\xi|^{-1} d\xi = i(\zeta\eta)^{-1} d\zeta$. Hence

$$V_{pq}(x) = i \int_{-i}^i \int_0^\infty p\Big(\frac{x}{y}\eta\Big) \bar q(xy\eta) e^{-\frac{1}{2}(\frac{1}{y}+y)\zeta x} \frac{dy d\zeta}{y\zeta\eta}. \tag{16.28}$$

Combining (16.24) and (16.25) we arrive at the following

PROPOSITION 16.1. *Let $m, n > 0$ and $p(y), q(y)$ be smooth, bounded functions on $\mathbb{R}^+$. Then we have*

(16.29)
$$\sum_{j=1}^{\infty} \bar{\rho}_j(m)\rho_j(n)\omega_p(t_j)\bar{\omega}_q(t_j) + \sum_{\mathfrak{a}} \frac{1}{4\pi}\int_{-\infty}^{\infty} \bar{\tau}_{\mathfrak{a}}(m,r)\tau_{\mathfrak{a}}(n,r)\omega_p(r)\bar{\omega}_q(r)dr$$
$$= \delta(m,n)(p,q) + \sum_{c\equiv 0 (\mathrm{mod}\ q)} c^{-1}S(m,n;c)V_{pq}\left(\frac{4\pi\sqrt{mn}}{c}\right).$$

Now let us choose $p(y) = q(y) = y^{\frac{1}{2}+i\nu}$ with $\nu \in \mathbb{R}$. For these test functions we get
$$\omega_p(t) = \pi^{\frac{1}{2}}\Gamma(\tfrac{1}{2}+it+i\nu)\Gamma(\tfrac{1}{2}-it+i\nu)\Gamma(1+i\nu)^{-1},$$
whence

$$\omega_p(t)\bar{\omega}_p(t) = \pi^2\sinh(\pi\nu)/\nu\cosh\pi(\nu-t)\cosh\pi(\nu+t). \tag{16.30}$$

On the other hand, $(p,q) = 1$ and $V_{pp}(x) = B_{2i\nu}(x)$, where

$$B_s(x) = 2ix\int_{-i}^{i} K_s(\zeta x)\zeta^{-1}d\zeta. \tag{16.31}$$

In this case (16.29) becomes (an auxiliary version of Kuznetsov's formula)

COROLLARY 16.2. *For $m, n > 0$ and $\nu \in \mathbb{R}$ we have*

$$\sum_{j=1}^{\infty} \frac{\bar{\rho}_j(m)\rho_j(n)}{\cosh\pi(\nu-t_j)\cosh\pi(\nu+t_j)} \tag{16.32}$$
$$+\sum_{\mathfrak{a}}\frac{1}{4\pi}\int_{-\infty}^{\infty}\frac{\bar{\tau}_{\mathfrak{a}}(m,r)\tau_{\mathfrak{a}}(n,r)}{\cosh\pi(\nu-r)\cosh\pi(\nu+r)}dr$$
$$= \frac{\pi^{-2}\nu}{\sinh(\pi\nu)}\Big(\delta(m,n) + \sum_{c\equiv 0(\mathrm{mod}\ q)} c^{-1}S(m,n;c)B_{2i\nu}\Big(\frac{4\pi\sqrt{mn}}{c}\Big)\Big).$$

The integral (16.31) can be expressed in terms of the Bessel function $J_s(y)$ of real positive argument. Indeed, by Cauchy's theorem
$$B_s(x) = 2ix\left(\int_{i}^{i\infty} + \int_{-i\infty}^{-i}\right)K_s(\zeta x)\zeta^{-1}d\zeta$$
$$= 2ix\int_{1}^{\infty}(K_s(ixy) - K_s(-ixy))y^{-1}dy.$$

For $y > 0$ we have (see (5.14.4), (5.22.5), (5.22.6) of [**Leb**])
$$K_s(iy) = \frac{\pi}{2\sin(\pi s)}(e^{-\pi is/2}J_{-s}(y) - e^{\pi is/2}J_s(y))$$
$$K_s(-iy) = \frac{\pi}{2\sin(\pi s)}(e^{\pi is/2}J_{-s}(y) - e^{-\pi is/2}J_s(y)).$$

Hence
$$K_s(iy) - K_s(-iy) = \frac{-\pi i}{2\cos(\pi s/2)}(J_s(y) + J_{-s}(y)).$$

Finally, this yields

$$B_s(x) = \frac{\pi x}{\cos(\pi s/2)} \int_x^\infty (J_s(y) + J_{-s}(y))y^{-1}dy. \tag{16.33}$$

With the real parameter ν one can generate quite a lot of functions $h(t)$ out of the particular one $h(t) = (\cosh \pi(\nu - t)\cosh \pi(\nu + t))^{-1}$. N. V. Kuznetsov [**Kuz**] produced from (16.32) the following

THEOREM 16.3. *Suppose $h(t)$ satisfies the conditions (15.20). Then for any $m, n > 0$ we have*

(16.34)
$$\sum_{j=1}^\infty \bar\rho_j(m)\rho_j(n)\frac{h(t_j)}{\cosh \pi t_j} + \sum_{\mathfrak{a}} \frac{1}{4\pi}\int_{-\infty}^\infty \bar\tau_{\mathfrak{a}}(m,r)\tau_{\mathfrak{a}}(n,r)\frac{h(r)dr}{\cosh \pi r}$$
$$= \delta(m,n)g_0 + \sum_{c\equiv 0 (\mathrm{mod}\ q)} c^{-1}S(m,n;c)g\left(\frac{4\pi\sqrt{mn}}{c}\right)$$

where

$$g_0 = \pi^{-2}\int_{-\infty}^\infty rh(r)\tanh(\pi r)dr \tag{16.35}$$

and

$$g(x) = \frac{2i}{\pi}\int_{-\infty}^\infty J_{2ir}(x)\frac{rh(r)}{\cosh(\pi r)}dr. \tag{16.36}$$

REMARKS. The formula (16.34) was also established independently by R. W. Bruggeman [**Bru**], but only for a certain type of test functions $h(t)$ and with $g(x)$ in less refined form than (16.36) of Kuznetsov. The result also holds (and the proof is similar) for $mn < 0$, except for the integral transform (16.36) which changes to

$$g^-(x) = \frac{4}{\pi^2}\int_0^\infty K_{2ir}(x)rh(r)\sinh(\pi r)dr.$$

The Kuznetsov formula (16.34) can be derived from (16.32) by means of the following

LEMMA 16.4. *Suppose $h(t)$ satisfies the conditions (15.20). Then*

$$\int_{-\infty}^\infty (h(r+\tfrac{i}{2}) + h(r-\tfrac{i}{2}))\frac{\cosh(\pi r)dr}{\cosh \pi(r-t)\cosh \pi(r+t)} = \frac{-2h(t)}{\cosh(\pi t)}. \tag{16.37}$$

PROOF (SKETCH). Move the integral of $h(r+\frac{i}{2})$ down to the line $\mathrm{Im}\,(r) = -\frac12$ excluding small half-circles centered at $r = t - \frac{i}{2}$, $r = -t - \frac{i}{2}$, and move the integral of $h(r-\frac{i}{2})$ up to the line $\mathrm{Im}\,(r) = \frac12$ excluding small half-circles centered at $r = t + \frac{i}{2}$, $r = t - \frac{i}{2}$. The horizontal integrals cancel out because $h(t)$ is even. The four integrals over the small half-circles yield in the limit $-2h(t)/\cosh(\pi t)$ by computing residues. □

By integrating (16.32) according to (16.37) one obtains immediately the spectral terms on the left side of (16.34). On the right side one obtains the diagonal term with

$$\begin{aligned} g_0 &= \frac{-1}{2\pi^2}\int_{-\infty}^{\infty}(h(r+\tfrac{i}{2})+h(r-\tfrac{i}{2}))\frac{rdr}{\tanh(\pi r)} \\ &= \frac{-1}{\pi^2}\int_{-\infty}^{\infty}h(r+\tfrac{i}{2})\frac{rdr}{\tanh(\pi r)} = \frac{1}{\pi^2}\int_{-\infty}^{\infty}h(r)\tanh(\pi r)(r-\tfrac{i}{2})dr \end{aligned}$$

which is exactly (16.35) because $h(r)$ is even. Moreover, one obtains the sum of Kloosterman sums with

$$g(x) = \frac{-1}{2\pi^2}\int_{-\infty}^{\infty}(h(r+\tfrac{i}{2})+h(r-\tfrac{i}{2}))B_{2ir}(x)\frac{rdr}{\tanh(\pi r)}.$$

Inserting (16.34) and interchanging the integration we get

$$\begin{aligned} g(x) &= \frac{-x}{2\pi}\int_{x}^{\infty}\int_{-\infty}^{\infty}(h(r+\tfrac{i}{2})+h(r-\tfrac{i}{2}))(J_{2ir}(y)+J_{-2ir}(y))\frac{rdr}{\sinh(\pi r)}\frac{dy}{y} \\ &= \frac{-x}{2\pi}\int_{x}^{\infty}\int_{-\infty}^{\infty}[(2ir+1)J_{2ir+1}(y)-(2ir-1)J_{2ir-1}(y)]\frac{h(r)dr}{\cosh(\pi r)}\frac{dy}{y}, \end{aligned}$$

because $h(r)$ is even. Next, by the recurrence formula

$$(s+1)J_{s+1}(y)-(s-1)J_{s-1}(y) = -2sy(J_s(y)/y)',$$

we get

$$g(x) = \frac{-2ix}{\pi}\int_{x}^{\infty}\int_{-\infty}^{\infty}\frac{rh(r)}{\cosh(\pi r)}(J_{2ir}(y)/y)'drdy$$

which coincides with (16.36).

The formula (16.34) finds its basic applications for estimation of the Fourier coefficients of cusp forms in spectral parameters and the level (see the next section). In the level aspect alone the preliminary formula (16.32) is often sufficient and easy to use. However, the extra flexibility offered by the large family of functions $h(t)$ in the Kuznetsov refined version (16.34) is helpful for tackling delicate issues, such as exceptional eigenvalues or the distribution of length of closed geodesics on the Riemann surface $\Gamma\backslash\mathbb{H}$; cf. [**I11**], [**I12**].

For problems in analytic number theory we need to treat the sums of Kloosterman sums rather than the Fourier coefficients of cusp forms. To this end we need an equation in which the test function attached to the Kloosterman sums is given a priori, and the ones attached to the spectral terms are obtained by integral transforms with explicit kernels (something like the Riemann explicit formula for primes). Unfortunately, the formula (16.34) is not sufficient because one cannot represent every requested function $f(x)$ by the integral (16.36). Moreover, even if $g(x)$ admits the integral representation (16.36), it is not easy to find its original function $h(r)$.

Historically speaking, it was E. C. Titchmarsh [**T3**] who first tried to invert the map $h \mapsto g$ given by (16.36) for g in a dense subspace of $L^2(\mathbb{R}^+, x^{-1}dx)$, but not in the context of automorphic forms. Then jointly with D. B. Sears [**ST**] they realized that the image of this map is not dense. Indeed, writing (16.36) as

$$g(x)=\frac{2i}{\pi}\int_0^\infty (J_{2it}(x)-J_{-2it})(x))\frac{h(t)t}{\cosh(\pi t)}dt$$

it is clear that the Bessel functions $J_{k-1}(x)$ of positive integral order $k-1$ with k even are missed, because they are orthogonal to $J_{2it}(x)-J_{-2it}(x)$. However, they showed that $\{J_\ell(x);\ \ell=1,3,5,\dots\}$ together with $\{J_{2it}(x)-J_{-2it}(x);\ 0<t<\infty\}$ form a complete, orthogonal system in $L^2(\mathbb{R}^+,x^{-1}dx)$, the latter in the sense of continuous spectral measure.

Precisely, let $f(x)$ be a function of class C^2 on $[0,\infty)$ such that

$$f(0)=0,\qquad f^{(a)}(x)\ll(x+1)^{-\alpha}\tag{16.38}$$

for $a=0,1,2$ with $\alpha>2$. Then we have

(16.39)
$$\begin{aligned}f(x)=&\int_0^\infty (J_{2it}(x)-J_{-2it}(x))\Big(\int_0^\infty (J_{2it}(h)-J_{-2it}(y))f(y)\frac{dy}{y}\Big)\frac{2tdt}{\sinh(2\pi t)}\\&+\sum_{k>0,k\text{ even}}2(k-1)J_{k-1}(x)\int_0^\infty J_{k-1}(y)f(y)y^{-1}dy.\end{aligned}$$

Here the integral component of $f(x)$ is in the image of the integral transform (16.36). Therefore, for completeness one needs, in addition to (16.34), a formula for sums of Kloosterman sums $S(m,n;c)$ twisted by the Bessel functions $J_{k-1}(4\pi\sqrt{mn}/c)$. These we already know for each k as the Petersson formula (14.15). Put

$$\mathcal{M}_f(t)=\frac{\pi i}{\sinh(2\pi t)}\int_0^\infty (J_{2it}(x)-J_{-2it}(x))f(x)x^{-1}dx\tag{16.40}$$

and

$$\mathcal{N}_f(k)=\frac{4(k-1)!}{(4\pi i)^k}\int_0^\infty J_{k-1}(x)f(x)x^{-1}dx.\tag{16.41}$$

Let $\rho_j(m)$ and $\tau_{\mathfrak{a}}(m,r)$ be the Fourier coefficients of a complete orthonormal system of cusp forms and the Eisenstein series respectively. In addition let $\psi_{jk}(m)$ for $1\leqslant j\leqslant\dim S_k(\Gamma)$ be the Fourier coefficients of a complete orthonormal system of holomorphic cusp forms

$$f_{jk}(z)=\sum_{m=1}^\infty \psi_{jk}(m)m^{\frac{k-1}{2}}e(mz)\tag{16.42}$$

of weight $k=2,4,6,\dots$ Adding Kuznetsov's formula (16.34) to Petersson's formula (14.15) in accordance to the decomposition (16.39) one derives

THEOREM 16.5. *Let f satisfy (16.38). For $mn > 0$ we have*

$$\sum_{c\equiv 0 (\mathrm{mod}\ q)} c^{-1} S(m,n;c) f\left(\frac{4\pi\sqrt{mn}}{c}\right) = \sum_{j=1}^{\infty} \mathcal{M}_f(t_j)\bar{\rho}_j(m)\rho_j(n)$$
$$+\sum_{\mathfrak{a}} \frac{1}{4\pi}\int_{-\infty}^{\infty} \mathcal{M}_f(r)\bar{\tau}_{\mathfrak{a}}(m,r)\tau_{\mathfrak{a}}(n,r)dr$$
$$+ \sum_{0<k\equiv 0(\mathrm{mod}\ 2)} \mathcal{N}_f(k) \sum_{1\leqslant j\leqslant \dim S_k(\Gamma)} \bar{\psi}_{jk}(m)\psi_{jk}(n). \tag{16.43}$$

Observe that the diagonal term on the spectral side is gone!

One also needs a formula for sums of Kloosterman sums $S(m,n;c)$ with m,n of different sign. This is a similar case, actually it is simpler because the holomorphic cusp forms do not appear (they have no Fourier coefficients at negative frequencies). Put

$$\mathcal{K}_f(t) = 4\int_0^{\infty} K_{2it}(x)f(x)x^{-1}dx. \tag{16.44}$$

THEOREM 16.6. *Let f satisfy (16.38). For $mn < 0$ we have*

(16.45)

$$\sum_{c\equiv 0 (\mathrm{mod}\ q)} c^{-1} S(m,n;c) f\left(\frac{4\pi\sqrt{|mn|}}{c}\right) = \sum_{j=1}^{\infty} \mathcal{K}_f(t_j)\bar{\rho}_j(m)\rho_j(n)$$
$$+\sum_{\mathfrak{a}} \frac{1}{4\pi}\int_{-\infty}^{\infty} \mathcal{K}_f(r)\bar{\tau}_{\mathfrak{a}}(m,r)\tau_{\mathfrak{a}}(n,r)dr.$$

PROOF. This follows from Theorem 16.3 (with $g(x)$ replaced by $g^{-}(x)$) and the following Kontorovich-Lebedev inversion (see (5.27.14) of [**Leb**])

$$f(x) = \frac{8}{\pi^2}\int_0^{\infty} K_{2it}(x)\left(\int_0^{\infty} K_{2it}(y)f(y)y^{-1}dy\right) t\sinh(2\pi t)dt. \tag{16.46}$$

□

We close this section with remarks on the Kloosterman sums zeta function. Applying the formula (16.43) for $f(x) = x^{2s-1}$ formally one derives the following spectral decomposition:

$$2\frac{(2\pi\sqrt{mn})^{2s-1}}{\sin(\pi s)} Z_{mn}(s) = \sum_{j=1}^{\infty} \frac{\Gamma(s-s_j)\Gamma(s-1+s_j)}{\cosh(\pi t_j)}\bar{\rho}_j(m)\rho_j(n)$$
$$+ \sum_{k>0, k\ \mathrm{even}} \pi^{-2}(4\pi)^{1-k}(k-1)!\Gamma(s-\tfrac{k}{2})\Gamma(s-1+\tfrac{k}{2}) \sum_{1\leqslant j\leqslant \dim S_K(\Gamma)} \bar{\psi}_{ji}(m)\psi_{jk}(n)$$
$$+\sum_{\mathfrak{a}} \frac{1}{4\pi}\int_{-\infty}^{\infty} \frac{\Gamma(s-\frac{1}{2}-ir)\Gamma(s-\frac{1}{2}+ir)}{\cosh(\pi r)}\bar{\tau}_{\mathfrak{a}}(m,r)\tau_{\mathfrak{a}}(n,r)dr. \tag{16.47}$$

Of course, the function $f(x) = x^{2s-1}$ does not satisfy the conditions (16.38), nevertheless one can prove rigorously that the formula (16.47) holds true.

The series

$$L_{mn}(s) = \sum_{c\equiv 0(\mathrm{mod}\ q)} c^{-1}S(m,n;c)J_s\Big(\frac{4\pi\sqrt{mn}}{c}\Big) \tag{16.48}$$

seems more natural than $Z_{mn}(s)$ because it satisfies a suitable functional equation

$$L_{mn}(s) - L_{mn}(-s) + \frac{1}{2s}\sum_{\mathfrak{a}} \tau_{\mathfrak{a}}\Big(m,\frac{is}{2}\Big)\tau_{\mathfrak{a}}\Big(n,-\frac{is}{2}\Big) = \pi^{-1}\sin\Big(\frac{\pi s}{2}\Big)\delta(m,n), \tag{16.49}$$

(see Theorem 9.2 of [**I5**]).

16.5. Estimates for the Fourier coefficients.

In this section we derive from the Petersson and the Kuznetsov formulas immediate estimates for the Fourier coefficients of basic modular forms. To this end we simply estimate the involved sums of Kloosterman sums by using Weil's bound for the individual sum. More advanced methods and stronger results have been established in [**I3**], [**I13**] and [**DI**].

Our problems reduce to estimating the series (16.48). First we show that

$$\sum_{c\equiv 0(\mathrm{mod}\ q)} c^{-1-\sigma}|S(m,n;c)| \leqslant 8(\sigma-\tfrac12)^{-2}\tau((m,n))(m,n,q)^{\frac12}\tau(q)q^{-\frac12-\sigma} \tag{16.50}$$

if $\frac12 < \sigma \leqslant 1$. Indeed, by applying (16.1) the left side of (16.50) is bounded by

$$(m,n,q)^{\frac12}q^{-\frac12-\sigma}\tau(q)\sum_{c=1}^{\infty}(m,n,c)^{\frac12}c^{-\frac12-\sigma}\tau(c),$$

and the last series is bounded by

$$\zeta^2(\sigma+\tfrac12)\sum_{d|(m,n)}\tau(d)d^{-\sigma}.$$

Hence (16.50) follows, because $\tau(d)\leqslant 2d^{\frac12}$ and $\zeta(\sigma+\frac12)\leqslant 2(\sigma-\frac12)^{-1}$.

For $k\geqslant 2$ we have

$$J_{k-1}(x) \ll \min\Big(1,\frac{x^{k-1}}{(k-1)!}\Big) \ll \Big(\frac{x}{k}\Big)^{\sigma}$$

for any $0\leqslant\sigma\leqslant 1$. Therefore (16.50) yields

$$L_{mn}(k-1) \ll (2\sigma-1)^{-2}\Big(\frac{\sqrt{mn}}{kq}\Big)^{\sigma}\frac{\tau(q)}{\sqrt{q}}\tau((m,n))(m,n,q)^{\frac12} \tag{16.51}$$

for any $\frac12<\sigma\leqslant 1$, where the implied constant is absolute.

Next for $s=\sigma+it$ with $\frac12<\sigma\leqslant 1$ we have (see (23.411.4) of [**GR**])

$$J_s(x) \ll \frac{x^{\sigma}}{|\Gamma(s+\frac12)|} \ll e^{\pi|s|/2}\Big(\frac{x}{|s|}\Big)^{\sigma}.$$

Therefore (16.50) yields

$$L_{mn}(s) \ll (2\sigma-1)^{-2}e^{\pi|s|/2}\Big(\frac{\sqrt{mn}}{|s|q}\Big)^{\sigma}\frac{\tau(q)}{\sqrt{q}}\tau((m,n))(m,n,q)^{\frac12} \tag{16.52}$$

where the implied constant is absolute.

Note the right side of the Petersson formula (14.14) equals

$$\delta(m,n)+2\pi i^k L_{mn}(k-1).$$

Hence, using (16.52) with $2\sigma-1=(\log 3mn)^{-1}$, we conclude

THEOREM 16.7. *Let $k\geqslant 2, k$ even. The Fourier coefficients of an orthonormal system of holomorphic cusp forms of weight k and level q satisfy*

$$\begin{aligned}(16.53)\quad \frac{\Gamma(k-1)}{(4\pi)^{k-1}}\sum_j \bar{\psi}_{jk}(m)\psi_{jk}(n) &= \delta(m,n)\\ &+O\Big(\frac{\tau(q)}{q\sqrt{k}}\tau((m,n))(m,n,q)^{\frac12}(mn)^{\frac14}\log^2 3mn\Big)\end{aligned}$$

where the implied constant is absolute.

Next notice that the right side of the Kuznetsov formula (16.34) equals

$$\delta(m,n)g_0+\frac{1}{2\pi i}\int_{(\sigma)} L_{mn}(s)\frac{h(is/2)s}{\cos(\pi s/2)}ds$$

for any $\frac12<\sigma<1$. Inserting (16.52) with $2\sigma-1\leqslant(\log 3mn)^{-1}$ we conclude

THEOREM 16.8. *Suppose $h(r)$ satisfies (15.20). Then for $m,n>0$*

$$\begin{aligned}(16.54)\quad \sum_{j=1}^{\infty}\bar{\rho}_j(m)\rho_j(n)\frac{h(t_j)}{\cosh(\pi t_j)}+\sum_{\mathfrak{a}}\frac{1}{4\pi}\int_{-\infty}^{\infty}\bar{\tau}_{\mathfrak{a}}(m,r)\tau_{\mathfrak{a}}(n,r)\frac{h(r)dr}{\cosh(\pi r)}\\ =\delta(m,n)g_0+O\Big(H\frac{\tau(q)}{q}\tau((m,n))(m,n,q)^{\frac12}(mn)^{\frac14}\Big)\end{aligned}$$

where

$$(16.55)\qquad H=\eta^{-2}\int_{\operatorname{Im} r=\frac14+\eta}|r^{\frac34}h(r)|dr$$

for any $0<\eta\leqslant(4\log 3mn)^{-1}$, and the implied constant is absolute.

In particular, Theorem 16.8 yields

$$\begin{aligned}(16.56)\quad \sum_{j=1}^{\infty}e^{-(t_j/T)^2}\frac{|\rho_j(n)|^2}{\cosh(\pi t_j)}+\sum_{\mathfrak{a}}\frac{1}{4\pi}\int_{-\infty}^{\infty}e^{-(t/T)^2}\frac{|\tau_{\mathfrak{a}}(n,t)|^2}{\cosh(\pi t)}dt\\ =g_0+O(T^{\frac74}\tau(q)q^{-1}\tau(n)(n,q)^{\frac12}n^{\frac12}\log^2 3n)\end{aligned}$$

where $g_0=\pi^{-2}T^2+O(1)$ and the implied constants are absolute.

Another interesting choice of the test function is

$$(16.57)\qquad h(t)=(X^{it}+X^{-it})^2(t^2+1)^{-2}$$

where $X\geqslant 1$. This is non-negative on the spectrum, i.e., for t or it real, and large for the exceptional points, precisely we have $h(t)\gg X^{2s-1}$ if $s=\frac12+it>\frac12$. Clearly $g_0\ll 1$ and the integral (16.55) satisfies

$$H\ll\eta^{-2}X^{\frac12+2\eta}\ll X^{\frac12}(\log 3mnX)^2$$

by choosing $\eta = (4\log 3mnX)^{-1}$. Therefore (16.54) for the test function (16.57) yields

$$\sum_{\frac{1}{2}<s_j<1} |\rho_j(n)|^2 X^{2s_j-1} \ll 1 + \frac{\tau(q)}{q}\tau(n)(n,q)^{\frac{1}{2}}(nX)^{\frac{1}{2}}(\log 3nX)^2. \tag{16.58}$$

According to Selberg's eigenvalue conjecture for the group $\Gamma_0(q)$ the sum on the left side of (16.58) is void. This has not yet been proved. However, the inequality (16.58) shows that the exceptional points appear infrequently. Specifically, taking $n=1$ and $X=q^2$ we get

$$\sum_{\frac{1}{2}<s_j<1} |\rho_j(1)|^2 q^{2(2s_j-1)} \ll \tau(q)(\log 2q)^2. \tag{16.59}$$

Here one can choose a basis (a Hecke-Pizer basis) such that $|\rho_j(1)| \gg q^{-\frac{1}{2}-\varepsilon}$, getting

$$\sum_{\frac{1}{2}<s_j<1} q^{2(2s_j-1)} \ll q^{1+\varepsilon}. \tag{16.60}$$

Hence we derive the following density estimate

$$|\{j;\ s_j > \alpha\}| \ll q^{3-4\alpha+\varepsilon} \tag{16.61}$$

for any $\alpha \geqslant \frac{1}{2}$ and $\varepsilon > 0$, the implied constant depending only on ε. If $\alpha > \frac{1}{2}$, this bound is quite small relative to the volume of $\Gamma_0(q)\backslash\mathbb{H}$.

16.6. Estimates for sums of Kloosterman sums.

Now we have everything ready for estimation of sums of Kloosterman sums. The original sum of Linnik given by (16.2) for $q=1$ was treated in 1977 by Kuznetsov [**Kuz**]. He applied his formula (16.43) for a suitable $f(x)$ and implemented Weil's bound for the Kloosterman sums near the boundary of the summation range getting

$$S_{mn}(X) \ll X^{\frac{1}{6}}(\log 2X)^{\frac{1}{3}} \tag{16.62}$$

for $X \geqslant 1$ and $m,n \geqslant 1$, where the implied constant depends on m,n. The exponent $\frac{1}{6}$ here is remarkable by comparison to the $\frac{1}{2}$ in (16.3), though it is not yet arbitrarily small as conjectured by Linnik. In practice one is satisfied with sums of $S(m,n;c)$ over the modulus c restricted smoothly in an interval rather than by the sharp cut-off. These smoothed sums, while not compromising real applications, can be given the true estimate (see (16.72)).

Fix $m,n,q \geqslant 1$ and a function f of class C^3 supported on $[\frac{1}{2},1]$. We shall estimate

$$S(X) = \sum_{c\equiv 0 (\mathrm{mod}\ q)} c^{-1} S(m,n;c) f(2\pi\sqrt{mn}X/c) \tag{16.63}$$

for all $X \geqslant 1$. By Theorem 16.5 the problem is transferred to estimation of the Fourier coefficients of basic modular forms and the integral transforms (16.40), (16.41) of $f(xX/2)$.

Introducing the power series expansion

$$J_\nu(2x) = \sum_{\ell=0}^{\infty} \frac{(-1)^\ell x^{2\ell+\nu}}{\ell!\Gamma(\ell+\nu+1)} \tag{16.64}$$

into

$$\mathcal{M}_f(t) = \frac{-\pi}{\sin(2\pi s)} \int_0^\infty (J_{2s-1}(2x) - J_{1-2s}(2x)) f(xX) x^{-1} dx$$

where $s = \frac{1}{2} + it$, we get

$$\mathcal{M}_f(t) = \frac{-\pi}{\sin(2\pi s)} \sum_{\ell=0}^{\infty} (-1)^\ell \frac{\hat{f}(2\ell+2s-1)}{\ell!\Gamma(\ell+2s)} X^{2s-1-2\ell}$$
$$+ \quad \text{the same expression with } s \text{ changed into } 1-s,$$

where $\hat{f}(s)$ is the Mellin transform of $f(x)$. All we need to know about f is that $|f^{(a)}| \leqslant \kappa$ for $a = 0, 1, 2, 3$ and some $\kappa > 0$. Integrating by parts up to three times we deduce that

$$\hat{f}(s) \ll \kappa(|s|+1)^{-3} \qquad \text{if } \mathrm{Re}\,(s) > 0, \tag{16.65}$$

where the implied constant is absolute. Moreover, we use the Stirling formula

$$\Gamma(s) = \left(\frac{2\pi}{s}\right)^{\frac{1}{2}} \left(\frac{s}{e}\right)^{s} \left(1 + O\left(\frac{1}{|s|}\right)\right) \qquad \text{if } \mathrm{Re}\,(s) > 0. \tag{16.66}$$

Hence, estimating all terms of the power series for $\mathcal{M}_f(t)$, except for $\ell = 0$, we deduce that

$$\mathcal{M}_f(t) = \omega(s, X) + O\left(\frac{\kappa X^{-1} \log 2X}{|s|^3 \cosh(\pi t)}\right) \tag{16.67}$$

where

$$\omega(s, X) = \frac{-\pi}{\sin(2\pi s)} \left(\frac{\hat{f}(2s-1)}{\Gamma(2s)} X^{2s-1} - \frac{\hat{f}(1-2s)}{\Gamma(2-2s)} X^{1-2s} \right).$$

Next by the functional equation $\Gamma(s)\Gamma(1-s) = \pi/\sin(\pi s)$ we write

$$\omega(s, X) = \Gamma(2s-1)\hat{f}(1-2s)X^{1-2s} - \Gamma(1-2s)\hat{f}(2s-1)X^{2s-1}. \tag{16.68}$$

We also estimate the leading term $\omega(s, X)$, but only for $s = \frac{1}{2} + it$ with t real, getting

$$\mathcal{M}_f(t) \ll \frac{\kappa \log 2X}{|s|^3 \cosh(\pi t)}. \tag{16.69}$$

Similarly for the integral transform (16.41) we deduce that

$$\mathcal{N}_f(k) \ll \kappa k^{-3} X^{-1}. \tag{16.70}$$

Finally introducing (16.69) and (16.70) into (16.43), except for the points $\frac{1}{2} < s_j < 1$, then applying Cauchy's inequality and the estimates (16.53), (16.56) we arrive at

THEOREM 16.9. *Let f be of class C^3 with compact support in $[\frac{1}{2}, 1]$ and $|f^{(a)}| \leqslant 1$ for $a = 0, 1, 2, 3$. Then for m n, q, $X \geqslant 1$ we have*

$$S(X) = \sum_{\frac{1}{2}<s_j<1} \omega(s_j, X)\bar{\rho}_j(m)\rho_j(n) + R(X) \tag{16.71}$$

where $\omega(s, X)$ is given by (16.68) and $R(X)$ satisfies

$$R(X) \ll \Big(1 + (m, q)\frac{m}{q^2}\Big)^{\frac{1}{4}} \Big(1 + (n, q)\frac{n}{q^2}\Big)^{\frac{1}{4}} \tau(q)\tau(mn)(\log 2mnX)^3 \tag{16.72}$$

with the implied constant being absolute.

Assuming Selberg's eigenvalue conjecture the sum over $\frac{1}{2} < s_j < 1$ is void, so $S(X) = R(X)$ and (16.72) becomes a pure bound for (16.63). In particular, it is known that the Selberg conjecture holds true for the modular group where $q = 1$ (for a proof, see e.g. [**DI**]; Hejhal computed that the smallest eigenvalue is $\lambda_1 = 91.14\ldots$). In this case we get the unconditional result

$$\sum_{c>0} c^{-1}S(m, n; c)f(2\pi\sqrt{mn}X/c) \ll \tau(mn)(mn)^{\frac{1}{4}}(\log 2mn)^2. \tag{16.73}$$

Notice we replaced $(\log 2mnX)^3$ by $(\log 2mn)^2$ because the factor $\log 2X$ in (16.69) appears for small t which is not in the spectrum for the modular group.

Denote the sum over the exceptional spectrum (hypothetically empty) by

$$E(X) = \sum_{\frac{1}{2}<s_j<1} \omega(s_j, X)\bar{\rho}_j(m)\rho_j(n). \tag{16.74}$$

Suppose $s_1 = \frac{1}{2} + \alpha$ is the largest point, i.e., $\lambda_1 = s_1(1 - s_1) = \frac{1}{4} - \alpha^2$ is the lowest eigenvalue in the space of cusp forms. Since $\omega(s_1, X) \gg X^{2\alpha}$ for sufficiently large X we deduce by comparing (16.71) with (16.3) for $m = n$ with $\rho_1(n) \neq 0$ that $\alpha \leqslant \frac{1}{4}$ which corresponds to the Selberg lower bound $\lambda_1 \geqslant \frac{3}{16}$. The best known estimate is $\alpha \leqslant \frac{7}{64}$ due to Kim and Sarnak [**KSa**] (see Section 5.11). The estimate $\alpha \leqslant \frac{3}{14}$ is obtained in [**I14**] by simpler arguments. Applying Cauchy's inequality and $\omega(s, X) \ll X^{2s-1} \log 2X$ uniformly for $\frac{1}{2} \leqslant s \leqslant 1$ we get

$$E(X) \ll E_m(X)^{\frac{1}{2}} E_n(X)^{\frac{1}{2}} \log 2X$$

where

$$E_m(X) = \sum_j |\rho_j(m)|^2 X^{2s_j - 1}.$$

For any $1 \leqslant Y \leqslant X$ we have $E_m(X) \leqslant (X/Y)^{2\alpha} E_m(Y)$, and (16.58) yields

$$E_m(Y) \ll 1 + \tau(q)\tau(m)q^{-1}(m, q)^{\frac{1}{2}}(mY)^{\frac{1}{2}}(\log 2mY)^2.$$

Choosing $Y = \min(X, \max(1, q^2/m(m, q)))$ we derive

$$E_m(X) \ll [1 + ((m, q)m/q^2)^{\frac{1}{2}} X^{2\alpha} + ((m, q)mX/q^2)^{2\alpha}]\tau(q)\tau(m)(\log 2mX)^2.$$

Combining the above estimates we obtain

PROPOSITION 16.10. *Suppose the lowest cuspidal eigenvalue for* $\Gamma_0(q)$ *is* $\lambda_1 = \frac{1}{4} - \alpha^2$ *with* $0 < \alpha \leqslant \frac{1}{4}$. *Then the exceptional sum (16.74) satisfies*

(16.75)
$$E(X) \ll \Big(1 + ((m,q)m/q^2)^{\frac{1}{4}}X^\alpha + ((m,q)m/q^2)^\alpha X^\alpha\Big)$$
$$\Big(1 + ((n,q)n/q^2)^{\frac{1}{4}}X^\alpha + ((n,q)n/q^2)^\alpha X^\alpha\Big)\tau(q)\tau(mn)(\log 2mnX)^3$$

where the implied constant is absolute.

Because the bound (16.72) is absorbed by (16.75), the latter constitutes our final estimate for the sum (16.63). Although the Selberg eigenvalue conjecture is not yet proved, our bound for $S(X)$ turns out to be strong enough in the level aspect to yield adequate results in the most important applications. Very often in practice it suffices to know that $\lambda_1 \geqslant \frac{3}{16}$. Putting $\alpha = \frac{1}{4}$ we get the following simple unconditional estimate

THEOREM 16.11. *For* $m, n, q, X \geqslant 1$ *we have*

$$S(X) \ll \Big(1 + (m,q)\frac{mX}{q^2}\Big)^{\frac{1}{4}}\Big(1 + (n,q)\frac{nX}{q^2}\Big)^{\frac{1}{4}}\tau(q)\tau(mn)(\log 2mnX)^3 \tag{16.76}$$

where the implied constant is absolute.

Similar arguments work and the same results hold true for the sum (16.63) if $S(m,n;c)$ is replaced by $S(-m,n;c)$ for any $m, n \geqslant 1$. To this end replace Theorem 16.5 by Theorem 16.6.

CHAPTER 17

PRIMES IN ARITHMETIC PROGRESSIONS

17.1. Introduction.

The Prime Number Theorem asserts that every residue class $a \pmod q$ with $(a,q)=1$ contains the same proportion of primes, i.e., for fixed q we have

$$\pi(x;q,a) \sim \frac{1}{\varphi(q)}\pi(x), \tag{17.1}$$

$$\psi(x;q,a) \sim \frac{1}{\varphi(q)}\psi(x) \tag{17.2}$$

as x tends to infinity. One even has reasonable estimates for the error terms in the asymptotics (17.1), (17.2). However, more important than the error term is the range of uniformity for the modulus q in terms of x. The Siegel-Walfisz Theorem (Corollary 5.29) asserts that

$$\psi(x;q,a) = \frac{x}{\varphi(q)} + O(x(\log x)^{-A}) \tag{17.3}$$

for any $q \geqslant 1$, $(a,q)=1$, $x \geqslant 2$ and $A \geqslant 0$, the implied constant depending only on A. Notice that this estimate is non-trivial only if $q \ll (\log x)^A$.

The Grand Riemann Hypothesis for $L(s,\chi)$ with $\chi \pmod q$ implies

$$\psi(x;q,a) = \frac{x}{\varphi(q)} + O(x^{\frac12}(\log x)^2) \tag{17.4}$$

where the implied constant is absolute, and this is non-trivial for $q \ll x^{\frac12}(\log x)^{-2}$. It is conjectured by H. Montgomery that

$$\psi(x;q,a) = \frac{x}{\varphi(q)} + O(q^{-\frac12}x^{\frac12+\varepsilon}) \tag{17.5}$$

where the implied constant depends only on ε. This implies that the asymptotic formula (17.2) holds true uniformly in $q \leqslant x^{1-\varepsilon}$. It was shown, however, by Friedlander and Granville ([**FG**], [**Gra**]), extending previous work of Maier [**Mai**] on primes in short intervals, that (17.1) and (17.2) cannot hold for $q \leqslant x(\log x)^{-A}$ for any $A > 0$.

The GRH may be out of reach for several generations of researchers, but we have already a satisfactory substitute of (17.4), or rather the extension of (17.3), in the following form.

THEOREM 17.1 (E. BOMBIERI, A. I. VINOGRADOV, 1965). *We have*

$$\sum_{q\leqslant Q}\max_{(a,q)=1}\left|\psi(x;q,a)-\frac{x}{\varphi(q)}\right|\ll x(\log x)^{-A} \tag{17.6}$$

for any $A\geqslant 0$, *where* $Q=x^{\frac{1}{2}}(\log x)^{-B}$ *with* $B=B(A)$, *the implied constant depending on* A.

Here are brief comments on the history of the problem. All treatments use in one way or another the idea of large sieve due to Linnik [**Li1**]. First, A. Renyi [**Re**] established (17.6) with $Q=x^{\theta-\varepsilon}$ for some $\theta>0$, then K. Roth [**Rot**] got this for $\theta=\frac{1}{3}$ while M. B. Barban [**Bar**] succeeded with $\theta=\frac{3}{8}$ (and died soon after). Finally A. I. Vinogradov [**Vi**] reached the level $\theta=\frac{1}{2}$ while E. Bombieri [**Bo5**] working independently in the same time proved a slightly stronger result as stated in Theorem 17.1. Both authors derived their results from new density theorems for zeros of L-functions which they established first. But Gallagher [**Ga3**] and Motohashi [**Mot2**] simplified the argument to avoid the use of zeros, instead feeding the main ideas directly into the estimation of bilinear forms over arithmetic progressions. These developments are related to the identities of Vinogradov, Linnik, Heath-Brown or Vaughan type for sums over primes discussed in Chapter 13. In this chapter we borrow from these innovations as much as possible to prove Theorem 17.1 with $B(A)=2A+6$ by employing minimal work.

EXERCISE 1. Prove the asymptotic formula for the Titchmarsh divisor problem:

$$\sum_{p\leqslant x}\tau(p-1)\sim cx \text{ with } c=\frac{\zeta(2)\zeta(3)}{\zeta(6)}=1.943596\dots$$

as $x\to+\infty$. This formula was proved first by Linnik [**Li7**]. [**Hint:** Use Dirichlet's hyperbola method to reduce to counting primes $\equiv 1 \pmod d$ with $d\leqslant\sqrt{x}$, then use Theorem 17.1 and the Brun-Titchmarsh inequality (6.95).]

See [**Ko1**] for a fairly natural analogue of this problem for elliptic curves.

The Bombieri-Vinogradov theorem is expected to be true with $Q=x^{1-\varepsilon}$ (the Elliott-Halberstam conjecture) which follows from (17.5), but not from the GRH. There are some unconditional results of type

$$\sum_{q\leqslant Q}\lambda(q)\Big(\psi(x;q,a)-\frac{x}{\varphi(q)}\Big)\ll x(\log x)^{-A} \tag{17.7}$$

with $Q=x^{\frac{4}{7}-\varepsilon}$, where $a\neq 0$ is fixed and $\lambda(q)$ is any well-factorable function in the sense of sieve theory (see [**BFI**]). Here $\varepsilon>0, A>0$ are arbitrary and the implied constant depends on a,ε,A. This result contains useful information about equidistribution of primes $p\leqslant x$ in a fixed residue class $a(\bmod q)$ of modulus q larger than $x^{\frac{1}{2}}$, in which range the GRH does not apply.

One can handle the larger moduli q if the averaging over the classes $a(\bmod q)$ is permitted. Thus, P. Turán [**Tu2**] derived from the GRH that

$$\sideset{}{^\star}\sum_{a(\bmod q)}\Big(\psi(x;q,a)-\frac{x}{\varphi(q)}\Big)^2\ll x(\log x)^4 \tag{17.8}$$

where the implied constant is absolute. This estimate is non-trivial for q as large as $x(\log x)^{-5}$. Introducing the additional averaging over the modulus M. B. Barban [**Bar**] and H. Davenport and H. Halberstam [**DH2**] established unconditional results of comparable strength.

THEOREM 17.2 (BARBAN, DAVENPORT AND HALBERSTAM, 1966). *We have*

$$\sum_{q\leqslant Q}\ \sideset{}{^\star}\sum_{a(\mathrm{mod}\ q)} \Big(\psi(x;q,a)-\frac{x}{\varphi(q)}\Big)^2 \ll x(\log x)^{-A} \tag{17.9}$$

for any $A>0$, where $Q=x(\log x)^{-B}$ with $B=B(A)$, the implied constant depending on A.

There are analogous results for the Möbius function in arithmetic progressions.

EXERCISE 2. Prove that

$$\sum_{q\leqslant Q}\max_a\Big|\sum_{\substack{m\leqslant x\\ m\equiv a(\mathrm{mod}\ q)}}\mu(m)\Big| \ll x(\log x)^{-A} \tag{17.10}$$

for any $A>0$, where $Q=x^{\frac12}(\log x)^{-B}$ with $B=B(A)$, the implied constant depending on A.

17.2. Bilinear forms in arithmetic progressions.

Given a reasonable arithmetic function f it is natural to expect its values to be equidistributed over primitive residue classes, i.e., one should have a non-trivial bound for

$$D_f(x;q,a)=\sum_{\substack{n\leqslant x\\ n\equiv a(\mathrm{mod}\ q)}} f(n)-\frac{1}{\varphi(q)}\sum_{\substack{n\leqslant x\\ (n,q)=1}} f(n) \tag{17.11}$$

for all $(a,q)=1$, provided q is not extremely large. In many cases one can show that

$$D_f(x;q,a)\ll \Big(\sum_{n\leqslant x}|f(n)|^2\Big)^{\frac12}x^{\frac12}(\log x)^{-A}. \tag{17.12}$$

This is non-trivial only for $q\leqslant(\log x)^A$. Having (17.12) one can prove by the large sieve inequality (7.31) a bound for $D_f(x;q,a)$ for some type of f which is good for almost all $q\leqslant x^{\frac12}(\log x)^{-B}$. It is surprising how little is required of f. Essentially f needs to be represented as a convolution of two sequences, say $f=\alpha\star\beta$, one of which, say β, is equidistributed over primitive residue classes uniformly with respect to the moduli $q\leqslant(\log x)^A$, or more generally, f should be well approximated by a linear combination of such convolutions.

We begin our investigations by examining a sequence $\beta=(\beta_n)$ of complex numbers supported on $1\leqslant n\leqslant N$. We assume that for some $0<\Delta\leqslant 1$ it satisfies

$$|D_\beta(N;q,a)|\leqslant \|\beta\|N^{\frac12}\Delta^9 \tag{17.13}$$

for all $(a,q)=1$, where

$$\|\beta\|=\Big(\sum_n|\beta_n|^2\Big)^{\frac12}.$$

First we show that (17.13) implies the following bound for character sums.

LEMMA 17.3. *For a non-principal character χ (mod r) and a positive integer s we have*

$$\Big|\sum_{(n,s)=1} \beta_n\chi(n)\Big| \leqslant \|\beta\| N^{\frac12}\Delta^3 r\tau(s). \tag{17.14}$$

PROOF. By Möbius' inversion we change the condition $(n,s)=1$ and then we split the sum as follows:

$$\begin{aligned}\sum_{(n,s)=1}\beta_n\chi(n) &= \sum_{k|s}\mu(k)\sum_{n\equiv 0(\text{mod } k)}\beta_n\chi(n)\\ &= \sum_{\substack{k|s\\ k\leqslant K}}\mu(k)\sum_{\ell|k}\mu(\ell)\sum_{(n,\ell)=1}\beta_n\chi(n) + \sum_{\substack{k|s\\ k>K}}\mu(k)\sum_{n\equiv 0(\text{mod } k)}\beta_n\chi(n).\end{aligned}$$

Next we estimate the sum of $\beta_n\chi(n)$ over $(n,\ell)=1$ by splitting into classes modulo ℓr, and for each class we apply the hypothesis (17.13). This gives us (the leading terms cancel out because χ is non-principal)

$$\|\beta\|N^{\frac12}\Delta^9\sum_{\substack{k|s\\ k\leqslant K}}\sum_{\ell|k}|\mu(\ell)|\varphi(\ell r) \leqslant \|\beta\|N^{\frac12}\Delta^9 K\varphi(r)\tau(s).$$

Then we estimate the sum of $\beta_n\chi(n)$ over $n\equiv 0(\text{mod } k)$ using Cauchy's inequality. This gives us

$$\|\beta\|N^{\frac12}\sum_{\substack{k|s\\ k>K}}k^{-\frac12} \leqslant \|\beta\|N^{\frac12}K^{-\frac12}\tau(s).$$

Adding both estimates we obtain (17.14) by taking $K=\Delta^{-6}$. □

Now we proceed to the main result

THEOREM 17.4. *Suppose $\beta=(\beta_n)$ is a sequence supported on $1\leqslant n\leqslant N$ which satisfies (17.13). Let $\alpha=(\alpha_m)$ be any sequence supported on $1\leqslant m\leqslant M$. Then we have*

(17.15)

$$\begin{aligned}&\sum_{q\leqslant Q}\max_{(a,q)=1}|D_{\alpha\star\beta}(MN;q,a)|\\ &\qquad\ll \|\alpha\|\,\|\beta\|(\Delta M^{\frac12}N^{\frac12}+M^{\frac12}+N^{\frac12}+Q)(\log Q)^2\end{aligned}$$

where the implied constant is absolute.

PROOF. Using Dirichlet characters we write

$$D_{\alpha\star\beta}(MN;q,a) = \frac{1}{\varphi(q)}\sum_{\substack{\chi(\text{mod } q)\\ \chi\neq\chi_0}}\bar\chi(a)\Big(\sum_m\alpha_m\chi(m)\Big)\Big(\sum_n\beta_n\chi(n)\Big).$$

Hence, reducing to primitive characters, the left side of (17.15) is bounded by

$$\sum_{s\leqslant Q}\frac{1}{\varphi(s)}\sum_{1<r\leqslant Q}\frac{1}{\varphi(r)}\sideset{}{^\star}\sum_{\chi(\text{mod } r)}\Big|\sum_{(m,s)=1}\alpha_m\chi(m)\Big|\Big|\sum_{(n,s)=1}\beta_n\chi(n)\Big|.$$

If r is small, say $r \leqslant R$, we apply (17.14) getting

$$\|\beta\| N^{\frac{1}{2}} \Delta^3 \sum_{s \leqslant Q} \frac{\tau(s)}{\varphi(s)} \sum_{r \leqslant R} r \ll \|\beta\| N^{\frac{1}{2}} \Delta^3 R^2 (\log Q)^2.$$

For the remaining $r > R$ we split the range into dyadic segments $P < r \leqslant 2P$ and for each of such reduced sums we apply the large sieve inequality (7.31) getting

$$\begin{aligned} &\|\alpha\| \, \|\beta\| \sum_{R<P<Q} P^{-1}(P^2+M)^{\frac{1}{2}}(P^2+N)^{\frac{1}{2}} \log Q \\ &\ll \|\alpha\| \, \|\beta\| (Q + M^{\frac{1}{2}} + N^{\frac{1}{2}} + M^{\frac{1}{2}} N^{\frac{1}{2}} R^{-1})(\log Q)^2. \end{aligned}$$

Adding both results for $R = \Delta^{-1}$ we get (17.15). □

17.3. Proof of the Bombieri-Vinogradov Theorem.

Suppose $x^{\frac{1}{5}} < n \leqslant x$. By (13.39) with $y = z = x^{\frac{1}{5}}$ we write $\Lambda(n) = \Lambda^{\#}(n) + \Lambda^{\flat}(n)$, where

$$\Lambda^{\sharp}(n) = \sum_{\substack{\ell m = n \\ m \leqslant x^{\frac{1}{5}}}} \lambda(\ell)\mu(m),$$

$$\Lambda^{\flat}(n) = \sum_{\substack{\ell m = n \\ x^{\frac{1}{5}} < m \leqslant x^{\frac{4}{5}}}} \lambda(\ell)\mu(m).$$

Here $\lambda(\ell)$ is the incomplete logarithm

$$\lambda(\ell) = \log \ell - \sum_{\substack{k \mid \ell \\ k \leqslant x^{\frac{1}{5}}}} \Lambda(k). \tag{17.16}$$

Accordingly we split

$$D_{\Lambda}(x; q, a) = D_{\Lambda^{\sharp}}(x; q, a) + D_{\Lambda^{\flat}}(x; q, a) + O(x^{\frac{1}{5}} \log x) \tag{17.17}$$

where the error term comes from the contribution of terms $n < x^{\frac{1}{5}}$ for which the above identities do not apply.

If f is continuous and monotonic on $[1, y]$, then by elementary arguments we derive

$$|D_f(y; q, a)| \leqslant 2|f(1)| + 2|f(y)|.$$

Applying this for $f(\ell) = \log \ell$ and $f(\ell) = 1$ we obtain $D_{\lambda}(y; q, a) \ll x^{\frac{1}{5}} \log x$. Hence $D_{\Lambda^{\sharp}}(x; q, a) \ll x^{\frac{2}{5}} \log x$ and summing over $q \leqslant Q$ we get

$$\sum_{q \leqslant Q} \max_{(a,q)=1} |D_{\Lambda^{\sharp}}(x; q, a)| \ll Q x^{\frac{2}{5}} \log x. \tag{17.18}$$

We shall estimate $D_{\Lambda^{\flat}}(x; q, a)$ by an appeal to Theorem 17.4. To this end we need $\Lambda^{\flat}(n)$ to be in a form of convolution of two sequences, and we almost have it except that the restriction $n = \ell m \leqslant x$ makes ℓ and m interconnected. To relax this and to keep hold of the sizes of ℓm, we are going to divide the interval $1 \leqslant n \leqslant x$

into short subintervals of type $y < n \leqslant (1+\delta)y$ with $x^{-\frac{1}{5}} < \delta \leqslant 1$. There are $O(\delta^{-1})$ such subintervals. We cover $\Lambda^\flat(n)$ by partial sums of type

$$\sum\sum_{\substack{\ell m=n\\ L<\ell\leqslant(1+\delta)L\\ M<m\leqslant(1+\delta)M}} \lambda(\ell)\mu(m) \tag{17.19}$$

with L, M taking values $(1+\delta)^j$ in the range $x^{\frac{1}{5}} < L, M < x^{\frac{4}{5}}$, $LM < x$, except for $n < x^{\frac{1}{5}}$ and $(1+\delta)^{-1}x < n < (1+\delta)x$ in which ranges the covering is not exactly with multiplicity one. Estimating the contribution in these excess areas trivially we obtain

$$D_{\Lambda^\flat}(x;q,a) = \sum_L\sum_M D(LM;q,a) + O(\delta q^{-1}x\log x) \tag{17.20}$$

where $D(LM;q,a)$ stands for

$$\sum\sum_{\substack{L<\ell\leqslant(1+\delta)L\\ M<m\leqslant(1+\delta)M\\ \ell m\equiv a(\mathrm{mod}\ q)}} \lambda(\ell)\mu(m) - \frac{1}{\varphi(q)}\sum\sum_{\substack{L<\ell\leqslant(1+\delta)L\\ M<m\leqslant(1+\delta)M\\ (\ell m,q)=1}} \lambda(\ell)\mu(m).$$

For each of $D(LM;q,a)$ we can apply Theorem 17.4 with $\Delta = (\log x)^{-A}$ because the sequence $\mu(m)$ satisfies the hypothesis (17.12) by the Siegel-Walfisz Theorem (also the sequence $\lambda(\ell)$ satisfies this hypothesis). This gives

$$\sum_{q\leqslant Q}\max_{(a,q)=1}|D(LM;q,a)| \ll \delta\Delta x(\log x)^3 \tag{17.21}$$

for $Q = \Delta x^{\frac{1}{2}}$. Summing over L, M (there are $O(\delta^{-2})$ such segments) we obtain by (17.20) and (17.21) that

$$\sum_{q\leqslant Q}\max_{(a,q)=1}|D_{\Lambda^\flat}(x;q,a)| \ll (\delta^{-1}\Delta+\delta)x(\log x)^3. \tag{17.22}$$

We choose $\delta = \Delta^{\frac{1}{2}}$ getting the bound $\Delta^{\frac{1}{2}}x(\log x)^3$. Adding (17.22) to (17.18) we obtain

$$\sum_{q\leqslant\Delta x^{\frac{1}{2}}}\max_{(a,q)=1}\left|\psi(x;q,a)-\frac{\psi(x)}{\varphi(q)}\right| \ll \Delta^{\frac{1}{2}}x(\log x)^3. \tag{17.23}$$

Here we can replace $\psi(x)$ by $x + O(\Delta x)$ by the Prime Number Theorem. Having done this, Theorem 17.1 follows with

$$B(A) = 2A + 6. \tag{17.24}$$

17.4. Proof of the Barban-Davenport-Halberstam Theorem.

We shall establish Theorem 17.2 for arithmetic functions f more general than Λ. It does not require f to be a convolution type as in Theorem 17.4, but only that f is well distributed over residue classes to quite small moduli.

THEOREM 17.5. *Suppose that for some $0 < \Delta \leqslant 1$ we have*

$$|D_f(x;q,a)| \leqslant \|f\| x^{\frac{1}{2}} \Delta^9 \tag{17.25}$$

for all $(a,q) = 1$. Then

$$\sum_{q\leqslant Q} \sum_{a(\bmod q)}^{\star} |D_f(x;q,a)|^2 \ll \|f\|^2 (\Delta x + Q)(\log Q)^2 \tag{17.26}$$

where the implied constant is absolute.

PROOF. By the orthogonality of characters

$$\sum_{a(\bmod q)}^{\star} |D_f(x;q,a)| = \frac{1}{\varphi(q)} \sum_{\substack{\chi(\bmod q)\\ \chi\neq\chi_0}} \Big|\sum_{n\leqslant x} f(n)\chi(n)\Big|^2.$$

Averaging over q, then reducing to the primitive characters, we see that the left side of (17.26) is bounded by

$$\sum_{s\leqslant Q} \frac{1}{\varphi(s)} \sum_{1<r\leqslant Q} \frac{1}{\varphi(r)} \sum_{\chi(\bmod r)}^{\star} \Big| \sum_{\substack{n\leqslant x\\ (n,s)=1}} f(n)\chi(n)\Big|^2.$$

At this point we apply the same arguments as in the proof of Theorem 17.4 with $\alpha = \beta = f$ getting the bound

$$\|f\|^2 \{x\Delta^3 R^2 + Q + x^{\frac{1}{2}} + xR^{-1}\}(\log Q)^2.$$

Choosing $R = \Delta^{-1}$ we arrive at (17.26). □

COROLLARY 17.6. *Let the conditions be as in Theorem 17.4. Let $ab \neq 0$. Then we have*

(17.27)

$$\sum_{\substack{q\leqslant Q\\ (q,ab)=1}} \Big| \sum\sum_{\substack{am\equiv bn(\bmod q)\\ (mn,q)=1}} \alpha_m\beta_n - \frac{1}{\varphi(q)}\Big(\sum_{(m,q)=1} \alpha_m\Big)\Big(\sum_{(n,q)=1} \beta_n\Big)\Big|$$
$$\ll \|\alpha\|\,\|\beta\|(M+Q)^{\frac{1}{2}}(\Delta N + Q)^{\frac{1}{2}}(\log Q)^2.$$

PROOF. The difference of the sums in (17.27) is also given by

$$\sum_{au\equiv bv(\bmod q)}^{\star}\sum^{\star} D_\alpha(M;q,u)D_\beta(N;q,v).$$

Hence the result follows by Cauchy's inequality from Theorem 17.5 applied for $f = \alpha$ with $\Delta = 1$ and for $f = \beta$ with Δ. □

CHAPTER 18

THE LEAST PRIME IN AN ARITHMETIC PROGRESSION

18.1. Introduction.

After having uniform distribution of primes in primitive residue classes to a given modulus q the natural question to ask is how big is the first prime in a given class? Denote

$$p(q,a) = \min\{p : p \equiv a (\mathrm{mod}\ q)\}. \tag{18.1}$$

Clearly $p(q,a)$ cannot be smaller than $(1+o(1))\varphi(q)\log q$ for some $a(\mathrm{mod}\ q)$. It is conjectured that

$$p(q,a) \ll q^{1+\varepsilon} \tag{18.2}$$

while the Riemann Hypothesis for Dirichlet L-functions of characters $\chi(\mathrm{mod}\ q)$ implies

$$p(q,a) \ll (\varphi(q)\log q)^2. \tag{18.3}$$

Unconditionally, it is easy to derive from the Siegel-Walfisz Theorem (see Corollary 5.29) that

$$p(q,a) \ll \exp(q^{\varepsilon})$$

for any $\varepsilon > 0$, the implied constant depending on ε (ineffectively). Moreover, if there is no exceptional character modulo q, then it follows from the Prime Number Theorem that

$$p(q,a) \ll q^{c \log q}$$

where c is a positive constant. The celebrated theorem of Yu. V. Linnik [**Li4**], [**Li5**], [**Li6**] of 1944 asserts

THEOREM 18.1 (LINNIK). *There exist absolute constants $c \geqslant 1$ and $L \geqslant 2$ such that*

$$p(q,a) \leqslant cq^L. \tag{18.4}$$

This beautiful theorem is one of the greatest achievements in analytic number theory. Originally Linnik did not give a numerical value of L, though his method was effective, i.e., both constants c and L could have been computed given the time and will. Here is a selected list of the Linnik constant produced by various researchers

$L = 10{,}000$	Pan (1957),	$L = 20$	Graham (1981),
$L = 777$	Chen (1965),	$L = 16$	Wang (1986),
$L = 80$	Jutila (1977),	$L = 5.5$	Heath-Brown (1992).

In this chapter we prove Linnik's theorem without determining the constant L. Our treatments borrow substantially from the magnificent work of S. Graham [**G1**], [**G2**]. For readers interested in learning other methods (the Turán power-sums), we recommend Chapter 6 of E. Bombieri [**Bo2**]. The sharpest tools are developed in Heath-Brown [**HB4**].

All proofs of Linnik's theorem use in one form or another the following three principles about the zeros of the Dirichlet L-functions. Throughout we denote

$$L_q(s) = \prod_{\chi(\mathrm{mod}\ q)} L(s,\chi). \tag{18.5}$$

A zero of $L(s,\chi)$ will be denoted by $\rho = \beta + i\gamma$, or by $\rho_\chi = \beta_\chi + i\gamma_\chi$ if the dependence on the character need be displayed. For $\frac{1}{2} \leqslant \alpha \leqslant 1$ and $T \geqslant 1$ we denote by $N(\alpha, T, \chi)$ the number of zeros ρ_χ counted with multiplicity in the rectangle

$$\alpha < \sigma \leqslant 1, \quad |t| \leqslant T. \tag{18.6}$$

Thus

$$N_q(\alpha, T) = \sum_{\chi(\mathrm{mod}\ q)} N(\alpha, T, \chi) \tag{18.7}$$

is the number of all zeros of $L_q(s)$ counted with multiplicity in the rectangle (18.6).

PRINCIPLE 1 (THE ZERO-FREE REGION). *There is a positive constant c_1 (effectively computable) such that $L_q(s)$ has at most one zero in the region*

$$\sigma \geqslant 1 - c_1/\log qT, \quad |t| \leqslant T. \tag{18.8}$$

The exceptional zero, if it exists, is real and simple and it is for a real, non-principal character.

PRINCIPLE 2 (THE LOG-FREE ZERO-DENSITY ESTIMATE). *There are positive constants c_1, c_2 (effectively computable) such that for any $\frac{1}{2} \leqslant \alpha \leqslant 1$ and $T \geqslant 1$,*

$$N_q(\alpha, T) \leqslant c(qT)^{c_2(1-\alpha)}. \tag{18.9}$$

PRINCIPLE 3 (THE EXCEPTIONAL ZERO REPULSION). *There is a positive constant c_3 (effectively computable) such that, if the exceptional zero β_1 exists, say $L(\beta_1, \chi_1) = 0$ with*

$$1 - c_1/\log qT \leqslant \beta_1 < 1, \tag{18.10}$$

then the function $L_q(s)$ has no other zeros in the region

$$\sigma \geqslant 1 - c_3 \frac{|\log(1-\beta_1)\log qT|}{\log qT}, \quad |t| \leqslant T. \tag{18.11}$$

The first principle is classical (due to Landau), and is proved in Section 5.9. The second principle is due to Linnik [**Li6**] and the third principle is a quantitative version of the Deuring-Heilbronn phenomenon (also due to Linnik [**Li5**]). Bombieri [**Bo2**] established a somewhat stronger version of Principle 3. He showed that if there is an exceptional zero, then Principle 2 can be made stronger as well. We extract from his result the following

PROPOSITION 18.2 (BOMBIERI). *There is a positive constant c_2 (effectively computable) such that, if the zero β_1 satisfying (18.10) exists, then*

$$N_q(\alpha, T) \ll (1-\beta_1)(\log qT)(qT)^{c_2(1-\alpha)} \tag{18.12}$$

for any $\frac{1}{2} \leqslant \alpha \leqslant 1$, where the implied constant is absolute (and effectively computable).

EXERCISE 1. Show that Proposition 18.2 implies Principle 3.

EXERCISE 2. Derive from Proposition 18.2 the Siegel bound (5.73) for real zeros of Dirichlet L-functions.

The numerical values of the constants c_1, c_2, c_3 and c yield the Linnik constant L, but from our estimates this will be quite large. Assuming (as we can) that q is sufficiently large and $T \leqslant \log q$ the above principles are known to hold for the constants $c_1 = \frac{1}{10}, c_2 = 3, c_3 = \frac{1}{2}$.

18.2. The log-free zero-density theorem.

Recall the Huxley density estimate (see Theorem 10.4)

$$N_q(\alpha, T) \ll (qT)^{\frac{12}{5}(1-\alpha)}(\log qT)^A \tag{18.13}$$

where A is an absolute constant. Therefore, (18.9) is new only for zeros near the line Re $s = 1$, namely, for α with

$$1 - \alpha \ll A\mathcal{L}^{-1}\log\mathcal{L} \tag{18.14}$$

where

$$\mathcal{L} = \log qT. \tag{18.15}$$

In this section we are going to prove Principle 2 with the constant $c_2 = 47$, however, the method is capable of giving a much smaller constant by direct modifications.

Note that for the principal character $N(\alpha, T, \chi_0)$ equals the number of zeros of the Riemann zeta function in the rectangle (18.6). Thus by the Vinogradov zero-free region (Corollary 8.28) and the Huxley density estimate (see Chapter 10) it follows that

$$N(\alpha, T, \chi_0) \ll T^{3(1-\alpha)}. \tag{18.16}$$

This could also be established without appealing to the zero-free region along the lines of this section, but we exclude the principal character from consideration here to simplify our exposition.

The main idea behind the proof of Principle 2 is the same as for the density estimates in Chapter 10, i.e., we construct a Dirichlet polynomial which serves as a zero-detector because it assumes enormous values at the zeros of $L(, s, \chi)$. Then, using the duality idea followed by estimation of resulting character sums, we obtain the desired bound for $N_q(\alpha, T)$. If we had applied this directly to $L(s, \chi)$ then a loss of logarithmic factors would occur, which is not acceptable. Therefore some refinements are necessary. We eliminate the logarithmic factors by reducing the coefficients of $L(s, \chi)$ with a device similar to that which gave us a positive proportion of zeros of $\zeta(s)$ on the critical line (see Chapter 24). There are other methods for establishing Principle 2, such as the Power Sum Method of P. Turán (used by Bombieri, Jutila, Montgomery).

We introduce two kinds of mollifications to $L(s,\chi)$. Put

$$K(s) = \sum_{1}^{\infty}\Big(\sum_{d|n}\lambda_d\Big)\Big(\sum_{b|n}\theta_b\Big)n^{-s} \tag{18.17}$$

where λ_d is a continuously cut-off Möbius function and θ_b is a kind of coefficient used in sieve methods. Though λ_d and θ_b appear to have similar structures they play here distinct roles. The first one is applied to annihilate the early terms of the series (except for $n = 1$) while the second one is applied to sift out (or rather to diminish the contribution) of the terms which have small prime factors. Precisely we choose

$$\lambda_d = \mu(d)\min\Big(1, \frac{\log z/d}{\log z/w}\Big) \tag{18.18}$$

for $1 \leqslant d \leqslant z$ where $1 < w < z$, and we set $\lambda_d = 0$ if $d > z$. Then we choose

$$\theta_b = \frac{\mu(b)b}{G\varphi(b)} \sum_{\substack{ab\leqslant y\\(a,bq)=1}} \frac{\mu^2(a)}{\varphi(a)} \tag{18.19}$$

where G is the normalization factor such that $\theta_1 = 1$, i.e.,

$$G = \sum_{\substack{a\leqslant y\\(a,q)=1}} \frac{\mu^2(a)}{\varphi(a)}. \tag{18.20}$$

The coefficients θ_b are coming from the Λ^2-sieve and have the following properties (see Section 6.5):

$$|\theta_b| \leqslant 1 \tag{18.21}$$

$$\sum\sum_{b_1,b_2} \frac{\theta_{b_1}\theta_{b_2}}{[b_1,b_2]} = G^{-1} \tag{18.22}$$

$$G \geqslant \frac{\varphi(q)}{q}\log y \tag{18.23}$$

(in fact one could choose θ_b from essentially any upper-bound sieve of level y).

Since λ_d agrees with $\mu(d)$ for $1 \leqslant d \leqslant w$ it follows by Möbius inversion that

$$\sum_{d|n}\lambda_d = \begin{cases} 1 & \text{if } n = 1,\\ 0 & \text{if } 1 < n \leqslant w.\end{cases} \tag{18.24}$$

For any $n > w$ we have the trivial bound

$$\Big|\sum_{d|n}\lambda_d\Big| \leqslant \tau(n), \tag{18.25}$$

however, it was shown by S. Graham [**G1**] that

$$\sum_{w<n\leqslant x}\Big(\sum_{d|n}\lambda_d\Big)^2 \leqslant \frac{x}{\log z/w}\Big(1+O\Big(\frac{1}{\log z/w}\Big)\Big). \tag{18.26}$$

Since we do not seek the best constants we make the comfortable choice of the levels y, w, z, namely

$$y=(qT)^2,\quad w=(qT)^7,\quad z=(qT)^8. \tag{18.27}$$

For any $\chi(\operatorname{mod} q)$ we have the twisted series

$$K(s,\chi)=\sum_1^\infty\Big(\sum_{d|n}\lambda_d\Big)\Big(\sum_{b|n}\theta_b\Big)\chi(n)n^{-s}. \tag{18.28}$$

This factors into

$$K(s,\chi)=L(s,\chi)M(s,\chi) \tag{18.29}$$

where

$$M(s,\chi)=\sum_m\Big(\sum\sum_{[b,d]=m}\lambda_d\theta_b\Big)\chi(m)m^{-s}. \tag{18.30}$$

Actually we shall take only a partial sum of $K(s,\chi)$,

$$K_x(s,\chi)=\sum_{1\leqslant n\leqslant x}\Big(\sum_{d|n}\lambda_d\Big)\Big(\sum_{b|n}\theta_b\Big)\chi(n)n^{-s} \tag{18.31}$$

with

$$x=(qT)^{23}. \tag{18.32}$$

Notice that $m=[b,d]\leqslant bd\leqslant yz$ and for $\chi\neq\chi_0$

$$\Big|\sum_{n>\frac{x}{m}}\chi(n)n^{-s}\Big|\leqslant 2q|s|\Big(\frac{m}{x}\Big)^\sigma. \tag{18.33}$$

Hence, for $s=\sigma+it$ in the rectangle (18.6) the tail of $K(s,\chi)$ satisfies

$$|K(s,\chi)-K_x(s,\chi)|\leqslant 2q|s|yzx^{-\sigma}\leqslant 4qTyzx^{-\alpha}\leqslant\tfrac12. \tag{18.34}$$

For $s=\rho$, a zero of $L(s,\chi)$, we have $K(\rho,y)=0$ by (18.29), so (18.34) becomes

$$|K_x(\rho,\chi)|\leqslant\tfrac12. \tag{18.35}$$

Extracting from $K_x(\rho,\chi)$ the first term, this inequality becomes the zero detector

$$\Big|\sum_{w<n\leqslant x}\Big(\sum_{d|n}\lambda_d\Big)\Big(\sum_{b|n}\theta_b\Big)\chi(n)n^{-\rho}\Big|\geqslant\tfrac12. \tag{18.36}$$

For the trivial character $\chi=\chi_0$ the detector of zeros is somewhat different from (18.36), so we shall count these zeros separately. Let $\mathcal{S}(\chi)$ denote the set of zeros of $L(s,\chi)$ (counted with multiplicity) in the rectangle (18.6). By (18.36) we deduce that the total number

$$R=\sum_{\chi\neq\chi_0}|\mathcal{S}(\chi)|=\sum_{\chi\neq\chi_0}N(\alpha,T,\chi) \tag{18.37}$$

satisfies

$$R \leqslant 2\sum_{\chi}\sum_{s\in\mathcal{S}(\chi)}\Big|\sum_{w<n\leqslant x}\Big(\sum_{d|n}\lambda_d\Big)\Big(\sum_{b|n}\theta_b\Big)\chi(n)n^{-s}\Big| \leqslant$$

$$2\sum_{w<n\leqslant x}\Big|\Big(\sum_{d|n}\lambda_d\Big)\Big(\sum_{b|n}\theta_b\Big)\Big|\,\Big|\sum_{\chi}\sum_{s\in\mathcal{S}(\chi)}c_\chi(s)\chi(n)n^{-s}\Big|$$

for some numbers $c_\chi(s)$ with $|c_\chi(s)| = 1$. By Cauchy's inequality we obtain

$$R^2 \leqslant 4UV$$

where

$$U = \sum_{w<n\leqslant x}\Big(\sum_{d|n}\lambda_d\Big)^2 n^{1-2\alpha} \tag{18.38}$$

and

$$V = \sum_{n} f(n)\Big(\sum_{b|n}\theta_b\Big)^2 n^{2\alpha-1}\Big|\sum_{\chi}\sum_{s\in\mathcal{S}(\chi)}c_\chi(s)\chi(n)n^{-s}\Big|^2. \tag{18.39}$$

Here f is any non-negative function such that $f(n) \geqslant 1$ for $w < n \leqslant x$.

Using (18.26) one derives by partial summation that for any $\frac{1}{2} \leqslant \alpha \leqslant 1$,

$$U \leqslant x^{2(1-\alpha)}\frac{\log x/w}{\log z/w}\Big\{1 + O\Big(\frac{1}{\log z/w}\Big)\Big\} \ll x^{2(1-\alpha)}. \tag{18.40}$$

To estimate V we square out and change the order of summation getting

$$V \leqslant \sum_{\chi_1}\sum_{\chi_2}\sum_{s_1}\sum_{s_2}|B(\chi_1\bar{\chi}_2, s_1 + \bar{s}_2 + 1 - 2\alpha)| \tag{18.41}$$

where

$$B(\chi, s) = \sum_{n} f(n)\Big(\sum_{b|n}\theta_b\Big)^2\chi(n)n^{-s}$$

for $\chi = \chi_1\bar{\chi}_2$ and $s = s_1 + \bar{s}_2 + 1 - 2\alpha$. Note that $\mathrm{Re}\,(s) \geqslant 1$. Take any f supported on $[w/v, xv]$ which is continuous, bounded and piecewise monotonic. Then we have

$$\sum_{n\equiv\alpha(\mathrm{mod}\ q)} f(dn)(dn)^{-s} = \frac{F(s)}{dq} + O\Big(|s|\frac{v}{w}\Big)$$

where

$$F(s) = \int f(\xi)\xi^{-s}d\xi \tag{18.42}$$

and the implied constant is absolute. Hence we derive that for any $(d, q) = 1$,

$$\sum_{n\equiv 0(\mathrm{mod}\ d)} f(n)\chi(n)n^{-s} = \delta(\chi)\frac{\varphi(q)}{dq}F(s) + O\Big(|s|q\frac{v}{w}\Big)$$

where $\delta(\chi_0) = 1$ and $\delta(\chi) = 0$ otherwise. This yields by (18.28) and (18.29)

$$B(\chi, s) = \delta(\chi)\frac{\varphi(q)}{q}\frac{F(s)}{G} + O\Big(|s|q\frac{v}{w}y^2\Big). \tag{18.43}$$

Introducing (18.43) into (18.41) we get by (18.30)

$$V \leqslant \sum_{\chi} \sum_{s_1,s_2\in\mathcal{S}(\chi)}\sum |F(s_1+\bar{s}_2+1-2\alpha)|(\log y)^{-1} + O(R^2Tqvw^{-1}y^2). \tag{18.44}$$

One can find f such that for $\mathrm{Re}\,(s) \geqslant 1$,

$$F(s) \ll (1+|s-1|\log v)^{-2}\log x. \tag{18.45}$$

For example, we may take

$$f(\xi) = \min\left\{1-\frac{\log w/\xi}{\log v}, 1, 1-\frac{\log \xi/x}{\log v}\right\}$$

for $w/v \leqslant \xi \leqslant xv$ and $f(\xi)=0$ otherwise. Indeed, writing

$$f(\xi)\log v = \log^+(xv/\xi) - \log^+(x/\xi) - \log^+(w/\xi) + \log^+(w/v\xi)$$

one verifies by the formula

$$\int_0^\infty (\log^+\xi)\xi^{s-1}d\xi = s^{-2}$$

that

$$\begin{aligned} F(1-s) = \hat{f}(s) = \int_0^\infty f(\xi)\xi^{s-1}d\xi &= \frac{(xv)^s - x^s - w^s + (w/v)^s}{s^2\log v} \\ &= \frac{(v^s-1)(x^s-w^2v^{-s})}{s^2\log v} \ll \min\Big(\log x, \frac{1}{|s|^2\log v}\Big). \end{aligned}$$

This gives (18.45).

For $s = s_1+\bar{s}_2+1-2\alpha$ we have $|s-1| = |\beta_1+\beta_2-2\alpha+i(\gamma_1-\gamma_2|) \geqslant |\gamma_1-\gamma_2|$. By (18.44) and (18.45) we get

$$V \ll \frac{\log x}{\log y}\sum_{\chi} \sum_{s_1,s_2\in\mathcal{S}(\chi)}\sum (1+|\gamma_1-\gamma_2|\log v)^{-2} + R^2qvw^{-1}y^2. \tag{18.46}$$

LEMMA 18.3. *Let χ (mod q) be a non-trivial character, $\frac{1}{2}\leqslant\alpha\leqslant 1$, $v\geqslant 2$ and t a real number. Then*

$$\sum_{\substack{L(\rho,\chi)=0\\ \beta\geqslant\alpha}} (1+|\gamma-t|\log v)^{-2} \leqslant \frac{1}{2}\Big(1-\alpha+\frac{1}{\log v}\Big)\log Avq(|t|+1) \tag{18.47}$$

where A is an absolute constant.

PROOF. We first suppose that χ is primitive. We have

$$\frac{L'}{L}(s,\chi) = -\frac{1}{2}\log\frac{q}{\pi} - \frac{1}{2}\frac{\Gamma'}{\Gamma}\Big(\frac{s}{2}+\frac{1}{4}(1+\chi(-1))\Big) + B(\chi) + \sum_\rho\Big(\frac{1}{s-\rho}+\frac{1}{\rho}\Big) \tag{18.48}$$

where the summation ranges over all zeros of $L(s,\chi)$ with $\beta>0$ and $B(\chi)$ is a constant with

$$\mathrm{Re}\, B(\chi) = -\sum_\rho \mathrm{Re}\,\frac{1}{\rho}$$

(see (5.29)). Hence it follows that for $1 < \operatorname{Re} s \leqslant 2$,

$$\sum_{\rho} \operatorname{Re} \frac{1}{s-\rho} = \frac{1}{2}\log(q|s|) + \operatorname{Re} \frac{L'}{L}(s,\chi) + O(1). \tag{18.49}$$

For $s = \sigma + it$ with $\sigma > 1$ we have

$$\operatorname{Re} \frac{1}{s-\rho} = \frac{(\sigma-\beta)}{(\sigma-\beta)^2 + (\gamma-t)^2} \geqslant \frac{1}{(\sigma-\beta)}\Big(1 + \frac{|\gamma - t|}{\sigma-1}\Big)^{-2},$$

$$\Big|\frac{L'}{L}(s,\chi)\Big| \leqslant \sum_{1}^{\infty} \Lambda(n) n^{-\sigma} = -\frac{\zeta'}{\zeta}(\sigma) = \frac{1}{\sigma-1} + O(1).$$

Hence (18.49) yields (drop the zeros with $\beta < \alpha$)

$$\sum_{\beta \geqslant \alpha} \Big(1 + \frac{|\gamma-t|}{\sigma-1}\Big)^{-2} \leqslant (\sigma-\alpha)(\tfrac{1}{2}\log q(|t|+1) + \frac{1}{\sigma-1} + O(1)).$$

Choosing $\sigma = 1 + 1/\log v$ we obtain (18.47). If $\chi (\bmod q)$ is induced by a primitive character $\chi^\star (\bmod q^\star)$ with $q^\star \mid q$, then the left side of (18.47) agrees with that for $\chi^\star$, so the result holds for any $\chi \neq \chi_0$.

Notice that $1 + (1-\alpha)\log v \leqslant v^{1-\alpha}$. Therefore Lemma 18.3 and (18.46) yield

$$V \ll Rv^{1-\alpha} \frac{\log x}{\log y} \frac{\log vqT}{\log v} + R^2 Tq\frac{v}{w} y^2 \ll Rv^{1-\alpha} \tag{18.50}$$

by the trivial bound $R \ll qT \log qT$ and by choosing

$$v = qT. \tag{18.51}$$

Multiplying (18.40) and (18.53) we obtain by (18.38) that

$$R \ll (x^2 v)^{1-\alpha} = (qT)^{47(1-\alpha)}. \tag{18.52}$$

Hence we get (18.9) by adding the number of zeros of $L(s, \chi_0)$ which is negligible by virtue of (18.16). □

18.3. The exceptional zero repulsion.

In this section we are going to prove Principle 3, however, our method is capable of giving (18.13). It begins by a suitable modification of the arguments which we used for the proof of Principle 2 in the previous section.

Suppose $\chi_1 (\bmod q)$ is the exceptional character and β_1 is the exceptional zero of $L(s, \chi_1)$ which satisfies

$$\delta_1 = 1 - \beta_1 \leqslant c_1 (\log qT)^{-1}. \tag{18.53}$$

Throughout we assume (as we can) that c_1 is a sufficiently small, absolute positive constant. Our goal is to prove there exists an absolute constant $c_0 \geqslant 2c_1$ such that the function $L_q(s)$ has no zero other than β_1 in the region

$$\sigma \geqslant 1 - \frac{\log(c_0/\delta_1 \log qT)}{92 \log qT}, \quad |t| \leqslant T. \tag{18.54}$$

Clearly this result implies Principle 3.

To take full advantage of the repulsion property of β_1 we apply the method of zero-detectors from the previous section to the function $\zeta(s)L(s+\delta_1,\chi_1)$ rather than to $\zeta(s)$. We have

$$\zeta(s)L(s+\delta_1,\chi_1)=\sum_1^\infty \rho(n)n^{-s} \tag{18.55}$$

where $\rho(n)$ is a positive, multiplicative function given by

$$\rho(n)=\sum_{a|n}\chi_1(a)a^{-\delta_1}. \tag{18.56}$$

REMARK. One could work as well with the product

$$\zeta(s)L(s,\chi_1)=\sum_1^\infty \tau(n,\chi_1)n^{-s} \tag{18.57}$$

where

$$\tau(n,\chi_1)=\sum_{a|n}\chi_1(a) \tag{18.58}$$

but the slight shift in $L(s+\delta_1,\chi_1)$ turns out to simplify some technical arguments.

As before we introduce two kinds of mollifications to $\zeta(s)L(s+\delta_1,\chi_1)$ getting

$$K(s)=\sum_1^\infty\Big(\sum_{d|n}\lambda_d\Big)\Big(\sum_{b|n}\theta_b\Big)\rho(n)n^{-s} \tag{18.59}$$

where λ_d are defined by (18.18) and θ_b are defined by (18.19) with the parameters y,w,z given by (18.27). Then for any non-principal character $\chi(\bmod q)$ we consider the twisted series

$$K(s,\chi)=\sum_1^\infty\Big(\sum_{d|n}\lambda_d\Big)\Big(\sum_{b|n}\theta_b\Big)\rho(n)\chi(n)n^{-s}. \tag{18.60}$$

This factors into

$$K(s,\chi)=L(s,\chi)L(s+\delta_1,\chi\chi_1)M(s;\chi) \tag{18.61}$$

where

$$M(s,\chi)=\sum_m\Big(\sum_{[b,d]=m}\sum\lambda_d\theta_b\Big)\prod_{p|m}(\rho(p)-\chi(p)p^{-s-2\delta_1})\chi(m)m^{-s}. \tag{18.62}$$

To see this we arrange $K(s,\chi)$ as follows:

$$K(s,\chi)=\sum_m\Big(\sum_{[b,d]=m}\sum\lambda_d\theta_b\Big)\chi(m)m^{-s}\sum_n\rho(mn)\chi(n)n^{-s}.$$

Opening $\rho(mn)$ by (18.56) we see that the innermost sum is

$$\begin{aligned}\sum_a \chi_1(a)a^{-\delta_1} \sum_{n\equiv 0(\text{mod } a/(a,m))} \chi(n)n^{-s} &= L(s,\chi)\sum_a \frac{\chi_1(a)}{a^{\delta_1}}\chi\Big(\frac{a}{(a,m)}\Big)\Big(\frac{(a,m)}{a}\Big)^s \\ &= L(s,\chi)\sum_{c|m}\chi_1(c)c^{-\delta_1}\sum_{(a,c)=1}\chi_1\chi(a)a^{-s-\delta_1} \\ &= L(s,\chi)L(s+\delta_1,\chi\chi_1)\sum_{c|m}\chi_1(c)c^{-\delta_1}\prod_{p|c}(1-\chi_1\chi(p))p^{-s-\delta_1}\end{aligned}$$

and this yields (18.61) and (18.62) because m is squarefree. By (18.61) it follows that $K(s,\chi)$ is holomorphic in the whole complex plane (if $\chi=\chi_1$, then the pole of $L(s+\delta_1,\chi_0)$ at $s=1-\delta_1=\beta_1$ cancels with the zero of $L(s,\chi_1)$).

As before we take only a partial sum of $K(s,\chi)$,

$$K_x(s,\chi) = \sum_{1\leqslant n\leqslant x}\Big(\sum_{d|n}\lambda_d\Big)\Big(\sum_{b|n}\theta_b\Big)\rho(n)\chi(n)n^{-s} \tag{18.63}$$

with x given by (18.32), and we show by contour integration that the tail of $K(s,\chi)$ satisfies

$$|K(s,\chi)-K_x(s,\chi)| \leqslant \tfrac{1}{2} \tag{18.64}$$

for $s=\sigma+it$ with $\sigma\geqslant\frac12$ and $|t|\leqslant T$. Then for $s=\rho$, a zero of $L(s,\chi)$ which is different from β_1 if $\chi=\chi_1$, we have $K(\rho,\chi)=0$ by (18.61), therefore (18.64) yields

$$\Big|\sum_{w<n\leqslant x}\Big(\sum_{d|n}\lambda_d\Big)\Big(\sum_{b|n}\theta_b\Big)\rho(n)\chi(n)n^{-\rho}\Big| \geqslant \tfrac12 \tag{18.65}$$

if $\rho=\beta+i\gamma$ with $\beta\geqslant\frac12$ and $|\gamma|\leqslant T$.

The inequality (18.65) is viewed as the zero-detector. This wouldn't be useful if not for the fact that $\rho(n)$ is very small quite frequently. One cannot exploit a possible cancellation of terms because of the complexity of the coefficients λ_d. Remember that it is due to the introduction of the factors

$$\omega(n) = \sum_{d|n}\lambda_d \tag{18.66}$$

that the early terms with $1<n\leqslant w$ are annihilated. Now, our strategy will be to remove these factors (by applying Hölder's inequality) to create a sum of $\nu(n)^2\rho(n)$, where

$$\nu(n) = \sum_{b|n}\theta_b \tag{18.67}$$

over n with $w<n\leqslant z$. When applying Hölder's inequality one could also remove the factor $\nu(n)^2$, but it would result in a loss of $\log q$ which is not acceptable. For this reason we retain $\nu(n)^2\rho(n)$. Still, the sum of $\nu(n)^2\rho(n)$ is manageable because the support of θ_b is relatively small by comparison to the range of n. In this way by taking advantage that $\rho(n)$ nearly vanishes quite often (depending on how close to one is the exceptional zero) we derive from the inequality (18.65) that β is not close to one, precisely this inequality prohibits $\rho=\beta+i\gamma$ to be in the region (18.54).

Having outlined the strategy we proceed to details. By (18.65) we get

$$\sum_{w<n\leqslant x}|\omega(n)\nu(n)|\rho(n)n^{-\beta}\geqslant \tfrac{1}{2}.$$

Hence applying Hölder's inequality

$$16U^2VW\geqslant 1 \tag{18.68}$$

where

$$U=\sum_{w<n\leqslant x}\omega^2(n)n^{1-2\beta},$$
$$V=\sum_{w<n\leqslant x}\nu^2(n)\rho^3(n)n^{-1},$$
$$W=\sum_{w<n\leqslant x}\nu^2(n)\rho(n)n^{-1}.$$

For U we have by (18.40)

$$U\ll x^{2(1-\beta)}. \tag{18.69}$$

In V we estimate $\rho^3(n)$ by $\tau^3(n)$ and apply a sieve of dimension 8 (see the Fundamental Lemma 6.3) getting

$$V\ll\left(\frac{\log x}{\log y}\right)^8\ll 1. \tag{18.70}$$

We shall show that

$$W\ll\delta_1\log x. \tag{18.71}$$

Multiplying these estimates we derive from (18.68) that

$$\delta_1(\log x)x^{4(1-\beta)}>23c_0, \tag{18.72}$$

where c_0 is a positive absolute constant. Assuming (18.53) holds with $2c_1\leqslant c_0$ we conclude by (18.72) with $x=(qT)^{23}$ that $\rho=\beta+i\gamma$ is not in the region (18.54) as claimed.

We now proceed to the proof of (18.71), it is here where the exceptional zero plays its role. Our treatment is essentially elementary, yet some arguments are quite subtle (we borrow these from the work of Bombieri [**Bo2**]). We begin by an asymptotic evaluation of W. To this end we examine the generating zeta function

$$W(s)=\sum_1^\infty\nu^2(n)\rho(n)n^{-s}. \tag{18.73}$$

This is just (18.60) with $\lambda_d=\theta_d$ and $\chi=\chi_0$, so by (18.61) and (18.62)

$$W(s)=L(s,\chi_0)L(s+\delta_1,\chi_1)M(s)$$

where

$$M(s)=\sum_{(m,q)=1}\sigma_m\prod_{p|m}(\rho(p)-p^{-s-2\delta_1})m^{-s}$$

and

$$\sigma_m = \sum\sum_{[b_1,b_2]=m} \theta_{b_1}\theta_{b_2}.$$

Hence $W(s)$ is holomorphic everywhere except for a simple pole at $s = 1$ with

$$\operatorname*{res}_{s=1} W(s) = \frac{\varphi(q)}{q} L(1+\delta_1, \chi_1) M(1).$$

By contour integration we derive that

$$W = \frac{\varphi(q)}{q} M(1) L(1+\delta_1, \chi_1) \log \frac{x}{w} + O\Big(\frac{1}{q}\Big). \tag{18.74}$$

Now we need estimates for $M(1)$ and $L(1+\delta_1, \chi_1)$. From sieve theory (see Chapter 6) we get

$$\begin{aligned} M(1) &= \sum_{(m,q)=1} \sigma_m \prod_{p|m} (\rho(p) - p^{-1-2\delta_1}) m^{-1} \\ &\ll \prod_{\substack{p\leqslant y \\ p\nmid q}} \Big(1 - \frac{\rho(p)}{p}\Big) = \frac{q}{\varphi(q)} \prod_{p\leqslant y} \Big(1 - \frac{\rho(p)}{p}\Big). \end{aligned} \tag{18.75}$$

To estimate $L(1+\delta_1, \chi_1)$ we first examine

$$P(x,y) = \sum_{y<p\leqslant x} (1+\chi_1(p)) p^{-1}. \tag{18.76}$$

LEMMA 18.4. *For $x > y \geqslant q^2$ and q sufficiently large we have*

$$P(x,y) \leqslant 4\delta_1 \log x. \tag{18.77}$$

PROOF. Consider

$$S(x) = \sum_{n\leqslant x} \tau(n, \chi_1) n^{-1}$$

where $\tau(n,\chi_1)$ is given by (18.58), so it is a multiplicative function with $\tau(p,\chi_1) = 1 + \chi_1(p)$. Multiplying $P(x,y)$ and $S(y)$ we deduce from the non-negativity of $\tau(n,\chi_1)$ that

$$P(x,y)S(y) \leqslant S(xy) - S(y).$$

Then by the elementary formula (see (22.11))

$$S(x) = L(1,\chi_1)(\log x + \gamma) + L'(1,\chi_1) + O(q^{\frac14} x^{-\frac12} \log x)$$

we get

$$S(xy) - S(x) = L(1,\chi_1)\log x + O(q^{\frac14} y^{-\frac12} \log y) \leqslant 2L(1,\chi_1)\log x$$

because $L(1,\chi_1) \gg q^{-\frac{1}{2}}$ and $y \geqslant q^2$. On the other hand, we derive a lower bound for $S(y)$ as follows

$$\begin{aligned} S(y) &\geqslant y^{-\delta_1} \sum_{n \leqslant y} \tau(n,\chi_1)\Big(1-\frac{n}{y}\Big) n^{-\beta_1} \\ &= \frac{2}{2\pi i}\int_{(1)} \zeta(s+\beta_1)L(s+\beta_1,\chi_1)\frac{y^{s-\delta_1}}{s(s+1)}ds \\ &= \frac{2L(1,\chi_1)}{\delta_1(\delta_1+1)} + O(y^{-\frac{1}{2}}q^{\frac{1}{4}}) \geqslant \frac{1}{2\delta_1}L(1,\chi_1). \end{aligned}$$

Combining these bounds we complete the proof of (18.77). □

REMARKS. Lemma 18.4 shows that if $\delta_1 = 1-\beta_1 = o(1/\log q)$, then $\chi_1(p) = -1$ for almost all primes in segments $q^2 < p \leqslant q^A$. For very small primes the above arguments are rather crude, but see Chapter 22, in particular (22.100).

Now we are ready to estimate $L(1+\delta_1,\chi_1)$. For any $1 < s \leqslant 2$ we have

$$\zeta(s)L(s,\chi_1) \asymp \prod_p (1+\tau(p,\chi_1)p^{-s})$$

and by Lemma 18.4 we obtain by partial summation

$$\prod_{p>y}(1+\tau(p,\chi_1)p^{-s}) \leqslant \exp\Big(\sum_{p>y}\tau(p,\chi_1)p^{-s}\Big) \ll \exp\Big(\frac{4\delta_1}{s-1}\Big).$$

Hence

$$L(s,\chi_1) \ll (s-1)\exp\left(\frac{4\delta_1}{s-1}\right)\prod_{p\leqslant y}(1+\tau(p,\chi_1)p^{-s}).$$

In particular, we get

$$L(1+\delta_1,\chi_1) \ll \delta_1 \prod_{p\leqslant y}\Big(1+\frac{\rho(p)}{p}\Big). \tag{18.78}$$

Combining (18.74), (18.75) and (18.78) we get $W \ll \delta_1 \log x + q^{-1}$ which completes the proof of (18.71), and of Principle 3.

18.4. Proof of Linnik's Theorem.

Throughout c_1, c_2, c_3 and c are the absolute constants from Principles 1, 2 and 3. We assume (as we can) that

$$c, c_2 > 1 > c_1, c_3 > 0. \tag{18.79}$$

Let

$$x \geqslant q^{4c_2} \tag{18.80}$$

and put

$$R = x^{1/2c_2}. \tag{18.81}$$

We use the truncated explicit formula (see (5.65))

$$\psi(x;q,a)=\frac{x}{\varphi(q)}-\frac{1}{\varphi(q)}\sum_{\chi(\mathrm{mod}\ q)}\bar{\chi}(a)\sum_{\rho_\chi}^{R}\frac{x^{\rho_\chi}}{\rho_\chi}+O\Big(\frac{x}{R}\log x\Big)$$

where $\sum^R$ restricts the summation to zeros $\rho_\chi=\beta_\chi+i\gamma_\chi$ of $L(s,\chi)$ in $\sigma\geqslant\frac{1}{2}, |t|\leqslant R$. First we estimate the sum over zeros ρ_χ with $\frac{1}{2}T<|\gamma_\chi|\leqslant T$ for $1\leqslant T\leqslant R$ by using Principle 2 as follows:

$$\begin{aligned}\Big|\sum_{\chi(\mathrm{mod}\ q)}\bar{\chi}(a)\sum_{\frac{1}{2}T<|\gamma_\chi|\leqslant T}x^{\rho_\chi}\rho_\chi^{-1}\Big|&\leqslant\frac{2}{T}\sum_{\chi}\sum_{|\gamma_\chi|\leqslant T}x^{\beta_\chi}=-\frac{2}{T}\int_{1/2}^{1}x^{\alpha}dN_q(\alpha,T)\\&=\frac{2}{T}x^{\frac{1}{2}}N_q(\tfrac{1}{2},T)+\frac{2}{T}(\log x)\int_{1/2}^{1}N_q(\alpha,T)x^{\alpha}d\alpha\\&\leqslant\frac{2c}{T}x^{\frac{1}{2}}(qT)^{\frac{c_2}{2}}+\frac{2c}{T}x(\log x)\int_{1/2}^{1}((qT)^{c_2}/x)^{1-\alpha}d\alpha\\&\leqslant\frac{2c}{T}\frac{x\log x}{\log(x(qT)^{-c_2})}\leqslant 2c(c_2+1)\frac{x}{T}\end{aligned}$$

provided $x\geqslant(qT)^{c_2+1}$ which is our case. Notice the absence of $\log x$ in the last bound. Hence the explicit formula reduces to

$$\psi(x;q,a)=\frac{x}{\varphi(q)}-\frac{1}{\varphi(q)}\sum_{\chi(\mathrm{mod}\ q)}\bar{\chi}(a)\sum_{\rho_\chi}^{T}\frac{x^{\rho_\chi}}{\rho_\chi}+O\Big(\frac{x}{\varphi(q)T}+\frac{x}{R}\log x\Big)\tag{18.82}$$

where $\sum^T$ restricts the summation to zeros ρ_χ with $|\gamma_\chi|\leqslant T$, and the implied constant is absolute. We choose

$$T=q,\tag{18.83}$$

so the error term in (18.82) is $O(x(\log q)/q\varphi(q))$.

Now we proceed to a more precise treatment of the sum over the zeros with $|\gamma_\chi|\leqslant T$. If there exists a real character $\chi_1(\mathrm{mod}\ q)$ for which $L(s,\chi_1)$ has a real zero $\beta_1=1-\delta_1$ with

$$\delta_1\leqslant c_1/2\log q,\tag{18.84}$$

we exclude it from $\sum^T$ and we estimate the remaining sum $\sum'$ by an appeal to Principle 1 as follows:

$$\begin{aligned}\Big|\sum_{\chi(\mathrm{mod}\ q}\bar{\chi}(a)\sum_{\rho_\chi}'\frac{x^{\rho_\chi}}{\rho_\chi}\Big|&\leqslant 2\sum_{\chi(\mathrm{mod}\ q)}\sum_{\rho_\chi}'x^{\beta_\chi}\\&=-2\int_{1/2}^{1-\eta}x^{\alpha}dN_q(\alpha,T)=2x^{\frac{1}{2}}N_q(\tfrac{1}{2},T)+2(\log x)\int_{\frac{1}{2}}^{1-\eta}N_q(\alpha,T)x^{\alpha}d\alpha,\end{aligned}$$

where

$$\eta=c_1/2\log q\tag{18.85}$$

if the zero β_1 does not exist, and

$$\eta=c_3|\log(2\delta_1\log q)|/2\log q\tag{18.86}$$

if β_1 exists. By Principle 2 we have

$$N_q(\alpha, T) \leqslant cq^{2c_2(1-\alpha)},$$

so our sum over the zeros ρ_χ other than β_χ is bounded by

$$\begin{aligned} 2cx^{\frac{1}{2}}q^{c_2} + 2cx(\log x)\int_{\frac{1}{2}}^{1-\eta}(q^{2c_2}/x)^{1-\alpha}d\alpha \\ \leqslant \frac{2cx\log x}{\log(xq^{-2c_2})}\left(\frac{q^{2c_2}}{x}\right)^{\eta} \leqslant 4cx^{1-\eta/2}. \end{aligned}$$

Therefore we have established the following

PROPOSITION 18.5. *Let c, c_1, c_2, c_3 be the absolute constants from Principles 1, 2, 3. Then for $x \geqslant q^{4c_2}$ we have*

$$\psi(x;q,a) = \frac{x}{\varphi(q)}\left\{1 - \chi_1(a)\frac{x^{\beta_1-1}}{\beta_1} + \theta cx^{-\eta/2} + O\Big(\frac{\log q}{q}\Big)\right\} \tag{18.87}$$

where the term involving β_1 does not exist if β_1 does not, η is given by (18.85) or (18.86) respectively, $|\theta| \leqslant 4$ and the implied constant is absolute.

In the first case of Proposition 18.5 we deduce

THEOREM 18.6. *Let c_1, c_2 and c be the absolute constants from Principle 1 and Principle 2 respectively. Suppose for any real character $\chi_1 \pmod q$ there is no real zero of $L(s, \chi_1)$ in*

$$\sigma > 1 - c_1/2\log q. \tag{18.88}$$

Then for $x \geqslant q^{4c_2}$ we have

$$\psi(x;q,a) = \frac{x}{\varphi(q)}\Big(1 + \theta c\exp\Big(-\frac{c_1}{4}\frac{\log x}{\log q}\Big) + O\Big(\frac{\log q}{q}\Big)\Big) \tag{18.89}$$

where $|\theta| \leqslant 4$, and the implied constant is absolute (effectively computable).

In the second case when β_1 exists, we assume that $x \geqslant q^\nu$ where

$$\nu = \max\Big(4c_2, \frac{4}{c_1}, \frac{4\log 8c}{c_3|\log c_1|}\Big) \tag{18.90}$$

and we estimate as follows:

$$\begin{aligned} 1 - \chi_1(a)\frac{x^{\beta_1-1}}{\beta_1} &\geqslant 1 - \frac{x^{-\delta_1}}{\beta_1} \geqslant \beta_1 - x^{-\delta_1} \geqslant \beta_1 - q^{-\nu\delta_1} \\ &= 1 - q^{-\nu\delta_1} - \delta_1 \geqslant \frac{\nu\delta_1\log q}{1+\nu\delta_1\log q} - \delta_1 \\ &\geqslant \frac{\nu\delta_1\log q}{1+\nu c_1/2} - \delta_1 \geqslant \frac{4\delta_1}{3c_1}\log q - \delta_1, \end{aligned}$$

$$x^{-\eta/2} \leqslant q^{-\nu\eta/2} = (2\delta_1\log q)^{\nu c_3/4} \leqslant (2\delta_1\log q)c_1^{\nu c_3/4-1} \leqslant \frac{\delta_1}{4cc_1}\log q.$$

Inserting these estimates into (18.87) we deduce

THEOREM 18.7. *Let c, c_1, c_2, c_3 be the absolute constants from Principles 1, 2 and 3. Suppose there exists a real zero β_1 of $L(s, \chi_1)$ for a real character $\chi_1 \pmod q$ such that $\delta_1 = 1 - \beta_1 \leqslant c_1/2\log q$. Then for $x \geqslant q^\nu$ with ν given by (18.90) we have*

$$\psi(x; q, a) \geqslant \frac{x}{\varphi(q)} \frac{\delta_1 \log q}{4c_1} \Big(1 + O\Big(\frac{1}{\sqrt{q}}\Big)\Big) \tag{18.91}$$

where the implied constant is absolute.

From Theorems 18.6 and 18.7 it follows that in any case we have (because $\delta_1 \log q \gg q^{-\frac{1}{2}}$)

COROLLARY 18.8. *If q is sufficiently large and $x \geqslant q^L$ with*

$$L = \max\Big\{4c_2, \frac{4}{c_1}\log 8c, \frac{4}{c_3}\frac{\log 8c}{|\log c_1|}\Big\}, \tag{18.92}$$

then

$$\psi(x; q, a) \gg \frac{x}{\varphi(q)\sqrt{q}} \tag{18.93}$$

where the implied constant is absolute.

This result is a quantitative version of Linnik's theorem.

CHAPTER 19

THE GOLDBACH PROBLEM

19.1. Introduction.

Christian Goldbach (1690-1764), a member of the Petersburg Academy, in a letter of 1742 to Leonard Euler posed the problem of proving that every number $N \geqslant 5$ can be represented as the sum of three primes. Euler answered with the equivalent hypothesis that every even number $N \geqslant 4$ is a sum of two primes. Inspired by many failures on these problems, Edmund Landau challenged (in 1912) to show that every number $N > 1$ is a sum of at most k primes with k sufficiently large but fixed. Landau's problem was solved in 1930 by L. G. Shnirelman (see e.g. [**Kh**]). The work of Shnirelman opened a new direction in general additive number theory.

In 1937 I. M. Vinogradov [**V6**] succeeded in solving the original Goldbach problem for all odd N which are sufficiently large. Before him, G. H. Hardy and J. E. Littlewood [**HL1**] made a serious attack by means of a then new circle method. Indeed, they succeeded conditionally under the Riemann Hypothesis for Dirichlet L-functions. Vinogradov followed their line of attack (with his own beautification of the circle method), and removed the use of GRH in the estimate of the exponential sum over primes

$$\sum_{p \leqslant N} e(\alpha p), \tag{19.1}$$

which is required by the circle method. His treatment of this particular sum (sieve and double sums methods) became a fundamental tool for estimating a large class of sums over primes (see Chapter 13).

In this chapter we derive Vinogradov's Three Primes Theorem by a mixture of his original and somewhat modern ideas. We replace the exponential sum over primes (19.1) by the exponential sum with the Möbius function for which we have a uniform bound (13.54) due to H. Davenport. In this way we minimize the role played by the circle method, and we do not appeal to the distribution of primes in residue classes (the use of these properties is accumulated in the proof of (13.54)). One can avoid completely the idea of circle method by using only the elementary dispersion method (see the Rutgers Lecture Notes of 1994, Chapter 6).

None of the current methods promises any chance to solve the Goldbach-Euler problem for two primes. However, it was established independently by N. G. Tchudakov [**Tchu**], van der Corput [**Cor3**] and T. Estermann [**Est**] that almost all (in the sense of density) even numbers are representable as the sum of two primes. This follows by a suitable generalization of the Vinogradov result (one of the three primes can be taken from any sequence of integers which is relatively dense). H.L. Montgomery and R.C. Vaughan [**MV3**] showed that the number of exceptional even integers (those which are not representable as the sum of two primes) is quite small.

For even numbers N we consider

$$G_2(N) = \sum_{n_1+n_2=N} \Lambda(n_1)\Lambda(n_2), \tag{19.2}$$

and for odd numbers N we consider

$$G_3(N) = \sum_{n_1+n_2+n_3=N} \Lambda(n_1)\Lambda(n_2)\Lambda(n_3). \tag{19.3}$$

Therefore

$$G_3(N) = \sum_{n<N} \Lambda(n)G_2(N-n). \tag{19.4}$$

Hence a good estimate of $G_2(N')$ for sufficiently many $N' < N$ will yield a correspondingly good estimate of $G_3(N)$.

The heuristic Möbius Randomness Law described in Section 13.1 justifies the following stronger form of the Goldbach problem (which implies its solution for sufficiently large N):

Conjecture. *For even numbers $N \geqslant 4$,*

$$G_2(N) = \mathfrak{G}_2(N)N + O(N(\log N)^{-A}) \tag{19.5}$$

where

$$\mathfrak{G}_2(N) = C_2 \prod_{\substack{p|N \\ p>2}} \frac{p-1}{p-2} \tag{19.6}$$

and

$$C_2 = 2\prod_{p>2}(1-(p-1)^{-2}). \tag{19.7}$$

Here A is any positive number and the implied constant in the error term depends only on A.

We shall prove the conjecture for almost all even numbers.

Theorem 19.1. *Let $A, B, X \geqslant 4$. The number of even integers N with $4 \leqslant N \leqslant X$ for which*

$$|G_2(N) - \mathfrak{G}_2(N)N| > BN(\log N)^{-A} \tag{19.8}$$

does not exceed $CX(\log X)^{-A}$ where C is a constant which depends only on A, B.

The celebrated result of Vinogradov for three primes follows immediately by (19.4) and Theorem 19.1.

THEOREM 19.2 (VINOGRADOV). *For odd numbers $N \geqslant 7$ we have*

$$G_3(N) = \mathfrak{S}_3(N)N^2 + O(N^2(\log N)^{-A}) \tag{19.9}$$

where $\mathfrak{S}_3(N) > 0$ is given by

$$\mathfrak{S}_3(N) = \frac{1}{2}\prod_{p|N}(1-(p-1)^{-2})\prod_{p\nmid N}(1+(p-1)^{-3}) > 0, \tag{19.10}$$

and A is any positive number, the implied constant depending on A.

19.2. Incomplete Λ-functions.

Given $z \geqslant 2$ we split $\Lambda(n)$ as follows:

$$\Lambda(n) = -\sum_{m|n}\mu(m)\log m = \Lambda^\sharp(n) + \Lambda^\flat(n) \tag{19.11}$$

where $\Lambda^\sharp(n)$ and $\Lambda^\flat(n)$ are the partial sums over the divisors $m \leqslant z$ and $m > z$ respectively. Then we write

$$G_2(N) = G_2^{\sharp\sharp}(N) + 2G_2^{\sharp\flat}(N) + G_2^{\flat\flat}(N) \tag{19.12}$$

where

$$G_2^{\sharp\sharp}(N) = \sum_{n_1+n_2=N}\Lambda^\sharp(n_1)\Lambda^\sharp(n_2)$$

and $G^{\sharp\flat}(N), G^{\flat\flat}(N)$ defined similarly.

LEMMA 19.3. *Let N be an even number $\geqslant 4$ and $z \geqslant 2$. We have*

$$G_2^{\sharp\sharp}(N) = \mathfrak{S}_2(N)N + O(N(\log z)^{-A} + \tau(N)Nz^{-\frac{1}{3}} + z^3) \tag{19.13}$$

for any $A \geqslant 0$, the implied constant depending only on A.

PROOF. We have

$$G_2^{\sharp\sharp}(N) = \sum_{m_1,m_2\leqslant z}\mu(m_1)\mu(m_2)(\log m_1)(\log m_2)\sum_{\ell_1 m_1+\ell_2 m_2=N} 1.$$

The equation $\ell_1 m_1 + \ell_2 m_2 = N$ is equivalent with the congruence conditions $(m_1, m_2)|N$ and

$$\ell m_1(m_1,m_2)^{-1} \equiv N(m_1,m_2)^{-1} \pmod{m_2(m_1,m_2)^{-1}}$$

with $1 \leqslant \ell < Nm_1^{-1}$. Therefore the number of solutions is $N[m_1,m_2]^{-1} + O(1)$ giving

$$G_2^{\sharp\sharp}(N) = N\mathfrak{S}_2(N;z) + O((z\log z)^2)$$

where

$$\mathfrak{S}_2(N;z) = \sum_{\substack{m_1,m_2\leqslant z\\(m_1,m_2)|N}}\frac{\mu(m_1)\mu(m_2)}{[m_1,m_2]}(\log m_1)(\log m_2).$$

It remains to estimate $\mathfrak{S}_2(N;z)$. We write

$$\mathfrak{S}_2(N;z)=\sum_{d|N}\frac{\mu(d)^2}{d}\sum_{\substack{m_1,m_2\leqslant z\\(m_1,m_2)=1\\(m_1m_2,d)=1}}\frac{\mu(m_1m_2)}{m_1m_2}(\log dm_1)(\log dm_2)$$

$$=\sum_{d|N}\frac{\mu(d)}{d}\sum_{c\leqslant z/d}\frac{\mu(cd)}{c^2}\Big(\sum_{\substack{m\leqslant z/cd\\(m,cd)=1}}\frac{\mu(m)}{m}\log cdm\Big)^2.$$

For $cd>z^{\frac12}$ we estimate trivially by

$$\sum_{d|N}\sum_{cd>z^{\frac12}}c^{-2}d^{-1}(\log z)^4\leqslant 2\tau(N)z^{-\frac12}(\log z)^4.$$

If $cd\leqslant z^{\frac12}$ we derive from the Prime Number Theorem that

$$-\sum_{\substack{m\leqslant z/cd\\(m,cd)=1}}\frac{\mu(m)}{m}\log cdm=\frac{cd}{\varphi(cd)}\{1+O(\tau(cd)(\log z)^{-A})\}.$$

Hence

$$\mathfrak{S}_2(N;z)=\sum_{d|N}\sum_{cd\leqslant z^{\frac12}}\mu(d)\mu(cd)d\varphi^{-2}(cd)+O((\log z)^{4-A}+\tau(N)z^{-\frac12}(\log z)^4).$$

Now extending the sum over $cd\leqslant z^{\frac12}$ to the infinite series (as we can with the error term which is already present) we obtain

$$\sum_{d|n}\frac{\mu^2(d)d}{\varphi^2(d)}\sum_{(c,d)=1}\frac{\mu(c)}{\varphi^2(c)}=\mathfrak{S}_2(n). \tag{19.14}$$

This completes the proof of Lemma 19.3. □

Lemma 19.3 reveals that the main contribution to the binary Goldbach-Euler equation comes from the incomplete function $\Lambda^\sharp(n)$. In other words, the complementary function $\Lambda^\flat(n)$ gives considerably less. This is due to the sign change of the Möbius function involved in $\Lambda^\flat(n)$ which ranges over a large segment. However, we do not have yet rigorous arguments to justify this heuristic. But it is a relatively simple matter in ternary additive problems.

19.3. A ternary additive problem with $\Lambda^\flat$.

In this section we give a bound for a general sum

$$T^\flat(N)=\sum_{\ell+m+n=N}u_\ell v_m\Lambda^\flat(n)$$

where u_ℓ and v_m are arbitrary complex numbers. To this end we need an estimate for the exponential sum

$$S^\flat(\alpha)=\sum_{n\leqslant N}\Lambda^\flat(n)e(\alpha n) \tag{19.15}$$

which is uniform in α. By Theorem 13.10 we derive by partial summation that for any $\alpha \in \mathbb{R}$ and $x \geqslant 2$,

$$\sum_{m \leqslant x} \mu(m)(\log m)e(\alpha m) \ll x(\log x)^{-A} \tag{19.16}$$

with any $A \geqslant 0$, the implied constant depending only on A. Since

$$|S^\flat(\alpha)| \leqslant \sum_{\ell \leqslant N/z} \Big| \sum_{z < m \leqslant N/\ell} \mu(m)(\log m)e(\alpha \ell m)\Big|$$

we get by (19.16) that

$$S^\flat(\alpha) \ll N(\log N)(\log z)^{-A}. \tag{19.17}$$

REMARK. Any sequence of numbers in place of $\Lambda^\flat(n)$ could be used provided the associated exponential sum (19.15) satisfies $S^\flat(\alpha) \ll N(\log N)^{-A}$ for any $A > 0$.

Now we are ready to prove the following

LEMMA 19.4. *For any complex numbers u_ℓ and v_m we have*

$$\sum_{\ell+m+n=N} u_\ell v_m \Lambda^\flat(n) \ll \|u\| \, \|v\| \, N(\log N)(\log z)^{-A} \tag{19.18}$$

with any $A \geqslant 0$, the implied constant depending only on A.

PROOF. We have

$$T^\flat(N) = \int_0^1 \Big(\sum_{\ell \leqslant N} u_\ell e(\alpha \ell)\Big)\Big(\sum_{m \leqslant N} v_m e(\alpha m)\Big) S^\flat(\alpha) e(-\alpha N) d\alpha.$$

Hence (19.18) follows by (19.17), Cauchy-Schwarz inequality and the Parseval formula

$$\int_0^1 \Big|\sum_{\ell \leqslant N} u_\ell e(\alpha \ell)\Big|^2 d\alpha = \sum_{\ell \leqslant N} |u_\ell|^2.$$

□

19.4. Proof of Vinogradov's three primes theorem.

Let c_N be any complex numbers for $N \leqslant X$, N odd. We write

$$\sum_{N \leqslant X} c_N G_2(N) = \sum_{N \leqslant X} c_N (G_2^{\sharp\sharp}(N) + 2G_2^{\sharp\flat}(N) + G_2^{\flat\flat}(N)).$$

Then we use Lemma 19.3 with $z = X^{\frac{1}{4}}$ getting

$$\sum_{N \leqslant X} c_N G_2^{\sharp\sharp}(N) = \sum_{N \leqslant X} c_N \mathfrak{G}_2(N) N + O(\|c\| X^{\frac{3}{2}} (\log X)^{-A}).$$

For the sums of $c_N G^{\sharp\flat}(N)$ and $c_N G_2^{\flat\flat}(N)$ we apply Lemma 19.4 getting the same bounds as the error term above. Hence

$$\sum_{N \leqslant X} c_N G_2(N) = \sum_{N \leqslant X} c_N \mathfrak{G}_2(N) N + O(\|c\| X^{\frac{3}{2}} (\log X)^{-A}). \tag{19.19}$$

For the special coefficients $c_N = G_2(N) - \mathfrak{S}_2(N)N$ this yields

PROPOSITION 19.5. *For any $X \geqslant 3$ we have*

$$\sum_{\substack{N\leqslant X\\ N \text{ even}}} (G_2(N) - \mathfrak{S}_2(N)N)^2 \ll X^3(\log X)^{-2A} \tag{19.20}$$

where A is any positive number and the implied constant depends only on A.

Clearly (19.20) implies Theorem 19.1. For the proof of Theorem 19.2 we apply (19.19) to (19.4) getting

$$G_3(N) = \sum_{n<N} \Lambda(n)\mathfrak{S}_2(N-n)(N-n) + O(N^3(\log N)^{-A}).$$

By (19.14) the main term is equal to

$$\sum_{(c,d)=1} \frac{\mu(c)\mu^2(d)d}{\varphi^2(c)\varphi^2(d)} \sum_{\substack{n<N\\ n\equiv N(\bmod d)}} \Lambda(n)(N-n).$$

The inner sum is equal to

$$\frac{N^2}{2\varphi(d)} + O(N^2(\log N)^{-A})$$

if $(d, N) = 1$ by the Prime Number Theorem. Since

$$\frac{1}{2} \sum_{(d,cN)=1} \frac{\mu(c)\mu^2(d)d}{\varphi^2(c)\varphi^3(d)} = \mathfrak{S}_3(N) \tag{19.21}$$

this completes the proof of (19.9).

CHAPTER 20

THE CIRCLE METHOD

20.1. The partition number.

The circle method originated in the paper of 1918 by G. H. Hardy and S. Ramanujan [**HR**] on partitions. Any decomposition $n = n_1 + n_2 + n_3 + \cdots$ into non-negative integers without regard to their order is called a partition of n. Let $p(n)$ denote the number of such (unrestricted) partitions. This is also equal to the number of ordered solutions $(x_1, x_2, \dots)$ in non-negative integers to the equation $n = x_1 + 2x_2 + 3x_3 + \cdots$. Algebraically, it is also the number of conjugacy classes in the symmetric group $\mathfrak{S}_n$.

L. Euler was fascinated by the remarkable properties of $p(n)$. He introduced the generating power series

$$F(z) = \sum_0^\infty p(n)z^n = \prod_1^\infty (1 - z^m)^{-1}$$

and showed that

$$\frac{1}{F(z)} = \prod_1^\infty (1 - z^m) = \sum_{-\infty}^\infty (-1)^\ell z^{(3\ell-1)\ell/2}$$

which is known as Euler's Pentagonal Numbers Theorem. Playing with other power series Euler derived many amazing formulas, such as

$$p(n) = \frac{1}{n} \sum_{0<h\leqslant n} \sigma(h)p(n-h)$$

where $\sigma(h)$ is the sum of divisors of h.

Ramanujan noticed, and later gave proofs of the congruence properties

$$\begin{aligned}
p(n) &\equiv 0(\text{mod } 5), &\quad \text{if} \quad n &\equiv 4(\text{mod } 5),\\
p(n) &\equiv 0(\text{mod } 7), &\quad \text{if} \quad n &\equiv 5(\text{mod } 7),\\
p(n) &\equiv 0(\text{mod } 11), &\quad \text{if} \quad n &\equiv 6(\text{mod } 11).
\end{aligned}$$

He was also fascinated by analytic expressions for $p(n)$ in terms of modular forms.

Changing the variable z into $e(z)$ one transforms the power series $F(z)$ into the Fourier series

$$f(z) = \sum_0^\infty p(n)e(nz). \tag{20.1}$$

We write $f(z) = e(z/24)/\eta(z)$ where

$$\eta(z) = e\Big(\frac{z}{24}\Big)\prod_1^\infty (1 - e(mz))$$

is the Dedekind eta function. This is a modular form on the upper half-plane $\mathbb{H}$ with respect to the modular group $SL_2(\mathbb{Z})$ of weight $\frac{1}{2}$ and a suitable multiplier system. Precisely Dedekind proved that

$$\eta(\gamma z) = \theta(\gamma)(cz+d)^{\frac{1}{2}}\eta(z) \tag{20.2}$$

for any $z \in \mathbb{H}$ and $\gamma = \left(\begin{smallmatrix} a & b \\ c & d \end{smallmatrix}\right) \in SL_2(\mathbb{Z})$, where the multiplier $\theta(\gamma)$ is defined by the properties $\theta(-\gamma) = e(\frac{1}{4})\theta(\gamma)$, $\theta(\left(\begin{smallmatrix} 1 & b \\ & 1 \end{smallmatrix}\right)) = e(\frac{b}{24})$ and

$$\theta\left(\begin{pmatrix} a & b \\ c & d \end{pmatrix}\right) = e\Big(\frac{a+d}{24} - \frac{c}{8} - \frac{1}{2}s(d,c)\Big),$$

if $c > 0$. Here $s(d,c)$ is the Dedekind sum

$$s(d,c) = \sum_{x \,(\mathrm{mod}\ c)} \psi\Big(\frac{x}{c}\Big)\psi\Big(\frac{dx}{c}\Big)$$

(recall that $\psi(x) = x - [x] - \frac{1}{2}$).

Note that the cusp ∞ is non-singular with respect to the Dedekind multiplier system (see **[I4]** for definition), so the Fourier expansion of $\eta(z)$ has no constant term; it is a lacunary series

$$\eta(z) = \sum_n a_n e(n^2 z/24)$$

with $a_n = 1$ if $n \equiv \pm 1 \pmod{12}$, $a_n = -1$ if $n \equiv \pm 5 \pmod{12}$ and $a_n = 0$ otherwise. However, the Fourier coefficients $p(n)$ of $f(z) = e(z/24)/\eta(z)$ are quite large. Note the trivial bound $p(n) \leqslant 2^n$ which follows from

$$\sum_0^\infty p(n)2^{-n} = \prod_1^\infty (1 - 2^{-m})^{-1}.$$

One can derive a better bound from the modular equation

$$f(z) = \Big(\frac{z}{i}\Big)^{\frac{1}{2}} e\Big(\frac{1}{24}\Big(z + \frac{1}{z}\Big)\Big) f\Big(-\frac{1}{z}\Big)$$

(see (20.2) for $\gamma = \left(\begin{smallmatrix} & -1 \\ 1 & \end{smallmatrix}\right)$). In particular, for $z = i/y$ with $y \geqslant 1$ this yields

$$p(n) \ll y^{-\frac{1}{2}} \exp\Big(2\pi\Big(\frac{y}{24} + \frac{n}{y}\Big)\Big).$$

Hence by taking $y = \pi\sqrt{n/6}$ one gets

$$p(n) \ll n^{-\frac{1}{4}} e^{B\sqrt{n}} \tag{20.3}$$

where $B = 2\pi/\sqrt{6}$. Hardy and Ramanujan proved the asymptotic formula

$$p(n) \sim (4\sqrt{3}n)^{-1} e^{B\sqrt{n}}.$$

They also gave an asymptotic expansion with an estimate for the error term.

Refining the method of Hardy-Ramanujan, H. Rademacher [**R**] established in 1937 the exact formula

THEOREM 20.1. *For $n \geqslant 1$ we have*

$$p(n) = \frac{1}{\pi\sqrt{2}} \sum_{c=1}^{\infty} c^{\frac{1}{2}} A_c(n) \frac{d}{dn} \frac{1}{\lambda_n} \sinh\Big(\frac{B}{c}\lambda_n\Big) \tag{20.4}$$

where $B = 2\pi/\sqrt{6}$, $\lambda_n = (n - \frac{1}{24})^{\frac{1}{2}}$ and

$$A_c(n) = \sum_{a (\mathrm{mod}\ c)}^{\star} e(\tfrac{1}{2}s(a,c) - an/c). \tag{20.5}$$

Notice that the Rademacher series (20.4) converges absolutely by virtue of the trivial estimates $|A_c(n)| \leqslant c$ and

$$\frac{d}{dn}\Big(\frac{1}{\lambda_n} \sinh \frac{B}{c}\lambda_n\Big) \ll c^{-3} \exp\Big(\frac{B}{c}\lambda_n\Big).$$

Taking only the first term in (20.4) one already gets a very good approximation

$$p(n) = \frac{e^{B\lambda_n}}{4\sqrt{3}\lambda_n^2}\Big(1 - \frac{1}{B\lambda_n} + O(e^{-\frac{1}{2}B\lambda_n})\Big). \tag{20.6}$$

The starting point of the Hardy-Ramanujan method is the Cauchy integral for the coefficients of a power series. In our case

$$p(n) = \frac{1}{2\pi i} \int_{|z|=r} F(z) z^{-n-1} dz \tag{20.7}$$

(this integration over a circle gave the method its name). Changing z into $e(z)$ we get by periodicity

$$p(n) = \int_{w}^{w+1} f(z) e(-nz) dz \tag{20.8}$$

where w is any point in the upper half-plane $\mathbb{H}$ and the path of integration is any continuous curve in $\mathbb{H}$ from w to $w+1$. We shall choose the path of integration as a continuous chain of certain circular arcs on which $f(z)$ can be analyzed quite precisely by exploiting its modularity.

To construct the arcs in question we need some well-known properties of the Farey sequence or series (see e.g. [**HW**]). Let C be a positive integer. The collection of all reduced fractions $\frac{a}{c}$ with $1 \leqslant c \leqslant C$ and $(a,c) = 1$ arranged in the increasing sequence

$$\cdots \frac{a'}{c'} < \frac{a}{c} < \frac{a''}{c''} \cdots$$

is called the Farey sequence of order C. Given a point $\frac{a}{c}$ in that sequence let $\frac{a'}{c'}$ and $\frac{a''}{c''}$ denote the adjacent points. The denominators c', c'' are determined by the conditions

$$\begin{cases} C - c < c' \leqslant C, & ac' \equiv 1 (\mathrm{mod}\ c), \\ C - c < c'' \leqslant C, & ac'' \equiv -1 (\mathrm{mod}\ c) \end{cases} \tag{20.9}$$

and the numerators are $a' = (ac' - 1)c^{-1}$, $a'' = (ac'' + 1)c^{-1}$, so the points are

$$\frac{a'}{c'} = \frac{a}{c} - \frac{1}{cc'}, \quad \frac{a''}{c''} = \frac{a}{c} + \frac{1}{cc''}. \tag{20.10}$$

Between a/c and $a'/c', a''/c''$ there are the mediants $(a'+a)/(c'+c)$, $(a+a'')/(c+c'')$ which are also reduced fractions but do not belong to the Farey sequence of order C. They have denominators in the segment $(C, c+C]$ and satisfy

$$\frac{a'+a}{c'+c} = \frac{a}{c} - \frac{1}{c(c+c')} = \frac{a'}{c'} + \frac{1}{c'(c+c')},$$
$$\frac{a''+a}{c''+c} = \frac{a}{c} + \frac{1}{c(c+c'')} = \frac{a''}{c''} - \frac{1}{c''(c+c'')}.$$

To each reduced fraction $\frac{a}{c}$ with $c \geqslant 1$, $(a,c) = 1$ we associate the circle

$$\left| z - \frac{a}{c} - \frac{i}{2c^2} \right| = \frac{1}{2c^2} \tag{20.11}$$

which is denoted by $\mathcal{C}(a/c)$ and is called a Ford circle; it is a circle in $\mathbb{H}$ tangent to the real line at $z = \frac{a}{c}$. Two different circles do not intersect; they are tangent if and only if they belong to consecutive fractions in the same Farey sequence. All the circles $\mathcal{C}(a/c)$ are obtained as the images of the horizontal line $\text{Im}\, z = 1$ under the action of $SL_2(\mathbb{Z})$. Indeed writing

$$\gamma z = \frac{az+b}{cz+d} = \frac{a}{c} - \frac{1}{c(cz+d)} \tag{20.12}$$

for $\gamma = \left(\begin{smallmatrix} a & b \\ c & d \end{smallmatrix}\right)$ with $c \geqslant 1$ it follows that

$$\gamma z = \frac{a}{c} + \frac{i}{2c^2} - \frac{i}{2c^2}\,\frac{\bar{z}+d/c}{z+d/c}$$

if $\text{Im}\,(z) = 1$, whence it is clear that γz satisfies (20.11).

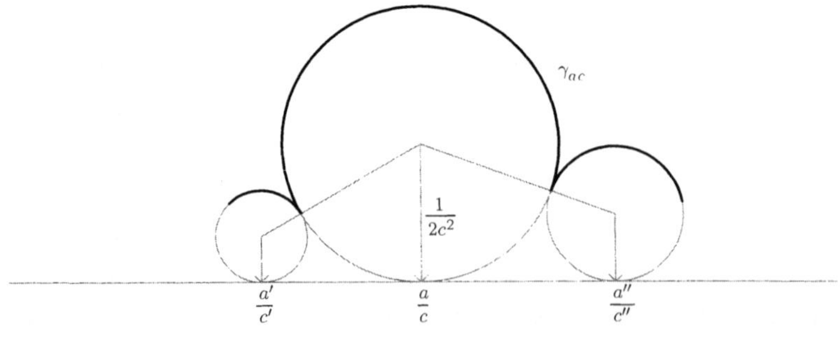

Now let $C \geqslant 1$ be given. Suppose $a'/c' < a/c < a''/c''$ are three consecutive points from the Farey sequence of order C. Then $\mathcal{C}(a/c)$ touches $\mathcal{C}(a'/c')$ and

$\mathcal{C}(a''/c'')$ at the points

$$\frac{a}{c} + \zeta'_{ac} \quad \text{with} \quad \zeta'_{ac} = \frac{-1}{c(c'+ic)}, \tag{20.13}$$

$$\frac{a}{c} + \zeta''_{ac} \quad \text{with} \quad \zeta''_{ac} = \frac{1}{c(c''-ic)}. \tag{20.14}$$

Indeed, we see that the point (20.13) is an appropriate mean of the centers of $\mathcal{C}(a/c)$ and $\mathcal{C}(a'/c')$

$$\left[\frac{1}{2c^2}\Big(\frac{a'}{c'} + \frac{i}{2c'^2}\Big) + \frac{1}{2c'^2}\Big(\frac{a}{c} + \frac{i}{2c^2}\Big)\right]\Big(\frac{1}{2c^2} + \frac{1}{2c'^2}\Big)^{-1} = \frac{a}{c} - \frac{1}{c(c'+ic)}$$

by (20.10), so this is the tangency point of the circles $\mathcal{C}(a/c)$ and $\mathcal{C}(a'/c')$. Similarly one verifies the point (20.14).

On each circle $\mathcal{C}(a/c)$ with $1 \leqslant c \leqslant C$ we choose the upper arc γ_{ac} which connects the tangency points (20.13) and (20.14) with the circles of the adjacent fractions. The chain of such arcs γ_{ac} gives us an infinite continuous curve which is periodic of period one. We choose as the path of integration in (20.8) the fragment of this curve which is contained in a vertical strip of width one beginning in a tangency point. We get

$$p(n) = \sum_{c \leqslant C} \sideset{}{^\star}\sum_{a(\mathrm{mod}\ c)} H_{ac}(n) \tag{20.15}$$

where

$$H_{ac}(n) = \int_{\gamma_{ac}} f(z) e(-nz) dz$$

by virtue of the periodicity of $f(z)$ and $e(-nz)$.

We shall evaluate each arc integral $H_{ac}(n)$ separately by applying specific transformation $\gamma \in SL_2(\mathbb{Z})$. First, changing the variable z into $a/c + iz/c^2$ (translation – magnification – rotation) we transform (20.11) onto the circle $|z - \frac{1}{2}| = \frac{1}{2}$ getting

$$H_{ac}(n) = \frac{i}{c^2} e\Big(-\frac{an}{c}\Big) \int_{z'_{ac}}^{z''_{ac}} f\Big(\frac{a}{c} + \frac{iz}{c^2}\Big) e\Big(-\frac{inz}{c^2}\Big) dz$$

where z is on the right arc of the circle between the points

$$z'_{ac} = -ic^2 \zeta'_{ac} = \frac{ic}{c'+ic}, \tag{20.16}$$

$$z''_{ac} = -ic^2 \zeta''_{ac} = \frac{-ic}{c''-ic}. \tag{20.17}$$

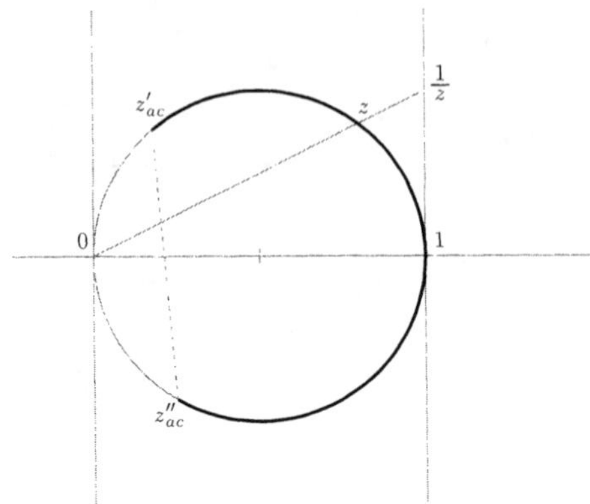

Given $c \geqslant 1$ and $(a, c) = 1$ the modular relation (20.2) with z replaced by $-d/c - 1/cz$ translates into

$$f\Big(\frac{a}{c} + \frac{z}{c}\Big) = \Big(\frac{z}{i}\Big)^{\frac{1}{2}} e\left(\tfrac{1}{2}s(d,c) + \frac{1}{24c}\Big(\frac{1}{z} + z\Big)\right) f\Big(\frac{-d}{c} - \frac{1}{cz}\Big)$$

where $ad \equiv 1 (\mathrm{mod}\ c)$. Changing z into iz/c we write this equation as

$$f\Big(\frac{a}{c} + \frac{iz}{c^2}\Big) = c^{-\frac{1}{2}} e(\tfrac{1}{2}s(d,c)) E_c(z) f\Big(-\frac{d}{c} + \frac{i}{z}\Big)$$

where $E_c(z) = z^{\frac{1}{2}} \exp(\frac{\pi}{12}(\frac{1}{z} - \frac{z}{c^2}))$. Applying this to the integral $H_{ac}(n)$ we get

$$H_{ac}(n) = ic^{-\frac{5}{2}} e(\tfrac{1}{2}s(a,c) - an/c) I_{ac}(n)$$

where

$$I_{ac}(n) = \int_{z'_{ac}}^{z''_{ac}} f\Big(-\frac{d}{c} + \frac{i}{z}\Big) E_c(z) e^{2\pi nz/c^2}\, dz.$$

(More precise but cumbersome notation would be $I_{a,c}(n)$).

To evaluate $I_{ac}(n)$ we replace f by its constant term in the Fourier expansion (20.1) which is $p(0) = 1$, and then we extend the arc of integration to the whole circle. This will give us the main term

$$I_c(n) = \int_{|z-\frac{1}{2}|=\frac{1}{2}} E_c(z) e^{2\pi nz/c^2}\, dz.$$

Changing z into $1/z$ we transform the circle $|z - \frac{1}{2}| = \frac{1}{2}$ into the vertical line $\mathrm{Re}\,(z) = 1$ getting

$$I_c(n) = \int_{1-i\infty}^{1+i\infty} \exp\left(\frac{\pi z}{12} + \frac{2\pi}{c^2 z}\Big(n - \frac{1}{24}\Big)\right) z^{-\frac{5}{2}}\, dz.$$

Therefore we have

$$I_c(n) = \frac{2\pi}{i}\Big(\frac{c}{12\lambda_n}\Big)^{\frac{3}{2}} I_{\frac{3}{2}}\Big(\frac{2\pi}{c\sqrt{6}}\lambda_n\Big) \tag{20.18}$$

where $I_\nu(y)$ is the Bessel function (see [**GR**], 8.4–8.5). (This unfortunate coincidence in notation will end shortly!)

It remains to estimate the difference $I_{ac}(n) - I_c(n) = I^\star_{ac}(n) - I'_{ac}(n) - I''_{ac}(n)$, say, where

$$I^\star_{ac}(n) = \int_{z'_{ac}}^{z''_{ac}} \Big(f\Big(-\frac{d}{c}+\frac{i}{z}\Big)-1\Big)E_c(z)e^{2\pi nz/c^2}\,dz,$$

$$I'_{ac}(n) = \int_0^{z'_{ac}} E_c(z)e^{2\pi nz/c^2}\,dz,$$

$$I''_{ac}(n) = \int_{z''_{ac}}^{0} E_c(z)e^{2\pi nz/c^2}\,dz.$$

For estimation of $I^\star_{ac}(n)$ we move the path of integration from the arc to the chord of the arc because on this chord i/z has large height so $f(-d/c+i/z)-1$ is small. Indeed on this chord we have $|z| \leqslant \min(|z'_{ac}|, |z''_{ac}|) \leqslant 2cC^{-1}$ and $\mathrm{Re}\, z \leqslant \max(\mathrm{Re}\, z'_{ac}, \mathrm{Re}\, z''_{ac}) \leqslant c^2C^{-2}$ by (20.16), (20.19) and (20.9). The length of the chord is bounded by $|z'_{ac}| + |z''_{ac}| \leqslant 4cC^{-1}$. Moreover, on the chord we have

$$\begin{aligned}\Big(f\Big(-\frac{d}{c}+\frac{i}{z}\Big)-1\Big)&E_c(z)e^{2\pi nz/c^2}\\ &= z^{\frac12}\sum_{m=1}^{\infty} p(m)e\Big(-\frac{dm}{c}\Big)\exp\Big(-\frac{2\pi}{z}\Big(m-\frac{1}{24}\Big)+\frac{2\pi z}{c^2}\Big(n-\frac{1}{24}\Big)\Big)\\ &\ll \Big(\frac{c}{C}\Big)^{\frac12}\exp\Big(\frac{2\pi n}{C^2}\Big)\end{aligned}$$

because $\mathrm{Re}\,(z^{-1}) = 1$, $\mathrm{Re}\,(z) \leqslant c^2C^{-2}$ and $p(m) \leqslant 2^m$. Therefore

$$I^\star_{ac}(n) \ll \Big(\frac{c}{C}\Big)^{\frac32}\exp\Big(\frac{2\pi n}{C^2}\Big). \tag{20.19}$$

Next we estimate $I'_{ac}(n)$. The length of the arc from 0 to z'_{ac} on the circle $|z-\frac12| = \frac12$ is bounded by $\frac{\pi}{2}|z'_{ac}| \leqslant \pi cC^{-1}$ and for z on that arc we have $|z| \leqslant cC^{-1}$. Moreover, we derive (as we did when estimating $I^\star_{ac}(n)$) that

$$E_c(z)e^{2\pi nz/c^2} = z^{\frac12}\exp\Big(\frac{\pi}{12z}+\frac{2\pi z}{c^2}\big(n-\frac{1}{24}\big)\Big) \ll \Big(\frac{c}{C}\Big)^{\frac12}\exp\Big(\frac{2\pi n}{C^2}\Big).$$

Therefore $I'_{ac}(n)$ satisfies the bound (20.19). Similarly we show that $I''_{ac}(n)$ satisfies the bound (20.19). Collecting these bounds and the formula (20.18) we get

$$H_{ac}(n) = \frac{2\pi}{c}e\Big(\tfrac12 s(a,c)-\frac{an}{c}\Big)(12\lambda_n)^{-\frac32}I_{\frac32}\Big(\frac{2\pi}{c\sqrt6}\lambda_n\Big) + O(c^{-1}C^{-\frac32}\exp(2\pi nC^{-2})).$$

Inserting this into (20.15) we get

$$p(n) = \sum_{c\leqslant C}\frac{2\pi}{c}A_c(n)(12\lambda_n)^{-\frac32}I_{\frac32}\Big(\frac{B}{c}\lambda_n\Big) + O(C^{-\frac12}\exp(2\pi nC^{-2})).$$

Letting C tend to infinity we obtain

$$p(n) = \frac{\pi}{12\sqrt3}\lambda_n^{-\frac32}\sum_{c=1}^{\infty}c^{-1}A_c(n)I_{\frac32}\Big(\frac{B}{c}\lambda_n\Big). \tag{20.20}$$

Using the formulas

$$I_{\frac{1}{2}}(y) = \left(\frac{2}{\pi y}\right)^{\frac{1}{2}} \sinh y, \quad I_{\frac{3}{2}}(y) = y^{\frac{1}{2}} \frac{d}{dy}(y^{-\frac{1}{2}} I_{\frac{1}{2}}(y))$$

we conclude (20.4).

REMARKS. Since $\eta(z)$ is essentially a theta function, the multiplier system $\theta(\gamma)$ can be expressed in terms of Gauss sums rather than the Dedekind sums. Consequently the coefficients $A_c(n)$ in the Rademacher formula (20.4) can be also expressed by simpler functions. Indeed A. Selberg showed (using directly the Euler Pentagonal Numbers Theorem) that

$$A_c(n) = \left(\frac{c}{3}\right)^{\frac{1}{2}} \sum_{\substack{\ell (\mathrm{mod}\ 2c) \\ \ell(3\ell-1) \equiv -2n (\mathrm{mod}\ 2\ell)}} (-1)^{\ell} \cos\left(\frac{\pi}{c}\left(\ell - \frac{1}{6}\right)\right). \tag{20.21}$$

Hence $|A_c(n)| \leqslant 2\tau(c) c^{\frac{1}{2}}$.

20.2. Diophantine equations.

After his historical paper with Ramanujan on the partition function Hardy continued working on the circle method jointly with J. E. Littlewood. From 1920 to 1928 they published a series of eight papers under the common title, "Some problems of 'Partitio Numerorum'". They realized that the modularity of the generating Fourier series is not an essential property for the method, but rather good estimates for the relevant exponential sums are sufficient. This observation opened numerous new applications, the most popular one being for the Waring problem.

This concerns the solvability of the equation

$$x_1^k + x_2^k + \cdots + x_n^k = N \tag{20.22}$$

in positive integers $x_1, \ldots, x_n$, where $k \geqslant 1$ and $N \geqslant 1$ are given numbers. In 1770 Edward Waring asserted that (20.22) does have a solution provided n is sufficiently large in terms of k alone. This was proved in 1909 by D. Hilbert [**H**]. By the circle method Hardy and Littlewood obtained an asymptotic formula for the number of representations (20.22). In this section we illustrate the method by showing a somewhat later result due to L-K. Hua [**H2**].

THEOREM 20.2. *Let $n > 2^k$. Then the number $\nu(N)$ of representations of N as the sum of n positive k-th powers satisfies*

$$\nu(N) = \mathfrak{S}(N) \frac{\Gamma(1+1/k)^n}{\Gamma(n/k)} N^{\frac{n}{k}-1} \{1 + O(N^{-\delta})\}. \tag{20.23}$$

Here $\mathfrak{S}(N)$ is given as the sum of infinite series of the multiplicative function

$$c(q, N) = \sum_{a (\mathrm{mod}\ q)}^{\star} \Big(\frac{1}{q} \sum_{x (\mathrm{mod}\ q)} e\Big(\frac{a x^k}{q}\Big)\Big)^n e\Big(-\frac{aN}{q}\Big). \tag{20.24}$$

The series

$$\mathfrak{S}(N) = \sum_{q=1}^{\infty} c(q, N) \tag{20.25}$$

converges absolutely and satisfies $c_1 \leqslant \mathfrak{S}(N) \leqslant c_2$, where c_1, c_2 are positive numbers depending on k and n, but not on N; also $\delta > 0$ and the implied constant depend at most on k, n.

We begin by applying the method to a general equation

$$f(x_1, \dots, x_n) = 0 \tag{20.26}$$

where f is a polynomial of degree $k \geqslant 1$ with integer coefficients. We seek integral solutions $x = (x_1, \dots, x_n)$ restricted to a box $\mathcal{B} \subset [-X, X]^n$ with $X \geqslant 1$. Thus

$$f(x) \ll X^k \quad \text{if } x \in \mathcal{B}. \tag{20.27}$$

The method works if the number of variables is much larger than the degree and the values of $f(x)$ vary considerably (so $f(x)$ cannot be constant in many variables after any unimodular transformation, see our analytic condition (20.35)). Therefore, by a statistical reasoning we anticipate that the number $\nu_f(\mathcal{B})$ of solutions to (20.26) in $x \in \mathcal{B} \cap \mathbb{Z}^n$ should satisfy

$$\nu_f(\mathcal{B}) \ll X^{n-k}.$$

Exactly, the number of such solutions is given by the integral (Vinogradov's variation on the Hardy-Littlewood power series setting)

$$\nu_f(\mathcal{B}) = \int_0^1 \Big(\sum_{x \in \mathcal{B} \cap \mathbb{Z}^n} e(\alpha f(x)) \Big) d\alpha. \tag{20.28}$$

As in the Hardy-Ramanujan work on the partition number we expect here that the main contribution comes from the integration in neighborhoods of rational points with small denominators.

According to the Dirichlet approximation theorem every α satisfies

$$\Big| \alpha - \frac{a}{q} \Big| \leqslant \frac{1}{qP}, \quad \text{with } 1 \leqslant q \leqslant P, \ \ (a, q) = 1, \tag{20.29}$$

where P is a fixed positive number. Let $P \geqslant 2Q \geqslant 2$. If α satisfies (20.29) with $q \leqslant Q$, we say α belongs to the "major arc"

$$\mathfrak{M} = \Big\{ \alpha; \ \Big| \alpha - \frac{a}{q} \Big| \leqslant \frac{1}{qP}, \quad \text{with } q \leqslant Q, (a, q) = 1 \Big\}. \tag{20.30}$$

Notice that the intervals of $\mathfrak{M}$ centered at different points a/q do not overlap because

$$\Big| \frac{a}{q} - \frac{a_1}{q_1} \Big| \geqslant \frac{1}{qq_1} \geqslant \frac{1}{qQ} \geqslant \frac{2}{qP}$$

if $a/q \neq a_1/q_1$ with $q_1 \leqslant Q$. The remaining part of the segment $\mathfrak{m} = [0, 1] \setminus \mathfrak{M}$ is called the "minor arc".

REMARK. In the Hardy-Ramanujan work the circle $\alpha \pmod 1$ was divided exactly by the mediants of the Farey sequence of order $C = P$, so there was nothing in the minor arc.

For $\alpha \in \mathfrak{M}$ we evaluate the exponential sum

$$S_f(\alpha) = \sum_{x \in \mathcal{B} \cap \mathbb{Z}^n} e(\alpha f(x))$$

by splitting into residue classes

$$S_f(\alpha) = \sum_{u \pmod q} e\Big(\frac{a}{q} f(u)\Big) \sum_{\substack{x \in \mathcal{B} \cap \mathbb{Z}^n \\ x \equiv u \pmod q}} e(\theta f(x))$$

where $\theta = \alpha - a/q$. Since $|\theta|$ is small, namely $|\theta| \leqslant (qP)^{-1}$ by (20.29), we can replace the sum over residue classes $u \pmod q$ by a corresponding integral with a reasonable error term. Suppose that $P \gg X^{k-1}$ with the implied constant large enough so that

$$\Big|\frac{\partial}{\partial x_\nu} f(x)\Big| \leqslant \frac{P}{2} \quad \text{for } 1 \leqslant \nu \leqslant n, x \in \mathcal{B}.$$

Hence $|\theta \partial f / \partial x_\nu| \leqslant 1/2q$, so Lemma 8.8 applied successively for each variable $x_\nu \equiv u_\nu \pmod q$ yields

$$\sum_{\substack{x \in \mathcal{B} \cap \mathbb{Z}^n \\ x \equiv u \pmod q}} e(\theta f(x)) = q^{-n} \mathcal{B}_f(\theta) + O\Big(\Big(1 + \frac{X}{q}\Big)^{n-1}\Big)$$

where $\mathcal{B}_f(\theta)$ is the integral in question

$$\mathcal{B}_f(\theta) = \int_{\mathcal{B}} e(\theta f(x)) dx. \tag{20.31}$$

A trivial error term would be $(1 + X/q)^n$, so we only got saving of the size of one variable of summation. Note that $\mathcal{B}_f(\theta)$ does not depend on the residue class $u \pmod q$ (in more sophisticated versions of the circle method the residue class does appear in the main term with a profound effect). Therefore summing over the classes $u \pmod q$ we get

$$S_f(\alpha) = \mathcal{B}_f(\theta) \mathcal{C}_f(a/q) + O(q(q+X)^{n-1}) \tag{20.32}$$

where $\mathcal{C}_f(a/q)$ is the normalized, complete exponential sum

$$\mathcal{C}_f(a/q) = \frac{1}{q^n} \sum_{u \pmod q} e\Big(\frac{a}{q} f(u)\Big). \tag{20.33}$$

Integrating (20.32) over the major arcs we get

$$\int_{\mathfrak{M}} S_f(\alpha) d\alpha = \sum_{q \leqslant Q} c_f(q) \int_{|\theta| \leqslant 1/qP} \mathcal{B}_f(\theta) d\theta + O(P^{-1} Q^2 (Q + X)^{n-1})$$

where

$$c_f(q) = \frac{1}{q^n} \sideset{}{^*}\sum_{a \pmod q} \sum_{u \pmod q} e\Big(\frac{a}{q} f(u)\Big). \tag{20.34}$$

Note that $c_f(q)$ is real and $|c_f(q)| \leqslant q$.

First we evaluate asymptotically the integral of $\mathcal{B}_f(\theta)$ over the segment $|\theta| \leqslant 1/qP$. Suppose that for any $\theta > 0$,

$$\mathcal{B}_f(\theta) \ll (\theta X^k)^{-1-\gamma} X^n \tag{20.35}$$

where $\gamma > 0$. Have in mind that X^n is the trivial bound and θX^k is about $\theta f(x)$, so we postulate a saving which is only slightly larger than the size of the amplitude. This condition is quite realistic for a smooth function $f(x)$ which changes fast in several variables. Observe that the bound (20.35) remains the same if a constant is added to f. For example, if

$$f(x) = N - x_1^k - \cdots - x_n^k \tag{20.36}$$

for $0 < x_\nu \leqslant X = N^{\frac{1}{k}}$, then

$$|\mathcal{B}_f(\theta)| = \Big|\int_0^X e(\theta x^k) dx\Big|^n \ll \theta^{-n/k},$$

so (20.35) holds with $\gamma = \frac{n}{k} - 1 > 0$, if $n > k$. Using (20.35) we can extend the integration over $|\theta| \leqslant 1/qP$ to the whole real line getting

$$\int_{|\theta| \leqslant 1/qP} \mathcal{B}_f(\theta) d\theta = V_f(\mathcal{B}) + O((qPX^{-k})^\gamma X^{n-k}) \tag{20.37}$$

where

$$V_f(\mathcal{B}) = \int_{-\infty}^{\infty} \mathcal{B}_f(\theta) d\theta. \tag{20.38}$$

The condition (20.35) also implies that

$$V_f(\mathcal{B}) \ll X^{n-k} \tag{20.39}$$

(use (20.35) if $|\theta| > X^{-k}$ and estimate trivially in $|\theta| \leqslant X^{-k}$).

Next we evaluate asymptotically the sum of $c_f(q)$ over the moduli $q \leqslant Q$. Suppose that for any $q \geqslant 1, (a, q) = 1$,

$$\mathcal{C}_f(a/q) \ll q^{-2-\eta} \tag{20.40}$$

where $\eta > 0$. Observe that this bound remains the same if a constant is added to f. Hence

$$c_f(q) \ll q^{-1-\eta}. \tag{20.41}$$

Using (20.41) we can extend the summation over $q \leqslant Q$ to all moduli getting

$$\sum_{q \leqslant Q} c_f(q) = \mathfrak{G}_f + O(Q^{-\eta}) \tag{20.42}$$

where

$$\mathfrak{G}_f = \sum_1^\infty c_f(q). \tag{20.43}$$

Now collecting the above results we obtain

$$\int_{\mathfrak{M}} S_f(\alpha) d\alpha = \mathfrak{G}_f V_f(\mathcal{B}) + O(\Delta X^{n-k}) \tag{20.44}$$

where

$$\Delta = Q^{-\eta} + Q^2(QPX^{-k})^{\gamma}. \tag{20.45}$$

Observe that we can take $\Delta \leqslant 2X^{-\delta\gamma\eta}$ with δ a small, positive number by choosing $Q = X^{\delta\gamma}$ and $P = X^{k-\delta(2+\gamma+\eta)}$, but some other choices of Q and P can be better, see the line above (20.62).

The infinite integral (20.38) and the infinite series (20.43), are called the "singular integral" and the "singular series" for the equation (20.26). These can be brought to expressions which allow meaningful interpretations in terms of local densities.

First we work out the singular integral. We have

$$V_f(\mathcal{B}) = \lim_{T\to\infty} \int_{-T}^{T} \Big(\int_{\mathcal{B}} e(\theta f(x))dx \Big) d\theta = \lim_{T\to\infty} \int_{B} \frac{\sin(2\pi f(x)T)}{\pi f(x)} dx.$$

Integrating over the level set $\{x \in \mathcal{B}; f(x) = t\}$ and using

$$\int_{-\infty}^{\infty} \frac{\sin 2\pi t}{\pi t} dt = 1$$

one can show that $V_f(\mathcal{B})$ is the $(n-1)$-dimensional volume of the set $\{x \in \mathcal{B}; f(x) = 0\}$ with respect to a suitable measure which depends on f. It is also approximately equal to the n-dimensional Lebesgue measure of the set $\{x \in \mathcal{B}; |f(x)| \leqslant \frac{1}{2}\}$. Precisely, we derive by the condition (20.35) that for any $V > 0$,

$$V_f(\mathcal{B}) = \frac{1}{2V} |\{x \in \mathcal{B}; |f(x)| \leqslant V\}| + O((VX^{-k})^{\gamma} X^{n-k}) \tag{20.46}$$

where $|\mathcal{A}|$ denotes the n-dimensional Lebesgue measure of $\mathcal{A}$. This is non-trivial for $V \ll X^k$, and it yields

$$V_f(\mathcal{B}) = \lim_{V\to 0} \frac{1}{2V} |\{x \in \mathcal{B}; |f(x)| \leqslant V\}|. \tag{20.47}$$

For the proof of (20.46) we use the formula

$$\int_{-\infty}^{\infty} \frac{\sin 2\pi\theta V}{\pi\theta} e(\theta y) d\theta = \begin{cases} 1 & \text{if } |y| < V, \\ 0, & \text{if } |y| > V. \end{cases}$$

and the estimates $\sin 2\pi\theta V = 2\pi\theta V + O(\theta^2 V^2)$, $\sin 2\pi\theta V \ll 1$. Hence by (20.35) we obtain

$$\begin{aligned} V_f(\mathcal{B}) &= \int_{-\infty}^{\infty} \mathcal{B}_f(\theta) d\theta = \int_{-\infty}^{\infty} \frac{\sin 2\pi\theta V}{2\pi\theta V} \mathcal{B}_f(\theta) d\theta \\ &\quad + O\Big(\int_0^T \theta V |\mathcal{B}_f(\theta)| d\theta + \int_T^{\infty} (\theta V)^{-1} |\mathcal{B}_f(\theta)| d\theta \Big) \\ &= \frac{1}{2V} |\{x \in \mathcal{B}; |f(x)| \leqslant V\}| + O((TV + (TV)^{-1})(TX^k)^{-\gamma} X^{n-k}). \end{aligned}$$

Choosing $TV = 1$ we get (20.46).

For example, if $f(x)$ is given by (20.36), then (20.47) yields

$$V(N) = \frac{1}{k} \int \cdots \int_{\substack{x_2,\dots,x_n>0 \\ x_2^k+\cdots+x_n^k<N}} (N - x_2^k - \cdots - x_n^k)^{\frac{1}{k}-1} dx_2 \cdots dx_n.$$

EXERCISE 1. Show that the above integral is

$$V(N) = \Gamma(1+\tfrac{1}{k})^n \Gamma(\tfrac{n}{k})^{-1} N^{\frac{n}{k}-1}. \tag{20.48}$$

Now we work out the singular series. The sum over $a(\bmod q)$ with $(a,q)=1$ in (20.34) is the Ramanujan sum

$$\sum_{a(\bmod q)}^{\star} e\Big(\frac{a}{q} f(u)\Big) = \sum_{\substack{d|q \\ d|f(u)}} \mu\Big(\frac{q}{d}\Big) d,$$

therefore

$$c_f(q) = \frac{1}{q^n} \sum_{d|q} \mu\Big(\frac{q}{d}\Big) d \, |\{u(\bmod q); f(u) \equiv 0(\bmod d)\}| = \sum_{d|q} \mu\Big(\frac{q}{d}\Big) \omega_f(d),$$

where $d^{n-1}\omega_f(d)$ is the number of solutions to the congruence

$$f(x) \equiv (\bmod d). \tag{20.49}$$

Since $\omega_f(d)$ is multiplicative, we have

$$c_f(q) = \prod_{p^\alpha \| q} (\omega_f(p^\alpha) - \omega_f(p^{\alpha-1})).$$

Hence the singular series is also given by the infinite product

$$\mathfrak{S}_f = \prod_p \delta_f(p) \tag{20.50}$$

where

$$\delta_f(p) = 1 + \sum_{\alpha=1}^{\infty} (\omega_f(p^\alpha) - \omega_f(p^{\alpha-1})). \tag{20.51}$$

Note that the series (20.51) and the product (20.50) converge absolutely by virtue of (20.41). The numbers $\delta_f(p)$ are called "local densities" for the reason that

$$\delta_f(p) = \lim_{\alpha\to\infty} \omega_f(p^\alpha), \tag{20.52}$$

so $\delta_f(p)$ is a density for the p-adic solutions of $f(x)=0$. By the analogous formula (20.47) one can interpret $V_f(\mathcal{B})$ as a density measure for the real solutions of $f(x) = 0$.

If $\omega_f(p^\alpha)$ stabilizes, i.e.,

$$\omega_f(p^{\alpha+1}) = \omega_f(p^\alpha), \qquad \text{for } \alpha \geqslant a, \tag{20.53}$$

then $\delta_f(p) = \omega_f(p^a)$ is the density of solutions of the congruence

$$f(x) \equiv 0 (\mathrm{mod}\ p^a). \tag{20.54}$$

For some polynomials f (such as (20.36)) the stability property (20.53) holds for all p with a common a which is sufficiently large depending on f. In this case

$$\mathfrak{S}_f = \prod_p \omega_f(p^a). \tag{20.55}$$

We now proceed to the estimation of the integral of the exponential sum $S_f(\alpha)$ over the minor arc $\mathfrak{m} = [0,1] \setminus \mathfrak{M}$. First Hardy and Littlewood estimated $S_f(\alpha)$ for each α in $\mathfrak{m}$ and then integrated trivially. For a polynomial in one variable we have

LEMMA 20.3 (WEYL). *Let $g(x) = cx^k + c_1 x^{k-1} + \cdots$ with c and k positive integers. Suppose*

$$\left|\alpha - \frac{a}{q}\right| \leqslant \frac{1}{q^2}, \qquad \text{with } (a,q) = 1. \tag{20.56}$$

Then

$$\sum_{0<n\leqslant X} e(\alpha g(n)) \ll X^{1+\varepsilon}\Big(\frac{c}{q} + \frac{c}{X} + \frac{q}{X^k}\Big)^{2^{1-k}} \tag{20.57}$$

for any $\varepsilon > 0$ and $X \geqslant 1$, the implied constant depending only on ε.

REMARK. The range of summation in (20.57) can be shifted to any segment of length X without affecting the bound.

PROOF. By Proposition 8.2 the exponential sum is bounded by

$$2X\Big\{X^{-k} \sum\cdots\sum_{-X<n_1,\dots,n_{k-1}<X} \min(X, \|\alpha ck! n_1 \cdots n_{k-1}\|^{-1})\Big\}^{2^{1-k}}$$

$$\ll X\Big\{X^{-1} + X^{-k}\sum_{1\leqslant m<M} \tau_{k-1}(m) \min(X, \|\alpha ck! m\|^{-1})\Big\}^{2^{1-k}}$$

where the first term comes from $n_1 \dots n_{k-1} = 0$ and $\tau_{k-1}(m)$ appears as a bound for the number of representations of $m = n_1 \cdots n_{k-1}$ with $1 \leqslant n_\nu < X$, so $1 \leqslant m < M = X^{k-1}$. After applying $\tau_{k-1}(m) \ll 2^{2^k} m^\varepsilon$ we extend the summation to

$$\sum_{1\leqslant \ell<L} \min(X, \|\alpha\ell\|^{-1}) \tag{20.58}$$

where $L = ck!M$ (i.e., we ignore the condition $\ell \equiv 0 (\mathrm{mod}\ ck!)$). If ℓ ranges over an interval of length $q/2$ the points $\alpha\ell (\mathrm{mod}\ 1)$ are spaced by q^{-1} by virtue of (20.56), therefore the sum (20.58) is bounded by

$$\Big(1 + \frac{2L}{q}\Big) \sum_{\ell (\mathrm{mod}\ q)} \min\Big(X, \Big\|\frac{\ell}{q}\Big\|^{-1}\Big) \ll \Big(1 + \frac{L}{q}\Big)(X + q\log q).$$

This estimate leads to (20.57). □

From Weyl's lemma for $\alpha = a/q$ and $X = q$ we obtain

COROLLARY 20.4. *Let $g(x)$ be a polynomial of degree $k \geqslant 2$ with integer coefficients. Let $(a,q)=1$. Then*

$$\sum_{x(\text{mod } q)} e\Big(\frac{a}{q}g(x)\Big) \ll (c,q)^{\frac{1}{2}} q^{1-2^{1-k}+\varepsilon} \tag{20.59}$$

where c is the leading coefficient of g, and the implied constant depends only on ε.

REMARK. By the Riemann Hypothesis for curves over finite fields (proved by A. Weil [**We1**]) one can obtain the bound $q^{\frac{1}{2}+\varepsilon}$, but with the implied constant depending on g. However, our exponent $1-2^{1-k}$ is more than sufficient for applications in this chapter.

For α in the minor arc $\mathfrak{m}$ we have (20.56) with $Q<q\leqslant P$, so the saving factor in the bound (20.57) is

$$\Big(\frac{1}{Q}+\frac{1}{X}+\frac{P}{X^k}\Big)^{2^{1-k}}$$

which is rather small. However, this can be multiplied if $f(x_1,\dots,x_n)$ has many summands in distinct single variables. For example let us consider

$$f(x_1,\dots,x_n)=f_1(x_1)+\cdots+f_n(x_n) \tag{20.60}$$

where $f_\ell(x_\ell)=c_\ell x^k+\cdots$ with $c_\ell>0$ for all $1\leqslant \ell\leqslant n$. Then, by Lemma 20.3

$$S_f(\alpha)=S_{f_1}(\alpha)\cdots S_{f_n}(\alpha)\ll X^{n+\varepsilon}\Big(\frac{1}{Q}+\frac{1}{X}+\frac{P}{X^k}\Big)^{n2^{1-k}} \tag{20.61}$$

for any $\alpha\in\mathfrak{m}$. The same bound holds for the integral of $S_f(\alpha)$ on $\mathfrak{m}$. We require this to be of smaller order than X^{n-k} which we accomplish by taking n sufficiently large in terms of k. Choosing $Q=X^{\frac{1}{4}}$ and $P=X^{k-\frac{1}{3}}$ we get

$$\int_{\mathfrak{m}} S_f(\alpha)d\alpha \ll X^{n-k-\delta} \tag{20.62}$$

with some $\delta>0$, if $n>k2^{k+1}$. Note that the condition (20.35) for the polynomial (20.60) holds with $\gamma=n/k-1$. So the above choice of Q and P also gives $\Delta\ll X^{-\delta}$, see (20.45). Moreover, we verify (20.40) using Corollary 20.4

$$\mathcal{C}_f(a/q)\ll q^{-n2^{1-k}+\varepsilon}\ll q^{-2-\eta}$$

if $n>2^k$. Therefore (20.44) becomes

$$\int_{\mathfrak{M}} S_f(\alpha)d\alpha=\mathfrak{S}_f V_f(\mathcal{B})+O(X^{n-k-\delta}) \tag{20.63}$$

where $\mathfrak{S}_f$ and $V_f(\mathcal{B})$ are the singular series and the singular integral for the polynomial (20.60). Adding (20.62) to (20.63) we conclude

THEOREM 20.5. *Let $f_1(x_1), \dots, f_n(x_n)$ be polynomials of degree $k \geqslant 2$ with integer coefficients. Suppose $n > k2^{k+1}$. Then the number of integral solutions to the equation*

$$f_1(x_1) + \cdots + f_n(x_n) = 0 \tag{20.64}$$

in a box $\mathcal{B} \subset [-X, X]^n$ with $X \geqslant 1$ satisfies

$$\nu_f(\mathcal{B}) = \mathfrak{S}_f V_f(\mathcal{B}) + O(X^{n-k-\delta}) \tag{20.65}$$

for some $\delta > 0$, where $\mathfrak{S}_f$ and $V_f(\mathcal{B})$ are the singular series and the singular integral for the polynomial (20.60).

The implied constant in the error term of (20.65) may depend on the polynomials $f_1, \dots, f_n$ but rather mildly; it does not depend on the constant terms of $f_1, \dots, f_n$. In particular, the asymptotic formula (20.65) applies nicely to the equation (by adding a constant term)

$$f_1(x_1) + \cdots + f_n(x_n) = N. \tag{20.66}$$

Let $c_1, \dots, c_n$ be the leading coefficients, and suppose all are positive. Let $\nu_f(N)$ be the number of solutions to (20.66) in positive integers. They are contained in the box $\mathcal{B} = [0, X]^n$ with $X = cN^{\frac{1}{k}}$ for some constant $c \geqslant 1$ depending on $f_1, \dots, f_n$. Therefore Theorem 20.5 yields

$$\nu_f(N) = \mathfrak{S}_f(N) V_f(N) + O(N^{\frac{n}{k}-1-\delta}) \tag{20.67}$$

for some $\delta > 0$. In this case we can compute the singular integral $V_f(N)$ asymptotically. First we know from (20.46) that

$$|\{0 \leqslant x_1, \dots, x_n \leqslant X;\ |f_1(x_1) + \cdots + f_n(x_n) - N| \leqslant U\}| \ll UX^{n-k}$$

for any $U > 0$. Hence and by (20.46) we derive using the approximations $f_\ell(x_\ell) = c_\ell x_\ell^k + O(X^{k-1})$ that

$$\begin{aligned} V_f(N) = \frac{1}{2U} &|\{0 \leqslant x_1, \dots, x_n \leqslant X; |c_1 x_1^k + \cdots + c_\ell x_\ell^k - N| \leqslant U\}| \\ &+ O(U^{-1}X^{n-1} + (UX^{-k})^\gamma X^{n-k}). \end{aligned}$$

In the error term we choose $U = X^{k-1/(1+\gamma)}$ getting $O(X^{n-k-\gamma/(1+\gamma)})$. In the main term we can reduce U to any smaller number V without changing the above error term by virtue of (20.46). Therefore

$$V_f(N) = (c_1 \cdots c_n)^{-\frac{1}{k}} V(N) + O(N^{\frac{n}{k}-1-\delta}) \tag{20.68}$$

where $V(N)$ is the same one as in (20.48). Inserting this to (20.67) we conclude that

$$\nu_f(N) = (c_1 \cdots c_n)^{-\frac{1}{k}} \Gamma(1 + \tfrac{1}{k})^n \Gamma(\tfrac{n}{k})^{-1} N^{\frac{n}{k}-1} (\mathfrak{S}_f(N) + O(N^{-\delta})). \tag{20.69}$$

In particular, we complete the proof of Theorem 20.2 if $n > k2^{k+1}$, except for the analysis of the singular series $\mathfrak{S}(N)$ about which we shall speak later.

In order to get results for smaller n we look for improvements in the treatment of the minor arc. Suppose f has the first variable separated,

$$f(x_1, \dots, x_n) = g(x_1) + h(x_2, \dots, x_n). \tag{20.70}$$

Then $S_f(\alpha) = S_g(\alpha)S_h(\alpha)$, so we can write

$$\left|\int_{\mathfrak{m}} S_f(\alpha)d\alpha\right| \leqslant (\max_{\alpha\in\mathfrak{m}} |S_g(\alpha)|) \int_0^1 |S_h(\alpha)|d\alpha. \tag{20.71}$$

In the minor arc we estimate $S_g(\alpha)$ by Weyl's lemma. Recall that for $\alpha \in \mathfrak{m}$ we have (20.44) with $Q < q \leqslant P$. We choose $Q = X^{\delta/3}$ and $P = X^{k-1+\delta}$, where δ is a small positive number. Assuming that g is a non-constant polynomial we get a non-trivial bound

$$S_g(\alpha) \ll X^{1-2\eta}, \quad \text{if } \alpha \in \mathfrak{m}. \tag{20.72}$$

Since η is small, we cannot waste much in the estimation of $S_h(\alpha)$. Our goal is

$$\int_0^1 |S_h(\alpha)|d\alpha \ll X^{n-1-k+\eta} \tag{20.73}$$

which is sufficient to yield

$$\int_{\mathfrak{m}} S_f(\alpha)d\alpha \ll X^{n-k-\eta}. \tag{20.74}$$

Moreover, the above choice of Q and P makes the error term in (20.44) of admissible order $X^{n-k-\eta}$. Adding (20.44) to (20.74) we get

$$\nu_f(\mathcal{B}) = \mathfrak{S}_f V_f(\mathcal{B}) + O(X^{n-k-\eta}). \tag{20.75}$$

It remains to prove (20.73). Note that this bound is close to the true order of magnitude. Indeed, by dropping the absolute value of $S_h(\alpha)$ we estimate from below by the integral which is exactly equal to the number of solutions to

$$h(x_2, \dots, x_n) = 0 \tag{20.76}$$

in the relevant box, and this number is expected to be of order X^{n-1-k}. In view of this remark the task of showing (20.73) may seem to be almost as difficult as the original problem, or even more difficult because we have one variable less in (20.76) than in the original equation (20.26). However, this situation is not as bad in practice, the reason being that it is easier to give an upper bound for the number of points on a variety than to count these with asymptotic precision.

In 1938 L-K. Hua [**H2**] succeeded in proving (20.73) with any $\eta > 0$ for the polynomial which is the sum of sufficiently many k-th powers.

LEMMA 20.6 (HUA). *Let $k \geqslant 1$ and*

$$S(\alpha) = \sum_{1\leqslant n\leqslant X} e(\alpha n^k). \tag{20.77}$$

Then for any $1 \leqslant \ell \leqslant k$ we have

$$\int_0^1 |S(\alpha)|^{2^\ell} d\alpha \ll X^{2^\ell-\ell+\varepsilon} \tag{20.78}$$

the implied constant depending on ε and k.

PROOF. For $\ell = 1$ we get

$$\int_0^1 |S(\alpha)|^2 d\alpha = X.$$

Suppose that (20.78) holds for ℓ with $1 \leqslant \ell < k$. We shall prove it holds for $\ell + 1$. By repeated application of the Weyl "differencing process" ℓ times (see Proposition 8.2) we arrive at the inequality

$$|S(\alpha)|^{2^\ell} \leqslant (2X)^{2^\ell-\ell-1} \sum_{-X<h_1,\dots,h_\ell<X}\cdots\sum \sum_{n\in I} e(\alpha h_1 \cdots h_\ell p(n; h_1, \dots, h_\ell))$$

where $I = I(h_1, \dots, h_\ell)$ is a subinterval of $[1, X]$ and $p(x; h_1, \dots, h_\ell)$ is a polynomial of degree $k - \ell$ with integer coefficients. We simplify this by writing

$$|S(\alpha)|^{2^\ell} \leqslant (2X)^{2^\ell-\ell-1} \sum_m a_m e(\alpha m)$$

where a_m is the number of solutions to the equation

$$h_1 \cdots h_\ell p(n; h_1, \dots, h_\ell) = m$$

in the above range. Therefore $a_0 \ll X^\ell$ and $a_m \ll X^\varepsilon$ if $m \neq 0$, because $p(x; h_1, \dots, h_\ell)$ is not a constant polynomial. We can also write

$$|S(\alpha)|^{2^\ell} = \sum_m b_m e(\alpha m)$$

where b_m is the number of solutions to

$$n_1^k + \cdots + n_s^k - n_{s+1}^k - \cdots - n_{2s}^k = m$$

with $s = 2^{\ell-1}$, $1 \leqslant n_1, \dots, n_{2s} \leqslant X$. Clearly

$$\sum_m b_m = |S(0)|^{2^\ell} = X^{2^\ell}$$

and

$$b_0 = \int_0^1 |S(\alpha)|^{2^\ell} d\alpha \ll X^{2^\ell-\ell+\varepsilon}$$

by the induction hypothesis. Now we derive by the Parseval identity

$$\int_0^1 |S(\alpha)|^{2^{\ell+1}} d\alpha \leqslant (2X)^{2^\ell-\ell-1} \sum_m a_m b_m \ll X^{2^\ell-\ell-1} X^{2^\ell+\varepsilon} = X^{2^{\ell+1}-\ell-1+\varepsilon}$$

completing the proof. □

From Hua's lemma with $\ell = k$ one gets (20.73) for any polynomial h in which at least 2^k variables appear exactly as k-th powers. This holds for the polynomial (20.36) provided $n - 1 \geqslant 2^k$ (we used one variable for the non-trivial estimation (20.72) in the minor arc). In particular, this completes the proof of Theorem 20.2 except for the analysis of the singular series.

In general this analysis of $\mathfrak{S}_f$ is not simple. In the special case of the equation (20.23) one can estimate the series (20.43) by elementary yet long arguments, see the comments below.

The asymptotic formula

$$\nu_f(\mathcal{B}) = \mathfrak{S}_f V_f(\mathcal{B}) + O(X^{n-k-\delta}) \tag{20.79}$$

for the equation $f(x) = 0$ in $x \in \mathcal{B} \cap \mathbb{Z}^n$ will be meaningful only if one knows that the singular series $\mathfrak{S}_f$ is bounded below by a positive quantity which does not or only barely depends on f, and that the box $\mathcal{B}$ is large enough so that $V_f(\mathcal{B})$ has order of magnitude X^{n-k}. The latter is essentially true if the equation $f(x) = 0$ has solutions in real variables laying deeply in the interior of $\mathcal{B}$. However, a good lower bound for $\mathfrak{S}_f$ is a more subtle question which we shall address on special occasions. The positivity of $\mathfrak{S}_f$ follows if the equation $f(x) = 0$ is solvable in every p-adic field by virtue of the product formula (20.50). Actually the local densities $\delta_f(p)$ are very close to one uniformly for all p sufficiently large, so the problem is to give good lower bounds for $\delta_f(p)$ at finite number of places which depend on f. If $f(x) = 0$ has local solutions and the number of variables n is considerably larger than the degree k, then (subject to some extra incidental conditions) there are plenty of local solutions so that $\mathfrak{S}_f \geqslant c(k, n) > 0$. For example, if $f(x) = N - x_1^k - \cdots - x_n^k$ (the case of the Waring problem) it is known that (see H. Davenport [**Da2**])

$$\mathfrak{S}(N) \geqslant c(k, n) > 0 \quad \text{if } n \geqslant 4k. \tag{20.80}$$

20.3. The circle method after Kloosterman.

The circle method is a tool which is suitable for detecting the equation $m = n$ by means of the additive characters $e(\alpha m)$, where m runs over a given set of integers and n is fixed. In a more general setting one has a sequence of complex numbers $\mathcal{A} = (\alpha_m)$ with

$$\sum_{m \in \mathbb{Z}} |\alpha_m| < \infty$$

and the problem is to evaluate asymptotically a selected term of $\mathcal{A}$. We have

$$\alpha_n = \int_0^1 S_{\mathcal{A}}(\alpha) e(-\alpha n) d\alpha$$

where

$$S_{\mathcal{A}}(\alpha) = \sum_m \alpha_m e(\alpha m).$$

Assuming we have an adequate knowledge about the distribution of $\mathcal{A}$ in residue classes to small moduli we can evaluate $S_{\mathcal{A}}(\alpha)$ at the points α which are close to rationals with small denominators. The set of such points is called the major arc because the main term in the asymptotic for α_n comes from this area. However, a harder work is needed for estimation of $S_{\mathcal{A}}(\alpha)$ for α in the remaining area which is called the minor arc. Here the treatment depends very much on the structure of the sequence $\mathcal{A}$. We are particularly pleased with $\mathcal{A}$ being an additive convolution of many independent sequences. In this case $S_{\mathcal{A}}(\alpha)$ factors into several exponential sums, so even a small saving in estimation of each of these can lead to a successful conclusion. It is intrinsic of the Hardy-Littlewood type arguments that one needs at least three summands to resolve the additive equation in question. The arguments certainly fail for binary additive problems (sometimes one summand can be partitioned into two so the additive problem under consideration seems to be of ternary type, but the range of resulting variables is reduced drastically so this rearrangement does not really change the problem).

Another characteristic of the equations $m = n$ for which the circle method as developed by Hardy-Littlewood works is that the expected number of solutions must be large relative to n, and it shouldn't depend essentially on n. For example, the method may work for a diophantine equation $f(x) = n$ if the number of variables is much larger than the degree.

A real challenge for the circle method is to detect the incident $m = n$ when it is expected to happen irregularly. Moreover, a new idea is needed to treat binary additive problems

$$\nu(n) = \sum_{m+\ell=n}\sum \alpha_m \beta_\ell \tag{20.81}$$

where $\mathcal{A} = (\alpha_m)$ and $\mathcal{B} = (\beta_\ell)$ are given sequences of some arithmetical nature. In 1926 H. D. Kloosterman [**Klo**] succeeded with the equation

$$a_1x_1^2 + a_2x_2^2 + a_3x_3^2 + a_4x_4^2 = n. \tag{20.82}$$

Actually his method works for the equation

$$Q(x) = n \tag{20.83}$$

where $Q(x)$ is any positive definite quadratic form in $r \geqslant 4$ variables and integral coefficients. (In particular, using it one can recover the four squares theorem of Lagrange, which the Hardy-Littlewood version could not). In the next section we shall treat the equation (20.83) by extending the original work of Kloosterman.

To explain the novelty of the Kloosterman method let us speculate on the general additive binary problem (20.81) (the quadric equation (20.82) is of this type, for one can take $m = a_1x_1^2 + a_2x_2^2$ and $\ell = a_3x_3^2 + a_4x_4^2$ as the summands with α_m, β_ℓ being the representation numbers by the corresponding binary forms). We have

$$\nu(n) = \int_0^1 S_{\mathcal{A}}(\alpha)S_{\mathcal{B}}(\alpha)e(-\alpha n)d\alpha$$

where $S_{\mathcal{A}}(\alpha)$ and $S_{\mathcal{B}}(\alpha)$ are the corresponding exponential sums. The best bounds for these exponential sums which one hopes to prove for almost all α are the ℓ_2-norms

$$\|\mathcal{A}\| = \Big(\sum_m |\alpha_m|^2\Big)^{\frac{1}{2}}$$

and $\|\mathcal{B}\|$ respectively. Even if the bounds $S_{\mathcal{A}}(\alpha) \ll \|\mathcal{A}\|$ and $S_{\mathcal{B}}(\alpha) \ll \|\mathcal{B}\|$ were accomplished for α in the minor arc (which is rarely the case), it is not sufficient because the expected main term does not exceed $\|\mathcal{A}\|\ \|\mathcal{B}\|$. Therefore one cannot rely only on estimates. For some sequences one can transform the corresponding exponential sum into another sum, say

$$S_{\mathcal{A}}(\alpha) = \sum_m \gamma_m(\alpha) + E_{\mathcal{A}}(\alpha), \tag{20.84}$$

where the new terms $\gamma_m(\alpha)$ have the status of being "dual" to $\alpha_m e(-\alpha m)$ and the error term $E_{\mathcal{A}}(\alpha)$ is significantly smaller than $\|\mathcal{A}\|$. Of course, the dual sum is essentially as large as the original one, but it may have fewer terms. Now, instead of estimating, one integrates the product of dual sums in α getting an extra saving from the cancellation which is due to the variation in the argument of each individual term $\gamma_m(\alpha)$. In this way one breaks the barrier $\|\mathcal{A}\|\ \|\mathcal{B}\|$ which couldn't be passed by absolute value integration. The main term emerges as before from the integration in small neighborhoods of rationals with small denominators. The diophantine nature of α plays a role in the transformation (20.84), so it is appropriate to divide the circle α (mod 1) by the Farey points $\frac{a}{c}$ with $1 \leqslant c \leqslant C$, $(a,c) = 1$. However, in contrast to the previous practice, neither gaps nor overlapping of these segments is allowed (because of the prohibition of employing estimates for the exponential sums $S_{\mathcal{A}}(\alpha)$ and $S_{\mathcal{B}}(\alpha)$). To cover the circle exactly we choose naturally the division points to be the mediants of the Farey sequence of order C. The resulting segments

$$\mathfrak{M}\Big(\frac{a}{c}\Big) = \Big(\frac{a'+a}{c'+c}, \frac{a+a''}{c+c''}\Big] = \Big(\frac{a}{c} - \frac{1}{c(c+c')}, \frac{a}{c} + \frac{1}{c(c+c'')}\Big]$$

have length which changes in arithmetic fashion (see (20.9)) and this property has to be considered when integrating different pieces of the dual exponential sum. Kloosterman controls the length of the segments $\mathfrak{M}(\frac{a}{c})$ by Fourier analysis, and consequently creates exponentials of type $e(\frac{du}{c})$ for various $u \in \mathbb{Z}$, where d is the multiplicative inverse of $a($ mod $c)$. By this analysis the integration in $\alpha($mod $1)$ amounts to the summation over the Farey points $\frac{a}{c}$ with factors $e(\frac{du}{c})$. At the same time a nice coincidence may occur that the dual terms $\gamma_m(\alpha)$ for α close to $\frac{a}{c}$ also have a phase of type $e(\frac{dv}{c})$ for various $v \in \mathbb{Z}$ (this happens for the sequence of values of a quadratic form). Multiplying these factors and summing over $a($mod $c)$ one arrives at the Kloosterman sum

$$S(m,n;c) = \sum_{ad\equiv 1 (\mathrm{mod}\ c)} e\Big(\frac{dm+an}{c}\Big)$$

with $m = -u - v$. Now any non-trivial bound for $S(m,n;c)$ corresponds to a cancellation in the integral of the dual sums giving a critical improvement of the direct bound $\|\mathcal{A}\|\ \|\mathcal{B}\|$. Kloosterman succeeded in showing that

$$S(m,n;c) \ll (m,n,c)^{\frac{1}{4}} c^{\frac{3}{4}} \tau(c),$$

while the Weil bound (Corollary 11.12) asserts that

$$|S(m,n;c)| \leqslant (m,n,c)^{\frac{1}{2}} c^{\frac{1}{2}} \tau(c). \tag{20.85}$$

Let us emphasize that if $Q(x)$ was not a quadratic polynomial, then the dual terms, if they exist, may have a phase other than $e(\frac{dv}{c})$, in which case another exponential sum to modulus c would appear in place of the Kloosterman sums, and the method still works by an appeal to appropriate results for the relevant sums.

We now proceed to a detailed presentation of Kloosterman's ideas in a somewhat abstract form. We begin by establishing a Fourier-Kloosterman expansion of the delta symbol

$$\delta(n) = \begin{cases} 1 & \text{if } n = 0, \\ 0 & \text{if } n \neq 0. \end{cases} \tag{20.86}$$

PROPOSITION 20.7. *Let C be a real number $\geqslant 1$. We have*

$$\delta(n) = 2\mathrm{Re} \int_0^1 \sum_{c\leqslant C<d\leqslant c+C}\sum{}^{\star} (cd)^{-1} e\Big(n\frac{\bar{d}}{c} - \frac{nx}{cd}\Big)dx \tag{20.87}$$

where $\star$ restricts the summation by $(c,d)=1$ and $\bar{d}$ is the multiplicative inverse to d modulo c.

REMARK. For $n=0$ the formula (20.86) gives a cute identity

$$\sum_{\substack{c\leqslant C<d\leqslant c+C\\(c,d)=1}}\sum (cd)^{-1} = \frac{1}{2}. \tag{20.88}$$

EXERCISE 2. Prove (20.88) directly.

For the proof of (20.87) we may assume that C is a positive integer. Let $f:\mathbb{R}\to\mathbb{C}$ be a periodic function of period 1. We shall evaluate the constant term of the Fourier series of f which is equal to the mean-value

$$\mu(f) = \int_0^1 f(x)dx.$$

Splitting into Farey segments $\mathfrak{M}(\frac{a}{c})$ we obtain by the periodicity of f,

$$\begin{aligned}
\mu(f) &= \sum_{\substack{0\leqslant a<c\leqslant C\\(a,c)=1}}\sum \int_{\mathfrak{M}(\frac{a}{c})} f(x)dx\\
&= \sum_{\substack{0\leqslant a<c\leqslant C\\(a,c)=1}}\sum \int_{-1/c(c+c')}^{1/c(c+c'')} f\Big(\frac{a}{c}+x\Big)dx\\
&= \sum_{\substack{c\leqslant C<d\leqslant c+C\\(c,d)=1}}\sum \bigg(\int_{-1/cd}^{0} f\Big(\frac{\bar{d}}{c}+x\Big)dx + \int_0^{1/cd} f\Big(-\frac{\bar{d}}{c}+x\Big)dx\bigg),
\end{aligned}$$

the last line following by (20.9). Suppose that f has the symmetry $f(-x)=\bar{f}(x)$ (i.e., f has real Fourier coefficients), then

$$\mu(f) = 2\mathrm{Re} \sum_{c\leqslant C}\int_0^{1/cC} \sum_{\substack{C<d\leqslant c+C\\cdx<1}}{}^{\star} f\Big(x-\frac{\bar{d}}{c}\Big)dx. \tag{20.89}$$

Applying this to $f(x)=e(nx)$ and changing x into x/cd one gets (20.87).

In the remaining part of this section we elaborate (20.89) further, not for immediate applications but rather to uncover more structure of the formula. We write (20.89) as

$$\mu(f) = 2\mathrm{Re} \int_C^\infty \Big(\sum_{c\leqslant C} c^{-1}K_f(x;c)\Big)x^{-2}dx \tag{20.90}$$

where $K_f(x;c)$ is a kind of incomplete Kloosterman sum associated with f

$$K_f(x;c) = \sideset{}{^\star}\sum_{C<d\leqslant\min(x,c+C)} f\Big(\frac{\bar{d}}{c} - \frac{1}{cx}\Big). \tag{20.91}$$

This can be expressed as a complete sum in $d(\mathrm{mod}\ c)$ by applying the formula

$$\min\Big(1, \Big[\frac{x-d}{c}\Big] - \Big[\frac{C-d}{c}\Big]\Big) = \begin{cases} 1 & \text{if } d \leqslant x, \\ 0 & \text{if } d > x \end{cases}$$

for $C < d \leqslant c + C$ and $x \geqslant C$. Hence we get

$$K_f(x;c) = \sideset{}{^\star}\sum_{d(\mathrm{mod}\ c)} \min\Big(1, \Big[\frac{x-d}{c}\Big] - \Big[\frac{C-d}{c}\Big]\Big) f\Big(\frac{\bar{d}}{c} - \frac{1}{cx}\Big). \tag{20.92}$$

Moreover, the "min" factor in (20.92) equals 1 if $x > c + C$, and it is equal to

$$\Big[\frac{x-d}{c}\Big] - \Big[\frac{C-d}{c}\Big] = \frac{x-C}{c} - \psi\Big(\frac{x-d}{c}\Big) + \psi\Big(\frac{C-d}{c}\Big)$$

if $C < x < c + C$, therefore

$$\begin{aligned} K_f(x;c) &= \min\Big(1, \frac{x-C}{c}\Big) \sideset{}{^\star}\sum_{d(\mathrm{mod}\ c)} f\Big(\frac{\bar{d}}{c} - \frac{1}{cx}\Big) \\ &\quad + \sideset{}{^\star}\sum_{d(\mathrm{mod}\ c)} \Big(\psi\Big(\frac{C-d}{c}\Big) - \psi\Big(\frac{x-d}{c}\Big)\Big) f\Big(\frac{\bar{d}}{c} - \frac{1}{cx}\Big) \end{aligned} \tag{20.93}$$

where the last sum is present only if $C < x < c + C$. Inserting this into (20.90) we obtain

PROPOSITION 20.8. *Let C be a positive integer and f a periodic function on $\mathbb{R}$ of period 1. Then the constant term of f in its Fourier series is given by*

(20.94)
$$\begin{aligned} \mu(f) = 2\mathrm{Re} \sum_{c\leqslant C} c^{-1} \int_C^\infty \min\Big(1, \frac{x-C}{c}\Big) \Big(\sideset{}{^\star}\sum_{d(\mathrm{mod}\ c)} f\Big(\frac{\bar{d}}{c} - \frac{1}{cx}\Big)\Big) x^{-2} dx \\ + 2\mathrm{Re} \sum_{c\leqslant C} c^{-1} \int_C^{c+C} \sideset{}{^\star}\sum_{d(\mathrm{mod}\ c)} \Big(\psi\Big(\frac{C-d}{c}\Big) - \psi\Big(\frac{x-d}{c}\Big)\Big) f\Big(\frac{\bar{d}}{c} - \frac{1}{cx}\Big) x^{-2} dx. \end{aligned}$$

REMARKS. Quite often in applications the first line of (20.94) yields the main term while the second line can be successfully estimated by means of exponential sums over finite fields. Expanding ψ and f into Fourier series the sum $K_f(x;c)$ can be expressed in terms of complete Kloosterman sums (20.85), but this option is not always practical. For $f(x) = e(-nx)$ one gets by (20.94) another expression for the delta-symbol

$$\begin{aligned} \delta(n) = 2\mathrm{Re} \sum_{c\leqslant C} c^{-1} R(n;c) \int_C^\infty e\Big(-\frac{n}{cx}\Big) \min\Big(1, \frac{x-C}{c}\Big) x^{-2} dx \\ + 2\mathrm{Re} \sum_{c\leqslant C} c^{-1} \int_C^{c+C} (R_C(n;c) - R_x(n;c)) e\Big(-\frac{n}{cx}\Big) x^{-2} dx \end{aligned}$$

where $R(n;c) = S(0,n;c)$ is the Ramanujan sum and

$$R_x(n;c) = \sum_{d(\bmod c)}^{\star} \psi\Big(\frac{x-d}{c}\Big) e\Big(\frac{n\bar{d}}{c}\Big).$$

Note that

$$R_x(n;c) \ll \Big(\sum_{d|(c,n)} d^{-\frac{1}{2}}\Big) c^{\frac{1}{2}} \tau(c) \log 2c$$

by applying the Fourier expansion

$$\psi\Big(\frac{x-d}{c}\Big) = \sum_{0<|\ell|<c} (2\pi i \ell)^{-1} e\Big(\ell \frac{x-d}{c}\Big) + O\Big(\Big(1 + c\Big\|\frac{x-d}{c}\Big\|\Big)^{-1}\Big)$$

and Weil's bound for the Kloosterman sums $S(\ell,n;c)$.

20.4. Representations by quadratic forms.

We proceed to treat the equation (20.83) by the Kloosterman circle method. Our goal is

THEOREM 20.9. *Let $Q(x)$ be a positive definite quadratic form in $r \geqslant 4$ variables and integer coefficients. Then for any $n > 0$ the number $r(n,Q)$ of integral solutions $m = (m_1, \dots, m_r)$ to $Q(m) = n$ satisfies*

$$r(n,Q) = \frac{(2\pi)^k n^{k-1}}{\Gamma(k)\sqrt{|A|}} \mathfrak{G}(n,Q) + O(n^{\frac{k}{2}-\frac{1}{4}+\varepsilon}) \tag{20.95}$$

where $k = \frac{r}{2}$, $|A|$ is the determinant of Q and $\mathfrak{G}(n,Q)$ is the singular series

$$\mathfrak{G}(n,Q) = \sum_{c=1}^{\infty} c^{-r} g_c(n,Q) \tag{20.96}$$

with

$$g_c(n,Q) = \sum_{d(\bmod c)}^{\star} \sum_{h(\bmod c)} e\Big(\frac{d}{c}(Q(h)-n)\Big), \tag{20.97}$$

the implied constant depends on ε and on the form Q.

The quadratic form $Q(x)$ in $x = (x_1, \dots, x_r)$ is associated with a symmetric matrix $A = (a_{ij})$ of rank r with $a_{ij} \in \mathbb{Z}$ and $a_{ii} \in 2\mathbb{Z}$ which is positive definite. In Siegel's notation

$$Q(x) = \tfrac{1}{2}A[x] = \tfrac{1}{2}{}^t x A x = \tfrac{1}{2}\sum_i a_{ii}x_i^2 + \sum_{i<j}\sum a_{ij}x_i x_j.$$

The discriminant of Q is defined by

$$\begin{cases} D = (-1)^{\frac{r}{2}}|A| & \text{if } r \text{ is even,} \\ D = \frac{1}{2}(-1)^{\frac{r+1}{2}}|A| & \text{if } r \text{ is odd} \end{cases}$$

where $|A|$ is the determinant (one shows that $D \equiv 0, 1 (\bmod 4)$ if r is even). Put

$$\theta(z) = \sum_{m\in\mathbb{Z}^r} e(Q(m)z) = \sum_{0}^{\infty} r(n,Q)e(nx) \tag{20.98}$$

where

$$r(n, Q) = |\{m \in \mathbb{Z}^r; Q(m) = n\}|.$$

This theta function is a modular form of weight $k = \frac{r}{2}$, level $2N$, where N is such that NA^{-1} is integral, and it transforms by a suitable multiplier system defined in terms of the Jacobi symbol (see Theorem 10.8 of [**I4**]), but we do not use this fact. All we need is the following Jacobi inversion formula

LEMMA 20.10. *Let $z \in \mathbb{H}$ and $x \in \mathbb{R}^r$. Then*

$$\sum_{m\in\mathbb{Z}^r} e\Big(\frac{z}{2}A[m+x]\Big) = |A|^{-\frac{1}{2}}\Big(\frac{i}{z}\Big)^k \sum_{m\in\mathbb{Z}^r} e\Big(\frac{-1}{2z}A^{-1}[m] + {}^t mx\Big). \tag{20.99}$$

SKETCH OF PROOF. This follows by the Poisson summation formula

$$\sum_{m\in\mathbb{Z}^r} f(m+x) = \sum_{m\in\mathbb{Z}^r} \hat{f}(m)e({}^t mx)$$

for $f(x) = e(\frac{z}{2}A[x])$. The computation of the Fourier transform $\hat{f}(m)$ is carried out by a linear change of variables which diagonalizes A, so it reduces to the familiar integral in one variable

$$\int_{\mathbb{R}} e\Big(\frac{z}{2}y^2 - uy\Big)dy = \Big(\frac{i}{z}\Big)^{\frac{1}{2}} e\Big(-\frac{u^2}{2z}\Big).$$

□

Let $ad \equiv 1(\text{mod } c)$. We shall use Lemma 20.10 to transform $\theta(z - \frac{a}{c})$ into a theta function for the adjoint quadratic form

$$Q^\star(x) = \tfrac{1}{2}A^{-1}[x]. \tag{20.100}$$

REMARK. The adjoint form $Q^\star(x)$ does not have integral coefficients. If N is a positive integer such that NA^{-1} is integral, then $NQ^\star(x)$ has integral coefficients. For example, $N = |A|$ satisfies this condition, yet a smaller number suffices quite often. Note that $N^r \equiv 0(\text{mod } |A|)$ so every prime factor of $|A|$ is in N.

Splitting the summation in (20.98) into residue classes $h = (h_1, \dots, h_r)$ of modulus c we get

$$\theta\Big(z - \frac{a}{c}\Big) = \sum_{h(\text{mod } c)} e\Big(-\frac{a}{c}Q(h)\Big) \sum_{m\equiv h(\text{mod } c)} e(Q(m)z).$$

Then by the Jacobi inversion formula (20.99) for $x = hc^{-1}$ we get

$$\sum_{m\equiv h(\text{mod } c)} e(Q(m)z) = |A|^{-\frac{1}{2}}c^{-r}\Big(\frac{i}{z}\Big)^k \sum_m e\Big(-\frac{Q^\star(m)}{c^2 z} - \frac{{}^t hm}{c}\Big).$$

Inserting this into the former sum and changing h into dh modulo c we arrive at

LEMMA 20.11. *Let $z \in \mathbb{H}$ and $ad \equiv 1 \pmod{c}$. Then*

$$\theta\Big(z - \frac{a}{c}\Big) = |A|^{-\frac{1}{2}} c^{-r} \Big(\frac{i}{z}\Big)^k \sum_{m \in \mathbb{Z}^r} G_m\Big(-\frac{d}{c}\Big) e\Big(-\frac{Q^\star(m)}{c^2 z}\Big) \tag{20.101}$$

where $Q^\star$ is the adjoint quadratic form and $G_m(d/c)$ is the Gauss sum

$$G_m\Big(\frac{d}{c}\Big) = \sum_{h \pmod{c}} e\Big(\frac{d}{c}(Q(h) + {}^t h m)\Big). \tag{20.102}$$

To pick up the coefficient $r(n, Q)$ in (20.98) we apply (20.89) for $f(x) = e(-nz)\theta(z)$, where $z = x + iy$ with $y > 0$ to be chosen later. We obtain

$$r(n, Q) = 2\mathrm{Re} \sum_{c \leqslant C} \int_0^{1/cC} T(c, n; x) e(-nz) dx \tag{20.103}$$

where

$$T(c, n; x) = \sum_{\substack{C < d \leqslant c + C \\ cdx < 1}}^\star e\Big(n\frac{\bar{d}}{c}\Big) \theta\Big(z - \frac{\bar{d}}{c}\Big). \tag{20.104}$$

Then by Lemma 20.11 we obtain

$$T(c, n; x) = |A|^{-\frac{1}{2}} c^{-r} \Big(\frac{i}{z}\Big)^k \sum_{m \in \mathbb{Z}^r} T_m(c, n; x) e(-Q^\star(m)/c^2 z) \tag{20.105}$$

where

$$T_m(c, n; x) = \sum_{\substack{C < d \leqslant c + C \\ cdx < 1}}^\star e\Big(n\frac{\bar{d}}{c}\Big) G_m\Big(-\frac{d}{c}\Big). \tag{20.106}$$

Now we are going to estimate $T_m(c, n; x)$. First we estimate the Gauss sum

LEMMA 20.12. *Let $(c, d) = 1$ and $m \in \mathbb{Z}^r$. We have*

$$G_m\Big(\frac{d}{c}\Big) \ll c^{\frac{r}{2}} \tag{20.107}$$

where the implied constant depends only on the quadratic form Q.

PROOF. We have

$$\begin{aligned}
\Big|G_m\Big(\frac{d}{c}\Big)\Big|^2 &= \sum_{x, y \pmod{c}}^\star e\Big(\frac{d}{c}(Q(x) - Q(y) + {}^t(x - y)m)\Big) \\
&= \sum_{y, z \pmod{c}}^\star e\Big(\frac{d}{c}({}^t yAz + Q(z) + {}^t zm)\Big) \\
&\leqslant c^r |\{z \pmod{c};\, Az \equiv 0 \pmod{c}\}|
\end{aligned}$$

which yields (20.107). □

Next we evaluate the Gauss sum precisely, but only if the modulus is odd and coprime with the discriminant.

LEMMA 20.13. *Let $(c, 2|A|d) = 1$ and $m \in \mathbb{Z}^r$. We have*

$$G_m\Big(\frac{d}{c}\Big) = \Big(\frac{|A|}{c}\Big)\Big(\varepsilon_c\Big(\frac{2d}{c}\Big)\sqrt{c}\Big)^r e\Big(-\frac{\bar{d}}{c}Q^\star(m)\Big). \tag{20.108}$$

PROOF. It reduces to the one-dimensional case by a local diagonalization and completing the square,

$$\sum_{y(\text{mod } c)} e\Big(\frac{ay^2}{c}\Big) = \varepsilon_c\Big(\frac{a}{c}\Big)\sqrt{c} \tag{20.109}$$

if $(c, 2a) = 1$ (see (3.38)). Recall that a nonsingular symmetric matrix over a field of characteristic $\neq 2$ is equivalent to a diagonal matrix. Hence one shows that there exist an integral matrix V and an integral diagonal matrix $B = \text{diag}\,(b_1, \dots, b_r)$ such that $(c, |V|) = 1$ and ${}^tVAV \equiv B(\text{mod } c)$. Changing x to Vy modulo c, we get

$$A[x] \equiv B[y] = \sum_\nu b_\nu y_\nu^2.$$

Moreover, letting $V^tm = [d_1, \dots, d_r]$, we get ${}^txm \equiv {}^tyV^tm = \sum_\nu d_\nu y_\nu$. Hence

$$Q(x) + {}^txm \equiv \tfrac{1}{2}\sum_{\nu=1}^r(b_\nu y_\nu^2 + 2d_\nu y_\nu) = \tfrac{1}{2}\sum_{\nu=1}^r b_\nu(y_\nu + \bar{b}_\nu d_\nu)^2 - \tfrac{1}{2}\sum_{\nu=1}^r \bar{b}_\nu d_\nu^2.$$

Here the last sum is congruent modulo c to

$$\sum_\nu b_\nu^{-1}d_\nu^2 = B^{-1}[{}^tVm] \equiv V^{-1}A^{-1}({}^tV)^{-1}[{}^tVm] = A^{-1}[m].$$

Moreover, we have $b_1 \cdots b_\nu = |B| \equiv |A|\,|V|^2(\text{mod } c)$. Therefore, applying (20.109) in each variable y_ν we conclude (20.108). □

To estimate the sum (20.106) we first complete it as follows

$$T_m(c, n; x) = \sum_{\ell(\text{mod } c)} \gamma(\ell)K(\ell, m, n; c) \tag{20.110}$$

where

$$K(\ell, m, n; c) = \sideset{}{^\star}\sum_{d(\text{mod } c)} e\Big(\frac{\ell d + n\bar{d}}{c}\Big)G_m\Big(-\frac{d}{c}\Big) \tag{20.111}$$

and

$$\gamma(\ell) = \frac{1}{c}\sum_{C<b\leqslant\min(c+C,1/cx)} e\Big(-\frac{b\ell}{c}\Big).$$

If $|\ell| \leqslant \frac{c}{2}$ we have $\gamma(\ell) \ll (1 + |\ell|)^{-1}$. Therefore the problem reduces to estimation of the complete sums $K(\ell, m, n; c)$.

The complete sum $K(\ell, m, n; c)$ is multiplicative in c. Precisely, if $c = c_0c_1$ with $(c_0, c_1) = 1$, then $K(\ell, m, n; c) = K^{(c_0)}(\ell, m, n; c_1)K^{(c_1)}(\ell, m, n; c_0)$ where the superscripts $(c_0), (c_1)$ indicate that the additive characters $e(\frac{\bar{c}_0}{c_1}x)$ and $e(\frac{\bar{c}_1}{c_0}x)$ are used in place of $e(\frac{x}{c_1})$ and $e(\frac{x}{c_0})$ respectively. Let c_0 denote the largest factor of c having all its prime divisors in $2|A|$ and c_1 be the remaining factor, so $(c_1, 2|A|) = 1$.

We estimate the sums (20.111) to moduli c_0 and c_1 separately. By (20.107) and trivial summation over $d_0 (\text{mod } c_0)$, we get

$$K^{(c_1)}(\ell, m, n; c_0) \ll c_0^{\frac{r}{2}+1} \tag{20.112}$$

where the implied constant depends only on the form Q. Then by (20.108) we get

$$K^{(c_0)}(\ell, m, n; c_1) = \Big(\frac{|A|}{c_1}\Big)\Big(\bar{\varepsilon}_{c_1}\Big(\frac{2}{c_1}\Big)\sqrt{c_1}\Big)^r S_r(\bar{c}_0\ell, \bar{c}_0(n + Q^\star(m))) \tag{20.113}$$

where

$$S_r(x,y) = \sum_{d (\text{mod } c_1)} \Big(\frac{d}{c_1}\Big)^r e\Big(\frac{xd + y\bar{d}}{c_1}\Big)$$

is either a Kloosterman sum if r is even, or a Salié sum if r is odd. In each case this sum satisfies the bound (20.85) giving

$$K^{(c_0)}(\ell, m, n; c_1) \ll (\ell, n + Q^\star(m), c_1)^{\frac{1}{2}} c_1^{\frac{r+1}{2}} \tau(c_1). \tag{20.114}$$

Inserting (20.113) and (20.114) into (20.110) and summing over $|\ell| \leqslant \frac{c}{2}$ we derive

LEMMA 20.14. *We have*

$$T_m(c, n; x) \ll (n + Q^\star(m), c_1)^{\frac{1}{2}} c_0^{\frac{1}{2}} c^{\frac{r+1}{2}} \tau(c) \log 2c \tag{20.115}$$

where the implied constant depends only on the form Q.

We shall apply the bound (20.115) to all terms in (20.105) except for $m = 0$ in the range $0 < x < 1/c(c + C)$ in which $T_0(c, n; x)$ is equal to

$$T(c, n) = \sum_{d (\text{mod } c)}^\star e\Big(n\frac{\bar{d}}{c}\Big) \sum_{h (\text{mod } c)} e\Big(-\frac{d}{c}Q(h)\Big). \tag{20.116}$$

We obtain

$$\begin{aligned} T(c, n; x) &= |A|^{-\frac{1}{2}} c^{-r} \Big(\frac{i}{z}\Big)^k T(c, n) \\ &\quad + O\Big((c_0 c)^{\frac{1}{2}} \tau(c)(\log 2c)(c|z|)^{-k} \sum_{m \in \mathbb{Z}^r}^\flat (n + Q^\star(m), c_1)^{\frac{1}{2}} \exp\Big(-\frac{2\pi y Q^\star(m)}{c^2|z|^2}\Big)\Big) \end{aligned}$$

where $\sum^\flat$ means that $m = 0$ is excluded from the summation if $0 < x < 1/c(c+C)$. We estimate this sum by

$$\sum^\flat \leqslant \Big(\sum_{\ell \geqslant 0} (nN + \ell, c_1)^{\frac{1}{2}} (1 + \ell)^{-2}\Big) \sum_{m \in \mathbb{Z}^r}^\flat (1 + NQ^\star(m))^2 \exp\Big(-\frac{2\pi y Q^\star(m)}{c^2|z|^2}\Big).$$

Recall that N is a positive integer such that $NQ^\star$ is integral and $(c_1, 2N) = 1$. We take

$$C = n^{\frac{1}{2}}, \quad \text{and} \quad y = C^{-2} = n^{-1}. \tag{20.117}$$

This choice implies that for $z = x + iy$ with $0 < x < (cC)^{-1}$,

$$c|z|y^{-\frac{1}{2}} \leqslant (C^{-1} + Cy)y^{-\frac{1}{2}} = 2,$$

and if $(c(c + C))^{-1} < x < (cC)^{-1}$, then we also have the lower bound

$$c|z|y^{-\frac{1}{2}} \geqslant (2C)^{-1} y^{-\frac{1}{2}} = \tfrac{1}{2}.$$

Moreover, we have $Q^\star(m) \gg |m|^2$. Hence, for any $0 < x < (cC)^{-1}$ we deduce

$$\sum_{m\in\mathbb{Z}^r}^{\flat}(1+NQ^\star(m))^2\exp\Big(-\frac{2\pi yQ^\star(m)}{c^2|z|^2}\Big) \ll (c|z|y^{-\frac12})^\kappa$$

for any $\kappa \geqslant 0$. Applying this with $\kappa = k$ we get

$$T(c,n;x) = |A|^{-\frac12}c^{-r}\Big(\frac{i}{z}\Big)^k T(c,n) + O(\xi(c_1)(c_0c)^{\frac12}\tau(c)(\log 2c)n^{\frac{k}{2}})$$

where

$$\xi(c_1) = \sum_{\ell\geqslant 0}(c_1,\ell+nN)^{\frac12}(\ell+1)^{-2}.$$

Inserting this into (20.103) we get

$$r(n,Q) = |A|^{-\frac12}\sum_{c\leqslant C}c^{-r}T(c,n)\int_{-1/c(c+C)}^{1/c(c+C)}(i/z)^k e(-nz)dx$$
$$+ O\Big(n^{\frac{k}{2}}C^{-1}\sum_{c\leqslant C}\xi(c_1)(c_0/c)^{\frac12}\tau(c)\log 2c\Big).$$

Here the error term is bounded by

$$n^{\frac{k}{2}}C^{\varepsilon-1}\sum_{c_0c_1\leqslant C}^{\star}\xi(c_1)c_1^{-\frac12} \ll n^{\frac{k}{2}}C^{\varepsilon-\frac12}\sum_{c_0}c_0^{-\frac12} \ll n^{\frac{k}{2}-\frac14+\varepsilon}.$$

On the other hand, the integral is equal to

$$\int_{-\infty}^{\infty}(i/z)^k e(-nz)dx = (2\pi)^k\Gamma(k)^{-1}n^{k-1}$$

up to an error term $O((cC)^{k-1})$. Summing this error term over $c \leqslant C$ we find that its total contribution is absorbed by the error term already present (use the bound (20.115) for $T(c,n)$), therefore

$$r(n,Q) = \frac{(2\pi)^k n^{k-1}}{\Gamma(k)\sqrt{|A|}}\sum_{c\leqslant C}c^{-r}T(c,n) + O(n^{\frac{k}{2}-\frac14+\varepsilon}).$$

Finally, extending the summation in the main term to all c we obtain (20.95) (notice that $T(c,n) = g_c(n,Q)$ by changing the variables h into dh and d into $-d$ in (20.116)).

Theorem 20.9 is also valid for ternary forms ($r = 2k = 3$), but the error term in (20.95) is just too large to yield an asymptotic formula. However, a slight improvement is possible by exploiting additional cancellation in sums of the sums (20.113) for various moduli of the relevant Farey sequence. In the case of $r = 3$ one obtains the Salié sums

$$K(a,b;c) = \sum_{d(\mathrm{mod}\ c)}\Big(\frac{d}{c}\Big)e\Big(\frac{ad+b\bar d}{c}\Big). \tag{20.118}$$

These can be computed explicitly in terms of quadratic roots modulo c. For example, if $(c,2a) = 1$, then (see Lemma 12.4)

$$K(a,b;c) = \varepsilon_c\Big(\frac{a}{c}\Big)c^{\frac12}\sum_{x^2\equiv ab(\mathrm{mod}\ c)}e\Big(\frac{2x}{c}\Big), \tag{20.119}$$

which easily yields $|K(a,b;c)| \leqslant c^{\frac{1}{2}}\tau(c)$. Clearly, this bound is best possible, yet one can derive a better estimate on average over the modulus c due to the variation of $e(\frac{2x}{c})$. Such a result belongs to the spectral theory of automorphic forms of weight $k = \frac{3}{2}$, and it leads to the following formula (see [**Du2**] and [**I6**])

THEOREM 20.15. *Let $Q(x)$ be a ternary, positive definite quadratic form with integral coefficients. Then for any $n > 0$ the number of integral representations $Q(m) = n$ satisfies*

$$r(n,Q) = 4\pi\left(\frac{2n}{|A|}\right)^{\frac{1}{2}} \mathfrak{G}(n,Q) + O(n^{\frac{1}{2}-\frac{1}{221}}) \tag{20.120}$$

where $\mathfrak{G}(n,Q)$ is the singular series given by (20.96)–(20.97), and the implied constant depends on Q.

The formulas (20.95) and (20.120) are true asymptotics only if the singular series $\mathfrak{G}(n,Q)$ does not vanish. We complete this section by highlighting a few features of $\mathfrak{G}(n,Q)$. We know from the general considerations in the previous section that the singular series (20.96) is the product of local densities (of p-adic solutions to $Q(x) = n$)

$$\mathfrak{G}(n,Q) = \prod_p \delta_p(n,Q). \tag{20.121}$$

Also the leading factor

$$\delta_\infty(n,Q) = \frac{(2\pi)^k n^{k-1}}{\Gamma(k)\sqrt{|A|}} \tag{20.122}$$

embodies the density of real solutions to $Q(x) = n$. We shall compute the local densities $\delta_p(n,Q)$ for $p \nmid 2|A|$ using the formula (20.108) for Gauss sums; it gives

$$g_c(n,Q) = \left(\frac{|A|}{c}\right)\varepsilon_c^{-r}c^{r/2} \sideset{}{^\star}\sum_{d(\mathrm{mod}\ c)} \left(\frac{d}{c}\right)^r e\Big(-n\frac{d}{c}\Big) \quad \text{if} \quad (c,2|A|) = 1. \tag{20.123}$$

Suppose r is even. Then the above sum is the Ramanujan sum giving

$$g_c(n,Q) = \chi_D(c)c^k \sum_{q|(c,n)} \mu\Big(\frac{c}{q}\Big)q$$

where $D = (-1)^{r/2}|A|$ is the discriminant of the quadratic form $Q(x)$ and $\chi_D(c) = (\frac{D}{c})$ is the Jacobi symbol. Hence for any P with $(P,2D) = 1$

$$\sum_{c|P} c^{-r}g_c(n,Q) = \sum_{q|(n,P)} \chi_D(q)q^{1-k}\prod_{p|\frac{P}{q}}(1-\chi_D(p)p^{-k}).$$

In particular, if every prime power factor of P is larger than that of n, this simplifies to

$$\sum_{c|P} c^{-r}g_c(n,Q) = \Big(\sum_{q|n}\chi_D(q)q^{1-k}\Big)\prod_{p|P}(1-\chi_D(p)p^{-k}).$$

Taking $P = p^{\nu+1}$, where $\nu = \mathrm{ord}_p n$, we get the local density

$$\delta_p(n,Q) = (1+\chi_D(p)p^{-k})^{-1}(1-\chi_D(p)p^{1-k})^{-1}(1-\chi_D(p^\nu)p^{\nu(1-k)}) \tag{20.124}$$

for any $p \nmid 2D$. Hence it is clear that

$$\delta_p(n,Q) \neq 0 \quad \text{if } p \nmid 2D. \tag{20.125}$$

Moreover, taking P to be the product of arbitrary large number of primes $p \nmid 2D$ we deduce that

$$\mathfrak{S}(n,Q) = \sigma_{1-k}(n,\chi_{4D})L(k,\chi_{4D})^{-1}\prod_{p|2D}\delta_p(n,Q) \tag{20.126}$$

where $L(s,\chi_{4D})$ denotes the Dirichlet L-function, and

$$\sigma_s(n,\chi) = \sum_{d|n}\chi(d)d^s. \tag{20.127}$$

The question when $\delta_p(n,Q) > 0$ for $p|2D$ is quite delicate. If $r > 4$ one can show that the non-vanishing of all the local densities $\delta_p(n,Q)$ with $p|2D$ is equivalent to the existence of $m \in \mathbb{Z}^r$ such that

$$Q(m) \equiv n \pmod{2^7|A|^3}. \tag{20.128}$$

Assuming this condition, one can infer by (20.96)–(20.97) that

$$\mathfrak{S}(n,Q) \asymp \prod_{p|n}(1+\chi_D(p)p^{1-k}) \tag{20.129}$$

if r is even, $r = 2k > 4$. For $r = 4$ a similar analysis applies, except for $\delta_2(n,Q)$.

If r is odd, $r \geqslant 5$, then the exact computation of the local densities of $\delta_p(n,Q)$ are more involved; nevertheless, one can derive directly by the trivial estimation

$$|g_c(n,Q)| \leqslant c^{\frac{r}{2}+1} \quad \text{if } (c,2|A|) = 1 \tag{20.130}$$

that the singular series $\mathfrak{S}(n,Q)$ is bounded from below and above by positive constants depending only on Q, provided the congruence (20.128) has integral solutions.

REMARK. If $r = 3$, we know only that $\mathfrak{S}(n,Q) \gg n^{-\varepsilon}$ for any $\varepsilon > 0$ where the implied constant depends on ε, Q, still assuming the solvability of (20.128) and that $\delta_2(n.Q) > 0$. At present the implied constant is not effectively computable, because the result uses Siegel's bound (5.76).

There is more to say about the formula (20.95) in terms of modular forms. Indeed, the representation number $r(n,Q)$ is the n-th Fourier coefficient of the theta function (20.98) which belongs to the space $M_k(2N,\vartheta)$ of modular forms of weight k, level $2N$, and a suitable multiplier ϑ. Suppose $r = 2k \geqslant 4$, so the whole space $M_k(2N,\vartheta)$ is spanned by the Eisenstein series (one for each singular cusp with respect to ϑ) and a finite number of cusp forms. Therefore

$$\theta(z,Q) = E(z,Q) + F(z,Q)$$

where $E(z,Q)$ is a unique combination of the standard Eisenstein series (called the Eisenstein series of the form Q) and $F(z,Q)$ is a cusp form uniquely determined by Q. From this decomposition

$$r(n,Q) = \rho(n,Q) + \tau(n,Q)$$

where $\rho(n,Q), \tau(n,Q)$ are the corresponding Fourier coefficients. It turns out that $\rho(n,Q)$ coincides with the main term in the formula (20.95)

$$\rho(n,Q) = \delta_\infty(n,Q)\prod_p \delta_p(n,q) = \frac{(2\pi)^k n^{k-1}}{\Gamma(k)\sqrt{|A|}}\mathfrak{S}(n,Q).$$

In view of these observations (due to Siegel) the error term in (20.95) is just a bound for the Fourier coefficient $\tau(n,Q)$ of the cusp form $F(z,Q)$. Having revealed this fact we could apply the Ramanujan-Petersson conjecture

$$\tau(n,Q) \ll \tau(n)n^{\frac{k-1}{2}}$$

(the implied constant depending on Q) which was proved by P. Deligne if k is an integer, $k \geqslant 2$, i.e., if $r = 2k$ is even, $r \geqslant 4$. Our result (20.95) corresponds to the Selberg bound for the Fourier coefficient $\tau(n,Q)$. If r is odd, the theory of Deligne is not applicable.

Siegel [**Sie3**] also observed that the Eisenstein series $E(z,Q)$ for a form Q is genus invariant. Actually he showed that

$$E(z,Q) = \sum_{Q_\nu \in \text{ gen } Q} w(Q_\nu)\theta(z,Q_\nu)$$

where Q_ν runs over inequivalent forms in the genus of Q, and $w(Q_\nu)$ are the genus masses. To define these numbers we first recall some definitions and basic facts from the theory of quadratic forms, in particular, genus theory.

Two quadratic forms Q_1, Q_2 (it is assumed throughout that all forms under consideration have the same rank $r \geqslant 2$) are equivalent if one can be obtained from the other by a unimodular (invertible over $\mathbb{Z}$) change of variables, i.e. if $A_1 = A_2U$ for some $U \in M_r(\mathbb{Z})$ with $|U| = \pm 1$. The equivalent forms form a class. The determinant is a class invariant.

A quadratic form $Q(x) = \frac{1}{2}A[x]$ is equivalent to itself in a number of ways. Denote by

$$O(Q) = \{U \in M_r(\mathbb{Z}); {}^tUAU = A\}$$

the group of automorphs of Q. This is finite and the same one for equivalent forms; we denote its order by $|O(Q)|$. If $r = 2$ the number $|O(Q)|$ depends only on the determinant $|A|$, but if $r > 2$ it varies from class to class.

Two forms Q_1, Q_2 are in the same genus if they are equivalent over every p-adic field and over the real numbers (the latter condition is redundant since both Q_1, Q_2 are assumed to be positive definite). The determinant is a genus invariant. One can show that two forms Q_1, Q_2 of the same determinant $|A_1| = |A_2|$ are in the same genus if and only if they are equivalent over $\mathbb{Z}$ to forms congruent modulo $8|A_1||A_2|$. The number of classes in a genus of a fixed determinant is finite.

Clearly, the number of representations $r(n,Q)$ is a class invariant, but it may vary within a given genus, because the number of automorphs may do so. Therefore, it is natural to add the following weight $w(Q)$ to the representations:

$$w(Q) = \frac{1}{|O(Q)|}\Big(\sum_{Q_\mu \in \text{ gen } Q}\frac{1}{|O(Q_\mu)|}\Big)^{-1}.$$

Note that the total mass of the forms in a genus is

$$\sum_{Q_\nu \in \text{ gen } Q} w(Q_\nu) = 1.$$

Accordingly, the average number of representation of n by forms of a given genus is defined as the weighted sum

$$r(n, \text{gen } Q) = \sum_{Q_\nu \in \text{ gen } Q} w(Q_\nu) r(n, Q_\nu).$$

Collecting the above notation and results we arrive at the Siegel mass formula,

$$r(n, \text{gen } Q) = \rho(n, Q) = \frac{(2\pi)^k n^{k-1}}{\Gamma(k)\sqrt{|A|}} \mathfrak{S}(n, Q). \tag{20.131}$$

20.5. Another decomposition of the delta-symbol.

Dissecting the unit circle by the Farey points $\frac{d}{c}$ of order C we developed a Fourier type expansion for $\delta(n)$ in terms of the additive characters $e(n\frac{\bar{d}}{c})$, where $\bar{d}d \equiv 1 (\text{mod } c)$. We call $\frac{\bar{d}}{c}$ the Kloosterman fractions. These appear in the expression (20.87) with d ranging over the interval $C < d \leqslant c + C$ of length exactly c, but not alone; they are escorted by the other factors involving d. This feature seems to be insignificant from an analytic viewpoint. In order to relax these "minor" factors one uses Fourier technique, and inevitably one is led to complete Kloosterman sums $S(m, n; c)$ in addition to the Ramanujan sums $S(0, n; c)$. One cannot avoid them in Kloosterman's treatment, even for small c.

In this section we give a different expression for $\delta(n)$ entirely in terms of the Ramanujan sums

$$S(0, n; c) = \sum_{d (\text{mod } c)}^{\star} e\Big(\frac{nd}{c}\Big) = \sum_{d | (c,m)} \mu\Big(\frac{c}{d}\Big) d.$$

We begin by a simple-minded treatment. An integer n is equal to zero if and only if it has a divisor larger than $|n|$. Therefore, for any $q > |n|$ we have

$$\delta(n) = \frac{1}{q} \sum_{a (\text{mod } q)} e\Big(\frac{an}{q}\Big).$$

Writing $\frac{a}{q}$ in its lowest terms we get

$$\delta(n) = \frac{1}{q} \sum_{c | q} S(0, n; c). \tag{20.132}$$

Next we average over q to gain an extra variable. Let $w(u)$ be a function supported in a segment $C \leqslant u \leqslant 2C$ and normalized by

$$\sum_{q=1}^{\infty} w(q) = 1. \tag{20.133}$$

Summing (20.132) over q we get

$$\delta(n)=\sum_{c=1}^{\infty}c^{-1}S(0,n;c)\sum_{r=1}^{\infty}r^{-1}w(cr). \tag{20.134}$$

This formula is valid only for $|n|<C$; consequently, it employs the characters $e(nd/c)$ to moduli as large as $|n|$, which is not a good feature in applications. In this respect a Kloosterman type expansion has the advantage of employing characters to much smaller moduli relative to $|n|$, precisely one can make it useful with $c\leqslant 2|n|^{\frac{1}{2}}$.

To refine the above idea we list here the following observations:

— every positive integer divides zero,

— a non-zero integer has few divisors,

— of the two complementary divisors of $n>0$ one does not exceed $n^{\frac{1}{2}}$.

Let $w(u)$ be a function of $u\geqslant 0$ which vanishes at $u=0$ and decays rapidly to zero as $u\to\infty$. We normalize $w(u)$ by the condition (20.133). Then by the above observations we deduce the following identity:

$$\delta(n)=\sum_{q|n}\Big(w(q)-w\Big(\frac{|n|}{q}\Big)\Big).$$

Detecting the condition $q|n$ by orthogonality of additive characters $e(an/q)$, and then reducing the fractions a/q we obtain

PROPOSITION 20.16. *Let $w(u)$ satisfy the above conditions. Then for any integer n we have*

$$\delta(n)=\sum_{c=1}^{\infty}S(0,n;c)\Delta_c(n) \tag{20.135}$$

where $\Delta_c(u)$ is a function on $\mathbb{R}$ given by

$$\Delta_c(u)=\sum_{r=1}^{\infty}(cr)^{-1}(w(cr)-w(|u|/cr)). \tag{20.136}$$

Since $\Delta_c(u)$ is not compactly supported (even if the test function $w(u)$ is chosen with compact support), we make a practical alteration by multiplying (20.135) with a nice function $f(u)$ supported in $|u|\leqslant 2N$ and normalized by $f(0)=1$. We get

$$\delta(n)=\sum_{c=1}^{\infty}S(0,n;c)\Delta_c(n)f(n). \tag{20.137}$$

Now suppose that $w(u)$ is also compactly supported, say in the dyadic interval $C\leqslant u\leqslant 2C$ (think of $w(u)$ as a bump function). Then the series (20.137) reduces to $c\leqslant 2\max(C,N/C)=X$, say. Clearly the optimal choice is

$$C=N^{\frac{1}{2}} \tag{20.138}$$

giving $X=2C=2N^{\frac{1}{2}}$ and

$$\delta(n)=\sum_{c\leqslant 2C}S(0,n;c)\Delta_c(n)f(n). \tag{20.139}$$

Therefore it takes characters of moduli $c \leqslant 2N^{\frac{1}{2}}$ to detect the event $n = 0$ within the integers $|n| \leqslant 2N$.

Of course, the optimal choice for the magnitude of moduli is not necessary, so we continue our investigation without the condition (20.138), because there are some technical benefits from having C independent of N. Yes, our new expressions (20.135) and (20.137) still contain only the Ramanujan sums (there are nowhere hidden Kloosterman fractions!), but what about the factor $\Delta_c(n)$? Assuming that the test function $w(u)$ is smooth one can regard $\Delta_c(u)$ as a continuous analog of the Ramanujan sum $S(0, n; c)$. To be practical we must be able to control the variation of $\Delta_c(u)$ in both variables c, u, and to separate c from u at a low cost. To this end we are going to prove

LEMMA 20.17. *Suppose $w(u)$ is smooth, supported in the segment $C \leqslant u \leqslant 2C$ with $C \geqslant 1$, normalized by (20.133), and has derivatives satisfying*

$$w^{(a)}(u) \ll C^{-a-1} \tag{20.140}$$

for $0 \leqslant a \leqslant A$. Then for any $c \geqslant 1$ and $u \in \mathbb{R}$ we have

$$\Delta_c(u) \ll \frac{1}{(c+C)C} + \frac{1}{|u| + cC}, \tag{20.141}$$

and for $1 \leqslant a \leqslant A$,

$$\Delta_c^{(a)}(u) \ll (cC)^{-1}(|u| + cC)^{-a}. \tag{20.142}$$

PROOF. We split

$$c\Delta_c(u) = V(c) - W\Big(\frac{|u|}{c}\Big) \tag{20.143}$$

where

$$V(y) = \sum_{r=1}^{\infty} r^{-1} w(yr) - \int_0^{\infty} r^{-1} w(r)dr, \tag{20.144}$$

$$W(y) = \sum_{r=1}^{\infty} r^{-1} w(y/r) - \int_0^{\infty} r^{-1} w(r)dr. \tag{20.145}$$

By the Euler-Maclaurin formula

$$\sum_r F(r) = \int (F(r) + \{r\}F'(r))dr,$$

where F is of class C^1 on $\mathbb{R}$ and $\{r\} = r - [r]$ is the fractional part of r, we get

$$V(y) = y \int_0^{\infty} \Big\{\frac{r}{y}\Big\} \Big(\frac{w(r)}{r}\Big)' dr. \tag{20.146}$$

Since $\{x\} \leqslant \min(1, x) \leqslant 2x(1+x)^{-1}$, we derive by (20.140) that

$$V(y) \ll C^{-1}(1 + C/y)^{-1}. \tag{20.147}$$

In the same way, by applying (20.147) for $w(u)$ changed into $w(1/u)$ and y into $1/y$, we derive that

$$W(y) \ll C^{-1}(1+y/C)^{-1}. \tag{20.148}$$

Inserting these estimates into (20.143) we obtain (20.141). For the proof of (20.142) observe that $\Delta_c^{(a)}(u)$ vanishes, unless $|u| > cC$ in which case

$$\Delta_c^{(a)}(u) = \sum_{r=1}^{\infty}(-cr)^{-a-1}w^{(a)}\Big(\frac{|u|}{cr}\Big) \ll (cC)^{-1}(|u|+cC)^{-a}.$$

□

Next we shall show that $\Delta_c(u)$ approximates to the Dirac distribution quite strongly on suitable test functions. To this end we first establish the following identity

LEMMA 20.18. *For any compactly supported function f of class C^a on $\mathbb{R}$ we have*

(20.149)
$$\int_{-\infty}^{\infty}\Delta_c(u)f(u)du = f(0)\hat{w}(0) + \hat{f}(0)c^{a-1}\int_0^{\infty}\beta_a\Big(\frac{r}{c}\Big)\Big(\frac{w(r)}{r}\Big)^{(a)}dr$$
$$- c^{a-1}\int_0^{\infty}\beta_a\Big(\frac{r}{c}\Big)\int_0^{\infty}w(u)u^a f^{(a)}(ru)dudr,$$

where $\beta_a(x) = \frac{1}{a!}B_a(\{x\})$ and $B_a(X)$ is the Bernoulli polynomial, so

$$\beta_a(x) = \sum_{m\neq 0}(-2\pi i m)^{-a}e(mx).$$

PROOF. We split the integral on the left side of (20.149) into two parts according to (20.143) and evaluate the sums over r by the Euler-Maclaurin formula (see Theorem 4.2)

$$\sum_r F(r) = \int (F(r) + \beta_a(r)F^{(a)}(r))dr. \tag{20.150}$$

First we get

$$V(Y) = y^a\int_0^{\infty}\beta_a\Big(\frac{r}{y}\Big)\Big(\frac{w(r)}{r}\Big)^{(a)}dr,$$

which leads to the second term on the right-hand side of (20.149). We are left with

$$\begin{aligned}\int_{-\infty}^{\infty}f(u)W\Big(\frac{|u|}{c}\Big)\frac{du}{c} &= \int_{-\infty}^{\infty}w(|u|)\sum_{r=1}^{\infty}f(cru)du - \frac{\hat{f}(0)}{c}\int_0^{\infty}\frac{w(r)}{r}dr\\ &= \int_0^{\infty}w(u)\sum_{r=-\infty}^{\infty}f(cru)du - f(0)\hat{w}(0) - \frac{\hat{f}(0)}{c}\int_0^{\infty}\frac{w(r)}{r}dr\\ &= \int_0^{\infty}w(u)\int_{-\infty}^{\infty}\{r\}\beta_a(r)\frac{\partial^a}{\partial r^a}f(cru)drdu - f(0)\hat{w}(0)\end{aligned}$$

by (20.150). Changing r into r/c we obtain the remaining terms of the right-hand side of (20.149). □

COROLLARY 20.19. *Let the conditions be as in Lemma 20.17. For any f compactly supported and of class C^a on $\mathbb{R}$, and for $c \geqslant 1$ we have*

$$\int_{-\infty}^{\infty} \Delta_c(u) f(u) du = f(0) + E_c(f) \tag{20.151}$$

where

$$E_c(f) \ll (cC)^{a-1} \int_{-\infty}^{\infty} (C^{-2a}|f(u)| + |f^{(a)}(u)|) du. \tag{20.152}$$

PROOF. This follows by $|\beta_a(x)| \leqslant 1$ and trivial estimation of the integrals in (20.149), but with the main term $f(0)\hat{w}(0)$ in place of $f(0)$. However, by the normalization condition (20.133) and the Euler-Maclaurin formula (20.150) we have

$$\hat{w}(0) = 1 + O(C^{-a-1}).$$

Here the error term is absorbed by (20.152) because

$$f(0) \ll \int_{-\infty}^{\infty} (|f(u)| + |f^{(a)}(u)|) du.$$

□

Suppose $f(u)$ is supported in $|u| \leqslant 2N$ and has derivatives

$$f^{(a)}(u) \ll N^{-a} \tag{20.153}$$

for $1 \leqslant a \leqslant A$. Then (20.152) gives

$$E_c(f) \ll \frac{N}{cC} \left(\frac{c}{C}\right)^a \left(1 + \frac{C^2}{N}\right)^a. \tag{20.154}$$

In particular, if $C = \sqrt{N}$, then this simplifies to $E_c(f) \ll (c/C)^a$ (change a into $a+1$). Recall that the series (20.137) terminates at $X = X(C, N) = 2\max(C, N/C) = 2\sqrt{N}$. The bound (20.154) is very small if $c \leqslant X^{1-\varepsilon}$ by letting a be sufficiently large.

When applying the formula (20.137) to solve an additive problem it is convenient to have $\Delta_c(n)f(n)$ expressed in additive characters $e(nv)$ with $|v|$ quite small. To this end we consider the Fourier transform

$$g_c(v) = \int_{-\infty}^{\infty} \Delta_c(u) f(u) e(-uv) du. \tag{20.155}$$

By Fourier inversion

$$\Delta_c(u) f(u) = \int_{-\infty}^{\infty} g_c(u) e(uv) du. \tag{20.156}$$

Inserting this to (20.137) we get

$$\delta(n) = \sum_{c \leqslant X} S(0, n; c) \int_{-\infty}^{\infty} g_c(v) e(nv) dv. \tag{20.157}$$

REMARKS. Here we keep the restriction $c \leqslant X$, because after applying the Fourier integral (20.156) for $\Delta_c(n)f(n)$ one loses sight of the support of the original test functions.

Now we need estimates for $g_c(v)$. By Corollary 20.19 for $f(u)e(-uv)$ in place of $f(u)$ we find that

$$g_c(v) = 1 + O\Big(\frac{1}{cC}\Big(\frac{c}{C} + \frac{cC}{N} + |v|cC\Big)^a\Big). \tag{20.158}$$

Moreover, we have by partial integration

$$g_c(v) = (-2\pi i v)^{-a} \int_{-\infty}^{\infty} (\Delta_c(u)f(u))^{(a)} e(-uv)du.$$

Hence we derive by (20.141), (20.142), (20.153) another estimate

$$g_c(v) \ll (|v|cC)^{-a} + (1 + NC^{-2})(|v|N)^{-a}. \tag{20.159}$$

The formula (20.158) is fine for small $|v|$ while the estimate (20.159) is for larger $|v|$, the $|v| = 1/cC$ being the transition value.

The test functions $f(u)$ and $w(u)$ do not need to be compactly supported, but only decaying rapidly enough outside the crucial ranges. We propose the following

EXERCISE 3. Give explicit computations for the special test functions

$$f(u) = \exp\Big(-\frac{u}{N}\Big), \qquad w(u) = \frac{\kappa}{C}\exp\Big(-\frac{u}{C} - \frac{C}{u}\Big), \tag{20.160}$$

where κ is a normalization constant.

It is understandable that the Fourier development of $\delta(n)$ of Kloosterman type (20.87) and our formula (20.135) have essentially the same power in applications, because they employ the additive characters to moduli of comparable size. However, the absence of Kloosterman fractions in (20.135) can simplify some treatments. For example, when solving the diophantine equation $f(x_1, \dots, x_n) = 0$ by means of (20.135) one is led to an exponential sum for a polynomial in $n+1$ variables over a finite field, whereas the Kloosterman method brings us a rational function.

CHAPTER 21

EQUIDISTRIBUTION

21.1. Weyl's criterion.

We begin by reviewing the general principles of equidistribution theory. Let $(X, \mathcal{B}, d\mu)$ be a probability space, where X is a topological space, $\mathcal{B}$ is the σ-algebra of Borel sets, and $d\mu$ is a measure normalized by

$$\int_X d\mu(x) = 1.$$

A sequence $\{x_n\} \subset X$ is said to be equidistributed (with respect to the measure $d\mu$) if

$$\lim_{N\to\infty} \frac{1}{N} |\{n \leqslant N; x_n \in B\}| = \mu(B)$$

for any open set $B \in \mathcal{B}$. Under some mild assumptions this property is essentially that

$$\lim_{N\to\infty} \frac{1}{N} \sum_{n\leqslant N} f(x_n) = \int_X f(x) d\mu(x) \tag{21.1}$$

for any compactly supported continuous function $f : X \to \mathbb{C}$.

Clearly the equidistribution implies that the sequence $\{x_n\}$ is dense in X, yet it is a more refined property. For showing the equidistribution it is sufficient to verify (21.1) for a system of test functions f which generates a dense subset of $C_0(X)$. The system of test functions can be simply the elements of an orthonormal basis of $L^2(X, d\mu)$. In particular, if $X = G$ is a compact group, the matrix coefficients of irreducible unitary representations of G form an orthonormal basis of $L^2(G)$ with respect to Haar measure (the Peter-Weyl theorem). If $X = G^\sharp$ is the space of conjugacy classes of a compact group, then the characters of G form an orthonormal basis of $L^2(G^\sharp, dg)$, with respect to the measure induced by Haar measure.

Since the formula (21.1) always holds for $f(x) = 1$ which is the character of the trivial representation, specializing the above discussion to this case gives the

WEYL CRITERION. *Let G be a compact group, $G^\sharp$ the set of conjugacy classes in G. Then a sequence of elements $x_n \in G^\sharp$ is equidistributed with respect to Haar measure if and only if for any non-trivial irreducible unitary representation $\rho : G \to GL(V)$, we have*

$$\sum_{n\leqslant N} \mathrm{Tr}(\rho(x_n)) = o(N), \qquad \text{as } N \to \infty. \tag{21.2}$$

A classical example in number theory (initiated by H. Weyl [**W1**] in 1916) is the circle $X = \mathbb{R}/\mathbb{Z}$ with the Lebesgue measure dx. In this case it is customary to look at a sequence of real numbers x_n, and for the purpose of equidistribution to take the fractional parts $\{x_n\}$ as representatives on the circle. If the fractional parts are equidistributed on the circle, then the original sequence is said to be equidistributed modulo one. The characters $e(hx)$ for $h \in \mathbb{Z}$, $h \neq 0$, are then the representations involved, and the Weyl criterion for equidistribution modulo 1 is

$$\sum_{n \leqslant N} e(hx_n) = o(N), \qquad \text{as } N \to \infty, \text{ for any } h \neq 0.$$

21.2. Selected equidistribution results.

Originally Weyl introduced his criterion to study the distribution modulo one of the sequences $x_n = f(n)$, where f is a polynomial with real coefficients.

PROPOSITION 21.1. *Let $f \in \mathbb{R}[X]$ be a real polynomial of degree $d \geqslant 1$. Assume that $f(x) = \alpha x^d + \cdots$ with $\alpha \notin \mathbb{Q}$. Then $x_n = f(n)$ is equidistributed modulo one.*

PROOF. We consider $f(x) = \alpha x^d$ for simplicity. Let $h \neq 0$ and $\beta = h\alpha$. For $Q = X^{d-\delta}$, with $0 < \delta < 1$, apply Lemma 20.3 (coming from Proposition 8.2) to an approximation $|\beta - a/q| \leqslant q^{-2}$ with $(a, q) = 1$ and $0 < q \leqslant Q$. This gives

$$\sum_{n \leqslant X} e(\beta n^d) \ll X^{1+\varepsilon}(q^{-1} + X^{-1} + qX^{-d})^{\gamma_d}$$

with $\gamma_d = 2^{1-d}$. If $q \geqslant X^d Q^{-1} = X^\delta$, this implies

$$\sum_{n \leqslant X} e(\beta n^d) \ll X^{1-\gamma} \text{ for some } \gamma > 0.$$

If $q < X^d Q^{-1}$, denoting $g(t) = (\alpha - \frac{a}{q})t^d$, we have by partial summation

$$S(X) = \sum_{n \leqslant X} e(\beta n^d) = \sum_{n \leqslant X} e\Big(\frac{an^d}{q}\Big) e(g(X)) - 2\pi i \int_1^X \Big(\sum_{n \leqslant t} e\Big(\frac{an^d}{q}\Big)\Big) g'(t) e(g(t)) dt.$$

By splitting into residue classes modulo q we get

$$\sum_{n \leqslant Y} e\Big(\frac{an^d}{q}\Big) = \frac{Y\mathfrak{S}}{q} + O(q) \text{ with } \mathfrak{S} = \sum_{x \pmod q} e\Big(\frac{ax^d}{q}\Big).$$

Hence by partial integration again we get

$$S(X) = \frac{\mathfrak{S}}{q} \int_1^X e(g(t)) dt + O\Big(q\Big(1 + \frac{X^d}{qQ}\Big)\Big) \ll \frac{X|\mathfrak{S}|}{q} + O(q).$$

To the complete sum $\mathfrak{S}$ we can apply either the bounds for Ramanujan sums (if $d = 1$), Gauss sums (if $d = 2$) or Weil's bound for higher degree (also Weyl's Lemma (20.59) would be sufficient), getting $\mathfrak{S} \ll q^{1/2+\varepsilon}$ and hence $S(X) \ll Xq^{-1/2+\varepsilon} + X^\delta$. Since $q \to +\infty$ because $\beta \notin \mathbb{Q}$, we have in any case

$$\sum_{n \leqslant X} e(\beta n^d) = o(X) \text{ as } X \to +\infty.$$

□

The equidistribution of arithmetical sequences built out of prime numbers are fascinating and most challenging. For primes themselves, if one takes as space

$G = (\mathbb{Z}/q\mathbb{Z})^\times$ with the counting measure, then equidistribution is equivalent with the Prime Number Theorem in Arithmetic Progressions (see Chapter 5). Similarly, if $K/\mathbb{Q}$ is a finite Galois extension, then one wishes to study the distribution of the Frobenius conjugacy classes σ_p in the Galois group of $K/\mathbb{Q}$. There the Chebotarev Density Theorem gives the result:

THEOREM 21.2. *Let $K/\mathbb{Q}$ be a finite Galois extension with Galois group G. For any set $C \subset G$ which is stable by conjugacy, we have*

$$\lim_{X\to+\infty} \frac{1}{X}|\{p \leqslant X \mid \sigma_p \in C\}| = \frac{|C|}{|G|}.$$

However, in this case as for arithmetic progressions, questions of the size of the error term and uniformity in terms of parameters is most important and quite different in perspective than the simple equidistribution statement.

One is naturally led to consider sequences $x_p = f(p)$, especially when $x_n = f(n)$ has already been proved to be equidistributed modulo one. There the methods of Chapter 13 are relevant. Indeed, Theorem 13.6 implies the following theorem of Vinogradov

THEOREM 21.3. *Let $\alpha \notin \mathbb{Q}$ be an irrational number. Then the sequence αp is equidistributed modulo one as $p \to +\infty$.*

PROOF. Let $\beta = \alpha h$ with $h \neq 0$. We must prove

$$\sum_{p\leqslant X} e(\beta p) = o(X(\log X)^{-1}) \text{ as } X \to +\infty. \tag{21.3}$$

Let $Q = x(\log x)^{-B}$ for some $B > 0$, and let a/q be a rational approximation to β such that $|\beta - a/q| \leqslant 1/(qQ) \leqslant q^{-2}$. If $q \geqslant xQ^{-1} = (\log x)^B$, applying Theorem 13.6 yields

$$\sum_{p\leqslant X} e(\beta p) \ll x(\log x)^{3-B/2}.$$

If $q < xQ^{-1}$, then by partial summation, splitting into arithmetic progressions modulo q and the Siegel-Walfisz Theorem, we have

$$\sum_{p\leqslant X} e(\beta p) \ll \frac{\mathrm{Li}(X)}{\varphi(q)} + X(\log X)^{B-A}$$

for any $A > 0$. If we take $B > 6$ and $A > B+1$, then (21.3) follows since $q \to +\infty$ as $X \to +\infty$ for β irrational. □

Similarly, (13.55) implies that $\alpha\sqrt{p}$ is equidistributed for any fixed real number $\alpha \neq 0$.

The techniques of exponential sums over finite fields, and particularly the powerful results based on the Riemann Hypothesis proved by Deligne (see Chapter 11), give very good tools to study many interesting equidistribution problems. We give an example of Fouvry and Katz [**FK**] using Theorem 11.43.

THEOREM 21.4. *Let $P_1(X),\dots, P_r(X)$ be polynomials in $\mathbb{Z}[X_1,\dots,X_n]$ such that the total degree of any non-trivial integral linear combination $a_1P_1+\cdots+a_rP_r$ is $\geqslant 2$. Let $\phi(x)\to+\infty$ as $x\to+\infty$. Then, for $p\to+\infty$, the sequence*

$$\Big\{\Big(\frac{P_1(x)}{p},\dots,\frac{P_r(x)}{p}\Big)\Big\}$$

where $0\leqslant x_1,\dots,x_n\leqslant \phi(p)\sqrt{p}\log p$, is equidistributed modulo one.

SKETCH OF PROOF. Let $w(p)=\phi(p)\sqrt{p}\log p$ and for $a=(a_1,\dots,a_r)\neq 0$, let

$$S=\sum_{\substack{x_1,\dots,x_r\\ 0\leqslant x_i\leqslant w(p)}}\cdots\sum e\Big(\frac{a_1P_1(x)+\cdots+a_rP_r(x)}{p}\Big).$$

By the Weyl Criterion of $(\mathbb{R}/\mathbb{Z})^r$, we must show that

$$S\ll w(p)^n \text{ as } p\to+\infty. \tag{21.4}$$

First assume $w(p)<p$. Using Fourier inversion on $\mathbb{Z}/p\mathbb{Z}$, we have

$$S=\frac{1}{p^n}\sum_h T(h_i,w(p))S_1(a,P,h)$$

where

$$T(h,x)=\prod_{1\leqslant i\leqslant n}\sum_{0\leqslant m\leqslant x}e\Big(\frac{h_im}{p}\Big)$$

$$S_1(a,P,h)=\sum_{x_1,\dots,x_n}e\Big(\frac{a\cdot P(x)-h\cdot x}{p}\Big)$$

(here $h\cdot x$ is the usual scalar product $h_1x_1+\cdots+h_nx_x$). We have

$$T(h,x)\ll\prod_{1\leqslant i\leqslant n}\min(x,\|h_i/p\|^{-1})$$

where $\|t\|$ is the distance to the nearest integer as usual. For the complete sums $S_1(a,P,h)$, we have two bounds: first, a result of Weil shows that

$$S_1(a,P,h)\ll p^{n-\frac{1}{2}} \tag{21.5}$$

(the implied constant depending only on the P_i) because $a\cdot P(x)-h\cdot x$ is never identically zero for $a\neq 0$ by assumption. This bound is insufficient because there are too many h. But by Theorem 11.43 with $f=a\cdot P$, $g=1$ and $V=\mathbb{A}^n_{\mathbb{Z}}$, there are varieties X_j of dimension $\leqslant n-j$ such that $S_1(a,P,h)\ll Cp^{n/2+(j-1)/2}$ for $h\notin X_j$. Weil's bound (21.5) shows that $X_n=\emptyset$. Denoting $X_0=\mathbb{A}^n$ we get

$$S\ll p^{-n}\sum_{1\leqslant j\leqslant n}p^{(n+j-1)/2}\sum_{h\in X_{j-1}}|T(h,w(p))|.$$

Lemma 9.5 of [**FK**] gives

$$\sum_{h\in X_{j-1}}|T(h,w(p))|\ll p^{n-(j-1)}w(p)^{j-1}(\log p)^{n-(j-1)}$$

from which

$$S \ll p^{1/2} w(p)^{n-1} (\log p)$$

hence (21.4) follows. If $w(p) > p$, one dissects the set of summation into cubes with sides $< p$ and apply the above bound for each of them. □

A different set of problems, still very natural, involves the "angles" of exponential sums to prime modulus, or the local roots of L-functions for varieties over finite fields. Quite similar is the question of distribution of eigenvalues of Hecke operators T_p.

Maybe the simplest non-trivial exponential sums are Gauss sums. Let χ be a primitive character modulo p and $\tau(\chi)$ the associated Gauss sum (3.10). By (3.14) one can write

$$\tau(\chi) = e(\alpha_p(\chi))\sqrt{p},$$

where $\alpha_p(\chi) \in \mathbb{R}/\mathbb{Z}$ is called the "angle" of the Gauss sum. The question is, how are those distributed?

This question admits at least two variants. One can consider Gauss sums for characters of a fixed order d modulo primes $p \equiv 1 \pmod{d}$, as $p \to +\infty$, or one can look at all characters modulo p. In the former case, when $d = 2$, Theorem 3.3 and the Prime Number Theorem show that $e(\alpha_p(\chi))$ is equidistributed among the two elements $\{1, i\}$ (note that in this case $e(\alpha_p(\chi)) \in \{\pm 1, \pm i\}$ follows from $\tau(\chi)^2 = \tau(\chi)\tau(\bar{\chi}) = \chi(-1)p$, so the angle is not equidistributed in the group $\{\pm 1, \pm i\}$).

Kummer studied the arguments of cubic Gauss sums. For $p \equiv 1 \pmod 3$, there exists a unique prime $\pi \in K = \mathbb{Q}(e(1/3)) \subset \mathbb{C}$ such that $\pi \equiv 1 \pmod 3$ and $N\pi = p$. The cubic residue symbol $\chi_\pi(x) = (\frac{x}{\pi})_3$ is defined as the cube root of unity such that

$$x^{(p-1)/3} \equiv \chi_\pi \pmod{\pi}, \text{ for } x \in \mathcal{O}/\pi\mathcal{O}$$

where $\mathcal{O}$ is the ring of integers in K. It is a character of order 3 on the multiplicative group of the residue field $\mathcal{O}/\pi\mathcal{O} = \mathbb{Z}/p\mathbb{Z}$, i.e. a (primitive) cubic Dirichlet character modulo p. Using Jacobi sums, one shows that the Gauss sum satisfies

$$\Big(\frac{\tau(\chi_\pi)}{\sqrt{p}}\Big)^3 = e(3\alpha_p(\chi_\pi)) = -\frac{\pi}{\bar{\pi}};$$

see [**IR**]. It is well-known that $\pi/\bar{\pi}$ is equidistributed on $\mathbb{R}/\mathbb{Z}$ (proved using Hecke characters of K, see Section 3.8), but to get $e(\alpha_p(\chi_\pi))$ a certain cube root of unity "interferes". Kummer and later Hasse conjectured a certain non-uniform distribution based on some numerical evidence. However R. Heath-Brown and S. Patterson [**HBP**] proved:

THEOREM 21.5. *The angles $e(\alpha_p(\chi_\chi))$ of cubic Gauss sums are equidistributed in $\mathbb{R}/\mathbb{Z}$ as $p \to +\infty$, $p \equiv 1 \pmod 3$.*

The proof is based on techniques for sums over primes, as in Chapter13, and properties of a (metaplectic) Eisenstein series which has the Gauss sums in its Fourier coefficients.

In the case of all characters modulo p, we have the following theorem of Deligne:

THEOREM 21.6. *The $p-2$ angles $(\alpha_p(\chi))$ of Gauss sums of primitive characters modulo p become equidistributed in $\mathbb{R}/\mathbb{Z}$ as p tends to ∞, i.e. we have*

$$\lim_{p\to+\infty}\frac{1}{p-2}\sum_{\chi \pmod p}^{\star} f(\alpha_p(\chi)) = \int_0^1 f(\theta)d\theta$$

as $p\to\infty$ for any continuous function $f:[0,1]\to\mathbb{C}$.

PROOF. By the Weyl Criterion, it is enough to show that

$$\lim_{p\to+\infty}\frac{1}{p-2}\sum_{\chi \pmod p}^{\star}\Big(\frac{\tau(\chi)}{\sqrt{p}}\Big)^n = 0 \tag{21.6}$$

for $n\neq 0$. One can reduce to $n\geqslant 1$ using the formula $\tau(\chi)\tau(\bar\chi)=\chi(-1)p$ (see (3.14)).

We have

$$\tau(\chi)^n = \sum_{x_1,\dots,x_n}\chi(x_1\cdots x_n)e\Big(\frac{x_1+\cdots+x_n}{p}\Big)$$

hence summing over all χ (including $\chi=1$) we get by orthogonality of characters

$$\sum_{\chi}\tau(\chi)^n = (p-1)K_n(1,p)$$

where $K_n(1,p)$ is the multiple Kloosterman sum defined in (11.55). Subtracting the contribution of $\chi=1$, we get

$$\sum_{\chi \pmod p}^{\star}\Big(\frac{\tau(\chi)}{\sqrt{p}}\Big)^n = p^{-n/2}((p-1)K_n(1,p)+(-1)^{n+1}).$$

As $p\to+\infty$, we have $K_n(1,p)\ll p^{(n-1)/2}$ by (11.58), where the implied constant depends only on n, hence (21.6) follows. □

Of course for analytic number theory a particular interest attaches to the classical Kloosterman sums (1.56). By Weil's bound, one can write

$$S(a,b;p) = 2\sqrt{p}\cos(2\pi\theta_p(a,b))$$

for some unique $\theta_p(a,b)\in[0,\pi]$. The problem of the distribution of the angles is fascinating. Using all $a\neq 0$ modulo p, N. Katz [**K4**] solved the problem.

THEOREM 21.7. *The $p-1$ angles $\theta_p(a,a)$ for $a\neq 0$ are equidistributed with respect to the Sato-Tate measure on $[0,\pi]$ given by*

$$d\mu_{ST} = 2\pi^{-1}(\sin\theta)d\theta$$

as $p\to+\infty$, i.e we have

$$\frac{1}{p-1}\sum_{a \pmod p}^{\star} f(\theta_p(a)) \sim \int_0^{\pi} f(\theta)d\mu.$$

as $p\to\infty$ for any continuous function $f:[0,\pi]\to\mathbb{C}$.

Moreover, he states the following conjecture where a is fixed:

THE SATO-TATE CONJECTURE FOR KLOOSTERMAN SUMS. *Let a, b be fixed non-zero integers. For p prime let $\theta_p \in [0, \pi]$ be such that*

$$S(a, b; p) = 2\sqrt{p}(\cos\theta_p).$$

Then as $X \to +\infty$, the angles θ_p, $p \leqslant X$, become equidistributed with respect to the Sato-Tate measure $d\mu_{ST}$.

It is not even known whether $S(a, b; p)$ changes sign infinitely often, or that the statement

$$(2 - \varepsilon) \leqslant \frac{S(a, b; p)}{\sqrt{p}} = 2\cos\theta_p \leqslant 2$$

(for all p large enough) is false, for any $\varepsilon \in]0, 2[$. See [**FoM**] for the best progress in this direction, combining sieve methods and cohomological studies of exponential sums as in Chapter 11.

In Corollary 21.9, on the other hand, we prove that the angles of Salié sums $T(a, b; p)$ (for fixed a, b with ab not a square) are equidistributed as $p \to +\infty$, but with respect to Lebesgue measure.

The same Sato-Tate distribution is conjectured to hold for the distribution of coefficients a_p associated to an elliptic curve $E/\mathbb{Q}$ without CM, or more generally for the eigenvalues $\lambda_f(p)$ of a non-dihedral primitive holomorphic cusp form f of weight $\geqslant 2$. If p is a prime not dividing the conductor of f, then by Deligne's bound we have

$$\lambda_f(p) = 2\cos\theta_p(f)$$

for some $\theta_p(f) \in [0, \pi]$. For E, this reads

$$a_E(p) = 2\sqrt{p}\cos\theta_p(E).$$

Then we have

THE SATO-TATE CONJECTURE FOR MODULAR FORMS. *Let f be a primitive holomorphic cusp form of weight $\geqslant 2$ which is not of dihedral type. With the above notation, the angles $\theta_p(f)$, $p \leqslant X$, become equidistributed as $X \to +\infty$ with respect to the Sato-Tate measure $d\mu_{ST}$.*

Using the modularity of f and the analytic properties of symmetric power L-functions, more progress has been made concerning this conjecture than about the analogue for Kloosterman sums (see Serre's letter at the end of [**Shah**]). In fact, there is a convincing argument in favor of this conjecture based on the expected functoriality of symmetric power L-functions and the analytic properties that would follow by the results of Chapter 5. We can sketch this argument, taking the case of the Ramanujan Δ function of weight 12 as an example – the level is 1 which eliminates complications arising from ramification. We have the Euler product factorizing as

$$L(\Delta, s) = \prod_p (1 - \alpha_p p^{-s})^{-1}(1 - \beta_p p^{-s})^{-1}$$

with $p^{11/2}(\alpha_p + \beta_p) = \tau(p)$ and $\alpha_p\beta_p = 1$. Here $\tau(p)$ is the Ramanujan τ-function, not the divisor function. For $n \geqslant 1$, the n-th symmetric power of Δ is the Euler product

$$L(\mathrm{Sym}^n \Delta, s) = \prod_p \prod_{0 \leqslant j \leqslant n} (1 - \alpha_p^j \beta_p^{n-j} p^{-s})^{-1}.$$

The functoriality principle of Langlands would imply that $L(\mathrm{Sym}^n \Delta, s)$ is an automorphic L-function of degree $n+1$ in the sense of Chapter 5, with conductor $q = 1$ and with no pole. Since $\alpha_p \beta_p = 1$, we have

$$\Lambda_{\mathrm{Sym}^n \Delta}(p) = \frac{\sin(n+1)\theta_p}{\sin \theta_p} = P_n(\cos \theta_p)$$

where $\theta_p = \theta_p(\Delta)$ and P_n is the Tchebyshev polynomial. Applying Theorem 5.10 we find that for $n \geqslant 1$,

$$\sum_{p \leqslant X} P_n(\cos \theta_p) = o(\pi(X))$$

as $X \to +\infty$, hence

$$\lim_{X \to +\infty} \frac{1}{\pi(X)} \sum_{p \leqslant X} P_n(\cos \theta_p) = 0 = \int_0^\pi P_n(\cos t) d\mu_{ST}(t).$$

It is easy to show that the functions $P_n(\cos t)$ span $C([0, \pi])$, so the Sato-Tate conjecture follows by the Weyl Criterion from these conjectured properties of the symmetric powers.

It is currently known that $L(\mathrm{Sym}^n \Delta, s)$ is automorphic and for $n \leqslant 4$: $n = 2$ is due to Gelbart and Jacquet [**GeJ**] and $n = 3$, 4 to Kim and Shahidi (see [**KSh**]), and that it is an L-function in the sense of Chapter 5, possibly with extra poles, for $n \leqslant 9$.

If f is dihedral, which is equivalent to E having CM in the elliptic curve case, the distribution is known (by work of Deuring) and is quite different. Notice that one can determine this distribution by the above argument if one knows the order of the pole of $L(\mathrm{Sym}^n f, s)$ at $s = 1$. Similarly for weight 1 forms where the Hecke eigenvalues can only take finitely many different values.

Finally we should mention that equidistribution problems in hyperbolic plane are also very interesting. There one has to use not only cusp forms as test functions in the Weyl Criterion, but also Eisenstein series. As an example, W. Duke [**Du2**] proved that Heegner points become equidistributed with respect to the Poincaré measure as the discriminant goes to infinity. Also we refer to [**Sa4**] for discussion of the problems of "quantum chaos" and the links between the quantum unique ergodicity conjecture and subconvexity bounds for certain high-degree L-functions.

21.3. Roots of quadratic congruences.

Let $f(X) = aX^2 + bX + c$ be a quadratic polynomial with integer coefficients. For a given prime $p \nmid 2a$ the congruence

$$f(\nu) \equiv 0 (\mathrm{mod}\ p) \tag{21.7}$$

has at most two solutions; precisely the number of solutions is

$$\rho(p) = 1 + \left(\frac{D}{p}\right)$$

where $D = b^2 - 4ac$ is the discriminant of f and $\left(\frac{D}{p}\right)$ is the Legendre symbol. Hence 50% of primes yield two roots and the other 50% yields none. Therefore $\rho(p)$ is one on average

$$\sum_{p\leqslant x}\rho(p)\sim\pi(x),\qquad \text{as } x\to\infty. \tag{21.8}$$

It is an interesting question how the two roots $\nu(\text{mod } p)$ are distributed as p runs over the primes with $p\nmid 2a$, $(\frac{D}{p})=1$. Obviously, if $f(X)$ factors over $\mathbb{Q}$, then the fractions $\{\nu/p\}$ are not dense in $[0,1]$; in fact the limit points can only be $\frac{u}{a}$ with $0\leqslant u\leqslant a$. Excluding this case W. Duke, J. Friedlander, H. Iwaniec [**DFI5**] and A. Toth [**Tot**] showed

THEOREM 21.8. *Suppose $f(X)$ is a quadratic polynomial with integer coefficients, irreducible over $\mathbb{Q}$. Then for any $0\leqslant\alpha<\beta\leqslant 1$ we have*

$$|\{\nu \ (\text{mod } p); p\leqslant x, f(\nu)\equiv 0 \ (\text{mod } p), \alpha\leqslant\left\{\frac{\nu}{p}\right\}\leqslant\beta\}\sim(\beta-\alpha)\pi(x) \tag{21.9}$$

as $x\to\infty$.

Equivalently, for any continuous, periodic function $F(t)$ of period one

$$\lim_{x\to\infty}\frac{1}{\pi(x)}\sum_{p\leqslant x}\ \sum_{f(\nu)\equiv 0(\text{mod } p)}F(\nu/p)=\int_0^1F(t)dt. \tag{21.10}$$

In other words, the sequence of numbers ν/p is equidistributed modulo one. Furthermore, by Weyl's criterion together with (21.8) this property is equivalent to

$$\sum_{p\leqslant x}\rho_h(p)=o(\pi(x)), \tag{21.11}$$

as $x\to\infty$, for any positive integer h, where

$$\rho_h(n)=\sum_{\substack{\nu(\text{mod } n)\\ f(\nu)\equiv 0(\text{mod } n)}}e\Big(\frac{h\nu}{n}\Big). \tag{21.12}$$

A consequence of Theorem 21.8 is the uniform distribution of angles of the Salié sums

$$T(m,n;p)=\sum_{d(\text{mod } p)}\Big(\frac{d}{p}\Big)\Big(\frac{\bar{d}m+dn}{p}\Big).$$

By Lemma 12.4 we have

$$T(m,n;p)=2\cos\left(\frac{2\pi\nu}{p}\right)T(0,n;p)$$

for $p\nmid 2mn$, where $T(0,n;p)=\varepsilon_p\sqrt{p}(\frac{n}{p})$ is the Gauss sum and ν is a root of $\nu^2\equiv 4mn(\text{mod } p)$. We think of $T(0,n;p)$ as the normalization factor and call $2\pi\nu/p$ the angle of the Salié sum. Therefore Theorem 21.8 implies

COROLLARY 21.9. *If mn is not a square, then the angles of the Salié sums $T(m,n;p)$ are uniformly distributed modulo 2π as p runs over primes.*

This contrasts with the conjectured Sato-Tate distribution for Kloosterman sums.

In the simplest case of the polynomial $f(X) = X^2+1$ the roots $\nu(\text{mod } p)$ with $p \equiv 1(\text{mod } 4)$ correspond to the representations of p as the sum of two squares $p = r^2+s^2$. By choosing $-s < r \leqslant s$ with s odd, we see that every such representation gives the unique root $\nu \equiv r\bar{s}(\text{mod } p)$. Hence

$$\frac{\nu}{p} \equiv -\frac{\bar{r}}{s} + \frac{1}{sp}(\text{mod } 1).$$

In view of this construction the assertion (21.11) becomes

$$\sum_{r^2+s^2=p\leqslant x}^{\star} e\Big(h\frac{\bar{r}}{s}\Big) = o(\pi(x))$$

after clearing the almost constant perturbation $e(h/sp) = 1+O(h/sp)$. This result is reminiscent to the following one:

$$\sum_{r^2+s^2=p\leqslant x}^{\star} \Big(\frac{r}{s}\Big) \ll x^{\frac{76}{77}}$$

which was established in [**FI**]. Other surprising applications of Theorem 21.8 are given in [**VdP**] and [**Ko2**].

In this chapter we present a proof of Theorem 21.8 for the polynomial

$$f(X) = X^2 - D \tag{21.13}$$

with $D < 0$ (so $f(X)$ is irreducible without further conditions). The general case of f with negative discriminant can be dealt with by completing the square and modifying slightly the arguments. However, the case of positive discriminant requires substantial changes (because of the infinite group of units in $\mathbb{Q}(\sqrt{D})$ if $D > 0$). This case was settled by A. Toth [**Tot**] using an alternative path passing directly through Kloosterman sums. Our approach also depends on Kloosterman sums, but indirectly via Section 16.5.

21.4. Linear and bilinear forms in quadratic roots.

By the Weyl Criterion we must show the inequality (21.11) for the polynomial (21.13) with $D < 0$ and $h \neq 0$. We can assume that D is squarefree because the Weyl sum (21.11) with frequency h for $D = m^2E$ is equal to the Weyl sum with frequency mh for E (up to bounded amount for primes dividing m).

According to the equidistribution criteria it suffices to prove that

$$\sum_p \rho_h(p) g\Big(\frac{p}{x}\Big) = o\Big(\frac{x}{\log x}\Big) \tag{21.14}$$

as $x \to \infty$, where h is a fixed positive integer and $g(y)$ is a fixed smooth function supported on $1 \leqslant y \leqslant 2$. To this end we apply Theorem 13.12 for the sequence

$\mathcal{A} = (a_n)$ with $a_n = \rho_h(n)g(n/x)$. We have to verify the conditions (13.67), (13.68) for $\Delta = x^{\varepsilon(x)}$ with $\varepsilon(x) \to 0$. The first condition is weaker than the estimate

$$(21.15)\qquad \sum_{d \leqslant x^{\frac{1}{2}}(\log x)^{-B}} \Big| \sum_{n \equiv 0 (\mathrm{mod}\ d)} \rho_h(n) g\Big(\frac{n}{x}\Big)\Big| \ll x(\log x)^{-A}$$

for any $A \geqslant 2$, where $B = B(A)$ and the implied constant depends on A, h and g. For the second condition it suffices to prove the following estimate

$$(21.16)\qquad \sum_m \Big| \sum_n \beta_n \rho_h(mn) g\Big(\frac{mn}{x}\Big)\Big| \ll x(\log x)^{-A}$$

for any complex numbers β_n supported in the interval

$$(21.17)\qquad (\log x)^B \leqslant n \leqslant x^{\frac{1}{3}}(\log x)$$

with $|\beta_n| \leqslant 1$. We can also restrict the support of β_n to primes (see (13.68)) which helps at technical points.

In this section we derive both (21.15) and (21.16) from one estimate for the linear forms

$$(21.18)\qquad \mathcal{L}_d(x) = \sum_{n \equiv 0 (\mathrm{mod}\ d)} \rho_h(n) g\Big(\frac{n}{x}\Big).$$

However, we need a very strong bound for $\mathcal{L}_d(x)$ which is uniform in large ranges of d and h.

PROPOSITION 21.10. *Let $1 \leqslant h \leqslant x$. Then*

$$(21.19)\qquad \mathcal{L}_d(x) \ll (x^{\frac{1}{2}} + d^{-\frac{1}{2}}(d,h)^{\frac{1}{4}} x^{\frac{3}{4}}) \tau^2(dh) \log^2 x$$

where the implied constant depends on the polynomial f and the test function g.

For $d = 1$ the spectral theorem for the modular group would give a stronger (best possible) result

$$\sum_n \rho_h(n) g\Big(\frac{n}{x}\Big) \ll x^{\frac{1}{2}} \log^2 x$$

where the implied constant depends on h. Without smoothing, the spectral method yields

$$\sum_{n \leqslant x} \rho_h(n) \ll x^{\frac{2}{3}} \log x$$

as shown by V. A. Bykovsky [**Byk**]. Our approach to $\mathcal{L}_d(N)$ was inspired by that of Bykovsky. We shall work with the group $\Gamma_0(d)$ so we have to deal with the exceptional eigenvalues that might be there. All these will be taken care of in estimation for a certain Poincaré series in the last section.

That (21.19) implies (21.15) is clear. Now we are going to derive (21.16) from (21.19). Let $\mathcal{B}(x)$ denote the double sum over m, n on the left side of (21.16). First we install the condition $(m, n) = 1$ and estimate the missing part by

$$\sum_m \sum_{n|m} \Big| \beta_n \rho(mn) g\Big(\frac{mn}{x}\Big)\Big| \ll x \sum_n |\beta_n| n^{-2} \ll x(\log x)^{-B}$$

because β_n are supported on primes in the interval (21.17). Let $\mathcal{B}^\star(x)$ denote the double sum reduced by $(m,n)=1$, so $\mathcal{B}(x)=\mathcal{B}^\star(x)+O(x(\log x)^{-B})$. We arrange $\mathcal{B}^\star(x)$ as follows

$$\mathcal{B}^\star(x)\leqslant\sum_m\sum_{f(\delta)\equiv 0(m)}\Big|\sum_{(n,m)=1}\beta_n g\Big(\frac{mn}{x}\Big)\sum_{\substack{f(\nu)\equiv 0(mn)\\ \nu\equiv\delta(m)}}e\Big(\frac{h\nu}{mn}\Big)\Big|.$$

Then by Cauchy's inequality $\mathcal{B}^\star(x)^2\ll\mathcal{C}^\star(x)\log x$, where

$$\mathcal{C}^\star(x)=\sum_m m\sum_{f(\delta)\equiv 0(m)}\Big|\sum_{(n,m)=1}\beta_n g\Big(\frac{mn}{x}\Big)\sum_{\substack{f(\nu)\equiv 0(mn)\\ \nu\equiv\delta(m)}}e\Big(\frac{h\nu}{mn}\Big)\Big|^2.$$

Squaring out and changing the order of summation we obtain

$$\mathcal{C}^\star(x)=\sum_{n_1}\sum_{n_2}\beta_{n_1}\bar{\beta}_{n_2}\mathcal{D}^\star(n_1,n_2)$$

where

$$\mathcal{D}^\star(n_1,n_2)=\sum_{(m,n_1n_2)=1}mg\Big(\frac{mn_1}{x}\Big)\bar{g}\Big(\frac{mn_2}{x}\Big)\sum_{\substack{f(\nu_j)\equiv 0(mn_j)\\ \nu_1\equiv\nu_2(m)}}^{\star}e\Big(\frac{h}{m}\Big(\frac{\nu_2}{n_1}-\frac{\nu_2}{n_2}\Big)\Big).$$

For $n_1=n_2$ we use the trivial bound $\mathcal{D}^\star(n,n)\ll n^{-2}x^2$. If $n_1\neq n_2$, then $(n_1,n_2)=1$ since we assumed they are primes, therefore

$$\mathcal{D}^\star(n_1,n_2)=\sum_{(m,n_1n_2)=1}mg\Big(\frac{mn_1}{x}\Big)\bar{g}\Big(\frac{mn_2}{x}\Big)\sum_{f(\nu)\equiv 0(mn_1n_2)}e\Big(\frac{h(n_2-n_1)\nu}{mn_1n_2}\Big).$$

Note that $\frac{1}{2}\leqslant\frac{n_1}{n_2}\leqslant 2$, or else the sum $\mathcal{D}^\star(n_1,n_2)$ is void. Here we remove the condition $(m,n_1n_2)=1$ and estimate the excess part by $O((n_1n_2)^{-3/2}x^2)$ again by taking advantage of n_1,n_2 being primes. With the condition $(m,n_1n_2)=1$ removed the resulting complete sum $\mathcal{D}(n_1,n_2)$ is just the linear form (21.18) for $d=n_1n_2$, h replaced by $h(n_2-n_1)$ and the test function $g(y)$ replaced by $\frac{y}{n_1n_2}g(\frac{y}{n_2})\bar{g}(\frac{y}{n_1})$. After rescaling, Proposition 21.10 yields

$$\mathcal{D}(n_1,n_2)\ll((n_1n_2)^{-\frac14}x^{\frac12}+(n_1n_2)^{-\frac58}x^{\frac34})x\log^2 x$$

where the implied constant depends only on f,g,h. Putting together the above estimates one can easily complete the proof of (21.16).

21.5. A Poincaré series for quadratic roots.

In this section we interpret the linear form $\mathcal{L}_d(x)$ given by (21.18) as a Poincaré series $P_h(z)$ for the group $\Gamma_0(d)$ of the kind considered in Sections 16.2 and 16.3. Throughout $z=x+iy$ denotes a point in the upper half-plane $\mathbb{H}$, so x is no longer the variable from the previous sections.

We begin by recalling various connections of positive definite binary quadratic forms with modular forms. A few relevant facts are also reviewed in Section 22.1. Fix a negative integer D. Consider all forms (with even middle coefficient)

$$\varphi(X,Y)=aX^2+2bXY+cY^2$$

of discriminant $b^2 - ac = D$. Multiplying by -1 (if necessary) we can assume that $a, c > 0$, so φ is positive definite. There are two zeros of φ which are complex conjugate. Choose the one in $\mathbb{H}$

$$z_\varphi = \frac{b + \sqrt{D}}{a} \in \mathbb{H}.$$

The modular group $\Gamma = SL_2(\mathbb{Z})$ acts on the quadratic forms of discriminant D by linear change of variables

$$\varphi^\sigma(X, Y) = \varphi(\alpha X + \gamma Y, \beta X + \delta Y)$$

if $\sigma = \begin{pmatrix} \alpha & \beta \\ \gamma & \delta \end{pmatrix} \in \Gamma$. This action is compatible with taking the zeros, i.e., $\sigma z_\varphi = z_\varphi \sigma$. Let F be the standard fundamental domain for Γ (see (22.10) and (22.11)). Every form of discriminant D is equivalent (with respect to the action of Γ) to the exactly one form with $z_\varphi \in F$. Denote the set of roots of representing classes by

$$\Lambda = \{z_\varphi \in F;\ \ \text{discr}(\varphi) = D\}.$$

Therefore, there exists a one-to-one correspondence between the solutions of $b^2 - ac = D$ with $a, b, c \in \mathbb{Z}$, $a, c > 0$ and the points of the orbits $\{\sigma z; \sigma \in \Gamma/\Gamma_z\}$ for $z \in \Lambda$, where Γ_z is the stability group of z. Hence we derive

$$\rho_h(n) = \sum_{b^2 \equiv D (\text{mod } n)} e\left(\frac{hb}{n}\right) = \sum_{z \in \Lambda} |\Gamma_z|^{-1} \sum_{\substack{\sigma \in \Gamma_\infty \backslash \Gamma \\ \text{Im } \sigma z = \sqrt{|D|}/n}} e(h \text{Re } \sigma z).$$

Next we split the inner sum into classes with respect to the subgroup $\Gamma_0(d)$ getting

$$\rho_h(n) = \sum_{z \in \Lambda} |\Gamma_z|^{-1} \sum_{\tau \in \Gamma_0(d) \backslash \Gamma} \sum_{\substack{\sigma \in \Gamma_\infty \backslash \Gamma_0(d) \\ \text{Im } \sigma \tau z = \sqrt{|D|}/n}} e(h \text{Re } \sigma \tau z). \tag{21.20}$$

We shall use this formula for numbers $n \equiv 0 (\text{mod } d)$. This congruence can be built into the summation conditions for z and τ (but not σ) as the following congruence

$$\sqrt{|D|}/\text{Im } \tau z \equiv 0 (\text{mod } d). \tag{21.21}$$

Note that this condition does not depend on the choice of τ in the coset $\Gamma_0(d)\tau$.

Now we are ready to evaluate the linear forms

$$\mathcal{L}_d = \sum_{n \equiv 0 (\text{mod } d)} \rho_h(n) F(2\pi h \sqrt{|D|}/n) \tag{21.22}$$

where $F(u)$ is a smooth function compactly supported on $\mathbb{R}^+$. Introducing (21.20) we arrange this into

$$\mathcal{L}_d = \sum_{z \in \Lambda} |\Gamma_z|^{-1} \sideset{}{^b}\sum_{\tau \in \Gamma_0(d) \backslash \Gamma} P_h(\tau z) \tag{21.23}$$

where $\sum^b$ restricts τ by the congruence (21.21) and $P_h(z)$ is the Poincaré series for the group $\Gamma_0(d)$

$$P_h(z) = \sum_{\sigma \in \Gamma_\infty \backslash \Gamma_0(d)} F(2\pi h \mathrm{Im}\, \sigma z) e(h \mathrm{Re}\, \sigma z). \tag{21.24}$$

The formula (21.23) truly represents the linear form $\mathcal{L}_d$ by the Poincaré series $P_h(\tau z)$ because there are relatively few points τz satisfying (21.21). Precisely the number of points $z \in \Lambda$ is the class number $h(D)$ and the number of cosets $\Gamma_0(d)\tau$ with $\tau \in \Gamma$ is the index

$$[\Gamma : \Gamma_0(d)] = d \prod_{p|d} \Big(1 + \frac{1}{p}\Big).$$

This alone is quite large, however, not every τ satisfies (21.21). For every $z \in \Lambda$, let $\nu(d, z)$ denote the number of cosets $\Gamma_0(d)\tau$ for which (21.21) holds. Let $\varphi = aX^2 + 2bXY + cY^2$ be the quadratic form corresponding to z. Then notice that $\sqrt{|D|}/\mathrm{Im}\,(z) = c = \varphi(0,1)$. Hence the condition (21.21) for

$$\tau = \begin{pmatrix} \alpha & \beta \\ \gamma & \delta \end{pmatrix}$$

is that

$$\varphi((0,1)\tau) = \varphi(\gamma, \delta) = a\gamma^2 + 2b\gamma\delta + c\delta^2 \equiv 0 \pmod{d}. \tag{21.25}$$

Moreover, counting the cosets $\Gamma_0(d)\tau$ means that we count the roots (γ, δ) of φ modulo d up to an invertible scalar modulo d (i.e. the projective solutions).

We claim that

$$\nu(d, z) \leqslant \tau(d). \tag{21.26}$$

To prove this, recall we assumed that D is squarefree, so by the Chinese Remainder Theorem we must prove that $\nu(p, z) \leqslant 2$ for $p \mid D$ prime. Since p^2 does not divide D, the quadratic form $aX^2 + 2bXY + cY^2$ is non-zero modulo p. The projective solutions to (21.25) for $d = p$ with $\delta \not\equiv 0 \pmod{p}$ correspond to roots of the non-zero quadratic polynomial $aX^2 + 2bX + c$, hence there are at most 2 of them. If $p \mid \delta$, the only possible projective root is the "point at infinity" $(1, p)$, which occurs if and only if $p \mid a$. In that case there was at most one solution in the first situation, so the total number of solutions if always $\leqslant 2$.

Note that (21.18) becomes (21.22) for the test function

$$F(u) = g(2\pi h \sqrt{|D|}/ux). \tag{21.27}$$

This particular function is supported in $Y^{-1} \leqslant u \leqslant 2Y^{-1}$ with

$$\pi h \sqrt{|D|} Y = x. \tag{21.28}$$

Therefore, by the above arrangements and observations, our problem of estimating the linear form $\mathcal{L}_d(x)$ reduces to that of the Poincaré series $P_h(z)$ for the test function $F(u)$ given by (21.27). In the next section we treat $P_h(z)$ by spectral methods.

21.6. Estimation of the Poincaré series.

We shall estimate $P_h(z)$ for all $z \in \mathbb{H}$ uniformly in h and the level d. Our goal is

LEMMA 21.11. *Let $P_h(z)$ be given by (21.24) with $F(u)$ supported in $Y^{-1} \leqslant u \leqslant 2Y^{-1}$ for $Y \geqslant 2$ such that $|F^{(j)}| \leqslant Y^j$ for $0 \leqslant j \leqslant 4$. Then for any $z \in \mathbb{H}$ and $\tau \in \Gamma$ we have*

$$P_h(\tau z) \ll (y + y^{-1})^{\frac{1}{2}} \left((hY)^{\frac{1}{2}} + d^{-\frac{1}{2}} (d,h)^{\frac{1}{4}} (hY)^{\frac{3}{4}} \right) \tau(dh) (\log Y)^2 \tag{21.29}$$

where $y = \operatorname{Im} z$ and the implied constant is absolute.

It is easy to see that the bound (21.29) for Y given by (21.28) together with (21.26) yield (21.19). Hence one completes the proof of Theorem 21.8 for the polynomial $f(X) = X^2 - D$.

We still have to prove Lemma 21.11. We start from the spectral decomposition (see Theorem 15.2)

$$P_h(z) = \sum_j \langle P_h, u_j \rangle u_j(z) + \sum_{\mathfrak{a}} \frac{1}{4\pi} \int_{\mathbb{R}} \langle f, E_{\mathfrak{a}}(\cdot, \tfrac{1}{2} + it) \rangle E_{\mathfrak{a}}(z, \tfrac{1}{2} + it) dt$$

where $(u_j(z))$ is an orthonormal basis of Maass cusp forms, together with the constant function for the zero eigenvalue, and $\mathfrak{a}$ runs over the cusps of $\Gamma_0(d)$.

By the Cauchy-Schwarz inequality

$$|P_h(z)|^2 \leqslant K_d(z) R_d(h) \tag{21.30}$$

where

$$K_d(z) = \sum_j h(t_j) |u_j(z)|^2 + \sum_{\mathfrak{a}} \frac{1}{4\pi} \int_{\mathbb{R}} h(t) |E_{\mathfrak{a}}(z, \tfrac{1}{2} + it)|^2 dt \tag{21.31}$$

and

$$R_d(h) = \sum_j \frac{1}{h(t_j)} |\langle P_h, u_j \rangle|^2 + \sum_{\mathfrak{a}} \frac{1}{4\pi} \int_{\mathbb{R}} \frac{1}{h(t)} |\langle P_h, E_{\mathfrak{a}}(\cdot, \tfrac{1}{2} + it) \rangle|^2 dt. \tag{21.32}$$

Here $h(t) > 0$ is a function decreasing fast enough at infinity, introduced for convergence. We choose

$$h(t) = (1 + t^2)^{-1} - (4 + t^2)^{-1} = 3(1 + t^2)^{-1} (4 + t^2)^{-1}. \tag{21.33}$$

First we estimate $K_d(z)$. To this end observe that (21.31) is the spectral decomposition of the automorphic kernel

$$K_d(z) = \sum_{\gamma \in \Gamma_0(d)} k(u(z, \gamma z))$$

where $k(u)$ is the Harish-Chandra/Selberg transform of $h(t)$ (see Theorem 15.7). Moreover, notice that $k(u)$ is non-negative because the Fourier transform of $h(t)$ is positive and decreasing on $\mathbb{R}^+$. These observations show that $K(z)$ will only increase if one replaces $\Gamma_0(d)$ by the modular group $\Gamma = \Gamma_0(1)$. Having done this the dependence on d is gone and the enlarged kernel is Γ-invariant, so we get

$K_d(\tau z) \leqslant K_1(\tau z) = K_1(z)$ for any $\tau \in \Gamma$ and $z \in \mathbb{H}$. Using crude estimates for cusp forms and Eisenstein series for the modular group we derive

$$K_d(\tau z) \ll y + y^{-1} \tag{21.34}$$

for any $\tau \in \Gamma$ and $z \in \mathbb{H}$, where the implied constant is absolute. Actually for our application we could stop at the inequality $K_d(\tau z) \leqslant K_1(z)$ because z runs over a fixed set of $h(D)$ points and we allow the implied constant to depend on D.

Now it remains to estimate $R_d(h)$. By (16.16) we have

$$\langle P_h, u_j\rangle = (2\pi h)^{\frac{1}{2}}\bar{\rho}_j(h)\tilde{F}(t_j)$$

where $\rho_j(h)$ is the Fourier coefficient of $u_j(z)$ and

$$\tilde{F}(t) = \int_0^\infty F(y)K_{it}(y)y^{-\frac{3}{2}}dy.$$

First we explain how to estimate $\tilde{F}(t)$. Since $F(y)$ is supported on the dyadic segment $Y^{-1} \leqslant y \leqslant 2Y^{-1}$ with $Y \geqslant 2$, it is useful to apply the power series expansion for $K_{it}(y)$ getting a rapidly convergent series of Mellin transform of $F(y)$ at the points $\frac{1}{2} \pm it + 2\ell$, for $\ell = 0, 1, 2, \ldots$. Then integrating the Mellin transform by parts four times we gain the factor $h(t)$ (see (21.33)). Next we estimate the resulting terms by Stirling's formula (5.112), and sum them up to deduce that $\tilde{F}(t)$ on the spectrum satisfies

$$\tilde{F}(t) \ll h(t)(\cosh \pi t)^{-\frac{1}{2}}Y^{\frac{1}{2}}(Y^{it} + Y^{-it} + 3\log Y).$$

Actually this result is a synthesis of three different bounds derived along the above lines separately in the ranges $0 < it < \frac{1}{2}$, $0 \leqslant t < \frac{1}{2}$, $\frac{1}{2} \leqslant t < \infty$. Hence we obtain

$$h(t_j)^{-1}|\langle P_h, u_j\rangle|^2 \ll hYH(t_j)|\rho_j(h)|^2,$$

where

$$H(t) = \frac{h(t)}{\cosh(\pi t)}(Y^{2it} + Y^{-2it} + 9\log^2 Y).$$

A similar estimate is deduced for the inner product of the Poincaré series P_h and the Eisenstein series $E_{\mathfrak{a}}$. From these estimates we get

$$R_d(h) \ll hY\Big(\sum_j H(t_j)|\rho_j(h)|^2 + \sum_{\mathfrak{a}} \frac{1}{4\pi}\int_{\mathbb{R}} H(t)|\tau_{\mathfrak{a}}(h,t)|^2 dt\Big),$$

where $\tau_{\mathfrak{a}}(h,t)$ is the Fourier coefficient of $E_{\mathfrak{a}}(z, \frac{1}{2}+it)$ (see (16.22)). Then applying (16.56) and (16.58) we arrive at

$$R_d(h) \ll hY\{1 + d^{-1}(d,h)^{\frac{1}{2}}(hY)^{\frac{1}{2}}\}\tau^2(dh)\log^4 Y. \tag{21.35}$$

Finally inserting (21.34) and (21.35) into (21.30) we finish the proof of Lemma 21.11.

CHAPTER 22

IMAGINARY QUADRATIC FIELDS

22.1. Binary quadratic forms.

We are primarily interested in the imaginary quadratic field $K = \mathbb{Q}(\sqrt{D})$, but to put our treatment into historical perspective we begin by reviewing the theory of binary quadratic forms

$$\varphi(X,Y) = aX^2 + bXY + cY^2 \tag{22.1}$$

of discriminant

$$D = b^2 - 4ac < 0. \tag{22.2}$$

We do not assume that D is fundamental, however, we only consider primitive forms, i.e.,

$$(a,b,c) = 1. \tag{22.3}$$

This is not a serious restriction because any form is a multiple of a primitive form. Changing the sign of all coefficients (if necessary) we can assume that

$$a > 0, \quad \text{and} \quad c > 0 \tag{22.4}$$

so φ is positive definite. We call a, b, c the first, the middle and the last coefficients of φ, respectively. The quadratic form $\varphi(X,Y)$ factors over complex numbers into linear forms

$$\varphi(X,Y) = a(X + z_a Y)(X + \bar{z}_a Y) = c(z_c X + Y)(\bar{z}_c X + Y) \tag{22.5}$$

with

$$z_a = \frac{b + \sqrt{D}}{2a} \quad \text{and} \quad z_c = \frac{b + \sqrt{D}}{2c}. \tag{22.6}$$

Actually the roots z_a, z_c depend on b but we do not display it. Throughout we choose the root so that z_a and z_c are in the upper half-plane $\mathbb{H}$. Note that $z_a \bar{z}_c = 1$.

The modular group $\Gamma = SL_2(\mathbb{Z})$ acts on the primitive forms of a given discriminant D by unimodular transformations

$$\begin{aligned}\varphi^\sigma(X,Y) &= \varphi(\alpha X + \gamma Y, \beta X + \delta Y), \quad \text{if } \sigma = \begin{pmatrix} \alpha & \beta \\ \gamma & \delta \end{pmatrix} \in \Gamma \\ &= a^\sigma X^2 + b^\sigma XY + c^\sigma Y^2,\end{aligned} \tag{22.7}$$

say, with

$$\begin{cases} a^\sigma &= a\alpha^2 + b\alpha\beta + c\beta^2 \\ b^\sigma &= 2a\alpha\gamma + b(\alpha\delta + \beta\gamma) + 2c\beta\delta \\ c^\sigma &= a\gamma^2 + b\gamma\delta + c\delta^2. \end{cases} \tag{22.8}$$

This action is compatible with action by linear fractional transformation on $\mathbb{H}$, and the roots are preserved.

Two forms φ, ψ are said to be equivalent if they belong to the same Γ-orbit, i.e., $\psi = \varphi^\sigma$ for some $\sigma \in \Gamma$; in this case we write $\varphi \sim \psi$. We denote the equivalence class which contains φ by

$$[\varphi] = \{\varphi^\sigma : \sigma \in \Gamma\}. \tag{22.9}$$

Obviously, every equivalence class is represented by a unique form $\varphi(X,Y) = aX^2 + bXY + cY^2$ whose root z_a is in the standard fundamental domain of the modular group $F = F^- \cup F^+$, where

$$F^- = \{z \in \mathbb{H} : |z| > 1, \quad -\frac{1}{2} < x < 0\}, \tag{22.10}$$

$$F^+ = \{z \in \mathbb{H} : |z| \geqslant 1, \quad 0 \leqslant x \leqslant \frac{1}{2}\}. \tag{22.11}$$

such a form φ is called reduced; it is characterized in terms of coefficients by

$$-a < b \leqslant a \leqslant c \quad \text{with} \quad b \geqslant 0 \quad \text{if} \quad a = c. \tag{22.12}$$

Since $\mathrm{Im}\,(z_a) = \sqrt{|D|}/2a \geqslant \sqrt{3}/2$, the first coefficient of a reduced form satisfies

$$a \leqslant \sqrt{|D|/3}. \tag{22.13}$$

Hence it follows that the number of reduced forms is finite. This number $h = h(D)$ is called the class number of discriminant D because it equals the number of distinct equivalence classes.

The transformations $\tau \in \Gamma$ which fix a form φ are called automorphs of φ; they form a finite cyclic group $\mathrm{Aut}(\varphi) = \{\tau \in \Gamma : \varphi^\tau = \varphi\}$ of order

$$w = \begin{cases} 6 & \text{if} \quad D = -3, \\ 4 & \text{if} \quad D = -4, \\ 2 & \text{if} \quad D < -4. \end{cases} \tag{22.14}$$

Equivalent forms have conjugate groups of automorphs, namely

$$\mathrm{Aut}(\varphi^\sigma) = \sigma \mathrm{Aut}(\varphi)\sigma^{-1}.$$

For reduced forms $\mathrm{Aut}(\varphi)$ is generated by

$$R = \begin{pmatrix} & -1 \\ 1 & 1 \end{pmatrix} \qquad \text{if } \varphi(X,Y) = X^2 + XY + Y^2,$$

$$S = \begin{pmatrix} & -1 \\ 1 & \end{pmatrix} \qquad \text{if } \varphi(X,Y) = X^2 + Y^2,$$

$$T = \begin{pmatrix} -1 & \\ & -1 \end{pmatrix} \qquad \text{if } D < -4.$$

Note that for all integers k the forms $\varphi(X+kY,Y)$ are equivalent; such forms are said to be parallel. The parallel forms have the same first coefficient a while the middle coefficients are in the same residue class modulo $2a$.

In "Disquisitiones Arithmeticae" Gauss introduced composition of forms of a given discriminant. His construction was subsequently simplified by Dirichlet but only for pairs of forms $\varphi_1(X,Y) = a_1X^2 + b_1XY + c_1Y^2$ and $\varphi_2(X,Y) = a_2X^2 + b_2XY + c_2Y^2$ satisfying

$$\left(a_1, \frac{b_1+b_2}{2}, a_2\right) = 1. \tag{22.15}$$

Note that b_1, b_2 have the same parity (namely $b_1 \equiv b_2 \equiv D(\mathrm{mod}\ 2)$) so $\frac{1}{2}(b_1+b_2)$ is an integer. If the Dirichlet condition (22.15) holds, then the forms φ_1, φ_2 are said to be united. Clearly, this relationship extends to the classes of parallel forms. For the united forms there exists a unique $b(\mathrm{mod}\ 2a_1a_2)$ such that

$$b \equiv b_1(\mathrm{mod}\ 2a_1), \quad b \equiv b_2(\mathrm{mod}\ 2a_2), \quad b^2 \equiv D(\mathrm{mod}\ 4a_1a_2). \tag{22.16}$$

Putting $b^2 - D = 4a_1a_2c$ we derive the form

$$\varphi(X,Y) = a_1a_2X^2 + bXY + cY^2 \tag{22.17}$$

which is primitive of discriminant D; it satisfies the identity

$$\varphi(x,y) = \varphi_1(x_1,y_1)\varphi_2(x_2,y_2) \tag{22.18}$$

with $x = x_1x_2 - cy_1y_2$ and $y = a_1x_1y_2 + by_1y_2 + a_2x_2y_1$. This is the Dirichlet formula for composition of united classes of parallel forms.

Now we use the Dirichlet formula to define composition of any two equivalence classes. We say that a positive integer m is properly represented by φ if

$$m = \varphi(\alpha,\gamma) \qquad \text{with } (\alpha,\gamma) = 1. \tag{22.19}$$

Since φ is positive definite the number of proper representations, denoted by $r_\varphi^\star(m)$, is finite. Equivalent forms represent properly the same numbers; these are exactly the numbers which appear as the first coefficients of forms in the equivalent class. Every primitive form represents properly a number m which is co-prime with any fixed positive integer. Therefore given two classes $\mathcal{A}_1, \mathcal{A}_2$ we can choose representatives φ_1, φ_2 with $(a_1, a_2) = 1$; such forms are united. Then (22.17) determines the third class $\mathcal{A} = \mathcal{A}_1\mathcal{A}_2 = [\varphi]$. One can show that $\mathcal{A}$ does not depend on the chosen representatives φ_1, φ_2, so the Dirichlet composition induces a well-defined multiplication of classes. The law of composition is commutative and also associative (the latter property is not obvious). This makes the set $\mathcal{H}$ of all distinct equivalence classes a finite abelian group of order h. The identity element of $\mathcal{H}$, which we denote by 1, is the class which contains the principal form

$$\begin{cases} \varphi = X^2 - \dfrac{D}{4}Y^2 & \text{if } D \equiv 0(\mathrm{mod}\ 4), \\[2ex] \varphi = X^2 + XY + \dfrac{1-D}{4}Y^2 & \text{if } D \equiv 1(\mathrm{mod}\ 4). \end{cases} \tag{22.20}$$

The inverse class of $\varphi(X,Y) = aX^2 + bXY + cY^2$ is the class which contains $\overline{\varphi}(X,Y) = aX^2 - bXY + cY^2$; these are called the opposite forms. Note the $\overline{\varphi}(X,Y)$ is equivalent to $\varphi(Y,X)$ (the form which is obtained by interchanging the first and the last coefficients). Therefore the sets of numbers properly represented by a class $\mathcal{A}$ and its inverse $\mathcal{A}^{-1}$ coincide.

The question of which numbers are represented properly by a given form, has no simple answer. However, there is a simple, sufficient and necessary condition for a positive integer m to be properly represented by some primitive form of discriminant D, namely it is the solvability of the congruence

$$b^2 \equiv D(\text{mod } 4m). \tag{22.21}$$

Let $R_D(m)$ denote the number of all representations of m by a complete system of inequivalent forms of discriminant D, thus

$$R_D(m) = \sum_{\varphi} \sum_{d^2|m} r_{\varphi}^{\star}(m/d^2)$$

where φ runs over inequivalent forms. If $(m, D) = 1$ we have

$$R_D(m) = w \sum_{d|m} \chi_D(d) \tag{22.22}$$

where χ_D is the Kronecker symbol. We shall see that for special discriminants (idoneal numbers) the above formula yields the number of representations for an individual form (because exactly one class of forms represents m). This "convenient" situation is explained in the genus theory due to Gauss, which we are going to present in more modern terms.

First we classify the binary quadratic forms of a given discriminant according to rational unimodular transformations rather than the integral ones. Two forms φ, ψ are said to be in the same genus if $\psi(X, Y) = \varphi(\alpha X + \gamma Y, \beta X + \delta Y)$ for some $\sigma = \begin{pmatrix} \alpha & \beta \\ \gamma & \delta \end{pmatrix} \in SL_2(\mathbb{Q})$. In fact it suffices to apply σ with all entries having the denominator $8D$. Obviously, equivalent forms are in the same genus but not conversely. Indeed, the inverse class $\mathcal{A}^{-1}$ is in the genus of $\mathcal{A}$, though $\mathcal{A}^{-1}$ does not always coincide with $\mathcal{A}$. Every genus has the same number of classes. The genus which contains the principal class is called the principal genus; we denote it by $\mathcal{G}$ and put

$$h_1 = |\mathcal{G}|. \tag{22.23}$$

A beautiful theorem of Gauss asserts that $\mathcal{G}$ consists of squares of classes,

$$\mathcal{G} = \{\mathcal{A}^2 : \mathcal{A} \in \mathcal{H}\}. \tag{22.24}$$

Thus $\mathcal{G}$ is a subgroup of $\mathcal{H}$. The factor group $\mathcal{F} = \mathcal{H}/\mathcal{G}$ is called the genus group.

We say that a form φ is ambiguous if it is equivalent to $\overline{\varphi}$; in other words, the class $\mathcal{A} = [\varphi]$ has exponent two in the class group $\mathcal{H}$. Such classes form the subgroup

$$\mathcal{E} = \{\mathcal{A} \in \mathcal{H} : \mathcal{A}^2 = 1\}. \tag{22.25}$$

Thus $\mathcal{E}$ is isomorphic to $\mathcal{F}$ because it is the kernel of the homomorphism $\mathcal{A} \to \mathcal{A}^2$. Putting

$$h_0 = |\mathcal{E}| \tag{22.26}$$

we have $h = h_0 h_1$.

The group of ambiguous classes $\mathcal{E}$ is the easier part of $\mathcal{H}$. An ambiguous form φ which is reduced is characterized by having its root z_a on the boundary of F^+,

$$z_a = \frac{b+\sqrt{D}}{2a} \in \partial F^+. \tag{22.27}$$

Hence a reduced form $\varphi(X,Y) = aX^2 + bXY + cY^2$ is ambiguous if and only if $a = b, a = c$ or $b = 0$ in which cases the discriminant factors into $D = a(a-2c), D = (b-2a)(b+2a)$ or $D = -4ac$, respectively. Using these characterizations one can compute h_0; we obtain

$$h_0 = 2^{r+s-1} \tag{22.28}$$

where r is the number of distinct, odd prime divisors of D and $s = 0, 1, 2$. Precisely, if $D \equiv 1(\text{mod } 4)$, then $s = 0$, and if $D = -4N$, then

$$\begin{cases} s = 0 & \text{if } N \equiv 3, 7(\text{mod } 8), \\ s = 1 & \text{if } N \equiv 1, 2, 4, 5, 6(\text{mod } 8), \\ s = 2 & \text{if } N \equiv 0(\text{mod } 8). \end{cases} \tag{22.29}$$

If a positive integer m prime to D is represented properly by a form of discriminant D (i.e., (22.21) is solvable), then it is represented by forms only of one genus, actually the residue class $m(\text{mod } D)$ alone determines this genus. In this connection Gauss assigned to a discriminant D a system of $r+s$ genus characters. These are the Legendre symbols χ_p for every $p \mid D, p > 2$, and if $D = -4N$, we have extra s characters modulo 8 listed below

$$\begin{cases} \text{none} & \text{if } N \equiv 3, 7(\text{mod } 8), \\ \chi_4 & \text{if } N \equiv 1, 4, 5(\text{mod } 8), \\ \chi_8 & \text{if } N \equiv 6(\text{mod } 8), \\ \chi_4\chi_8 & \text{if } N \equiv 2(\text{mod } 8), \\ \chi_4 \text{ and } \chi_8 & \text{if } N \equiv 0(\text{mod } 8). \end{cases}$$

Let $\mathcal{S}$ denote the system of genus characters. We have

$$\prod_{\chi \in \mathcal{S}} \chi = \chi_D. \tag{22.30}$$

Hence if m is represented by a form of discriminant D and $(m, D) = 1$, then by (22.30) and (22.21) we obtain $\prod_{\chi \in \mathcal{S}} \chi(m) = \chi_D(m) = 1$. Two such numbers m, n are represented by forms of the same genus if and only if $\chi(m) = \chi(n)$ for all $\chi \in \mathcal{S}$. Therefore a genus of forms is characterized by a collection of signs $\varepsilon_\chi = \pm 1$ such that

$$\prod_{\chi \in \mathcal{S}} \varepsilon_\chi = 1, \tag{22.31}$$

and we have exactly $2^{|\mathcal{S}|-1}$ possibilities.

Now we return to the formula (22.22) which, in view of the genus theory, gives the number of representations of m by forms representing the genus whose invariants are

$$\varepsilon_\chi = \chi(m) \qquad \text{for all } \chi \in \mathcal{S}. \tag{22.32}$$

If there is only one class in each genus, then $R_D(m)$ reduces to the number of representations by a single form. Hence a question: which discriminants D satisfy $h_1 = |\mathcal{G}| = 1$, or equivalently

$$\mathcal{A}^2 = 1 \qquad \text{for every } \mathcal{A} \in \mathcal{H}. \tag{22.33}$$

After Euler we call such a discriminant "numerus idoneus", (in English it is called "idoneal number", or "convenient number"; in French "nombre convenable"). As a matter of fact Euler considered discriminants of type $D = -4N$ and called N a numerus idoneus rather than D. Gauss gave a list of 65 such numbers:

$$\begin{cases} N = 1, 2, 3, 4, 7 & \text{with } h(D) = 1, \\ N = 5, 6, 8, 9, 10 \text{ (plus ten more)} & \text{with } h(D) = 2, \\ N = 21, 24, 30, 33 \text{ (plus twenty more)} & \text{with } h(D) = 4, \\ N = 105, 120 \text{ (plus fifteen more)} & \text{with } h(D) = 8, \\ N = 840, 1320, 1365, 1848 & \text{with } h(D) = 16. \end{cases}$$

The last number in Gauss' list $N = 1848$ is the largest known numerus idoneus. W. E. Briggs and S. Chowla [**BC**] showed that there is at most one such number beyond 10^{65}.

Idoneal numbers N have the property that the form $x^2 + Ny^2$ represents primes essentially in no more than one way, whereas a composite number has several representations if any (for details see [**Cox**]). This property offers an algorithm for primality testing which was an inspiration for Euler to study the idoneal numbers in the first place.

22.2. The class group.

From now on we assume that D is a negative fundamental discriminant. A fundamental discriminant is characterized by the property that every form $\varphi(X, Y) = aX^2 + bXY - cY^2$ with $b^2 - 4ac = D$ is primitive, i.e., $(a, b, c) = 1$. Such negative discriminants are of type $D \equiv 1 (\text{mod } 4)$, D squarefree, or $D = -4N$, N squarefree with

$$N \equiv 1, 2, 5, 6 \pmod 8. \tag{22.34}$$

The associated Kronecker symbol $\chi_D(m) = (\frac{D}{m})$ (see (3.43)) is a real, primitive character of conductor $-D$. It turns out that the fundamental discriminants are exactly the discriminants of quadratic fields.

Throughout we use the notation from the theory of the imaginary quadratic field $K = \mathbb{Q}(\sqrt{D})$ as described in Section 3.8. The quadratic forms $\varphi(X, Y)$ (with $a > 0$, $c > 0$, $b^2 - 4ac = D < 0$) correspond in a one-to-one way to the primitive ideals

$$\mathfrak{a} = \left[a, \frac{b + \sqrt{D}}{2}\right] = a[1, z_\mathfrak{a}] \subset \mathcal{O}. \tag{22.35}$$

Hence the terminology and results for quadratic forms which were used in the previous section translate naturally for ideals. In this section we go through this transition.

Recall the following:

I - the group of non-zero fractional ideals,
P - the subgroup of principal ideals,
$\mathcal{H} = I/P$ - the class group,
$h = |\mathcal{H}|$ - the class number,
$\mathfrak{d} = (\sqrt{D})$ - the different.

Every class $\mathcal{A} \in \mathcal{H}$ has a unique, primitive ideal (22.35) with $z_{\mathfrak{a}}$ in the fundamental domain $F = F^- \cup F^+$ of the modular group $\Gamma = SL_2(\mathbb{Z})$; such ideal is said to be reduced. Thus the number of reduced ideals is h. The subgroup

$$\mathcal{G} = \{\mathcal{A}^2 : \mathcal{A} \in \mathcal{H}\}$$

is called the principal genus and the factor group $\mathcal{F} = \mathcal{H}/\mathcal{G}$ is called the genus group. Hence two non-zero ideals $\mathfrak{a}, \mathfrak{b}$ belong to the same genus if and only if $\mathfrak{a} = \mathfrak{b}\mathfrak{c}^2$ for some $\mathfrak{c} \in I$. One can also show that $\mathfrak{a}, \mathfrak{b}$ are in the same genus if and only if

$$N\mathfrak{a} = N\mathfrak{b}N\gamma \qquad \text{for some } \gamma \in K^\star. \tag{22.36}$$

A class $\mathcal{A} \in \mathcal{H}$ is said to be ambiguous if $\mathcal{A} = \mathcal{A}^{-1}$, i.e., $\mathcal{A}^2 = 1$. The group of ambiguous classes

$$\mathcal{E} = \{\mathcal{A} \in \mathcal{H} : \mathcal{A}^2 = 1\},$$

as the kernel of the homomorphism $\mathcal{A} \to \mathcal{A}^2$, is isomorphic to the genus group $\mathcal{F}$. The correspondence between quadratic forms and primitive ideals casts ambiguous forms into ambiguous ideals ($\mathfrak{a}$ is ambiguous if $\mathfrak{a}^2$ is principal); it proves that every ambiguous class $\mathcal{A}$ has its reduced primitive ideal $\mathfrak{a}$ with $z_{\mathfrak{a}}$ on the boundary of F^+. Recall that a reduced ambiguous form $\varphi(X,Y) = aX^2 + bXY = cY^2$ has $a = b, a = c$ or $b = 0$, hence it yields a factorization of the discriminant into

$$-D = d_1 d_2 = a(4c - a), \quad (2a - b)(2a + b), \quad 4ac,$$

respectively, with $d_2 > d_1 > 0$ and $(d_1, d_2) = 1$ (the coprimality of these factors results from D being fundamental). Conversely, a factorization $-D = d_1 d_2$ with $d_2 > d_1 > 0$, $(d_1, d_2) = 1$ yields a reduced ambiguous form. For example. if $D \equiv 1 (\mathrm{mod}\ 4)$ we obtain

$$\begin{cases} \varphi(X,Y) = d_1 X^2 + d_1 XY + \dfrac{d_1 + d_2}{4} Y^2, & \text{if } d_2 > 3d_1, \\ \varphi(X,Y) = \frac{d_1 + d_2}{4} X^2 + \dfrac{d_2 - d_1}{2} XY + \dfrac{d_1 + d_2}{4} Y^2, & \text{if } d_2 < 3d_1. \end{cases}$$

Correspondingly, a reduced primitive ambiguous ideal yields a factorization of the different $\mathfrak{d} = (\sqrt{D})$ into $\mathfrak{d} = \mathfrak{d}_1\mathfrak{d}_2$ with $(\mathfrak{d}_1, \mathfrak{d}_2) = 1$; note that $\mathcal{C}\ell(\mathfrak{d}_1) = \mathcal{C}\ell(\mathfrak{d}_2)$. Distinct such factorizations (up to the order of factors $\mathfrak{d}_1, \mathfrak{d}_2$) yield distinct ambiguous classes, therefore

$$h_0 = |\mathcal{E}| = 2^{t-1} \tag{22.37}$$

where $t = \omega(|D|)$ is the number of distinct prime divisors of D (note that (22.37) agrees with (22.28) in case of fundamental discriminants).

The genus of an ideal can be determined by values of real characters of the class group. These characters are assigned to every factorization $D = D_1 D_2$ into two fundamental discriminants D_1, D_2. Note that D_1, D_2 have different signs and are coprime, so we have exactly 2^{t-1} distinct factorizations up to the order. We define χ_{D_1,D_2} first on prime ideals by

$$\chi_{D_1,D_2}(\mathfrak{p}) = \begin{cases} \chi_{D_1}(N\mathfrak{p}) & \text{if } \mathfrak{p} \nmid D_1, \\ \chi_{D_2}(N\mathfrak{p}) & \text{if } \mathfrak{p} \nmid D_2 \end{cases} \tag{22.38}$$

(this is well defined because $\chi_D(N\mathfrak{a}) = 1$ if $(\mathfrak{a}, D) = 1$), and we extend χ_{D_1,D_2} to all non-zero fractional ideals by multiplicativity. We obtain a character $\chi_{D_1,D_2} : I \to \{\pm 1\}$ such that $\chi_{D_1,D_2}(\mathfrak{a}) = 1$ for all $\mathfrak{a} \in P$. Indeed, for $\mathfrak{a} = (\alpha)$ with $\alpha = \frac{1}{2}(m + n\sqrt{D})$, $(\alpha, D) = 1$ we have $\chi_{D_1,D_2}(\mathfrak{a}) = \chi_{D_1}(\mathfrak{a}) = \chi_{D_1}(\frac{1}{4}(m^2 - n^2 D)) = \chi_{D_1}(\frac{m^2}{4}) = 1$ (similarly one can check the other cases). Therefore $\chi_{D_1,D_2} \in \mathcal{H}$; these are exactly all real characters of the class group called the genus characters. Two ideals $\mathfrak{a}, \mathfrak{b} \in I$ are in the same genus if and only if

$$\chi(\mathfrak{a}) = \chi(\mathfrak{b}) \qquad \text{for all genus characters.}$$

The genus theory of Gauss with its explicit characterization of ambiguous classes proved to be very useful for primality testing and factorization techniques. A method for factoring $-D$ which exploits the class group structure, as described by D. Shanks [**Sha**], requires only $O(|D|^{\frac{1}{4}})$ operations.

If $-D$ is prime, then the whole class group is the principal genus $\mathcal{H} = \mathcal{G}$. In this case $\mathcal{H}$ tends to be cyclic more often than not. After Gauss, a discriminant D (not necessarily negative prime) is said to be regular if the principal genus is cyclic (this is not the same as the regular primes occurring in the theory of cyclotomic fields). We still do not know if there are infinitely many regular discriminants. However, the numerical evidence supports the existence of a large proportion of these. It has been conjectured in [**Ge**] that the proportion of regular to all discriminants (both negative) is

$$\Big(\zeta(6) \prod_{n=4}^{\infty} \zeta(n)\Big)^{-1} = 0.8469\ldots.$$

Gauss genus theory gives us a full description of the 2-Sylow subgroups of the class group. There are some interesting, however, incomplete developments for other p-Sylow subgroups of $\mathcal{H}$, in particular, for $p = 3$ by Davenport and Heilbronn [**DH**]. According to the Cohen-Lenstra heuristics the probability that an imaginary quadratic field has in its ideal class group an element of order p is

$$1 - \prod_{n=1}^{\infty} (1 - p^{-n}).$$

22.3. The class group L-Functions.

Let $\chi : \mathcal{H} \to \mathbb{C}^\star$ be a character of the class group. Define

$$L_K(s,\chi) = \sum_{\mathfrak{a}} \chi(\mathfrak{a})(N\mathfrak{a})^{-s} \tag{22.39}$$

for $\mathrm{Re}\,(s) > 1$ where the series converges absolutely. Notice that $L_K(s,\overline{\chi}) = L_K(s,\chi)$. For any primitive ideal $\mathfrak{a} = [a, \frac{b+\sqrt{D}}{2}]$ we have

$$\sum_{\chi\in\hat{\mathcal{H}}} \chi(\mathfrak{a}) L_K(s,\chi) = h\sum_{\mathfrak{b}\sim\mathfrak{a}} (N\mathfrak{b})^{-s} = \frac{h}{w} a^{-s} \sum_{0\neq\alpha\in\mathfrak{a}^{-1}} |\alpha|^{-2s}$$

by writing $\mathfrak{b} = (\alpha)\mathfrak{a}$. Since $\mathfrak{a}^{-1} = \mathbb{Z} + \overline{z}_{\mathfrak{a}}\mathbb{Z}$, we derive

$$\begin{aligned}\sum_{0\neq\alpha\in\mathfrak{a}^{-1}} |\alpha|^{-2s} &= \sum\sum_{(m,n)\neq(0,0)} |m+nz_{\mathfrak{a}}|^{-2s}\\ &= \zeta(2s)\sum\sum_{(m,n)=1} |m+nz_{\mathfrak{a}}|^{-2s} = 2\zeta(2s)\Big(\frac{\sqrt{|D|}}{2a}\Big)^{-s} E(z_{\mathfrak{a}}, s)\end{aligned}$$

where

$$E(z,s) = \sum_{\gamma\in\Gamma_\infty\backslash\Gamma} (\mathrm{Im}\gamma z)^s = \frac{1}{2}\sum\sum_{(m,n)=1} y^s|m+nz|^{-2s}$$

is the Eisenstein series for the modular group. This gives us

$$\sum_{\chi\in\hat{\mathcal{H}}} \chi(\mathfrak{a}) L_K(s,\chi) = \frac{2h}{w}\Big(\frac{\sqrt{|D|}}{2}\Big)^{-s} \zeta(2s) E(z_{\mathfrak{a}}, s). \tag{22.40}$$

We put

$$\theta(s) = \pi^{-s}\Gamma(s)\zeta(2s), \tag{22.41}$$

$$E^\star(z,s) = \theta(s)E(z,s), \tag{22.42}$$

$$\Lambda_K(s,\chi) = (2\pi)^{-s}\Gamma(s)|D|^{\frac{s}{2}} L_K(s,\chi). \tag{22.43}$$

In this notation (22.40) becomes

$$\sum_{\chi\in\hat{\mathcal{H}}} \chi(\mathfrak{a})\Lambda_K(s,\chi) = \frac{2h}{w} E^\star(z_{\mathfrak{a}}, s). \tag{22.44}$$

By Fourier inversion (the orthogonality of characters) we obtain

$$\Lambda_K(s,\chi) = \frac{2}{w}\sum_{\mathfrak{a}} \chi(\mathfrak{a}) E^\star(z_{\mathfrak{a}}, s) \tag{22.45}$$

where $\mathfrak{a}$ runs over a set of primitive, inequivalent ideals (a good choice being the set of reduced ideals). By virtue of this formula the Hecke L-function inherits analytic properties of the Eisenstein series.

It is well known (see (15.13)) that $E^\star(z,s)$ has the Fourier expansion

$$\begin{aligned}E^\star(z,s) = \theta(s)y^s + \theta(1-s)y^{1-s}\\ + 4\sqrt{y}\sum_1^\infty \tau_{s-1/2}(n) K_{s-1/2}(2\pi n y)\cos(2\pi n x)\end{aligned} \tag{22.46}$$

where

$$\tau_\nu(n) = \sum_{ad=n} \Big(\frac{a}{d}\Big)^\nu. \tag{22.47}$$

The Fourier expansion yields analytic continuation of $E(z,s)$ to the whole complex s-plane, it shows that $E(z,s)$ in $\mathrm{Re}\,(s) \geqslant \frac{1}{2}$ has only a simple pole at $s=1$ with constant residue $\frac{3}{\pi}$, and it satisfies the functional equation

$$E^\star(z,s) = E^\star(z,1-s). \tag{22.48}$$

Inserting (22.46) into (22.45) we obtain the Fourier expansion

$$\begin{aligned}\Lambda_K(s,\chi) = \frac{2}{w}\sum_{\mathfrak{a}} \chi(\mathfrak{a})\Big\{\theta(s)\Big(\frac{\sqrt{|D|}}{2a}\Big)^s + \theta(1-s)\Big(\frac{\sqrt{|D|}}{2a}\Big)^{1-s} \\ + 4\Big(\frac{\sqrt{|D|}}{2a}\Big)^{1/2} \sum_1^\infty \tau_{s-1/2}(n) K_{s-1/2}\Big(\pi n \frac{\sqrt{|D|}}{a}\Big)\cos\Big(\pi n \frac{b}{a}\Big)\Big\}\end{aligned} \tag{22.49}$$

where $\mathfrak{a}$ runs over primitive representatives of ideal classes. Note that the Fourier series converges rapidly since the Bessel function $K_\nu(y)$ has exponential decay

$$K_\nu(y) = \Big(\frac{\pi}{2y}\Big)^{1/2} e^{-y}\Big(1 + \frac{\theta}{2y}\Big).$$

for $y > 0, \nu \in \mathbb{C}$ where $|\theta| \leqslant |\nu^2 - \frac{1}{4}|$ (see (23.451.6) of [**GR**]).

We deduce by (22.45) or (22.49) that $L_K(s,\chi)$ has analytic continuation to the whole complex s-plane, it is entire except for a simple pole at $s=1$ if χ is trivial with

$$\operatorname*{res}_{s=1} \zeta_K(s) = \frac{2\pi h}{w\sqrt{|D|}}, \tag{22.50}$$

and it satisfies the functional equation

$$\Lambda_K(s,\chi) = \Lambda_K(1-s,\chi). \tag{22.51}$$

All the above properties of $L_K(s,\chi)$ can be also deduced from the integral representation (à la Riemann, see (4.77)) due to Hecke

$$\Lambda_K(s,\chi) = \frac{h\delta(\chi)}{ws(s-1)} + \int_1^\infty (y^{-s} + y^{s-1}) f_\chi\Big(\frac{iy}{\sqrt{|D|}}\Big)dy \tag{22.52}$$

where $\delta(\chi) = 1$ if χ is trivial, or else $\delta(\chi) = 0$, and

$$f_\chi(z) = \sum_{\mathfrak{a}} \chi(\mathfrak{a}) e(zN\mathfrak{a}) \tag{22.53}$$

where $\mathfrak{a}$ runs over all non-zero integral ideals. Integrating termwise we get

$$\begin{aligned}\Lambda_K(s,\chi) = \frac{h\delta(\chi)}{ws(s-1)} + \sum_{\mathfrak{a}} \chi(\mathfrak{a}) \Big(\frac{\sqrt{|D|}}{2\pi a}\Big)^s \Gamma\Big(s, \frac{2\pi a}{\sqrt{|D|}}\Big) \\ + \sum_{\mathfrak{a}} \chi(\mathfrak{a}) \Big(\frac{\sqrt{|D|}}{2\pi a}\Big)^{1-s} \Gamma\Big(1-s, \frac{2\pi a}{\sqrt{|D|}}\Big)\end{aligned} \tag{22.54}$$

where $a = N\mathfrak{a}$ and $\Gamma(s, x)$ is the incomplete gamma function

$$\Gamma(s, x) = \int_x^\infty e^{-y} y^{s-1} dy.$$

Hecke derived the formula (22.52) by splitting $L_K(s, \chi)$ into a sum of Epstein zeta functions (apparently introduced first by Dirichlet)

$$Z_\mathfrak{a}(s) = \sum\sum_{(m,n)\neq(0,0)} \varphi_\mathfrak{a}(m, n)^{-s} \tag{22.55}$$

where $\varphi_\mathfrak{a}(X, Y) = aX^2 + bXY + cY^2$ is the quadratic form corresponding to the ideal $\mathfrak{a}$. This in turn is the Mellin transform of the theta function

$$\theta_\mathfrak{a}(z) = \sum_m \sum_n e(\varphi_\mathfrak{a}(m, n)z).$$

Both $\theta_\mathfrak{a}(z)$ and $f_\chi(z)$ are modular forms of weight one for the group $\Gamma_0(|D|)$ and character χ_D. If χ is not real, then $f_\chi(z)$ is a primitive cusp form with Hecke eigenvalues

$$\lambda_\chi(n) = \sum_{N\mathfrak{a}=n} \chi(\mathfrak{a}). \tag{22.56}$$

If χ is real, say $\chi = \chi_{D_1,D_2}$ with $D_1D_2 = D$, then the Hecke L-function factors into Dirichlet L-functions. Precisely, it can be verified by comparing local factors of Euler products and using (22.38) together with the factorization law of ideals in $K = \mathbb{Q}(\sqrt{D})$ that

KRONECKER FACTORIZATION FORMULA.

$$L_K(s, \chi_{D_1,D_2}) = L(s, \chi_{D_1})L(s, \chi_{D_2}). \tag{22.57}$$

For the trivial character we have the Dedekind zeta function

$$\zeta_K(s) = \sum_\mathfrak{a} (N\mathfrak{a})^{-s} = \zeta(s)L(s, \chi_D) = \sum_1^\infty \tau(n, \chi_D)n^{-s}.$$

In this case, (22.45) becomes

$$\Lambda_K(s) = \frac{2}{w} \sum_\mathfrak{a} E^\star(z_\mathfrak{a}, s). \tag{22.58}$$

Comparing the residues at $s = 1$ on both sides we obtain the celebrated

DIRICHLET CLASS NUMBER FORMULA.

$$L(1, \chi_D) = \frac{2\pi h(D)}{w\sqrt{|D|}}. \tag{22.59}$$

EXERCISE 1. Prove that for $D < -4$,

$$(2 - \chi_D(2))h(D) = \sum_{0<n<\frac{|D|}{2}} \chi_D(n).$$

Another interesting formula comes from (22.45) at $s = \frac{1}{2}$. By the Fourier expansion (22.46) we see that $E(z, \frac{1}{2}) \equiv 0$ and

$$E'(z, \tfrac{1}{2}) = \sqrt{y} \log y + 4\sqrt{y} \sum_{1}^{\infty} \tau(n) K_0(2\pi n y) \cos(2\pi n x).$$

Hence the central value of $L_K(s, \chi)$ is given by

$$L_K(\tfrac{1}{2}, \chi) = \frac{\sqrt{2}}{w} |D|^{-1/4} \sum_{\mathfrak{a}} \chi(\mathfrak{a}) E'(z_{\mathfrak{a}}, \tfrac{1}{2}).$$

Inserting the Fourier expansion (22.61) we arrive at

$$(22.60)\quad L_K(\tfrac{1}{2}, \chi) = \frac{1}{w} \sum_{\mathfrak{a}} \frac{\chi(\mathfrak{a})}{\sqrt{a}} \Big\{ \log \frac{\sqrt{|D|}}{2a} + 4 \sum_{1}^{\infty} \tau(n) K_0\Big(\pi n \frac{\sqrt{|D|}}{a}\Big) \cos\Big(\frac{\pi n b}{a}\Big) \Big\}.$$

Using $K_0(y) \ll y^{-\frac{1}{2}} e^{-y}$ one derives by trivial estimation that

$$L_K(\tfrac{1}{2}, \chi) = \frac{1}{2} \sum_{\mathfrak{a}} \frac{\chi(a)}{\sqrt{a}} \log \frac{\sqrt{|D|}}{2a} + O(h(D)|D|^{-\frac{1}{4}}).$$

This is interesting for the trivial class character χ_0. Assuming $L(\frac{1}{2}, \chi_D) \geqslant 0$ (which follows from the Riemann Hypothesis for $L(s, \chi_D)$, but may conceivably be established sometime without recourse to the GRH) the left side is $L_K(\frac{1}{2}, \chi_0) = \zeta(\frac{1}{2}) L \frac{1}{2}, \chi_D) \leqslant 0$ so we derive

$$h(D) \gg \sum_{\mathfrak{a}} \Big(\frac{\sqrt{|D|}}{a}\Big)^{\frac{1}{2}} \log \frac{\sqrt{|D|}}{a}$$

where $\mathfrak{a}$ runs over the complete system of primitive, reduced ideals. Note that we dropped the factor 2 in the logarithm (find out why this is permissible!). Recall that $a = N\mathfrak{a} \leqslant \sqrt{|D|/3}$ so all terms above are positive. This implies that $h(D)$ cannot be very small, namely

$$h(D) \gg |D|^{\frac{1}{4}} \log |D|,$$

where the implied constant is effectively computable. Moreover, this implies that a positive proportion of the primitive, reduced ideals have the norm $N\mathfrak{a} \asymp \sqrt{|D|}$. Duke [**Du2**] has shown unconditionally that the points $z_{\mathfrak{a}}$ are equidistributed in the fundamental domain F with respect to the hyperbolic measure $d\mu = y^{-2} dx dy$. His result is ineffective because it uses Siegel's bound $h(D) \gg |D|^{\frac{1}{2}-\varepsilon}$ (see (5.76)).

The Class Number Formula (22.59) reveals the algebraic nature of the special value $L(1, \chi_D)$. One can derive infinitely many formulas of that kind by comparing coefficients in the Taylor expansion at $s=1$ on both sides of (22.45). Of particular interest are the constant terms. On the left side we have

$$\Lambda_K(s,\chi) \sim \frac{\sqrt{|D|}}{2\pi} L_K(1,\chi) \tag{22.61}$$

if χ is non-trivial, and

$$\Lambda_K(s,\chi) \sim \frac{h}{w}\Big(\frac{1}{s-1} + \log\frac{\sqrt{|D|}}{2\pi} + \frac{L'}{L}(1,\chi_D)\Big) \tag{22.62}$$

if χ is trivial by the following expansions:

$$\begin{aligned}\Big(\frac{\sqrt{|D|}}{2\pi}\Big)^{s-1} &= 1 + (s-1)\log\tfrac{\sqrt{|D|}}{2\pi} + \cdots\\ \Gamma(s) &= 1 - (s-1)\gamma + \cdots\\ \zeta(s) &= \frac{1}{s-1} + \gamma + \cdots\\ L(s,\chi_D) &= L(1,\chi_D) + (s-1)L'(1,\chi_D) + \cdots\end{aligned}$$

and the Class Number Formula (22.59). On the right side of (22.45) we need an expansion of $E^\star(z,s)$ which we shall derive from the Fourier series (22.46). First notice that $2\theta(s)y^s \sim -s^{-1} + \gamma - \log 4\pi y$ as $s \to 0$ by the following expansions:

$$\begin{aligned}\Big(\frac{y}{\pi}\Big)^s &= 1 + s\log\frac{y}{\pi} + \cdots\\ \Gamma(s) &= \frac{1}{s} - \gamma + \cdots\\ \zeta(2s) &= -\frac{1}{2} - s\log 2\pi + \cdots\end{aligned}$$

using the values $\zeta(0) = -\frac{1}{2}$ and $\zeta'(0) = -\frac{1}{2}\log 2\pi$. Moreover, $2\theta(1-s)y^{1-2} \sim \frac{\pi}{3}$ by using the value $\zeta(2) = \frac{\pi^2}{6}$. Hence

$$2E^\star(z,s) \sim \frac{1}{s-1} + \gamma - \log 4\pi y + \frac{\pi}{3}y + 2D(z) \tag{22.63}$$

where $D(z)$ is the tail in the Fourier expansion (22.46) at $s=1$, precisely

$$D(z) = 4\sqrt{y}\sum_1^\infty \tau_{1/2}(n)K_{1/2}(2\pi n y)\cos(2\pi n x).$$

Since $K_{\frac{1}{2}}(y) = (\pi/2y)^{\frac{1}{2}}e^{-y}$ and $\tau_{1/2}(n) = \sigma_{-1}(n)n^{\frac{1}{2}}$, this becomes

$$D(z) = 2\mathrm{Re}\sum_1^\infty \sigma_{-1}(n)e(nz). \tag{22.64}$$

If we execute the summation as follows

$$D(z) = 2\mathrm{Re}\sum_1^\infty\sum_1^\infty m^{-1}e(mnz) = -2\sum_1^\infty \log|1-e(nz)|$$

we find that

$$D(z) = -\frac{\pi}{6}y - 2\log|\eta(z)|, \tag{22.65}$$

where $\eta(z)$ is the Dedekind eta function

$$\eta(z) = e\Big(\frac{z}{24}\Big)\prod_{1}^{\infty}(1 - e(nz)). \tag{22.66}$$

The eta function is a modular form of weight $\frac{1}{2}$ on the group $\Gamma = SL_2(\mathbb{Z})$ with a suitable multiplier system involving Dedekind sums (see e.g. [**I4**]). Then $\eta(z)^{24} = \Delta(z)$ is the famous Ramanujan Δ function, the primitive cusp form of weight 12. Therefore the function

$$F(z) = \Big(\frac{2y}{\sqrt{|D|}}\Big)^{\frac{1}{2}}|\eta(z)|^2 \tag{22.67}$$

is Γ-invariant (here the factor $(2/\sqrt{|D|})^{1/2}$ is introduced for normalization so that $F(z_{\mathfrak{a}}) = a^{-\frac{1}{2}}|\eta(z_{\mathfrak{a}})|^2$). In this notation (22.65) becomes

$$D(z) = -\frac{\pi}{6}y - \frac{1}{2}\log\frac{\sqrt{|D|}}{2y} - \log F(z). \tag{22.68}$$

Inserting (22.68) into (22.63) we obtain

$$2E^\star(z,s) \sim \frac{1}{s-1} + \gamma - \log 2\pi\sqrt{|D|} - 2\log F(z). \tag{22.69}$$

Next inserting (22.69) into (22.45) we obtain

$$\Lambda_K(s,\chi) \sim \delta(\chi)\frac{h}{w}\Big(\frac{1}{s-1} + \gamma - \log 2\pi\sqrt{|D|}\Big) - \frac{2}{w}\sum_{\mathfrak{a}}\chi(\mathfrak{a})\log F(z_{\mathfrak{a}}). \tag{22.70}$$

Now matching (22.70) with (22.61), or (22.62) in case χ is trivial, we derive an exact equation which is called the

Kronecker Limit Formula. *If χ is a non-trivial class group character, then*

$$L_K(1,\chi) = \frac{-4\pi}{w\sqrt{|D|}}\sum_{\mathfrak{a}}\chi(\mathfrak{a})\log F(z_{\mathfrak{a}}). \tag{22.71}$$

Moreover,

$$-\frac{L'}{L}(1,\chi_D) = \log|D| - \gamma + \frac{2}{h}\sum_{\mathfrak{a}}\log F(z_{\mathfrak{a}}). \tag{22.72}$$

Consider (22.71) for a non-trivial genus character $\chi = \chi_{B,C}$ which corresponds to the factorization $D = BC$ into non-trivial fundamental discriminants B, C. One of these is negative and the other is positive, say $B < 0 < C$. By the Kronecker formula (22.57) we have $L_K(1,\chi_{B,C}) = L(1,\chi_B)L(1,\chi_C)$ and by the Dirichlet Class Number Formula (22.59) we have $L(1,\chi_B) = 2\pi h(B)/w(B)\sqrt{|B|}$. Dirichlet also established the class number formula for positive discriminants (see (2.31)), namely

$$L(1,\chi_C) = 2h(C)|C|^{-\frac{1}{2}}\log\varepsilon(C) \tag{22.73}$$

where $h(C)$ is the class number of the real quadratic field $\mathbb{Q}(\sqrt{C})$ and $\varepsilon(C)$ is the fundamental unit. Combining these formulas we infer from (22.71) that

$$\varepsilon(C)^{2h(B)h(C)/w(B)} = \prod_{\mathfrak{a}} F(\mathfrak{a})^{-\chi(\mathfrak{a})}. \tag{22.74}$$

Besides their significance for algebraic numbers the Kronecker limit formulas can be used to estimate derivatives of L-functions. For this purpose it is more convenient to operate with $D(z)$ rather than $F(z)$, so we return to (22.68) getting by (22.72) and (22.59)

$$L'(1,\chi_D) = \frac{2\pi}{w}\sum_{\mathfrak{a}}\Big(\frac{\pi}{6a} + \frac{1}{\sqrt{|D|}}\Big(\gamma + \log\frac{a}{|D|} + 2D(z_{\mathfrak{a}})\Big)\Big) \tag{22.75}$$

where $\mathfrak{a}$ runs over inequivalent primitive ideals and $a = N\mathfrak{a}$. It is not easily seen that the right side of (22.75) does not depend on the choice of representatives of the ideal classes while this is clear in (22.72). A good choice in practice is the system of primitive, reduced ideals because the points $z_{\mathfrak{a}}$ have the largest height making $D(z_{\mathfrak{a}})$ easy to estimate. Estimating trivially $D(z_{\mathfrak{a}}) \ll 1$ we obtain

$$L'(1,\chi_D) = \pi\sum_{\mathfrak{a}}\Big(\frac{\pi}{6a} - \frac{1}{\sqrt{|D|}}\log\frac{|D|}{a}\Big) + O\Big(\frac{h(D)}{\sqrt{|D|}}\Big) \tag{22.76}$$

where the implied constant is absolute.

REMARKS. A different formula for $L'(1,\chi_D)$ was derived by S. Chowla and A. Selberg [**CS**], namely

$$|D|^{\frac{1}{2}}L'(1,\chi_D) = -\pi\sum_{0<n<|D|}\chi_D(n)\log\Gamma\Big(\frac{n}{|D|}\Big) + \frac{2\pi h}{w}(\gamma + \log 2\pi).$$

22.4. The class number problems.

Estimation of the order of the class group of a number field is one of the most intricate problems in arithmetic. In particular, one wants to know which fields have class number one since it means that the ring of integers of such a field has the unique factorization property. A small class number (relative to the discriminant) is not a common feature. As with many stories in arithmetic this one starts with Gauss. In his Disquisitiones Arithmeticae, Gauss is seriously concerned about the class number $h(D)$ of the imaginary quadratic field $K = \mathbb{Q}(\sqrt{D})$.

GAUSS CONJECTURE. *We have $h(D) \to \infty$ as $-D \to \infty$.*

Even if this conjecture is proved the real issue in practice is

GAUSS CLASS NUMBER PROBLEM. *Find an effective algorithm for determining all imaginary quadratic fields $K = \mathbb{Q}(\sqrt{D})$ with a given class number $h = h(D)$.*

Gauss knew that $h(D) = 1$ for the following nine discriminants

$$-D = 3, 4, 7, 8, 11, 19, 43, 67, 163. \tag{22.77}$$

Here is a remarkable characterization of these discriminants

PROPOSITION 22.1 (Rabinovitch, 1913). *For $D \equiv 1(mod\ 4)$ we have $h(D) = 1$ if and only if the polynomial*

$$f(x) = x^2 - x + \tfrac{1-D}{4}$$

represents primes for all natural numbers $x < \frac{1-D}{4}$.

The question becomes:

GAUSS CLASS NUMBER ONE PROBLEM. *Prove that the above list of nine negative discriminants D with $h(D) = 1$ is complete.*

Recall that $h(D)$ is linked to $L(1, \chi_D)$ by the Dirichlet formula (22.59) but it does not help much. In fact this formula served Dirichlet to estimate $L(1, \chi_D)$ rather than $h(D)$ giving

$$L(1, \chi_D) \geqslant \frac{2\pi}{w\sqrt{|D|}} \tag{22.78}$$

since $h(D) \geqslant 1$. The class number one problem boils down to improving the lower bound (22.78) for $|D| > 163$.

Next, recall that the effective lower bound

$$h(D) \gg |D|^{\frac{1}{4}} \log |D|, \tag{22.79}$$

follows from the formula (22.60) subject to the hypothesis that $L(\frac{1}{2}, \chi_D) \geqslant 0$. This can solve the CNOP assuming GRH (if the constant are made explicit).

There are much stronger connections between $h(D)$ and zeros of $L(s, \chi_D)$, for example

PROPOSITION 22.2 (Hecke-Landau, 1918). *If $L(s, \chi_D)$ does not vanish in the region $s > 1 - a/\log |D|$, then*

$$h(D) > b\frac{\sqrt{|D|}}{\log |D|} \tag{22.80}$$

where a, b are some positive constants, effectively computable.

In modern times the above result is routine, and one can push this implication much deeper when powered by new technologies (Linnik's density theorem and Turán's power-sums method). However, the following implication was quite surprising in its time.

PROPOSITION 22.3 (Deuring, 1933). *If the Riemann Hypothesis for the zeta function $\zeta(s)$ is false, then $h(D) > 1$ for all sufficiently large $|D|$.*

A year later Mordell improved Deuring's result to the effect that $h(D) \to \infty$ as $-D \to \infty$ under the same assumption. In the same time Heilbronn expanded the assumption to the falsity of the Riemann hypothesis for any Dirichlet L-series.

PROPOSITION 22.4 (Heilbronn, 1934). *If GRH is false, then $h(D) \to \infty$ as $-D \to \infty$.*

Combining both Proposition 22.2 and 22.4 one completes an unconditional solution of the Gauss Conjecture that $h(D) \to \infty$ as $-D \to \infty$. Of course, the result is not effective because we do not know which one of the two propositions is needed. The problem is that if some $L(s, \chi)$ has a zero off the line $\mathrm{Re}\,(s) = \frac{1}{2}$ the resulting lower bound for $h(D)$ depends on this hypothetical point which has no numerical value. Therefore the CNOP still cannot be reduced to a finite number of computations. Some degree of effectivization was achieved in the following fashion.

PROPOSITION 22.5 (Heilbronn and Linfoot, 1934). *There is at most one imaginary quadratic field $K = \mathbb{Q}(\sqrt{D})$ with $h = h(D) = 1$ which is not in the list of Gauss.*

For the above achievements some attention should be given to E. Landau for his early ideas in [**La2**]. Promptly after Heilbronn and Linfoot, he was able to prove the following remarkable estimate.

PROPOSITION 22.6 (Landau, 1935). *There exists an absolute, effective constant $c > 0$ such that for every $h \geqslant 1$ all but one of the discriminants $D < 0$ with $h(D) = h$ satisfy*

$$|D| \leqslant ch^8(\log 3h)^6. \tag{22.81}$$

Next C. L. Siegel [**Sie1**] used similar arguments to establish his famous

THEOREM 22.7 (Siegel, 1935). *For any $\varepsilon > 0$ there exists a constant $c(\varepsilon) > 0$ (not computable) such that*

$$L(1, \chi_D) > c(\varepsilon)|D|^{-\varepsilon}. \tag{22.82}$$

Hence

$$h(D) > c(\varepsilon)|D|^{1/2-\varepsilon}. \tag{22.83}$$

Both papers of Landau and Siegel appeared in the first volume of Acta Arithmetica, 1935. A simple proof of Siegel's theorem was given by D. Goldfeld [**Gol**] (see Chapter 5). The next interesting result is

THEOREM 22.8 (Tatuzawa, 1951). *Siegel's bound is effective except for one possible D for each $\varepsilon > 0$. Precisely, if $0 < \varepsilon \leqslant \frac{1}{12}$ and $\varepsilon \log |D| \geqslant 1$, then with at most one exception*

$$L(1, \chi_D) \geqslant \tfrac{3}{5}\varepsilon|D|^{-\varepsilon}. \tag{22.84}$$

In 1980 J. Hoffstein [**Hof**] established further numerical improvements of Tatuzawa's estimate using Goldfeld's technique.

The CNOP was solved for the first time by Kurt Heegner [**Hee**] in 1952, but his arguments (modular forms, complex multiplication) were understood and recognized only after his death in 1968. An additional note of tragedy was that this first recognition came two years after another solution was found by A. Baker. Baker's method [**B**] is very different: it uses an effective lower bound for linear forms of three logarithms of algebraic numbers. In the meantime, H. Stark [**St1**] gave a third solution which turned out to be similar to Heegner's. Then Stark looked into Baker's proof and realized that it sufficed to use linear forms only in two logarithms, so that the problem could have been solved already in 1949 by Gelfand and Linnik. Retrospectively, it seems that H. Weber was capable of solving the CNOP long before. See [**Cox**] for a presentation of the Heegner-Stark arguments.

Next was the problem of class number two. This has been solved independently in 1971 by A. Baker [**B**] and H. Stark [**St2**]. They established that there are exactly eighteen negative discriminants D with $h(D) = 2$, namely

$$-D = 15, 20, 24, 35, 40, 51, 52, 88, 91, 115, 123, 148, 187, 232, 235, 267, 403, 427.$$

Our story culminates at two remarkable achievements by Goldfeld (1976) and Gross and Zagier (1983). Goldfeld [**Go2**] gave an effective lower bound for $h(D)$, the quality of which depends on the rank of an auxiliary elliptic curve, and its satisfying the Birch and Swinnerton-Dyer Conjecture. Gross and Zagier ([**GZ1**], [**GZ2**]) produced a curve with the required properties. The combined results of Goldfeld, Gross and Zagier solves the Class Number Problem for any fixed h, up to a finite amount of computation. J. Oesterlé [**Oe**] reduced the implied constants in Goldfeld's work so much as to make the result practical for computers. He succeeded in showing the following neat estimate:

$$h(D) > \frac{1}{55}(\log|D|)\prod_{p|D}\Big(1-\frac{[2\sqrt{p}]}{p+1}\Big). \tag{22.85}$$

We shall present Goldfeld's approach in Section 22.7 with some variations of our own, and we survey the construction of Gross and Zagier in the Appendix to Chapter 23.

The best effective lower bounds for $h(D)$ which current technology allows us to hope for are of type $h(D) \geqslant c_g(\log|D|)^g$ for any $g > 0$ and some computable constant $c_g > 0$, yet it is too short to solve some other popular problems such as

EULER IDONEAL NUMBER PROBLEM. *Find all discriminants D for which the class group of $\mathbb{Q}(\sqrt{D})$ has one class in each genus.*

If D is an idoneal discriminant, then $h(D) = 2^{t-1}$ where $t = \omega(|D|)$ is the number of distinct prime divisors of D. Since $\omega(|D|)$ can be as large as $\log|D|/\log\log|D|$ the problem requires an effective lower bound

$$h(D) \gg |D|^{c/\log\log|D|} \quad \text{with } c > \log 2. \tag{22.86}$$

We know by Landau's estimate (22.81) that there are only finitely many idoneal discriminants but we cannot list them all because of ineffective constants.

22.5. Splitting primes in $\mathbb{Q}(\sqrt{D})$.

If the class number $h = h(D)$ is small, then there are only few prime ideals $\mathfrak{p}$ of degree one with small norm. Indeed, if $p = \mathfrak{p}\bar{\mathfrak{p}}$ with $(\mathfrak{p},\bar{\mathfrak{p}}) = 1$, then $\mathfrak{p}^h$ is a principal ideal generated by $\frac{1}{2}(m+n\sqrt{D})$ with $n \neq 0$, whence $p^h = \frac{1}{4}(m^2-n^2D) \geqslant \Delta$ where

$$\Delta = \frac{|D|}{4}. \tag{22.87}$$

Therefore the least prime $p_1 = p_1(D)$ with $\chi_D(p_1) = 1$ satisfies

$$p_1 \geqslant \sqrt[h]{\Delta}. \tag{22.88}$$

Hence $\chi_D(n)$ agrees with $\mu(n)$ on all squarefree numbers $n \leqslant \sqrt[h]{\Delta}$ with $(n,\Delta) = 1$. This property is not likely to hold in long segments (because χ_D is periodic while μ is not), therefore (22.88) suggests that h is rather large.

In this section we establish several estimates which hold for any $D < -4$ but are interesting only when h is relatively small. Let $\rho_D(a)$ be the number of solutions to

$$b^2 \equiv D(\text{mod } 4a) \tag{22.89}$$

in $b(\bmod 2a)$. This is the multiplicative function with $\rho_D(p^\alpha) = 1 + \chi_D(p)$ if $p \nmid D$, $\rho_D(p) = 1$ if $p \mid D$ and $\rho_D(p^\alpha) = 0$ if $p \mid D, \alpha > 1$. Thus the generating Dirichlet series for $\rho_D(a)$ has the Euler product

$$\zeta_K^\star(s) = \sum_a \rho_D(a)a^{-s} = \prod_{p|D}\Big(1 + \frac{1}{p^s}\Big) \prod_{\chi_D(p)=1}\Big(1 + \frac{1}{p^s}\Big)\Big(1 - \frac{1}{p^s}\Big)^{-1}.$$

Since $\rho_D(a)$ is the number of primitive ideals of norm a we have

$$\zeta_K^\star(s) = \zeta(2s)^{-1}\zeta_K(s) = \zeta(2s)^{-1}\zeta(s)L(s, \chi_D)$$

whence

(22.90)
$$\rho_D(a) = \sum_{bc^2|a} \chi_D(b)\mu(c).$$

First we show the following basic inequalities

(22.91)
$$\sum_{a\leqslant\sqrt{\Delta}} \rho_D(a) \leqslant h \leqslant \sum_{a\leqslant\sqrt{|D|/3}} \rho_D(a).$$

This follows because the class number h is equal to the number of primitive, reduced ideals $\mathfrak{a} = [a, \frac{b+\sqrt{D}}{2}]$ with b satisfying (22.89). For $a \leqslant \sqrt{\Delta}$ the points

(22.92)
$$z_\mathfrak{a} = \frac{b+\sqrt{D}}{2a}$$

have height $\mathrm{Im} z_\mathfrak{a} \geqslant 1$, so choosing b with $-a < b \leqslant a$, these points lie in the standard fundamental domain of $\Gamma = SL_2(\mathbb{Z})$ proving the first inequality of (22.91). The second inequality follows along the same lines since every reduced ideal $\mathfrak{a}$ has $a = N\mathfrak{a} \leqslant \sqrt{|D|/3}$.

Now suppose $p_1 < \cdots < p_r$ are the first r primes which split completely in $K = \mathbb{Q}(\sqrt{D})$. By (22.91) we have

$$\sum_{p_1^{\alpha_1}\cdots p_r^{\alpha_r}\leqslant\sqrt{\Delta}} 2^{r'} \leqslant h,$$

where $\alpha_1, \ldots, \alpha_r$ run over non-negative integers and r' is the number of positive exponents. Hence $\nu_r(\log\Delta/2\log p_r) \leqslant h$ where

$$\nu_r(\alpha) = \sum_{\alpha_1+\cdots+\alpha_r\leqslant\alpha} 2^{r'}.$$

Since $\nu_r(\alpha) \geqslant \frac{(2\alpha)^r}{2r!}$, this implies $\log\Delta \leqslant (2hr!)^{1/r}(\log p_r)$ ($\leqslant rh^{1/r}\log p_r$ if $r \geqslant 2$), i.e.,

(22.93)
$$\log p_r \geqslant \frac{\log\Delta}{\sqrt[r]{2hr!}}.$$

In particular, the second prime which splits completely satisfies $p_2 \geqslant \Delta^{\frac{1}{2\sqrt{h}}}$. If $h \leqslant (\log\Delta)^g$, then

(22.94)
$$|\{p \leqslant \exp(\sqrt{\log\Delta}) : \chi_D(p) = 1\}| \leqslant 2g$$

provided Δ is sufficiently large in terms of g, namely $\log \Delta > (4g)^{4g}$. Moreover, (22.91) yields

$$\frac{1}{r!}\Big(\sum_{\substack{p\leqslant \sqrt[2r]{\Delta}\\ \chi(p)=1}} 2\Big)^r \leqslant h.$$

Hence for any positive integer r we obtain

$$|\{p \leqslant \Delta^{\frac{1}{2r}} : \chi_D(p) = 1\}| \leqslant r h^{\frac{1}{r}}.$$

Choosing $r = [\frac{1}{2}\log 8h]$ this gives

$$|\{p \leqslant \Delta^{1/\log 8h} : \chi(p) = 1\}| \leqslant \log 8h. \tag{22.95}$$

Next we extend the basic inequality (22.91) to larger ranges. To this end we use the following elementary estimate (see Proposition 15.10):

$$|\{\gamma \in \Gamma_\infty\backslash\Gamma : \operatorname{Im}\gamma z > Y\}| \leqslant 1 + \frac{10}{Y}$$

which holds for any $Y > 0$ and $z \in \mathbb{H}$. Applying this for the points $z_{\mathfrak{a}}$ in the standard fundamental domain of $\Gamma = SL_2(\mathbb{Z})$ with $Y = \sqrt{\Delta}/A$ we deduce that

$$\sum_{a\leqslant A} \rho_D(a) \leqslant h\Big(1 + \frac{10A}{\sqrt{\Delta}}\Big) \tag{22.96}$$

for any $A > 0$. As before this implies

$$\frac{1}{r!}\Big(\sum_{\substack{p\leqslant A\\ \chi(p)=1}} 2\Big)^r \leqslant h\Big(1 + \frac{10A^r}{\sqrt{\Delta}}\Big).$$

Hence for any positive integer r we obtain

$$|\{p \leqslant A : \chi_D(p) = 1\}| < r h^{\frac{1}{r}} + 5rA\Big(\frac{h}{\sqrt{\Delta}}\Big)^{\frac{1}{r}}.$$

Choosing $r = [\frac{1}{8}\log 2h] + 1$ we conclude that

$$|\{p \leqslant A : \chi_D(p) = 1\}| \ll (1 + A|D|^{-4/\log 2h}) \log 2h \tag{22.97}$$

for any $A > 0$, where the implied constant is absolute.

In Chapter 23 we shall need the above estimates in various forms, therefore we end this section by deriving the required results. First restricting the summation in (22.96) to squarefree numbers we obtain

$$\sideset{}{^\flat}\sum_{a\leqslant A} \tau(a,\chi_D) \leqslant h\Big(1 + \frac{10A}{\sqrt{\Delta}}\Big). \tag{22.98}$$

Hence for any positive A, B we derive by partial summation

$$\sideset{}{^\flat}\sum_{B<a\leqslant A} \tau(a,\chi_D) a^{-1/2} \ll h\Big(\frac{1}{B} + \frac{A}{\Delta}\Big)^{1/2} \tag{22.99}$$

$$\sideset{}{^\flat}\sum_{B<a\leqslant A} \tau(a,\chi_D) a^{-1} \ll h\Big(\frac{1}{B} + \frac{\log A}{\sqrt{\Delta}}\Big). \tag{22.100}$$

Next we are going to estimate the sum

$$S(A,B)=\sideset{}{^\flat}\sum_{\substack{1<a\leqslant A\\(a,C)=1}}\tau(a)\tau(a,\chi_D)a^{-1/2} \tag{22.101}$$

where C stands for the product of primes in D and all the primes $p\leqslant B$ with $\chi_D(p)=1$.

We have

$$\begin{aligned}S(A,B)&\leqslant\sideset{}{^\flat}\sum\sideset{}{^\flat}\sum_{\substack{1<a_1a_2\leqslant A\\(a_1a_2,C)=1}}\tau(a_1,\chi_D)\tau(a_2,\chi_D)(a_1a_2)^{-1/2}\\&\leqslant 2\sideset{}{^\flat}\sum_{B<a\leqslant A}\tau(a,\chi_D)a^{-1/2}+2\sideset{}{^\flat}\sum\sideset{}{^\flat}\sum_{\substack{a_1a_2\leqslant A\\a_1,a_2>B}}\tau(a_1,\chi_D)\tau(a_2,\chi_D)(a_1a_2)^{-1/2}.\end{aligned}$$

Applying (22.99) and (22.100) we get

$$\begin{aligned}S(A,B)&\ll h\Big(\frac{1}{B}+\frac{A}{\Delta}\Big)^{1/2}+h\sideset{}{^\flat}\sum_{B<a\leqslant A}\frac{\tau(a,\chi_D)}{\sqrt{a}}\Big(\frac{1}{B}+\frac{A}{a\Delta}\Big)^{1/2}\\&\ll h\Big(\frac{1}{B}+\frac{A}{\Delta}\Big)^{1/2}+\frac{h^2}{\sqrt{B}}\Big(\frac{1}{B}+\frac{A}{\Delta}\Big)^{1/2}+h^2\Big(\frac{1}{B}+\frac{\log A}{\sqrt{\Delta}}\Big)\Big(\frac{A}{\Delta}\Big)^{1/2}.\end{aligned}$$

Rearranging terms this bound simplifies to

$$S(A,B)\ll h\Big(1+\frac{h}{\sqrt{B}}\Big)\Big(\frac{A}{\Delta}+\frac{1}{B}\Big)^{1/2}+h^2\Delta^{-1}\sqrt{A}\log A \tag{22.102}$$

where the implied constant is absolute, and effective.

22.6. Estimations for derivatives $L^{(k)}(1,\chi_D)$.

If the class number of the imaginary field $K=\mathbb{Q}(\sqrt{D})$ is unusually small, say

$$h(D)=o\Big(\frac{\sqrt{|D|}}{\log|D|}\Big), \tag{22.103}$$

then so is the value of the Dirichlet series $L(s,\chi_D)$ at $s=1$, namely

$$L(1,\chi_D)=o\Big(\frac{1}{\log|D|}\Big) \tag{22.104}$$

by virtue of the Class Number Formula (22.59). Though there is no direct relation between the class number and derivatives of $L(s,\chi_D)$ at $s=1$ all of these values are also affected by the improbable hypothesis (22.103).

Truncating the Dirichlet series for $L^{(k)}(s,\chi_D)$ we get

$$L^{(k)}(1,\chi_D)=\sum_{n\leqslant x}\frac{\chi_D(n)}{n}(-\log n)^k+O(|D|^{\frac12}x^{-1}(\log x)^{k+1}) \tag{22.105}$$

by the Polyá-Vinogradov inequality (12.50)

$$\sum_{y<n\leqslant x}\chi_D(n)\ll|D|^{\frac12}\log|D|. \tag{22.106}$$

Hence by absolute summation

$$L^{(k)}(1,\chi_D) \ll (\log|D|)^{k+1}. \tag{22.107}$$

It is amazing that this trivial bound is the best known so far (apart from some improvements of the implied constant). By the Riemann Hypothesis for $L(s,\chi_D)$ one can show that

$$L^{(k)}(1,\chi_D) \ll (\log\log|D|)^{k+1}. \tag{22.108}$$

A slight improvement on (22.107) can be derived from the rather subtle results of Graham and Ringrose [**GRi**] for special D having only small prime divisors (see Chapter 12). In this section we give estimates which are better than (22.107) if the class number $h(D)$ is relatively small.

First we examine $L'(1,\chi_D)$. We know by Landau's analytic method (see Chapter 5) that $L(s,\chi_D)$ cannot have two real zeros (counted with multiplicity) close to $s=1$, therefore $L(1,\chi_D)$ being small should imply $L'(1,\chi_D)$ is not small. This can be shown quickly by elementary arguments. To this end we evaluate the sum

$$\begin{aligned}\sum_{n\leqslant x} n^{-1}\tau(n,\chi_D) &= \sum_{m\leqslant y}\frac{\chi_D(m)}{m}\Big(\log\frac{x}{m}+\gamma+O\Big(\frac{m}{x}\Big)\Big)+O\Big(\frac{\sqrt{|D|}}{y}\log^2 x\Big)\\ &= L(1,\chi_D)(\log x+\gamma)+L'(1,\chi_D)+O\Big(\frac{y}{x}+\frac{\sqrt{|D|}}{y}\log^2 x\Big)\end{aligned}$$

for any $2\leqslant y\leqslant x$, where the error terms above came from estimation of the relevant character sums with $y<m\leqslant x$ by means of the Polyá-Vinogradov inequality. Hence we deduce the approximate formula

$$\sum_{n\leqslant x} n^{-1}\tau(n,\chi_D) = L(1,\chi_D)(\log x+\gamma)+L'(1,\chi_D)+O(|D|^{\frac14}x^{-\frac12}\log x). \tag{22.109}$$

Notice that on the left side all terms are non-negative and for $n=dm^2$ with $d\mid D$ we have $\tau(n,\chi_D)\geqslant 1$. Summing over these numbers (with d squarefree to ensure the uniqueness of the representation $n=dm^2$) we infer that (by choosing $x=e^{-\gamma}D$)

$$L(1,\chi_D)\log|D|+L'(1,\chi_D) > \Big(\frac{\pi^2}{6}+O\big(\frac{1}{\log|D|}\big)\Big)\nu(D),$$

where $\nu(D)$ is almost constant, precisely

$$\nu(D)=\prod_{p\mid D}\Big(1+\frac{1}{p}\Big). \tag{22.110}$$

Assuming (22.104) this yields $L'(1,\chi_D) > (\frac{\pi^2}{6}+o(1))\nu(D)$.

Now we derive a precise asymptotic for $L'(1,\chi_D)$ by appealing to the Kronecker Limit Formula (22.75) which allows us to take full advantage of the small class number condition. One could derive the same things from (22.109) but we choose the alternative path since it works for any class character and reveals other features. First, estimating trivially, we get

$$L'(1,\chi_D)=\frac{\pi^2}{6}\ell(D)+O\Big(\frac{h(D)}{\sqrt{|D|}}\log|D|\Big) \tag{22.111}$$

where $\ell(D)$ is the sum over the norms of primitive reduced ideals

$$\ell(D) = \sum_{\mathfrak{a}} a^{-1}. \tag{22.112}$$

We shall approximate $\ell(D)$ by the product

$$P(D) = \prod_{p|D}\Big(1+\frac{1}{p}\Big) \prod_{\substack{p\leqslant|D|\\ \chi_D(p)=1}} \Big(1+\frac{1}{p}\Big)\Big(1-\frac{1}{p}\Big)^{-1}. \tag{22.113}$$

Clearly we have the upper bound

$$\ell(D) < \sum_{a\leqslant|D|} \rho_D(a)a^{-1} \leqslant P(D).$$

For a lower bound we apply Rankin's trick (recall that $\Delta = |D|/4$):

$$\ell(D) \geqslant \sum_{a\leqslant\sqrt{\Delta}} \rho_D(a)a^{-1} > \zeta_K^\star(s) - \sum_{a>\sqrt{\Delta}} \rho_D(a)a^{-s}$$

where $\zeta_K^\star(s) = \zeta(2s)^{-1}\zeta_K(s)$ is the zeta function of $K = \mathbb{Q}(\sqrt{D})$ (reduced to the primitive ideals)

$$\zeta_K^\star(s) = \sum_a \rho_D(a)a^{-s} = \prod_{p|D}(1+p^{-s}) \prod_{\chi_D(p)=1} (1+p^{-s})(1-p^{-s})^{-1},$$

and s is any real number > 1 to be chosen at the end to optimize the obtained results. Applying (22.88) by partial summation we get

$$\sum_{a>\sqrt{\Delta}} \rho_D(a)a^{-s} < \frac{11sh}{(s-1)\sqrt{\Delta}}.$$

Here we can replace $11s$ by 33 because if $s \geqslant 3$ we have also the trivial bound $3/s\sqrt{\Delta}$ derived by using $\rho_D(a) \leqslant a$ and $h \geqslant 1$. Then we estimate $\zeta_K^\star(s)$ from below by the partial Euler product restricted to primes $p \leqslant |D|$, say $P_s(D)$, and replace $P_s(D)$ by $P(D)$ with admissible correction. Precisely, using the inequality $\prod(1-x_p) \geqslant 1 - \sum x_p$ with

$$\begin{aligned} x_p &= 1 - \Big(1+\frac{1}{p^s}\Big)\Big(1-\frac{1}{p}\Big)\Big(1+\frac{1}{p}\Big)^{-1}\Big(1-\frac{1}{p^s}\Big)^{-1} \\ &= \frac{2}{p+1}\,\frac{1-p^{1-s}}{1-p^{-s}} \leqslant 2(s-1)\frac{p\log p}{p^2-1} \end{aligned}$$

we get $P_s(D)/P(D) > 1 - 2(s-1)\eta(D)$ where $\eta(D)$ is given by

$$\eta(D) = \sum_{\substack{p\leqslant|D|\\ \chi_D(p)\neq-1}} \frac{p\log p}{p^2-1}. \tag{22.114}$$

Hence $\ell(D) > P(D)\{1-2(s-1)\eta(D)\} - 33h(D)/(s-1)\sqrt{\Delta}$. Choosing s close to 1, namely $s = 1 + (33h(D)/\eta(D)P(D)\sqrt{|D|})^{\frac{1}{2}}$, we derive the unconditional formula

$$\ell(D) = P(D) - \theta(\eta(D)P(D)h(D)/\sqrt{|D|})^{1/2} \tag{22.115}$$

where $0 < \theta < 3\sqrt{33}$. Note that $\eta(D) \ll \log|D|$. Hence we obtain

Proposition 22.9. *Put $\varepsilon(D) = h(D)|D|^{-\frac{1}{2}}\log|D|$. We have*

$$\ell(D) = P(D)\{1 + O(\varepsilon(D)^{\frac{1}{2}})\} \tag{22.116}$$

where the implied constant is absolute and effective.

Inserting (22.116) into (22.111) we come up with

Proposition 22.10. *Suppose $\varepsilon(D) = h(D)|D|^{-\frac{1}{2}}\log|D| \leqslant 1$. Then we have*

$$L'(1,\chi_D) = \Big(\frac{\pi^2}{6} + O(\varepsilon(D)^{\frac{1}{2}})\Big)P(D). \tag{22.117}$$

The product over splitting primes in $P(D)$ can be reduced considerably to that over a very few small primes by means of (22.97). Indeed (22.97) gives

$$\sum_{\substack{B<p\leqslant|D|\\ \chi_D(p)=1}} p^{-1} \ll (B^{-1} + |D|^{-4/\log 2h}\log|D|)\log 2h. \tag{22.118}$$

Applying this bound with $B = z\log 2h$ we deduce from (22.117) the following

Proposition 22.11. *Suppose*

$$h \leqslant |D|^{1/\log\log|D|}, \tag{22.119}$$

$$2 \leqslant z \leqslant \log|D|. \tag{22.120}$$

Then we have

$$L'(1,\chi_D) = \frac{\pi^2}{6}\nu(D)\prod_{\substack{p\leqslant z\log 2h\\ \chi_D(p)=1}}\Big(1+\frac{1}{p}\Big)\Big(1+\frac{1}{p}\Big)^{-1}\Big(1+O\Big(\frac{1}{z}\Big)\Big). \tag{22.121}$$

where $\nu(D)$ is defined by (22.110).

Now we estimate higher derivatives. We start from the sum

$$S_k(x) = \sum_{n\leqslant x}\chi_D(n)n^{-1}(\log n)^k$$

which approximates to $(-1)^k L^{(k)}(1,\chi_D)$ with an error term given by (22.105). On the other hand, we have

$$S_k(x) = \sum\sum_{mn\leqslant x}\tau(n,\chi_D)\frac{\mu(m)}{mn}(\log mn)^k.$$

By the convergence of $\sum\mu(m)m^{-1}(\log m)^\ell$ for $0 \leqslant \ell \leqslant k$, and in case of $\ell = 0$ using

$$\sum_{m\leqslant z}\mu(m)m^{-1} \ll (\log 2z)^{-1},$$

we obtain

$$S_k(x) \ll \sum_{n\leqslant x}\frac{\tau(n,\chi_D)}{n}(\log 2n)^{k-1}\frac{\log x}{\log(2x/n)}.$$

Estimating the terms with $y < n \leqslant x$ by means of (22.100) we get

$$S_k(x) \ll \sum_{n \leqslant y} \frac{\tau(n, \chi_D)}{n}(\log 2n)^{k-1} + \frac{h}{y}(\log y)^{k-1} + \frac{h}{\sqrt{\Delta}}(\log x)^k.$$

Here we estimate $\log n$ by $\log y$, then extend the summation to $n \leqslant D^2$ and apply (22.109) getting

$$S_k(x) \ll L'(1, \chi_D)(\log y)^{k-1} + \frac{h}{y}(\log y)^{k-1} + \frac{h}{\sqrt{\Delta}}(\log x)^k.$$

We choose $x = D^2$ and $y = 2h^2$ to arrive at

$$L^{(k)}(1, \chi_D) \ll |L'(1, \chi_D)|(\log 2h)^{k-1} + h|D|^{-\frac{1}{2}}(\log |D|)^k. \tag{22.122}$$

Combining (22.122) and (22.117) we obtain

PROPOSITION 22.12. *Suppose $h(D) \log |D| \leqslant \sqrt{|D|}$, then for any $k \geqslant 1$,*

$$L^{(k)}(1, \chi_D) \ll P(D)(\log 2h(D))^{k-1} \tag{22.123}$$

where $P(D)$ is the product given by (22.113) and the implied constant depends only on k.

REMARKS. Notice that (22.122) implies the trivial bound (22.107), and it becomes stronger when the class number is relatively small. In the derivation of (22.122) (which is unconditional) we made an appeal to the Prime Number Theorem in the form

$$\sum_{m \leqslant z} \frac{\mu(m)}{m}(\log m)^\ell \ll 1. \tag{22.124}$$

One can avoid these not so simple results altogether by examining a special combination of the truncated derivatives $S_\ell(x)$ with $0 \leqslant \ell \leqslant k$.

EXERCISE 2. Show that for $x \geqslant 2$ and a primitive character $\chi(\text{mod } |D|)$,

$$2\sum_{n \leqslant x} \tau(n, \chi)\frac{\log n}{n} = (\log^2 x - \gamma_1)L(1, \chi) - 2\gamma L'(1, \chi) - L''(1, \chi) + O(|D|^{\frac{1}{4}} x^{-\frac{1}{2}} \log^2 x).$$

Derive from this that

$$L''(1, \chi_D) < -2\sum_{d|D} \frac{\log 2d}{d} < 0$$

provided $L(1, \chi_D) < (\log |D|)^{-2}$ and $|D|$ is sufficiently large.

CHAPTER 23

EFFECTIVE BOUNDS FOR THE CLASS NUMBER

We have already seen how the assertion of $h(D)$ being small relatively to the discriminant D leads to simple, yet fake, approximations to the special values $L(\frac{1}{2}, \chi_D)$ and $L^{(k)}(1, \chi_D)$. In this section we consider central values of derivatives of certain L-functions associated with automorphic forms twisted by the character χ_D. In this case the level of the twisted form is much larger than the conductor of the character (it is about D^2), so we shall be able to observe positive effects only for derivatives of order $g \geqslant 2$ and only when the class number is extremely small (essentially $h(D) \ll (\log|D|)^{g-1}$). As in the previous cases our results are effective but we do not dwell on computing all the implied constants. Our motivation is to illustrate the masterwork of D. Goldfeld [**Go2**] which after Gross and Zagier [**GZ1**] has applications to give effective lower bounds for $h(D)$. Although for this application Goldfeld (and J. Oesterlé [**Oe**]) considered only the L-function of an elliptic curve, we take any primitive cusp form of a fixed level and a central character (possibly the trivial central character). At the end we derive the lower bound for $h(D)$ in question, and we make suggestions for further search of the L-functions which can be employed for estimating the class number.

23.1. Landau's plot of automorphic L-functions.

Throughout, f is a primitive cusp form of weight $k \geqslant 1$ and character $\varepsilon(\text{mod } N)$ on the group $\Gamma_0(N)$ such that $\varepsilon(-1) = (-1)^k$. Therefore (see Chapter 14) f has the Fourier expansion

$$f(z) = \sum_1^\infty \lambda(n) n^{\frac{k-1}{2}} e(nz)$$

with coefficients $\lambda(n)$ which are the eigenvalues of Hecke operators T_n for all n. Notice we normalized these so that the associated L-function

$$L(s, f) = \sum_1^\infty \lambda(n) n^{-s}$$

has an Euler product of type

$$L(s, f) = \prod_p (1 - \lambda(p)p^{-s} + \varepsilon(p)p^{-2s})^{-1} \tag{23.1}$$

and the complete L-function

$$\Lambda(s, f) = \Big(\frac{\sqrt{N}}{2\pi}\Big)^s \Gamma(s + \tfrac{k-1}{2}) L(s, f)$$

(which is entire) satisfies the functional equation

$$\Lambda(s,f)=w(f)\Lambda(1-s,\overline{f}). \tag{23.2}$$

Here $w(f)$ is a complex number depending on f with $|w(f)|=1$ and $\overline{f}$ is the cusp form whose Fourier coefficients are $\overline{\lambda}(n)$.

Let χ be the character associated with the field $K=\mathbb{Q}(\sqrt{D})$, that is the Kronecker symbol $\chi(n)=(\frac{D}{n})$; it is a real, primitive character of conductor $|D|$. We create the form $f\otimes\chi$ by twisting the coefficients with χ,

$$(f\otimes\chi)(z)=\sum_1^\infty\lambda(n)\chi(n)n^{\frac{k-1}{2}}e(nz).$$

The twisted form $f\otimes\chi$ is a cusp form of weight k, character ε and level ND^2, but it is not always primitive. However, there exists a unique primitive cusp form

$$f_\chi(z)=\sum_1^\infty\lambda_\chi(n)n^{\frac{k-1}{2}}e(nz)$$

of level $M|ND^2$ and character ε_χ such that $\varepsilon_\chi(n)=\varepsilon(n)$ and $\lambda_\chi(n)=\lambda(n)\chi(n)$ for all $(n,ND)=1$. Therefore the associated L-function

$$L(s,f_\chi)=\sum_1^\infty\lambda_\chi(n)n^{-s}$$

has the Euler product

$$\begin{aligned}L(s,f_\chi)=&\prod_{p|ND}(1-\lambda_\chi(p)p^{-s}+\varepsilon_\chi(p)p^{-2s})^{-1}\\&\prod_{p\nmid ND}(1-\lambda(p)\chi(p)p^{-s}+\varepsilon(p)p^{-2s})^{-1}.\end{aligned} \tag{23.3}$$

The complete L-function

$$\Lambda(s,f_\chi)=\Big(\frac{\sqrt{M}}{2\pi}\Big)^s\Gamma(s+\tfrac{k-1}{2})L(s,f_\chi)$$

is entire and it satisfies the functional equation

$$\Lambda(s,f_\chi)=w(f_\chi)\Lambda(1-s,\overline{f}_\chi) \tag{23.4}$$

with $|w(f_\chi)|=1$.

Given f and χ we consider the product L-function (in Landau's style)

$$L(s)=L(s,f)L(s,f_\chi)=\sum_1^\infty a_n n^{-s}. \tag{23.5}$$

Note that $L(s)$ has an Euler product of degree four.

Now we have all the information about $L(s,f)$ and $L(s,f_\chi)$ that is needed to work with $L(s)$. The complete product L-function defined by

$$\Lambda(s)=Q^s\Gamma^2(s+\tfrac{k-1}{2})L(s,f)L(s,f_\chi)\text{ with }Q=\sqrt{MN}/4\pi^2$$

satisfies the functional equation

$$\Lambda(s) = w\overline{\Lambda}(1-\overline{s}) \tag{23.6}$$

with the root number $w = w(f)w(f_\chi)$. Our first aim is to evaluate asymptotically the derivative $\Lambda^{(g)}(s)$ at $s = \frac{1}{2}$ of order $g \geqslant 0$. We make no further hypothesis until the time comes to examine the main terms in our formula.

23.2. A partition of $\Lambda^{(g)}(\frac{1}{2})$.

We begin by expressing $\Lambda^{(g)}(\frac{1}{2})$ as a sum of two rapidly converging conjugate series in coefficients of $L(s)$, similar to the formula (5.12). To this end we compute the contour integral

$$I = \frac{g!}{2\pi i}\int_{(1)} \Lambda(s+\tfrac{1}{2})s^{-g-1}ds$$

in two ways. First moving to the line $\mathrm{Re}\,(s) = -1$ we pass a pole of order $g+1$ at $s = \frac{1}{2}$ with residue $\Lambda^{(g)}(\frac{1}{2})$. Then we observe that the new integral on the line $\mathrm{Re}\,(s) = -1$ is equal to $-(-1)^g w\overline{I}$ by the functional equation (23.6), therefore we obtain $\Lambda^{(g)}(\frac{1}{2}) = I + (-1)^g w\overline{I}$. Here $\overline{I}$ denotes the complex conjugate of I. Next we compute I by integrating termwise the Dirichlet series (23.5). We obtain $I = Q^{1/2}S$, so

$$Q^{-1/2}\Lambda^{(g)}(\tfrac{1}{2}) = S + (-1)^g w\bar{S} \tag{23.7}$$

where

$$S = \sum_1^\infty \frac{a(n)}{\sqrt{n}} V\Big(\frac{n}{Q}\Big) \tag{23.8}$$

and $V(y)$ is the inverse Mellin transform of $g!\Gamma^2(s+\frac{k}{2})s^{-g-1}$,

$$V(y) = \frac{g!}{2\pi i}\int_{(1)} y^{-s}\Gamma^2(s+\tfrac{k}{2})s^{-g-1}ds. \tag{23.9}$$

Note that (move the integration to $\mathrm{Re}\,(s) = \frac{k}{2}$)

$$V(y) = \sum_{0\leqslant j\leqslant g} c_j(\log\tfrac{1}{y})^j + O(y^{\frac{k}{2}}\log\tfrac{2}{y}) \tag{23.10}$$

if $0 < y \leqslant 1$ with the leading coefficient $c_g = \Gamma^2(\frac{k}{2})$. Moreover, $V(y) \ll y^k e^{-2\sqrt{y}}$ if $y \geqslant 1$ (move the integration to $\mathrm{Re}\,(s) = \sqrt{y}$ and estimate by Stirling's formula (5.113)). Combining both cases we deduce the bound for the cut-off function

$$V(y) \ll e^{-\sqrt{y}}\log^g(1+y^{-1}) \tag{23.11}$$

which holds for all $y > 0$.

By virtue of (23.11) the series (23.8) begins to decay exponentially with $n \gg Q$ so we are left essentially with Q terms, which is still a lot. However, if the class number h is very small, then many coefficients $a(n)$ vanish. Under this fictitious condition a main term for S will emerge from a lacunary subsequence of the coefficients (essentially from $a(m^2)$). This main term is proportional to the value at $s = 1$ of the symmetric square L-function attached to f (which does not vanish, see Theorem 5.44 and (5.101)).

Before estimating S we pull out all the ramified places and a few small ones which split in the field $K = \mathbb{Q}(\sqrt{D})$. Precisely, let C be the product of primes in ND and all primes $p \leqslant B$ with $\chi(p) = 1$. We shall choose B later, it will be quite small but for now we assume only $B \leqslant |D|$. We write

$$S = \sum_{c|C^\infty} \frac{a(c)}{\sqrt{c}} \sum_{(m,C)=1} \frac{a(m)}{\sqrt{m}} V\Big(\frac{cm}{Q}\Big). \tag{23.12}$$

By the Euler product (23.1) the coefficients $\lambda(n)$ satisfy

$$\lambda(m)\lambda(n) = \sum_{d|(m,n)} \varepsilon(d)\lambda(mnd^{-2}). \tag{23.13}$$

Hence for $(m, ND) = 1$ we have

$$a(m) = \sum_{d^2n=m} \psi(d)\tau(n,\chi)\lambda(n)$$

where $\psi = \varepsilon\chi$. Accordingly S is partitioned into

$$S = \sum_{c|C^\infty} \frac{a(c)}{\sqrt{c}} \sum_{(d,C)=1} \frac{\psi(d)}{d} \sum_{(n,C)=1} \tau(n,\chi)\frac{\lambda(n)}{\sqrt{n}} V\Big(\frac{cd^2n}{Q}\Big).$$

We expect that the main contribution to S comes from $n = m^2$ and with $\tau(m^2, \chi)$ deleted, which is

$$S_1 = \sum_{c|C^\infty} \frac{a(c)}{\sqrt{c}} \sum_{(d,C)=1} \frac{\psi(d)}{d} \sum_{(n,C)=1} \frac{\lambda(m^2)}{m} V\Big(\frac{cd^2m^2}{Q}\Big).$$

The remaining terms yield

$$S_2 = \sum_{c|C^\infty} \frac{a(c)}{\sqrt{c}} \sum_{(d,C)=1} \frac{\psi(d)}{d} \sum_{(m,C)=1} (\tau(m^2,\chi) - 1)\frac{\lambda(m^2)}{m} V\Big(\frac{cd^2m^2}{Q}\Big),$$

$$S_3 = \sum_{c|C^\infty} \frac{a(c)}{\sqrt{c}} \sum_{(d,C)=1} \frac{\psi(d)}{d} \sum_{\substack{(n,C)=1\\ n\neq m^2}} \tau(n,\chi)\frac{\lambda(n)}{\sqrt{n}} V\Big(\frac{cd^2n}{Q}\Big).$$

Thus we have $S = S_1 + S_2 + S_3$. We shall treat each of these three sums separately. The first sum S_1 will be evaluated asymptotically by an appeal to analytic properties of the symmetric square L-function attached to the cusp form f, and the other two sums will be estimated by means of (22.86) and (22.88) respectively. We estimate S_3 essentially by the class number while S_2 turns out to be negligible. For the coefficients of the cusp form f we use Deligne's bound

$$|\lambda(n)| \leqslant \tau(n). \tag{23.14}$$

This rather advanced result is not essential, however, it is useful to simplify exposition.

23.3. Estimation of S_3 and S_2.

We have

$$\tau(n,\chi)=\prod_{p^{2\alpha-1}\|n}\alpha(1+\chi(p))\prod_{p^{2\alpha}\|n}(1+\alpha(1+\chi(p))).$$

Hence, writing $n=am^2$, where a is squarefree we see that $\tau(n,\chi)\leqslant\tau(a,\chi)\tau(m^2)$. Moreover, by Deligne's bound $|\lambda(n)|\leqslant\tau(n)\leqslant\tau(a)\tau(m^2)$. Therefore

$$|S_3|\leqslant\sum_{c|C^\infty}\frac{|a(c)|}{\sqrt{c}}\sum_d\sum_m\frac{\tau^2(m^2)}{dm}\sideset{}{^\flat}\sum_{\substack{a>1\\(a,C)=1}}\tau(a,\chi)\frac{\tau(a)}{\sqrt{a}}\Big|V\Big(\frac{acd^2m^2}{Q}\Big)\Big|.$$

Applying (22.88) and (23.11) we deduce that

$$S_3\ll h\Big(1+\frac{h}{\sqrt{B}}\Big)\{\nu_C(1)+\nu_C(\tfrac12)B^{-\frac12}(\log\Delta)^{g+10}\}$$

where

$$\nu_C(s)=\sum_{c|C^\infty}|a(c)|c^{-s}.$$

We have $a(p^\ell)=\sum_0^\ell\lambda(p^k)\lambda_\chi(p^{\ell-k})$, whence for any $s>0$ by (23.14)

$$\nu_C(s)\leqslant\prod_{p|C}\Big(\sum_0^\infty\tau(p^\ell)p^{-\ell s}\Big)^2=\prod_{p|C}(1-p^{-s})^{-4}=\xi_C(s),$$

say. For $\nu_D(1)$ we may have a more precise estimate

$$\nu_D(1)\asymp\prod_{p|D}\Big(1+\frac{|\lambda(p)|}{p}\Big)\ll\nu^2(D)\tag{23.15}$$

where $\nu(D)$ is defined by (22.110).

To estimate S_2 first notice that $\tau(m^2,\chi)=1$ unless m has a prime divisor with $\chi(p)=1$, indeed it is clear from the formula

$$\tau(m^2,\chi)=\prod_{p^\alpha\|m}(1+\alpha(1+\chi(p))).$$

Therefore by (23.11) and Deligne's bound

$$\begin{aligned}S_2&\ll\nu_C(\tfrac12)(\log\Delta)^{g+1}\sum_{(m,C)=1}\frac{\tau(m^2)}{m}(\tau(m^2,\chi)-1)e^{-m/\sqrt{Q}}\\&\ll\nu_C(\tfrac12)(\log\Delta)^{g+1}\sum_m\frac{\tau^2(m^2)}{m}\sum_{\substack{p\nmid C\\\chi(p)=1}}p^{-1}e^{-mp/\sqrt{Q}}.\end{aligned}$$

Applying (22.86) we infer that

$$S_2\ll h\Big(1+\frac{h}{\sqrt{B}}\Big)\nu_C(\tfrac12)B^{-1}(\log\Delta)^{g+10}.\tag{23.16}$$

We simplify both results by taking B in the following range

$$h^2(\log\Delta)^{2g+20}\leqslant B\leqslant\Delta^{1/\log 2h}.\tag{23.17}$$

For $s = \frac{1}{2}, 1$ we have $\nu_C^2(s) \leqslant \xi_C^2(s) \ll h\xi_D(s) \ll h\tau(|D|) \leqslant h^2$, whence we conclude that

$$S_2 + S_3 \ll h\nu_C(1). \tag{23.18}$$

23.4. Evaluation of S_1.

To this end we use complex integration and analytic properties of the series which come out. Inserting (23.9) for $V(cd^2m^2/Q)$ into (23.12) we get

$$S_1 = \frac{g!}{2\pi i}\int_{(1)} Q^s\Gamma^2(s+\tfrac{k}{2})Z(2s+1)s^{-g-1}ds \tag{23.19}$$

where $Z(s)$ is the corresponding Dirichlet series

$$Z(s) = \Big(\sum_{c|C^\infty} a(c)c^{-s/2}\Big)\Big(\sum_{(d,C)=1} \psi(d)d^{-s}\Big)\Big(\sum_{(m,C)=1} \lambda(m^2)m^{-s}\Big).$$

Recall that $\psi = \varepsilon\chi$. We write $Z(s) = L(s,\psi)M(s,f)P(s,f)$ where

$$P(s,f) = \Big(\sum_{c|C^\infty} a(c)c^{-s/2}\Big)\Big(\sum_{d|C^\infty} \psi(d)d^{-s}\Big)^{-1}\Big(\sum_{m|C^\infty} \lambda(m^2)m^{-s}\Big)^{-1},$$

$L(s,\psi)$ is the Dirichlet L-function for the character $\psi = \varepsilon\chi$ and

$$M(s,f) = \sum_1^\infty \lambda(m^2)m^{-s}. \tag{23.20}$$

Every Dirichlet series above has an Euler product. To compute them we factor the Hecke polynomial for $L(s,f)$ into

$$1 - \lambda(p)p^{-s} + \varepsilon(p)p^{-2s} = (1-\alpha(p)p^{-s})(1-\beta(p)p^{-s}) \tag{23.21}$$

with $\alpha(p)+\beta(p) = \lambda(p)$ and $\alpha(p)\beta(p) = \varepsilon(p)$. Similarly, this holds for the local factors of $L(s,f_\chi)$ with $\alpha_\chi(p) = \alpha(p)\chi(p)$, $\beta_\chi(p) = \beta(p)\chi(p)$ in place of $\alpha(p)$, $\beta(p)$. This yields

$$\lambda(p^\ell) = \frac{\alpha^{\ell+1}-\beta^{\ell+1}}{\alpha-\beta}. \tag{23.22}$$

Here and hereafter we drop the argument p in the roots $\alpha, \beta, \alpha_\chi, \beta_\chi$ for simplicity. By (23.22) we derive

$$\begin{aligned}\sum_0^\infty \lambda(p^{2\ell})p^{-\ell s} &= \frac{\alpha}{\alpha-\beta}(1-\alpha^2p^{-s})^{-1} - \frac{\beta}{\alpha-\beta}(1-\beta^2p^{-s})^{-1}\\ &= (1-\alpha^2p^{-s})^{-1}(1+\alpha\beta p^{-s})(1-\beta^2p^{-s})^{-1}\\ &= (1-\alpha^2p^{-s})^{-1}(1-\alpha\beta p^{-s})^{-1}(1-\beta^2p^{-s})^{-1}(1-\varepsilon^2p^{-2s}).\end{aligned}$$

Hence the C-part is equal to

$$\begin{aligned}P(s,f) = &\prod_{p|C}(1-\alpha p^{-s/2})^{-1}(1-\beta p^{-s/2})^{-1}(1-\alpha_\chi p^{-s/2})^{-1}(1-\beta_\chi p^{-s/2})^{-1}\\ &\prod_{p|C}(1-\psi p^{-s})(1-\alpha^2p^{-s})(1+\varepsilon p^{-s})^{-1}(1-\beta^2p^{-s}).\end{aligned}$$

After factoring $1-\alpha^2p^{-s}$ and $1-\beta^2p^{-s}$ we arrange this product as

(23.23)

$$P(s,f)=\prod_{p|C}(1+\varepsilon p^{-s})^{-1}(1-\psi p^{-s})$$
$$\prod_{p|C}(1+\alpha p^{-s/2})(1+\beta p^{-s/2})(1-\alpha_\chi p^{-s/2})^{-1}(1-\beta_\chi p^{-s/2})^{-1}.$$

Moreover, we get $M(s,f)=L(2s,\varepsilon^2)^{-1}L(s,\mathrm{Sym}^2 f)$ where

$$L(s,\mathrm{Sym}^2 f)=\prod_p(1-\alpha^2p^{-s})^{-1}(1-\alpha\beta p^{-s})^{-1}(1-\beta^2p^{-s})^{-1}$$

is the symmetric square L-function attached to f; see Section 5.12. This L-function appears as a factor in the Rankin-Selberg L-function

$$L(s,f\otimes f)=\sum_1^\infty\lambda^2(n)n^{-s}. \tag{23.24}$$

Indeed, by (23.12) for $m=n$ we deduce that

$$L(s,f\otimes f)=L(s,\varepsilon)M(s,f)=\frac{L(s,\varepsilon)}{L(2s,\varepsilon^2)}L(s,\mathrm{Sym}^2 f). \tag{23.25}$$

The L-function $L(s,\mathrm{Sym}^2 f)$ admits meromorphic continuation to the whole complex s-plane and has a functional equation. As proved by Shimura [**Sh2**], it is holomorphic except possibly for simple poles at $s=0$ and $s=1$.

The cases for which $L(s,\mathrm{Sym}^2 f)$ has a pole at $s=1$ are completely understood (the dihedral forms) and very rare. We shall assume that our function $L(s,\mathrm{Sym}^2 f)$ has no pole at $s=1$. By (5.101) or Theorem 5.44, we have

$$L(1,\mathrm{Sym}^2 f)\neq 0. \tag{23.26}$$

Besides the above arithmetic properties of the zeta functions we need crude estimates on the line $\mathrm{Re}\,(s)=\frac{1}{2}$. By (23.23) we get

$$|P(s,f)|\leqslant \xi_C(\tfrac14)\ll h. \tag{23.27}$$

For the L-function of the character $\psi=\varepsilon\chi$ we use the convexity bound (see Section 5.9)

$$L(s,\psi)\ll|s||D|^{\frac14}. \tag{23.28}$$

Hence we arrive at

$$Z(s)\ll|s|^4|D|^{\frac14}h \tag{23.29}$$

where the implied constant depends on the cusp form f.

Having gathered the arithmetic and analytic properties of the relevant zeta functions we are now ready to evaluate the series S_1. Moving in (23.19) to the line $\mathrm{Re} s=-1/4$ we obtain

$$S_1=S_0+O(h) \tag{23.30}$$

where S_0 is the residue at $s = 0$, or simply

$$S_0 = \frac{d^g}{ds^g} Q^s \Gamma^2(s + \tfrac{k}{2}) Z(2s + 1), \qquad \text{at } s = 0. \tag{23.31}$$

REMARK. The error term in (23.30) could be improved by factor $|D|^{-1/16}$ using Burgess' bound in place of (23.28) but we do not need anything better than (23.30).

The dependence of S_0 on the character $\psi = \varepsilon\chi$ needs to be exposed. To this end we set

$$R(s) = Q^s \Gamma^2(s + \tfrac{k}{2}) M(2s + 1, f) P(2s + 1, f) \tag{23.32}$$

and differentiate $L(2s + 1, \psi)R(s)$ by Leibniz rule getting

$$S_0 = \sum_{u+v=g} 2^u \binom{g}{u} L^{(u)}(1, \psi) R^{(v)}(0). \tag{23.33}$$

Next, when computing the derivatives $R^{(v)}(0)$ we think of Q^s as a dominant factor in $R(s)$. Recalling that all primes in C are bounded by Q, we derive by repeated application of Leibniz rule and the product representation (23.23) that

$$R^{(v)}(0) = R(0)(\log Q)^v \Big(1 + O\Big(\frac{t(C)}{\log Q}\Big)\Big) \tag{23.34}$$

where

$$t(C) = \sum_{p|C} p^{-\frac{1}{2}} \log p. \tag{23.35}$$

This approximation would be useless if the error term was not small, but thanks to (23.17) we get by (22.81) that $t(C) = t(D) + O(\log 2h)$. We have also $t(D) \ll \omega(|D|) \ll \log 2h$, therefore $t(C) \ll \log 2h$.

Finally by (23.30), (23.34) and (23.36) we obtain

$$S_1 = R(0) \sum_{u+v=g} 2^u \binom{g}{u} L^{(u)}(1, \psi)(\log |D|)^v \Big(1 + O\Big(\frac{\log 2h}{\log |D|}\Big)\Big) + O(h). \tag{23.36}$$

23.5. An asymptotic formula for $\Lambda^{(g)}(\frac{1}{2})$.

Adding (23.18) to (23.37) we obtain

$$S = R(0) \sum_{u+v=g} 2^u \binom{g}{u} L^{(u)}(1, \psi)(\log |D|)^v \Big(1 + O\Big(\frac{\log 2h}{\log |D|}\Big)\Big) + O(h\nu_C(1)) \tag{23.37}$$

where $R(0) = \Gamma^2(\frac{k}{2}) M(1, f) P(1, f)$. Recall that

$$M(1, f) = L(2, \varepsilon^2)^{-1} L(1, \mathrm{Sym}^2 f) \tag{23.38}$$

and $P(1,f)$ is given by (23.23). Have in mind that $P(1,f)$ depends on D but rather mildly, precisely

$$P(1,f)=\prod_{p|C}\Big(1+\frac{\varepsilon(p)}{p}\Big)^{-1}\Big(1-\frac{\varepsilon(p)\chi(p)}{p}\Big)\Big(1+\frac{\alpha(p)}{\sqrt{p}}\Big)\Big(1+\frac{\beta(p)}{\sqrt{p}}\Big)\Big(1-\frac{\alpha(p)\chi(p)}{\sqrt{p}}\Big)^{-1}\Big(1-\frac{\beta(p)\chi(p)}{\sqrt{p}}\Big)^{-1} \tag{23.39}$$

where the product is over primes $p \mid D$ and $p \leqslant B$ with $\chi(p)=1$. Inserting (23.37) to (23.7) we obtain the desired expression for $\Lambda^{(g)}(\frac{1}{2})$ in terms of derivatives of the L-function for the character $\psi=\varepsilon\chi$ at the point $s=1$.

From now on we assume that ε is a real character, the symmetric square $\mathrm{Sym}^2 f$ has real coefficients (though f may have both real and imaginary coefficients) and $P(1,f)$ is also real. Then the approximate formula (23.37) for S and $\overline{S}$ is the same. Recall the formula (23.7). Since we do not wish the main terms canceled we must match the parity of g with the sign of the functional equation (23.6) for the product L-function $L(s)=L(s,f)L(s,f_\chi)$, i.e., we require

$$w=w(f)w(f_\chi)=(-1)^g. \tag{23.40}$$

Now inserting (23.37) into (23.7) we arrive at

$$Q^{-\frac{1}{2}}\Lambda^{(g)}(\tfrac{1}{2})=2\Gamma^2(\tfrac{k}{2})M(1,f)P(1,f)\sum_{u+v=g}2^u\binom{g}{u}L^{(u)}(1,\psi)(\log|D|)^v\Big(1+O\Big(\frac{\log 2h}{\log|D|}\Big)\Big)+O(h\nu_C(1)). \tag{23.41}$$

We shall refine this rudimentary formula in special cases.

Suppose $\varepsilon=1$, so $\psi=\chi_D$. We can use the estimates for $L^{(u)}(1,\chi_D)$ established in Section 22.6 under the minor assumption $h(D)\log|D|\leqslant\sqrt{|D|}$ (see (22.122)) reducing (23.41) to

$$Q^{-\frac{1}{2}}\Lambda^{(g)}(\tfrac{1}{2})=2g\Gamma^2(\tfrac{k}{2})M(1,f)P(1,f)L'(1,\chi_D)(\log|D|)^{g-1}\Big(1+O\Big(\frac{\log 2h}{\log|D|}\Big)\Big)+O(h\nu_C(1)). \tag{23.42}$$

Also we could replace $L'(1,\chi_D)$ by $\frac{\pi^2}{6}P(D)$ using (22.117). Though we made so far rather minor restrictions for the class number (exactly such that the segment (23.17) is not void), our asymptotic formula (23.42) is interesting only if $h(D)\ll(\log|D|)^g$, or else the error term exceeds the main term. Having assumed the class number is that small, notice that there are at most $2g$ primes $p\leqslant\exp(\sqrt{\log|D|})$ which split in $K=\mathbb{Q}(\sqrt{D})$ (see (22.95)). Choosing $B=(\log|D|)^{4g+20}$ (which is within (23.17)) we find that $\nu_C(1)\asymp\nu_D(1)\ll\nu(D)^2$ by (23.15), then $L'(1,\chi_D)\asymp\nu(D)$ by (22.119), and

$$P(1,f)\asymp\prod_{p|D}\Big(1+\frac{1}{p}\Big)^{-1}\Big(1+\frac{\lambda(p)}{\sqrt{p}}+\frac{1}{p}\Big)=\prod_{p|D}\Big(1+\frac{\lambda(p)\sqrt{p}}{p+1}\Big) \tag{23.43}$$

by (23.39).

23.6. A lower bound for the class number.

As before $\chi = \chi_D$ is the character of the imaginary quadratic field $K = \mathbb{Q}(\sqrt{D})$. If the left side of (23.42) vanishes, we are left with a lower bound for the error term which contains the class number $h = h(D)$. We put this observation as

PROPOSITION 23.1. *Suppose f is a primitive cusp form of weight $k \geqslant 1$, level N and trivial central character. Suppose the product L-function vanishes at $s = \frac{1}{2}$ to order*

$$m = \operatorname*{ord}_{s=\frac{1}{2}} L(s,f)L(s,f_\chi) \geqslant 3. \tag{23.44}$$

Let $g = m-1$ or $g = m-2$ be such that $(-1)^g = w = w(f)w(f_\chi)$. Then

$$h(D) \gg \theta(D)(\log|D|)^{g-1} \tag{23.45}$$

where

$$\theta(D) = \prod_{p|D}\Big(1+\frac{1}{p}\Big)^{-3}\Big(1+\frac{\lambda(p)\sqrt{p}}{p+1}\Big)^{-1}. \tag{23.46}$$

The implied constant depends only on f and m and it is effectively computable.

The theory of elliptic curves provides plenty of examples of cusp forms $f \in S_2(\Gamma_0(N))$ for which the associated L-function $L_E(s+\frac{1}{2}) = L(s,f)$ should have large order of vanishing at $s = \frac{1}{2}$, precisely equal to the rank of the group of rational points on E, by the Birch-Swinnerton-Dyer Conjecture. Besides the rank we need to know the root numbers $w(f), w(f_\chi)$ to control the parity of g. In general, for a primitive form $f \in S_k(\Gamma_0(N))$ of squarefree level N we have (see e.g. Proposition 14.16 or [**I4**])

$$w(f) = i^k\mu(N)\lambda(N). \tag{23.47}$$

There is no simple formula for $w(f)$ if N is not squarefree. If the greatest common divisor (N, D^2) is squarefree, then the twisted form is primitive, i.e., $f_\chi = f\otimes\chi$, has level $M = [N, D^2]$ and the root number is

$$w(f_\chi) = \chi(-N_1)\mu(N_2)\lambda(N_2)w(f) \tag{23.48}$$

where $N = N_1N_2$ with $(N_1, D) = 1$ and $N_2|D$. This simplifies to $w(f_\chi) = \chi(-N)w(f)$ if $(N, D) = 1$. Exploiting (23.47) and (23.48) we shall derive from Proposition 23.1 our final result

THEOREM 23.2. *There exists an absolute, effectively computable constant $c > 0$ such that for any fundamental discriminant $D < 0$ we have*

$$h(D) \geqslant c\theta(D)\log|D| \tag{23.49}$$

where

$$\theta(D) = \prod_{p|D}\Big(1+\frac{1}{p}\Big)^{-3}\Big(1+\frac{2\sqrt{p}}{p+1}\Big)^{-1}. \tag{23.50}$$

PROOF. First we handle the case when $p = 37$ splits in $\mathbb{Q}(\sqrt{D})$; in this case (22.72) implies $h(D)\log 37 \geqslant \log|D/4|$ which is better than (23.49). Therefore, from now on we assume that

$$\chi_D(37) \neq 1. \tag{23.51}$$

We begin with the elliptic curve

$$E_0 : y^2 = x^3 + 10x^2 - 20x + 8 \tag{23.52}$$

which was considered by Gross and Zagier [**GZ1**]; this is a modular curve of conductor $N_0 = 37$ and rank $r_0 = 0$. The corresponding primitive cusp form is

$$f_0(z) = \sum_1^\infty \lambda_0(n) n^{\frac{1}{2}} e(nz) \in S_2(\Gamma_0(N_0))$$

and the associated Hasse-Weil zeta function is $L_{E_0}(s) = L(s - \frac{1}{2}, f_0)$ where

$$L(s, f_0) = \sum_1^\infty \lambda_0(n) n^{-s}.$$

This satisfies the appropriate functional equation with sign $w(f_0) = 1$. For the proof of Theorem 23.2 we take the twisted curve

$$E : -139y^2 = x^3 + 10x^2 - 20x + 8. \tag{23.53}$$

This is a modular curve of conductor $N = 37 \cdot 139^2$ and rank $r = 3$. The corresponding primitive cusp form is

$$f(z) = \sum_1^\infty \lambda(n) n^{\frac{1}{2}} e(nz) \in S_2(\Gamma_0(N))$$

with $\lambda(n) = \lambda_0(n)\chi_{-139}(n)$, i.e., $f = f_0 \otimes \chi_{-139}$, and the associated Hasse-Weil zeta function is $L_E(s) = L(s - \frac{1}{2}, f)$ where

$$L(s, f) = \sum_1^\infty \lambda(n) n^{-s}.$$

The sign of the functional equation for f is

$$w(f) = \chi_{-139}(-1) w(f_0) = \chi_{-139}(-1) = -1$$

according to (23.48). Hence $L(\frac{1}{2}, f) = 0$, but Gross and Zagier showed that also $L'(\frac{1}{2}, f) = 0$, therefore $L(s, f)$ vanishes at $s = \frac{1}{2}$ to the order at least three. (In Appendix 23.A, we sketch the proof of the result of Gross and Zagier).

Now we examine the twisted form $f \otimes \chi_D \in S_2(\Gamma_0(ND^2))$ and its corresponding primitive form $f_{\chi_D} \in S_2(\Gamma_0(M))$ of level $M | ND^2$.

CASE $139 \nmid D$. Then $M = ND^2$, $f_{\chi_D} = f \otimes \chi_D = f_0 \otimes \chi_{-139D}$, and by virtue of (23.48)

$$w(f_{\chi_D}) = \chi_{-139D}(-N_1)\mu(N_2)\lambda_0(N_2)$$

where $N_1 N_2 = 37$ with $(N_1, D) = 1$ and $N_2 | D$. If $N_1 = 37$ and $N_2 = 1$, we get $w(f_{\chi_D}) = \chi_{-139D}(-37) = \chi_{-139}(-37) = -w(f_0) = -1$. If $N_1 = 1$ and $N_2 = 37$, we get $w(f_{\chi_D}) = -\chi_{-139D}(-1)\lambda_0(37) = -\lambda_0(37) = -w(f_0) = -1$ by virtue of (23.47).

CASE $139|D$. We write $D = -139C$ where C is a positive fundamental discriminant coprime with 139. Accordingly the field character factors into $\chi_D = \chi_{-139}\chi_C$ and $f \otimes \chi_D = f_0 \otimes \chi_{-139}^2 \otimes \chi_C$. Here the form $f_0 \otimes \chi_{-139}^2$ is not primitive because the trivial character χ_{-139}^2 simply kills the place $p = 139$ in its L-function, however, it is induced by the primitive form f_0. Therefore the primitive form which induces $f \otimes \chi_D$ is $f_{\chi_D} = f_0 \otimes \chi_C$, it has the level $M = [37, C^2] = 37C^2/(37, C)$ and the root number

$$w(f_{\chi_D}) = \chi_C(-N_1)\mu(N_2)\lambda_0(N_2)$$

by virtue of (23.48), where $N_1N_2 = 37$ with $(N_1, C) = 1$ and $N_2|C$. If $N_1 = 37$ and $N_2 = 1$, we get $w(f_{\chi_D}) = \chi_C(-37) = \chi_{-139D}(-37) = -1$, as in the former case. If $N_1 = 1$ and $N_2 = 37$, we get $w(f_{\chi_D}) = -\lambda_0(37) = -1$.

Now we conclude that in all cases $w(f) = w(f_{\chi_D}) = -1$, whence the L-function $L(s) = L(s, f)L(s, f_\chi)$ has root number $w = w(f)w(f_{\chi_D}) = 1$. Since $L(s)$ vanishes at $s = \frac{1}{2}$ to order $m \geqslant 3$ (actually $m \geqslant 4$) we can apply Proposition 23.1 with $g = 2$ proving (23.49). □

EXERCISE. Derive from (23.49) the following effective bound

$$h(D) \gg \exp(-(\log\log|D|)^{\frac{1}{2}})(\log|D|). \tag{23.54}$$

[**Hint:** Use genus theory if D has many small prime factors.]

23.7. Concluding notes.

Our formula (23.41) is not restricted to the L-functions associated with elliptic curves, and one hopes to use it more efficiently by employing other L-functions which vanish to high order at the central point. An interesting proposition is to try the L-function of a cusp form with non-trivial central character, the point being that the leading term in (23.41) has the order of magnitude $L(1, \varepsilon\chi_D)(\log|D|)^g$ which for $\varepsilon \neq 1$ is better than the bound $L'(1, \chi_D)(\log|D|)^{g-1}$ deduced from (23.42) when $\varepsilon = 1$.

There are several examples of cuspidal L-functions other than the Hasse-Weil zeta functions of elliptic curves which vanish at the central point $s = \frac{1}{2}$ to the order at least two (but still with the trivial central character, unfortunately) found by F. Rodriguez Villegas [**RV**]. These are the L-functions for Hecke characters of an imaginary quadratic field which are described in Chapter 3 by formulas (3.92)–(3.95). Suppose ξ is such a character of the field $\mathbb{Q}(\sqrt{-q})$ with $q > 3, q \equiv 3(\text{mod } 4)$ and of weight $\ell \geqslant 1, \ell \equiv 1(\text{mod } 2)$, i.e., ξ is a character which on principal ideals takes

$$\xi((\alpha)) = \left(\frac{2m}{q}\right)\left(\frac{\alpha}{|\alpha|}\right)^\ell$$

for $\alpha = \frac{1}{2}(m + n\sqrt{-q})$. There are exactly $h(-q)$ characters of this type, each of which yields a primitive cusp form

$$f_\xi(z) = \sum_{\mathfrak{a}} \xi(\mathfrak{a})(N\mathfrak{a})^{\frac{\ell}{2}} e(zN\mathfrak{a}) \in S_{\ell+1}(\Gamma_0(q))$$

(see (3.89)), and the associated L-function

$$L(s, \xi) = \sum_{\mathfrak{a}} \xi(\mathfrak{a})(N\mathfrak{a})^{-s}$$

satisfies the following equation (self-dual)

$$\Big(\frac{\sqrt{q}}{2\pi}\Big)^s\Gamma\Big(s+\frac{\ell}{2}\Big)L(s,\xi)=w\Big(\frac{\sqrt{q}}{2\pi}\Big)^{1-s}\Gamma\Big(1-s+\frac{\ell}{2}\Big)L(1-s,\xi)$$

(see Theorem 3.8) with the root number $w=(-1)^{\frac{\ell-1}{2}}(\frac{2}{q})$. In the Main Theorem of Rodriguez-Villegas [**RV**] the central value $L(\frac{1}{2},\xi)$ is expressed as a square of a linear combination of certain theta functions, showing that $L(\frac{1}{2},\xi)\geqslant 0$. It is also proved that $L(\frac{1}{2},\xi)$ is positive if $\ell=1$ (because it is true on average over all characters ξ and these characters are Galois conjugates of each other). For $\ell=3$ the non-vanishing of $L(\frac{1}{2},\xi)$ is guaranteed only if the class number $h(-q)$ is not a multiple of 3 (this condition is required to show there is only one Galois orbit). Consistently, for $q=59$ we have $h(-59)=3$ and $L(\frac{1}{2},\xi)=0$. Since the root number is $w=-(\frac{2}{59})=1$ we get three distinct functions $L(s,\xi)$ each of which vanishes at $s=\frac{1}{2}$ to order at least 2, and the corresponding cusp forms have weight $k=\ell+1=4$.

23.A Appendix: The Gross-Zagier L-function vanishes to order 3.

We will sketch the proof by Gross and Zagier that the Hasse-Weil zeta function of the curve (23.53)

$$-139y^2=x^3+10x^2-20x+8$$

vanishes to order $\geqslant 3$ at the central critical point. Precisely, we explain the proof of:

THEOREM 23.3. *For the above L-function, we have* $L'(f,\frac{1}{2})=0$.

Since the sign of the function equation for $L(f,s)$ is -1, this implies that the order of vanishing is at least 3 as desired. The proof is based on a formula of Gross and Zagier for the special value of the derivative of the L-function of E (over a suitable auxiliary imaginary quadratic field) in terms of the height of a special point. To state this, we recall some notation.

Let k be a number field and M_k the set of (normalized) absolute values of k, denoted $|\cdot|_v$ for $v\in M_k$; those include archimedean as well as non-archimedean absolute values and we have the product formula

$$\prod_{v\in M_k}|x|_v=1 \text{ for any } x\in k^{\star}.$$

For any $x=[x_0:x_1:x_2]$ in the projective plane $\mathbb{P}^2(k)$, the absolute logarithmic height of x is

$$h(x)=\frac{1}{[k:\mathbb{Q}]}\log\prod_{v\in M_k}\max\{|x_0|_v,|x_1|_v,|x_2|_v\} \tag{23.55}$$

(this is well-defined because of the product formula). If $E\subset\mathbb{P}^2$ is an elliptic over k, then the function $x\mapsto h(x)$ is almost a quadratic form (with respect to the group structure). Denoting

$$h_E(x)=\lim_{n\to+\infty}\frac{h(2^nx)}{2^{2n}},$$

Tate and Néron showed that h_E is a quadratic form with $h_E(x)=h(x)+O(1)$, called the canonical height on E. We only need the obvious fact that $h_E(0)=0$, which follows because $h(\infty)=1$; see [**Sil**] for more details.

Let $E/\mathbb{Q}$ be an elliptic curve with conductor N. It is modular and a (non-obvious) consequence is that there exists non-constant maps $\pi : X_0(N) \to E$. Among these, there is a unique one such that $\pi(\infty) = 0$ and $\pi^*(dz) = cf(z)dz$ for some constant $c > 0$, where dz is the translation-invariant differential form on E and f is the primitive weight 2 cusp form associated to E. The map π is then called the modular parameterization of E and in fact it is induced by the holomorphic function $\mathbb{H} \to \mathbb{C}$ defined by

$$\pi(z) = 2\pi i c \int_\infty^z f(w)dw.$$

Let $K/\mathbb{Q}$ be any imaginary quadratic field with discriminant $D < 0$ such that every prime p dividing N splits in K (in particular, is not ramified). Note that there are infinitely many such fields because the conditions are $\chi_K(p) = 1$ for $p \mid N$, which are congruence conditions for D. There exist $2^{\omega(N)}$ ideals $\mathfrak{n} \subset \mathcal{O}$ such that $\mathcal{O}/\mathfrak{n} \simeq \mathbb{Z}/N\mathbb{Z}$ (where $\mathcal{O} \subset K$ is the ring of integers), those being in one-to-one correspondence with solutions $\beta \pmod{2N}$ to the congruence $\beta^2 \equiv D \pmod{4N}$.

To each ideal class $\mathfrak{a}$ of K is associated the corresponding complex point $z_\mathfrak{a} \in \mathbb{H}$ (see Section 22.1). One easily shows that there is an $SL(2,\mathbb{Z})$-equivalent point $\tilde{z}_\mathfrak{a}$ such that

$$\tilde{z}_\mathfrak{a} = \frac{-B + \sqrt{-D}}{2A}$$

with $N \mid A$ and $B \equiv \beta \pmod{2N}$. This point is unique modulo $\Gamma_0(N)$. The Heegner point on E associated to K and $\mathfrak{n}$ is defined to be

$$y_K = \sum_\mathfrak{a} \pi(\tilde{z}_\mathfrak{a}) \in E(\mathbb{C}),$$

where the sum runs over all ideal classes of K.

THEOREM 23.4 (THE GROSS-ZAGIER FORMULA). *Let $E/\mathbb{Q}$ be an elliptic curve with associated cusp form f, K an imaginary quadratic field as above with discriminant D, χ the Kronecker character associated to K. Let*

$$L_K(f,s) = L(f,s)L(f \otimes \chi, s).$$

Then $L_K(f, \frac{1}{2}) = 0$ and

$$L'_K(f, \tfrac{1}{2}) = \frac{8\pi^2 \|f\|^2}{u^2(\deg \pi)|D|^{1/2}} h(y_K) \tag{23.56}$$

where $\|f\|$ is the Petersson norm of $f \in S_2(N)$, $u \in \{1,2,3\}$ is half the number of units of K, $\deg(\pi) \geqslant 1$ is the degree of the map $\pi : X_0(N) \to E$.

REMARKS. (1) In the application below, the modularity of E and the existence of π will be obvious.

(2) What is implicit (and already quite deep) is that y_K is an algebraic point in $E(\mathbb{C})$, so that its canonical height is defined. This comes from the modular interpretation of $X_0(N)$: points $\tau \in \Gamma_0(N)\backslash\mathbb{H}$ are in one-to-one correspondence with pairs (E, H), where E is an elliptic curve over $\mathbb{C}$ and $H \subset E(\mathbb{C})$ is a cyclic subgroup of order N. This bijection is induced by the map $\tau \mapsto (\mathbb{C}/(\mathbb{Z} \oplus \tau\mathbb{Z}), \langle N^{-1}\rangle)$, and the pair corresponding to $z_\mathfrak{a}$ is $(\mathbb{C}/\mathfrak{a}, \mathfrak{q}^{-1}\mathfrak{a}/\mathfrak{a})$, where $N_{K/\mathbb{Q}}\mathfrak{q} = N$. This curve has complex multiplication, hence finitely many Galois conjugates, which implies that $z_\mathfrak{a}$ is defined over $\bar{\mathbb{Q}}$.

More precisely, one shows that $y_K \in E(K)$. The $z_{\mathfrak{a}}$, where $\mathfrak{a}$ runs over the ideal classes in K, are in fact a single Galois orbit under the Galois group of K.

SKETCH OF PROOF OF THEOREM 23.3. The principle is that if $E/\mathbb{Q}$ is an elliptic curve, with associated cusp form f, such that the root number of E is $+1$, and if $D < 0$ is a fundamental discriminant of an imaginary quadratic field $K/\mathbb{Q}$ such that the twisted curve $E_D/\mathbb{Q}$ has root number -1, then one has $L(f \otimes \chi, \frac{1}{2}) = 0$. If in addition the Heegner point $y_K \in E(\mathbb{C})$ turns out to have $h(y_K) = 0$, then

$$L'_K(f, \tfrac{1}{2}) = L(f, \tfrac{1}{2}) L'(f \otimes \chi, \tfrac{1}{2}) = 0.$$

If $L(f, \frac{1}{2}) \neq 0$ (which can be rigorously checked numerically), then it follows that $L'(f \otimes \chi, \frac{1}{2}) = 0$.

We now apply this following [**GZ2**] to the curve

$$E : y^2 = x^3 + 10x^2 - 20x + 8$$

with the imaginary quadratic field $K = \mathbb{Q}(\sqrt{-139})$, so that E_D is indeed the curve (23.53). This curves satisfies the assumptions of Theorem 23.4 since we have $\chi_{-139}(-37) = -1$. We choose the ideal $\mathfrak{n} = (1 + \bar{\omega})$ (where $\omega = (1 + \sqrt{-139})/2$).

The discriminant of E is $\Delta = 2^{12} \cdot 37$ and the j-invariant is $j = \frac{2^{15} \cdot 5^3}{37} = \frac{4096000}{37}$. Thus the given Weierstrass equation is not minimal. A corresponding minimal model is the curve

$$E' : y^2 + y = x^3 + x^2 - 3x + 1,$$

the coordinate change bringing E' to E being $(x, y) \mapsto (4x - 2, 8y + 4)$. The curve E' has discriminant 37, and since it is squarefree, the conductor of E is 37. We denote by f the weight 2 primitive cusp form of level 37 corresponding to E. Write $E(\mathbb{C}) = \mathbb{C}/(\mathbb{Z} \oplus \tau\mathbb{Z})$ for some $\tau \in \mathbb{H}$, as given by the theory of elliptic functions.

Since the conductor $N = 37$ is squarefree, the sign of the functional equation is given by $\varepsilon_E = a_{E'}(37)$ which gives $\varepsilon_E = 1$, and numerical checking gives

$$L(f, \tfrac{1}{2}) = L(E, 1) \approx 0.72568106193615278\ldots \neq 0$$

as required. In fact, one can check that $E(\mathbb{Q}) \simeq \mathbb{Z}/3\mathbb{Z}$ is generated by $(x, y) = (2, 4)$.

The ideal class group of K is of order three, generated by $\mathfrak{a} = [5, \frac{1+\sqrt{-139}}{2}]$. The three Heegner points for the chosen $\mathfrak{n}$ are given by

$$\tilde{z}_{\mathfrak{a}} = \frac{151 + \sqrt{-139}}{370}, \quad \tilde{z}_{-\mathfrak{a}} = \frac{-71 + \sqrt{-139}}{370}, \quad \tilde{z}_{\mathcal{O}} = \frac{3 + \sqrt{-139}}{74}.$$

Denoting by π the modular parameterization $X_0(37) \to E$, we need to prove that $\pi(\tilde{z}_{\mathfrak{a}}) + \pi(\tilde{z}_{-\mathfrak{a}}) + \pi(\tilde{z}_{\mathcal{O}}) = 0 \in E(\mathbb{C})$, or equivalently, that

$$\pi(\tilde{z}_{\mathfrak{a}}) + \pi(\tilde{z}_{-\mathfrak{a}}) + \pi(\tilde{z}_{\mathcal{O}}) \in \mathbb{Z} \oplus \tau\mathbb{Z}. \tag{23.57}$$

Notice that this is a problem stated in analytic terms. To prove it, one uses the following lemma from the theory of elliptic functions:

LEMMA 23.5. *Let $\Lambda \subset \mathbb{C}$ be a lattice, and let $f \neq 0$ be a Λ-periodic meromorphic function on $\mathbb{C}$. Let $s_1, \ldots, s_d$ be a set of representatives of Λ-equivalence classes of poles and zeros of f, with multiplicity. Then we have $s_1 + \cdots + s_d \in \Lambda$.*

This is the complex-analytic analogue of part of Proposition 11.33, which we proved there for elliptic curves over finite fields. It is in fact only a special case of the Abel-Jacobi Theorem (see [**Rey**] for instance).

PROOF. Choose a fundamental parallelogram $P \subset \mathbb{C}$ for $\mathbb{C}/\Lambda$ and representatives for the s_i such that s_i is in the interior of P for all i. Then consider the integral

$$\frac{1}{2\pi i}\int_{\partial P} z\frac{f'(z)}{f(z)}dz = \sum_i s_i$$

by the residue theorem. If γ and $\gamma+\lambda_0$ are parallel sides of P, with $\lambda_0 \in \Lambda$, then (taking into account orientation) the substitution $z \mapsto z - z_0$ on the second one yields by Λ-periodicity

$$\frac{1}{2\pi i}\int_{\gamma} z\frac{f'(z)}{f(z)}dz + \frac{1}{2\pi i}\int_{\gamma+\lambda_0} z\frac{f'(z)}{f(z)}dz = \frac{\lambda_0}{2\pi i}\int_{\gamma}\frac{f'(z)}{f(z)}dz \in \lambda_0\mathbb{Z}$$

since the integral is the winding number around 0 of the closed loop $t \mapsto f(\gamma(t))$. Hence the result follows. □

We will exhibit an elliptic function on $\mathbb{C}/(\mathbb{Z}\oplus\tau\mathbb{Z})$ with a pole of order 3 at 0 and simple zeros at the three points $\pi(\tilde{z}_{\mathfrak{b}})$, thereby proving (23.57) by Lemma 23.5. Start with the meromorphic function u on $X_0(37)$ given by

$$u(z) = \frac{\eta(z)^2}{\eta(37z)^2}$$

where η is the Dedekind eta function given by

$$\eta(z) = e\Big(\frac{z}{24}\Big)\prod_{k\geqslant 1}(1 - e(kz))$$

(see (22.66)).

Letting $q = e(z)$, we have at infinity the expansion

$$u(z) = \frac{q^{\frac{1}{12}}\prod_k(1-q^k)^2}{q^{\frac{37}{12}}\prod_k(1-q^{37k})^2} = q^{-3}(1+\dots),$$

so u has a pole of order 3 at ∞. Since η doesn't vanish except at infinity, this is the only pole of u. Moreover, because η is of level 1, the substitution $z \mapsto -1/z$ shows that u has a zero of order 3 at the cusp 0, and no other zeros. Now consider $v(z) = u(z) - u(\tilde{z}_{\mathcal{O}})$. It has still a pole of order 3 at infinity, and a zero at $\tilde{z}_{\mathcal{O}}$. This is a simple zero, and v as two other simple zeros at $\tilde{z}_{\mathfrak{a}}$ and $\tilde{z}_{-\mathfrak{a}}$. Indeed, because there are as many zeros as poles, it suffices to prove $v(\tilde{z}_{\pm\mathfrak{a}}) = 0$, or $u(\tilde{z}_{\pm\mathfrak{a}}) = u(\tilde{z}_{\mathcal{O}})$, to show this.

Notice that $u(z)^{12} = \Delta(z)\Delta(37z)^{-1}$ where Δ is the Ramanujan function. For a point $z \in X_0(37)$ corresponding to a pair (E,H) of an elliptic curve and a cyclic subgroup of order $N = 37$, as in the remark above, one has $\Delta(z) = \Delta(E)$ and $\Delta(37z) = \Delta(E/H)$. For a Heegner point $\tilde{z}_{\mathfrak{b}}$, this pair is $(\mathbb{C}/\mathfrak{b}, \mathfrak{n}^{-1}\mathfrak{b}/\mathfrak{b})$. It follows that $u(\tilde{z}_{\mathcal{O}})^{12} = u(\tilde{z}_{\pm\mathfrak{a}})^{12}$. Then one can check (even numerically) that $u(\tilde{z}_{\mathcal{O}}) = u(\tilde{z}_{\pm\mathfrak{a}})$ or, more precisely, prove that $u(\tilde{z}_{\mathcal{O}}) = 1+\omega \in K$, where $37 = N(1+\omega)$, and use the fact from Complex Multiplication theory, that if σ is the generator of the Galois group of the Hilbert Class Field of K corresponding to $\mathfrak{a}$, then $u(\tilde{z}_{\pm\mathfrak{a}}) = u(\tilde{z}_{\mathcal{O}})^{\sigma_{\pm\mathfrak{a}}} = u(\tilde{z}_{\mathcal{O}})$.

To conclude, we need to push the function v to $E(\mathbb{C})$. To do this, put

$$w(z) = \prod_{\pi(z')=z} v(z')$$

for $z \in E(\mathbb{C})$. It is easy to check that this is a well-defined meromorphic function on $E(\mathbb{C})$, and it has a pole of order 3 at $\pi(\infty) = 0$ and simple zeros at $\pi(\tilde{z}_{\mathfrak{b}})$. This finishes the proof. □

REMARK. Using programs such as PARI/GP, one can nowadays find explicit equations and formulas for all objects in the above constructions. Some are stated in [**MSD**], Sec. 5. We thank C. Delaunay for computing the Heegner points. It is easier to deal with the curve

$$F \,:\, y^2 + y = x^3 + x^2 - 23x - 50$$

instead of E. By [**MSD**], we have $E \simeq F/\mu_3$, where μ_3 is the group of 3-roots of unity (one has $F[3] \simeq \mu_3 \times \mathbb{Z}/3\mathbb{Z}$ as Galois-module), so F is isogenous to E, in particular, E and F (and their twists) have identical L-functions.

The curve $X_0(37)$ is the nonsingular model of the plane curve with equations

$$X_0(37) \,:\, y^2 = -x^6 - 9x^4 - 11x + 37,$$

and the modular parameterization of F is the map

$$\pi(x,y) = (\tfrac{1}{4}(37x^{-2} - 5), \tfrac{1}{8}(37yx^{-3} - 4)).$$

The Hilbert Class Field H of $K = \mathbb{Q}(\sqrt{-139})$ is the splitting field over K of $P = X^3 + 4X^2 + 6X + 1$. Denote by $\theta_{-\mathfrak{a}}$ the real root of P, by $\theta_{\mathcal{O}}$ the root in $\mathbb{H}$ and $\theta_{\mathfrak{a}} = \bar{\theta}_{\mathcal{O}}$. Then $\pi(\tilde{z}_{\mathfrak{b}}) \in F(H)$ is given by

$$\pi(\tilde{z}_{\mathfrak{b}}) = (A_1(\theta_{\mathfrak{b}}) + A_2(\theta_{\mathfrak{b}})\sqrt{-139}, B_1(\theta_{\mathfrak{b}}) + B_2(\theta_{\mathfrak{b}})\sqrt{-139})$$

with

$$\begin{aligned} A_1 &= -\frac{1}{41^2}(1152X^2 + 6398X + \tfrac{13469}{2}) \\ A_2 &= \frac{1}{139 \cdot 41^2}(32579X^2 + 53474X + \tfrac{13881}{2}) \\ B_1 &= \frac{1}{41^3}(-39658X^2 + 225512X + 84810) \\ B_2 &= -\frac{1}{139 \cdot 41^3}(2662280X^2 + 6561700X + 1270577). \end{aligned}$$

One can then check "by hand" that $\pi(\tilde{z}_{\mathfrak{a}}) + \pi(\tilde{z}_{-\mathfrak{a}}) + \pi(\tilde{z}_{\mathcal{O}}) = 0$ in $F(\mathbb{C})$. (This means that the three points $\pi(\tilde{z}_{\mathfrak{b}})$ are on a line in $\mathbb{P}^2$). Note, however, that $\pi(\tilde{z}_{\mathcal{O}})$ is of infinite order (in $F(H)$) because the class number of $\mathbb{Q}(\sqrt{-139})$ is larger than the degree ($= 2$) of the modular parameterization π (this is a theorem of Nakazato [**Nak**]).

CHAPTER 24

THE CRITICAL ZEROS OF THE RIEMANN ZETA FUNCTION

The zeros $\rho = \frac{1}{2} + i\gamma$ of $\zeta(s)$ on the critical line are called the critical zeros. Let $N_0(T)$ denote the number of critical zeros with $0 < \gamma \leqslant T$. Recall that $N(T)$ denotes the number of all zeros $\rho = \beta + i\gamma$ with $0 < \beta < 1$ and $0 < \gamma \leqslant T$, and that we have (see Theorem 5.24)

$$N(T) = \frac{T}{2\pi} \log \frac{T}{2\pi e} + O(\log T). \tag{24.1}$$

The Riemann Hypothesis asserts that $N_0(T) = N(T)$. By the density theorem (Theorem 10.1) we know that almost all zeros are near the critical line. Riemann showed by hand computations that there are zeros exactly on the line Re $s = \frac{1}{2}$, the first critical zero being $\rho_1 = \frac{1}{2} + i\gamma_1$ with $\gamma_1 = 14.13\ldots$. Using modern computers it has been checked that millions of the first zeros of $\zeta(s)$ are on the line Re $s = \frac{1}{2}$ and they all are simple (see e.g. [**LRW**]).

24.1. A lower bound for $N_0(T)$.

In 1914, G. H. Hardy proved that $\zeta(s)$ has infinitely many zeros on the line Re $s = \frac{1}{2}$, and seven years later he showed jointly with J. E. Littlewood [**HL2**] that

$$N_0(T) \gg T \tag{24.2}$$

for all large T. In this section we give a proof of this result. Then in the next section we improve this bound to the following estimate of A. Selberg,

$$N_0(T) \gg T \log T, \tag{24.3}$$

which says that a positive proportion of zeros lies on the critical line.

We set

$$f(u) = \frac{g(\frac{1}{2} + iu)}{|g(\frac{1}{2} + iu)|} \zeta(\tfrac{1}{2} + iu) \tag{24.4}$$

where $g(s) = \pi^{-s/2}\Gamma(s/2)$. By the functional equation $g(s)\zeta(s) = g(1-s)\zeta(1-s)$ it follows that $f(u)$ is real and even. We consider the two integrals

$$I(t) = \int_t^{t+\Delta} f(u)du, \tag{24.5}$$

$$J(t) = \int_t^{t+\Delta} |f(u)|du \tag{24.6}$$

where Δ is a fixed positive number. Clearly $|I(t)| \leqslant J(t)$. If

$$|I(t)| < J(t), \tag{24.7}$$

then $f(u)$ must change sign in the interval $(t, t+\Delta)$, hence $f(u)$ must have a zero in $(t, t+\Delta)$, so has $\zeta(\frac{1}{2}+iu)$ because $g(s)$ does not vanish anywhere. Therefore our problem reduces to showing that (24.7) occurs quite often. To this end we are going to prove a lower bound for $I(t)$ and an upper bound for $J(t)$ on average over the segment $[T, 2T]$.

LEMMA 24.1. *Let $\Delta \geqslant 1$. There exists a function $K(t)$, which also depends on Δ, such that*

$$J(t) \geqslant \Delta - K(t), \tag{24.8}$$

$$\int_T^{2T} |K(t)|^2 dt \ll T, \tag{24.9}$$

if $T \geqslant \Delta^2$, where the implied constant is absolute.

LEMMA 24.2. *Let $\Delta \geqslant 1$. We have*

$$\int_T^{2T} |I(t)|^2 dt \ll \Delta T \tag{24.10}$$

if $T \geqslant \Delta^6$, where the implied constant is absolute.

By Lemmas 24.1 and 24.2 the estimate of Hardy-Littlewood can be deduced as follows. Let $\mathcal{T}$ be the subset of $[T, 2T]$ where $|I(t)| = J(t)$, thus

$$\int_{\mathcal{T}} |I(t)| dt = \int_{\mathcal{T}} J(t) dt.$$

We get by Cauchy's inequality and by Lemma 24.2

$$\int_{\mathcal{T}} |I(t)| dt \leqslant \int_T^{2T} |I(t)| dt \ll \Delta^{\frac{1}{2}} T.$$

On the other hand, we get by Lemma 24.1 and Cauchy's inequality

$$\int_{\mathcal{T}} J(t) dt \geqslant \Delta|\mathcal{T}| - \int_{\mathcal{T}} K(t) dt = \Delta|\mathcal{T}| + O(|\mathcal{T}|^{\frac{1}{2}} T^{\frac{1}{2}}).$$

Combining the above bounds we deduce that the measure of $\mathcal{T}$ satisfies $|\mathcal{T}| \ll \Delta^{-\frac{1}{2}} T$ where the implied constant is absolute. For Δ sufficiently large this gives $|\mathcal{T}| \leqslant \frac{1}{2} T$. In other words, the set $\mathcal{S} = [T, 2T] \backslash \mathcal{T}$ on which (24.7) holds has measure $|\mathcal{S}| \geqslant \frac{1}{2} T$. Clearly $\mathcal{S}$ contains a sequence $\{t_1, \dots, t_R\}$ of Δ-spaced points of length $R \geqslant \Delta^{-1} |\mathcal{S}| \geqslant T/2\Delta$. For each $t_r, 1 \leqslant r \leqslant R$, there is a sign change of $f(u)$ in $t_r < u < t_r + \Delta$, hence there is a critical zero $\rho_r = \frac{1}{2} + i\gamma_r$ with $t_r < \gamma_r < t_r + \Delta$. This proves that $N_0(T) \gg T/\Delta$ which is (24.2).

It remains to prove the two lemmas.

PROOF OF LEMMA 24.1. We have

$$\begin{aligned} J(t) = \int_0^\Delta |\zeta(\tfrac{1}{2} + it + iu)| du &\geqslant \Big| \int_0^\Delta \zeta(\tfrac{1}{2} + it + iu) du \Big| \\ &\geqslant \Delta - \Big| \int_0^\Delta (\zeta(\tfrac{1}{2} + it + iu) - 1) du \Big| = \Delta - K(t), \end{aligned}$$

say. To estimate $K(t)$ we use the approximation

$$\zeta(s) = \sum_{1\leqslant n\leqslant T} n^{-s} + O(T^{-\frac{1}{2}}) \tag{24.11}$$

which is valid for $s = \frac{1}{2} + it$ with $T < t < 3T$ (see (8.3)). Hence

$$K(t) = \Big| \sum_{1<n\leqslant T} n^{-\frac{1}{2}-it} \frac{1-n^{-i\Delta}}{\log n} \Big| + O(\Delta T^{-\frac{1}{2}}),$$

and by Theorem 9.1 we conclude that

$$\int_T^{2T} |K(t)|^2 dt \ll T \sum_{1<n\leqslant T} n^{-1}(\log n)^{-2} + \Delta^2 \ll T.$$

□

PROOF OF LEMMA 24.2. Using the convexity bound $\zeta(s) \ll |s|^{\frac{1}{4}}$ we arrange the integral of (24.10) as follows

$$\begin{aligned} I &= \int_T^{2T} |I(t)|^2 dt = \int_T^{2T} \Big| \int_0^\Delta f(t+u) du \Big|^2 dt \\ &= \int_0^\Delta \int_0^\Delta \int_T^{2T} f(t+u_1)\bar{f}(t+u_2) dt\, du_1 du_2 \\ &= \int_0^\Delta \int_0^\Delta \int_T^{2T} f(t)\bar{f}(t+u_2-u_1) dt\, du_1 du_2 + O(\Delta^3 T^{\frac{1}{2}}) \\ &= \int_{-\Delta}^\Delta (\Delta - |u|) \int_T^{2T} f(t)\bar{f}(t+u) dt\, du + O(\Delta^3 T^{\frac{1}{2}}). \end{aligned}$$

Now we approximate $f(t)\bar{f}(t+u)$ by a Dirichlet polynomial. First by Stirling's formula

$$\Gamma(\sigma+it) = (2\pi)^{\frac{1}{2}} (it)^{\sigma-\frac{1}{2}} (t/e)^{it} e^{-\frac{\pi}{2}t} (1 + O(t^{-1}))$$

for $\sigma > 0, t > 0$ we find that

$$\frac{g(\frac{1}{2}+it)\bar{g}(\frac{1}{2}+it+iu)}{|g(\frac{1}{2}+it)\bar{g}(\frac{1}{2}+it+iu)|} = \Big(\frac{2\pi}{t}\Big)^{\frac{iu}{2}} \Big(1 + O\Big(\frac{u^2+1}{t}\Big)\Big).$$

Then by (24.11) and the convexity bound $\zeta(s) \ll |s|^{\frac{1}{4}}$ we get

$$f(t)\bar{f}(t+u) = \sum\sum_{1\leqslant m,n\leqslant T} (mn)^{-\frac{1}{2}} (\frac{m}{n})^{it} \Big(\frac{2\pi m^2}{t}\Big)^{\frac{iu}{2}} + O(\Delta^2 T^{-\frac{1}{2}}).$$

Hence

$$I = \sum\sum_{1\leqslant m,n\leqslant T} c(m,n)(mn)^{-\frac{1}{2}} + O(\Delta^4 T^{\frac{1}{2}})$$

where

$$\begin{aligned} c(m,n) &= \int_{-\Delta}^\Delta (\Delta - |u|) \int_T^{2T} \Big(\frac{m}{n}\Big)^{it} \Big(\frac{2\pi m^2}{t}\Big)^{\frac{iu}{2}} dt du \\ &= \Delta^2 \int_T^{2T} \Big(\frac{m}{n}\Big)^{it} \chi\Big(\frac{\Delta}{4} \log \frac{2\pi m^2}{t}\Big) dt \end{aligned}$$

with $\chi(x) = x^{-2}(\sin x)^2$. If $m \neq n$, we integrate by parts getting

$$\begin{aligned} c(m,n) \log \frac{m}{n} &\ll \left(\frac{\sin \frac{1}{4}\Delta \log(\pi m^2/T)}{\log(\pi m^2/T)}\right)^2 + \left(\frac{\sin \frac{1}{4}\Delta \log(2\pi m^2/T)}{\log(2\pi m^2/T)}\right)^2 \\ &+ \int_T^{2T} \left| d\left(\frac{\sin \frac{1}{4}\Delta \log(2\pi m^2/t)}{\log(2\pi m^2/t)}\right)^2 \right| \\ &\ll \min\left\{\Delta^2, \left|\log \frac{\pi m^2}{T}\right|^{-2} + \left|\log \frac{2\pi m^2}{T}\right|^{-2}\right\}. \end{aligned}$$

If $m = n$, it is easier to integrate over t first and then over u getting

$$\begin{aligned} c(m,m) &= 2T \int_{-\Delta}^{\Delta} (\Delta - |u|) \frac{2 - 2^{\frac{iu}{2}}}{2 - iu} \left(\frac{\pi m^2}{T}\right)^{\frac{iu}{2}} du \\ &\ll \Delta T \min\left\{1, \left|\log \frac{\pi m^2}{T}\right|^{-1} + \left|\log \frac{2\pi m^2}{T}\right|^{-1}\right\}. \end{aligned}$$

Using these estimates one derives that

$$\sum\sum_{1 \leqslant m,n \leqslant T} c(m,n)(mn)^{-\frac{1}{2}} \ll \Delta T$$

which completes the proof of (24.10). □

24.2. A positive proportion of critical zeros.

In 1942, A. Selberg modified the arguments of Hardy and Littlewood showing that a positive proportion of zeros of $\zeta(s)$ lies on the line Re $s = \frac{1}{2}$. He did not give the percentage number. A very strong result is due to B. Conrey [**Co2**], namely that

$$N_0(T) > \tfrac{2}{5} N(T) \tag{24.12}$$

for sufficiently large T. Conrey's approach is based on Levinson's method (see [**Lev**]) which is quite different from the one used by Selberg.

Selberg's refinement of the Hardy-Littlewood arguments appears in mollification of $\zeta(s)$ by a Dirichlet polynomial, its role being to diminish the contribution of large values of $\zeta(s)$. This idea was first exercised by Bohr and Landau [**BL**], however, less effectively than in the hands of A. Selberg [**S8**]. Selberg considers

$$f(u) = \frac{g(\frac{1}{2} + iu)}{|g(\frac{1}{2} + iu)|} \zeta(\tfrac{1}{2} + iu) |\varphi(\tfrac{1}{2} + iu)|^2 \tag{24.13}$$

where $\varphi(s)$ is a Dirichlet polynomial of type

$$\varphi(s) = \sum_{d \leqslant D} h\left(\frac{\log d}{\log D}\right) \gamma_d d^{-s} \tag{24.14}$$

and $h(x)$ is a real, continuous function on $[0, 1]$ such that

$$h(x) = 1 + O(x) \tag{24.15}$$

$$h(x) \ll 1 - x \tag{24.16}$$

while the coefficients γ_d have arithmetical nature (they are also real and bounded). Clearly $f(u)$ is real and even. The reason that $|\varphi(\frac{1}{2}+iu)|^2$ is applied rather than $\varphi(\frac{1}{2}+iu)$ is to ensure that any sign change of $f(u)$ at $u=\gamma$ comes from the sign change of $\zeta(\frac{1}{2}+iu)$ because $|\varphi(\frac{1}{2}+iu)|^2 \geqslant 0$. The mollifier $|\varphi(\frac{1}{2}+iu)|^2$ gives some extra zeros to $f(u)$, but they are not counted by the sign changes!

A reasonable guess of the coefficients γ_d comes by trying $\varphi(s)$ to approximate $\zeta(s)^{-\frac{1}{2}}$. Indeed Selberg chooses γ_d to be exactly the coefficients in the Dirichlet series representation of the Euler product

$$\zeta(s)^{-\frac{1}{2}} = \prod_p \Big(1-\frac{1}{p^s}\Big)^{\frac{1}{2}} = \sum_d \gamma_d d^{-s}. \tag{24.17}$$

We have for $s>1$,

$$\Big(1-\frac{1}{p^s}\Big)^{\frac{1}{2}} \leqslant \Big(1-\frac{1}{p^s}\Big)^{-\frac{1}{2}} \leqslant \Big(1-\frac{1}{p^s}\Big)^{-1},$$

whence $|\gamma_d| \leqslant |\tilde{\gamma}_d| \leqslant 1$, where $\tilde{\gamma}_d$ are the coefficients of

$$\zeta(s)^{\frac{1}{2}} = \prod_p \Big(1-\frac{1}{p^s}\Big)^{-\frac{1}{2}} = \sum_d \tilde{\gamma}_d d^{-s}. \tag{24.18}$$

As in the Hardy-Littlewood proof we shall compare the integrals

$$I(t) = \int_t^{t+\Delta} f(u)du, \tag{24.19}$$

$$J(t) = \int_t^{t+\Delta} |f(u)|du \tag{24.20}$$

in the range $0 \leqslant t \leqslant T$. Due to the mollification, we shall be able to work with smaller Δ,

$$\Delta \asymp (\log T)^{-1} \tag{24.21}$$

where the implied constant is absolute. We shall prove three estimates

LEMMA 24.3. *Let* $(\log T)^{-1} \leqslant \Delta \leqslant 1$ *and* $D = T^\theta$ *with* $0<\theta\leqslant\frac{1}{80}$. *Then we have*

$$\int_0^T |f(t)|dt \gg T, \tag{24.22}$$

$$\int_0^T |f(t)|^2 dt \ll T, \tag{24.23}$$

$$\int_0^T |I(t)|^2 dt \ll \Delta T(\log T)^{-1}, \tag{24.24}$$

where the implied constant depends only on θ.

From these estimates we derive Selberg's bound (24.3) as follows. Let $\mathcal{E}$ be the subset of $[0,T]$ in which $|I(t)| < J(t)$. We obtain

$$A = \int_{\mathcal{E}} J(t)dt \geqslant \int_{\mathcal{E}} (J(t)-|I(t)|)dt = \int_0^T (J(t)-|I(t)|)dt = B - C$$

say. By the Cauchy-Schwarz inequality and by (24.23) we derive

$$A^2 \leqslant |\mathcal{E}| \int_0^T J^2(t)dt = |\mathcal{E}| \int_0^T \Big(\int_0^\Delta |f(t+v)|dv \Big)^2 dt$$
$$\leqslant \Delta |\mathcal{E}| \int_0^T \int_0^\Delta |f(t+v)|^2 dv dt \leqslant 2\Delta^2 |\mathcal{E}| \int_0^{T+\Delta} |f(t)|^2 dt \ll \Delta^2 |\mathcal{E}| T.$$

On the other hand, we have by (24.22)

$$B = \int_0^T \int_t^{t+\Delta} |f(u)| du dt \geqslant \Delta \int_\Delta^T |f(u)| du \gg \Delta T,$$

and by (24.24)

$$C^2 \leqslant T \int_0^T |I(t)|^2 dt \ll \Delta T^2 (\log T)^{-1}.$$

Combining the above estimates we obtain $\Delta T \ll \Delta^{\frac{1}{2}} T (\log T)^{-\frac{1}{2}} + \Delta |\mathcal{E}|^{\frac{1}{2}} T^{\frac{1}{2}}$, whence $|\mathcal{E}| \gg T$, provided $\Delta \log T$ is sufficiently large. The set $\mathcal{E}$ contains a sequence of points $\{t_1, \dots, t_R\}$ which are Δ-spaced of length $R \geqslant \Delta^{-1} |\mathcal{E}| \gg T \log T$. For each t_r there is a sign change of $f(u)$ in $(t_r, t_r + \Delta)$, so there is a zero of $\zeta(\frac{1}{2} + iu)$ in this interval because $|\varphi(\frac{1}{2} + iu)|^2$ is non-negative. Hence we conclude (24.3).

It remains to prove the estimates for the three integrals. The first one (24.22) is easy. We have

$$\int_0^T |f(t)| dt \geqslant \Big| \int_{T/2}^T \zeta(\tfrac{1}{2} + it) \varphi^2(\tfrac{1}{2} + it) dt \Big|.$$

By the approximation (24.11) and the trivial bound $\varphi(s) \ll D^{\frac{1}{2}}$ for $\mathrm{Re}\,(s) = \frac{1}{2}$ we get

$$\zeta(s)\varphi^2(s) = \sum_{n \leqslant N} a_n n^{-s} + O(DT^{-\frac{1}{2}})$$

where $a_1 = 1$ and $|a_n| \leqslant \tau_3(n)$ for $n \leqslant N = D^2 T$. Assuming $D \leqslant T^{\frac{1}{2}} (\log T)^{-1}$ we obtain

$$\int_{T/2}^T \zeta(s)\varphi^2(s) dt \geqslant \frac{T}{2} - 2 \sum_{2 \leqslant n \leqslant N} |a_n| n^{-\frac{1}{2}} (\log n)^{-1} + O\Big(\frac{T}{\log T}\Big) \geqslant \frac{T}{3}$$

if T is sufficiently large, which proves (24.22).

The other two estimates (24.23) and (24.24) are much harder to establish because we have to appeal to the nature of the mollifier $\varphi^2(s)$. We split the arguments into several groups and evaluate a few integrals of the zeta function which are of independent interest.

THEOREM 24.4. *Let $s_1 = \frac{1}{2} + iv_1, s_2 = \frac{1}{2} + iv_2$ with $|v_1| \leqslant 1, |v_2| \leqslant 1$. Let a, b be positive integers with $(a, b) = 1$. We have*

$$\int_0^T \zeta(s_1 + it) \bar{\zeta}(s_2 + it) a^{s_1 + it} b^{\bar{s}_2 - it} dt = T P_v \Big(\frac{T}{2\pi ab} \Big) + O((ab)^{\frac{3}{2}} T^{\frac{8}{9}} (\log T)^6), \tag{24.25}$$

where $v = v_2 - v_1$ and $P_v(X)$ is defined by

$$P_v(X) = \zeta(1 - iv) + \zeta(1 + iv) \frac{X^{iv}}{1 + iv}. \tag{24.26}$$

The implied constant in (24.25) is absolute.

Notice that by the Laurent expansion of $\zeta(1+iv)$ we get

$$P_v(X) = \frac{X^{iv}-1}{iv} + 2\gamma - 1 + O(|v|) \tag{24.27}$$

which is quite good if $|v|$ is small. In particular, we have

$$P_0(X) = \log X + 2\gamma - 1. \tag{24.28}$$

COROLLARY 24.5. *If $(a,b)=1$, we have*

$$\int_0^T |\zeta(\tfrac{1}{2}+it)|^2 \Big(\frac{a}{b}\Big)^{it} dt = \frac{T}{\sqrt{ab}}\Big(\log\frac{T}{2\pi ab} + 2\gamma - 1\Big) + O(abT^{\frac{8}{9}}(\log T)^6) \tag{24.29}$$

where the implied constant is absolute.

We begin the proof of Theorem 24.4 by evaluating the following partial integrals

$$S = \int_{T_1}^{T_2} \zeta(s_1+it)\bar{\zeta}(s_2+it)a^{-s_1-it}b^{-\bar{s}_2+it}dt \tag{24.30}$$

for $T \leqslant T_1 \leqslant T_2 \leqslant 2T$. Changing the variable $t \to t - v_1$ we get

$$S = (ab)^{-\frac{1}{2}}b^{iv}\int_{T_1}^{T_2} \zeta(\tfrac{1}{2}+it)\bar{\zeta}(\tfrac{1}{2}+iv+it)\Big(\frac{b}{a}\Big)^{it}dt + E,$$

where E is the non-overlapping part of the integral, and it satisfies

$$E \ll (ab)^{-\frac{1}{2}}T^{\frac{1}{2}} \tag{24.31}$$

by applying the convexity bound $\zeta(s) \ll |s|^{\frac{1}{4}}$ on the line Re $s = \frac{1}{2}$.

Next we approximate $\zeta(s)$ by its partial sum

$$\zeta_X(s) = \sum_{n\leqslant X} n^{-s}. \tag{24.32}$$

Recall that for Re $s = \sigma \geqslant \frac{1}{2}$ and $|s| \leqslant \pi X$ we have

$$\zeta(s) = \zeta_X(s) - \frac{X^{1-s}}{1-s} + O(X^{-\sigma}). \tag{24.33}$$

In particular, for s with Re $s = \frac{1}{2}$ and $T \ll |s| \leqslant \pi T$ we have

$$\zeta(s) = \zeta_T(s) + O(T^{-\frac{1}{2}}). \tag{24.34}$$

Hence

$$S = (ab)^{-\frac{1}{2}}b^{iv}\int_{T_1}^{T_2} \zeta_T(\tfrac{1}{2}+it)\zeta_T(\tfrac{1}{2}-iv-it)\Big(\frac{b}{a}\Big)^{it}dt + E + E_\infty$$

where E_∞ accounts the contribution of the error term in (24.34), thus it satisfies

$$E_\infty \ll (ab)^{-\frac{1}{2}}(T\log T)^{\frac{1}{2}} \tag{24.35}$$

by virtue of the following estimate (obtained from (7.52) by Cauchy's inequality)

$$\int_0^{3T} |\zeta(\tfrac{1}{2}+it)|dt \ll T(\log T)^{\frac{1}{2}}.$$

Inserting the Dirichlet polynomials (24.32) we get

$$S = \int_{T_1}^{T_2} \sum_{m,n \leqslant T}\sum (am)^{-\frac{1}{2}-it}(bn)^{-\frac{1}{2}+iv+it}\,dt + E + E_\infty. \tag{24.36}$$

First we pull out the terms on the diagonal $am = bn$. Since a, b are co-prime it follows that $m = b\ell$ and $n = a\ell$ with $\ell \leqslant L = Tc^{-1}$ where $c = \max(a,b)$. Hence these terms contribute

$$S_0 = (T_2 - T_1)(ab)^{-1+iv}\zeta_L(1 - iv).$$

Introducing the approximation (24.33) we arrive at

$$S_0 = (T_2 - T_1)(ab)^{-1+iv}\Big(\zeta(1 - iv) + \frac{L^{iv}}{iv}\Big) + O(1). \tag{24.37}$$

Now we consider the contribution to the integral (24.36) of the terms $am \neq bn$, say $\mathcal{S}^\star$. By integration we get

$$\mathcal{S}^\star = \mathcal{S}(T_2) - \mathcal{S}(T_1), \tag{24.38}$$

where

$$\mathcal{S}(t) = i\sum_{am\neq bn}\sum (am)^{-\frac{1}{2}-it}(bn)^{-\frac{1}{2}+iv+it}\Big(\log\frac{am}{bn}\Big)^{-1}. \tag{24.39}$$

We shall show that the main contribution to $\mathcal{S}(t)$ comes from the terms near the diagonal, i.e., from m, n with $1 - \varepsilon < \frac{am}{bn} < 1 + \varepsilon$ for some small ε to be chosen later. Put

$$\mathcal{S}_\varepsilon(t) = i \sum_{0<|am-bn|<\varepsilon bn}\sum (am)^{-\frac{1}{2}-it}(bn)^{-\frac{1}{2}+iv+it}\Big(\log\frac{am}{bn}\Big)^{-1}. \tag{24.40}$$

For the remaining terms we have $|\log\frac{am}{bn}| > \frac{\varepsilon}{2}$, and since $\log x$ is monotonic, we can apply partial summation to estimate the contribution of these terms by

$$\varepsilon^{-1}(ab)^{-\frac{1}{2}}(\log T)^2|\zeta_x(\tfrac{1}{2} + it)\zeta_y(\tfrac{1}{2} - iv - it)|$$

for some x, y with $1 \leqslant x, y \leqslant T$. Now by Weyl's subconvexity bound $\zeta_x(s) \ll |s|^{\frac{1}{6}}(\log 3|s|)^2$, which is valid for s with $\mathrm{Re}\ s = \frac{1}{2}$ and $x \leqslant 4|s|^2$ (see (8.22)), it follows that

$$\mathcal{S}(t) = \mathcal{S}_\varepsilon(t) + O(\varepsilon^{-1}(ab)^{-\frac{1}{2}}T^{\frac{1}{3}}(\log T)^6). \tag{24.41}$$

For the terms in $\mathcal{S}_\varepsilon(t)$ we write $am = bn + h$ with $O < |h| < \varepsilon bn$, and use the approximations

$$(abmn)^{-\frac{1}{2}}\Big(\log\frac{am}{bn}\Big)^{-1} = \frac{1}{h} + O\Big(\frac{1}{bn}\Big)$$

$$\Big(\frac{bm}{am}\Big)^{it} = \Big(1 + \frac{h}{bn}\Big)^{-it} = e^{-ith/bn} + O\Big(\frac{th^2}{b^2n^2}\Big)$$

getting

$$\mathcal{S}_\varepsilon(t) = \sum_{\substack{am-bn=h\\0<|h|<\varepsilon bn}}\sum ih^{-1}(bn)^{iv}e^{-ith/bn} + E_1$$

where E_1 accounts for the contribution of the above error terms. We have

$$\begin{aligned} E_1 &\ll (1+\varepsilon t) \sum\sum_{|am-bn|<\varepsilon bn} (bn)^{-1} \\ &\leqslant (1+\varepsilon t)\sum_n (\varepsilon nba^{-1}+1)(bn)^{-1} \ll (1+\varepsilon T)(\varepsilon a^{-1}T + b^{-1}\log T). \end{aligned}$$

Clearly we can interchange a and b (see the first inequality), so combining both estimates we derive a symmetric bound

$$E_1 \ll (ab)^{-\frac{1}{2}}(\varepsilon T)^2 \tag{24.42}$$

provided $\varepsilon T \geqslant \log T$, which condition we henceforth assume.

Now we write $\mathcal{S}_\varepsilon(t)$ as

$$\mathcal{S}_\varepsilon(t) = \sum_{\substack{H^-<h<H^+ \\ h\neq 0}} ih^{-1} \sum_{\substack{N_1<n<N_2 \\ n\equiv -h\bar{b} (\mathrm{mod}\ a)}} (bn)^{iv} e^{-ith/bn} + E_1$$

where

$$\begin{aligned} H^+ &= \min\{\varepsilon bT,\ \varepsilon(1+\varepsilon)^{-1}aT\} \\ H^- &= -\min\{\varepsilon bT,\ \varepsilon(1-\varepsilon)^{-1}aT\} \\ N_1 &= |h|/\varepsilon b, \quad N_2 = \min\{T,\ (aT-h)b^{-1}\}. \end{aligned}$$

For the inner sum we apply the following formula

$$\sum_{\substack{N_1<n<N_2 \\ n\equiv\alpha (\mathrm{mod}\ a)}} e(f(n)) = \frac{1}{a}\int_{N_1}^{N_2} e(f(x))dx + O(1)$$

which holds true for any smooth function f such that $|f'| \leqslant (2a)^{-1}$ and $f'' \neq 0$ in the interval $[N_1, N_2]$ with $N_1 < N_2$ (this follows from Proposition 8.7). Assuming

$$2\varepsilon^2 abT \leqslant 1 \tag{24.43}$$

we obtain

$$\sum_{\substack{N_1<n<N_2 \\ n\equiv -h\bar{b} (\mathrm{mod}\ a)}} (bn)^{iv} e^{-ith/bn} = \frac{1}{a}\int_{N_1}^{N_2} (bx)^{iv} e^{-ith/bx} dx + O(1).$$

Here we replace N_2 by $T\min\{1, a/b\}$ making an error $O(|h|/ab)$. Then we change the variable x into $xt/2\pi b$ getting

$$\sum_{\substack{N_1<n<N_2 \\ n\equiv -h\bar{b} (\mathrm{mod}\ a)}} (bn)^{iv} e^{-ith/bn} = \frac{1}{ab}\Big(\frac{t}{2\pi}\Big)^{1+iv} \int_{t/2\pi dT}^{\varepsilon t/2\pi|h|} e(-hx)x^{-iv-2}dx + O\Big(\frac{|h|}{ab}+1\Big)$$

where $d = \min\{a, b\}$. Inserting this into $\mathcal{S}_\varepsilon(t)$ we get

$$\mathcal{S}_\varepsilon(t) = \frac{t}{ab}\Big(\frac{t}{2\pi}\Big)^{iv} \int_{t/2\pi dT}^{\infty} x^{-2-iv}\Big(\sum_{\substack{0<|h|<\varepsilon t/2\pi x \\ H^-<h<H^+}} (2\pi ih)^{-1} e(-hx)\Big)dx + E_1 + E_2$$

where

$$E_2 \ll \sum_{0<h<2\varepsilon dT} \left(\frac{1}{ab}+\frac{1}{h}\right) \ll (ab)^{-\frac{1}{2}}\varepsilon T + \log T. \tag{24.44}$$

The inner sum over h in the above integral is a partial sum in the Fourier expansion for $-\psi(x) = [x] - x + \frac{1}{2}$. Applying the approximation

$$-\psi(x) = \sum_{\substack{-H_1<h<H_2\\ h\neq 0}} (2\pi i h)^{-1} e(hx) + O((1+\|x\|H)^{-1})$$

where $H = \min\{H_1, H_2\}$ (compare this with (4.18)) we get

$$\mathcal{S}_\varepsilon(t) = \frac{t}{ab}\left(\frac{t}{2\pi}\right)^{iv} \int_{t/2\pi dT}^{\infty} x^{-2-iv}\psi(x)dx + E_1 + E_2 + E_3$$

where

$$E_3 \ll \frac{t}{ab}\int_{t/2\pi dT}^{\infty}\left(1+\varepsilon t\frac{\|x\|}{x}\right)^{-1} x^{-2}dx = \frac{1}{\varepsilon ab}\int_{t/2\pi dT}^{\infty}\left(\|x\| + \frac{x}{\varepsilon t}\right)^{-1}\frac{dx}{x}$$

$$\ll \frac{1}{\varepsilon ab}\left(\int_{t/2\pi dT}^{1} x^{-2}dx + \sum_{1}^{\infty} k^{-1}\log\left(1+\frac{\varepsilon t}{k}\right)\right) \ll \frac{1}{\varepsilon ab}(d+\log T).$$

We simplify this bound to

$$E_3 \ll \varepsilon^{-1}(ab)^{-\frac{1}{2}}\log T. \tag{24.45}$$

Finally we use the formula

$$\int_y^{\infty} \psi(x)x^{-2-w}dx = \frac{\zeta(1+w)}{1+w} - \frac{y^{-w}}{w} + \frac{1}{2}\frac{y^{-1-w}}{1+w} \tag{24.46}$$

which holds for $0 < y < 1$ and $\mathrm{Re}\, w > -1$, getting

$$\mathcal{S}_\varepsilon(t) = \frac{t}{ab}\left(\frac{t}{2\pi}\right)^{iv}\frac{\zeta(1+iv)}{1+iv} - \frac{t}{ab}\frac{(dT)^{iv}}{iv} + \frac{\pi}{ab}\frac{(dT)^{1+iv}}{1+iv} + E_1 + E_2 + E_3.$$

Now adding up the relevant formulas we obtain

$$\mathcal{S} = (ab)^{-1+iv}\left[T_2 P_v\left(\frac{T_2}{2\pi ab}\right) - T_1 P_v\left(\frac{T_1}{2\pi ab}\right)\right] + R$$

where $P_v(X)$ is defined by (24.26) and R is the total error term. By (24.31), (24.35), (24.37), (24.42)–(24.45) we get

$$R \ll (ab)^{-\frac{1}{2}}(T^{\frac{1}{2}} + \varepsilon^{-1}T^{\frac{1}{3}} + \varepsilon^2 T^2)(\log T)^6 + (\log T)^2$$

provided $\varepsilon T \geqslant \log T$ and $2\varepsilon^2 abT \leqslant 1$. We choose $\varepsilon = T^{-\frac{5}{9}}$ getting

$$R \ll (ab)^{-\frac{1}{2}}T^{\frac{8}{9}}(\log T)^6$$

provided $2ab \leqslant T^{\frac{1}{9}}$. We also have the trivial bound

$$\mathcal{S} \ll (ab)^{-\frac{1}{2}}\int_0^T |\zeta(\tfrac{1}{2}+it)|^2 dt \ll (ab)^{-\frac{1}{2}} T\log T.$$

Combining both results we get rid of the condition $2ab \leqslant T^{\frac{1}{9}}$ claiming a weaker result

$$\mathcal{S} = (ab)^{-1+iv}\left[T_2 P_v\Big(\frac{T_2}{2\pi ab}\Big) - T_1 P_v\Big(\frac{T_1}{2\pi ab}\Big)\right] + O((ab)^{\frac{1}{2}} T^{\frac{8}{9}} (\log T)^6).$$

This yields (24.25) by adding the results for dyadic points T_j.

REMARKS. The error terms in (24.25) and (24.29) could be improved substantially. For example, if $a = b = 1$, our Corollary 24.5 yields

$$\int_0^T |\zeta(\tfrac{1}{2}+it)|^2 dt = T\Big(\log\frac{T}{2\pi} + 2\gamma - 1\Big) + E(T) \tag{24.47}$$

with $E(T) \ll T^{\frac{8}{9}} \log^6 T$, while it is known that $E(T) \ll T^{\frac{7}{22}+\varepsilon}$. It is conjectured that (24.47) holds with $E(T) \ll T^{\frac{1}{4}+\varepsilon}$ (see [**Iv**]).

Theorem 24.4 can be easily generalized to integrals of type

$$G_v(T) = \int_0^T \zeta(\tfrac{1}{2}+it)\zeta(\tfrac{1}{2}-it-iv)\Big(\frac{a}{b}\Big)^{it} g(t)dt$$

where $g(t)$ is a smooth function on $[0,T]$. For $g(t) = 1$ we get by (24.25)

$$G_v(T) = \frac{Tb^{iv}}{\sqrt{ab}}\left\{\zeta(1-iv) + \frac{\zeta(1+iv)}{1+iv}\Big(\frac{T}{2\pi ab}\Big)^{iv}\right\} + O(abT^{\frac{8}{9}}(\log T)^6)$$

if $(a,b) = 1$, $|v| \leqslant 1$ and $T \geqslant 2$, the implied constant being absolute. Hence we derive by partial integration that in general

$$G_v(T) = \frac{b^{iv}}{\sqrt{ab}}\int_0^T g(t)\Big(\zeta(1-iv) + \zeta(1+iv)\Big(\frac{t}{2\pi ab}\Big)^{iv}\Big)dt + O(abGT^{\frac{8}{9}}(\log T)^6)$$

where

$$G = |g(T)| + \int_0^T |g'(t)|\frac{tdt}{t+1}.$$

For $g(t) = (2\pi/t)^{iv/2}$ this yields

COROLLARY 24.6. *Let $(a,b) = 1$, $|v| \leqslant 1$ and $T \geqslant 2$. Then we have*

$$\begin{aligned} &\int_0^T \zeta(\tfrac{1}{2}+it)\zeta(\tfrac{1}{2}-it-iv)\Big(\frac{a}{b}\Big)^{it}\Big(\frac{2\pi}{t}\Big)^{\frac{iv}{2}} dt \\ &= \frac{2T}{\sqrt{ab}}\left\{\frac{\zeta(1+iv)}{2+iv}\Big(\frac{T}{2\pi a^2}\Big)^{\frac{iv}{2}} + \frac{\zeta(1-iv)}{2-iv}\Big(\frac{T}{2\pi b^2}\Big)^{\frac{iv}{2}}\right\} + O(abT^{\frac{8}{9}}(\log T)^7) \end{aligned} \tag{24.48}$$

where the implied constant is absolute.

Now we proceed to the proof of (24.23). Let α_ℓ be the coefficients of

$$\varphi^2(s) = \Big(\sum_{d\leqslant D}\beta_d d^{-s}\Big)^2 = \sum_{\ell\leqslant D^2}\alpha_\ell \ell^{-s}$$

where $\beta_d = \gamma_d h(\log d/\log D)$, thus $\alpha_\ell \ll \tau(\ell)$. We have by Corollary 24.5

$$\begin{aligned}\int_0^T |f(t)|^2 dt &= \int_0^T |\zeta(\tfrac{1}{2}+it)\varphi^2(\tfrac{1}{2}+it)|^2 dt\\ &= \sum_a\sum_b \alpha_a\alpha_b(ab)^{-\frac{1}{2}}\int_0^T |\zeta(\tfrac{1}{2}+it)|^2(a/b)^{it}dt\\ &= AT\Big(\log\frac{T}{2\pi}+2\gamma-1\Big) - BT + O(D^6T^{\frac{8}{9}}(\log T)^8)\end{aligned}$$

where

$$A = \sum_d \sum_{(a,b)=1}\sum \alpha_{ad}\alpha_{bd}(abd)^{-1}, \tag{24.49}$$

$$B = \sum_d \sum_{(a,b)=1}\sum \alpha_{ad}\alpha_{bd}(abd)^{-1}\log ab. \tag{24.50}$$

Therefore it suffices to prove that

$$A \ll (\log D)^{-1}, \tag{24.51}$$
$$B \ll 1. \tag{24.52}$$

First we treat A in considerable detail and then we modify the arguments to reduce the case of B to that of A.

We begin by diagonalizing the quadratic form A as follows

$$\begin{aligned}A &= \sum_d d^{-1}\sum_\delta \mu(\delta)\delta^{-2}\Big(\sum_a \alpha_{a\delta d}a^{-1}\Big)^2\\ &= \sum_d d^{-1}\Big(\sum_{\delta|d}\mu(\delta)\delta^{-1}\Big)\Big(\sum_a \alpha_{ad}a^{-1}\Big)^2 = \sum_d \varphi(d)A_d^2,\end{aligned}$$

where

$$A_d = \sum_{a\equiv 0(\text{mod } d)} \alpha_a a^{-1} = \sum_{a\equiv 0(\text{mod } d)} \beta_{d_1}\beta_{d_2}(d_1d_2)^{-1}$$

for $d \leqslant D^2$. Set $d_1 = \delta_1 k$ and $\delta_2 = \delta_2\ell$ with $\delta_1\delta_2 \mid d^\infty$, $d \mid \delta_1\delta_2$, $(k\ell, d) = 1$ getting

$$A_d = \sum_{\substack{\delta_1\delta_2|d^\infty\\ d|\delta_1\delta_2}}\sum (\delta_1\delta_2)^{-1}A_d(\delta_1)A_d(\delta_2),$$

where

$$A_d(\delta) = \sum_{(k,d)=1} \beta_{\delta k}k^{-1}.$$

From now on we take $h(x) = 1 - x$, so that

$$\beta_d = \gamma_d\frac{\log^+ D/d}{\log D}. \tag{24.53}$$

These are the original coefficients of Selberg, however one could choose $h(x)$ almost arbitrarily subject to the conditions (24.15) and (24.16). In this particular case we have a simple Mellin integral representation

(24.54) $$\log^+ x = \frac{1}{2\pi i}\int_{(\varepsilon)} x^s s^{-2} ds.$$

Now

$$A_d(\delta)\log D = \sum_{(k,d)=1} \gamma_{\delta k} k^{-1} \log^+ \frac{D}{\delta k} = \gamma_\delta \frac{1}{2\pi i}\int_{(\varepsilon)} Z(s+1)(D/\delta)^s s^{-2} ds$$

where $Z(s)$ is the corresponding zeta function

$$Z(s) = \sum_{(k,d)=1} \gamma_k k^{-2} = \prod_{p\nmid d}(1-p^{-s})^{\frac{1}{2}} = \zeta_d(s)^{\frac{1}{2}}\zeta(s)^{-\frac{1}{2}}$$

with

$$\zeta_d(s) = \prod_{p|d}(1-p^{-s})^{-1}.$$

We have $\zeta_d(s+1) \ll \zeta_d(1) = d/\varphi(d)$, $\zeta(s+1) \gg |s|^{-1}$ and $(D/\delta)^s \ll 1$ on the line $\mathrm{Re}\, s = \varepsilon$ with $\varepsilon = 1/\log D$. We also have

$$\int_{(\varepsilon)} |s|^{-3/2} ds \ll \varepsilon^{-1/2} = (\log D)^{\frac{1}{2}}.$$

Hence we derive $A_d(\delta) \ll |\gamma_\delta|(d/\varphi(d)\log D)^{\frac{1}{2}}$. This yields $A_d \ll \lambda(d) d/\varphi(d)\log D$, where

$$\lambda(d) = \sum_{\substack{\delta_1\delta_2|d^\infty \\ d|\delta_1\delta_2}}\sum |\gamma_{\delta_1}\gamma_{\delta_2}|(\delta_1\delta_2)^{-1} \leqslant \sum_{\substack{r|d^\infty \\ d|r}} r^{-1} \sum_{\delta_1\delta_2=r} \tilde{\gamma}_{\delta_1}\tilde{\gamma}_{\delta_2} = \frac{1}{\varphi(d)}$$

because $|\gamma_\delta| \leqslant \tilde{\gamma}_\delta$ and $\tilde{\gamma} \star \tilde{\gamma} = 1$. Therefore we have

(24.55) $$A_d \ll d(\varphi(d))^{-2}(\log D)^{-1}.$$

Finally inserting this to the diagonal form of A we argue as follows:

$$\begin{aligned} A(\log D)^2 &\ll \sum_{d\leqslant D^2} d^2\varphi^{-3}(d) = \sum_{d\leqslant D^2} d^{-1}\prod_{p|d}\Big(1-\frac{1}{p}\Big)^{-3} \\ &\ll \sum_{d\leqslant D^2} d^{-1}\prod_{p|d}(1+p^{-\frac{1}{2}}) \leqslant \sum_{d\leqslant D^2} d^{-1}\sum_{w|d} w^{-\frac{1}{2}} \\ &\leqslant \zeta(\tfrac{3}{2})\sum_{d\leqslant D^2} d^{-1} \ll \log D \end{aligned}$$

which completes the proof of (24.51).

For the proof of (24.52) we write $\log ab = \log abd^2 - 2\log d$ and we split B accordingly, say $B = B' - 2B''$. By the arguments which led us to the diagonalization of A we derive similar expressions for B' and B''. Thus we have

$$B' = \sum_d \varphi(d) \sum_{a,b\equiv 0 (\mathrm{mod}\ d)}\sum \alpha_a\alpha_b(ab)^{-1}\log ab = 2\sum_d \varphi(d)A_d B_d$$

where

$$B_d = \sum_{a\equiv 0 (\mathrm{mod}\ d)} \alpha_a a^{-1}\log a = \sum_q \Lambda(q) A_{[d,q]}$$

by the formula $L = \Lambda \star 1$. Applying (24.55) we derive

$$B_d \ll \Big(\sum_{q\leqslant D^2}(d,q)\Lambda(q)q^{-1}\Big) d\varphi^{-2}(d)(\log D)^{-1} \ll d\varphi^{-2}(d).$$

This bound is larger than that for A_d by factor $\log D$, hence the former arguments yield $B' \ll 1$ as required for (24.52). Moreover, we have

$$B'' = \sum_d \sum\sum_{(a,b)=1} \alpha_{ad}\alpha_{bd}(abd)^{-1}\log d = \sum_d \psi(d) A_d^2,$$

where

$$\begin{aligned}\psi(d) &= \sum_{\delta|d}\mu(\delta)\frac{d}{\delta}\log\frac{d}{\delta} = \sum_{\delta|d}\mu(\delta)\frac{d}{\delta}\sum_{\delta q|d}\Lambda(q)\\ &= \sum_{q|d}\Lambda(q)q\varphi(d/q) \leqslant 2\varphi(d)\sum_{q|d}\Lambda(q) = 2\varphi(d)\log d.\end{aligned}$$

Hence $B'' \leqslant 4A\log D \ll 1$ by (24.51). This completes the proof of (24.52).

Finally we sketch a proof of (24.24). We have

$$\begin{aligned} I &= \int_0^T |I(t)|^2 dt = \int_0^T \Big|\int_0^\Delta f(t+u)du\Big|^2 dt\\ &= \int_{-\Delta}^{\Delta}(\Delta - |u|)\int_0^T f(t)\bar{f}(t+u)dtdu + O(D^2T^{\frac12})\end{aligned}$$

and by the argument (Stirling's formula) which was applied in the proof of Lemma 24.2 we get

$$\begin{aligned} f(t)\bar{f}(t+u) &= \zeta(\tfrac12+it)\zeta(\tfrac12-it-iu)\varphi^2(\tfrac12+it)\varphi^2(\tfrac12-it-iu)\Big(\frac{2\pi}{t}\Big)^{\frac{iu}{2}} + O\Big(\frac{D^2}{T^{\frac12}}\Big)\\ &= \sum_a\sum_b \frac{\alpha_a\alpha_b}{\sqrt{ab}}\zeta(\tfrac12+it)\zeta(\tfrac12-it-iu)\Big(\frac{a}{b}\Big)^{it}\Big(\frac{2\pi a^2}{t}\Big)^{\frac{iu}{2}} + O\Big(\frac{D^2}{T^{\frac12}}\Big).\end{aligned}$$

Integrating over t by Corollary 24.6 we get

$$\begin{aligned}\int_0^T f(t)\bar{f}(t+u)dt &= 2T\sum_d\sum\sum_{(a,b)=1}\alpha_{ad}\alpha_{bd}(abd)^{-1}\\ &\Big\{\frac{\zeta(1+iu)}{2+iu}\Big(d\sqrt{\frac{T}{2\pi}}\Big)^{iu} + \frac{\zeta(1-iu)}{2-iu}\Big(abd\sqrt{\frac{2\pi}{T}}\Big)^{iu}\Big\} + O(D^4T^{\frac89}(\log T)^7).\end{aligned}$$

Now integrating over u we get

$$\text{(24.56)}\quad I = 2T\sum_d\sum\sum_{(a,b)=1}\alpha_{ad}\alpha_{bd}(abd)^{-1}\Big\{\Phi\Big(d\sqrt{\frac{T}{2\pi}}\Big) + \Phi\Big(\frac{\sqrt{T/2\pi}}{abd}\Big)\Big\} + O(D^4T^{\frac89}(\log T)^7)$$

where

$$\begin{aligned}\Phi(X) &= \int_{-\Delta}^{\Delta}(\Delta-|u|)\frac{\zeta(1+iu)}{2+iu}X^{iu}du\\ &= \int_{0}^{\Delta}(\Delta-u)(X^{iu}-X^{-iu})\frac{du}{2iu}+\int_{-\Delta}^{\Delta}(\Delta-|u|)X^{iu}\Big(\frac{\zeta(1+u)}{2+iu}-\frac{1}{2iu}\Big)du\\ &= \Delta\int_{0}^{\Delta\log X}\Big(\frac{\sin u}{u}\Big)^2du+O\Big(\frac{\Delta}{\log X}\Big).\end{aligned}$$

Putting

$$\omega(z)=\int_{z}^{\infty}\Big(\frac{\sin u}{u}\Big)^2du \tag{24.57}$$

we write $\Phi(X)=\frac{\pi}{2}\Delta-\Delta\omega(\Delta\log X)+O(\Delta/\log X)$. Inserting this for $X=(T/2\pi)^{\frac12}(abd)^{-1}$ into (24.56) we arrive at

$$I=-2\Delta TA'+O(\Delta TA''+\Delta TA'''(\log X)^{-1}+D^4T^{\frac89}(\log T)^7)$$

where

$$\begin{aligned}A' &= \sum_d\sum\sum_{(a,b)=1}\alpha_{ad}\alpha_{bd}(abd)^{-1}\omega\Big(\Delta\log\frac{\sqrt{T/2\pi}}{abd}\Big),\\ A'' &= \sum_d\Big|\sum\sum_{(a,b)=1}\alpha_{ad}\alpha_{bd}(abd)^{-1}\Big|,\\ A''' &= \sum_d\sum\sum_{(a,b)=1}|\alpha_{ad}\alpha_{bd}|(abd)^{-1}.\end{aligned}$$

The sum A'' can be estimated in the same way as A,

$$A''\leqslant\sum_d d\prod_{p|d}\Big(1+\frac1p\Big)A_d^2\ll(\log D)^{-2}\sum_d d^{-1}\prod_{p|d}\Big(1+\frac1p\Big)^5\ll(\log D)^{-1}.$$

The sum A''' looks like A except that its terms are taken with absolute value. Consequently, in the analytic argument applied for A''' the Riemann zeta function occurs as $\zeta(s)^{1/2}$ rather than $\zeta(s)^{-1/2}$ resulting in a loss of factor $\log D$ in the bound for A''' by comparison to A, i.e., one gets $A'''\ll 1$.

For the sum A' one derives the same bound as for A, namely $A'\ll(\log D)^{-1}$, since the function $\omega(\Delta\log X)$ is bounded and it has nice derivatives. For example, one can use the Mellin transform of $\omega(\Delta\log X)$ in X which brings a slight shift in the argument of the involved Dirichlet series, and this makes no essential change in the rest of the proof. Another approach would be to reduce A' to A by arithmetic arguments as we reduced B' to B. To this end write the power series

$$\omega(z)=\sum_{k=0}^{\infty}\frac{(-1)^{k+1}}{(2k+2)!}\,\frac{(2z)^{2k+1}}{2k+1}.$$

For $z = \Delta \log X = \frac{1}{2}\Delta \log(T/2\pi) - \Delta \log(abd)$ this reduces the problem to estimating sums of type

$$A^{(k)} = \sum_d \sum_{(a,b)=1} \sum \alpha_{ad}\alpha_{bd}(abd)^{-1}(\log abd)^k.$$

Writing

$$(\log abd)^k = \sum_{q|abd} \Lambda_k(q)$$

where Λ_k is the von Mangoldt function of degree k we arrange $A^{(k)}$ in terms of A_d and use the bound (24.55).

Now collecting the above estimates for A', A'', A''' we get

$$I \ll \Delta T(\log D)^{-1} + D^4 T^{\frac{8}{9}}(\log T)^7.$$

Taking $D = T^{\frac{1}{40}}$ we complete the proof of (24.24).

REMARKS. Selberg's mollification method works well for counting zeros on and near the critical line. He showed that

$$N(\tfrac{1}{2} + 4\delta, T) \ll T^{1-\delta} \log T \tag{24.58}$$

uniformly for $\delta \geqslant 0$. Hence it follows that almost all zeros of $\zeta(s)$ lie in the region

$$|\sigma - \tfrac{1}{2}| \leqslant \frac{\eta(t)}{\log(|t| + 3)} \tag{24.59}$$

where $\eta(t)$ is any positive function which increases to infinity.

CHAPTER 25

THE SPACING OF ZEROS OF THE RIEMANN ZETA-FUNCTION

25.1. Introduction.

Throughout this chapter we assume the validity of the Riemann Hypothesis for $\zeta(s)$. This allows us to put the critical zeros $\rho = \frac{1}{2} + i\gamma$ in an increasing sequence according to their ordinates

$$\ldots \leqslant \gamma_{-2} \leqslant \gamma_{-1} < 0 < \gamma_1 \leqslant \gamma_2 \leqslant \ldots \tag{25.1}$$

where $\gamma_{-n} = -\gamma_n$. Recall that the counting function $N(T) = |\{n \mid 0 < \gamma_n \leqslant T\}|$ satisfies

$$N(T) = \frac{T}{2\pi} \log \frac{T}{2\pi e} + O(\log T) \tag{25.2}$$

for $T \geqslant 2$ (see Theorem 5.24). Hence it follows that

$$\gamma_n \sim 2\pi n (\log n)^{-1}, \quad \text{as } n \to \infty, \tag{25.3}$$

so the numbers

$$\zeta_n = \frac{1}{2\pi} \gamma_n \log |\gamma_n| \tag{25.4}$$

have unit mean spacing, i.e., $\zeta_n \sim n$ as $|n| \to \infty$. Actually (25.2) is much more precise than (25.4), still it is not satisfactory for understanding the subtle aspects of the distribution of γ_n's, in particular, it tells very little about gaps between consecutive zeros. The Riemann Hypothesis alone does not shed enough light on the problem (it does improve the error term in (25.2) only by factor $\log\log T$) .

Our first goal would be to understand the statistic of the sequence of numbers ζ_n. One expects that the consecutive spacings

$$\delta_n = \zeta_{n+1} - \zeta_n \tag{25.5}$$

are not purely random (i.e. are not poissonian), as expected for the normalized distances between primes, or for the normalized distances between eigenvalues of the Laplace operator for the modular group (see the final remarks of Chapter 10 and [**Sa4**] respectively), but rather they follow the Gaussian Unitary Ensemble distribution from random matrix theory. We say that a sequence $0 < \zeta_1 \leqslant \zeta_2 \leqslant \ldots$ follows GUE if

$$\frac{1}{N} \sum_{1 \leqslant n \leqslant N} f(\delta_n) \sim \int_0^\infty f(s) P(s) ds \tag{25.6}$$

for any nice function $f : \mathbb{R}^+ \to \mathbb{C}$ (say of Schwarz class), where $P(s)$ is the limit density distribution of consecutive spacing of eigenvalues of random unitary matrices (the limit with respect to the rank). Explicitly this distribution was determined by Gaudin and Mehta [**GM**], and is given by $P(s) = \det(I - Q_s)$, where Q_s is the integral operator on $L^2([-1,1])$ whose kernel is

$$Q_s(x,y) = \frac{\sin \frac{\pi s}{2}(x-y)}{\pi(x-y)}. \tag{25.7}$$

Specifically the density can be expressed by the following infinite product

$$P(s) = \prod_0^\infty (1 - \lambda_j(s)) \tag{25.8}$$

where $1 \geqslant \lambda_0(s) \geqslant \lambda_1(s) \geqslant \dots$ are the eigenvalues of the integral operator.

In general the GUE law as stated above is very hard to establish, because harmonic analysis cannot control consecutive points of a sequence, but it can locate the points in small domains. Therefore more tractable problems are about correlation of sets of points in question. We begin by presenting the original work of H. Montgomery [**Mo4**] on the pair correlation of zeros of the Riemann zeta function, and then we state more general results and conjectures for the n-level correlation (see definitions in Section 25.3).

25.2. The pair correlation of zeros.

The main tool for dealing with the correlation problems is the explicit formula of Riemann (or its variations) which connects a sum over the zeros of $\zeta(s)$ with a sum over prime numbers. General versions are proved in Chapter 5, but here we use a formula in somewhat customized form. Let $g \in C_c^\infty(\mathbb{R})$ and let

$$h(r) = \int_{-\infty}^{\infty} g(u) e^{iur} du. \tag{25.9}$$

Put $\Gamma_R(s) = \pi^{-s/2}\Gamma(s/2)$, the local factor at the infinite place for the Euler product of $\zeta(s)$. Then

$$\begin{aligned} \sum_\gamma h(\gamma) = h\Big(\frac{i}{2}\Big) + h\Big(-\frac{i}{2}\Big) + \frac{1}{2\pi}\int_{-\infty}^{\infty} h(r)\left\{\frac{\Gamma_R'}{\Gamma_R}(\tfrac12 + ir) + \frac{\Gamma_R'}{\Gamma_R}(\tfrac12 - ir)\right\} dr \\ - \sum_1^\infty \frac{\Lambda(n)}{\sqrt{n}}(g(\log n) + g(-\log n)). \end{aligned} \tag{25.10}$$

This is the case of Theorem 5.12 for the Riemann zeta function.

The strategy goes as follows. First we localize the γ-terms, say γ near t, by choosing a suitable test function $h(r) = h(r,t)$, where t is a parameter at our disposal ($h(r)$ has to be an entire function). Of course, by the uncertainty principle of harmonic analysis, such a localization cannot be exact; at best one can see r near t at the distance $c/\log t$, where c is a positive constant. Computing a properly weighted L_2-norm of such γ-sum with respect to the parameter t involved in the test function $h(\gamma,t)$ one picks up terms which are close to the diagonal, i.e., one gets a sum over pairs of zeros γ, γ' with $\gamma - \gamma'$ being quite small. Then, by an approximation argument, or by Fourier inversion in another parameter which is built in the test function, one creates a sum whose terms are the desired function of the difference $\gamma - \gamma'$.

The original approach of Montgomery was somewhat special, and it is still being used by many researchers. Since his choice of test functions is quite natural we use them first. Montgomery [**Mo4**] introduced the function (a Fourier transform of the gaps)

$$F(\alpha, T) = \frac{2\pi}{T \log T} \sum\sum_{0<\gamma,\gamma' \leqslant T} w(\gamma - \gamma') T^{i\alpha(\gamma-\gamma')} \tag{25.11}$$

for any real α and $T \geqslant 2$, where $w(u)$ is a suitable localizing function. Specifically he takes

$$w(u) = 4(4 + u^2)^{-1}. \tag{25.12}$$

Note that $F(\alpha, T)$ is real and $F(\alpha, T) = F(-\alpha, T) \geqslant 0$, because the Fourier transform of w is positive. Precisely, $\hat{w}(v) = 2\pi e^{-4\pi|v|}$, so

$$F(\alpha, T) = \frac{2\pi}{T \log T} \int_{-\infty}^{\infty} e^{-2|v|} \Big| \sum_{0<\gamma\leqslant T} e^{i\gamma(v+\alpha \log T)} \Big|^2 dv. \tag{25.13}$$

By trivial estimation, using (25.2), one derives

$$F(\alpha, T) \leqslant F(0, T) \ll \log T. \tag{25.14}$$

Our goal is to give an asymptotic formula for $F(\alpha, T)$ as $T \to \infty$ which is uniform in α in ranges as large as possible.

THEOREM 25.1 (MONTGOMERY). *For $0 \leqslant \alpha \leqslant 1$ and $T \geqslant 2$ we have*

$$F(\alpha, T) = \alpha + T^{-2\alpha} \log T + O(\alpha T^{\alpha-1} + T^{-\alpha} \log 2T^{\alpha} + (\log T)^{-1}) \tag{25.15}$$

where the implied constant is absolute.

REMARKS. For α with $2(\log\log T)/\log T \leqslant \alpha \leqslant 1 - 1/(\log\log T)\log T$ we have

$$F(\alpha, T) \sim \alpha, \quad \text{as } T \to \infty. \tag{25.16}$$

The asymptotic formula (25.15) holds for all $\alpha \geqslant 0$ but it loses its meaning if $\alpha \geqslant 1$, because the error term $O(\alpha T^{\alpha-1})$ exceeds the leading term α.

In proving Theorem 25.1, Montgomery appealed to the explicit formula

$$L(x, t) = R(x, t) \tag{25.17}$$

where $L(x, t)$ is a special sum over the zeros of $\zeta(s)$ localized near t, namely

$$L(x, t) = 2 \sum_{\gamma} \frac{x^{i\gamma}}{1 + (t - \gamma)^2}, \tag{25.18}$$

and $R(x, t)$ is the corresponding sum over primes, including the infinite place terms; see Section 5.5. We have

$$R(x, t) = -\sum_{1}^{\infty} \frac{\Lambda(n)}{\sqrt{n}} \Big(\frac{x}{n}\Big)^{it} \min\Big(\frac{n}{x}, \frac{x}{n}\Big) + x^{-1+it} \log(t + 2) + E(x, t) \tag{25.19}$$

where the last term satisfies

$$E(x,t) \ll \frac{1}{x} + \frac{\sqrt{x}}{t+1}, \tag{25.20}$$

if $x \geqslant 1$ and $t > 0$, the implied constant being absolute. This particular equation (25.17) can be deduced from the formula

$$\sum_{n \leqslant x} \Lambda(n) n^{-s} = -\frac{\zeta'}{\zeta}(s) = \frac{x^{1-s}}{1-s} - \sum_{\rho} \frac{x^{\rho - s}}{\rho - s} + \sum_{1}^{\infty} \frac{x^{-2n-s}}{2n+s} \tag{25.21}$$

which holds for $x > 1$, $x \neq p^n$ and $s \neq 1, \rho, -2n$, or directly by a contour integration of

$$-\frac{\zeta'}{\zeta}(s) = \sum_{1}^{\infty} \Lambda(n) n^{-s} \tag{25.22}$$

and the functional equation

$$-\frac{\zeta'}{\zeta}(s) - \frac{\zeta'}{\zeta}(1-s) = \frac{\Gamma_R'}{\Gamma_R}(s) + \frac{\Gamma_R'}{\Gamma_R}(1-s) = \log(|s|+3) + O(1). \tag{25.23}$$

We square the equation (25.17) and integrate (this is a way of picking small differences $\gamma - \gamma'$), getting

$$\int_0^T |L(x,t)|^2 dt = \int_0^T |R(x,t)|^2 dt. \tag{25.24}$$

The left side is equal to

$$\int_0^T |L(x,t)|^2 dt = 4 \sum_{\gamma} \sum_{\gamma'} x^{i(\gamma - \gamma')} \int_0^T (1 + (t-\gamma)^2)^{-1} (1 + (t-\gamma')^2)^{-1} dt$$

where the summation is over all the critical zeros $\rho = \frac{1}{2} + i\gamma$, $\rho' = \frac{1}{2} + i\gamma'$ (with γ, γ' positive and negative). Using (25.2) we reduce the above expression as follows:

$$4 \sum_{0<\gamma,\gamma' \leqslant T}\sum x^{i(\gamma - \gamma')} \int_0^T (1 + (t-\gamma)^2)^{-1} (1 + (t-\gamma')^2)^{-1} dt + O(\log^3 T)$$

$$\begin{aligned} &= 4 \sum_{0<\gamma,\gamma' \leqslant T}\sum x^{i(\gamma - \gamma')} \int_{-\infty}^{\infty} (1 + (t-\gamma)^2)^{-1} (1 + (t-\gamma')^2)^{-1} dt + O(\log^3 T) \\ &= 2\pi \sum_{0<\gamma,\gamma' \leqslant T}\sum x^{i(\gamma - \gamma')} w(\gamma - \gamma') + O(\log^3 T). \end{aligned}$$

Putting $x = T^\alpha$ with $\alpha \geqslant 0$ we arrive at

$$\int_0^T |L(T^\alpha, t)|^2 dt = F(\alpha, T) T \log T + O(\log^3 T) \tag{25.25}$$

where the implied constant is absolute.

Next we evaluate the other side of (25.24). To this end we use the following form of Parseval's formula for Dirichlet series (see [**MV1**], or refine the proof of Theorem 9.1)

$$\int_0^T \Big|\sum_1^\infty a_n n^{-it}\Big|^2 dt = \sum_1^\infty (T + O(n))|a_n|^2. \tag{25.26}$$

Write (25.20) as $R(x,t) = C(t) + D(t) + E(t)$, respectively. Then

$$|R(x,t)|^2 = |C(t)|^2 + |D(t)|^2 + |E(t)|^2 + O(|C(t)D(t)| + |D(t)E(t)| + |E(t)C(t)|).$$

Integrating in $0 < t \leqslant T$ we derive by the Cauchy-Schwarz inequality

$$\int_0^T |R(x,t)|^2 dt = C^2 + D^2 + E^2 + O(CD + DE + EC), \tag{25.27}$$

with $C^2 = \displaystyle\int_0^T |C(t)|^2 dt$ and D^2, E^2 defined similarly. By (25.26) we obtain

$$C^2 = \sum_1^\infty (T + O(n)) \frac{\Lambda^2(n)}{n} \min{}^2\Big(\frac{n}{x}, \frac{x}{n}\Big).$$

By the Prime Number Theorem $\psi(x) = x + O(x/\log 2x)$ this is

$$C^2 = T\log x + O(T + x\log x). \tag{25.28}$$

For D^2 we get simply

$$D^2 = x^{-2}\int_0^T \log^2(t+2)dt = x^{-2}T\log^2 T + O(x^{-2}T\log T), \tag{25.29}$$

and from the error term (25.20) we get

$$E^2 \ll x^{-2}T + x. \tag{25.30}$$

Inserting (25.28), (25.29), (25.30) into (25.27) we arrive at

$$\int_0^T |R(x,t)|^2 dt = T\log x + x^{-2}T\log^2 T + B(x,T) \tag{25.31}$$

where

$$B(x,T) \ll T + x\log x + x^{-1}(\log 2x)T\log T. \tag{25.32}$$

Comparing (25.31) for $x = T^\alpha$ with (25.25) we complete the proof of (25.15).

Now we derive a few consequences of Theorem 25.1.

THEOREM 25.2. *Let $f(x)$ be a Schwartz function on $\mathbb{R}$ such that its Fourier transform*

$$\hat{f}(y) = \int_{-\infty}^{\infty} f(x)e(-xy)dx \tag{25.33}$$

is of class C^1 with $\operatorname{supp} \hat{f} \subset (-1,1)$. *Then*

$$\begin{aligned} &\sum\sum_{0<\gamma,\gamma'\leqslant T} w(\gamma-\gamma')f((\gamma-\gamma')\frac{\log T}{2\pi}) \\ &\qquad = \left\{\int_{-1}^{1} \hat{f}(\alpha)|\alpha|d\alpha + \hat{f}(0)\right\}\frac{T}{2\pi}\log T + O(T) \end{aligned} \tag{25.34}$$

where the implied constant depends only on f.

PROOF. For any f of Schwartz class we have

$$\sum\sum_{0<\gamma,\gamma'\leqslant T} w(\gamma-\gamma')f\Big((\gamma-\gamma')\frac{\log T}{2\pi}\Big) = \frac{T}{2\pi}(\log T)\int_{-\infty}^{\infty} f(\alpha)F(\alpha,T)d\alpha. \tag{25.35}$$

Here the integral equals

$$\int_0^1 (\hat{f}(\alpha)+\hat{f}(-\alpha))\left\{\alpha + T^{-2\alpha}\log T + O(T^{\alpha-1} + T^{-\alpha}\log 2T^{\alpha} + (\log T)^{-1})\right\}d\alpha.$$

The first term agrees with that on the right-hand side of (25.34). The second term is

$$2(\log T)\int_0^1 (\hat{f}(0)+O(\alpha))T^{-2\alpha}d\alpha = \hat{f}(0) + O((\log T)^{-1}),$$

and the error term is trivially estimated by $O(1/\log T)$. This completes the proof of (25.34). □

In particular, choosing $f(x) = ((\sin \pi\alpha x)/\pi\alpha x)^2$ we get

COROLLARY 25.3. *If $0<\alpha<1$ is fixed, then as $T\to\infty$,*

$$\sum\sum_{0<\gamma,\gamma'\leqslant T} w(\gamma-\gamma')\left(\frac{\sin((\gamma-\gamma')\frac{\alpha}{2}\log T)}{(\gamma-\gamma')\frac{\alpha}{2}\log T}\right)^2 \sim \left(\frac{1}{\alpha}+\frac{\alpha}{3}\right)\frac{T}{2\pi}\log T. \tag{25.36}$$

From this Montgomery deduced a lower bound for the number of simple zeros of $\zeta(s)$,

$$N_1(T) = |\{0<\gamma\leqslant T;\quad \rho=\tfrac{1}{2}+i\gamma \quad \text{is simple}\}|. \tag{25.37}$$

(recall we assume the Riemann Hypothesis).

COROLLARY 25.4 (MONTGOMERY). *As $T\to\infty$, we have*

$$N_1(T) > \Big(\frac{2}{3}+o(1)\Big)\frac{T}{2\pi}\log T, \tag{25.38}$$

i.e., at least 2/3 of the zeros are simple.

PROOF. Discarding all terms in (25.36) except for $\gamma = \gamma'$ we get

$$\sum_{0<\gamma\leqslant T} m_\gamma^2 < \Big(\frac{1}{\alpha} + \frac{\alpha}{3} + o(1)\Big)\frac{T}{2\pi}\log T,$$

where m_γ is the multiplicity of $\rho = \frac{1}{2}+i\gamma$. This holds for any fixed α with $0 < \alpha < 1$, the best bound being $\frac{4}{3} + o(1)$ for $\alpha \to 1-$. Hence we derive that

$$N_1(T) \geqslant \sum_{0<\gamma\leqslant T} (2 - m_\gamma)m_\gamma > (2 - \tfrac{4}{3} + o(1))\frac{T}{2\pi}\log T$$

completing the proof of (25.38). □

REMARK. The test function which was used in (25.36) is not optimal for the purpose of estimating $N_1(T)$ along the above lines, so the percentage 2/3 of simple zeros can be slightly increased.

Another consequence of Theorem 25.2 (which we do not draw here) is that $\gamma_{n+1} - \gamma_n$ can be smaller than its average $2\pi/\log\gamma_n$ by a factor $\lambda < 1$ infinitely often, i.e., $\liminf \delta_n \leqslant \lambda$. Montgomery succeeded in showing this with $\lambda = 0.68$. The best published result is $\lambda = 0.5171$ which is due to Conrey, Ghosh and Gonek [**CGG**], while a new refinement points to $\lambda = 0.5169\ldots$. Recently it was shown that if $\delta_n \leqslant \frac{1}{2} - \varepsilon$ sufficiently often, then the class number $h(-D)$ of the imaginary quadratic field $\mathbb{Q}(\sqrt{-D})$ is bounded below by $c(\varepsilon)D^{\frac{1}{2}}(\log D)^{-2}$ for some constant $c(\varepsilon)$ effectively computable in terms of ε; see [**CI2**].

The formula (25.16) holds uniformly for $\varepsilon \leqslant \alpha \leqslant 1-\varepsilon$. With a little more effort one could show that (25.16) holds uniformly for $\varepsilon \leqslant \alpha \leqslant 1$. One may wonder what is the true behavior of $F(\alpha, T)$ for $\alpha \geqslant 1$? Montgomery showed that (25.16) cannot hold for $\alpha > 1$. Moreover, he described heuristic arguments which suggest that

$$F(\alpha, T) \sim 1, \quad \text{as} \quad T \to \infty \tag{25.39}$$

uniformly in bounded intervals $1 \leqslant \alpha \leqslant A$. By applying this to (25.35) with appropriate test functions f of Schwartz class, Montgomery was led to the following

PAIR CORRELATION CONJECTURE. *For $\alpha < \beta$ put*

$$N(\alpha, \beta; T) = \Big|\Big\{m \neq n;\ 0 < \gamma_m, \gamma_n \leqslant T,\quad \tfrac{2\pi\alpha}{\log T} < \gamma_m - \gamma_n < \tfrac{2\pi\beta}{\log T}\Big\}\Big|.$$

Then, as $T \to \infty$ we have

$$N(\alpha, \beta; T) \sim N(T)\int_\alpha^\beta \Big(1 - \Big(\frac{\sin \pi u}{\pi u}\Big)^2\Big)du. \tag{25.40}$$

REMARKS. The assertions (25.39) for $\alpha \geqslant 1$ together with (25.16) for $0 < \alpha \leqslant 1$ are essentially equivalent to (25.40). By (25.39) the asymptotic formula (25.36) extends to all $\alpha \geqslant 1$ with the function $\alpha^{-1} + \alpha/3$ replaced by $1 + 1/3\alpha^2$. Hence letting $\alpha \to \infty$ it follows by the former argument that almost all zeros of $\zeta(s)$ are simple.

25.3. The n-level correlation function for consecutive spacing.

The Pair Correlation Conjecture of Montgomery indicates that the zeros of $\zeta(s)$ behave much like the eigenvalues of large complex Hermitian random matrices whose statistical distribution is known as the Gaussian Unitary Ensemble. Further support for this behavior was provided by Z. Rudnick and P. Sarnak [**RS**] who extended Montgomery's results to the n-level correlation sums for the normalized zeros of $\zeta(s)$. In this section we formulate the main results of Rudnick and Sarnak and we give a few comments about their arguments.

Recall that ζ_m denote the normalized ordinates of the zeros $\rho_m = \frac{1}{2} + i\gamma_m$ of $\zeta(s)$ given by (25.4), they are ordered in increasing sequence $\ldots \leqslant \zeta_{-2} \leqslant \zeta_{-1} < 0 < \zeta_1 \leqslant \zeta_2 \leqslant \ldots$, and ζ_n have unit mean spacing. The Pair Correlation Conjecture answers the question of how often the differences $\zeta_{m_1} - \zeta_{m_2}$ fall in a given interval. One may inquire about pairs of differences $\zeta_{m_1} - \zeta_{m_2}, \zeta_{m_2} - \zeta_{m_3}$ falling into a given rectangle, the triple of differences in a box, and so on. To control the problem we restrict our considerations to the first M points ζ_m, $1 \leqslant m \leqslant M$, where M is a large number. Let $n \geqslant 2$ and $B \subset \mathbb{R}^{n-1}$ be a box of dimension $n-1$. Put

$$R_n(M;B) = \frac{1}{M}|\{1 \leqslant m_1, \ldots, m_n \leqslant M \quad \text{all distinct}; \\ (\zeta_{m_1} - \zeta_{m_2}, \zeta_{m_2} - \zeta_{m_3} \cdots, \zeta_{m_{n-1}} - \zeta_{m_n}) \in B\}|.$$

For $n = 2$ and $B = (\alpha, \beta)$ this is simply

$$R_2(M;B) = \frac{1}{M}|\{1 \leqslant m_1 \neq m_2 \leqslant M; \quad \alpha < \zeta_{m_1} - \zeta_{m_2} < \beta\}|.$$

For $M = N(T)$ this is very close to $N(\alpha, \beta; T)/N(T)$. As in (25.41) we expect that there is a function $R_n(B)$ such that

$$R_m(M;B) \sim R_n(B), \quad \text{as } M \to \infty. \tag{25.41}$$

Our goal is to determine $R_n(B)$ for any $n \geqslant 2$.

To simplify the Fourier analysis here we introduce test functions $f(x_1, \ldots, x_n)$ of the following type:

TF1. $f(x_1, \ldots, x_n)$ is symmetric,

TF2. $f(x_1 + t, \ldots, x_n + t) = f(x_1, \ldots, x_n)$ for any $t \in \mathbb{R}$,

TF3. $f(x_1, \ldots, x_n) \to 0$ rapidly as $|x| \to 0$ on the planes $x_1 + \cdots + x_n = 0$.

REMARKS. The condition TF2 says that $f(x_1, \ldots, x_n)$ depends only on the successive differences. This together with TF3 implies that $f(x_1, \ldots, x_n)$ decays rapidly on any hyperplane $x_1 + \cdots + x_n = t$. For $n = 2$ such a function is given by $f(x_1, x_2) = f(x_1 - x_2)$ where $f(x)$ is an even function on $\mathbb{R}$ which decays rapidly to zero as $|x| \to \infty$.

For any test function as above we put

$$R_n(M;f) = \frac{n!}{M} \sum_{\substack{1 \leqslant m_1, \ldots, m_n \leqslant M \\ \text{all distinct}}} \cdots \sum f(\zeta_{m_1}, \ldots, \zeta_{m_n}), \tag{25.42}$$

and seek an asymptotic formula

$$R_n(M; f) \sim R_n(f), \quad \text{as } M \to \infty. \tag{25.43}$$

The multiple sum $R_n(M; f)$ can be smoothed further by replacing the range of summation $1 \leqslant m_1, \dots, m_n \leqslant M$ by a smooth (rapidly decreasing) cut-off function. Thus we also consider

$$R_n(T; f, h) = \sum_{\substack{m_1, \dots, m_n \\ \text{all distinct}}} \cdots \sum h(\gamma_{m_1} T^{-1}) \dots h(\gamma_{m_n} T^{-1}) f(\gamma_{m_1} L, \dots, \gamma_{m_n} L) \tag{25.44}$$

where $h(y)$ is the cut-off function in question and

$$L = \frac{1}{2\pi} \log T. \tag{25.45}$$

REMARK. There is no reason to use a cut-off function $h(y_1, \dots, y_n)$ more general than the product $h(y_1) \dots h(y_n)$ because we wish the points $\gamma_{m_1}, \dots, \gamma_{m_n}$ to be close to each other.

In 1962, Dyson [**D**] considered a statistical distribution of energy levels of some complex systems in which context he determined the n-level correlation density $W_n(x_1, \dots, x_n)$ for the GUE model. He showed that

$$W_n(x_1, \dots, x_n) = \det(K(x_i - x_j)), \tag{25.46}$$

where

$$K(x) = \frac{\sin \pi x}{\pi x}. \tag{25.47}$$

One can show the following properties of $W_n(x)$:

$$0 \leqslant W_n(x) \leqslant 1, \tag{25.48}$$

$$W_n(x) = 0 \text{ if and only if } x_i = x_j \text{ for some } i \neq j, \tag{25.49}$$

$$W_n(x) = 1 \text{ if and only if } x \in \mathbb{Z}^n \text{ and } x_i \neq x_j \text{ for } i \neq j. \tag{25.50}$$

Let us compute the density function for $n = 2$. We get

$$W_2(x_1, x_2) = \det \begin{pmatrix} K(0) & K(x_1 - x_2) \\ K(x_1 - x_2) & K(0) \end{pmatrix} = 1 - \left(\frac{\sin \pi(x_1 - x_2)}{\pi(x_1 - x_2)} \right)^2,$$

which agrees with that in (25.40).

THEOREM 25.5 (RUDNICK-SARNAK). *Suppose that the cut-off function $h(r)$ is given by the Fourier integral (25.9) with some smooth compactly supported function $g(u)$. Suppose the Fourier transform of the test function $f(x)$,*

$$\hat{f}(\xi) = \int_{\mathbb{R}^n} f(x) e(-\xi \cdot x) dx$$

is supported in the polyhedral domain $|\xi_1| + \dots + |\xi_n| \leqslant 2$. Then, as $T \to \infty$ we have

$$R_n(T; f, h) \sim N(T) \Big(\int_{-\infty}^{\infty} h(r)^n dr \Big) \int_{\mathbb{R}^n} f(x) W_n(x) \delta \Big(\frac{1}{n}(x_1 + \dots + x_n) \Big) dx \tag{25.51}$$

where $\delta(x)$ is the Dirac distribution at point 0.

By making an appropriate approximation to the cut-off function one can derive from (25.51) the following

COROLLARY 25.6. *Suppose the test function $f(x)$ has its Fourier transform $\hat{f}(\xi)$ supported in $|\xi_1| + \cdots + |\xi_n| \leqslant 2$. Then we have (25.43) with*

$$R_n(f) = \int_{\mathbb{R}^n} f(x) W_n(x) \delta\Big(\frac{1}{n}(x_1 + \cdots + x_n)\Big) dx \tag{25.52}$$

where $x = (x_1, \ldots x_n)$. This can be written as

$$R_n(f) = n \int_{\mathbb{R}^{n-1}} f(x_1, \ldots, x_n) W_n(x_1, \ldots, x_n) dx_1 \ldots dx_{n-1} \tag{25.53}$$

where $x_n = -x_1 - \ldots - x_n$.

REMARKS. In particular, we have

$$R_2(f) = 2 \int_{\mathbb{R}} f(x, -x) W_2(x, -x) dx = \int_{\mathbb{R}} f(x) \Big(1 - \Big(\frac{\sin \pi x}{\pi x}\Big)^2\Big) dx$$

where $f(x) = f(x, 0)$, which agrees with the Montgomery pair correlation result.

25.4. Low-lying zeros of L-functions.

A somewhat different problem from the consecutive spacing of zeros of an L-function concerns the zeros near the central point, i.e., in diminishing distance to $s = \frac{1}{2}$. Such a problem, of course, depends on the scaling of zeros. Consider an L-function of degree d (still satisfying GRH) as in Chapter 5, given by

$$L(f, s) = \sum_1^\infty \lambda_f(n) n^{-s}, \tag{25.54}$$

with conductor $c_f > 1$ (see (5.7), (5.8)). The number of zeros of $L(f, s)$

$$\rho_f = \tfrac{1}{2} + i\gamma_f \tag{25.55}$$

with $|\gamma_f| \leqslant T$ satisfies

$$N(T, f) \sim \frac{dT}{2\pi} \log c_f T \tag{25.56}$$

as $T \to \infty$ by Theorem 5.8. This suggests that the proper scaling factor for the low-lying zeros of $L(f, s)$ should be $\frac{1}{2\pi} \log c_f$. Now the question is: how does $N(T, f)$ behave for small T?

To make the question more interesting we take a test function ϕ in the Schwartz class on $\mathbb{R}$ and define the low-zeros sum

$$D(f; \phi) = \sum_{\gamma_f} \phi\Big(\gamma_f \frac{\log c_f}{2\pi}\Big), \tag{25.57}$$

where γ_f runs over the ordinates of zeros counted with multiplicity. For ϕ with rapid decay this sum concentrates on zeros which are within $O(1/\log c_f)$ of the point $s = \frac{1}{2}$.

Because, for a single L-function, we are effectively counting very few zeros, classical harmonic analysis cannot detect their distribution. Therefore we must take a sufficiently large family of L-functions into consideration. Suppose $\mathcal{F}$ is an infinite family ordered by the conductor. Put

$$\mathcal{F}(Q)=\{f\in\mathcal{F};\ c_f\leqslant Q\}. \tag{25.58}$$

Now we want to know the asymptotic behavior of the average density

$$AD(\mathcal{F};\phi,Q)=\frac{1}{|\mathcal{F}(Q)|}\sum_{f\in\mathcal{F}(Q)}D(f,\phi). \tag{25.59}$$

If $\mathcal{F}$ is a complete family in a certain spectral sense, one should expect that

$$AD(\mathcal{F};\phi,Q)\sim\int_{\mathbb{R}}\phi(x)W(\mathcal{F})(x)dx \tag{25.60}$$

where $W(\mathcal{F})(x)$ is a density function characterized by $\mathcal{F}$.

Some investigations in this direction were undertaken in the 1970's (see M. Jutila [**Ju5**], H. L. Montgomery [**Mo4**]) for the family of Dirichlet L-functions $L(s,\chi)$. However, the recent studies by N. Katz and P. Sarnak [**KS1**], [**KS2**] expanded far beyond this family. By examining various families $\mathcal{F}$ they revealed that the low-lying zeros are always governed by a symmetry (or monodromy) group $G(\mathcal{F})$ associated with $\mathcal{F}$. Their assertions are particularly convincing for zeta and L-functions over finite fields, not only because the Riemann Hypothesis is known to be true, but also due to connections with geometry which provides additional intuition (the zeros are then eigenvalues of the Frobenius operator, see Section 11.11 and [**KS1**]).

There are also attractive results for families of global L-functions, but progress is not yet as deep as in the case of local L-functions. In this section we present a few such results.

First we inspect the family of Dirichlet L-functions with real primitive characters χ_d. Here $L(s,\chi_d)$ has the conductor $c_d=|d|$. Since $L(s,\chi_d)$ have even functional equations, we may put their zeros $\rho_d^{(\ell)}=\frac{1}{2}+i\gamma_d^{(\ell)}$ into the sequence

$$\ldots\leqslant\gamma_d^{(-2)}\leqslant\gamma_d^{(-1)}\leqslant 0\leqslant\gamma_d^{(1)}\leqslant\gamma_d^{(2)}\leqslant\ldots \tag{25.61}$$

with $\gamma_d^{(-\ell)}=\gamma_d^{(\ell)}$ for $\ell\in\mathbb{Z}\backslash\{0\}$ (it is believed that $L(\frac{1}{2},\chi_d)$ is never zero). Let

$$\mathcal{F}(Q)=\{d\text{ fundamental discriminant};\ |d|\leqslant Q\}. \tag{25.62}$$

CONJECTURE (KATZ-SARNAK). *As $Q\to\infty$, we have*

$$\frac{1}{|\mathcal{F}(Q)|}\sum_{d\in\mathcal{F}(Q)}\sum_{\ell>0}\phi\Big(\gamma_d^{(\ell)}\frac{\log|d|}{2\pi}\Big)\sim\int_{\mathbb{R}}\phi(x)W_{sp}(x)dx \tag{25.63}$$

where the density function is given by

$$W_{sp}(x)=1-\frac{\sin 2\pi x}{2\pi x}. \tag{25.64}$$

REMARKS. Katz and Sarnak [**KS2**] proved this conjecture for any test function ϕ whose Fourier transform

$$\hat{\phi}(\xi) = \int_{\mathbb{R}} \phi(x)e(-\xi x)dx \tag{25.65}$$

is supported in $(-2,2)$ (of course, subject to the validity of the GRH).

The case of $L(s,\chi_d)$ is obtained by twisting $\zeta(s)$ with χ_d. One may generalize this case by starting with any $L(f,s)$ in place of $\zeta(s)$ and taking the family of L-functions $L(f\otimes\chi_d,s)$ twisted by the real characters χ_d. For example, choose an elliptic curve $E/\mathbb{Q}$, and let $E^{(d)}/\mathbb{Q}$ be the corresponding quadratic twists. Then the associated L-functions $L(E^{(d)},s)$ are obtained from $L(E,s)$ by twisting with χ_d. The corresponding conjecture of Katz and Sarnak predicts that the density function for the low-lying zeros of $L(E^{(d)},s)$ is given by

$$W^+(x) = 1 + \frac{\sin 2\pi x}{2\pi x}. \tag{25.66}$$

Many interesting examples of families of L-functions are offered by the theory of automorphic forms, and a variety of these are studied by Iwaniec, Luo and Sarnak [**ILS**]. We are going to describe some of their results. Let $\Gamma = SL_2(\mathbb{Z})$ and $k \geqslant 2$ be an even integer. Let $S_k(\Gamma)$ be the linear space of cusp forms for Γ of weight k. Let $H_k(\Gamma)$ be the Hecke basis of primitive forms of $S_k(\Gamma)$. To any

$$f(z) = \sum_1^\infty \lambda_f(n)n^{\frac{k-1}{2}}e(nz) \in H_k(\Gamma)$$

we associate the Hecke L-function

$$L(f,s) = \sum_1^\infty \lambda_f(n)n^{-s} = \prod_p (1-\lambda_f(p)p^{-s}+p^{-2s})^{-1}.$$

As described in Chapter 14, the completed function

$$\Lambda(f,s) = (2\pi)^{-s}\Gamma(s+\tfrac{k-1}{2})L(f,s)$$

is entire and satisfies the functional equation $\Lambda(f,s) = \varepsilon_f\Lambda(f,1-s)$ with root number given by $\varepsilon_f = i^k = \pm 1$ (note ε_f depends only on k but not on f). Because the sign of the functional equation has some impact on the distribution of low-lying zeros we separate the families of even and odd L-functions. Put

$$M^+(K) = \sum_{\substack{k\leqslant K\\ k\equiv 0 (\mathrm{mod}\ 4)}} |H_k(\Gamma)|, \tag{25.67}$$

$$M^-(K) = \sum_{\substack{k\leqslant K\\ k\equiv 2 (\mathrm{mod}\ 4)}} |H_k(\Gamma)|. \tag{25.68}$$

The corresponding densities turn out to be

$$W^+(x) = 1 + \frac{\sin 2\pi x}{2\pi x}, \tag{25.69}$$

$$W^-(x) = 1 - \frac{\sin 2\pi x}{2\pi x} + \delta_0(x). \tag{25.70}$$

THEOREM 25.7 (IWANIEC-LUO-SARNAK). *Suppose ϕ is a smooth function on $\mathbb{R}$ compactly supported in $]-2,2[$. Then, as $K \to \infty$, we have*

$$\frac{1}{M^+(K)} \sum_{\substack{k \leqslant K \\ k \equiv 0 (\mathrm{mod}\ 4)}} \sum_{f \in H_k(\Gamma)} D(f;\phi) \sim \int_{\mathbb{R}} \phi(x) W^+(x) dx, \tag{25.71}$$

$$\frac{1}{M^-(K)} \sum_{\substack{k \leqslant K \\ k \equiv 2 (\mathrm{mod}\ 4)}} \sum_{f \in H_k(\Gamma)} D(f;\phi) \sim \int_{\mathbb{R}} \phi(x) W^-(x) dx. \tag{25.72}$$

Let $L(\mathrm{Sym}^2 f, s)$ be the symmetric square L-function associated with f. Recall from Section 5.12 that it is the L-function of degree 3 given by

$$\begin{aligned} L(\mathrm{Sym}^2 f, s) &= \zeta(2s) \sum_1^\infty \lambda_f(n^2) n^{-s} \\ &= \prod_p (1 - \lambda_f(p^2) p^{-s} + \lambda_f(p^2) p^{-2s} - p^{-3s})^{-1}. \end{aligned} \tag{25.73}$$

As shown by Shimura [**Sh2**], the completed function

$$\Lambda(\mathrm{Sym}^2 f, s) = \pi^{-\frac{3s}{2}} \Gamma\left(\tfrac{s+1}{2}\right) \Gamma\left(\tfrac{s+k-1}{2}\right) \Gamma\left(\tfrac{s+k}{2}\right) L(\mathrm{Sym}^2 f, s) \tag{25.74}$$

is entire and satisfies the functional equation $\Lambda(\mathrm{sym}^2 f, s) = \Lambda(\mathrm{Sym}^2 f, 1-s)$ (note that the root number is always one). In this case we have

THEOREM 25.8 (IWANIEC-LUO-SARNAK). *Suppose ϕ is a smooth function on $\mathbb{R}$ compactly supported in $]-\frac{4}{3}, \frac{4}{3}[$. Then, as $K \to \infty$,*

$$\frac{1}{M(K)} \sum_{\substack{k \leqslant K \\ k \equiv 0 (\mathrm{mod}\ 2)}} \sum_{f \in H_k(\Gamma)} D(Sym^2 f;\phi) \sim \int_{\mathbb{R}} \phi(x) W_{sp}(x) dx \tag{25.75}$$

where $M(K) = M^+(K) + M^-(K)$ and $W_{sp}(x)$ is given by (25.64).

REMARKS. We emphasize that the conditions on the test function ϕ in Theorem 25.7 and Theorem 25.8 allow the support of the Fourier transform $\hat{\phi}$ (see (25.65)) to be larger than $[-1,1]$. To appreciate this fact we write

$$\int_{\mathbb{R}} \phi(x) W(x) dx = \int_{\mathbb{R}} \hat{\phi}(y) \hat{W}(y) dy$$

by Plancherel's theorem. Here the Fourier transforms of the corresponding density distributions are:

$$\begin{aligned} \hat{W}^+(y) &= \delta_0(y) + \tfrac{1}{2}\eta(y), \\ \hat{W}^-(y) &= \delta_0(y) - \tfrac{1}{2}\eta(y) + 1, \\ \hat{W}_{sp}(y) &= \delta_0(y) - \tfrac{1}{2}\eta(y) \end{aligned}$$

where $\delta_0(y)$ is the Dirac distribution and $\eta(y)$ is the characteristic function of the segment $[-1,1]$. These Fourier transforms have discontinuity at $y=1$ and $y=-1$, therefore the results are distinguishable only if the support of $\hat{\phi}$ covers the points $y=1$ and $y=-1$.

CHAPTER 26

CENTRAL VALUES OF L-FUNCTIONS

26.1. Introduction.

We have already mentioned many times the Grand Riemann Hypothesis for L-functions and discussed many aspects of the fascinating zeros of L-functions (see Chapters 5 or 24 for instance) in various aspects.

Recently, much of the analytic theory of L-functions focused on estimates for the values of L-functions at the central critical point $s = \frac{1}{2}$. These special values appear in various contexts and vanishing or non-vanishing are the main questions. For example, algebraic techniques developed by Mazur to study rational points on certain modular curves, with applications to diophantine equations, require for full effect to prove that there exists a modular form of a certain type which does not vanish at the central point (see [**Ell**] for a recent example). Another very different situation is the surprising link discovered by Iwaniec and Sarnak between the proportion of non-vanishing of various families of L-functions and the Gauss Class Number Problem (see Theorem 26.1 below).

Here is a short panorama of selected accomplishments in recent years, emphasizing the strongest results from analytic number theory.

The first is the Iwaniec-Sarnak result mentioned above [**IS2**]. Among different variants, we choose the following:

THEOREM 26.1. *For $k \equiv 0 \pmod 4$, let H_k be the set of weight k Hecke forms on $SL(2,\mathbb{Z})$. For any real primitive character ψ modulo q with $q \leqslant k^{\vartheta}$, we have*

$$|\{f \in H_k \mid |L(f \otimes \psi, \tfrac{1}{2})| \geqslant \frac{1}{(\log k)^2}\}| \geqslant \Big(\frac{1}{2} - o(1)\Big)|H_k| \tag{26.1}$$

as $k \to +\infty$ and $\vartheta \to 0$.

They also show that

$$\sum_{f \in H_k} L(f, \tfrac{1}{2})L(f \otimes \chi, \tfrac{1}{2}) \sim L(\chi, 1)|H_k|$$

for any real primitive character χ modulo D, uniformly for $D \leqslant k^{\vartheta}$ as $k \to +\infty$ and $\vartheta \to 0$. Hence, if the constant $\frac{1}{2}$ in (26.1) for $\psi = 1$ could be replaced by $\frac{1}{2} + \delta$ for any $\delta > 0$, with an effective implied constant, it would follow that

$$L(1, \chi) \gg \frac{1}{(\log D)^2}$$

for any real primitive character χ modulo D with an absolute and effective implied constant.

Note it is important to have a lower bound as in (26.1) instead of simply $L(f,\frac{1}{2}) \neq 0$, so it is conceivable that all special values are non-zero, yet too small to solve the Class Number Problem along these lines. From the point of view of analytic number theory, a lower bound as in (26.1) and simple non-vanishing are usually indistinguishable.

In part because of the relationship to the Birch and Swinnerton-Dyer Conjecture, the order of vanishing of weight 2 cusp forms at $s=\frac{1}{2}$ has attracted particular attention. Recall that in one instance it is easy to guarantee that the central point is a zero, namely if $L(f,s)$ is a self-dual L-function and the sign of the functional equation is -1. It is reasonable to assume that the order of vanishing will be the smallest compatible with this restriction, namely a "generic" even form will have $L(f,\frac{1}{2}) \neq 0$ and a generic odd form $L'(f,\frac{1}{2}) \neq 0$. Brumer (and independently Murty) conjectured this and used GRH to give some evidence.

Iwaniec and Sarnak proved that half the special values of even forms do not vanish (and are not too small, as in (26.1)). Kowalski and Michel [**KM1**], and independently VanderKam [**vdK**] considered higher derivatives. We quote two results:

THEOREM 26.2 (Kowalski-Michel). *Let $S_2(q)^*$ be the set of primitive weight 2 forms of level q. We have*

$$|\{f \in S_2(q)^* \mid \varepsilon_f = -1 \text{ and } L'(f,\tfrac{1}{2}) \neq 0\}| \geqslant \Big(\frac{7}{16} - o(1)\Big)|S_2(q)^*|$$

for primes $q \to +\infty$.

THEOREM 26.3 (Kowalski-Michel-VanderKam). *For any fixed $k \geqslant 0$ we have*

$$|\{f \in S_2(q)^* \mid \varepsilon_f = (-1)^k \text{ and } L^{(k)}(f,\tfrac{1}{2}) \neq 0\}| \geqslant \tfrac{1}{2}(p_k - o(1))|S_2(q)^*|$$

for primes $q \to +\infty$, where $p_0 = 1/4$, $p_1 = 7/16$, $p_2 = 0.4825$, $p_3 = 0.495$, and moreover, $p_k = \frac{1}{2} - \frac{1}{32k^2} + O(\frac{1}{k^3})$.

Although the error term depends on k, it is possible to combine this result with an average bound

$$\sum_{f \in S_2(q)^*} (\operatorname*{ord}_{s=\frac{1}{2}} L(f,s))^2 \ll 1$$

for q prime and derive (see [**KMV3**] for both results)

COROLLARY 26.4. *We have*

$$\sum_{f \in S_2(q)^*} \operatorname*{ord}_{s=\frac{1}{2}} L(f,s) \leqslant (c + o(1))|S_2(q)^*|$$

for primes $q \to +\infty$ with $c = 1.1891$.

This is particularly interesting because $1.1891 < \frac{3}{2}$, which was the value obtained by Brumer on GRH.

In the following sections we will give considerable details of the proof of Theorem 26.2. This result expands an earlier one of B. Duke [**Du3**] where the proportion was $q/(\log q)^4$.

Those non-vanishing results have nice interpretation in terms of arithmetic geometry, as already mentioned in Section 5.14. Indeed associated to the curve $X_0(q) = \Gamma_0(q)\backslash\bar{\mathbb{H}}$ is an abelian variety, its jacobian variety, denoted $J_0(q)$, which is of dimension $\dim J_0(q) = \dim S_2(q) = |S_2(q)^*|$ (the latter only for q prime). The jacobian is defined over $\mathbb{Q}$ and its group of rational points is finitely generated by the Mordell-Weil theorem. Eichler and Shimura computed the Hasse-Weil zeta function of $J_0(q)$ in terms of modular forms: for q prime it takes the form

$$L(J_0(q), s) = \prod_{f \in S_2(q)^*} L(f, s).$$

If the Birch and Swinnerton-Dyer Conjecture for abelian varieties holds for $J_0(q)$, we have

$$\operatorname{rank} J_0(q)(\mathbb{Q}) = \operatorname*{ord}_{s=1/2} L(J_0(q), s) = \sum_{f \in S_2(q)^*} \operatorname*{ord}_{s=1/2} L(f, s),$$

so lower bounds for the order of vanishing give lower bounds for the rank of $J_0(q)$. In fact, by the general form of the Gross-Zagier formula (Theorem 23.4) for an arbitrary weight 2 cusp form, one shows that if the order of vanishing is exactly 1, then f contributes 1 to the rank (by conjugates of Heegner points). So Theorem 26.2 implies

THEOREM 26.5. *For q prime, $q \to +\infty$, we have*

$$\operatorname{rank} J_0(q)(\mathbb{Q}) \geqslant \Big(\frac{7}{16} - o(1)\Big) \dim J_0(q).$$

On the Birch and Swinnerton-Dyer Conjecture, Corollary 26.4 gives a corresponding fairly good upper bound for $\operatorname{rank} J_0(q)$. There is currently no comparable unconditional bound.

Recall from Proposition 5.21 that, on GRH, the maximal order of vanishing of an L-function of degree d with conductor q at the central critical point is about $(\log q)(\log \frac{3}{d} \log q)^{-1}$. Assuming the Birch and Swinnerton-Dyer Conjecture, this translates to a conjectural upper bound for the maximum rank of an abelian variety over $\mathbb{Q}$. One can see it is best possible by taking powers of an elliptic curve with positive rank. However, more convincing is the case of $J_0(q)$ (which is conjecturally "almost" irreducible). For q prime one gets by GRH and the Petersson formula the slightly better estimate (5.94). The conductor is $N_q = q^{\dim J_0(q)}$, hence by $\dim J_0(q) \sim q/12$ we have $\dim J_0(q) \sim (\log N_q)(\log\log N_q)^{-1}$. So Theorem 26.5 shows that the upper bound on GRH cannot be improved (except for the implied constant).

Of course one can also study classical Dirichlet L-functions. Michel and VanderKam [**MvdK**] studied the analogue of Theorem 26.3 for the family of primitive characters modulo q. A different challenge is to consider only real characters modulo $q \leqslant Q$. There Soundararajan [**Sou**] has proved

THEOREM 26.6. *Let $\mathcal{Q}$ be the set of odd squarefree integers. For $q \in \mathcal{Q}$, let $\chi_{8q}(n) = (8q/n)$ be the even quadratic character modulo $8q$. Then*

$$|\{q \leqslant Q \mid q \in \mathcal{Q} \text{ and } L(\tfrac{1}{2}, \chi_{8q}) \neq 0\}| \geqslant \Big(\frac{7}{8} - o(1)\Big) |\{q \leqslant Q \mid q \in \mathcal{Q}\}|$$

as $Q \to +\infty$.

Notice the coincidence of the proportion 7/8 with Theorem 26.2 (where a factor $\frac{1}{2}$ really comes from consideration of odd forms). In fact, Soundararajan's proof uses a mollifier $M(\chi)$ as in Theorem 26.2 (see Section 26.2 and following), and remarkably for a mollifier of length Q^{Δ}, the proportion obtained is $1-(2\Delta+1)^{-3}$ exactly as in Theorem 26.9! The Katz-Sarnak philosophy of monodromy groups of a family of L-functions can nicely explain this phenomenon.

This concept of monodromy group of families, although lacking a formal definition yet, has already been used to explain very striking results. For example, Conrey and Soundararajan [**CoSo**] have been able to prove for the first time that there are infinitely many L-functions of real characters with no real zeros in the critical strip (not only at $s=\frac{1}{2}$). The success of their approach (which depends on a constant turning out to be positive) was "predicted" by the conjectured symmetry of the family of real characters.

THEOREM 26.7. *Let $\mathcal{Q}$ be as in Theorem 26.6. We have*

$$|\{q \leqslant Q \mid L(\chi_{-8q}, s) \neq 0 \text{ for } s \in [0,1]\}| \geqslant \Big(\frac{1}{5} - o(1)\Big)|\{q \leqslant Q \mid q \in \mathcal{Q}\}|$$

as $Q \to +\infty$.

We conclude with a still challenging problem: prove that, given a modular form f (say of weight 2 and level q), there exists a positive proportion of quadratic characters χ such that $L(f\otimes\chi, \frac{1}{2}) \neq 0$. This is a conjecture of Goldfeld that still resists proof. The best known result is due to Ono [**Ono**], using algebraic methods (congruences and Fourier coefficients of half-integral weight modular forms), losing a factor $(\log Q)^{\delta}$ for some $\delta \in (0,1)$. Analytically, using Heath-Brown's large sieve inequality for real characters (Theorem 7.20), Perelli and Pomykala [**PP**] have the best result, with a factor $Q^{-\varepsilon}$ lost.

REMARK. Other special points on the critical line are those $s_j = \frac{1}{2}+it_j$ associated to an eigenvalue of the Laplace operator of a Maass cusp form φ_j for a congruence subgroup. The deformation theory of Phillips and Sarnak shows that the vanishing of the special value of the Rankin-Selberg L-function $L(f\otimes\varphi_j, \frac{1}{2}+it_j)$ (where f is a weight 4 holomorphic cusp form) has an interpretation for the question of the existence of cusp forms in non-arithmetic groups. W. Luo [**Luo**] has shown that for fixed $f \in S_4(p)^*$, where p is prime, a positive proportion of special values $L(f\otimes\varphi_j, \frac{1}{2}+it_j)$ are non-zero.

26.2. Principle of the proof of Theorem 26.2.

Since q is prime and $k=2<12$, we can apply the Petersson formula to the basis $S_2(q)^*$ (properly normalized) of primitive forms of $S_2(q)$, as described in Corollary 14.23 and afterwards. Let $f \in S_2(q)^*$. We denote its Fourier expansion

$$f(z) = \sum_{n\geqslant 1} \lambda_f(n)\sqrt{n}e(nz)$$

and recall its multiplicative properties (14.50), (14.51). To prove Theorem 26.2, we first derive the weighted analogue:

THEOREM 26.8. *We have*

$$\sideset{}{^h}\sum_{\substack{f\in S_2(q)^* \text{ odd} \\ L'(f,1/2)\neq 0}} 1 \geqslant \Big(\frac{7}{16} - o(1)\Big)$$

for primes $q \to +\infty$.

Recall from (14.60) that the h in superscript indicates that the normalizing factor $(4\pi)^{-1}\|f\|^{-2}$ is inserted.

The method of proof is based on comparison of the first and second moments of special values $L'(f, 1/2)$. Since "large values" of $L'(f, 1/2)$ have a large effect on the second moment, we use mollification, as in the proof of Selberg's theorem in Chapter 24. Let

$$M_1 = \sum_{f \in S_2(q)^*}^{h} \tfrac{1}{2}(1-\varepsilon_f)M(f)L'(f, \tfrac{1}{2})$$

and

$$M_2 = \sum_{f \in S_2(q)^*}^{h} \tfrac{1}{2}(1-\varepsilon_f)|M(f)L'(f, \tfrac{1}{2})|^2$$

for some $M(f) \in \mathbb{C}$. Comparison of an upper bound for M_2 and a lower bound for M_1 yields by Cauchy's inequality a lower bound for the quantity we study

$$\sum_{\substack{\varepsilon_f = -1 \\ L'(f,\frac{1}{2}) \neq 0}}^{h} 1 \geqslant \frac{M_1^2}{M_2}. \tag{26.2}$$

In order to achieve the best possible estimate, we will find asymptotics for M_1 and M_2. Note that if the mollifier is ignored (take $M(f) = 1$), a factor $\log q$ is lost in the final estimate.

To get sums amenable to Petersson's formula, we look for mollifiers as linear forms in $\lambda_f(m)$:

$$M(f) = \sum_{m \leqslant M} \frac{x_m}{\sqrt{m}} \lambda_f(m) \tag{26.3}$$

for some $M = q^{\Delta}$ with $0 \leqslant \Delta < \frac{1}{2}$ and real coefficients x_m supported on squarefree numbers $\leqslant M$, such that

$$x_m \ll \tau(m)(\log q)^3 \tag{26.4}$$

for some absolute implied constant.

Theorem 26.8 follows by letting $\Delta \to \frac{1}{2}$ in the following

THEOREM 26.9. *For* $0 \leqslant \Delta < 1/2$, *we have*

$$\sum_{\substack{f \in S_2(q)^* \text{ odd} \\ L'(f,1/2) \neq 0}}^{h} 1 \geqslant \frac{1}{2}\Big(1 - \frac{1}{(2\Delta+1)^3}\Big)$$

for all primes q *large enough in terms of* Δ.

26.3. Formulas for the first and the second moment.

The moments M_1 and M_2 can both be reduced to combinations of averages of $L'(f,\frac{1}{2})$ or $L'(f,\frac{1}{2})^2$ twisted by Hecke eigenvalues. For $m \geqslant 1$, let

$$D(m) = \sum_{f\in S_2(q)^*}^{h} \tfrac{1}{2}(1-\varepsilon_f)\lambda_f(m)L'(f,\tfrac{1}{2}), \tag{26.5}$$

$$H(m) = \sum_{f\in S_2(q)^*}^{h} \tfrac{1}{2}(1-\varepsilon_f)\lambda_f(m)L'(f,\tfrac{1}{2})^2. \tag{26.6}$$

We have

$$M_1 = \sum_m \frac{x_m}{\sqrt{m}} D(m) \tag{26.7}$$

and by (26.3)

$$\begin{aligned} M_2 &= \sum_{\varepsilon_f=-1}^{h} L'(f,\tfrac{1}{2})^2 \sum_{m_1,m_2} \frac{x_{m_1}x_{m_2}}{\sqrt{m_1m_2}}\lambda_f(m_1)\lambda_f(m_2) \\ &= \sum_b \frac{1}{b} \sum_{m_1,m_2} \frac{x_{bm_1}x_{bm_2}}{\sqrt{m_1m_2}} H(m_1m_2). \end{aligned} \tag{26.8}$$

We will obtain asymptotic formulas for $D(m)$ and $H(m)$ valid if m is small enough with respect to q.

We first use contour integration and the functional equation to express $L'(f,\frac{1}{2})$ and $L'(f,\frac{1}{2})^2$ as sums of rapidly convergent series. This could be derived from (5.12) by differentiation, but we redo the computations anyway.

Let $N \geqslant 2$ and G be a fixed polynomial such that $G(-s) = G(s)$, $G(-N) = \cdots = G(-1) = 0$ and $G(0) = 1$. In particular, $G'(0) = G^{(3)}(0) = 0$. Consider

$$I = \frac{1}{2\pi i}\int_{(2)} \Lambda(f, s+\tfrac{1}{2})G(s)\frac{ds}{s^2}.$$

We can shift the contour of integration to the line $\mathrm{Re}\,(s) = -2$, picking up a double pole at $s = 0$, hence

$$I = \operatorname*{res}_{s=0} \Lambda(f,s+\tfrac{1}{2})\frac{G(s)}{s^2} + \frac{1}{2\pi i}\int_{(-2)} \Lambda(f,s+\tfrac{1}{2})G(s)\frac{ds}{s^2}.$$

By the functional equation (Theorem 14.17), with sign $-\bar{\eta} = \varepsilon_f$, the integral on $\mathrm{Re}\,(s) = -2$ is equal to $\varepsilon_f I$, so

$$(1-\varepsilon_f)I = \operatorname*{res}_{s=0} \Lambda(f,s+\tfrac{1}{2})\frac{G(s)}{s^2}.$$

Computing the residue by writing the Taylor expansions around 0, we obtain

$$2(1-\varepsilon_f)I = (1-\varepsilon_f)\Big(\frac{\sqrt{q}}{2\pi}\Big)^{1/2} L'(f,\tfrac{1}{2}).$$

On the other hand, we can compute I by expanding the L-function in an absolutely convergent Dirichlet series on the line $\mathrm{Re}\,(s) = 2$, which gives

$$I = \Big(\frac{\sqrt{q}}{2\pi}\Big)^{1/2} \sum_{l\geqslant 1} \frac{\lambda_f(l)}{\sqrt{l}} V\Big(\frac{2\pi l}{\sqrt{q}}\Big)$$

where

$$V(y) = \frac{1}{2\pi i} \int_{(3/2)} \Gamma(s+1)G(s)y^{-s}\frac{ds}{s^2}.$$

Hence, comparing both expressions, we have

$$(1-\varepsilon_f)L'(f,\tfrac{1}{2}) = 2(1-\varepsilon_f)\sum_{l\geqslant 1} \frac{\lambda_f(l)}{\sqrt{l}} V\Big(\frac{2\pi l}{\sqrt{q}}\Big). \tag{26.9}$$

We estimate V easily by shifting the contour to the left, or right: we have

$$V(y) = -\log y - \gamma + O(y), \tag{26.10}$$

$$V(y) \ll y^{-j} \quad \text{for all } j \geqslant 1. \tag{26.11}$$

Similarly, we consider the integral

$$J = \frac{1}{2\pi i} \int_{(2)} \Lambda(f, s+\tfrac{1}{2})^2 G(s)\frac{ds}{s^3}$$

and proceed to evaluate it in the same fashion as for I. Shifting the contour to $\mathrm{Re}\,(s) = -2$ and applying the functional equation for the square of the L-function

$$\Lambda(f,s)^2 = \Lambda(f,1-s)^2$$

(where the sign is $\varepsilon_f^2 = 1$), we get

$$2J = \operatorname*{res}_{s=0} \Lambda(f,s+\tfrac{1}{2})^2\frac{G(s)}{s^3}.$$

Further, from the multiplicativity of the Hecke eigenvalues (see (14.50)) we derive the Dirichlet series expansion

$$L(f,s)^2 = \zeta_q(2s)\sum_{n\geqslant 1}\tau(n)\lambda_f(n)n^{-s},$$

so the term by term integration yields

$$J = \frac{\sqrt{q}}{2\pi}\sum_{n\geqslant 1}\frac{\lambda_f(n)}{\sqrt{n}}\tau(n)W\Big(\frac{4\pi^2 n}{q}\Big)$$

where

$$W(y) = \frac{1}{2\pi i}\int_{(1/2)} \zeta_q(1+2s)\Gamma(s)^2G(s)y^{-s}\frac{ds}{s}. \tag{26.12}$$

Hence

$$\frac{\sqrt{q}}{\pi}\sum_{n\geqslant 1}\frac{\lambda_f(n)}{\sqrt{n}}\tau(n)W\Big(\frac{4\pi^2 n}{q}\Big) = \operatorname*{res}_{s=0}\Lambda(f,s+\tfrac{1}{2})^2\frac{G(s)}{s^3}.$$

For odd f, we have $L(f,\frac{1}{2})=\Lambda(f,\frac{1}{2})=0$ so the above residue equals $\Lambda'(f,\frac{1}{2})^2=\frac{\sqrt{q}}{2\pi}L'(f,\frac{1}{2})^2$ from the Taylor expansions of $\Lambda(f,s+\frac{1}{2})$ and $G(s)$. Therefore for odd forms we have

$$L'(f,\tfrac{1}{2})^2=2\sum_{n\geqslant 1}\frac{\lambda_f(n)}{\sqrt{n}}\tau(n)W\Big(\frac{4\pi^2 n}{q}\Big). \tag{26.13}$$

The function W satisfies the bound

$$y^iW^{(j)}(y)\ll\log^3(y+y^{-1}),\quad \text{for all } i\geqslant j\geqslant 0, \tag{26.14}$$

$$y^iW^{(i)}(y)\ll y^{-j},\quad \text{for all } i\geqslant 0,\ j\geqslant 1. \tag{26.15}$$

(the implied constant depending on i and j) and there exists a polynomial P, independent of q, of degree at most 2, such that

$$W(y)=-\frac{1}{12}(\log y)^3+P(\log y)+O(q^{-1}(\log y)^2+y). \tag{26.16}$$

This last fact follows by writing

$$W(y)=\operatorname*{res}_{s=0}\frac{G(s)\Gamma(s)^2\zeta_q(1+2s)}{sy^s}+O(y)$$

by shifting the contour, and computing the residue.

Using (26.5) and (26.9) we get

$$D(m)=\sum_{\ell}\frac{1}{\sqrt{\ell}}V\Big(\frac{2\pi\ell}{\sqrt{q}}\Big)\Delta'(\ell,m)$$

where

$$\Delta'(l,m)=\sum_f{}^{h}(1-\varepsilon_f)\lambda_f(l)\lambda_f(m).$$

We now appeal to (14.65) of Corollary 14.26 which shows that Δ' is a strong approximation to the Kronecker delta

$$\Delta'(l,m)=\delta(l,m)+O\Big(\frac{\tau_3((l,m))}{\sqrt{q}}\Big(\frac{lm}{q}\Big)^{\frac{1}{4}}(\log q)^{\frac{1}{4}}\Big).$$

One could use also (14.64), keeping the Kloosterman sums, but this will not be necessary for the application to M_1 and Theorem 26.2. Hence we have by (26.10) and (26.11)

$$D(m)=\frac{1}{\sqrt{m}}V\Big(\frac{2\pi m}{\sqrt{q}}\Big)+O(q^{-3/8}m^{1/4}(\log q)^2)$$

with an absolute implied constant. By (26.7), (26.10) and (26.4) we get

PROPOSITION 26.10. *We have for $m\geqslant 1$,*

$$D(m)=\frac{1}{\sqrt{m}}\Big(\log\frac{\hat{q}}{m}\Big)+O(q^{-3/8}m^{1/4}(\log q)^2+mq^{-1/2}) \tag{26.17}$$

where $\hat{q}=\sqrt{q}/(2\pi e^{\gamma})$, and if $M=q^{\Delta}$, then

$$M_1=\sum_{m\leqslant M}\frac{x_m}{m}\Big(\log\frac{\hat{q}}{m}\Big)+O(q^{-\delta}(\log q)^7). \tag{26.18}$$

where $\delta = \frac{3}{4}(\frac{1}{2} - \Delta)$.

In the following, when we write an error term of the form $O(q^{-\delta})$, it is implied that $\delta > 0$, δ depends only on Δ, and the value of δ may change from line to line.

We now proceed to get an expression for M_2 as a quadratic form in the x_m. The idea is the same as above, but we cannot simply estimate the contribution of the sum of Kloosterman sums in the Petersson formula by Weil's bound: we open the Kloosterman sums and transform further to derive the required expression.

Let $m \geqslant 1$ be fixed, $m \leqslant q^{2\Delta} < q$. We are now ready to compute M_2. By (26.6) and (26.13) we have

$$\begin{aligned} H(m) &= \sum_{n\geqslant 1} \frac{\tau(n)}{\sqrt{n}} W\Big(\frac{4\pi^2 n}{q}\Big) \sum_f{}^h (1-\varepsilon_f)\lambda_f(m)\lambda_f(n) \\ &= \sum_{n\geqslant 1} \frac{\tau(n)}{\sqrt{n}} W\Big(\frac{4\pi^2 n}{q}\Big) \Delta'(m,n). \end{aligned}$$

Our previous approximation to $\Delta'(m,n)$ by $\delta(m,n)$ is not sufficiently strong. From Corollary 14.26 we have the more precise formula

$$\Delta'(m,n) = \delta(m,n) - \frac{2\pi}{\sqrt{q}} \sum_{(r,q)=1} \frac{1}{r} S(m\bar{q},n;r) J_1\Big(\frac{4\pi}{r}\sqrt{\frac{mn}{q}}\Big) + O\Big(\tau_3((m,n))\frac{(mn)^{\frac{1}{4}}}{q}(\log q)\Big)$$

Hence by (26.14), (26.15)

$$H(m) = \frac{\tau(m)}{\sqrt{m}} W\Big(\frac{4\pi^2 m}{q}\Big) + X(m) + O\Big(\Big(\frac{m}{q}\Big)^{\frac{1}{4}} (\log q)^5\Big) \tag{26.19}$$

where $X(m)$ is a sum of series of Kloosterman sums

$$X(m) = \frac{2\pi}{\sqrt{q}} \sum_{(r,q)=1} r^{-1} X_r(m), \tag{26.20}$$

$$X_r(m) = -\sum_{n\geqslant 1} \frac{\tau(n)}{\sqrt{n}} S(m\bar{q},n;r) J_1\Big(\frac{4\pi}{r}\sqrt{\frac{mn}{q}}\Big) W\Big(\frac{4\pi^2 n}{q}\Big)\xi(n). \tag{26.21}$$

For technical reasons we have inserted in the summation the factor $\xi(n)$ where ξ is a fixed function $\xi : \mathbb{R}^+ \longrightarrow [0,1]$, which is C^∞ and satisfies

$$\xi(x) = 0,\ 0 \leqslant x \leqslant \tfrac{1}{2}, \quad \xi(x) = 1,\ x \geqslant 1.$$

Obviously this extra factor doesn't affect the sum (all positive integers are at least 1!), but it will be useful to gain convergence in some series appearing later.

We first estimate the contribution of those terms with $r > R$ in (26.20), where $R > 0$ will be chosen later ($R = q^2$). Using $J_1(x) \ll x$, Weil's bound for Kloosterman sums, (26.14) and (26.15), we get

$$\frac{2\pi}{\sqrt{q}} \sum_{\substack{r>R \\ (r,q)=1}} \frac{1}{r} X_r(m) \ll \sqrt{\frac{m}{R}} (\log q)^{11}.$$

We denote now $X'(m)$ the remaining part of the sum in $X(m)$, namely

$$X'(m) = \frac{2\pi}{\sqrt{q}} \sum_{\substack{r\leqslant R \\ (r,q)=1}} r^{-1} X_r(m). \tag{26.22}$$

Let $r \leqslant R$. The expression $X_r(m)$ can be computed using the summation formula (4.56) for sums of Kloosterman sums. This yields

$$\begin{aligned} X_r(m) = &-\frac{2}{r} S(m,0;r) \int_0^\infty (\log \frac{\sqrt{x}}{r} + \gamma) t(x) dx \\ &+ \frac{2\pi}{r} \sum_{h\geqslant 1} \tau(h) S(hq-m,0;r) \int_0^{+\infty} Y_0\Big(\frac{4\pi\sqrt{hx}}{r}\Big) t(x) dx \\ &- \frac{4}{r} \sum_{h\geqslant 1} \tau(h) S(hq+m,0;r) \int_0^{+\infty} K_0\Big(\frac{4\pi\sqrt{hx}}{r}\Big) t(x) dx \end{aligned}$$

where

$$t(x) = J_1\Big(\frac{4\pi}{r}\sqrt{\frac{mx}{q}}\Big) W\Big(\frac{4\pi^2 x}{q}\Big) \frac{\xi(x)}{\sqrt{x}}. \tag{26.23}$$

Hence

$$\begin{aligned} X'(m) = &\frac{4\pi}{\sqrt{q}} \sum_{\substack{r\leqslant R \\ (r,q)=1}} \frac{1}{r^2} S(m,0;r) L(r) \\ &+ \frac{4\pi^2}{\sqrt{q}} \sum_{\substack{r\leqslant R \\ (r,q)=1}} \frac{1}{r^2} \sum_{h\geqslant 1} \tau(h) S(hq-m,0;r) y(h) \\ &- \frac{8\pi}{\sqrt{q}} \sum_{\substack{r\leqslant R \\ (r,q)=1}} \frac{1}{r^2} \sum_{h\geqslant 1} \tau(h) S(hq+m,0;r) k(h) + O\Big(\frac{(\log q)^5}{\sqrt{q}}\Big) \end{aligned} \tag{26.24}$$

with

$$L(r) = \int_0^\infty (\log \frac{\sqrt{x}}{r} + \gamma) J_1\Big(\frac{4\pi}{r}\sqrt{\frac{mx}{q}}\Big) W\Big(\frac{4\pi^2 x}{q}\Big) \frac{dx}{\sqrt{x}}, \tag{26.25}$$

$$y(h) = \int_0^{+\infty} Y_0\Big(\frac{4\pi\sqrt{hx}}{r}\Big) t(x) dx, \tag{26.26}$$

$$k(h) = \int_0^{+\infty} K_0\Big(\frac{4\pi\sqrt{hx}}{r}\Big) t(x) dx, \tag{26.27}$$

the error term coming from removing the function $\xi(x)$ from the first term using $t(x) \ll (\log q)^3 x^{-1/2}$ for $0 \leqslant x \leqslant 1$, and $|S(m,0;r)| \leqslant r$.

Let $X''(m)$ be the first term in $X'(m)$. We have thus by changing the integration variable x into $\frac{r^2qy}{4\pi^2}$,

$$X''(m) = -2 \sum_{\substack{r\leqslant R\\(r,q)=1}} \frac{1}{r} S(m,0;r) \int_0^\infty (\log \frac{\sqrt{qx}}{2\pi} + \gamma) J_1(2\sqrt{mx}) W(r^2x) \frac{dx}{\sqrt{x}}$$

$$= \frac{1}{2\pi i} \int_{(1/2)} Z_m^R(1+2s)\zeta_q(1+2s)s^{-1}\Gamma(s)^2 G(s) L(s) ds,$$

by (26.12), with

$$Z_m^R(s) = \sum_{\substack{r\leqslant R\\(r,q)=1}} S(m,0;r)r^{-s},$$

$$L(s) = -2 \int_0^{+\infty} (\log \frac{\sqrt{qx}}{2\pi} + \gamma) J_1(2\sqrt{mx}) x^{-s-1/2} dx.$$

Extending the sum over r to all $r \geqslant 1$, we get

$$Z_m^R(s) = \zeta_q(s)^{-1} \sum_{d|m} d^{1-s} + O(\tau(m)R^{-1})$$

if $\mathrm{Re}\,(s) = 2$ by the formula (3.2) for the Ramanujan sum. Moreover, for all s with $\frac{1}{4} < \mathrm{Re}\,(s) < 1$, we have

$$L(s) = m^{s-1/2}\Gamma(-s)\Gamma(s)^{-1}\Big(\log \frac{\hat{Q}}{m} + 2\gamma + \psi(1+s) + \psi(1-s)\Big) \tag{26.28}$$

where $\psi = \Gamma'/\Gamma$ and $\hat{Q} = q/4\pi^2$. Indeed formula 14.561.14 of [**GR**] gives (for $-2 < \mathrm{Re}\,(s) < -\frac{1}{2}$)

$$\int_0^{+\infty} J_1(x)x^s dx = 2^s \frac{\Gamma(1+\frac{s}{2})}{\Gamma(1-\frac{s}{2})}$$

$$\int_0^{+\infty} (\log x) J_1(x) x^s dx = 2^s \frac{\Gamma(1+\frac{s}{2})}{\Gamma(1-\frac{s}{2})} \Big(\log 2 + \frac{1}{2}\psi\Big(1+\frac{s}{2}\Big) + \frac{1}{2}\psi\Big(1-\frac{s}{2}\Big)\Big)$$

and a simple computation gives (26.28). Consequently we have

$$X''(m) = \frac{1}{2\pi i} \int_{(1/2)} (-2)\sigma_{-2s}(m)s^{-1}\Gamma(s)^2 G(s) L(s) ds + O\Big(\frac{\tau(m)}{R} \log q\Big)$$

since (on $\mathrm{Re}\,(s) = \frac{1}{2}$)

$$\zeta_q(1+2s)\Gamma(s)^2 G(s) L(s) \ll |\Gamma(s)\Gamma(-s)s^{-1}|(\log q + |\psi(1+s)| + |\psi(1-s)|).$$

The function $F(s)$ being integrated can be written

$$F(s) = m^{-1/2} s\eta_s(m)^{-1} G(s)\Gamma(s)\Gamma(-s)\Big(\log \frac{\hat{Q}}{m} + 2\gamma + \psi(1+s) + \psi(1-s)\Big)$$

where

$$\eta_s(m) = \sum_{ab=m} \Big(\frac{a}{b}\Big)^s$$

(Fourier coefficients of the non-holomorphic Eisenstein series for $SL(2,\mathbb{Z})$). Thus, $F(s)$ is an odd function of s. Moreover, it is holomorphic in the strip $|\mathrm{Re}\,(s)| < 1$, except for a triple pole at $s = 0$, and decreases exponentially in vertical strips. Shifting the contour to $\mathrm{Re}\,(s) = -\frac{1}{2}$ and then changing s into $-s$, we conclude that

$$X''(m) = \tfrac{1}{2} \operatorname*{res}_{s=0} F(s) + O\Big(\frac{\tau(m)}{R} \log q\Big).$$

Expanding around $s = 0$ we have

$$\begin{aligned} s^{-1}\Gamma(s)\Gamma(-s) &= -s^{-3} + (\gamma^2 - \Gamma''(1))s^{-1} + O(s), \\ G(s) &= 1 + \tfrac{1}{2}G''(0)s^2 + O(s^3), \\ 2\gamma + \psi(1+s) + \psi(1-s) &= \psi''(0)s^2 + O(s^4), \\ \eta_s(m) &= \tau(m) + \tfrac{1}{2}T(m)s^2 + O(s^3) \end{aligned}$$

where

$$T(m) = \sum_{ab=m} \Big(\log \frac{a}{b}\Big)^2. \tag{26.29}$$

Combining those, we obtain therefore

$$X''(m) = -\frac{T(m)}{4\sqrt{m}}\Big(\log \frac{\hat{Q}}{m}\Big) + \alpha \frac{\tau(m)}{\sqrt{m}}\Big(\log \frac{\hat{Q}}{m}\Big) + O\Big(\frac{\tau(m)}{R} \log q\Big), \tag{26.30}$$

where $\alpha = \frac{1}{2}(\gamma^2 - \Gamma''(1) - \frac{1}{2}G''(0) - \psi''(0))$.

Coming to the other two terms in (26.24), we claim that for $q < m$ and $R \leqslant q^2$

$$\frac{1}{\sqrt{q}} \sum_{r \leqslant R}{}^{\star} \frac{1}{r^2} \sum_{h \geqslant 1} \tau(h) S(hq - m, 0; r) y(h) \ll m^{1/2} q^{-1+\varepsilon}, \tag{26.31}$$

$$\frac{1}{\sqrt{q}} \sum_{r \leqslant R}{}^{\star} \frac{1}{r^2} \sum_{h \geqslant 1} \tau(h) S(hq + m, 0; r) k(h) \ll (m^{1/2} q^{-1} + q^{-1/2}) q^{\varepsilon} \tag{26.32}$$

for any $\varepsilon > 0$, the implied constant depending only on ε.

We choose $R = q^2$. By (26.20), (26.22), (26.24), (26.30), (26.31), (26.32) together, we deduce

$$X(m) = -\frac{T(m)}{4\sqrt{m}}\Big(\log \frac{\hat{Q}}{m}\Big) + \alpha \frac{\tau(m)}{\sqrt{m}}\Big(\log \frac{\hat{Q}}{m}\Big) + O\Big(\Big(\frac{\sqrt{m}}{q} + \frac{1}{\sqrt{q}}\Big) q^{\varepsilon}\Big). \tag{26.33}$$

Inserting this in (26.19) and appealing to (26.16), we obtain

PROPOSITION 26.11. *The twisted mean-value (26.6) is given by*

(26.34)

$$H(m) = \frac{1}{12} \frac{\tau(m)}{\sqrt{m}} \Big(\log \frac{\hat{Q}}{m}\Big)^3 - \frac{1}{4} \frac{T(m)}{\sqrt{m}} \Big(\log \frac{\hat{Q}}{m}\Big) + \frac{\tau(m)}{\sqrt{m}} P_1\Big(\log \frac{\hat{Q}}{m}\Big) + O(m^{1/4} q^{-1/4+\varepsilon})$$

where $\hat{Q} = q/4\pi^2$ and $P_1(X) = P(X) + \alpha X$ for $1 \leqslant m \leqslant q^{2\Delta}$ with $\Delta < 1/2$. The implied constant depends only on ε and Δ.

REMARK. In particular, for $m = 1$, (26.17) and (26.34) give asymptotic formulas for the first and second moments of the special values of the derivatives $L'(f, \frac{1}{2})$:

$$\sum_{f \in S_2(q)^*}^{h} \tfrac{1}{2}(1-\varepsilon_f)L'(f,\tfrac{1}{2}) = \tfrac{1}{2}\log q - \gamma \log 2\pi + O(q^{-1/2}), \tag{26.35}$$

$$\sum_{f \in S_2(q)^*}^{h} \tfrac{1}{2}(1-\varepsilon_f)L'(f,\tfrac{1}{2})^2 = \frac{1}{12}\Big(\log\frac{q}{4\pi^2}\Big)^3 + P_1\Big(\log\frac{q}{4\pi^2}\Big) + O(q^{-1/4+\varepsilon}). \tag{26.36}$$

Finally, the expression for the second moment of special values as quadratic form in the mollifier coefficients follows from (26.8) and (26.34).

THEOREM 26.12. *Assume $M = q^{\Delta}$ with $\Delta < \frac{1}{2}$. Then*

$$M_2 = \frac{1}{12}M_{21} - \frac{1}{4}M_{22} + M_{23} + O(q^{-\delta})$$

where M_{21}, M_{22} and M_{23} are quadratic forms in the variables x_m given by

$$M_{21} = \sum_b \frac{1}{b} \sum_{m_1,m_2} \frac{\tau(m_1m_2)}{m_1m_2} x_{bm_1}x_{bm_2}\Big(\log\frac{\hat{Q}}{m_1m_2}\Big)^3, \tag{26.37}$$

$$M_{22} = \sum_b \frac{1}{b} \sum_{m_1,m_2} \frac{T(m_1m_2)}{m_1m_2} x_{bm_1}x_{bm_2}\Big(\log\frac{\hat{Q}}{m_1m_2}\Big), \tag{26.38}$$

$$M_{23} = \sum_b \frac{1}{b} \sum_{m_1,m_2} \frac{\tau(m_1m_2)}{m_1m_2} x_{bm_1}x_{bm_2}P_1\Big(\log\frac{\hat{Q}}{m_1m_2}\Big), \tag{26.39}$$

and δ is a positive number depending only on Δ. Recall that $\hat{Q} = q/(4\pi^2)$.

REMARK. In fact our proof is complete only for $\Delta < \frac{1}{4}$. This limitation comes from the error term in (26.19), all other parts being adequate for $\Delta < \frac{1}{2}$. The full extension to this case can be established by arguments essentially the same as in the other parts, so we omit the details.

The proofs of (26.31) and (26.32) require only upper bounds for the integrals $y(h)$ and $k(h)$. These are estimated by standard integration by parts (as in Section 4.4). Combining with the bound for Ramanujan sums $|S(hq \pm m, 0; r)| \leqslant (hq \pm m, r)$ one derives the results (for details see the original paper [**KM1**]).

26.4. Optimizing the mollifier.

Now (26.18) expresses M_1 as a linear form, and Theorem 26.12 expresses M_2 as a quadratic form, in the coefficients x_m. The optimal choice of x_m to maximize M_1^2/M_2 (see (26.2)) is not easy to express explicitly because the quadratic form is not diagonal. Our strategy will be to decompose M_2 as a sum of quadratic forms, one of which, say M_2', is easily diagonalized. We then choose x_m to maximize M_1^2/M_2', and simply evaluate the other quadratic forms for this chosen vector. One could justify afterwards that this is asymptotically best possible even for the full quadratic form. It turns out that both M_{21} and M_{22} in Theorem 26.12 have contributions of the same order of magnitude, whereas M_{23} is smaller.

We separate m_1 and m_2 in (26.37) by means of the formula

$$\tau(m_1m_2) = \sum_{a|(m_1,m_2)} \mu(a)\tau\Big(\frac{m_1}{a}\Big)\tau\Big(\frac{m_2}{a}\Big)$$

getting (after relaxing the resulting coprimality condition by means of the Möbius function)

$$M_{21} = \sum_b \frac{1}{b}\sum_a \frac{\mu(a)}{a^2}\sum_{m_1,m_2}\frac{\tau(m_1)\tau(m_2)}{m_1m_2}x_{abm_1}x_{abm_2}\Big(\log\frac{\hat{Q}}{a^2m_1m_2}\Big)^3.$$

Expanding the logarithm, M_{21} is a linear combination of the quadratic forms

$$\Pi(t,u,v,w) = (\log\hat{Q})^u\sum_k \nu_t(k)y_k^{(v)}y_k^{(w)}$$

where

$$y_k^{(i)} = \sum_m \frac{\tau(m)}{m}(\log m)^i x_{km},$$

$$\nu_t(k) = \frac{1}{k}\sum_{ab=k}\frac{\mu(a)}{a}(\log a)^t,$$

and t, u, v and w are non-negative integers such that $t+u+v+w=3$. In fact M_{23} is also such a linear combination with parameters such that $t+u+v+w\leqslant 2$, however.

For the chosen vector (x_m) it will be clear that

$$\Pi(t,u,v,w) \ll \Pi(0,u,v,w)\frac{(\log\log q)^{t+2}}{\log q}$$

by direct estimation, so we can restrict our attention to $\Pi(u,v,w)=\Pi(0,u,v,w)$. Accordingly we write $\nu(k)$ for $\nu_0(k)$. Note that for $k\leqslant M$, we have $\nu(k)=\varphi(k)k^{-2}$.

The contribution to M_{21} of those $\Pi(u,v,w)$ is the quadratic form

$$m_{21} = \Pi(3,0,0) - 6\Pi(2,1,0) + 6\Pi(1,1,1) + 6\Pi(1,2,0) - 6\Pi(0,1,2) - 2\Pi(0,0,3)$$

(using the obvious symmetry $\Pi(u,v,w)=\Pi(u,w,v)$). We select the one quadratic form $\Pi=\Pi(3,0,0)$ as reference and we choose (x_m) which maximizes M_1^2/Π. Then we evaluate the other quadratic forms $\Pi(u,v,w)$ and M_{22} for that vector.

By definition, Π is already diagonalized

$$\Pi = (\log\hat{Q})^3\sum_k \nu(k)y_k^2.$$

The x_m are recovered by the formula

$$x_m = \sum_k \frac{g(k)}{k}y_{km} \tag{26.40}$$

where $g=\mu\star\mu$ is the Dirichlet convolution inverse of τ. Hence the linear form M_1 can be written as

$$M_1 = \sum_m \frac{x_m}{m}\log\frac{\hat{q}}{m} = \sum_k j(k)y_k$$

where

$$j(k) = \frac{1}{k} \sum_{ab=k} g(a)\Big(\log \frac{\hat{q}}{b}\Big).$$

Since

$$\sum_{k \geqslant 1} g(k) k^{-s} = \zeta(s)^{-2},$$

we have

$$\begin{aligned}\sum_{k \geqslant 1} j(k) k^{-s} &= \zeta(s+1)^{-2}\Big((\log \hat{q})\zeta(s+1) + \zeta'(s+1)\Big) \\ &= \frac{\log \hat{q}}{\zeta(s+1)} - (\zeta^{-1})'(s+1)\end{aligned}$$

which shows that

$$j(k) = \frac{\mu(k)}{k}(\log \hat{q}k). \tag{26.41}$$

By Cauchy's inequality we have

$$M_1^2 \leqslant \Big(\sum_k \frac{j(k)^2}{\nu(k)}\Big)\Big(\sum_k \nu(k) y_k^2\Big)$$

with equality if $y_k = \frac{\nu(k)}{j(k)}$ for $k \leqslant M$. Hence the best choice of mollifier is

$$y_k = \begin{cases} \frac{\mu(k)k}{\varphi(k)} \log \hat{q}k, \text{ if } k \leqslant M, \\ 0, \text{ if } k > M. \end{cases}$$

In particular y_k, and so also x_m (see (26.40)), is supported on squarefree numbers and the growth condition (26.4) is immediately verified. Since $j(k)$ is about $(\log k)/k$ and ν is about k^{-1}, it is quite clear from looking at the various expressions involved that we will get a positive (harmonic) proportion if $M = q^{\Delta}$ for any $\Delta > 0$, with both M_1^2 and M_2 being of size $(\log q)^6$.

The results of evaluating M_1 and M_{21}, M_{22} are as follows

PROPOSITION 26.13. *Let $M = q^{\Delta}$ and x_m be as above. We have*

$$\begin{aligned}&M_1 = \Delta\Big(\frac{\Delta^2}{3} + \frac{\Delta}{2} + \frac{1}{4}\Big)(\log q)^3 + O\big((\log q)^2\big), \\ &M_{21} = \Big(\frac{4}{3}\Delta^6 + \frac{28}{5}\Delta^5 + \frac{33}{4}\Delta^4 + \frac{35}{6}\Delta^3 + 2\Delta^2 + \frac{1}{4}\Delta\Big)(\log q)^6 + O\big((\log q)^5\big), \\ &M_{23} \ll (\log q)^5 (\log\log q)^4.\end{aligned}$$

PROOF. For M_1, by definition of y_k and (26.41), we have

$$M_1 = (\log \hat{Q})^{-3}\Pi = \sum_k \frac{j(k)^2}{\nu(k)} = \sum_k \frac{\mu(k)^2}{\varphi(k)}(\log \hat{q}k)^2$$

and the result follows by partial summation from

$$\sum_{k \leqslant K} \frac{\mu(k)^2}{\varphi(k)} = \log K + O(1).$$

For M_{21}, we have

$$\Pi(u,v,w)=(\log\hat{Q})^u\sum_k\nu(k)y_k^{(v)}y_k^{(w)}$$

and we express $y_k^{(i)}$ in terms of y_k using the higher von Mangoldt function $\Lambda_i = \mu\star(\log)^i$. Precisely we have

$$y_k^{(i)}=\sum_{\ell\leqslant M/k}\frac{\tau(\ell)}{\ell}\Lambda_i(\ell)y_{k\ell}. \tag{26.42}$$

See Section 1.4 ((1.43) and after) for the properties of Λ_i that we need.

The sum in (26.42) is thus a sum over squarefree ℓ having at most i prime factors ($i\leqslant 3$). To evaluate we can find the associated Dirichlet series or more elementarily separate the sum into the parts with a fixed number of prime factors. Those are multiple sums over primes, and they are of Mertens type since $\tau(\ell)\ell^{-1}=2^j\ell^{-1}$ for such ℓ with $\omega(\ell)=j$ prime factors, hence easy to evaluate asymptotically. We quote the result (see [**KM1**], 2.5.2, for details): one finds

$$y_k^{(i)}=c_i\frac{k\mu(k)}{\varphi(k)}\Bigl(\log\frac{M}{k}\Bigr)^i(\log\hat{q}^{i+1}M^ik)+O\bigl(\frac{k}{\varphi(k)}(\log q)^i(\log\log q)\bigr)$$

for $i=1,2,3$ with $c_1=-1$, $c_2=\frac{1}{3}$ and $c_3=0$. Then the computation of the various $\Pi(u,v,w)$ is a simple matter of summation by parts. For instance we have

$$\begin{aligned}\Pi(0,1,2)&=-\frac{1}{3}\sum_{k\leqslant M}\frac{\mu(k)^2}{\varphi(k)}\Bigl(\log\frac{M}{k}\Bigr)^3(\log\hat{q}^3M^2k)(\log\hat{q}^2Mk)\\&\qquad\qquad\qquad+O((\log q)^5(\log\log q)^3)\\&=\int_1^M\Bigl(\log\frac{M}{x}\Bigr)^3(\log\hat{q}^3M^2x)(\log\hat{q}^2Mx)\frac{dx}{x}+O((\log q)^5(\log\log q)^3)\\&=\int_0^{\log M}y^3(3\log\hat{q}M-y)(2\log\hat{q}M-y)dy+O((\log q)^5(\log\log q)^3)\\&=\Bigl(-\frac{2}{9}\Delta^6-\frac{1}{3}\Delta^5-\frac{1}{8}\Delta^4\Bigr)(\log q)^6+O((\log q)^5(\log\log q)^3)\end{aligned}$$

(recall $\log\hat{q}=\log\sqrt{q}+O(1)$). When everything is computed, the value of M_{21} is obtained as claimed. □

There remains to treat the quadratic form M_{22} defined by (26.38). Recall that $T(n)$ is defined by (26.29). It follows that

$$T(m_1m_2)=\sum_{d|(m_1,m_2)}\mu(d)\Bigl(\tau\Bigl(\frac{m_1}{d}\Bigr)T\Bigl(\frac{m_2}{d}\Bigr)+\tau\Bigl(\frac{m_2}{d}\Bigr)T\Bigl(\frac{m_1}{d}\Bigr)\Bigr). \tag{26.43}$$

We use this to separate m_1 and m_2, which yields the expression

$$\begin{aligned} M_{22} &= 2\sum_b \frac{1}{b}\sum_a \frac{\mu(a)}{a^2}\sum_{m_1,m_2}\frac{\tau(m_1)T(m_2)}{m_1m_2}x_{abm_1}x_{abm_2}\Big(\log\frac{\hat{Q}}{a^2m_1m_2}\Big) \\ &= 2\sum_k \nu(k)\sum_{m_1,m_2}\frac{\tau(m_1)T(m_2)}{m_1m_2}x_{km_1}x_{km_2}\Big(\log\frac{\hat{Q}}{m_1m_2}\Big) \\ &- 4\sum_k \nu_1(k)\sum_{m_1,m_2}\frac{\tau(m_1)T(m_2)}{m_1m_2}x_{km_1}x_{km_2}. \end{aligned}$$

This is rewritten as

$$M_{22} = 2\Big(\tilde{\Pi}(1,0,0) - \tilde{\Pi}(0,1,0) - \tilde{\Pi}(0,0,1)\Big) - 4\sum_k \nu_1(k)y_kz_k \tag{26.44}$$

where

$$\begin{aligned} z_k = z_k^{(0)} &= \sum_m \frac{T(m)}{m}x_{km}, \\ z_k^{(1)} &= \sum_m \frac{T(m)}{m}(\log m)x_{km} \end{aligned}$$

and

$$\tilde{\Pi}(a,b,c) = (\log\hat{Q})^a\sum_k \nu(k)y_k^{(b)}z_k^{(c)}.$$

As before we express the new variables in terms of y_k by

$$z_k = 2\sum_{\ell\leqslant M/k}\frac{\Lambda(\ell)\log\ell}{\ell}y_{k\ell}, \tag{26.45}$$

$$z_k^{(1)} = \sum_{\ell\leqslant M/k}\frac{\tau(\ell)\Lambda(\ell)}{\ell}z_{k\ell} + \sum_{\ell\leqslant M/k}\frac{T(\ell)\Lambda(\ell)}{\ell}y_{k\ell}. \tag{26.46}$$

To see this, notice that by (26.40) we have

$$z_k = \sum_\ell\Big(\sum_{mn=\ell}\frac{T(m)}{m}g(n)\Big)y_{k\ell}.$$

The Dirichlet generating series for the coefficient of ℓ is $L(s+1)$ where

$$L(s) = \zeta(s)^{-2}\sum_n T(n)n^{-s}.$$

It is easy to see that

$$\sum_n T(n)n^{-s} = 4\zeta\zeta'' - 2(\zeta\zeta')' = 2(\zeta\zeta'' - (\zeta')^2),$$

so $L(s) = 2(\zeta'\zeta^{-1})'$, which gives (26.45). Then we derive (26.46) as follows

$$\begin{aligned} z_k^{(1)} &= \sum_m \frac{T(m)}{m} \sum_{\ell b = m} \Lambda(\ell) x_{km} = \sum_\ell \frac{\Lambda(\ell)}{\ell} \sum_{m \leqslant M/\ell} \frac{T(\ell m)}{m} x_{k\ell m} \\ &= \sum_{\ell \leqslant M/k} \frac{\tau(\ell)\Lambda(\ell)}{\ell} z_{k\ell} + \sum_{\ell \leqslant M/k} \frac{T(\ell)\Lambda(\ell)}{\ell} y_{k\ell} \end{aligned}$$

by (26.43).

By (26.45) and (26.46) we can evaluate z_k and $z_k^{(1)}$ much as before with $y_k^{(i)}$, and we get

$$\begin{aligned} z_k &= -\frac{1}{3}\frac{k\mu(k)}{\varphi(k)}\Big(\log\frac{M}{k}\Big)^2 (\log \hat{q}^3 M^2 k) + O(\frac{k}{\varphi(k)}(\log q)^2) \\ &= -y_k^{(2)} + O\Big(\frac{k}{\varphi(k)}(\log q)^2 (\log\log q)\Big), \\ z_k^{(1)} &\ll \frac{k}{\varphi(k)}(\log q)^3. \end{aligned}$$

From this we easily evaluate the quadratic forms that occur in (26.44)

$$\begin{aligned} &\tilde{\Pi}(1,0,0) = -(\log \hat{Q}) \sum_k \nu(k) y_k y_k^{(2)} + O((\log q)^5) = -\Pi(1,2,0) + O((\log q)^5), \\ &\tilde{\Pi}(0,1,0) = -\sum_k \nu(k) y_k^{(1)} y_k^{(2)} + O((\log q)^5) = -\Pi(0,1,2) + O((\log q)^5), \\ &\tilde{\Pi}(0,0,1) \ll (\log q)^5. \end{aligned}$$

Moreover, by direct estimation

$$\sum_k \nu_1(k) y_k z_k \ll (\log q)^5 (\log\log q)^3.$$

Now (26.44) yields:

PROPOSITION 26.14. *Let $m = q^\Delta$. For x_m as above, we have*

$$M_{22} = \Big(-\frac{4}{9}\Delta^6 - \frac{4}{5}\Delta^5 - \frac{7}{12}\Delta^4 - \frac{1}{6}\Delta^3\Big)(\log q)^6 + O((\log q)^5 (\log\log q)^3).$$

From Theorem 26.12, Proposition 26.13 and Proposition 26.14, we can compute that for $\Delta < \frac{1}{2}$,

$$\frac{M_1^2}{M_2} = \frac{1}{2}\Big(1 - \frac{1}{(2\Delta+1)^3}\Big) + O\Big(\frac{(\log\log q)^4}{\log q}\Big)$$

(where the very simple fraction appears quite miraculously by partial fraction expansion). Inequality (26.2) then implies Theorem 26.9, and therefore Theorem 26.8.

EXERCISE. Show by similar methods that

$$\sum_{\substack{f\in S_2(q)^*\\ L(f,\frac{1}{2})\neq 0}}^h 1 \geqslant \Big(\frac{1}{6}-o(1)\Big)$$

for primes $q \to +\infty$.

26.5. Proof of theorem 26.2.

Passing from Theorem 26.8 (non-vanishing in "harmonic" average) to Theorem 26.2 (non-vanishing in "natural" average) can be done in different ways. We use Shimura's formula

$$4\pi\|f\|^2 = \frac{|S_2(q)^*|}{\zeta(2)} L(\mathrm{Sym}^2 f, 1) + O((\log q)^3) \quad \text{(for } q \text{ prime)} \tag{26.47}$$

relating the Petersson norm to the special value of the symmetric square L-function of f at $s = 1$. The latter is given by

$$L(\mathrm{Sym}^2 f, s) = \sum_{n\geqslant 1} \rho_f(n) n^{-s} = \zeta_q(2s) \sum_{n\geqslant 1} \lambda_f(n^2) n^{-s}$$

for $\mathrm{Re}\,(s) > 1$; see Section 5.12. At $s = 1$ the series barely fails to converge absolutely, but on average it can be represented by a very short sum. From [**KM2**], we quote a general useful result based on the large sieve type estimate for symmetric square L-functions (see Theorem 7.28).

PROPOSITION 26.15. *Let α_f, for $f \in S_2(q)^*$, be complex numbers such that*

$$\alpha_f \ll q^{1-\delta} \text{ for some } \delta > 0, \tag{26.48}$$

$$\sum_f^h |\alpha|_f \ll (\log q)^A \text{ for some } A > 0, \tag{26.49}$$

Then we have

$$\sum_f \alpha_f = \frac{|S_2(q)^*|}{\zeta(2)} \sum_{n\leqslant x} \sum_f^h \alpha_f \frac{\rho_f(n)}{n} + O(q^{1-\gamma})$$

with $x = q^\kappa$, where κ is any positive number and γ is a positive number which depends only on δ and κ. The implied constant depends only on κ, δ and A.

We use a mollifier $M(f)$ as in (26.3) of length $M = q^\Delta$ with $\Delta < \frac{1}{2}$, defining

$$N_1 = \sum_{\varepsilon_f=-1} M(f) L'(f, 1/2),$$

$$N_2 = \sum_{\varepsilon_f=-1} |M(f) L'(f, 1/2)|^2$$

and apply Proposition 26.15 with $\alpha_f = M(f)L'(f, \frac{1}{2})$ and $\alpha = |M(f)L'(f, \frac{1}{2})|^2$ respectively. The convexity bound $L'(f, \frac{1}{2}) \ll q^{1/4}(\log q)^2$ (use (26.9) to prove it again) yields easily (26.48) and (26.49).

It follows that

$$N_1 = \frac{|S_2(q)^*|}{\zeta(2)} \sum_{\varepsilon_f=-1}^{h} \sum_{\substack{d\ell^2 \leqslant x \\ (\ell,q)=1}} \frac{1}{d\ell^2} \lambda_f(d^2) M(f) L'(f, \tfrac{1}{2}) + O(q^{1-\delta}),$$

$$N_2 = \frac{|S_2(q)^*|}{\zeta(2)} \sum_{\varepsilon_f=-1}^{h} \sum_{\substack{d\ell^2 \leqslant x \\ (\ell,q)=1}} \frac{1}{d\ell^2} \lambda_f(d^2) |M(f) L'(f, \tfrac{1}{2})|^2 + O(q^{1-\delta})$$

with $x = q^\kappa$ where $\kappa > 0$ can be chosen arbitrarily.

Inserting the formulas (26.9) and (26.13), and using results of Section 26.3, we are led to a linear form expression for N_1 and a quadratic form for N_2 which are very similar to those of Theorem 26.12. It is important to be able to take κ arbitrarily small to retain an asymptotic formula. The same principle as in Section 26.4 applies to find the mollifier and evaluate N_1 and N_2, but there are many technical complications due to lack of complete multiplicativity of the coefficients involved.

REMARK. If one is willing to accept a slightly weaker result

$$|\{f \in S_2(q)^* \mid \varepsilon_f = -1 \text{ and } L'(f, \tfrac{1}{2}) \neq 0\}| \gg |S_2(q)^*| \tag{26.50}$$

for q prime, with no explicit constant (the analogue of Selberg's theorem on critical zeros of $\zeta(s)$), some shortcuts are available. First, there is no reason to optimize the mollifier as in Section 26.4, and the computations there can be simplified quite a bit. Then instead of appealing to Proposition 26.15 and performing the computation of N_1 and N_2 one can argue by means of Cauchy's inequality

$$\sum_{\substack{f \in S_2(q)^* \text{ odd} \\ L'(f,1/2) \neq 0}}^{h} 1 = \sum_{\substack{f \in S_2(q)^* \text{ odd} \\ L'(f,1/2) \neq 0}} \frac{1}{4\pi \|f\|^2} \leqslant \Big(\sum_f \frac{1}{16\pi^2 \|f\|^4} \Big)^{1/2} \Big(\sum_{\substack{f \in S_2(q)^* \text{ odd} \\ L'(f,1/2) \neq 0}} 1 \Big)^{1/2}.$$

Then the asymptotic formula

$$\sum_f \frac{1}{16\pi^2 \|f\|^4} \sim cq^2 \tag{26.51}$$

as $q \to +\infty$ for some constant $c > 0$ together with Theorem 26.8 imply (26.50). Formula (26.51) is a special case of a result of E. Royer [**Roy**] about moments of $L(\mathrm{Sym}^2 f, 1)$ over $f \in S_2(q)^*$. Note that by (26.47), this amounts to the negative second moment of $L(\mathrm{Sym}^2 f, 1)$. The proof of (26.51) uses a density theorem for zeros of families of L-functions to restrict the summation to f satisfying a "quasi-Riemann Hypothesis".

REMARK. Besides being used to build mollifiers, as we have done, the twisted mean-values of type (26.5), (26.6) for families of L-functions have a powerful use for building an "amplifier", i.e. a sum of the type

$$A_k = \sum_{f \in \mathcal{F}} A(f) |L(f, \tfrac{1}{2})|^k$$

where $A(f) \geqslant 0$, say

$$A(f) = \Big| \sum_{\ell \leqslant L} c_\ell \lambda_f(\ell) \Big|^2$$

but this time $A(f)$ is meant to be large for $f = f_0 \in \mathcal{F}$, some fixed form, so that by positivity one deduces from an upper bound for the mean-value A_k an individual upper bound

$$|L(f_0, \tfrac{1}{2})| \leqslant A_k^{1/k} A(f_0)^{-1}. \tag{26.52}$$

In the best circumstances, the bound for A_k can be achieved if L is a small power of $|\mathcal{F}|$, and matches that from Lindelöf Hypothesis and orthogonality

$$A_k \ll |\mathcal{F}|^{1+\varepsilon} \Big(\sum_{\ell} |c_\ell|^2\Big)^{1/2}$$

so that (26.52) is a gain if the mean-value of c_ℓ is comparable in size to the peak value $A(f_0)$. Think of the case of primitive Dirichlet characters of conductor q where one can take $c_\ell = \overline{\chi}(\ell)$ (see Proposition 14.22 for a suitable choice for classical modular forms).

This amplification method has been the most successful for the proof of subconvexity bounds for L-functions of degree $\geqslant 2$, particularly in conductor aspect where other techniques (Weyl shifts, reducing the averaging family) are inoperant. The implementation is, however, quite delicate in all cases (requiring off-diagonal analysis) so we refer to papers like [**DFI2**], [**DFI3**], [**KMV2**] and the survey [**M2**] for more information on this important subject.

Bibliography

[AS1] A. Adolphson and S. Sperber, *On twisted exponential sums*, Math. Ann. **290** (1991), 713–726.

[AS2] A. Adolphson and S. Sperber, *Character sums in finite fields*, Compositio Math. **52** (1984), 325–354.

[AS3] A. Adolphson and S. Sperber, *Newton polyhedra and the total degree of the L-function associated to an exponential sum*, Invent. math. **88** (1987), 555–569.

[Ahl] L. Ahlfors, *Complex Analysis*, McGraw Hill, 1978.

[AM] M. Atiyah and I.G. MacDonald, *Introduction to commutative algebra*, Addison-Wesley, 1969.

[AL] A.O.L Atkin and J. Lehner, *Hecke operators on* $\Gamma_0(m)$, Math. Ann. **185** (1970), 134–160.

[At] F. V. Atkinson, *The mean value of the zeta-function on the critical line*, Proc. London Math. Soc. (2) **47** (1941), 174–200.

[B] A. Baker, *Transcendental Number Theory*, Cambridge Univ. Press, 1990.

[BH] R. Baker and G. Harman, *The difference between consecutive primes*, Proc. London Math. Soc. **72** (1996), 261–280.

[Ba] W. Banks, *Twisted symmetric-square L-functions and the nonexistence of Siegel zeros on* $GL(3)$, Duke Math. J. **87** (1997), 343–353.

[Bar] M. B. Barban, *The "large sieve" method and its application to number theory*, Uspehi Mat. Nauk **21** (1966), 51–102; English transl. in Russian Math. Surveys **21** (1966), 49–103.

[BG] J. Bernstein and S. Gelbart, *An introduction to the Langlands program*, Birkhaüser, 2003.

[BL] H. Bohr and E. Landau, *Sur les zéros de la fonction* $\zeta(s)$ *de Riemann*, Compte Rendus de l'Acad. des Sciences (Paris) **158** (1914), 106–110.

[Bo1] E. Bombieri, *Maggiorazione del resto nel "Primzahlsatz" col metodo di Erdös–Selberg.*, Ist. Lombardo Accad. Sci. Lett. Rend. A **96** (1962), 343–350.

[Bo2] E. Bombieri, *Le grand crible dans la théorie analytique des nombres*, S.M.F, 1974.

[Bo3] E. Bombieri, *Counting points on curves over finite fields (d'après S. A. Stepanov)*, Lecture Notes in Math. **383** (1974), 234–241.

[Bo4] E. Bombieri, *On exponential sums in finite fields, II*, Invent. math. **47** (1978), 29–39.

[Bo5] E. Bombieri, *On the large sieve*, Mathematika **12** (1965), 201–225.

[BD] E. Bombieri and H. Davenport, *Some inequalities involving trigonometrical polynomials*, Ann. Scuola Norm. Sup. Pisa (3) **23** (1969), 223–241.

[BFI] E. Bombieri, J. Friedlander and H. Iwaniec, *Primes in arithmetic progressions to large moduli*, Acta Math. **156** (1986), 203–251.

[BI] E. Bombieri and H. Iwaniec, *Some mean-value theorems for exponential sums*, Ann. Scuola Norm. Sup. Pisa Cl. Sci. **13** (1986), 473–486.

[Bou] J. Bourgain, *Remarks on Montgomery's conjectures on Dirichlet sums*, Lecture Notes in Math. **1469** (1991), 153–165.

[BDCT] C. Breuil, B. Conrad, F. Diamond and R. Taylor, *On the modularity of elliptic curves over* $\mathbb{Q}$*: wild 3-adic exercises*, J. Amer. Math. Soc. **14** (2001), 843–939.

[BC] W. E. Briggs and S. Chowla, *On discriminants of binary quadratic forms with a single class in each genus*, Canad. J. Math. **6** (1954), 463–470.

[Bru] R. W. Bruggeman, *Fourier coefficients of cusp forms*, Invent. math. **45** (1978), 1–18.

[Br1] V. Brun, *Über das Goldbachsche Gesetz und die Anzahl der Primzahlpaare*, Archiv for Math.og Naturvid. **B34** (1915), no. 8.

[Br2] V. Brun, *Le crible d'Eratosthène et le théorème de Goldbach*, C. R. Acad. Sci. Paris **168** (1919), 544–546.

[Bu] D. Bump, *Automorphic forms and representations*, Cambridge Univ. Press, 1996.

[BDHI] D. Bump, W. Duke, J. Hoffstein and H. Iwaniec, *An estimate for the Hecke eigenvalues of Maass forms*, Internat. Math. Res. Notices **4** (1992), 75–81.

[Bur1] D. A. Burgess, *On character sums and L-series, I*, Proc. London Math. Soc. (3) **12** (1962), 193–206.

[Bur2] D. A. Burgess, *On character sums and L-series, II*, Proc. London Math. Soc. (3) **13** (1963), 524–536.

[BH] C. J. Bushnell and G. Henniart, *An upper bound on conductors for pairs*, J. Number Theory **65** (1997), 183–196.

[Byk] V. A. Bykovsky, *Spectral expansion of certain automorphic functions and its number-theoretical applications*, Proc. Steklov Inst. (LOMI) **134** (1984), 15–33; English transl. in J. Soviet Math. **36** (1987), 8–21.

[Car] F. Carlson, *Über die Nullstellen der Dirichletschen Reihen und der Riemannschen ζ-Funktion*, Arkiv. für Mat. Astr. och Fysik **15** (1920).

[CF] J.W.S Cassels and A. Frölich, *Algebraic Number Theory*, Academic Press, 1990.

[ChI] Chamizo and H. Iwaniec, *On the sphere problem*, Rev. Mat. Iberoamericana **11** (1995), 417–429.

[Ch] J. R. Chen, *On the representation of a larger even integer as the sum of a prime and the product of at most two primes*, Sci. Sinica **16** (1973), 157–176.

[CS] S. Chowla and A. Selberg, *On Epstein's zeta-function*, J. Reine Angew. Math. **227** (1967), 86–110.

[Co1] B. Conrey, *Zeros of derivatives of Riemann's ξ-function on the critical line*, J. Number Theory **16** (1983), 49–74.

[Co2] B. Conrey, *More than two fifths of the zeros of the Riemann zeta function are on the critical line*, J. Reine Angew. Math. **399** (1989), 1–26.

[CGG] B. Conrey, A. Ghosh and S. M. Gonek, *A note on gaps between zeros of the zeta function*, Bull. London Math. Soc. **16** (1984), 421–424.

[CI1] B. Conrey and H. Iwaniec, *The cubic moment of central values of automorphic L-functions*, Ann. of Math. (2) **151** (2000), 1175–1216.

[CI2] B. Conrey and H. Iwaniec, *Spacing of zeros of Hecke L-functions and the class number problem*, Acta Arithmetica **103** (2002), 259–312.

[CoSo] B. Conrey and K. Soundararajan, *Real zeros of quadratic Dirichlet L-functions*, Invent. math. **150** (2002), 1–44.

[Cor1] J.G. van der Corput, *Zahlentheoretische Abschätzungen*, Math. Ann. **84** (1921), 53–79.

[Cor2] J.G. van der Corput, *Verscharfung der Abschätzungen beim Teilerproblem*, Math. Ann. **87** (1922), 39–65.

[Cor3] J. G. van der Corput, *Sur l'hypothèse de Goldbach pour presque tous les nombres premiers*, Acta Arithmetica **2** (1937), 266–290.

[Cox] D. Cox, *Primes of the form $x^2 + ny^2$*, Wiley, 1989.

[Cra] H. Cramér, *On the order of magnitude of the difference between consecutive prime numbers*, Prace Mat.-Fiz. **45** (1937), 51–74.

[Da1] H. Davenport, *On some infinite series involving arithmetical functions. II*, Quart. J. Math. Oxf. **8** (1937), 313–320.

[Da2] H. Davenport, *Analytic methods for Diophantine equations and Diophantine inequalities*, Ann Arbor Publishers, 1963.

[DH1] H. Davenport and H. Halberstam, *The values of a trigonometrical polynomial at well spaced points*, Mathematika **13** (1966), 91–96.

[DH2] H. Davenport and H. Halberstam, *Primes in arithmetic progressions*, Michigan Math. J. **13** (1966), 485–489.

[DHe] H. Davenport and Heilbronn, *On the class-number of binary cubic forms, I, II*, J. London Math. Soc. **26** (1951), 183–192, 192–198.

[De1] P. Deligne, *La conjecture de Weil, I*, Inst. Hautes Études Sci. Publ. Math. **43** (1972), 206–226.

[De2] P. Deligne, *La conjecture de Weil, II*, Inst. Hautes Études Sci. Publ. Math. **52** (1980), 137–252.

[De3] P. Deligne, *Cohomologie étale, SGA $4\frac{1}{2}$*, Lecture Notes. Math. 569, Springer Verlag, 1977.

[DeSe] P. Deligne and J-P. Serre, *Formes modulaires de poids* 1, Ann. Sci. École Norm. Sup. (4) **7** (1974), 507–530.

[DI] J.M. Deshouillers and H. Iwaniec, *Kloosterman sums and Fourier coefficients of cusp forms*, Invent. math. **70** (1982/83), 219–288.

[DS] H. Diamond and J. Steinig, *An elementary proof of the prime number theorem with a remainder term*, Invent. math. **11** (1970), 199–258.

[Dir] P.G. Dirichlet, *Démonstration d'une propriété analogue à la loi de réciprocité qui existe entre deux nombres premiers quelconques*, J. Reine Angew. Math. **9** (1832), 379–389.

[Du1] W. Duke, *The dimension of the space of cusp forms of weight one*, Internat. Math. Res. Notices (1995), 99–109.

[Du2] W. Duke, *Hyperbolic distribution problems and half-integral weight Maass forms*, Invent. math. **92** (1988), 73–90.

[Du3] W. Duke, *The critical order of vanishing of automorphic L-functions with large level*, Invent. math. **119** (1995), 165–174.

[Du4] W. Duke, *Some problems in multidimensional analytic number theory*, Acta Arithmetica **52** (1989), 203–228.

[Du5] W. Duke, *Elliptic curves with no exceptional primes*, C. R. Acad. Sci. Paris Sér. I Math. **325** (1997), 813–818.

[DFI1] W. Duke, J. Friedlander and H. Iwaniec, *A quadratic divisor problem*, Invent. math. **115** (1994), 209–217.

[DFI2] W. Duke, J. Friedlander and H. Iwaniec, *Bounds for automorphic L-functions, II*, Invent. math. **115** (1994), 219–239.

[DFI3] W. Duke, J. Friedlander and H. Iwaniec, *The subconvexity problem for Artin L-functions*, Invent. math. **149** (2002), 489–577.

[DFI4] W. Duke, J. Friedlander and H. Iwaniec, *Bilinear forms with Kloosterman fractions*, Invent. math. **128** (1997), 23–43.

[DFI5] W. Duke, J. Friedlander, H. Iwaniec, *Equidistribution of roots of a quadratic congruence to prime moduli*, Ann. of Math. (2) **141** (1995), 423–441.

[DuI1] W. Duke and H. Iwaniec, *Bilinear forms in the Fourier coefficients of half-integral weight cusp forms and sums over primes*, Math. Ann. **286** (1990), 783–802.

[DuI2] W. Duke and H. Iwaniec, *Estimates for coefficients of L-functions, I*, Automorphic forms and analytic number theory (Montreal, PQ, 1989), CRM, 1989, pp. 43–47.

[DuI3] W. Duke and H. Iwaniec, *Estimates for coefficients of L-functions, II*, Proceedings of the Amalfi Conference on Analytic Number Theory, Univ. Salerno, Salerno, 1992, pp. 71–82.

[DuI4] W. Duke and H. Iwaniec, *Estimates for coefficients of L-functions, IV*, Amer. J. Math. **116** (1994), 207–217.

[DK] W. Duke and E. Kowalski, *A problem of Linnik for elliptic curves and mean-value estimates for automorphic representations*, Invent. math. **139** (2000), 1–39.

[D] F. J. Dyson, *Statistical theory of the energy levels of complex systems, I*, J. Mathematical Phys. **3** (1962), 140–156.

[Ell] J. Ellenberg, *Galois representations attached to $\mathbb{Q}$-curves and the generalized Fermat equation $A^4 + B^2 = C^p$*, Amer. J. Math. (to appear).

[El] P.D.T.A. Elliott, *On inequalities of large sieve type*, Acta Arithmetica **18** (1971), 405–422.

[EK] P. Erdös and M. Kac, *The Gaussian law of errors in the theory of additive number theoretic functions*, Amer. J. Math. **62** (1940), 738–742.

[Est] T. Estermann, *On Goldbach's Problem: Proof that Almost All Even Positive Integers are Sums of Two Primes*, Proc. London Math. Soc. (2) **44** (1938), 307–314.

[Fon] J.M. Fontaine, *Il n'y a pas de variétés abéliennes sur $\mathbb{Z}$*, Invent. math. **81** (1985), 515–538.

[FK] É. Fouvry and N. Katz, *A general stratification theorem for exponential sums, and applications*, J. Reine Angew. Math. **540** (2001), 115–166.

[FI] É. Fouvry and H. Iwaniec, *Gaussian Primes*, Acta. Arithmetica **79** (1997), 249–287.

[FoM] É. Fouvry and P. Michel, *Sur le changement de signe des sommes de Kloosterman* (preprint).

[Fri] J. Friedlander, *Primes in arithmetic progressions and related topics*, Analytic Number Theory and Diophantine problems, Birkhaüser, 1987, pp. 125–134.

[FG] J. Friedlander and A. Granville, *Limitations to the equi-distribution of primes, III*, Compositio Math. **81** (1992), 19–32.

[FI1] J. Friedlander and H. Iwaniec, *The polynomial $X^2 + Y^4$ captures its primes*, Ann. of Math. (2) **148** (1998), 945–1040.

[FI2] J. Friedlander and H. Iwaniec, *Summation formulae for coefficients of L-functions* (to appear).

[FI3] J. Friedlander and H. Iwaniec, *A Note on Character Sums*, Contemporary Math. **166** (1994), 295–299.

[FI4] J. Friedlander and H. Iwaniec, *Incomplete Kloosterman sums and a divisor problem*, Ann. of Math. **121** (1985), 319–350.

[Ga1] P.X. Gallagher, *A large sieve density estimate near $\sigma = 1$*, Invent. math. **11** (1970), 329–339.

[Ga2] P.X. Gallagher, *Primes in progressions to prime-power modulus*, Invent. math. **16** (1972), 191–201.

[Ga3] P. X. Gallagher, *Bombieri's mean value theorem*, Mathematika **15** (1968), 1–6.

[Ga4] P. X. Gallagher, *The large sieve and probabilistic Galois theory*, Proc. Sympos. Pure Math., Vol. XXIV, Amer. Math. Soc., 1973, pp. 91–101.

[GM] M. L. Gaudin and M. Mehta, *On the density of eigenvalues of a random matrix*, Nuclear Phys. **18** (1960), 420–427.

[Ge] F. Gerth III, *Extension of conjectures of Cohen and Lenstra*, Expositiones Math. **5** (1987), 181–184.

[GeJ] S. Gelbart and H. Jacquet, *A relation between automorphic representations of $GL(2)$ and $GL(3)$*, Ann. Sci. École Norm. Sup. (4) **11** (1978), 471–542.

[GJ] R. Godement and H. Jacquet, *Zeta functions of simple algebras*, Lecture Notes Math. 260, Springer Verlag, 1972.

[Go1] D. Goldfeld, *A simple proof of Siegel's theorem*, Proc. Nat. Acad. Sci. U.S.A. **71** (1974), 1055.

[Go2] D. Goldfeld, *The class number of quadratic fields and the conjectures of Birch and Swinnerton-Dyer*, Ann. Scuola Norm. Sup. Pisa Cl. Sci. (4) **3** (1976), 624–663.

[GS] D. Goldfeld and P. Sarnak, *Sums of Kloosterman sums*, Invent. math. **71** (1983), 243–250.

[GHL] D. Goldfeld, J. Hoffstein and D. Lieman, Annals of Math. (2) **140** (1994), 161–181.

[GHB] D. Goldston and D. R. Heath-Brown, *A note on the differences between consecutive primes*, Math. Ann. **266** (1984), 317–320.

[GR] I.S. Gradhsteyn and I.M. Rizhik, *Table of integrals, series and products*, 6th Edition, Academic Press, 2000.

[G1] S. Graham, *An asymptotic estimate related to Selberg's sieve*, J. Number Theory **10** (1978), 83–94.

[G2] S. Graham, *On Linnik's constant*, Acta Arithmetica **39** (1981), 163–179.

[GK] S.W Graham and G. Kolesnik, *van der Corput's method of exponential sums*, Cambridge Univ. Press, 1991.

[GRi] S. W. Graham and C. J. Ringrose, *Lower bounds for least quadratic nonresidues*, Analytic number theory (Allerton Park, IL, 1989), Birkhäuser, 1990, pp. 269–309.

[Gra] A. Granville, *Unexpected irregularities in the distribution of prime numbers*, Proceedings of the International Congress of Mathematicians, Vol. 1, 2 (Zürich, 1994), Birkhäuser, Basel, 1995, pp. 388–399.

[Gr] G. Greaves, *Sieves in number theory*, Ergebnisse der Mathematik und ihrer Grenzgebiete (3), vol. 43, Springer Verlag, 2001.

[GZ1] B. Gross and D. Zagier, *Heegner points and derivatives of L-series*, Invent. math. **84** (1986), 225–320.

[GZ2] B. Gross and D. Zagier, *Points de Heegner et dérivées de fonctions L*, C. R. Acad. Sci. Paris Sér. I Math. **297** (1983), 85–87.

[HaTu] G. Halász and P.Turán, *On the distribution of roots of Riemann zeta and allied functions, I*, J. Number Theory **1** (1969), 121–137.

[Hal] H. Halberstam, *On the distribution of additive number-theoretic functions, II, III*, J. London Math. Soc. **31** (1956), 1–14, 14–27.

[HaRi] H. Halberstam and H. E. Richert, *Sieve methods*, Academic Press, 1974.

[HaLa] G. H. Hardy and E. Landau, *The lattice points of a circle*, Proc. Royal Soc. A **105** (1924), 244–258.

[HL1] G. H. Hardy and J. E. Littlewood, *Some Problems of 'Partitio Numerorum.' III. On the Expression of a Number as a Sum of Primes.*, Acta Math. **44** (1922), 1–70.

[HL2] G. H. Hardy and J. E. Littlewood, *The zeros of Riemann's zeta function on the critical line*, Math. Z. **10** (1921), 283–317.

[HR] G.H. Hardy and S. Ramanujan, *Asymptotic Formulae in Combinatory Analysis*, Proc. London Math. Soc. **17** (1918), 75–115.

[HW] G.H. Hardy and E.M. Wright, *An introduction to the theory of numbers*, 5th edition, Oxford University Press, 1979.

[HT] M. Harris and R. Taylor, *The geometry and cohomology of some simple Shimura varieties*, Princeton Univ. Press, 2002.

[Ha] R. Hartshorne, *Algebraic Geometry*, Grad. Texts in Math. 52, Springer Verlag, 1977.

[HB1] D.R. Heath-Brown, *A mean value estimate for real character sums*, Acta Arithmetica **72** (1995), 235–275.

[HB2] D.R. Heath-Brown, *An estimate for Heilbronn's exponential sum*, Analytic number theory, Vol. 2 (Allerton Park, IL, 1995), Birkhaüser, 1995, pp. 451–463.

[HB3] D. R. Heath-Brown, *Prime numbers in short intervals and a generalized Vaughan identity*, Canad. J. Math. **34** (1982), 1365–1377.

[HB4] D. R. Heath-Brown, *Zero-free regions for Dirichlet L-functions, and the least prime in an arithmetic progression*, Proc. London Math. Soc. (3) **64** (1992), 265–338.

[HB5] D. R. Heath-Brown, *Lattice points in the sphere*, Number theory in progress, Vol. 2, de Gruyter, Berlin, 1999, pp. 883–892.

[HB6] D. R. Heath-Brown, *Cubic forms in ten variables*, Proc. London Math. Soc. (3) **47** (1983), 225–257.

[HBP] D.R. Heath-Brown and S.J. Patterson, *The distribution of Kummer sums at prime arguments*, J. Reine Angew. Math. **310** (1979), 111–130.

[Hec1] E. Hecke, *Über eine neue Art von Zetafunktionen*, Math. Zeit. **6** (1920), 11–51.

[Hee] K. Heegner, *Diophantische Analysis und Modulfunktionen*, Math. Z. **56** (1952), 227–253.

[H] D. Hilbert, *Beweis für die Darstellbarkeit der ganzen Zahlen durch eine feste Anzahl n-ter Potenzen (Waringsches Problem)*, Math. Annalen **67** (1909), 281–305.

[Hi1] A. Hildebrand, *An asymptotic formula for the variance of an additive function*, Math. Z. **183** (1983), 145–170.

[Hi2] A. Hildebrand, *On the constant in the Pólya-Vinogradov inequality*, Canad. Math. Bull. **31** (1988), 347–352.

[Hof] J. Hoffstein, *On the Siegel-Tatuzawa theorem*, Acta Arithmetica **38** (1980/81), 167–174.

[HL] J. Hoffstein and P. Lockhart, *Coefficients of Maass forms and the Siegel zero*, Annals of Math. **140** (1994), 161–181.

[Ho] G. Hoheisel, *Primzahlprobleme in der Analysis*, S.-B. Preuss. Akad. Wiss. Phys.-Math. Kl. (1930), 580–588.

[H1] Loo-Keng Hua, *Introduction to number theory*, Springer, 1982.

[H2] Loo Keng Hua, *On Waring's problem*, Quart. J. Math. Oxford **9** (1938), 199–202.

[Hub] H. Huber, *Zur analytischen Theorie hyperbolischen Raumformen und Bewegungsgruppen*, Math. Ann. **138** (1959), 1–26.

[Hu1] M. N. Huxley, *The large sieve inequality for algebraic number fields*, Mathematika **15** (1968), 178–187.

[Hu2] M. N. Huxley, *Large values of Dirichlet polynomials*, Acta Arithmetica **24** (1973), 329–346.

[Hu3] M. N. Huxley, *On the differences between consecutive primes*, Invent. math. **15** (1972), 164–170.

[Hu4] M. N. Huxley, *Area, lattice points, and exponential sums*, The Clarendon Press, 1996.

[HuW] M. N. Huxley and N. Watt, *Exponential sums and the Riemann zeta function*, Proc. London Math. Soc. **57** (1988), 1–24.

[Ing] A. E. Ingham, *On the estimation of* $N(\sigma, T)$, Quart. J. Math. **11** (1940), 291–292.

[IR] K. Ireland and M. Rosen, *A Classical Introduction to Modern Number Theory*, 2nd Edition, Grad. Texts in Math. 84, Springer-Verlag, 1990.

[Iv] A. Ivić, *The Riemann zeta-function, Theory and applications*, Dover Publications, 2003.

[I1] Iwaniec, *Character sums and small eigenvalues for* $\Gamma_0(p)$, Glasgow Math. J. **27** (1985), 99–116.

[I2] H. Iwaniec, *On zeros of Dirichlet's L series*, Invent. math. **23** (1974), 97–104.

[I3] H. Iwaniec, *Spectral theory of automorphic functions and recent developments in analytic number theory*, Proceedings of the ICM Berkeley 1986, Amer. Math. Soc., 1987, pp. 444–456.

[I4] H. Iwaniec, *Topics in Classical Automorphic Forms*, A.M.S, 1997.

[I5] H. Iwaniec, *Introduction to the spectral theory of automorphic forms*, 2nd edition, A.M.S and R.M.I, 2002.

[I6] H. Iwaniec, *Fourier coefficients of modular forms of half-integral weight*, Invent. math. **87** (1987), 385–401.

[I7] H. Iwaniec, *The spectral growth of automorphic L-functions*, J. Reine Angew. Math. **428** (1992), 139–159.

[I8] H. Iwaniec, *The half-dimensional sieve*, Acta Arithmetica **29** (1976), 69–95.

[I9] H. Iwaniec, *A new form of the error term in the linear sieve*, Acta Arithmetica **37** (1980), 307–320.

[I10] H. Iwaniec, *Almost primes represented by quadratic polynomials*, Invent. math. **47** (1978), 171–188.

[I11] H. Iwaniec, *Small eigenvalues of Laplacian for* $\Gamma_0(N)$, Acta Arithmetica **56** (1990), 65–82.

[I12] H. Iwaniec, *Prime geodesic theorem*, Journal Reine Angew. Math. **349** (1984), 136–159.

[I13] H. Iwaniec, *Nonholomorphic modular forms and their applications*, Modular forms (Durham, 1983), Horwood, 1984, pp. 157–196.

[I14] H. Iwaniec, *The lowest eigenvalue for congruence groups*, Topics in geometry, Birkhäuser, 1996, pp. 203–212.

[ILS] H. Iwaniec, W. Luo and P. Sarnak, *Low-lying zeros of families of L-functions*, Inst. Hautes Études Sci. Publ. Math. **91** (2001), 55–131.

[IM] H. Iwaniec and P. Michel, *The second moment of the symmetric square L-functions*, Ann. Acad. Sci. Fenn. Math. **26** (2001), 465–482.

[IS1] H. Iwaniec and P. Sarnak, *Perspectives on the analytic theory of L-functions*, GAFA **Special Volume GAFA2000** (2000), 705–741.

[IS2] H. Iwaniec and P. Sarnak, *The non-vanishing of central values of automorphic L-functions and Landau-Siegel zeros*, Israel J. Math. **120** (2000), 155–177.

[JS] H. Jacquet and J. Shalika, *On Euler products and the classification of automorphic representations, I and II*, Amer. J. Math. **103** (1981), 499–588, 777–815.

[JPS] H. Jacquet, I. Piatetskii-Shapiro and J. Shalika, *Rankin-Selberg convolutions*, Amer. J. Math. **105** (1983), 367–464.

[Ju1] M. Jutila, *Lectures on a method in the theory of exponential sums*, Springer Verlag, 1987.

[Ju2] M. Jutila, *On character sums and class numbers*, J. Number Theory **5** (1973), 203–214.

[Ju3] M. Jutila, *On large values of Dirichlet polynomials*, Topics in number theory (Proc. Colloq., Debrecen, 1974), North-Holland, 1976, pp. 129–140.

[Ju4] M. Jutila, *Zero-density estimates for L-functions*, Acta Arithmetica **32** (1977), 55–62.

[Ju5] M. Jutila, *Statistical Deuring-Heilbronn phenomenon*, Acta Arithmetica **37** (1980), 221–231.

[K1] N. Katz, *Twisted L-functions and monodromy*, Princeton Univ. Press, 2002.

[K2] N. Katz, *Exponential sums over finite fields and differential equations over the complex numbers: some interactions*, Bull. Amer. Math. Soc. (N.S.) **23** (1990), 269–309.

[K3] N. Katz, *Gauss sums, Kloosterman sums and monodromy groups*, Princeton Univ. Press, 1988.

[K4] N. Katz, *Sommes exponentielles*, S.M.F, 1980.

[K5] N. Katz, *Sums of Betti numbers in arbitrary characteristic*, Finite Fields Appl. **7** (2001), 29–44.

[KS1] N. Katz and P. Sarnak, *Random matrices, Frobenius eigenvalues, and monodromy*, A.M.S, 1999.

[KS2] N. Katz and P. Sarnak, *Zeroes of zeta functions and symmetry*, Bull. Amer. Math. Soc. **36** (1999), 1–26.

[Kh] A. Y. Khinchine, *Three pearls of number theory*, Dover Publications, 1998.

[KSa] H. Kim and P. Sarnak, *Refined estimates towards the Ramanujan and Selberg conjectures*, J. Amer. Math. Soc. **16** (2003), 175–181.

[KSh] H. Kim and F. Shahidi, *Cuspidality of symmetric powers with applications*, Duke Math. J. **112** (2002), 177–197.

[Klo] H.D. Kloosterman, *On the representation of numbers in the form* $ax^2+by^2+cz^3+dt^2$, Acta Math. **49** (1926), 407–464.

[Ko1] E. Kowalski, *Analytic problems for elliptic curves* (2001) (preprint).

[Ko2] E. Kowalski, *On the "reducibility" of arctangents of integers*, Amer. Math. Monthly **111** (2004), 351–354.

[KM1] E. Kowalski and P. Michel, *A lower bound for the rank of* $J_0(q)$, Acta Arithmetica **94** (2000), 303–343.

[KM2] E. Kowalski and P. Michel, *The analytic rank of* $J_0(q)$ *and zeros of automorphic L-function*, Duke Math. J. **100** (1999), 503–542.

[KMV1] E. Kowalski, P. Michel and J. VanderKam, *Mollification of the fourth moment of automorphic L-functions and arithmetic applications*, Invent. math. **142** (2000), 95–151.

[KMV2] E. Kowalski, P. Michel and J. VanderKam, *Rankin-Selberg L-functions in the level aspect*, Duke Math. J **114** (2002), 123–191.

[KMV3] E. Kowalski, P. Michel and J. VanderKam, *Non-vanishing of high derivatives of automorphic L-functions at the center of the critical strip*, J. für die Reine und Angew. Math. **526** (2000), 1–34.

[Kor] N. M. Korobov, *Estimates of trigonometric sums and their applications*, Uspehi Mat. Nauk. **13** (1958), 185–192.

[Kub1] J. Kubilius, *Probability methods in number theory*, Usp. Mat. Nauk. **68** (1956), 31–66.

[Kub2] J. Kubilius, *Sharpening of the estimate of the second central moment for additive arithmetical functions*, Litovsk. Mat. Sb. **25** (1985), 104–110.

[Kuz] N. V. Kuznetsov, *The Petersson conjecture for cusp forms of weight zero and the Linnik conjecture*, Mat. Sb. (N.S.) **111** (1980), 334–383; Math. USSR-Sb **39** (1981), 299–342.

[LO] J. Lagarias and A. Odlyzko, *Effective versions of the Chebotarev Density Theorem*, Algebraic number fields: L-functions and Galois properties (Proc. Sympos., Univ. Durham, Durham, 1975), Academic Press, 1977, pp. 409–464.

[La1] E. Landau, *Über die Einteilung der positiven ganzen Zahlen in vier Klassen nach der Mindestzahl der zu ihrer additiven Zusammensetzung erforderlichen Quadrate*, Arch. der Math. u. Phys. (3) **13** (1908), 305–312.

[La2] E. Landau, *Bemerkungen zum Heilbronnschen Satz*, Acta Arithmetica **1** (1936), 1–18.

[La3] E. Landau, *Über die Nullstellen der Dirichletschen Reihen und der Riemannschen* ζ*-Funktion*, Arkiv. für Mat. Astr. och Fysik **16** (1921).

[La] S. Lang, *Algebraic Number Theory*, 2nd edition, Grad. Texts in Math. 110, Springer-Verlag, 1994.

[LT] S. Lang and H. Trotter, *Frobenius distribution in* GL_2 *extensions*, Lecture Notes in Math. 504, Springer Verlag, 1976.

[Lau] G. Laumon, *Exponential sums and l-adic cohomology: a survey*, Israel J. Math. **120** (2000), 225–257.

[Lav] A. F. Lavrik, *Approximate functional equations of Dirichlet functions*, Izv. Akad. Nauk SSSR Ser. Mat. **32** (1968), 134–185.

[Leb] N. N. Lebedev, *Special functions and their applications*, Dover Publications, 1972.

[Lev] N. Levinson, *More than one-third of the zeros of the Riemann zetafunction are on* $\sigma = 1/2$, Adv. Math. **13** (1974), 383–436.

[L] W. Li, *L-series of Rankin type and their functional equations*, Math. Ann. **244** (1979), 135–166.

[Li1] Yu. V. Linnik, *The large sieve*, Dokl. Akad. Nauk SSSR **30** (1941), 292–294. (in Russian)

[Li2] Yu. V. Linnik, *The dispersion method in binary additive problems*, AMS, 1963.

[Li3] Yu V. Linnik, *Additive problems and eigenvalues of the modular operators*, Proc. Internat. Congr. Mathematicians (Stockholm, 1962), pp. 270–284.

[Li4] Yu. V. Linnik, *On the least prime in an arithmetic progression, I. The basic theorem*, Rec. Math. [Mat. Sbornik] N.S. **15(57)** (1944), 139–178.

[Li5] Yu. V. Linnik, *On the least prime in an arithmetic progression, II. The Deuring-Heilbronn phenomenon*, Rec. Math. [Mat. Sbornik] N.S. **15(57)** (1944), 347–368.

[Li6] Yu. V. Linnik, *On Dirichlet's L-series and prime-number sums*, Rec. Math. [Mat. Sbornik] N.S. **15(57)** (1944), 3–12.

[Li7] Yu. V. Linnik, *New versions and new uses of the dispersion methods in binary additive problems*, Dokl. Akad. Nauk SSSR **137** (1961), 1299–1302.

[LR] J. H. van Lint and H. -E. Richert, *On primes in arithmetic progressions*. Acta Arithmetica **11** (1965), 209–216.

[Lit] J. E. Littlewood, *On the zeros of the Riemann Zeta-function*, Cambridge Phil. Soc. Proc. **22** (1924), 295–318.

[LRW] J. van de Lune, H.J.J te Riele, D.T. Winter, *On the zeros of the Riemann zeta function in the critical strip, IV*, Math. Comp. **174** (1986), 667–681.

[Luo] W. Luo, *Nonvanishing of L-values and the Weyl law*, Ann. of Math. (2) **154** (2001), 477–502.

[LRS] W. Luo, Z. Rudnick and P. Sarnak, *On Selberg's eigenvalue conjecture*, Geom. Funct. Anal. **5** (1995), 387–401.

[Ma] H. Maass, *Über eine neue Art von nichtanalytischen automorphen Funktionen und die Bestimmung Dirichletscher Reihen durch Funktionalgleichungen*. Math. Ann. **121** (1949), 141–183.

[Mai] H. Maier, *Primes in short intervals*, Michigan Math. J. **32** (1985), 221–225.

[Mar] G. Margulis, *Discrete Subgroups of Semisimple Lie Groups*, Ergebnisse der Math. und ihrer Grenzgebiete 68, Springer Verlag, 1991.

[MSD] B. Mazur and P. Swinnerton-Dyer, *Arithmetic of Weil curves*, Invent. math. **25** (1974), 1–61.

[Mes] J-F. Mestre, *Formules explicites et minorations de conducteurs de variétés algébriques*, Compositio Math. **58** (1986), 209–232.

[M1] P. Michel, *The subconvexity problem for Rankin-Selberg L-functions and equidistribution of Heegner points*, Annals. of Math. (to appear).

[M2] P. Michel, *Analytic number theory and families of automorphic L-functions* (to appear).

[M3] P. Michel, *Autour de la conjecture de Sato-Tate pour les sommes de Kloosterman, I*, Invent. Math. **121** (1995), 61–78.

[MvdK] P. Michel and J. VanderKam, *Non-vanishing of high derivatives of Dirichlet L-functions at the central point*, J. Number Theory **81** (2000), 130–148.

[Mi] T. Miyake, *Modular forms*, Springer Verlag, 1989.

[MW] C. Moeglin and J-L. Waldspurger, *Pôles des fonctions L de paires pour $GL(N)$, app. to Le spectre résiduel de $GL(N)$*, Ann. Sci. ENS (4ème série) **22** (1989), 605–674.

[Mol] G. Molteni, *Upper and lower bounds at $s = 1$ for certain Dirichlet series with Euler product*, Duke Math. J. **111** (2002), 133–158.

[Mo1] H.L. Montgomery, *The analytic principle of the large sieve*, Bull. Amer. Math. Soc. **84** (1978), 547–567.

[Mo2] H.L. Montgomery, *Topics in multiplicative number theory*, Lecture Notes in Math. 227, Springer Verlag, 1971.

[Mo3] H. Montgomery, *Zeros of L-functions*, Invent. math. **8** (1969), 346–354.

[Mo4] H. L. Montgomery, *The pair correlation of zeros of the zeta function*, Analytic number theory (Proc. Sympos. Pure Math., Vol. XXIV), Amer. Math. Soc., 1972, pp. 181–193.

[MV1] H. L. Montgomery and R. C., Vaughan, *The large sieve*, Mathematikia **20** (1973), 119–134.

[MV2] H.L. Montgomery and R.C. Vaughan, *Hilbert's inequality*, J. London Math.Soc. (2) **8** (1974), 73–82.

[MV3] H.L. Montgomery and R.C. Vaughan, *The exceptional set in Goldbach's problem*, Acta Arithmetica **27** (1975), 353–370.

[Mor1] C. Moreno, *Prime number theorems for the coefficients of modular forms*, Bull. Amer. Math. Soc. **78** (1972), 796–798.

[Mor2] C. Moreno, *Algebraic curves over finite fields*, Cambridge Univ. Press, 1991.

[Mor3] C. Moreno, *Analytic proof of the strong multiplicity one theorem*, Amer. J. Math. **107** (1985), 163–206.

[Mot1] Y. Motohashi, *Spectral theory of the Riemann zeta-function*, Camdridge Univ. Press, 1997.

[Mot2] Y. Motohashi, *An induction principle for the generalization of Bombieri's prime number theorem*, Proc. Japan Acad. **52** (1976), 273–275.

[Nak] H. Nakazato, *Heegner points on modular elliptic curves*, Proc. Japan Acad. Ser. A Math. Sci. **72** (1996), 223–225.

[Ne] J. Nekovàr, *On the parity of ranks of Selmer groups, II*, C. R. Acad. Sci. Paris Sér. I Math. **332** (2001), 99–104.

[Od] A. M. Odlyzko, *Some analytic estimates of class numbers and discriminants*, Invent. math. **29** (1975), 275–286.

[Oe] J. Oesterlé, *Nombres de classes des corps quadratiques imaginaires*, Astérisque **121–122** (1985), 309–323.

[Ono] K. Ono, *Nonvanishing of quadratic twists of modular L-functions and applications to elliptic curves*, J. Reine Angew. Math. **533** (2001), 81–97.

[PP] A. Perelli and J. Pomykala, *Averages over twisted elliptic L-functions*, Acta Arithmetica **80** (1997), 149–163.

[PSa] Y. Petridis and P. Sarnak, *Quantum unique ergodicity for $SL_2(\mathcal{O})\backslash\mathbb{H}_3$ and estimates for L-functions*. Journal of Evolution Equations **1** (2001), 277–290.

[Ph] E. Phillips, *The zeta-function of Riemann; further developments of van der Corput's method*, Quart. J. Math. **4** (1933), 209–225.

[PS] R. Phillips and P. Sarnak, *On cusp forms for co-finite subgroups of $PSL(2,\mathbb{R})$*, Invent. math. **80** (1985), 339–364.

[Pi] N. Pitt, *On shifted convolution of $\zeta^3(s)$ with automorphic L-functions*, Duke Math. J. **77** (1995), 383–406.

[Poi] G. Poitou, *Sur les petits discriminants*, Séminaire Delange-Pisot-Poitou, 18e année (1976-77).

[Pol] G. Pólya, *Über die Verteilung der quadratischen Reste und Nichtreste*, Nachr. Königl. Gesell. Wissensch. Göttingen, Math.-phys. Klasse (1918), 21–29.

[Pos] A. G. Postnikov, *On Dirichlet L-series with the character modulus equal to the power of a prime number*, J. Indian Math. Soc. (N.S.) **20** (1956), 217–226.

[PR] A. G. Postnikov and N. P. Romanov, *A simplification of A. Selberg's elementary proof of the asymptotic law of distribution of prime numbers*, Uspehi Mat. Nauk (N.S.) **10** (1955), 75–87.

[R] H. Rademacher, *On the Partition Function $p(n)$*, Proc. London Math. Soc. **43** (1937), 241–254.

[Ra] D. Ramakrishnan, *Modularity of the Rankin-Selberg L-series, and multiplicity one for $SL(2)$*, Annals of Math. (2) **152** (2000), 45–111.

[Ra1] R. A. Rankin, *Van der Corput's method and the theory of exponent pairs*, Quart. J. Math. (2) **6** (1955), 147–153.

[Ra2] R. A. Rankin, *The difference between consecutive prime numbers, V*, Proc. Edinburgh Math. Soc. (2) **13** (1962/1963), 331–332.

[Ra3] R. A. Rankin, *Contributions to the theory of Ramanujan's τ function and similar arithmetical functions, II*, Proc. Camb. Phil. Soc. **35** (1939), 351–372.

[Re] A. Renyi, *On the representation of an even number as the sum of a single prime and single almost-prime number*, Izvestiya Akad. Nauk SSSR. Ser. Mat. **12** (1948), 57–78.

[Rey] É. Reyssat, *Quelques aspects des surfaces de Riemann*, Progress in Math. 77, Birkhaüser, 1989.

[Rie] B. Riemann, *Über die Anzahl der Primzahlen unter einer gegebenen Grösse*, Monatsber. Berlin. Akad. (1859), 671–680.

[RV] F. Rodriguez-Villegas, *Square root formulas for central values of Hecke L-series, II*, Duke Math. J. **72** (1993), 431–440.

[Rot] K.F. Roth, *On the large sieves of Linnik and Rényi*, Mathematatika **12** (1965), 1–9.

[Roy] E. Royer, *Statistique de la variable aléatoire* $L(Sym^2 f, 1)$, Math. Ann. **321** (2001), 667–687.

[Ru] W. Rudin, *Real and complex analysis*, McGraw-Hill, 1987.

[RS] Z. Rudnick and P. Sarnak, *Zeros of principal L-functions and random matrix theory*, Duke Math. J. **81** (1996), 269–322.

[Sal] H. Salié, *Über die Kloostermanschen Summen* $S(u,v,q)$, Math. Z. **34** (1931), 91–109.

[Sa1] P. Sarnak, *Estimates for Rankin-Selberg L-Functions and Quantum Unique Ergodicity*, J. Funct. Anal. **184** (2001), 419–453.

[Sa2] P. Sarnak, *Class Numbers of indefinite binary quadratic forms*, Journal of Number Theory **15** (1982), 229–247.

[Sa3] P. Sarnak, *Some applications of modular forms*, Cambridge Univ. Press, 1990.

[Sa4] P. Sarnak, *Arithmetic quantum chaos*, Bar-Ilan Univ., 1995.

[Sch] W. Schmidt, *Equations over finite fields. An elementary approach*, Lecture Notes in Math. 536, Springer Verlag, 1976.

[ST] D. B. Sears and E.C. Titchmarsh, *Some eigenfunction formulae*, Quart. J. Math. Oxford **1** (1950), 165–175.

[S1] A. Selberg, *The general sieve method and its place in prime number theory*, Proc. ICM, vol. 1, Cambridge, MA., 1950, pp. 286–292.

[S2] A. Selberg, *On the estimation of Fourier coefficients of modular forms*, Proc. Sympos. Pure Math., Vol. VIII, Amer. Math. Soc., 1965, pp. 1–15.

[S3] A. Selberg, *Lectures on sieves*, Collected Papers Vol. II, Springer Verlag, 1991, pp. 66–247.

[S4] A. Selberg, *On the zeros of Riemann's zeta-function*, Skr. Norske Vid. Akad. Oslo I (1942).

[S5] A. Selberg, *Bemerkungen über eine Dirichletsche Reihe, die mit der Theorie der Modulformen nahe verbunden ist*, Arch. Math. Naturvid. **43** (1940), 47–50.

[S6] A. Selberg, *Harmonic analysis and discontinuous groups in weakly symmetric Riemannian spaces with applications to Dirichlet series*, J. Indian Math. Soc. (N.S.) **20** (1956), 47–87.

[S7] A. Selberg, *On the estimation of Fourier coefficients of modular forms*, Proc. Symp. Pure Math. 8, 1965, pp. 1–8.

[S8] A. Selberg, *On the zeros of the zeta function of Riemann*, Der Kong. Norske Vidensk. Selsk. Forhand. **15** (1942), 59–62.

[Se1] J-P. Serre, *Cours d'arithmétique*, 2nd edition, P.U.F, 1977.

[Se2] J-P. Serre, *Modular forms of weight one and Galois representations*, Algebraic number fields: L-functions and Galois properties (Proc. Sympos., Univ. Durham, Durham, 1975), Academic Press, 1977, pp. 193–268.

[Se3] J-P. Serre, *Congruences et formes modulaires (d'après H. P. F. Swinnerton-Dyer)*, Séminaire Bourbaki, 24e année (1971/1972), Exp. No. 416, Lecture Notes in Math. 317, Springer Verlag, 1973, pp. 319–338.

[Se4] J-P. Serre, *Corps locaux*, Hermann, 1968.

[Se5] J-P. Serre, *Représentations linéaires des corps finis*, Hermann, 1971.

[Se6] J-P. Serre, *Quelques applications du théorème de densité de Chebotarev*, Publ. Math. IHES **54** (1981), 123–201.

[Se7] J-P. Serre, *Minorations de discriminants*, Oeuvres, Vol. III, Springer-Verlag, 1986, pp. 240–243.

[Se8] J-P. Serre, *Letter to J.M. Deshouillers*.

[Shah] F. Shahidi, *Symmetric power L-functions for* $GL(2)$, Elliptic curves and related topics, CRM Proc. Lecture Notes 4, Amer. Math. Soc., 1994, pp. 159–182.

[Sha] D. Shanks, *Class number, a theory of factorization, and genera*, 1969 Number Theory Institute (Proc. Sympos. Pure Math., Vol. XX), Amer. Math. Soc., 1971, pp. 415–440.

[Sh1] G. Shimura, *On modular forms of half-integral weight*, Annals of Math. **97** (1973), 440–481.

[Sh2] G. Shimura, *On the holomorphy of certain Dirichlet series*, Proc. London Math. Soc. (3) **31** (1975), 79–98.

[Sh3] G. Shimura, *Introduction to the arithmetic theory of automorphic functions*, Princeton Univ. Press, 1971.

[Sie1] C. L. Siegel, *Über die Classenzahl quadratischer Zahlkörper*, Acta Arithmetica **1** (1936), 83–86.

[Sie2] C.L. Siegel, *On the theory of indefinite quadratic forms*, Ann. of Math. **45** (1944), 577–622.

[Sie3] C. L. Siegel, *Lectures on quadratic forms*, Tata Institute, 1967.

[Sil] J. Silverman, *The arithmetic of elliptic curves*, Grad. Texts in Math. 106, Springer-Verlag, 1986.

[Sou] K. Soundararajan, *Nonvanishing of quadratic Dirichlet L-functions at $s = \frac{1}{2}$*, Ann. of Math. (2) **152** (2000), 447–488.

[St1] H. Stark, *A complete determination of the complex quadratic fields with class-number one*, Michigan Math. J. **14** (1967), 1–27.

[St2] H. Stark, *A transcendence theorem for class-number problems. II*, Annals of Math. (2) **96** (1972), 174–209.

[St3] H. Stark, *Some effective cases of the Brauer-Siegel theorem*, Invent. math. **23** (1974), 135–152.

[Ste] S. A. Stepanov, *The number of points of a hyperelliptic curve over a finite prime field*, Izv. Akad. Nauk SSSR Ser. Mat. **33** (1969), 1171–1181.

[Ta1] J. Tate, *Fourier analysis in number fields and Hecke's zeta functions*, Algebraic Number Theory, Academic Press, 1990, pp. 305–347.

[Tat] J. Tate, *Number theoretic preliminaries*, Proceedings of Symposia in Pure Math. 33, vol 2, A.M.S, 1979, pp. 3–26.

[TW] R. Taylor and A. Wiles, *Ring-theoretic properties of certain Hecke algebras*, Ann. of Math. (2) **141** (1995), 553–572.

[Tchu] N. G. Tchudakov, *Sur le problème de Goldbach*, C. R. (Dokl.) Acad. Sci. URSS, n. Ser. **17** (1937), 335–338.

[Th] A. Thue, *Über Annäherungswerte algebraischer Zahlen*, J. Reine Angew. Math. **135** (1909), 284–305.

[T1] E.C. Titchmarsh, *The theory of functions*, 2nd Edition, Oxford Univ. Press, 1939.

[T2] E. C. Titchmarsh, *The theory of the Riemann zeta-function*, 2nd edition, Oxford Univ. Press, 1986.

[T3] E. C. Titchmarsh, *Eigenfunction expansions associated with second-order differential equations*, Clarendon Press, 1962.

[Tot] A. Toth, *Roots of quadratic congruences*, Internat. Math. Res. Notices (2000), 719–739.

[Tu1] P. Turan, *Über einige Verallgemeinerungen eines Satzes von Hardy und Ramanujan*, J. Lond. Math. Soc. **11** (1936), 125–133.

[Tu2] P. Turán, *Über die Primzahlen der arithmetischen Progression*, Acta Litt. Sci. Szeged **8** (1937), 226–235.

[vdK] J. VanderKam, *The rank of quotients of $J_0(N)$*, Duke Math. J. **97** (1999), 545–577.

[VdP] M. van der Put, *Grothendieck's conjecture for the Risch equation $y' = ay + b$*, Indag. Mathem. N.S. **12** (2001), 113–124.

[Va] R. C. Vaughan, *Sommes trigonométriques sur les nombres premiers*, C. R. Acad. Sci. Paris Sér. A-B **285** (1977), A981–A983.

[V1] I. M. Vinogradov, *A new estimate for $\zeta(1 + it)$*, Izv. Akad. Nauk SSSR, Ser. Mat. **22** (1958), 161–164.

[V2] I. M. Vinogradov, *On Weyl's sums*, Mat. Sbornik **42** (1935), 521–530.

[V3] I. M. Vinogradov, *A new method of estimation of trigonometrical sums*, Mat. Sbornik (1) **43** (1936), 175–188.

[V4] I. M. Vinogradov, Perm. Univ. Fiz.-Mat. ob.-vo Zh **1** (1918), 18–24.

[V5] I. M.Vinogradov, *Some theorems concerning the theory of primes*, Math. Sb. 2 **44** (1937), 179–195.

[V6] I. M. Vinogradov, *Representation of an odd number as a sum of three primes*, Dokl. Akad. Nauk SSSR **15** (1937), 291–294.

[Vi] A. I. Vinogradov, *The density hypothesis for Dirichet L-series*, Izv. Akad. Nauk SSSR Ser. Mat. **29** (1965), 903–934.

[Vor] G. Voronoi, *Sur une fonction transcendante et ses applications à la sommation de quelques séries*, Ann. Sci. École Norm. Sup. (3) **21** (1904), 207–267, 459–533..

[W] L. Washington, *Introduction to cyclotomic fields*, Grad. Texts in Math. 83, 2nd edition, Springer Verlag, 1997.

[Wa] T. Watson, *Rankin triple products and quantum chaos, PhD thesis, Princeton University*, 2001.

[We1] A. Weil, *On some exponential sums*, Proc. Nat. Acad. Sci. U. S. A. **34** (1948), 204–207.

[We2] A. Weil, *Über die Bestimmung Dirichletscher Reihen durch Funktionalgleichungen*, Math. Ann. **168** (1967), 149–156.

[W1] H. Weyl, *Über die Gleichverteilung von Zahlen mod. Eins*, Math. Ann. **77** (1916), 313–352.

[W2] H. Weyl, *Zur Abschätzung von* $\zeta(1+ti)$, Math. Zeit. **10** (1921), 88–101.

[W] A. Wiles, *Modular elliptic curves and Fermat's last theorem*, Ann. of Math. (2) **141** (1995), 443–551.

[Wi1] E. Wirsing, *Das asymptotische Verhalten von Summen über multiplikative Funktionen, II*, Acta Math. Acad. Sci. Hungar. **18** (1967), 411–467.

[Wi2] E. Wirsing, *Elementare Beweise des Primzahlsatzes mit Restglied, II*, J. Reine Angew. Math. **214/215** (1964), 1–18.

[Wi3] E. Wirsing, *Growth and differences of additive arithmetic functions*, Topics in classical number theory, Vol. I, II (Budapest, 1981), Colloq. Math. Soc. János Bolyai, 34, North-Holland, Amsterdam, 1984, pp. 1651–1661.

Index

Titles in This Series

53 **Henryk Iwaniec and Emmanuel Kowalski,** Analytic number theory, 2004

52 **Dusa McDuff and Dietmar Salamon,** J-holomorphic curves and symplectic topology, 2004

51 **Alexander Beilinson and Vladimir Drinfeld,** Chiral algebras, 2004

50 **E. B. Dynkin,** Diffusions, superdiffusions and partial differential equations, 2002

49 **Vladimir V. Chepyzhov and Mark I. Vishik,** Attractors for equations of mathematical physics, 2002

48 **Yoav Benyamini and Joram Lindenstrauss,** Geometric nonlinear functional analysis, Volume 1, 2000

47 **Yuri I. Manin,** Frobenius manifolds, quantum cohomology, and moduli spaces, 1999

46 **J. Bourgain,** Global solutions of nonlinear Schrödinger equations, 1999

45 **Nicholas M. Katz and Peter Sarnak,** Random matrices, Frobenius eigenvalues, and monodromy, 1999

44 **Max-Albert Knus, Alexander Merkurjev, and Markus Rost,** The book of involutions, 1998

43 **Luis A. Caffarelli and Xavier Cabré,** Fully nonlinear elliptic equations, 1995

42 **Victor Guillemin and Shlomo Sternberg,** Variations on a theme by Kepler, 1990

41 **Alfred Tarski and Steven Givant,** A formalization of set theory without variables, 1987

40 **R. H. Bing,** The geometric topology of 3-manifolds, 1983

39 **N. Jacobson,** Structure and representations of Jordan algebras, 1968

38 **O. Ore,** Theory of graphs, 1962

37 **N. Jacobson,** Structure of rings, 1956

36 **W. H. Gottschalk and G. A. Hedlund,** Topological dynamics, 1955

35 **A. C. Schaeffer and D. C. Spencer,** Coefficient regions for Schlicht functions, 1950

34 **J. L. Walsh,** The location of critical points of analytic and harmonic functions, 1950

33 **J. F. Ritt,** Differential algebra, 1950

32 **R. L. Wilder,** Topology of manifolds, 1949

31 **E. Hille and R. S. Phillips,** Functional analysis and semigroups, 1957

30 **T. Radó,** Length and area, 1948

29 **A. Weil,** Foundations of algebraic geometry, 1946

28 **G. T. Whyburn,** Analytic topology, 1942

27 **S. Lefschetz,** Algebraic topology, 1942

26 **N. Levinson,** Gap and density theorems, 1940

25 **Garrett Birkhoff,** Lattice theory, 1940

24 **A. A. Albert,** Structure of algebras, 1939

23 **G. Szegö,** Orthogonal polynomials, 1939

22 **C. N. Moore,** Summable series and convergence factors, 1938

21 **J. M. Thomas,** Differential systems, 1937

20 **J. L. Walsh,** Interpolation and approximation by rational functions in the complex domain, 1935

19 **R. E. A. C. Paley and N. Wiener,** Fourier transforms in the complex domain, 1934

18 **M. Morse,** The calculus of variations in the large, 1934

17 **J. M. Wedderburn,** Lectures on matrices, 1934

16 **G. A. Bliss,** Algebraic functions, 1933

15 **M. H. Stone,** Linear transformations in Hilbert space and their applications to analysis, 1932

14 **J. F. Ritt,** Differential equations from the algebraic standpoint, 1932

TITLES IN THIS SERIES

13 **R. L. Moore,** Foundations of point set theory, 1932

12 **S. Lefschetz,** Topology, 1930

11 **D. Jackson,** The theory of approximation, 1930

10 **A. B. Coble,** Algebraic geometry and theta functions, 1929

9 **G. D. Birkhoff,** Dynamical systems, 1927

8 **L. P. Eisenhart,** Non-Riemannian geometry, 1927

7 **E. T. Bell,** Algebraic arithmetic, 1927

6 **G. C. Evans,** The logarithmic potential, discontinuous Dirichlet and Neumann problems, 1927

5.1 **G. C. Evans,** Functionals and their applications; selected topics, including integral equations, 1918

5.2 **O. Veblen,** Analysis situs, 1922

4 **L. E. Dickson,** On invariants and the theory of numbers

W. F. Osgood, Topics in the theory of functions of several complex variables, 1914

3.1 **G. A. Bliss,** Fundamental existence theorems, 1913

3.2 **E. Kasner,** Differential-geometric aspects of dynamics, 1913

2 **E. H. Moore,** Introduction to a form of general analysis

M. Mason, Selected topics in the theory of boundary value problems of differential equations

E. J. Wilczyński, Projective differential geometry, 1910

1 **H. S. White,** Linear systems of curves on algebraic surfaces

F. S. Woods, Forms on noneuclidean space

E. B. Van Vleck, Selected topics in the theory of divergent series and of continued fractions, 1905